STUDENT SOLUTIONS MANUAL

Chemistry: The Science in Context

Fourth Edition

STUDENT SOLUTIONS MANUAL

Chemistry: The Science in Context

Fourth Edition

Thomas R. Gilbert, Rein V. Kirss, Natalie Foster, Geoffrey Davies

Bradley M. Wile
OHIO NORTHERN UNIVERSITY

Karen S. Brewer
HAMILTON COLLEGE

W·W·NORTON

NEW YORK · LONDON

W. W. Norton & Company has been independent since its founding in 1923, when William Warder Norton and Mary D. Herter Norton first published lectures delivered at the People's Institute, the adult education division of New York City's Cooper Union. The Nortons soon expanded their program beyond the Institute, publishing books by celebrated academics from America and abroad. By mid-century, the two major pillars of Norton's publishing program—trade books and college texts—were firmly established. In the 1950s, the Norton family transferred control of the company to its employees, and today—with a staff of four hundred and a comparable number of trade, college, and professional titles published each year—W. W. Norton & Company stands as the largest and oldest publishing house owned wholly by its employees.

Media Editor: Rob Bellinger
Associate Media Editor: Jennifer Barnhardt
Assistant Media Editor: Paula F. Iborra
Project Editor: Carla Talmadge
Production Manager: Sean Mintus
Composition by Bradley M. Wile
Manufacturing by Courier

ISBN 978-0-393-93647-6

W. W. Norton & Company, Inc., 500 Fifth Avenue, New York, NY 10110
www.wwnorton.com

W. W. Norton & Company, Ltd., Castle House, 75/76 Wells Street, London W1T 3QT
1 2 3 4 5 6 7 8 9 0

Contents

Preface

Students often start an introductory course in chemistry with a little trepidation and want to know the secret formula for success. The simple answer (admittedly hard to put into practice) is for you to engage fully with the course during lectures, in laboratories, and in working problems. We hope that this *Solutions Manual* will help you with the last of these.

In the textbook, you have been introduced to the COAST method as an approach to solving problems and answering conceptual questions. This method encourages you to assemble relevant information and to plan an approach for any new problem you might encounter—say, on an exam—before you start calculating an answer. After solving the problem, this method guides you to think further about the problem and to extend your knowledge. The COAST method, as summarized below, is used throughout this *Manual*:

COLLECT AND ORGANIZE Restatement of the problem to delineate exactly what information has been provided and what is being asked.

ANALYZE Strategy to solve the problem including which formulas are relevant and which unit conversions are necessary.

SOLVE Application of the strategy to solve a problem numerically or to answer a conceptual question.

THINK ABOUT IT Reminders to check the answer or consider whether the answer makes sense. In this *Manual*, this step is sometimes used to add factual information extending the context of the question.

In this *Manual*, you will see the worked-out solutions to all the odd-numbered problems at the end of each chapter of the textbook. Generally speaking, the conventions for significant figures as outlined in Chapter 1 of the textbook are used, except when an additional significant figure in the answer would clarify the difference that you might see on your calculator compared to what strict adherence to the significant figure rules would give as an answer. For most of the calculations presented, all of the significant digits were kept in the calculator for intermediate values in a multistep calculation.

As you use this *Solutions Manual*, keep in mind that there are other valid approaches to solving a particular problem in chemistry. Neither your professor nor you will always solve the problem in exactly the same way as it is presented here. Also, in class, you may be shown additional approaches to solving the many problems that you will encounter in introductory chemistry. That's okay as long as those methods are applicable to the range of problems of that type.

When you sit down to solve a set of problems from the textbook, make sure you have all the relevant tools—textbook, periodic table, and calculator—and are ready to dig in. Sometimes working alone is best, but sometimes you might find that you learn more when discussing solutions with a study group. Also, you will be more successful in the course if you work the end-of-chapter problems on a regular basis, and not just on weekends or right before a test. Doing 8–10 problems every other day is less overwhelming (so you'll accomplish more) than trying to do 30 a week or over 100 right before a test. If you solve problems steadily and conscientiously throughout the course, you will be amazed by how much you are learning.

You will find this *Manual* most useful if you first try to solve the problems on your own. If you get stuck on a particular question, open the *Manual* and consult the solution shown, read it carefully and compare it to your approach, and then try to solve the problem again. Most of the odd-numbered problems are paired with an even-numbered problem that is similar. So you can hone your problem-solving approach by using the solutions in this *Manual* as a model for solving those even-numbered problems.

Finally, as you consider the connections to biology, physics, geology, materials science, environmental science, medicine, and astronomy that are presented in the text and used as a basis for many of the end-of-chapter problems, we hope that you will see the contributions of the science of chemistry to these fields.

CHAPTER 1 | Matter and Energy: The Origin of the Universe

1.1. Collect and Organize

In Figure P1.1(a) we are shown "molecules" each consisting of one red sphere and one blue sphere and in Figure P1.1(b) we have separate blue spheres and red spheres. In each figure we are to identify whether the substance(s) depicted is a solid, liquid, or gas and if the figures show pure elements or compounds.

Analyze

A pure substance is composed of all the same type of element or compound, not a mixture of two kinds. An element is composed of all the same type of atom and a compound is composed of two or more types of atoms. Solids have a definite volume and a highly ordered arrangement where the particles are close together, liquids also have a definite volume but have a disordered arrangement of particles that are close together, and gases have disordered particles that fill the volume of the container and are far apart from each other.

Solve

(a) Because the particles each consist of one red sphere and one blue sphere, all the particles are the same—this is a pure compound. The particles fill the container and are disordered, so these particles are in the gas phase.
(b) Because Figure P1.1(b) shows a mixture of red and blue spheres, this is depicting a mixture of blue element atoms and red element atoms. The blue spheres fill the container and are disordered, so these particles are in the gas phase. The red spheres have a definite volume and are slightly disordered, so these particles are in the liquid phase.

Think About It

Remember that both elements and compounds may be either pure or present in a mixture.

1.3. Collect and Organize

In this question we are to consider whether the reactants as depicted undergo a chemical reaction and/or phase change.

Analyze

Chemical reactions involve the breaking and making of bonds in which atoms are combined differently in the products compared to that of the reactants. In considering a possible phase change, solids have a definite volume and a highly ordered arrangement where the particles are close together, liquids also have a definite volume but have a disordered arrangement of particles that are close together, and gases have disordered particles that fill the volume of the container and are far apart from each other. It may help to visualize each scenario, as shown below

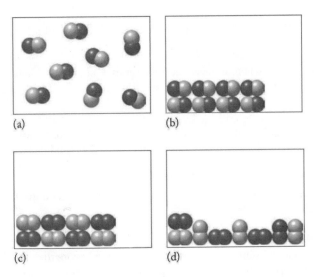

(a)   (b)

(c)   (d)

Solve

In Figure P1.3 two pure elements (red–red and blue–blue) in the gas phase recombine to form a compound (red–blue) in the solid phase (ordered array of molecules). Therefore, answer (b) describes the reaction shown.

Think About It

A phase change does not necessarily accompany a chemical reaction. We will learn later that the polarity of the product will determine whether or not a substance will be in the solid, liquid, or gaseous state at a given temperature.

1.5. Collect and Organize

From the space-filling model shown, we are to write the formula for the chemical represented.

Analyze

Based on the Atomic Color Palette (see the inside back cover of the textbook) we see that this model contains three hydrogen atoms bonded to a carbon atom that is bonded to an O–H unit.

Solve

H_3COH or CH_4O

Think About It

This model represents methanol, and the presence of the O–H unit classifies it as an alcohol. Sometimes methanol is called "wood alcohol."

1.7. Collect and Organize

This question considers *if* and *how* matter and energy are related. In particular, we consider whether the sun is all mass or all energy.

Analyze

Einstein showed that matter and energy are interconvertible through $E = mc^2$.

Solve

The sun is an example where matter is being changed into energy through nuclear fusion reactions. Therefore, both students are correct.

Think About It

Through Einstein's equation we see that a little bit of mass contains a great deal of energy locked in the nuclei of the atoms.

1.9. Collect and Organize

For this question we are to list some chemical and physical properties of gold.

Analyze

A chemical property is seen when a substance undergoes a chemical reaction thereby becoming a different substance. A physical property can occur without any transformation of one substance into another.

Solve

One chemical property of gold is its resistance to corrosion (oxidation). Gold's physical properties include its density, color, melting point, metallic luster, malleability, ductility, and electrical and thermal conductivity.

Think About It

Another metal that does not corrode (or rust) is platinum. Platinum and gold, along with palladium, are often called "noble metals."

1.11. Collect and Organize

This question asks us to use differences in physical or chemical properties to separate a mixture of substances. Using filtration, we are to propose a method to separate salt (NaCl) from sand in a mixture.

Analyze
The difference between sand and salt is salt's solubility in water versus sand's insolubility in water.

Solve
Adding water to the salt–sand mixture will dissolve the salt. After putting the sand–salt-solution mixture through a filter, the sand will remain on the filter. The salt can be recovered by evaporating the water from the solution that passed through the filter.

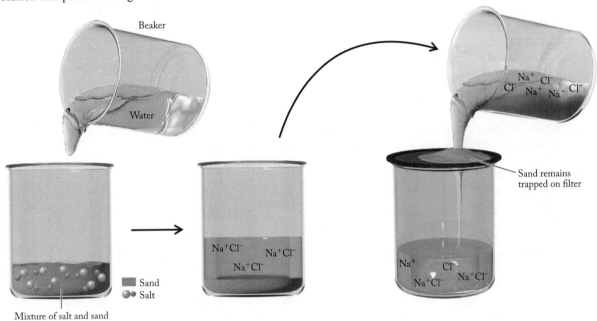

Think About It
The filtration method works very well for a mixture of two substances when one is soluble and the other is insoluble in a particular solvent. It is used often to separate and purify the desired product of a chemical reaction.

1.13. **Collect and Organize**
For the four processes named, we are to determine which involve a chemical change.

Analyze
Chemical changes involve transforming one substance into another to give that substance different physical and chemical properties.

Solve
(a) Distillation purifies a substance—not a chemical change.
(b) Combustion transforms the fuel (such as wood) into carbon dioxide and water—a chemical change.
(c) Filtration separates substances from each other—not a chemical change.
(d) Condensation changes a vapor into a liquid—not a chemical change.

Think About It
Distillation, filtration, and condensation all involve physical changes, not chemical changes.

1.15. **Collect and Organize**
For the foods listed, we are to determine which are heterogeneous.

Analyze
A heterogeneous mixture has visible regions of different compositions.

Solve

A Snickers bar and an uncooked hamburger (b, d) are heterogeneous—they have clear regions of differing composition, but solid butter and grape juice (a, c) are homogeneous.

Think About It

When butter is melted, you notice that there are milk solids and clear regions that are definitely discernible. Therefore, homogeneous solid butter becomes heterogeneous when heated.

1.17. Collect and Organize

For the foods listed, we are to determine which are heterogeneous.

Analyze

A heterogeneous mixture has visible regions of different compositions.

Solve

Orange juice (with pulp; d) is heterogeneous—there are clear regions of differing composition. Apple juice, cooking oil, solid butter, and tomato juice (a–c, e) are homogeneous.

Think About It

When butter is melted, you notice that there are milk solids and clear regions that are definitely discernible. Therefore, homogeneous solid butter becomes heterogeneous when heated.

1.19. Collect and Organize

We are asked to name three properties to distinguish between table sugar, water, and oxygen.

Analyze

We can distinguish between substances using either physical properties (color, melting point, density, etc.) or chemical properties (chemical reactions, corrosion, flammability, etc.).

Solve

We can distinguish between table sugar, water, and oxygen by examining their physical states (at room temperature and standard pressure, sugar is a solid, water is a liquid, and oxygen is a gas), densities, melting points, and boiling points.

Think About It

These three substances are also very different at the atomic level. Oxygen is a pure element made up of diatomic molecules, water is a liquid compound made up of discrete molecules of hydrogen and oxygen (H_2O), and table sugar is a solid compound made up of carbon, hydrogen, and oxygen atoms ($C_{12}H_{22}O_{11}$).

1.21. Collect and Organize

From the list of properties of sodium, we are to determine which are physical and which are chemical properties.

Analyze

Physical properties are those that can be observed without transforming the substance into another substance. Chemical properties are only observed when one substance reacts with another and therefore is transformed into another substance.

Solve

Density, melting point, thermal and electrical conductivity, and softness (a–d) are all physical properties, whereas tarnishing and reaction with water (e, f) are both chemical properties.

Think About It

Because the density of sodium is less than that of water, a piece of sodium will float on water as it reacts.

1.23. **Collect and Organize**

We are asked to determine which of the given species is *not* a pure substance.

Analyze

All matter may be classified as either a pure substance or a mixture. If a substance may be separated by physical means, it is not a pure substance. In addition to sodium chloride, table salt includes anti-caking agents and sodium iodide.

Solve

Both air and table salt may be separated into simpler components, so they cannot be pure substances. Nitrogen gas, oxygen gas, and argon gas are all pure elements.

Think About It

Air is a mixture of nitrogen gas, oxygen gas, argon gas, and other pure substances.

1.25. **Collect and Organize**

From the given list, we are asked to identify the elements.

Analyze

An element is a pure substance that cannot be broken down into simpler components by any process.

Solve

Only Cl_2 is a pure element. The other species listed are compounds of more than one type of element.

Think About It

The elements are listed in the periodic table of the elements. Some elements, such as chlorine, form diatomic molecules (i.e., Cl_2) in their pure form.

1.27. **Collect and Organize**

We are asked to determine which of the given mixtures is homogeneous.

Analyze

A homogeneous mixture has no visible boundaries and contains a uniform distribution of components. We should picture each of these items, and decide if their components are uniformly distributed.

Solve

Only filtered water is uniformly distributed, and so it is the only homogeneous mixture.

Think About It

All of the other species listed are heterogeneous mixtures, since they have components that are visually different, or are not uniformly distributed. Clouds may take many shapes and densities, so they do not exhibit a uniform distribution.

1.29. **Collect and Organize**

We are to determine which mixtures can be separated into their components by filtering.

Analyze

Filtration is used to separate suspended solids from a liquid or gas; this technique will only work if our sample is a heterogeneous mixture containing a solid suspended in a liquid or a gas.

Solve

Sand and water may be separated using filtration. The sand will be trapped on the filter, and the water will pass through.

Think About It

The other mixtures listed in the question are homogeneous and cannot be separated by filtration.

1.31. **Collect and Organize**

From the list of properties of formaldehyde, we are asked to determine which one is a chemical property.

Analyze

Chemical properties are observed when a chemical reaction takes place. We should identify which property relates to a chemical reaction between formaldehyde and another substance.

Solve

Formaldehyde must react with the air in order to burn. Flammability (c) is a chemical property.

Think About It

Smell, solubility, phase of matter, and color are all physical properties. We can observe these properties without destroying (reacting) the formaldehyde.

1.33. **Collect and Organize**

We are to explain if an extensive property can be used to identify a substance.

Analyze

An extensive property is one that, like mass, length, and volume, is determined by size or amount.

Solve

Extensive properties will change with the size of the sample and therefore cannot be used to identify a substance.

Think About It

We could, for example, have the same mass of feathers and lead, but their mass alone will not tell us which mass measurement belongs to which—the feathers or the lead.

1.35. **Collect and Organize**

In this question we think about the information needed to formulate a hypothesis.

Analyze

A hypothesis is a tentative explanation for an observation.

Solve

To form a hypothesis we need at least one observation, experiment, or idea (from examining nature).

Think About It

A hypothesis that is tested and shown to be valid can become a theory.

1.37. **Collect and Organize**

We are to consider whether we can disprove a hypothesis.

Analyze

A hypothesis is a tentative explanation for an observation.

Solve

It is possible to disprove a scientific hypothesis. In fact, many experiments are designed to do just that as the best test of the hypothesis's validity.

Think About It

It is even possible to disprove a theory (albeit harder to do so) or cause a theory to be modified when new evidence, a new experimental technique, or new data from a new instrument give observations that are counter to the explanation stated by the theory.

1.39. Collect and Organize
We are to define *theory* as used in conversation.

Analyze
Theory in everyday conversation has a different meaning than it does in science.

Solve
Theory in normal conversation is someone's idea or opinion or speculation that can easily be changed and may not have much evidence or many arguments to support it.

Think About It
A *theory* in science is a generally accepted and highly tested explanation of observed facts.

1.41. Collect and Organize
We are to compare SI units to English units.

Analyze
SI units are based on a decimal system to describe basic units of mass, length, temperature, energy, etc. whereas English units vary.

Solve
SI units, which were based on the original metric system, can be easily converted into a larger or smaller unit by multiplying or dividing by multiples of 10. English units are more complicated to manipulate. For example, to convert miles to feet you have to know that there are 5280 feet in 1 mile and to convert gallons to quarts you have to know that 4 quarts are in 1 gallon.

Think About It
Once you can visualize a meter, a gram, and a liter, using the SI system is quite convenient.

1.43. Collect and Organize
This problem asks which value from the given set contains the fewest number of significant figures.

Analyze
The listed values are a mix of standard and scientific notation. For values expressed using scientific notation, only the values prior to the exponent are counted as significant. By expressing all the quantities using scientific notation, we can better compare the number of significant figures.
   (a) 5.45×10^2
   (b) 6.4×10^{-3}
   (c) 6.50×10^0
   (d) 1.346×10^2

Solve
(b) 6.4×10^{-3} contains two significant figures, which is the smallest number of significant figures for this set.

Think About It
The value 6.50 has three significant figures, as the zero trailing the decimal is significant.

1.45. Collect and Organize
This problem asks which value from the given set contains the fewest number of significant figures.

Analyze
These values are all fractions. We should evaluate each fraction, and express it using scientific notation to help determine the number of significant figures.
   (a) $1/545 = 0.00183 = 1.83 \times 10^{-3}$
   (b) $1/6.4 \times 10^{-3} = 156.25 = 1.6 \times 10^2$

(c) $1/6.50 = 0.1538 = 1.54 \times 10^{-1}$
(d) $1/1.346 \times 10^2 = 0.007429 = 7.429 \times 10^{-3}$

Solve

The smallest number of significant figures present is two. The quantity with two significant figures is (b) $1/6.4 \times 10^{-3}$.

Think About It

The number of significant figures is derived from the denominator of the fraction. Despite the fact that $1/6.4 \times 10^{-3}$ gives us a number that appears to contain at least three digits before the decimal, we cannot be more confident than we would be in 6.4×10^{-3}, which contains two significant figures.

1.47. Collect and Organize

From the values given, we must identify those that contain four significant figures.

Analyze

Writing all the quantities in scientific notation will help determine the number of significant figures in each.
(a) $0.0592 = 5.92 \times 10^{-2}$
(b) $0.08206 = 8.206 \times 10^{-2}$
(c) 8.314
(d) $5420 = 5.42 \times 10^3$ or 5.420×10^3 (if the 0 is significant)
(e) 5.4×10^3

Solve

The quantities that have four significant figures are (b) 0.08206, (c) 8.314, and maybe (d) 5420, if the 0 is significant.

Think About It

Remember that zeros at the end of the number may be significant or they may simply be acting as placeholders.

1.49. Collect and Organize

We are to express the result of each calculation to the correct number of significant figures.

Analyze

The rules regarding the significant figures that carry over in calculations are given in Section 1.8 in the textbook. Remember to operate on the weak-link principle.

Solve

(a) The least well-known value has three significant figures so the calculator result of 17.363 is reported as 17.4 with rounding up the tenths place.
(b) The least well-known value has only one significant figure so the calculator result of 1.044×10^{-13} is reported as 1×10^{-13}.
(c) The least well-known value has three significant figures so the calculator result of 5.701×10^{-23} is reported as 5.70×10^{-23}.
(d) The least well-known value has three significant figures so the calculator result of 3.5837×10^{-3} is reported as 3.58×10^{-3}.

Think About It

Indicating the correct number of significant figures for a calculated value indicates the level of confidence we have in our calculated value. Reporting too many significant figures would indicate a higher level of precision in our number than we actually have.

1.51. **Collect and Organize**

To compute the runner's speed we have to use the definition: speed = change in distance/change in time. In the marathon runner's case we are given distance in miles and time in hours plus additional minutes. The first calculation of speed, therefore, in miles per hour will not require any unit conversion. That result will be used to compute the runner's speed in meters per second using conversions for miles to meters and hours to seconds.

Analyze

The equation to compute speed is given by

$$\text{Speed} = \frac{\Delta \text{ distance}}{\Delta \text{ time}}$$

Because the time is given as 3 hours 40 minutes, we will have to convert the 40 minutes into a part of an hour using the fact that 1 hour = 60 minutes. We can then divide the marathon distance by this time in hours to get the speed in miles per hour.

To convert this speed to meters per second, we can use the following conversions:

$$\frac{1 \text{ km}}{1000 \text{ m}} \text{ and } \frac{0.6214 \text{ mi}}{1 \text{ km}}$$

$$\frac{1 \text{ min}}{60 \text{ s}} \text{ and } \frac{1 \text{ hr}}{60 \text{ min}}$$

Solve

First, the number of hours for the runner to complete the marathon is

$$3 \text{ hr} + \left(40 \text{ min} \times \frac{1 \text{ hr}}{60 \text{ min}} \right) = 3.67 \text{ hr}$$

(a) The speed in miles per hour is

$$\text{speed} = \frac{26.2 \text{ mi}}{3.67 \text{ hr}} = 7.14 \text{ mi/hr}$$

(b) Converting this speed to meters per second gives

$$7.14 \frac{\text{mi}}{\text{hr}} \times \frac{1 \text{ km}}{0.6214 \text{ mi}} \times \frac{1000 \text{ m}}{1 \text{ km}} \times \frac{1 \text{ hr}}{60 \text{ min}} \times \frac{1 \text{ min}}{60 \text{ s}} = 3.19 \text{ m/s}$$

Think About It

Both of these values seem reasonable. A walking pace is about 3 miles per hour, so running could be imagined to be twice that fast. Also, 3 meters per second can be run easily by a fast runner.

1.53. **Collect and Organize**

We are asked to compare the volume of an imperial gallon and a U.S. gallon. Though both of these are "gallons," it may be helpful to consider the volume contained in each using another unit, like the liter. Since the U.S. gallon and the imperial gallon are similar in volume, the values should be approximately equal.

Analyze

We are told that one imperial gallon contains 4.546 L. From the text, we know that one U.S. gallon contains 3.7854 L. These may be expressed as fractions:

$$\frac{4.546 \text{ L}}{1 \text{ imperial gal}}, \quad \frac{1 \text{ U.S. gal}}{3.7854 \text{ L}}$$

By canceling units common to both, we can determine the number of U.S. gallons in one imperial gallon.

Solve

$$\frac{4.546 \text{ L}}{1 \text{ imperial gal}} \times \frac{1 \text{ U.S. gal}}{3.7854 \text{ L}} = 1.201 \text{ U.S. gal/imperial gal}$$

Think About It

The imperial gallon is a slightly larger unit of volume. We could also calculate the number of imperial gallons contained in one U.S. gallon by flipping both conversion factors. This would give us a value of 0.8327 imperial gallons in one U.S. gallon. Each "gallon" is approximately, though not exactly, equal to the other.

1.55. Collect and Organize

This question asks us to convert a mass in pounds to grams. There are approximately two pounds in a kilogram, so the mass should be just under 1000 grams.

Analyze

We may convert this mass using the following conversion

$$\frac{453.4 \text{ g}}{1 \text{ lb}}$$

Solve

$$1.65 \text{ lb} \times \frac{453.4 \text{ g}}{1 \text{ lb}} = 748 \text{ g or } 7.48 \times 10^2 \text{ g}$$

Think About It

A gram is a relatively small unit of mass, so it makes sense that 1.65 pounds would contain 748 grams (just under one kilogram).

1.57. Collect and Organize

This question asks us to convert a volume in gallons to milliliters. Remember that the prefix *milli-* can be read as 10^{-3}. There are approximately 4 liters in a gallon, so our value (just under 2.5 gallons) should be approximately 10 liters, or 10,000 milliliters.

Analyze

We can convert gallons to liters, then liters to milliliters, using the following conversions:

$$\frac{3.7854 \text{ L}}{1 \text{ gal}}, \frac{1000 \text{ mL}}{1 \text{ L}}$$

Solve

$$2.44 \text{ gal} \times \frac{3.7854 \text{ L}}{1 \text{ gal}} \times \frac{1000 \text{ mL}}{1 \text{ L}} = 9236.4 \text{ mL} = 9.24 \times 10^3 \text{ mL}$$

Think About It

Since the milliliter is a relatively small unit of volume, it makes sense that a large unit like a gallon would contain many milliliters. This volume is just under our estimate of 10 liters. You may also have arrived at the same answer using the conversion

$$\frac{1 \text{ mL}}{1 \times 10^{-3} \text{ L}}$$

1.59. **Collect and Organize**

We are asked to compare two lengths, expressed using different units. We should express these values using the same units to find the larger number.

Analyze

We can convert feet to inches, and then add this value to the 11 remaining inches to determine Peter's height in inches. We may then convert inches to centimeters and compare Peter and Paul's heights on the same scale. The conversions we will use are

$$\frac{12\ \text{in}}{1\ \text{ft}}, \frac{2.54\ \text{cm}}{1\ \text{in}}$$

Solve

First, we will convert Peter's height to inches

$$\left(5\ \text{ft} \times \frac{12\ \text{in}}{1\ \text{ft}}\right) + 11\ \text{in} = 71\ \text{in}$$

Peter's height in centimeters is

$$71\ \text{in} \times \frac{2.54\ \text{cm}}{1\ \text{in}} = 180\ \text{cm} = 1.8 \times 10^2\ \text{cm}$$

Peter is slightly taller than Paul.

Think About It

Because their heights are so similar, we can only tell who is taller because of the number of significant figures in Paul's height. If we had measured Paul's height with only two significant figures, the two would not appear different. You might also convert Paul's height into inches, and find that it is 69.3 in, which is also slightly smaller than Peter's height.

1.61. **Collect and Organize**

To find the Calories burned by the wheelchair marathoner in a race, we can first find the number of hours the race will be for the marathoner at the pace of 13.1 miles per hour. The Calories burned can then be calculated from that value and the rate at which the marathoner burns Calories.

Analyze

The time it takes for the marathoner to complete the race will be given by

$$\text{Time to complete the marathon} = \frac{\text{distance of the marathon}}{\text{pace of the marathoner}}$$

The Calories burned will be computed by

$$\text{Calories burned} = \frac{\text{Calories burned}}{\text{hour}} \times \text{time of the marathon race}$$

Solve

$$\text{Time to complete the marathon} = 26.2\ \text{mi} \times \frac{1\ \text{hr}}{13.1\ \text{mi}} = 2.00\ \text{hr}$$

$$\text{Calories burned} = \frac{665\ \text{Cal}}{\text{hr}} \times 2.00\ \text{hr} = 1.33 \times 10^3\ \text{Cal}$$

Think About It

This problem could be solved without touching a calculator! Because it takes the marathoner 2.00 hr to complete the race, the Calories she burns is simply twice the number of Calories she burns in one hour.

1.63. Collect and Organize

This problem asks for a simple conversion of length from meters to miles.

Analyze

The conversions that we need include meters to kilometers and kilometers to miles.

$$\frac{1 \text{ km}}{1000 \text{ m}}, \frac{0.6214 \text{ mi}}{1 \text{ km}}$$

Solve

$$4.0 \times 10^3 \text{ m} \times \frac{1 \text{ km}}{1000 \text{ m}} \times \frac{0.6214 \text{ mi}}{1 \text{ km}} = 2.5 \text{ mi}$$

Think About It

The answer is reasonable because 4000 meters would be a little over 2 miles when estimated. It is surprising, though, for a natural piece of silk to be that long.

1.65. Collect and Organize

In this problem we need to use the density of magnesium to find the mass of a specific size block of the metal.

Analyze

Density is defined as the mass of a substance per unit volume. The density of magnesium is given in Appendix 3 as 1.738 g/cm^3. We have to find the volume of the block of magnesium by multiplying the length by the height by the depth (the value will be in cm^3).

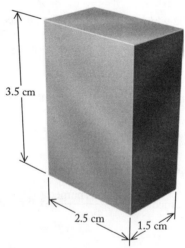

We can then find mass using the following formula:

$$\text{Mass (g)} = \text{density (g/cm}^3) \times \text{volume (cm}^3)$$

Solve

The volume of the block of magnesium is

$$2.5 \text{ cm} \times 3.5 \text{ cm} \times 1.5 \text{ cm} = 13 \text{ cm}^3$$

Therefore, the mass of the block is

$$13 \text{ cm}^3 \times \frac{1.738 \text{ g}}{1 \text{ cm}^3} = 23 \text{ g}$$

Think About It

The mass of a sample depends on how much there is of a substance. In this case, we have about 23 grams. As a quick estimate, a block of magnesium of about 10 cm³ would weigh 10 times that of 1 cm³, or about 17 grams. Because we have more than 10 cm³ of this sample and the density is a little greater than 1.7 g/cm³, our answer of 23 grams is reasonable.

1.67. Collect and Organize

To answer this question we need to compute the mass of a copper sample that is 125 cm³ in volume using the density of copper. Next, we use that mass to find out what volume (in cm³) that mass of gold would occupy.

Analyze

We need the density both of copper and of gold from Appendix 3 to make the conversions from volume to mass (for copper) and then from mass to volume (for gold). These densities are 8.96 g/mL for copper and 19.3 g/mL for gold. One milliliter is equivalent to 1 cm³, so the densities are 8.96 g/cm³ and 19.3 g/cm³, respectively. The density formulas that we need are

$$\text{Mass of copper} = \text{density of copper} \times \text{volume}$$

and

$$\text{Volume of gold} = \frac{\text{mass}}{\text{density of gold}}$$

Solve

$$\text{Mass of copper} = 8.96 \text{ g/cm}^3 \times 125 \text{ cm}^3 = 1120 \text{ g}$$

$$\text{Volume of gold} = \frac{1120 \text{ g}}{19.3 \text{ g/cm}^3} = 58.0 \text{ cm}^3$$

Think About It

Because gold is more than twice as dense as copper, we would expect the volume of a gold sample to have about half the volume of that of the same mass of copper.

1.69. Collect and Organize

Using the density of mercury, we can find the volume of 1.00 kg of mercury.

Analyze

The density of mercury is given in Appendix 3 as 13.546 g /mL. Because this property is expressed in grams per milliliter, not kilograms per milliliter, we have to convert kilograms into grams using the conversion

$$\frac{1000 \text{ g}}{1 \text{ kg}}$$

Once we have the mass in grams, we can use the rearranged formula for density to find volume

$$\text{Volume of mercury (mL)} = \frac{\text{mass of mercury (g)}}{\text{density of mercury (g/mL)}}$$

Solve

$$1.00 \text{ kg } \times \frac{1000 \text{ g}}{1 \text{ kg}} = 1.00 \times 10^3 \text{ g}$$

$$\text{Volume of mercury} = \frac{1.00 \times 10^3 \text{ g}}{13.546 \text{ g/mL}} = 73.8 \text{ mL}$$

Think About It

This is a fairly small amount that weighs one kilogram. This is due to the relatively high density of mercury.

1.71. Collect and Organize

To determine whether the HDPE will float on water we need to compare the density of the HDPE with that of water. If its density is less than water, then HDPE will float.

Analyze

To compare the densities we need to have the densities of the two substances (water and HDPE) in the same units. We can approach this in either of two ways—convert the seawater density to kg/m^3 or convert the HDPE density to g/cm^3. Let's do the latter using the following conversions:

$$\frac{100 \text{ cm}}{1 \text{ m}} \quad \text{and} \quad \frac{1000 \text{ g}}{1 \text{ kg}}$$

To calculate the density of the HDPE sample, we must divide the mass of the cube of HDPE in grams by the volume in cm^3.

Solve

$$\text{Volume of the HDPE cube} = \left(1.20 \times 10^{-3} \text{ m}\right)^3 \times \left(\frac{100 \text{ cm}}{1 \text{ m}}\right)^3 = 1.728 \text{ cm}^3$$

$$\text{Mass of the HDPE cube in grams} = 1.70 \times 10^{-3} \text{ kg} \times \frac{1000 \text{ g}}{1 \text{ kg}} = 1.70 \text{ g}$$

$$\text{Density of the HDPE cube} = \frac{1.70 \text{ g}}{1.728 \text{ cm}^3} = 0.984 \text{ g/cm}^3$$

This density is less than the density of the seawater (1.03 g/cm^3), so the cube of HDPE will float on water.

Think About It

Certainly boats are made out of materials (like iron) that are more dense than water. These boats float because the mass of the water they displace is greater than the mass of the boat.

1.73. Collect and Organize

In this problem we use the mass of a carat (the unit of weight for diamonds) to find the mass of a large diamond and then use the density to calculate the volume of that large diamond.

Analyze

We need the fact that 1 carat = 0.200 g and that the density is defined as mass per volume. To find the volume of the diamond we can rearrange the density equation to read

$$\text{Volume} = \frac{\text{mass}}{\text{density}}$$

Solve

The mass of the 5.0 carat diamond is

$$5.0 \text{ carat } \times \frac{0.200 \text{ g}}{1 \text{ carat}} = 1.0 \text{ g}$$

The volume of the diamond is then

$$\frac{1.0 \text{ g}}{3.51 \text{ g/cm}^3} = 0.28 \text{ cm}^3$$

Think About It

For this relatively large diamond in terms of carats, the mass is fairly small (one gram is about a fifth of the mass of a nickel); in this case, even though the density is relatively low, the volume is also quite small.

1.75. **Collect and Organize**

Given the data from three different manufacturers of circuit boards for copper line widths, we can determine which manufacturers were precise and which were accurate.

Analyze

By first calculating the range of values for the width of the copper lines from each manufacturer, we can see which manufacturer has the highest precision. By comparing measured line widths with the specified width of 0.500 μm, we can assess each manufacturer's accuracy.

Solve

(a) Manufacturer 1 has a range of 0.516 − 0.504 = 0.012 μm; Manufacturer 2 has a range of 0.514 − 0.512 = 0.002 μm; and Manufacturer 3 has a range of 0.502 − 0.500 = 0.002 μm.
(b) Yes, Manufacturers 2 and 3 can claim "high precision." They print the circuits with very little variability in the width of the copper lines.
(c) In the case of Manufacturer 2, the lines are printed at widths wider than the ones specified—a problem of accuracy. The claim of precision, without claiming accuracy, often misleads buyers.

Think About It

In electronic circuit boards, the specifications must be very strictly adhered to, and Manufacturer 3, who prints boards with the highest precision *and* accuracy, will win the contract.

1.77. **Collect and Organize**

This question asks if a temperature in Celsius would ever equal the temperature in Fahrenheit. We have to make use of the conversion equation between Celsius and Fahrenheit degrees.

Analyze

The equation converting between the temperatures is given as

$$^\circ\text{C} = \frac{5}{9}\left(^\circ\text{F} - 32\right)$$

To find the temperature at which these temperature scales meet (°C = °F in the above equation) substitute °C for °F:

$$^\circ\text{C} = \frac{5}{9}\left(^\circ\text{C} - 32\right)$$

Solve

Rearranging this equation and solving for °C gives

$$°C = \frac{5}{9}\left(°C - 32\right)$$

$$\frac{9}{5}\left(°C\right) = \left(°C - 32\right)$$

$$\frac{9}{5}\left(°C\right) - °C = -32$$

$$\frac{4}{5}\left(°C\right) = -32$$

$$°C = -40°C = -40°F$$

Think About It

Because the intervals between degrees on the Celsius scale are larger than the degrees on the Fahrenheit scale, the two scales will eventually meet at one temperature. This solution shows that they meet at $-40°$.

1.79. Collect and Organize

We are asked in this problem to convert from Kelvin to degrees Celsius.

Analyze

The relationship between the Kelvin temperature scale and the Celsius temperature scale is given by

$$K = °C + 273.15$$

Rearranging this gives the equation to convert Kelvin temperatures to Celsius:

$$°C = K - 273.15$$

Solve

$$°C = 4.2 \text{ K} - 273.15 = -269.0°C$$

Think About It

Because 4.2 K is very cold, we would expect that the Celsius temperature would be very negative. It should not, however, be lower than –273.15°C, since that is the lowest temperature possible.

1.81. Collect and Organize

This question asks us to convert a temperature in Fahrenheit degrees to a temperature in Celsius degrees.

Analyze

The relationship between the Celsius and Fahrenheit temperature scales is given by

$$°C = \frac{5}{9}\left(°F - 32\right)$$

Solve

$$°C = \frac{5}{9}\left(102.5°F - 32\right) = 39.2°C$$

Think About It

This temperature makes sense because it is higher than the normal body temperature, 37°C.

1.83. **Collect and Organize**

This question asks us to convert the coldest temperature recorded on Earth from Fahrenheit to Celsius and Kelvin.

Analyze

Since the Celsius and Kelvin scales are similar (offset by 273.15 degrees), once we convert from Fahrenheit to Celsius, finding the Kelvin temperature will be straightforward. The equations we need are

$$^{\circ}C = \frac{5}{9}\left(^{\circ}F - 32\right)$$

$$K = {^{\circ}C} + 273.15$$

Solve

$$^{\circ}C = \frac{5}{9}\left(-128.6\,^{\circ}F - 32\right) = -89.2\,^{\circ}C$$

$$K = -89.2\,^{\circ}C + 273.15 = 183.9\ K$$

Think About It

This temperature is cold on any scale!

1.85. **Collect and Organize**

Given the freezing and boiling point of a radiator coolant, we are to convert these temperatures from the Celsius scale to the Fahrenheit scale.

Analyze

The relationship between Celsius and Fahrenheit temperature scales is given by

$$^{\circ}C = \frac{5}{9}\left(^{\circ}F - 32\right)$$

This will have to be rearranged to find °F from °C.

$$^{\circ}F = \frac{9}{5}\left(^{\circ}C\right) + 32$$

Solve

The freezing point of this coolant in degrees Fahrenheit is

$$^{\circ}F = \frac{9}{5}\left(-39.0\,^{\circ}C\right) + 32 = -38\,^{\circ}F$$

The boiling point of this coolant in degrees Fahrenheit is

$$^{\circ}F = \frac{9}{5}\left(110\,^{\circ}C\right) + 32 = 230\,^{\circ}F$$

Think About It

In computing the freezing point of this coolant in degrees Fahrenheit, notice that the result is nearly the same as the freezing point in degrees Celsius. This is because the two temperature scales do share a temperature value (see P1.80).

1.87. Collect and Organize

We are asked to compare the critical temperature (T_c) of three superconductors. The critical temperatures, however, are given in three different temperature scales, so for the comparison, we will need to convert them to a single scale.

Analyze

It does not matter which temperature scale we use as the common one, but since the critical temperatures are low, it might be easiest to express all of the temperatures in Kelvin. The equations we will need are

$$K = {}^\circ C + 273.15 \quad \text{and} \quad {}^\circ C = \frac{5}{9}\left({}^\circ F - 32\right)$$

Solve

The T_c for $YBa_2Cu_3O_7$ is already expressed in Kelvin, $T_c = 93.0$ K.
The T_c of Nb_3Ge is expressed in $^\circ C$ and can be converted to K by

$$K = -250.0\,{}^\circ C + 273.15 = 23.2 \text{ K}$$

The T_c of $HgBa_2CaCu_2O_6$ is expressed in Fahrenheit degrees. To get this temperature in Kelvin, first convert to Celsius degrees, then to Kelvin:

$$^\circ C = \frac{5}{9}\left(-231.1\,{}^\circ F - 32\right) = -146.2\,{}^\circ C$$

$$K = -146.2\,{}^\circ C + 273.15 = 127.0 \text{ K}$$

The superconductor with the highest T_c is $HgBa_2CaCu_2O_6$ with a T_c of 127.0 K.

Think About It

The superconductor with the lowest T_c is Nb_3Ge with a T_c of 23.2 K, more than 100 K lower than the T_c of $HgBa_2CaCu_2O_6$.

1.89. Collect and Organize

This question considers the runoff of nitrogen every year into a stream resulting from a farmer's application of fertilizer. We must consider that not all of the fertilizer contains nitrogen and not all of the fertilizer runs off into the stream. We must also account for the flow of the stream in taking up the nitrogen runoff.

Analyze

First, we have to determine the amount of nitrogen that is in the fertilizer (10% of 1500 kg). Then, we need to find how much of that nitrogen gets washed into the stream (15% of the mass of N in the fertilizer). Our final answer must be in milligrams of N, so we can convert the mass of N that gets washed into the stream from kilograms to milligrams.

$$\text{Mass of fertilizer in kg} \times 0.10 = \text{mass of N in fertilizer in kg}$$

$$\text{Mass of N in fertilizer in kg} \times 0.15 = \text{mass of N washed into the stream in kg}$$

$$\text{Mass of N that washes into the stream in kg} \times \frac{1000 \text{ g}}{1 \text{ kg}} \times \frac{1000 \text{ mg}}{1 \text{ g}} = \text{mass of N that washes into the stream in mg}$$

Next, we need to know how much water flows through the farm each year via the stream. To find this, we must convert the rate of flow in cubic meters per minute to liters per year. We can convert this using dimensional analysis with the following conversions:

$$\frac{1000 \text{ L}}{1 \text{ m}^3}, \quad \frac{1 \text{ hr}}{60 \text{ min}}, \quad \frac{1 \text{ d}}{24 \text{ hr}}, \quad \frac{1 \text{ yr}}{365.25 \text{ d}}$$

Solve

The amount of N washed into the stream each year is

$$1500 \text{ kg} \times 0.10 = 150 \text{ kg N in the fertilizer}$$

$$150 \text{ kg} \times 0.15 = 22.5 \text{ kg N washed into the stream in one year}$$

$$22.5 \text{ kg} \times \frac{1000 \text{ g}}{1 \text{ kg}} \times \frac{1000 \text{ mg}}{1 \text{ g}} = 2.25 \times 10^7 \text{ mg of N washed into the stream in one year}$$

The amount of stream water flowing through the field each year is

$$\frac{1.4 \text{ m}^3}{1 \text{ min}} \times \frac{1000 \text{ L}}{1 \text{ m}^3} \times \frac{60 \text{ min}}{1 \text{ hr}} \times \frac{24 \text{ hr}}{1 \text{ d}} \times \frac{365.25 \text{ d}}{1 \text{ yr}} = 7.36 \times 10^8 \text{ L/yr}$$

The additional concentration of N that is added to the stream by the fertilizer is

$$\frac{2.25 \times 10^7 \text{ mg N/yr}}{7.36 \times 10^8 \text{ L/yr}} = 0.031 \text{ mg/L}$$

Think About It

The calculated amount of nitrogen that is added to the stream seems reasonable. The concentration is relatively low because the stream is moving fairly swiftly and the total amount of nitrogen that washes into the stream over the course of the year is not too great. This analysis, however, does not tell us if this amount would cause harm to the plant and animal life in the stream.

1.91. **Collect and Organize**

In this problem we need to express each mixture of chlorine and sodium as a ratio. The mixture that is closest to the ratio for chlorine to sodium will be the one with the desired product, leaving neither sodium nor chlorine left over.

Analyze

First, we must calculate the ratio of chlorine to sodium in sodium chloride. This is a simple ratio of the masses of these two substances:

$$\frac{\text{mass of chlorine}}{\text{mass of sodium}} = \text{ratio of the two components}$$

We can compare the ratios of the other mixtures by making the same calculations.

Solve

In sodium chloride the mass ratio of chlorine to sodium is

$$\frac{1.54 \text{ g of chlorine}}{1.00 \text{ g of sodium}} = 1.54$$

Repeating this calculation for the four mixtures, we obtain each ratio of chlorine to sodium:

$$\frac{17.0 \text{ g}}{11.0 \text{ g}} = 1.55 \text{ for mixture a} \qquad \frac{12.0 \text{ g}}{6.5 \text{ g}} = 1.8 \text{ for mixture c}$$

$$\frac{10.0 \text{ g}}{6.5 \text{ g}} = 1.5 \text{ for mixture b} \qquad \frac{8.0 \text{ g}}{6.5 \text{ g}} = 1.2 \text{ for mixture d}$$

Both mixtures a and b react so that there is neither sodium nor chlorine left over.

Think About It

Mixture c has leftover chlorine and mixture d has leftover sodium after the reaction is complete.

1.93. **Collect and Organize**

Given the amounts of recognition molecule, capture molecule, and detector molecule along with the amounts of each needed for one HIV assay plate, we are to determine whether we have sufficient amounts for 96 assay plates.

Analyze

To calculate the amount of each molecule required for the 96 assay plates, we first multiply the required amounts of each molecule by 96. We then subtract this amount from the quantity of each molecule that is available to determine if there are sufficient amounts of each molecule.

Solve

(a) Amount of recognition molecule needed:

$$0.550 \text{ mg} \times 96 = 52.8 \text{ mg}$$

This is less than the amount of recognition molecule available.

$$100.00 \text{ mg} - 52.8 \text{ mg} = 47.2 \text{ mg left over}$$

Amount of capture molecule needed:

$$1.200 \text{ mg} \times 96 = 115.2 \text{ mg}$$

This is more than the amount of recognition molecule available.

$$100.00 \text{ mg} - 115.2 \text{ mg} = -15.2 \text{ mg (deficit)}$$

Amount of detector molecule needed:

$$0.450 \text{ mg} \times 96 = 43.2 \text{ mg}$$

This is less than the amount of detector molecule available.

$$50.00 \text{ mg} - 43.2 \text{ mg} = 6.8 \text{ mg left over}$$

(b) We can make only the number of plates that use up the amount of capture molecules.

$$\frac{100.00 \text{ mg}}{1.200 \text{ mg}} = 83 \text{ plates}$$

Think About It

At most we could prepare 83 plates. Practically speaking, however, we might expect that we would be able to prepare slightly fewer than 83 plates because it is likely that some material would be lost during weighing and transfer of the capture molecule sample.

1.95. **Collect and Organize**

To make a complete bicycle we need all the parts. Here, we have to look for the maximum number of bicycles that can be built from the number of parts available.

Analyze

For each part, calculate the number of bicycles that can be built. This depends on the number of parts available *and* the number of those parts needed for each bicycle. For example, each bike needs two pedals so 112 pedals can build 56 bicycles. The part that is available in the lowest amount limits the number of bicycles that can be built. All the other parts will be left over.

Solve

For each bicycle part, the bikes that can be built are as follows:

111 frames will build 111 bicycles
81 front wheels will build 81 bicycles
95 rear wheels will build 95 bicycles

112 pedals will build 56 bicycles
47 sets of handlebars will build 47 bicycles
38 bike chains will build 38 bicycles
17 front brakes will build 17 bicycles
35 rear brakes will build 35 bicycles

We can build only as many bicycles as there are complete sets of parts. The part that is limiting the number of bicycles built in this problem is the front brakes. Therefore, only 17 complete bicycles can be built.

Think About It
After building 17 bicycles, we would have 94 frames, 64 front wheels, 78 rear wheels, 78 pedals, 30 handlebars, 21 bike chains, and 18 rear brakes left over.

1.97. **Collect and Organize**
This problem asks us to compute the percentages of the two ingredients in trail mix as manufactured on different days.

Analyze
Because we compare each day's percentage of peanuts in the trail mix bags to the ideal range of 65–69%, each day's percentage of peanuts has to be computed from the data given.
Each day has a total of 82 (peanuts plus raisins), so the percentage of the mix in peanuts for each day is calculated by the equation

$$\% \text{ peanuts} = \frac{\text{number of peanuts in mix}}{82} \times 100$$

Solve
For each day, the percent peanuts is

$$\text{Day 1: } \frac{50}{82} \times 100 = 61\% \text{ peanuts}$$

$$\text{Day 11: } \frac{56}{82} \times 100 = 68\% \text{ peanuts}$$

$$\text{Day 21: } \frac{48}{82} \times 100 = 59\% \text{ peanuts}$$

$$\text{Day 31: } \frac{52}{82} \times 100 = 63\% \text{ peanuts}$$

The only day that met the specifications for the percentage of peanuts in the trail mix was Day 11.

Think About It
On Days 1, 21, and 31 too few peanuts were in the trail mix.

1.99. **Collect and Organize**
We are told that the same force is used to stretch two springs. The stronger spring will stretch less than the weaker spring, so we are looking for which spring will experience a smaller percentage increase in length.

Analyze
We can find the percentage length increase in spring A by considering the length when stretched ($l_{\text{stretched}}$) and the natural length (l_{natural}). We can determine the percentage increase in length using

$$\% \text{ length increase} = \frac{l_{\text{stretched}} - l_{\text{natural}}}{l_{\text{natural}}} \times 100\%$$

Solve

$$\% \text{ length increase} = \frac{5.4\text{ cm} - 4.0\text{ cm}}{4.0\text{ cm}} \times 100\% = 35\%$$

Spring A's length increases by 35%, while spring B's length only increases by 15%. Spring B is stronger.

Think About It

If we assume that spring B was also 4.0 cm in its natural state, a 15% increase in length corresponds to a final length of 4.6 cm. This is much shorter than spring A when stretched, supporting our answer above.

CHAPTER 2 | Atoms, Ions, and Molecules: Matter Starts Here

2.1. Collect and Organize
This question asks us to look at the connectivity of the atoms of nitrogen (blue spheres) and oxygen (red spheres) to decide which species are represented in the figure.

Analyze
Figure P2.1 shows seven molecules of red and blue spheres. In some cases the spheres are in a group of three; in others they are in a group of two. In all cases, nitrogen and oxygen are present in the molecules; there are no all-red or all-blue molecules. Therefore, the answer will be some mixture of different nitrogen–oxygen species.

Solve
Looking specifically at the molecules made up of three atoms, we see that each contains two oxygen atoms and one nitrogen atom; this must be NO_2. For the two-atom molecules depicted, each is composed of one nitrogen atom and one oxygen atom; this must be written as NO. Therefore, the answer is (c) a mixture of NO_2 and NO.

Think About It
Even though there are 11 red spheres depicted with 7 blue spheres, the answer cannot be (b) N_7O_{11} because that formula implies that all 18 atoms of nitrogen and oxygen are bonded together in one molecule. Answer (a), N_2O_3, does not have the nitrogen-to-oxygen ratio correct, and it indicates that only one type of molecule, composed of two nitrogen and three oxygen atoms, is shown in the figure, whereas two different molecules are actually depicted. Finally, (d)—a mixture of N_2 and O_3—cannot be correct because there are no blue–blue (N_2) or red–red–red (O_3) molecules depicted; all molecules shown contain both nitrogen and oxygen.

2.3. Collect and Organize
This question asks us to determine which of the given diagrams is the best representation for $^{13}_{6}C^+$.

Analyze
Before identifying the correct depiction, we must interpret the symbol associated with the given nuclide.

Solve
The element may be identified as carbon by noting the symbol (C), as well as the atomic number ($Z = 6$). The atomic number also tells us that carbon contains 6 protons. This eliminates (a), which contains 7 protons, and (c), which contains only 5 protons. The mass number ($A = 13$) for this nuclide tells us that there are 7 neutrons ($A - Z = 13 - 6 = 7$ neutrons). Both (b) and (d) contain 7 neutrons, so we cannot eliminate either based on the number of neutrons depicted. Since there are 6 protons, and a charge of +1 for the ion, the ion must contain 5 electrons, $(+6) + (-5) = +1$, so (b) is the best representation of this ion.

Think About It
It will be helpful to consider atomic- or molecular-scale diagrams such as these as you read this textbook.

2.5. Collect and Organize
For this question we need to correlate properties of the elements with their positions in the periodic table. We need to access the definitions of *reactive* and *inert* and know the general regions of the periodic table that have gaseous, metallic, and nonmetallic elements.

Analyze
Reactive means that the species readily combines with other elements to form compounds, and *inert* means that the chemical species does not combine with other species. On the periodic table, metallic elements tend to be on the left side; nonmetallic elements are on the right side. Under standard conditions, the gases tend to be in groups 18 (noble gases) and 17 (the halogens: fluorine and chlorine) along with oxygen, nitrogen, and hydrogen. Of these gases, the noble gases are monatomic, but the others are all diatomic (N_2, O_2, H_2, F_2, Cl_2).

Solve

In the periodic table shown in Figure P2.5, elements Na, Ne, Cl, Au, and Lr are highlighted.
(a) Chlorine (Cl_2) is a reactive, diatomic gas at room temperature (yellow). It is a nonmetal.
(b) Neon (Ne) is a chemically inert gas (red).
(c) Sodium (Na) is a reactive metal (dark blue).

Think About It

Gold and lawrencium are also metals. However, gold is not very reactive (it can be found as an element in nature), and lawrencium is radioactive and synthesized in small amounts, meaning that its chemistry is relatively unexplored.

2.7. Collect and Organize

In this question we have to consider which elements combine with oxygen in a certain ratio to give a neutral compound (i.e., a compound with no overall charge). We must first determine the charge on the unknown element in each molecule. We are then asked to find the element highlighted in the periodic table in Figure P2.7 that corresponds to that charge.

Analyze

Oxygen has a 2– charge when combined with elements to form compounds. To balance out the charge, an element in a 1:1 ratio with O would have to carry a 2+ charge to form a neutral compound. Likewise, an element with in 2:1 ratio with O would have to carry a 1+ charge. The other ratios can be found similarly. Once the charges of the unknown elements are determined, we can use the information in Figure 2.17 to identify the element for each oxygen compound.

Solve

The highlighted elements in the periodic table in Figure P2.7 are K, Mg, Ti, and Al.
    (a) The ratio of X to O in XO is 1:1; therefore the charge on X is 2+. Elements in group 2 typically have this charge in compounds. Mg (green) will form MgO.
    (b) The ratio of X to O in X_2O is 2:1; therefore the charge on X is 1+. Elements in group 1 typically have this charge in compounds. K (red) will form K_2O.
    (c) The ratio of X to O in XO_2 is 1:2; therefore the charge on X is 4+. Elements in group 4 typically have this charge in compounds. Ti (yellow) will form TiO_2.
    (d) The ratio of X to O in X_2O_3 is 2:3; therefore the charge on X is 3+ (three O ions have a total charge of 6–, so two cations of 3+ charge will form a neutral compound). Elements in group 13 can have this charge in compounds. Al (dark blue) will form Al_2O_3.

Think About It

Elements from across the periodic table, from the alkali metals to the main group elements, combine with oxygen to form compounds.

2.9. Collect and Organize

In this question we are asked to explain how Rutherford's gold-foil experiment changed the plum-pudding model of the atom.

Analyze

The plum-pudding model of the atom viewed the electrons as small particles in a diffuse, positively charged "pudding." In Rutherford's experiment, most of the α particles (positively charged particles) directed at the gold foil went straight through, but a few of them bounced back toward the source of the α particles.

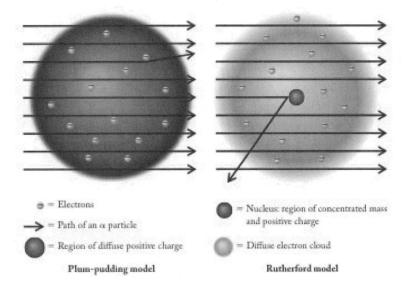

= Electrons

= Path of an α particle

= Region of diffuse positive charge

= Nucleus: region of concentrated mass and positive charge

= Diffuse electron cloud

Plum-pudding model   **Rutherford model**

Solve

From his experiments, Rutherford concluded that the positive charge in the atom could not be spread out (the pudding) in the atom, but must result from a concentration of charge in the center of the atom (the nucleus). Most of the α particles were deflected only slightly or passed directly through the gold foil; so he reasoned that the nucleus must be small compared to the size of the entire atom. The negatively charged electrons do not deflect the α particles, and Rutherford reasoned that the electrons took up the remainder of the space of the atom outside the nucleus.

Think About It

The nucleus is about 10^{-15} m in diameter, whereas the atom is about 10^{-10} m. This size difference has often been compared to "a fly in a cathedral."

2.11. Collect and Organize

In this question we are to explain how J. J. Thomson discovered that cathode rays were not pure energy, but actually particles.

Analyze

Thomson's experiment directed the cathode ray through a magnetic field, and he discovered that the ray was deflected.

Solve

When Thomson observed cathode rays being deflected by a magnetic field, he reasoned that the rays were streams of charged particles because only moving charged particles would interact with a magnetic field. Pure energy rays would not.

Think About It

Thomson's discovery of the electron in cathode rays did not completely eliminate the use of the term *cathode ray*. CRTs (cathode-ray tubes) are the traditional (not LCD) television and computer screens.

2.13. Collect and Organize

We need to define *weighted average* for this question.

Analyze

An average is a number that expresses the middle of the data (in this case of various masses of atoms or isotopes).

Solve

A *weighted average* takes into account the proportion of each value in the group of values to be averaged. For example, the average of 2 and 5 is $(2 + 5)/2 = 3.5$, while the weighted average of 2, 2, 2, and 5 would be computed as $(2 + 2 + 2 + 5)/4 = 2.75$. This average shows the heavier weighting toward the 2 values.

Think About It

Because isotopes for any element are not equally present but have a range of natural abundances, all the masses of the elements in the periodic table are calculated weighted averages.

2.15. Collect and Organize

This question asks us to consider the ratio of neutrons to protons in an element where we are given the fact that the mass number is more than twice the atomic number.

Analyze

We can find the number of neutrons for an isotope by relating the number of protons to the mass number. From that result we can then determine the neutron-to-proton ratio.

Solve

We are given an isotope in which the mass number is more than twice the number of protons. With m being the mass number and p the number of protons, we can express this relationship as

$$m > 2p$$

The mass number is also equal to the number of protons plus the number of neutrons (n),

$$m = p + n$$

Combining these expressions

$$p + n > 2p$$

and solving for n gives

$$n > 2p - p$$
$$n > p$$

Therefore, the number of neutrons in this isotope is greater than the number of protons and the neutron-to-proton ratio is greater than 1.

Think About It

We would not have had to express the relationships between the nuclear particles mathematically if the isotope had a mass number equal to twice the number of protons. In that case, the number of neutrons would have to be exactly the same as the number of protons, giving a neutron-to-proton ratio of 1 to 1.

2.17. Collect and Organize

We have to consider the concept of weighted average atomic mass to answer this question.

Analyze

We are asked to compare two isotopes and their weighted average mass. If the lighter isotope is more abundant, then the average atomic mass will be less than the average if both isotopes are equally abundant. If the heavier isotope is more abundant, the average atomic mass will be greater than the simple average of the two isotopes. We are given the mass number for the isotopes as part of the isotope symbol, and we will take that as the mass of that isotope in atomic mass units.

Solve

(a) The simple average atomic mass for ^{10}B and ^{11}B would be 10.5 amu. The actual weighted average mass (10.81 amu) is greater than this; ^{11}B is more abundant.
(b) The simple average atomic mass for ^{6}Li and ^{7}Li would be 6.5 amu. The actual weighted average mass (6.941 amu) is greater than this; ^{7}Li is more abundant.

(c) The simple average atomic mass for ^{14}N and ^{15}N would be 14.5 amu. The actual weighted average mass (14.01 amu) is less than this; ^{14}N is more abundant.
(d) The simple average atomic mass for ^{20}Ne and ^{22}Ne would be 21 amu. The actual weighted average mass (20.18 amu) is less than this; ^{20}Ne is more abundant.
Therefore, for both (a) and (b) the heavier isotope is more abundant.

Think About It
This is a quick question to answer for elements like boron, lithium, nitrogen, and neon that have the dominance of only two isotopes in terms of their abundance. It is a little harder to answer the same question for elements with more than two stable isotopes in relatively high abundances.

2.19. **Collect and Organize**
In this question we are provided with the abundances of the two naturally occurring isotopes of chlorine. From this information and the isotope masses from Appendix 3, we can calculate the weighted average atomic mass of chlorine.

Analyze
To calculate the weighted average atomic mass, we have to consider the relative abundances according to the following formula:

$$m_x = a_1m_1 + a_2m_2 + a_3m_3 + \ldots$$

where a_n refers to the abundance of isotope n and m_n refers to the mass of isotope n. If the relative abundances are given as percentages, the value we use for a_n in the formula is the percentage divided by 100.

Solve
For the average atomic mass of chlorine

$$m_{Cl} = (0.7578 \times 34.968852 \text{ amu}) + (0.2422 \times 36.965903 \text{ amu}) = 35.45 \text{ amu}$$

Think About It
Because chlorine-35 is more abundant than chlorine-37, we expect that the average atomic mass for chlorine would be below the simple average of 36 amu.

2.21. **Collect and Organize**
Here we are asked to find out if the weighted average atomic mass of magnesium on Mars is the same as here on Earth. For this we have to work backward from the exact masses and abundances of MgO measured by the *Sojourner* robot to the weighted average mass of magnesium.

Analyze
To find the mass of magnesium in each isotope of MgO, we need to subtract the mass of oxygen-16, given as 15.9949 amu, from the exact masses of the MgO isotopes. Because the oxygen is assumed not to be present in any other isotope, the percent abundances given must be the same as the percent abundances for the magnesium isotopes for the Mars rocks. From this we can calculate the weighted average from the masses of the Mg isotopes and their corresponding abundances.

Solve
The mass of Mg in each of the isotopes is the exact mass given minus the exact mass for oxygen-16. The weighted mass is the sum of the exact masses of Mg isotopes multiplied by their abundances (percentage divided by 100). The results of these calculations follow:

Exact Mass of MgO (amu)	Mass of Mg in MgO Isotope (amu)	Abundance	Weighted Mass of Mg Isotope (amu)
39.9872	23.9893	0.7870	18.88
40.9886	24.9937	0.1013	2.532
41.9846	25.9897	0.1117	2.903

Average atomic mass of Mg in Mars MgO samples = 24.32 amu

The average mass of Mg on Mars is about the same as here on Earth.

Think About It
There would be no reason why the mass of Mg on Mars should not be close to the same value as on Earth; the magnesium on both planets arrived in the solar system via the same ancient stardust.

2.23. Collect and Organize
In this problem, we again use the concept of weighted average atomic mass, but in this case we are asked to work backward from the average mass to find the exact mass of the ^{48}Ti isotope.

Analyze
We can use the formula for finding the weighted average atomic mass, but this time our unknown quantity is one of the isotope masses. In this case,

$$m_{Ti} = a_{46_{Ti}} m_{46_{Ti}} + a_{47_{Ti}} m_{47_{Ti}} + a_{48_{Ti}} m_{48_{Ti}} + a_{49_{Ti}} m_{49_{Ti}} + a_{50_{Ti}} m_{50_{Ti}}$$

Solve

$$47.87 \text{ amu} = (0.0825 \times 45.95263 \text{ amu}) + (0.0744 \times 46.9518 \text{ amu}) + (0.7372 \times m_{48_{Ti}})$$
$$+ (0.0541 \times 48.94787 \text{ amu}) + (0.0518 \times 49.9448 \text{ amu})$$

$$m_{48_{Ti}} = 47.95 \text{ amu}$$

Think About It
This answer makes sense since the exact mass of ^{48}Ti should be close to 48 amu.

2.25. Collect and Organize
We are asked why Mendeleev arranged his periodic table based on atomic masses instead of atomic numbers.

Analyze
Mendeleev announced his periodic table in 1869. At this point, atoms were considered to be indivisible and electrons, neutrons, and (especially) protons (which distinguish one element from another today,) were not yet discovered.

Solve
Mendeleev knew only the masses of the elements when he arranged the elements into his periodic table. At the time, there was no idea that an element distinguished itself from another element by the number of protons in its nucleus.

Think About It
Mendeleev, however, also looked for patterns in chemical reactivity and physical properties to arrange the known elements in his periodic table. Remarkably, he left open spots in his periodic table for yet undiscovered elements.

2.27. Collect and Organize
We are asked why some of the elements in Mendeleev's chart are out of order when compared to the modern periodic table.

Analyze

Mendeleev organized his periodic table using the average atomic masses for the known elements. The modern version of the periodic table organizes the elements based on the atomic number (Z).

Solve

When Mendeleev put together his periodic table, he was unaware of the presence of subatomic particles. Because of this, he did not know that the atomic masses he used to sort the elements were a weighted average of the naturally occurring isotopes of a given element. Elements with a large natural abundance of heavier isotopes exhibit a higher average atomic mass than might be predicted by atomic number alone, hence creating a different order.

Think About It

Because the identity of an element changes with the number of protons in the nucleus, it is a reliable and consistent way to organize the elements in the modern periodic table.

2.29. Collect and Organize

For each element in this question, we must look at the relationship of the neutrons, protons, and electrons. We need to determine what the element's atomic number is from the periodic table and, from the mass number given for the isotope, compute the number of neutrons in that isotope.

Analyze

An isotope is given by the symbol $_Z^A$X, where X is the element symbol from the periodic table, Z is the atomic number (the number of protons in the nucleus), and A is the mass number (the number of protons and neutrons in the nucleus). Often, Z is omitted because the element symbol gives us the same information about the identity of the element. To determine the number of neutrons in the nucleus for each of the named isotopes, we subtract Z (number of protons) from A (mass number). If the elements are neutral (no charge), the number of electrons equals the number of protons in the nucleus.

Solve

	Atom	Mass Number	Number of Protons = Atomic Number	Number of Neutrons = Mass Number – Atomic Number	Number of Electrons = Number of Protons
(a)	^{14}C	14	6	8	6
(b)	^{59}Fe	59	26	33	26
(c)	^{90}Sr	90	38	52	38
(d)	^{210}Pb	210	82	128	82

Think About It

Isotopes of an element contain the same number of protons but a different number of neutrons. Thus, isotopes have different masses.

2.31. Collect and Organize

To fill in the table, we have to consider how the numbers of nuclear particles relate to each other. We also need to recall how the symbols for the isotopes are written. In looking at the table, it is apparent that in some cases we have to work backward from the number of electrons or protons and mass number for the element symbol.

Analyze

An isotope is given by the symbol $_Z^A$X, where X is the element symbol from the periodic table, Z is the atomic number (the number of protons in the nucleus), and A is the mass number (the number of protons and neutrons in the nucleus). We can determine the number of neutrons in the nucleus for the isotopes by subtracting Z (number of protons) from A (mass number). If the elements are neutral (no charge), the number of electrons equals the number of protons in the nucleus.

Solve

Symbol	^{23}Na	^{89}Y	^{118}Sn	^{197}Au
Number of Protons	11	39	50	79
Number of Neutrons	12	50	68	118
Number of Electrons	11	39	50	79
Mass Number	23	89	118	197

Think About It

Because the nuclear particles are all related to each other, we can work from either the isotope symbol to find the number of protons, neutrons, and electrons for a particular isotope, or we can work from the mass number and the number of electrons or protons to determine the number of neutrons and write the element symbol.

2.33. **Collect and Organize**

An isotope is given by the symbol $^A_Z X^n$, where X is the element symbol from the periodic table, Z is the atomic number (the number of protons in the nucleus), A is the mass number (the number of protons and neutrons in the nucleus), and n is the charge on the species.

Analyze

If we are given the number of protons in the nucleus, the element can be identified from the periodic table. The mass number can be determined by adding the protons to the neutrons in the nucleus for the isotope. We can determine the number of neutrons or protons in the nucleus for the isotopes by subtracting Z (number of protons) or the number of neutrons from A (mass number), respectively. We can account for the charge on the species by adding electrons (to form a negatively charged ion) or by subtracting electrons (to form a positively charged ion).

Solve

Symbol	^{37}Cl$^-$	^{23}Na$^+$	^{81}Br$^-$	^{226}Ra^{2+}
Number of Protons	17	11	35	88
Number of Neutrons	20	12	46	138
Number of Electrons	18	10	36	86
Mass Number	37	23	81	226

Think About It

To form a singly charged ion, there has to be one electron more (for a negative charge) or one electron less (for a positive charge) compared to the number of protons in the nucleus. For a doubly charged ion, we add or take away two electrons.

2.35. **Collect and Organize**

We need to refer to the periodic table to determine the group that typically forms cations with charges of 2+.

Analyze

According to the periodic table, the elements of group 2 often combine with nonmetals to form salts in which the cation has a charge of 2+. To answer the question we need only to match up the elements that are in group 2.

Solve

Only (c) Be is likely to form Be^{2+}.

Think About It

The only other element that is likely to form a cation in the question is (d) Al because it is a metal. Both (a) S and (b) P are nonmetals and are more likely to form anions.

2.37. **Collect and Organize**

We need to refer to the periodic table to determine the atomic number of the element or anion and then count up the electrons for each species.

Analyze

According to the periodic table, neutral fluorine has 9 electrons, oxygen with 2 extra electrons (the oxide anion) has 10 electrons, sulfur with 2 extra electrons (the sulfide anion) has 18 electrons, and a neutral chlorine atom has 17 electrons.

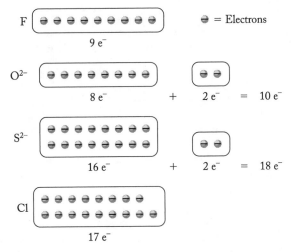

Solve

The species with the most electrons is (c) S^{2-}.

Think About It

What determines the greatest number of electrons for this series of atoms and ions is not only position in the periodic table (larger elements have more electrons), but also the charge on the species (a chlorine atom has fewer electrons than a sulfide anion).

2.39. **Collect and Organize**

We need to refer to the periodic table to determine the atomic number of the element or ion and then count up the electrons for each species. Here we are looking for species that have the same number of electrons as an argon atom.

Analyze

According to the periodic table, argon has 18 electrons. From the periodic table and by adding or subtracting electrons as necessary to make anions and cations as needed, we see that S^{2-} has 18 electrons, P^{3-} has 18 electrons, Be^{2+} has 2 electrons, and Ca^{2+} has 18 electrons.

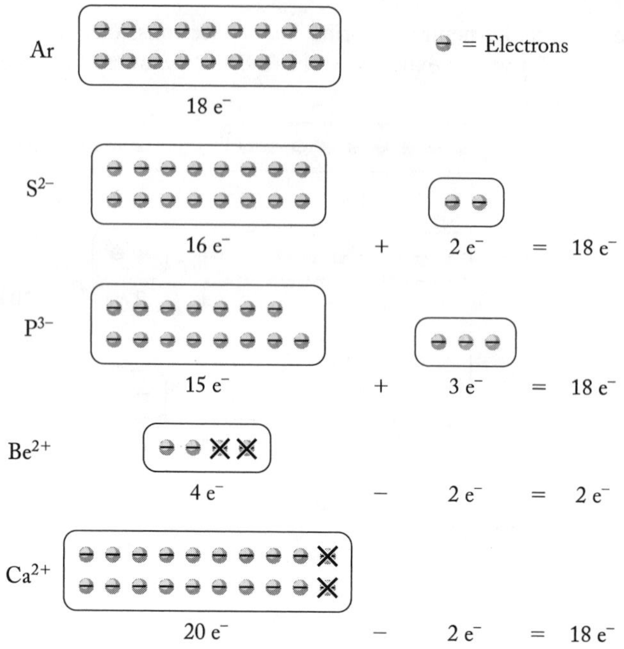

Ar 18 e⁻

⊖ = Electrons

S²⁻ 16 e⁻ + 2 e⁻ = 18 e⁻

P³⁻ 15 e⁻ + 3 e⁻ = 18 e⁻

Be²⁺ 4 e⁻ − 2 e⁻ = 2 e⁻

Ca²⁺ 20 e⁻ − 2 e⁻ = 18 e⁻

Solve
The species with the same number of electrons as Ar are (a) S^{2-}, (b) P^{3-}, and (d) Ca^{2+}.

Think About It
These species are *isoelectronic* with each other, meaning that they have the same number of electrons.

2.41. Collect and Organize
We need to refer to the periodic table to determine which element is classified as a nonmetal.

Analyze
Nonmetals are to the right side of the stair–step line on the periodic table, metals are on the left side, and metalloids (semimetals) are between the two.

Solve
Only (b) Br is clearly a nonmetal. (c) Ca and (d) Ru are clearly metals, and (a) Si is classified as a metalloid.

Think About It
There are more metallic elements in the periodic table than nonmetals.

2.43. Collect and Organize
We must refer to the periodic table and calculate which of the listed ions contains 10 electrons.

Analyze
All four of the listed ions are formed from a halogen (group 17 element) gaining one electron. We must consider the number of electrons already present in the neutral atom, as well as the additional electron gained to form the anion. For neutral atoms, the number of electrons is equal to the atomic number.

Solve
The number of electrons in each ion are
(a) $F^- = 9\,e^- + 1\,e^- = 10\,e^-$
(b) $Cl^- = 17\,e^- + 1\,e^- = 18\,e^-$

(c) $Br^- = 35\ e^- + 1\ e^- = 36\ e^-$
(d) $I^- = 53\ e^- + 1\ e^- = 54\ e^-$
The fluoride ion, F^-, is the only species in this list that contains 10 electrons.

Think About It
Remember to consider all of the electrons in the neutral atom, as well as the additional electron gained when the anion is formed.

2.45. **Collect and Organize**
We will want to classify the listed elements to determine which is a halogen.

Analyze
Halogens are elements located in group 17 of the periodic table.

Solve
Nitrogen (N_2) is in group 15; xenon (Xe) is a noble gas (group 18); hydrogen (H_2) is located in group 1. From this list, only chlorine (Cl_2) is a halogen.

Think About It
Some chemists have suggested that because it may form an ion with a 1– charge (the hydride ion, H^-), hydrogen may be classified as a halogen. However, we will focus on the more common H^+ ion, so it makes sense to group hydrogen with other elements that form 1+ ions.

2.47. **Collect and Organize**
We will want to classify the listed elements to determine which is an alkali metal.

Analyze
Alkali metals are elements located in group 1, at the far left of the modern periodic table.

Solve
Both chlorine (Cl_2) and bromine (Br_2) are a halogens (group 17); xenon (Xe) is a noble gas (group 18). From this list, only sodium (Na) is an alkali metal (group 1).

Think About It
Alkali metals form cations with a 1+ charge in 1:1 complexes with halogens, such as sodium chloride, NaCl.

2.49. **Collect and Organize**
We are to write the chemical formulas for the chloride and sulfate salts of the sodium, potassium, calcium, and magnesium cations.

Analyze
The chemical symbols and charges of the cations are: Na^+, K^+, Ca^{2+}, and Mg^{2+}. The chemical symbols and charges of the anions are: Cl^- and SO_4^{2-}. In writing the formula for the neutral salts, we must balance the cation charge with the anion charge.

Solve
NaCl and Na_2SO_4
KCl and K_2SO_4
$CaCl_2$ and $CaSO_4$
$MgCl_2$ and $MgSO_4$

Think About It
One formula unit of the salts Na_2SO_4 and K_2SO_4 contains three ions (2 cations + 1 anion). One formula unit of the salts $MgSO_4$ and $CaSO_4$ contains two ions (1 cation + 1 anion).

2.51. **Collect and Organize**
We are asked how Dalton's atomic theory applies to the decomposition of water into hydrogen and oxygen.

Analyze
Dalton's atomic theory states that the ratio of atoms of different elements in a compound is always in small whole numbers.

Solve
Dalton's atomic theory states that, because atoms are indivisible, the ratios of the atoms (elements) in a compound will occur as whole-number ratios. Thus, the ratio of hydrogen gas to oxygen gas is 2:1, a whole-number ratio, because the atoms in water are in the ratio of 2:1.

Think About It
Another way of thinking about this part of Dalton's atomic theory is that a compound cannot have a fraction of an atom in its formula.

2.53. **Collect and Organize**
We are asked to describe the types of elements that form ionic and molecular compounds.

Analyze
Molecular compounds are held together by the sharing of electrons (covalent interactions) between atoms. Ionic compounds are held together by an electrostatic attraction between cations and anions. These differences in bonding may be attributed to the types of elements that comprise each compound.

Solve
In a molecular compound, two nonmetals share electrons in a covalent fashion. In an ionic compound, electrons are transferred between a metal and a nonmetal to form discrete cations and anions.

Think About It
Both types of bonding involve the sharing of electrons, but if a metal is involved, the compound is likely to be ionic rather than molecular.

2.55. **Collect and Organize**
This question asks how the masses of cobalt and sulfur in two different compounds are related. To answer this question, we apply Dalton's law of multiple proportions.

Analyze
The law of multiple proportions states that, if more than one compound is formed from two elements, the ratios of the masses of the two elements in each compound will be whole-number ratios. In this problem we consider the mass of sulfur that would combine with a certain mass of cobalt to form CoS and Co_2S_3. We need only look at the ratio of the two elements in the compounds to answer this.

Solve
Because the ratio of Co to S in CoS is 1:1 and the ratio is 2:3 in Co_2S_3, applying Dalton's law of multiple proportions means that, if we compare the mass of sulfur needed to react with one gram of cobalt to form CoS in one reaction to the amount of sulfur needed to form Co_2S_3 in a second reaction, we would find the mass of sulfur required in the second reaction to be 1.5 times that needed for the first reaction. Adjusting that to a whole-number ratio would give us a 2:3 ratio for Co to S for the second compound.

Think About It
Notice here that the masses of the elements are compared, not the actual number of atoms in the molecules themselves; that idea came later. Dalton's link is with the indivisible atom, which means that the mass ratios also must be in whole-number ratios when making comparisons between different compounds containing the same elements.

2.57. **Collect and Organize**

Given the mass ratio of sulfur to oxygen needed to make SO_2, we are asked in this problem to use Dalton's law of multiple proportions to determine the amount of oxygen required to prepare SO_3 from a given mass of sulfur.

Analyze

The masses of elements in compounds of the same elements occur in whole-number ratios when compared to each other. If 5 g of sulfur combine with 5 g of oxygen to give SO_2 (a 1:2 ratio), we can use that mass ratio to determine the mass of oxygen to combine with 5 g of sulfur to give SO_3 (a 1:3 ratio).

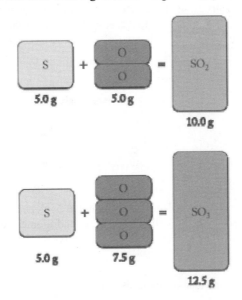

Solve

By comparing the mass ratios (1:2 versus 1:3), we see that we need half again as much of oxygen to make SO_3 compared to SO_2. Therefore, we will need $5.0 \times 1.5 = 7.5$ g of oxygen to react with 5.0 g of sulfur to make SO_3.

Think About It

This amount of oxygen makes sense because it is still less than twice the amount of oxygen (10 g), which would create a 1:4 ratio of S:O, giving a formula of SO_4.

2.59. **Collect and Organize**

We are asked to write formulas for chloride and sulfate salts in this question. We have to keep in mind the charges on the cations and anions and be sure to balance the charges to obtain neutral salts.

Analyze

The chloride ion has a charge of 1– and is written as Cl^-. The sulfate polyatomic anion has a charge of 2– and is written as SO_4^{2-}. Group 1 elements typically form cations with a 1+ charge (Na^+ and K^+) and those of group 2 form cations with a 2+ charge (Ca^{2+}, Mg^{2+}, and Sr^{2+}) in salts. In writing the formulas for the salts, we balance the charges between the anion and cation so as to obtain a neutral salt.

Solve

Combining the chloride anion with both sodium and potassium cations will give a 1:1 cation-to-anion ratio to form the neutral salts NaCl and KCl. For magnesium, calcium, and strontium ions, two chloride ions are needed in the formula to obtain neutral salts: $MgCl_2$, $CaCl_2$, and $SrCl_2$. In combination with the sulfate anion, the group 2 cations have formulas with a 1:1 cation-to-anion ratio: $MgSO_4$, $CaSO_4$, and $SrSO_4$. Sulfate salts of the group 1 cations will require a 2:1 ratio of cations to anion to form neutral salts: Na_2SO_4 and K_2SO_4.

Think About It

The dissolved salts in seawater come mainly from the dissolution of rocks and minerals through erosion.

2.61. **Collect and Organize**

We are asked to determine the chemical name for $Mg(OH)_2$. Before naming the compound, we must classify the compound as ionic or molecular.

Analyze

Magnesium is an alkaline earth metal, and OH^- is a polyatomic anion. We must follow the rules for naming a binary ionic compound.

Solve

Ionic compounds are named starting with the cation, followed by the name of the anion. The hydroxide anion, OH^-, is a polyatomic anion (see Table 2.3) and already has an *–ide* ending. The correct, unambiguous name for this compound is *magnesium hydroxide*, as it can only describe a compound of one magnesium cation and two hydroxide anions.

Think About It

Because magnesium forms a cation with a 2+ charge exclusively, it is not necessary to specify a charge, as we would for a transition metal.

2.63. **Collect and Organize**

We are asked to write the names for the listed molecular compounds.

Analyze

All of these compounds are binary molecular compounds. Molecular compounds are named by listing the elements in the order written, with the ending of the second element changed to *–ide*. Finally, prefixes are added to each name to indicate the number of atoms of each type present in the molecule.

Solve

The correct names for these molecular compounds are:
(a) hydrogen bromide
(b) carbon disulfide
(c) nitrogen trifluoride
(d) phosphorus pentafluoride

Think About It

Remember not to use the prefix *mono-* with the first element in a molecular compound.

2.65. **Collect and Organize**

In this question we are asked to distinguish between molecular and ionic substances.

Analyze

Ionic substances usually contain a metal (left side of the periodic table) and a nonmetal (right side of the periodic table), whereas covalent substances usually are a combination of nonmetallic elements.

Solve

(a) Both phosphorus and oxygen are nonmetals. This compound consists of molecules.
(b) Strontium is a metal and chlorine is a nonmetal. $SrCl_2$ consists of ions.
(c) Magnesium is a metal, and both carbon and oxygen are nonmetals (they are in combination to form the polyatomic anion carbonate, CO_3^{2-}). $MgCO_3$ consists of ions.
(d) All the elements in H_2SO_4 are nonmetals, though hydrogen may act like a metal in some cases. Sulfuric acid consists of molecules in the gas phase, and ions (H^+ and HSO_4^-) in aqueous solution.

Think About It

The formation of ionic compounds versus the formation of molecular (covalent) compounds can often be predicted by the difference in electronegativity between the bonding elements. Later you will learn the definition of electronegativity (Chapter 8).

2.67. Collect and Organize
We are asked to determine the total number of atoms in the formula unit for the listed ionic compounds.

Analyze
Subscripts denote the number of atoms in a compound or ion.

Solve
(a) 4 atoms   (b) 5 atoms   (c) 8 atoms   (d) 15 atoms

Think About It
It is important to remember that subscripts apply to all atoms within a set of brackets. In (d), you may treat $(H_2PO_4)_2$ as $2\ H_2PO_4^-$

$$2\ \text{ions} \times \frac{7\ \text{atoms}}{1\ \text{ion}} = 14\ \text{atoms}$$

2.69. Collect and Organize
For the oxoanions of element X, XO_2^{2-} and XO_3^{2-}, we are to assign one as *-ite*.

Analyze
Compound names that end in *-ite* represent oxoanions that have one fewer oxygen atom than compounds ending in *-ate*.

Solve
Between XO_2^{2-} and XO_3^{2-}, the oxoanion XO_2^{2-} would have a name ending in *-ite*.

Think About It
In Table 2.4 we can see this pattern for oxoanions of chlorine. ClO_3^- is *chlorate* and ClO_2^- is *chlorite*.

2.71. Collect and Organize
When writing names for transition metal compounds, we include Roman numerals in the name. We are asked in this question what purpose these Roman numerals serve.

Analyze
Transition metals often take on more than one oxidation state (or charge).

Solve
Roman numerals are used in the names of compounds of transition elements to indicate the charge on the transition metal cation.

Think About It
The indication of charge on the transition metal cation in the name of the compounds makes it easy to write the formula for a compound. For example, without the *(III)* in iron(III) chloride, we would not be sure whether to write $FeCl_3$ or $FeCl_2$ because both compounds exist and exhibit different chemical and physical properties.

2.73. Collect and Organize
All the compounds here are oxides of nitrogen. These are all molecular compounds composed of two nonmetallic elements. We name these using the rules for binary compounds.

Analyze
We will use prefixes (Table 2.2) to indicate the number of oxygen atoms in these compounds. The nitrogen atom is always first in the formula; so it is named first. If there is only one nitrogen atom in the formula, we do not need to use the prefix *mono-* for the nitrogen. If there is more than one nitrogen atom, however, we will indicate the number with the appropriate prefix. Also, since *oxide* begins with a vowel, it would be awkward to say *pentaoxide;* so we shorten the double vowel in this part of the chemical name to *pentoxide*.

Solve

(a) NO_3, nitrogen trioxide
(b) N_2O_5, dinitrogen pentoxide
(c) N_2O_4, dinitrogen tetroxide
(d) NO_2, nitrogen dioxide
(e) N_2O_3, dinitrogen trioxide
(f) NO, nitrogen monoxide
(g) N_2O, dinitrogen monoxide
(h) N_4O, tetranitrogen monoxide

Think About It

All of these many binary compounds of nitrogen and oxygen are uniquely named.

2.75. **Collect and Organize**

To predict the formula for the binary ionic compounds formed from the elements listed in the problem, we first have to decide what charges the metal and nonmetal typically have in ionic compounds. To name the compounds, we use the naming rules for ionic compounds.

Analyze

Metallic elements in group 1 of the periodic table (Na and Li) have a 1+ charge in ionic compounds, those in group 2 (Sr) have a 2+ charge, and those in group 13 (Al) have a 3+ charge. Nonmetals in group 16 (S and O) have a 2– charge and those in group 17 (Cl) have a 1– charge. Hydrogen here has a 1– charge because it is combining with a metal and is thus a hydride. To write the formulas of the neutral salts, the charges of the anion must be balanced with the charges of the cation. In naming binary ionic compounds, the cation is named first as the element, and the anion is named second with the ending *-ide* added.

Solve

(a) sodium (Na^+) and sulfur (S^{2-}): Na_2S, sodium sulfide
(b) strontium (Sr^{2+}) and chlorine (Cl^-): $SrCl_2$, strontium chloride
(c) aluminum (Al^{3+}) and oxygen (O^{2-}): Al_2O_3, aluminum oxide
(d) lithium (Li^+) and hydrogen (H^-): LiH, lithium hydride

Think About It

In naming binary ionic salts of the main group elements, we do not need to indicate the numbers of anions or cations in the formula with prefixes, making the naming of these compounds very direct.

2.77. **Collect and Organize**

We are asked which of the chemical formulas for the listed ionic compounds is incorrect. We must assume that all of the given chemical names are correct.

Analyze

You may find it helpful to predict the formula for each ionic compound from the chemical name given and compare these with the formula given in the question. In each case, consider the charge based on location in the periodic table or from the list of common polyatomic ions (Table 2.3).

Solve

(a) Calcium forms a cation with a 2+ charge, Ca^{2+}; oxygen forms an anion with a 2– charge, O^{2-}. CaO is correct.
(b) Lithium forms a cation with a 1+ charge, Li^+; sulfate is a polyatomic anion with a 2– charge, SO_4^{2-}. The correct chemical formula would be Li_2SO_4.
(c) Barium forms a cation with a 2+ charge, Ba^{2+}; sulfur forms an anion with a 2– charge, S^{2-}. BaS is correct.
(d) Potassium forms a cation with a 1+ charge, K^+; oxygen forms an anion with a 2– charge, O^{2-}. K_2O is correct.
The only incorrect chemical formula is (b).

Think About It

You may be tempted to use the charge on each ion as the subscript for the other ion. This shortcut would suggest that the chemical formula for barium sulfide is Ba_2S_2, which is incorrect. If the charge on the cation and anion are the same, only one of each ion is required to balance charge.

2.79. Collect and Organize
We are asked to identify the oxoanion from the name of a salt and write its formula with associated charge.

Analyze
The oxoanions are polyatomic ions. The element other than oxygen appears first in the name, and the ending depends on the number of oxygen atoms in the anion. Oxoanions with -*ate* as an ending have one more oxygen in their structure than those ending in -*ite*. Prefixes such as *per*- and *hypo*- can indicate the largest and smallest number of oxygens, respectively. We can use these rules and the examples in the text for chlorine (Table 2.4) as well as the polyatomic ions listed in Table 2.3 to help us write the formulas for the oxoanions in this question.

Solve
(a) hypobromite, BrO^- in analogy with hypochlorite
(b) sulfate, SO_4^{2-}
(c) iodate, IO_3^-
(d) nitrite, NO_2^-

Think About It
The names here do not really help us write the formulas; we have to just remember them. Learning them well for oxoanions containing chlorine can help because we can name the other halogen oxoanions by analogy with chlorine oxoanions.

2.81. Collect and Organize
Each of these compounds contains a metal in combination with a polyatomic anion. These compounds are ionic and follow the naming rules for ionic compounds.

Analyze
For these compounds, we name the metal cation first as the element name, then the anion.

Solve
(a) K_2CO_3, potassium carbonate
(b) NaCN, sodium cyanide
(c) $LiHCO_3$, lithium bicarbonate or lithium hydrogen carbonate
(d) $Ca(ClO)_2$, calcium hypochlorite

Think About It
These are named much like the binary ionic compounds. The anion name often ends in -*ide*, but can end in -*ate* or -*ite*, depending on the name of the polyatomic anion.

2.83. Collect and Organize
We need to name or write the formula for each of the compounds according to the rules for naming acids.

Analyze
Binary acids are named by placing *hydro*- in front of the element name other than hydrogen along with replacing the last syllable with -*ic* and adding *acid*. For acids containing oxoanions that end in -*ate*, the acid name becomes -*ic acid*. For acids containing oxoanions that end in -*ite*, the acid name becomes -*ous acid*.

Solve
(a) HF(*aq*), a binary acid, hydrofluoric acid
(b) $HBrO_3$(*aq*), an acid of the bromate anion, bromic acid
(c) phosphoric acid, acid of the phosphate anion, H_3PO_4(*aq*)
(d) nitrous acid, acid of the nitrite anion, HNO_2(*aq*)

Think About It
The rules are somewhat systematic but have to be learned and practiced.

2.85. Collect and Organize
All the compounds listed are salts of sodium. Some are composed of polyatomic ions, and others are binary salts.

Analyze
For ionic compounds, name the cation as the element first, then name the anion. If the anion is an element, the suffix -*ide* is used; if the anion is a polyatomic oxoanion, we simply add the name of that anion.

Solve
(a) Na_2O, sodium oxide
(b) Na_2S, sodium sulfide
(c) Na_2SO_4, sodium sulfate
(d) $NaNO_3$, sodium nitrate
(e) $NaNO_2$, sodium nitrite

Think About It
All of these names uniquely describe the compounds. Once we are familiar with the names of the polyatomic anions, there is no ambiguity in the identity of the compound when named.

2.87. Collect and Organize
We are asked to write the formula of an ionic salt from the name given.

Analyze
We use the rules for naming ionic salts. The first element in the name is the cation in the formula. In binary ionic salts, the anion is the element name with the ending -*ide*. If the ion is a polyatomic ion, the name of that polyatomic anion follows the name of the metal. In all cases when writing the formula, we have to balance the charges of the anion and the cation to give a neutral salt.

Solve
(a) potassium sulfide, K^+ with S^{2-} gives K_2S
(b) potassium selenide, K^+ with Se^{2-} gives K_2Se
(c) rubidium sulfate, Rb^+ with SO_4^{2-} gives Rb_2SO_4
(d) rubidium nitrite, Rb^+ with NO_2^- gives $RbNO_2$
(e) magnesium sulfate, Mg^{2+} with SO_4^{2-} gives $MgSO_4$

Think About It
Most of the anions in these salts are dianions with a 2– charge. When combining with cations with a 1+ charge, we have to balance the charge by having two cations for every anion in the formula.

2.89. Collect and Organize
We are asked to identify a compound by name from a list of formulas.

Analyze
Sodium sulfite would have the sodium cation (Na^+) and the sulfite anion (SO_3^{2-}) in its formula. To balance out the charge, there would be two Na^+ cations for every one SO_3^{2-} anion.

Solve
The answer is (b) Na_2SO_3.

Think About It
To write a formula, we have to be very familiar with the names of the cations and anions. It would be easy here to confuse sulfite with sulfate (SO_4^{2-}) or sulfide (S^{2-}).

2.91. Collect and Organize

These compounds are all binary ionic compounds of cobalt and oxygen. Because cobalt is a transition metal and thus has more than one available ionic charge, we use the naming rules that incorporate Roman numerals into the name to indicate the charge on the cobalt ion in the compound.

Analyze

In these compounds oxygen has a charge of 2–. The charge on the cobalt ion must balance out the charge on the oxide ion to give the neutral species listed. In naming these compounds the metal is named first, followed by the charge in Roman numerals in parentheses. The anion is named as a separate word with the ending *-ide*.

Solve

(a) CoO: cobalt ion has a 2+ charge, cobalt(II) oxide
(b) Co_2O_3: cobalt ion has a 3+ charge, cobalt(III) oxide
(c) CoO_2: cobalt ion has a 4+ charge, cobalt(IV) oxide

Think About It

Because the charges of the cations and anions must balance to give a neutral species, we do not have to indicate the number of oxide anions in these compounds; the cation charge dictates the number of oxide ions that must be present in the formula.

2.93. Collect and Organize

The formulas for the compounds to be named all contain transition metal ions with variable charges. In the name, therefore, we must be sure to indicate the charge of the cation.

Analyze

The naming of these compounds is the same as naming other ionic compounds. All that is needed is to add Roman numerals to indicate the charge on the transition metal cation.

Solve

(a) MnS, manganese(II) sulfide
(b) V_3N_2, vanadium(II) nitride
(c) $Cr_2(SO_4)_3$, chromium(III) sulfate
(d) $Co(NO_3)_2$, cobalt(II) nitrate
(e) Fe_2O_3, iron(III) oxide

Think About It

The addition of the Roman numerals to these names clearly indicates the charge of the cation. If the charges were not indicated, it would be unclear from the names of these compounds how to write their formulas.

2.95. Collect and Organize

We are asked which of the chemical formulas for the listed ionic compounds is incorrect. Each of these compounds contains a transition metal cation, so we must be sure to use the correct charge for the metal cation.

Analyze

The naming of these compounds is the same as naming other ionic compounds, except that Roman numerals are required to indicate the charge on the transition metal cation. You may find it helpful to predict the formula for each ionic compound from the chemical name given and compare these with the formula given in the question.

Solve

(a) Iron(II) refers to the 2+ cation Fe^{2+}; oxide refers to a 2– anion, O^{2-}. FeO is correct.
(b) Titanium(IV) refers to the 4+ cation Ti^{4+}; sulfate refers to a 2– polyatomic anion SO_4^{2-}. $Ti(SO_4)_2$ is correct.
(c) Cobalt(II) refers to the 2+ cation Co^{2+}; chloride refers to a 1– anion, Cl^-. $CoCl_2$ is correct.
(d) Vanadium(IV) refers to the 4+ cation V^{4+}; oxide refers to a 2– anion, O^{2-}. The correct chemical formula would be VO_2.
Of the listed compounds, only (d) is incorrect.

Think About It

It is important to consider the charge of the transition metal cations, as indicated by the Roman numerals. Without the charges, it would be unclear if the chemical formulas are correct.

2.97. Collect and Organize

We must define *chemistry* and *cosmology* and give examples of how these two sciences are related.

Analyze

Chemistry looks at the properties of matter at the atomic or molecular level. Cosmology looks at the properties of the universe.

Solve

Chemistry is the study of the composition, structure, properties, and reactivity of matter. Cosmology is the study of the history, structure, and dynamics of the universe. A few of the ways that these two sciences are related might be: (1) Because the universe is composed of matter and the study of matter is chemistry, the study of the universe is really chemistry; (2) The changing universe is driven by chemical and atomic or nuclear reactions, which are also studied in chemistry; (3) Cosmology often asks what the universe (including stars, black holes, etc.) are made of at the atomic level.

Think About It

Cosmologists, even though they study the vast universe, must understand chemistry to understand the dynamic processes that formed and continue to shape our universe.

2.99. Collect and Organize

We are asked why chemists do not include quarks in their study of subatomic particles.

Analyze

Chemists are interested in the reactivity and properties of atoms and molecules.

Solve

Chemists do not include quarks in the category of subatomic particles because the very small quarks combine to make up the three larger subatomic particles that are important to the properties and reactivity of atoms: protons, neutrons, and electrons.

Think About It

The quarks make up the neutron and protons, each containing three quarks. There are six "flavors" of quarks, named: up, down, charm, strange, top, and bottom. The word *quark* was first used by physicist Murray Gell-Mann, but later he found the word was also used previously by James Joyce in *Finnegan's Wake*.

2.101. Collect and Organize

We consider how the density of the universe is changing as it expands.

Analyze

Density is simply defined as mass per volume.

Solve

If the universe's volume is expanding with time and the amount of matter in the universe is constant, then its density is decreasing.

Think About It

The density of the universe has been calculated to be about 2.11×10^{-29} g/cm^3.

2.103. Collect and Organize

We are asked to explain why it takes twice the energy to remove an electron from He as it does from H.

Analyze
The electrons in He must be held more tightly to the nucleus so that they are harder to remove.

Solve
The electron is twice as hard to remove from a helium atom compared to a hydrogen atom because it is being held by a nucleus with a 2+ charge rather than one with a 1+ charge.

Think About It
As we will see later in the textbook, it does not take three times the energy to remove an electron from Li because the electron that we remove is higher in energy in the atom and the positive charge of the nucleus is being "shielded" by the other electrons.

2.105. **Collect and Organize**
We are told that an alpha particle and a neon-21 nucleus combine to produce the nucleus of another element and a neutron. We are asked to write an equation describing this nuclear reaction. Recall that an alpha particle contains two protons and two neutrons, and may be depicted as $_2^4\alpha$, and a neutron may be depicted as $_0^1n$. You should also recall that the 21 in neon-21 represents the mass number and that the identity of an element is determined by the number of protons, (the atomic number).

Analyze
When balancing nuclear equations, both atomic mass and atomic number are conserved (i.e., the sum of the atomic masses on the reactant side must equal the sum of the atomic masses on the product side, and similarly for atomic number). It may be easier to rephrase the question as a nuclear equation that we must balance by determining the identity of the particle formed. Write each reactant or product showing both the mass number and the atomic number:

$$_{10}^{21}Ne + _2^4\alpha \rightarrow _0^1n + \underline{\qquad}$$

Here the blank space represents the new element whose identity we are to determine. The new element is likely to have a mass slightly higher than that of neon, but it should not be significantly higher in mass.

Solve
There are 12 protons on the reactant side, so our new element must contain $12 - 0 = 12$ protons. This identifies the new element as magnesium. Using the same process, we may determine which isotope of magnesium is formed. The atomic mass of the magnesium nuclide is $21 + 4 - 1 = 24$. The equation describing this process is

$$_{10}^{21}Ne + _2^4\alpha \rightarrow _0^1n + _{12}^{24}Mg$$

Think About It
Our estimate was reasonable—the mass number and atomic number for the new element are very close to those of neon-21. It is easier to see that the number of subatomic particles are conserved when the nuclear equation is written out.

2.107. **Collect and Organize**
This question considers Thomson's cathode-ray tube experiment when using a radioactive source in place of electricity to generate a beam.

Analyze
The particles that are coming from the radioactive source include the β particle (an electron, with 1– charge), an α particle (a helium nucleus with 2+ charge), and rays of energy with no charge (perhaps these are γ rays). Each of these is deflected differently when the beam passes through two charged metal plates.

Solve

(a) Because each of the beams of particles or energy is deflected differently due to their different charges, there will be three spots on the phosphorescent screen.

(b) The rays of energy with no charge are unaffected by the electric field, so these would appear in the center of the screen. The β particles are deflected toward the positively charged plate. The α particles, with a higher positive charge (2+), would be deflected toward the negatively charged plate.

Think About It

This simple experiment could discriminate not only differently charged particles (negative, positive, and neutral), but could also discriminate between two positively or negatively charged particles of different masses (e.g., protons versus α particles) or two species with the same mass but with different charges.

2.109. Collect and Organize

We consider that the early universe was composed of 75% ^{1}H and 25% ^{4}He by mass.

Analyze

We must first convert the percentage by mass into percentage of atoms by considering that the helium nucleus weighs four times that of hydrogen. We can then compare that value to the 10:1 hydrogen-to-helium composition in our solar system and propose a way that any difference might have occurred.

Solve

(a) Because helium's mass is four times that of hydrogen, the atom ratio of these two elements would be given by

$$\frac{(25 \text{ g He/4 g})}{(75 \text{ g H/1 g})} = \frac{6.25}{75} = \frac{1}{12}$$

This means that in the early universe there were 12 H atoms for every 1 He atom.

(b) Comparing the ratio of hydrogen to helium in the solar system (10:1) to the ratio in the early universe (12:1), we see that there is proportionately more helium present in our solar system than at the beginning of the universe.

(c) Stars burn hydrogen fuel, which gives off energy. One product of this fusion process is helium. As a star burns away its hydrogen, more helium is produced.

(d) We could propose to look at the composition of galaxies that have older stars. Are more helium and heavier elements present as the stars burn up their hydrogen?

Think About It

If the fusion process continues, with lighter elements fusing to form heavier elements, it could be imagined that stars might run out of hydrogen fuel and then have to rely on helium fusion to produce energy.

2.111. Collect and Organize

Stainless steel contains chromium and nickel at 19% and 9% by mass, respectively. We are asked to name the two compounds that form on the surface of stainless steel to protect it from corrosion, Cr_2O_3 and NiO, and determine the charges on the transition metal cations in those ionic compounds.

Analyze

In naming the compounds, we must indicate the charge on the metal. To determine the charge on the Cr and Ni cations in these compounds, we use the common charge for oxygen (2–).

Solve

(a) Cr_2O_3, chromium(III) oxide

NiO, nickel(II) oxide

(b) Each Cr ion must bear a charge of 3+, for a total charge of 6+ on the two Cr atoms, to balance the $3 \times -2 = 6-$ charge of the three oxygen atoms. The charge on the Ni atom must be 2+ to balance the 2– charge on the oxygen atom in NiO.

Think About It
Stainless steel, because it contains a homogeneous mixture of metals, is an alloy.

2.113. Collect and Organize
We are to discern from the formulas for the composition of the element with hydrogen (MH_3) and oxygen (M_2O_5) the Roman numeral Mendeleev assigned to the group containing these elements.

Analyze
The Roman numerals that Mendeleev used are closely related to the charge on the ion for the element. In compounds, hydrogen can be 1+ or 1– and oxygen is usually 2–.

Solve
Looking first at M_2O_5, we would assign an oxidation number of 5+ to M so that the 10+ from two M ions in the compound balances the 10– charge from the five O atoms. The compound MH_3 fits if we consider compounds such as NH_3 and PH_3 that also have oxides of N_2O_5 and P_2O_5. Mendeleev would have designated this group as V.

Think About It
With close inspection of Mendeleev's table in Figure 2.9, you can see that indeed his Roman numerals often correspond to oxidation numbers typical for those elements.

2.115. Collect and Organize
We are asked to give the systematic names and chemical formulas for some commonly named chemicals.

Analyze
Any Internet source or search engine could help us find the names and chemical formulas for these. One such source is Wikipedia (http://en.wikipedia.org/wiki/Main_Page).

Solve
(a) magnesia = magnesium oxide, MgO
(b) Epsom salt = magnesium sulfate, $MgSO_4$
(c) K-Dur = potassium chloride, KCl
(d) lime = calcium oxide, CaO
(e) baking soda = sodium bicarbonate, $NaHCO_3$
(f) caustic soda = sodium hydroxide, $NaOH$
(g) muriatic acid = hydrogen chloride in aqueous solution, $HCl(aq)$
(h) zirconia = zirconium dioxide, ZrO_2

Think About It
Epsom salt and lime are often in hydrated form, $[MgSO_4 \cdot 7H_2O]$ and $Ca(OH)_2$.

2.117. Collect and Organize
Given that *thio-* in a chemical name means that a sulfur atom has replaced one of the oxygen atoms, we are asked to write the formulas for the thiosulfate ion and for its salt with sodium.

Analyze
The sulfate ion is SO_4^{2-}.

Solve
(a) If we replace an oxygen atom with a sulfur in SO_4^{2-}, we get the thiosulfate ion, $S_2O_3^{2-}$.
(b) To balance the charges to form a neutral salt, we need two sodium ions in the formula for sodium thiosulfate, $Na_2S_2O_3$.

Think About It
Other *thio-* compounds are also similarly named. Thiocyanate is SCN⁻ from cyanate (OCN⁻), and thiourea is $(NH_2)_2CS$ from urea, $(NH_2)_2CO$.

2.119. Collect and Organize
One isotope of bromine weighs 78.9183 amu (^{79}Br), and the other weighs 80.9163 amu (^{81}Br). From the average atomic mass of 79.9091 amu, we are to calculate the abundance of ^{81}Br.

Analyze
The average atomic mass is derived from a weighted average of the isotopes' atomic masses. If x = abundance of ^{79}Br and y = abundance of ^{81}Br, then the weighted average can be expressed as

$$78.9183 \text{ amu } (x) + 80.9163 \text{ amu } (y) = 79.9091 \text{ amu}$$

Because there are only two isotopes, the sum of their abundances (in decimal form) must equal 1.00.
$$x + y = 1.00$$
Thus
$$x = 1.00 - y$$

Substituting this expression for x into the preceding weighted average equation gives

$$78.9183 \text{ amu } (1-y) + 80.9163 \text{ amu } (y) = 79.9091 \text{ amu}$$

Solve
$$78.9183 \text{ amu} - 78.9183 \text{ amu } (y) + 80.9163 \text{ amu } (y) = 79.9091 \text{ amu}$$
$$1.9980 \text{ amu } (y) = 0.9908 \text{ amu}$$
$$y = 0.4959$$

The abundance of ^{81}Br is $y \times 100 = 49.59\%$.

Think About It
Because there are only two isotopes for bromine, it follows that the abundance of ^{79}Br is $100 - 49.59 = 50.41\%$.

2.121. Collect and Organize
In this problem we are provided with the masses of the three isotopes of magnesium (^{24}Mg = 23.9850 amu, ^{25}Mg = 24.9858 amu, and ^{26}Mg = 25.9826 amu) and given that the abundance for ^{24}Mg is 78.99%. From this information and the average (weighted) atomic mass units of magnesium (24.3050 amu), we must calculate the abundances of the other two isotopes, ^{25}Mg and ^{26}Mg.

Analyze
The average atomic mass is derived from a weighted average of the isotopes' atomic masses. If x = abundance of ^{25}Mg and y = abundance of ^{26}Mg, the weighted average of magnesium is

$$(0.7899 \times 23.9850 \text{ amu}) + 24.9858 \text{ amu } (x) + 25.9826 \text{ amu } (y) = 24.3050 \text{ amu}$$

Because the sum of the abundances of the isotopes must add up to 1.00

$$0.7899 + x + y = 1.00$$
So
$$x = 1.00 - 0.7899 - y = 0.2101 - y$$

Substituting this expression for x in the weighted average mass equation gives

$$(0.7899 \times 23.9850 \text{ amu}) + 24.9858 \text{ amu } (0.2101 - y) + 25.9826 \text{ amu } (y) = 24.3050 \text{ amu}$$

Solve

$$18.94575 + 5.249517 - 24.9858y + 25.9826y = 24.3050$$
$$0.9968y = 0.1097$$
$$y = 0.1101$$

So

$$x = 0.2101 - 0.1101 = 0.1000$$

The abundance of ^{25}Mg is $x \times 100 = 10.00\%$, and the abundance of ^{26}Mg is $y \times 100 = 11.01\%$.

Think About It

Although the abundances of ^{25}Mg and ^{26}Mg are nearly equal to each other at the end of this calculation, we cannot assume that when setting up the equation. We have to solve this problem algebraically by setting up two equations with two unknowns.

2.123. **Collect and Organize**

We are asked to write a chemical symbol based on the information given about each nuclide.

Analyze

Chemical symbols may include information about the atomic number (number of protons), mass number (sum of the number of protons and the number of neutrons), elemental symbol, and charge. Knowing something about the number of electrons, atomic number, or mass number, we may deduce the other values.

Solve

(a) Atomic number 12 identifies the element as magnesium (Mg). Incorporating the information about the charge and mass number, the chemical symbol is $^{24}_{12}Mg^{2+}$.

(b) An ion with a charge of 3+ and 48 electrons, must have $48 + 3 = 51$ protons. This identifies the element as antimony (Sb). With 70 neutrons, the mass number for this nuclide is $51 + 70 = 121$, and the chemical symbol for the nuclide is $^{121}_{51}Sb^{3+}$.

(c) A noble gas with 48 neutrons is likely to have a similar number of protons. The likely candidates are xenon (Xe, $Z = 54$) and krypton (Kr, $Z = 36$). If the element were xenon, the mass number for this nuclide would be 102, which is significantly smaller than the average atomic mass for xenon (131.293 amu). Krypton gives us a much more realistic mass number of $36 + 48 = 84$ (close to the average atomic mass for krypton of 83.798 amu). Since no charge is mentioned, we can assume it is a neutral atom, as is often the case for the noble gases. The chemical symbol for the nuclide is $^{84}_{36}Kr$.

Think About It

Remember that electrons have a negative charge. A cation is formed by losing one or more electrons, and an anion is formed by gaining one or more electrons. Forgetting this small detail could lead you to the wrong element!

2.125. **Collect and Organize**

We are asked to predict some physical and chemical properties of the element radium, Ra. We are also asked to predict the melting points for $RaCl_2$ and RaO by comparing the melting points for other alkaline earth metal chlorides and oxides.

Analyze

Radium is the heaviest known member of group 2, the alkaline earth metals. Elements in the same group may be assumed to have similar physical and chemical properties.

Solve

Like other alkaline earth metals, it is reasonable to expect that radium would adopt a 2+ charge to form Ra^{2+} compounds. As a metallic element, it would likely be malleable, relatively dense, conduct heat and electric current, and melt at a fairly high temperature. The melting points of various alkaline earth metal chlorides and oxides are listed below, along with estimates for $RaCl_2$ and RaO.

	Melting Point (°C)		Melting Point (°C)
$CaCl_2$	772	CaO	2572
$SrCl_2$	874	SrO	2531
$BaCl_2$	962	BaO	1923
$RaCl_2$	(950 to 1050)	RaO	(1700 to 2000)

Think About It

Mendeleev used the periodicity of the chemical and physical properties of the elements to predict the location of several undiscovered elements. The similarity of elements in a group is one of the defining characteristics of the modern periodic table.

2.127. Collect and Organize

We are asked to explain why argon is placed before potassium in the modern periodic table, despite having a larger average atomic mass.

Analyze

The modern periodic table is sorted by atomic number rather than by average atomic mass, as Mendeleev's periodic table was.

Solve

Despite being heavier (on average), argon contains 18 protons, whereas potassium contains 19 protons. Since the modern periodic table is organized by increasing atomic number, argon is placed before potassium.

Think About It

The argon nuclide $^{40}_{18}Ar$ is 99.6% abundant in nature; the potassium nuclide $^{39}_{19}K$ is 93.3% abundant.

CHAPTER 3 | Stoichiometry: Mass, Formulas, and Reactions

3.1. Collect and Organize

This exercise asks us to interpret diagrams (Figure P3.1) drawn from a molecular perspective in order to write a chemical reaction, including the state of the substances (solid, liquid, or gas).

Analyze

When the atoms are isolated, they are written as atomic species. If the atoms are bound to each other, they are in the form of molecular species. Any representation of the substances that has a high degree of order (all molecules arranged) and a shape independent of the container represents the solid phase. Representations with less order and that conform to the shape of the container indicate the liquid phase (molecules lined up with some order). Representations with a random distribution of the substances and that fill the container indicate the gas phase.

Solve

(a) There are four atoms of X (red spheres) and four atoms of Y (blue spheres), both in the gas phase, on the reactant side of the equation (left of the arrow). On the product side (right of the arrow), there are four gaseous molecules of XY (red–blue). Therefore, the chemical equation reads

$$4\ X(g) + 4\ Y(g) \rightarrow 4\ XY(g)$$

(b) There are four atoms of X (red spheres) and four atoms of Y (blue spheres), both in the gas phase, on the reactant side of the equation. On the product side there are four solid molecules of XY (red–blue). Therefore, the chemical equation reads

$$4\ X(g) + 4\ Y(g) \rightarrow 4\ XY(s)$$

(c) There are four atoms of X (red spheres) and four atoms of Y (blue spheres), both in the gas phase, on the reactant side of the equation. On the product side there are four gaseous substances: two molecules of XY_2 and two atoms of X. Therefore, the chemical equation reads

$$4\ X(g) + 4\ Y(g) \rightarrow 2\ XY_2(g) + 2\ X(g)$$

(d) There are four molecules of X_2 (red spheres bonded together) and four molecules of Y_2 (blue spheres bonded together), both in the gas phase, on the reactant side of the equation. On the product side there are eight gaseous molecules of XY (red–blue). Therefore, the chemical equation reads

$$4\ X_2(g) + 4\ Y_2(g) \rightarrow 8\ XY(g)$$

Think About It

In parts a, b, and d when the reactants react, there are no leftover atoms of the reactants; however, in reaction c two atoms of X are left over on the product side after making XY_2. In this case, Y is the limiting reactant, and the amount of Y atoms determines how many molecules of XY_2 can form from the reactant mixture.

3.3. Collect and Organize

By examining Figure P3.3, we are to determine which diagram (drawn in the molecular perspective) best depicts the reaction between N_2 and O_2 to produce N_2O.

Analyze

We are told that the red spheres represent oxygen, and the blue spheres represent nitrogen. The best depiction will be one that produces only the desired product, and ideally, depicts reactants in the correct ratio. The desired product (N_2O) should contain two blue spheres and one red sphere.

Solve

Diagrams (b), (c), and (d) all depict the formation of N_2O, but (c) and (d) also suggest the formation of other molecules, such as NO_2 and NO. The best depiction of this reaction is (b).

Think About It

Diagram (a) correctly depicts the formation of NO_2, not N_2O.

3.5. **Collect and Organize**

We are asked to identify molecules with the same empirical formula.

Analyze

An empirical formula is the simplest ratio of atoms present in a molecule. We can identify the molecular formulas for all species from molecular perspective diagrams, and determine the empirical formulas by finding the whole number ratio of atoms in each.

Solve

The molecular and empirical formulas for each species are:

	Molecular Formula	Empirical Formula
(a)	NO_2	NO_2
(b)	N_2O	N_2O
(c)	N_2O_4	NO_2
(d)	N_2O	N_2O

The molecules in (a) and (c) have the same empirical formula, as do those in (b) and (d).

Think About It

Despite the difference in connectivity between (b) (NNO) and (d) (NON), both have the same ratio of atoms, and thus the same empirical formula.

3.7. **Collect and Organize**

We are asked to write balanced equations for the unbalanced equations in Problem 3.6.

Analyze

To balance an unbalanced equation, we adjust the stoichiometric coefficients, but not the molecular formulas. By consulting the tables created when solving Problem 3.6, we can determine which coefficients must be increased or decreased. The unbalanced equations from Problem 3.6 are (b) and (d).

Solve

(b) As originally written:

$$2\,SO_2 + 2\,O_2 \rightarrow 2\,SO_3$$

Element	Left Side	Right Side
S	2	2
O	8	6

We need fewer O atoms on the reactant side. This can be accomplished by changing the O_2 coefficient to 1.

(b)

$$2\,SO_2 + O_2 \rightarrow 2\,SO_3$$

Element	Left Side	Right Side
S	2	2
O	6	6

The reaction is now balanced.

(d) As originally written:

$$2\,CS_2 + 4\,O_2 \rightarrow 2\,CO_2 + 2\,SO_2$$

Element	Left Side	Right Side
C	2	2
S	4	2
O	8	8

We need more S atoms on the product side. This can be accomplished by changing the SO_2 coefficient to 4.

$$2\,CS_2 + 4\,O_2 \rightarrow 2\,CO_2 + 4\,SO_2$$

Element	Left Side	Right Side
C	2	2
S	4	4
O	8	12

Now we need more O atoms on the reactant side. This can be accomplished by changing the O_2 coefficient to 6.

$$2\,CS_2 + 6\,O_2 \rightarrow 2\,CO_2 + 4\,SO_2$$

Element	Left Side	Right Side
C	2	2
S	4	4
O	12	12

All the coefficients are even, so we can divide by 2 to get the smallest whole number ratio of coefficients.

$$CS_2 + 3\,O_2 \rightarrow CO_2 + 2\,SO_2$$

Element	Left Side	Right Side
C	1	1
S	2	2
O	6	6

The reaction is now balanced.

Think About It
Remember to make sure the stoichiometric coefficients are in the smallest whole number ratio at the end.

3.9. **Collect and Organize**
By examining Figure 3.1, we are to determine which substance (SiO_2, Al_2O_3, or Fe) is the most dense.

Analyze
In Figure 3.1, iron is found in Earth's core; silicon and aluminum are found in the mantle and crust.

Solve
Because the most dense material would sink to the core, and iron is found at the core, $Fe\,(\ell)$ is the most dense material of $SiO_2(s)$, $Al_2O_3(s)$, and $Fe\,(\ell)$.

Think About It
Earth's core is believed to have temperatures of between 5000°C and 6000°C and a pressure of about 3,000,000 atm.

3.11. **Collect and Organize**
From the definition of a combination reaction, we are to determine how the number of different products compares to the number of different reactants.

Analyze
A combination reaction is one in which reactants combine to form a product.

Solve
Because reactants combine in a combination reaction, the number of different products formed is less than the number of different reactants in the reaction.

Think About It
The opposite of a combination reaction is a decomposition reaction in which a substance breaks down into simpler products.

3.13. Collect and Organize

When thinking about a collection of atoms or molecules, we are asked why it might not be a good idea to use the unit *dozen* to express quantities of atoms or molecules.

Analyze

A dozen is 12 objects—a relatively small group of objects.

Solve

A dozen is a convenient and recognizable unit for donuts and eggs, but it is too small a unit to express the very large number of atoms, ions, or molecules present in a mole.

$$\frac{6.022 \times 10^{23} \text{ atoms}}{1 \text{ mole}} \times \frac{1 \text{ dozen}}{12 \text{ atoms}} = 5.02 \times 10^{22} \text{ dozen/mole}$$

Think About It

The mole (6.022×10^{23}) is a much more convenient unit to express the number of atoms or molecules in a sample.

3.15. Collect and Organize

We are asked to compare the molar mass of a molecular compound with three atoms to that of a molecular compound with two atoms.

Analyze

The molar mass is calculated by adding up the individual molar masses of the elements that make up the molecular compound. The molar masses of the naturally occurring elements in the periodic table range from 1 g/mol for hydrogen to over 238 g/mol for uranium.

Solve

No, the molar mass of a substance does not directly correlate to the number of atoms in a molecular compound. This statement would be true only if the two compounds were composed of the same element: for example, O_2 and O_3. For all others, the mass of the individual atoms composing the compound matters greatly. For example, I_2 has a molar mass of $126.904 \times 2 = 253.808$ g/mol, whereas O_3 has a molar mass of $15.999 \times 3 = 47.997$ g/mol.

Think About It

Remember that 1 mole of a substance contains the same number of atoms or molecules as in 1 mole of another substance. Therefore, 18.01 g of H_2O contains the same number of molecules as the number of atoms in 190.23 g of osmium.

3.17. Collect and Organize

In this exercise, we must convert the given number of atoms or molecules of each gas to moles.

Analyze

To convert the number of atoms or molecules to moles, we divide by Avogadro's number.

Solve

(a) $\dfrac{4.4 \times 10^{14} \text{ atoms of Ne}}{6.022 \times 10^{23} \text{ atoms/mol}} = 7.3 \times 10^{-10} \text{ mol Ne}$

(b) $\dfrac{4.2 \times 10^{13} \text{ molecules of CH}_4}{6.022 \times 10^{23} \text{ molecules/mol}} = 7.0 \times 10^{-11} \text{ mol CH}_4$

(c) $\dfrac{2.5 \times 10^{12} \text{ molecules of O}_3}{6.022 \times 10^{23} \text{ molecules/mol}} = 4.2 \times 10^{-12} \text{ mol O}_3$

(d) $\dfrac{4.9 \times 10^{9} \text{ molecules of NO}_2}{6.022 \times 10^{23} \text{ molecules/mol}} = 8.1 \times 10^{-15}$ mol NO$_2$

Think About It

The trace gas that has the highest number of atoms or molecules present also has the highest number of moles present. In this sample of air, the amount of the trace gases decreases in the order Ne > CH$_4$ > O$_3$ > NO$_2$.

3.19. **Collect and Organize**

In this exercise we convert from the moles of titanium contained in a substance to the number of atoms present.

Analyze

For each substance we need to take into account the number of moles of titanium *atoms* present in *1 mole* of the substance. For 0.125 mol of substance, then, a substance that contains two atoms of titanium in its formula contains 0.125 × 2 = 0.250 mol of titanium. We can then use Avogadro's number to convert the moles of titanium to the number of atoms present in the sample.

Solve

(a) Ilmenite, FeTiO$_3$, contains one atom of Ti per formula unit; so 0.125 mol of ilmenite contains 0.125 mol of Ti.

$$0.125 \text{ mol Ti} \times \frac{6.022 \times 10^{23} \text{ Ti atoms}}{1 \text{ mol Ti}} = 7.53 \times 10^{22} \text{ Ti atoms}$$

(b) The formula for titanium(IV) chloride is TiCl$_4$. This formula contains only one Ti atom per formula unit as well; so the answer is identical to that calculated in (a).

$$0.125 \text{ mol Ti} \times \frac{6.022 \times 10^{23} \text{ Ti atoms}}{1 \text{ mol Ti}} = 7.53 \times 10^{22} \text{ Ti atoms}$$

(c) Ti$_2$O$_3$ contains two titanium atoms in its formula; so 0.125 mol of Ti$_2$O$_3$ contains 0.125 × 2 = 0.250 mol of titanium.

$$0.250 \text{ mol Ti} \times \frac{6.022 \times 10^{23} \text{ Ti atoms}}{1 \text{ mol Ti}} = 1.51 \times 10^{23} \text{ Ti atoms}$$

(d) Ti$_3$O$_5$ contains three titanium atoms in its formula; so 0.125 mol of Ti$_3$O$_5$ contains 0.125 × 3 = 0.375 mol of titanium.

$$0.375 \text{ mol Ti} \times \frac{6.022 \times 10^{23} \text{ Ti atoms}}{1 \text{ mol Ti}} = 2.26 \times 10^{23} \text{ Ti atoms}$$

Think About It

The number of atoms of titanium in 0.125 mol of each compound reflects the number of atoms of Ti in the chemical formula. Ti$_2$O$_3$ has twice the number of Ti atoms, and Ti$_3$O$_5$ has three times the number of Ti atoms compared to the number of Ti atoms in the same number of moles of FeTiO$_3$ or TiCl$_4$.

3.21. **Collect and Organize**

Given the formulas and the moles of each substance in a pair, we are asked to decide which compound of the pair contains more moles of oxygen.

Analyze

To answer this question we have to take into account the moles of oxygen present in the substance formulas as well as the number of moles specified for each substance.

Solve

(a) One mole of Al_2O_3 contains three moles of oxygen, and one mole of Fe_2O_3 also contains three moles of oxygen. These compounds contain the same number of moles of oxygen.

(b) One mole of SiO_2 contains two moles of oxygen, and one mole of N_2O_4 contains four moles of oxygen. Therefore, N_2O_4 contains more moles of oxygen (twice as much, in fact).

(c) Three moles of CO contains three moles of oxygen, and two moles of CO_2 contains four moles of oxygen. Therefore, two moles of CO_2 contains more oxygen than three moles of CO.

Think About It

Notice that we cannot decide which substance has more moles of oxygen by comparing only the amounts of the compounds. If that were the case, we would have concluded incorrectly that three moles of CO contains more moles of oxygen than two moles of CO_2.

3.23. **Collect and Organize**

For each aluminosilicate, we are given the chemical formula. From that formula we are asked to deduce the number of moles of aluminum in 1.50 mol of each substance.

Analyze

The number of moles of aluminum in one mole of each substance is reflected in its chemical formula. We need next to take into account that we are starting with 1.50 mol of each substance.

Solve

(a) Each mole of pyrophyllite, $Al_2Si_4O_{10}(OH)_2$, contains two moles of Al atoms. Therefore, 1.50 mol of pyrophyllite contains $1.50 \text{ mol} \times 2 = 3.00$ mol Al.

(b) Each mole of mica, $KAl_3Si_3O_{10}(OH)_2$, contains three moles of Al. Therefore, 1.50 mol of mica contains $1.50 \text{ mol} \times 3 = 4.50$ mol Al.

(c) Each mole of albite, $NaAlSi_3O_8$, contains one mole of Al. Therefore, 1.50 mol of albite contains 1.50 mol of Al.

Think About It

These minerals could all be distinguished by analysis of the amount of aluminum present in equimolar samples of each substance.

3.25. **Collect and Organize**

We are asked to convert a mass of carbon in grams to moles.

Analyze

We need the mass of 1 mole of carbon to compute the number of moles of carbon in the 500.0 g sample. From the periodic table, we see that the molar mass of carbon is 12.011 g/mol.

Solve

$$500.0 \text{ g C} \times \frac{1 \text{ mol C}}{12.011 \text{ g C}} = 41.63 \text{ mol C}$$

Think About It

Because carbon's molar mass is relatively low, at 12 g/mol, 500 g of this substance contains a fairly substantial number of moles.

3.27. **Collect and Organize**

Given the dimensions of an iridium rod, we are asked to determine the number of iridium atoms in the sample. We are provided with the diameter and length of the rod, as well as the density of iridium metal.

Analyze

Using the dimensions given ($d = 0.6$ mm, $l = 3.5$ mm), we can determine the volume of the rod in cm^3. With the volume, we can use the given density of iridium (22.42 g/cm^3) to determine the mass of the rod, and from this, the number of moles of iridium. Lastly, we may determine the number of atoms in the rod using Avogadro's number (6.022×10^{23}). Since there are a large number of atoms in a mole (6.022×10^{23}), we expect the total number of atoms to be quite large, even for a small sample.

Solve
We will need the radius and length in centimeters in order to determine the volume:

$$r = \frac{d}{2} = \frac{0.6 \text{ mm}}{2} \times \frac{1 \text{ cm}}{10 \text{ mm}} = 0.03 \text{ cm}$$

$$l = 3.5 \text{ mm} \times \frac{1 \text{ cm}}{10 \text{ mm}} = 0.35 \text{ cm}$$

The volume of the iridium rod is

$$V = l\pi r^2 = (0.35 \text{ cm})(\pi)(0.03 \text{ cm})^2 = 9.9 \times 10^{-4} \text{ cm}^3$$

This volume corresponds to a mass of

$$9.9 \times 10^{-4} \text{ cm}^3 \times \frac{22.42 \text{g}}{1 \text{ cm}^3} = 0.022 \text{ g}$$

Solving for the number of atoms

$$0.022 \text{ g} \times \frac{1 \text{ mol}}{192.217 \text{ g}} \times \frac{6.022 \times 10^{23} \text{ atoms}}{1 \text{ mol}} = 7 \times 10^{19} \text{ atoms}$$

Think About It
Our answer makes sense. Despite the small mass of the rod, there are still a large number of iridium atoms in the sample.

3.29. **Collect and Organize**
From the chemical formulas for various iron compounds with oxygen, we are asked to determine how many moles of iron are in 1 mole of each substance.

Analyze
The chemical formula reflects the molar ratios of the elements in the compound. If there is one atom of iron in the compound's chemical formula, then there is 1 mole of iron in 1 mole of the compound. Likewise, if there are three atoms of iron in the chemical formula, there are 3 moles of iron present in 1 mole of the substance.

Solve
(a) There is one atom of iron in FeO; therefore, 1 mole of FeO contains 1 mole of iron.
(b) There are two atoms of iron in Fe_2O_3; therefore, 1 mole of Fe_2O_3 contains 2 moles of iron.
(c) There is one atom of iron in $Fe(OH)_3$; therefore, 1 mole of $Fe(OH)_3$ contains 1 mole of iron.
(d) There are three atoms of iron in Fe_3O_4; therefore, 1 mole of Fe_3O_4 contains 3 moles of iron.

Think About It
The parentheses used in $Fe(OH)_3$ show that there are three OH units in this compound. If the question had asked how many moles of oxygen were present in 1 mole of this substance, the answer would be 3 moles of oxygen.

3.31. **Collect and Organize**
This exercise asks us to compute the molar mass of various molecular compounds of oxygen.

Analyze

The molar mass of each of the compounds can be found by adding the molar mass of each element from the periodic table, taking into account the number of moles of each atom present in 1 mole of the substance.

Solve

(a) SO_2: $32.06 + 2(16.00) = 64.06$ g/mol
(b) O_3: $3(16.00) = 48.00$ g/mol
(c) CO_2: $12.01 + 2(16.00) = 44.01$ g/mol
(d) N_2O_5: $2(14.01) + 5(16.00) = 108.02$ g/mol

Think About It

Notice that three compounds, SO_2, O_3, and CO_2, have three atoms in their chemical formula, but each has a different molar mass.

3.33. **Collect and Organize**

This exercise asks us to compute the molar mass of various flavorings.

Analyze

The molar mass of each flavoring can be found by adding the molar mass of each element from the periodic table, taking into account the number of moles of each atom present in 1 mole of the flavoring. Each of these flavorings contains only carbon (12.01 g/mol), hydrogen (1.01 g/mol), and oxygen (16.00 g/mol).

Solve

(a) Vanillin, $C_8H_8O_3$: $8(12.01) + 8(1.01) + 3(16.00) = 152.16$ g/mol
(b) Oil of cloves, $C_{10}H_{12}O_2$: $10(12.01) + 12(1.01) + 2(16.00) = 164.22$ g/mol
(c) Anise oil, $C_{10}H_{12}O$: $10(12.01) + 12(1.01) + 16.00 = 148.22$ g/mol
(d) Oil of cinnamon, C_9H_8O: $9(12.01) + 8(1.01) + 16.00 = 132.17$ g/mol

Think About It

Each of these flavorings has a distinctive odor and flavor, due in part to its different chemical formula. Another factor, however, in differentiating these flavorings is their chemical structure, or the arrangement in which the atoms are attached, as shown by the structures of these flavorings:

| Vanillin | Clove oil | Anise | Cinnamon |

3.35. **Collect and Organize**

Between two balloons filled with 10.0 g of different gases, we are to choose which balloon in the pair has the greater number of particles.

Analyze

The balloon with the greater number of particles has the greater number of moles. The greater number of moles in 10.0 g of a gas will be for the gas with the lowest molar mass. A gas with a lower molar mass contains more moles in a 10.0 g sample and, therefore, has a greater number of moles than a 10.0 g mass of a higher molar mass gas.

Solve
(a) The molar mass of CO_2 is 44 g/mol, and the molar mass of NO is 30 g/mol. Therefore, the balloon containing NO has the greater number of particles.
(b) The molar mass of CO_2 is 44 g/mol, and the molar mass of SO_2 is 64 g/mol. Therefore, the balloon containing CO_2 has the greater number of particles.
(c) The molar mass of O_2 is 32 g/mol, and the molar mass of Ar is 40 g/mol. Therefore, the balloon containing O_2 has the greater number of particles.

Think About It
We could numerically determine the number of moles of gas in each balloon to make the comparisons in this problem, but that is unnecessary because we know the relationship between moles and molar mass.

3.37. Collect and Organize
Given a mass of quartz, we are to determine the moles of SiO_2 present.

Analyze
To convert from mass to moles, we divide the mass given by the molar mass of SiO_2 [28.09 + 2(16.00) = 60.09 g/mol].

Solve

$$45.2 \text{ g SiO}_2 \times \frac{1 \text{ mol SiO}_2}{60.084 \text{ g SiO}_2} = 0.752 \text{ mol SiO}_2$$

Think About It
Because the initial mass is less than the molar mass, we would expect there to be less than 1 mole of SiO_2 in the quartz sample.

3.39. Collect and Organize
We are to calculate the mass of a given number of moles of magnesium carbonate.

Analyze
To convert from moles to mass, multiply the number of moles by the molar mass of the substance. The molar mass of $MgCO_3$ is 24.30 + 12.01 + 3(16.00) = 84.31 g/mol.

Solve

$$0.122 \text{ mol MgCO}_3 \times \frac{84.31 \text{ g MgCO}_3}{1 \text{ mol MgCO}_3} = 10.3 \text{ g}$$

Think About It
Moles in chemistry are like a common currency in exchanging money. From moles we can calculate mass; from mass we can calculate moles.

3.41. Collect and Organize
This exercise asks us to compute the moles of uranium and carbon (diamond) atoms in a 1 cm^3 block of each element and compare them.

Analyze
Starting with the 1 cm^3 block of each element, we can obtain the mass of the block by multiplying the volume by the density of the element. Dividing that result by the molar mass of the element gives us the moles of atoms in that block. The block with the larger number of moles of atoms must have the larger number of

atoms. We can compute the actual number of atoms by multiplying the moles of atoms for each element by Avogadro's number.

Solve

$$1 \text{ cm}^3 \text{ C} \times \frac{3.514 \text{ g}}{1 \text{ cm}^3} \times \frac{1 \text{ mol}}{12.01 \text{ g}} \times \frac{6.022 \times 10^{23} \text{ C atoms}}{1 \text{ mol C}} = 1.762 \times 10^{23} \text{ atoms of C}$$

$$1 \text{ cm}^3 \text{ U} \times \frac{19.05 \text{ g}}{1 \text{ cm}^3} \times \frac{1 \text{ mol}}{238.03 \text{ g}} \times \frac{6.022 \times 10^{23} \text{ U atoms}}{1 \text{ mol U}} = 4.820 \times 10^{22} \text{ atoms of U}$$

Therefore, the 1 cm^3 block of diamond contains more atoms.

Think About It
We might expect that, because the block of uranium weighs so much more than the diamond block (over five times as much), the uranium block would contain more atoms. However, we also have to take into account the very large molar mass of uranium. The result is that the diamond block has about 3.7 times more atoms in it than the same-sized block of uranium.

3.43. **Collect and Organize**
We are asked whether the number of moles of reactants in a balanced chemical equation must equal the number of moles of the products.

Analyze
A balanced chemical reaction follows the law of conservation of mass. This means that for each and every atom present in a reaction, we balance the number of a particular kind of atom (element) in reactants and products. This definition does not include any relationship between the moles of reactants and products.

Solve
The number of moles of reactants in a balanced chemical equation *does not always* equal the number of moles of the products. Elements in compounds are rearranged in chemical equations, and we may have fewer or more moles present after the completion of a reaction.

Think About It
For example, in the balanced equation for the production of ammonia

$$N_2(g) \ + \ 3 \, H_2(g) \ \rightarrow \ 2 \, NH_3(g)$$

4 moles of reactants produce 2 moles of products.

3.45. **Collect and Organize**
We are asked whether the sum of the masses of gaseous reactants in a balanced chemical equation must equal the sum of the masses of the gaseous products.

Analyze
A balanced chemical reaction follows the law of conservation of mass. This means that for all types of atoms present, we balance the number of a particular kind of atom (element) in reactants and products. It also means that the combined mass of all the reactants must be equal to the combined mass of all the products.

Solve
The sum of the masses of the gaseous reactants in a balanced chemical equation *does not always* equal the sum of the masses of the gaseous products. Compounds can react to form products other than gases; so this statement is inaccurate. What is true is that the sum of the masses of all the reactants is equal to the sum of the masses of all the products, regardless of their physical states.

Think About It
We must apply the law of conservation of mass to all of the reactants and products.

3.47. Collect and Organize
We are asked to sketch a reaction between five C atoms and some number of O_2 molecules to produce a 50% mixture of CO and CO_2.

Analyze
The term "50% mixture" describes a mixture with an equal number of molecules of CO and CO_2. Since this will require an even number of C atoms, one will be left over. Writing an equation to describe this reaction will help ensure that we sketch the correct number of each atom or molecule.

Solve
The unbalanced equation is

$$5\,C + O_2 \rightarrow CO + CO_2 + C$$

Element	Left Side	Right Side
C	5	3
O	2	3

The species are all present, but there are too few C atoms on the product side. We can change the coefficients on CO and CO_2 to 2; they must be the same to maintain a "50% mixture."

$$5\,C + O_2 \rightarrow 2\,CO + 2\,CO_2 + C$$

Element	Left Side	Right Side
C	5	5
O	2	6

There are not enough O atoms on the reactant side. We can change the coefficient on O_2 to 3 to fix this.
$$5\,C + 3\,O_2 \rightarrow 2\,CO + 2\,CO_2 + C$$

Element	Left Side	Right Side
C	5	5
O	6	6

The equation is now balanced. We should sketch five carbon atoms and three molecules of O_2 on the reactant side, and two molecules of CO, two molecules of CO_2, and one C atom on the product side.

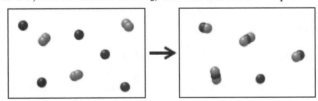

Think About It
It is important to ensure the reaction is balanced before trying to draw the diagram.

3.49. Collect and Organize
To balance the chemical equations, we use the three steps described in the textbook.

Analyze
To balance each equation we first write the unbalanced equation, using the chemical formulas of the reactants and products. Next, we balance an element that is present in only one reactant and product. Finally, we balance the other elements present by placing coefficients in front of the species in the reaction so that the number of the atoms for each element is equal on both sides of the equation. If there are any fractional coefficients, we multiply the entire equation to eliminate all fractions.

Solve

(a) 1. The unbalanced reaction is

$$FeSiO_3(s) + H_2O(\ell) \rightarrow Fe_3Si_2O_5(OH)_4(s) + H_4SiO_4(aq)$$

Element	Left Side	Right Side
Fe	1	3
Si	1	2 + 1 = 3
O	3 + 1 = 4	9 + 4 = 13
H	2	4 + 4 = 8

None of the atoms, in comparing the right and left sides, are equal. This reaction is not balanced.

2. We begin by balancing the Fe atoms, as they appear in only one reactant and one product. We can do this by placing a 3 in front of $FeSiO_3$ on the left side.

$$3 \, FeSiO_3(s) + H_2O(\ell) \rightarrow Fe_3Si_2O_5(OH)_4(s) + H_4SiO_4(aq)$$

Element	Left Side	Right Side
Fe	3 × 1 = 3	3
Si	3 × 1 = 3	2 + 1 = 3
O	3(3) + 1 = 10	9 + 4 = 13
H	2	4 + 4 = 8

3. This also balances the silicon atoms. The number of H atoms on the right side is four times the number of that on the left side. To balance the H atoms, therefore, we place 4 as the coefficient before H_2O on the left side of the equation.

$$3 \, FeSiO_3(s) + 4 \, H_2O(\ell) \rightarrow Fe_3Si_2O_5(OH)_4(s) + H_4SiO_4(aq)$$

Element	Left Side	Right Side
Fe	3 × 1 = 3	3
Si	3 × 1 = 3	2 + 1 = 3
O	3(3) + 4(1) = 13	9 + 4 = 13
H	4 × 2 = 8	4 + 4 = 8

The equation is now balanced.

(b) 1. The unbalanced reaction is

$$Fe_2SiO_4(s) + CO_2(g) + H_2O(\ell) \rightarrow FeCO_3(s) + H_4SiO_4(aq)$$

Element	Left Side	Right Side
Fe	2	1
Si	1	1
C	1	1
O	4 + 2 + 1 = 7	3 + 4 = 7
H	2	4

Though the number of silicon, carbon, and oxygen atoms on both sides of the equation are equal, the number of iron and hydrogen atoms are not. This reaction is not balanced.

2. We begin by balancing the Fe atoms, since they appear in only one reactant and one product. We can do this by placing a 2 in front of $FeCO_3$ on the right side.

$$Fe_2SiO_4(s) + CO_2(g) + H_2O(\ell) \rightarrow 2\,FeCO_3(s) + H_4SiO_4(aq)$$

Element	Left Side	Right Side
Fe	2	$2 \times 1 = 2$
Si	1	1
C	1	$2 \times 1 = 2$
O	$4 + 2 + 1 = 7$	$2(3) + 4 = 10$
H	2	4

3. The number of C atoms on the right side is twice the number of that on the left side. To balance the C atoms, therefore, we place 2 as the coefficient before CO_2 on the left side of the equation.

$$Fe_2SiO_4(s) + 2\,CO_2(g) + H_2O(\ell) \rightarrow 2\,FeCO_3(s) + H_4SiO_4(aq)$$

Element	Left Side	Right Side
Fe	2	$2 \times 1 = 2$
Si	1	1
C	$2 \times 1 = 2$	$2 \times 1 = 2$
O	$4 + 2(2) + 1 = 9$	$2(3) + 4 = 10$
H	2	4

To balance the H atoms, we place a coefficient of 2 in front of H_2O on the left side. This also balances the O atoms on each side.

$$Fe_2SiO_4(s) + 2\,CO_2(g) + 2\,H_2O(\ell) \rightarrow 2\,FeCO_3(s) + H_4SiO_4(aq)$$

Element	Left Side	Right Side
Fe	2	$2 \times 1 = 2$
Si	1	1
C	$2 \times 1 = 2$	$2 \times 1 = 2$
O	$4 + 2(2) + 2(1) = 10$	$2(3) + 4 = 10$
H	$2 \times 2 = 4$	4

The equation is now balanced.
(c) 1. The unbalanced reaction is

$$Fe_3Si_2O_5(OH)_4(s) + CO_2(g) + H_2O(\ell) \rightarrow FeCO_3(s) + H_4SiO_4(aq)$$

Element	Left Side	Right Side
Fe	3	1
Si	2	1
C	1	1
O	$9 + 2 + 1 = 12$	$3 + 4 = 7$
H	$4 + 2 = 6$	4

Though the number of C atoms on both sides of the equation is equal, the number of Fe, Si, O, and H atoms are not. This reaction is not balanced.
2. We begin by balancing the Fe and Si atoms, as they appear in only one reactant and one product. We can do this by placing a 3 in front of $FeCO_3$ and a 2 in front of H_4SiO_4 on the right side.

$$Fe_3Si_2O_5(OH)_4(s) + CO_2(g) + H_2O(\ell) \rightarrow 3\,FeCO_3(s) + 2\,H_4SiO_4(aq)$$

Element	Left Side	Right Side
Fe	3	$3 \times 1 = 3$
Si	2	$2 \times 1 = 2$
C	1	$3 \times 1 = 3$
O	$9 + 2 + 1 = 12$	$3(3) + 2(4) = 17$
H	$4 + 2 = 6$	$2 \times 4 = 8$

3. The number of C atoms on the right side is three times the number of that on the left side. To balance the C atoms, therefore, we place 3 as the coefficient before CO_2 on the left side of the equation.

$$Fe_3Si_2O_5(OH)_4(s) + 3\,CO_2(g) + H_2O(\ell) \rightarrow 3\,FeCO_3(s) + 2\,H_4SiO_4(aq)$$

Element	Left Side	Right Side
Fe	3	$3 \times 1 = 3$
Si	2	$2 \times 1 = 2$
C	$3 \times 1 = 3$	$3 \times 1 = 3$
O	$9 + 3(2) + 1 = 16$	$3(3) + 2(4) = 17$
H	$4 + 2 = 6$	$2 \times 4 = 8$

To balance the H atoms, we place a coefficient of 2 in front of H_2O on the left side.

$$Fe_3Si_2O_5(OH)_4(s) + 3\,CO_2(g) + 2\,H_2O(\ell) \rightarrow 3\,FeCO_3(s) + 2\,H_4SiO_4(aq)$$

Element	Left Side	Right Side
Fe	3	$3 \times 1 = 3$
Si	2	$2 \times 1 = 2$
C	$3 \times 1 = 3$	$3 \times 1 = 3$
O	$9 + 3(2) + 2(1) = 17$	$3(3) + 2(4) = 17$
H	$4 + 2(2) = 8$	$2 \times 4 = 8$

The equation is now balanced.

Think About It

These complex formulas looked difficult at first glance to balance. A good strategy here is to first balance all atoms except O and H atoms, then finish by balancing the O and H with a coefficient in front of H_2O.

3.51. **Collect and Organize**

To balance these chemical reactions we use the three steps described in the textbook.

Analyze

To balance each equation we first write the unbalanced equation using the chemical formulas of the reactants and products. Next, we balance an element that is present in only one reactant and one product. Finally, we balance the other elements present by placing coefficients in front of the species in the reaction so that the number of the atoms for each element is equal on both sides of the equation. If there are any fractional coefficients, we multiply the entire equation to eliminate all fractions.

Solve

(a) 1. The unbalanced reaction is

$$N_2(g) + O_2(g) \rightarrow NO(g)$$

Element	Left Side	Right Side
N	2	1
O	2	1

Neither the N nor the O atoms are balanced in this equation. This reaction is not balanced.

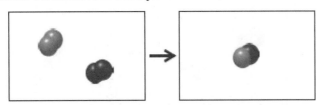

2. We begin by balancing the N atoms. We can do this by placing a 2 in front of NO on the right side.

$$N_2(g) + O_2(g) \rightarrow 2\,NO(g)$$

Element	Left Side	Right Side
N	2	$2 \times 1 = 2$
O	2	$2 \times 1 = 2$

The equation is now balanced.
(b) 1. The unbalanced reaction is

$$NO(g) + O_2(g) \rightarrow NO_2(g)$$

Element	Left Side	Right Side
N	1	1
O	$1 + 2 = 3$	2

Though the number of N atoms on both sides of the equation is equal, the number of O atoms is not. This reaction is not balanced.

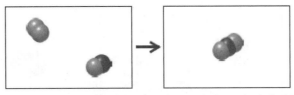

2. We can start by balancing the O atoms by placing a $\frac{1}{2}$ in front of O_2 on the left side.

$$NO(g) + \tfrac{1}{2}O_2(g) \rightarrow NO_2(g)$$

Element	Left Side	Right Side
N	1	1
O	$1 + \frac{1}{2}(2) = 2$	2

3. To eliminate the fractional coefficients, we multiply all of the coefficients by 2.

$$2\,NO(g) + O_2(g) \rightarrow 2\,NO_2(g)$$

Element	Left Side	Right Side
N	$2(1) = 2$	$2(1) = 2$
O	$2(1) + 2 = 4$	$2(2) = 4$

The equation is now balanced.

(c) 1. The unbalanced reaction is

$$NO(g) + NO_3(g) \rightarrow NO_2(g)$$

Element	Left Side	Right Side
N	$1 + 1 = 2$	1
O	$1 + 3 = 4$	2

Neither the N nor the O atoms are equal on both sides of the equation. This reaction is not balanced.

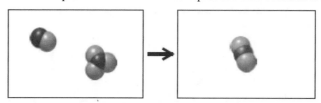

(2) We can start by balancing the N atoms. We can do this by placing a 2 in front of NO_2 on the right side.

$$NO(g) + NO_3(g) \rightarrow 2\,NO_2(g)$$

Element	Left Side	Right Side
N	$1 + 1 = 2$	$2(1) = 2$
O	$1 + 3 = 4$	$2(2) = 4$

The equation is now balanced.
(d) 1. The unbalanced reaction is

$$N_2(g) + O_2(g) \rightarrow N_2O(g)$$

Element	Left Side	Right Side
N	2	2
O	2	1

The number of N atoms on both sides of the equation is equal, the number of O atoms is not. This reaction is not balanced.

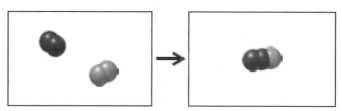

2. We can start by balancing the O atoms. We can do this by placing a coefficient of $\frac{1}{2}$ in front of O_2 on the left-hand side.

$$N_2(g) + \tfrac{1}{2}O_2(g) \rightarrow N_2O(g)$$

Element	Left Side	Right Side
N	2	2
O	$\frac{1}{2}(2) = 1$	1

3. To eliminate the fractional coefficients we multiply all of the coefficients by 2.

$$2\,N_2(g) + O_2(g) \rightarrow 2\,N_2O(g)$$

Element	Left Side	Right Side
N	$2(2) = 4$	$2(2) = 4$
O	$2(\frac{1}{2}(2)) = 2$	$2(1) = 2$

The equation is now balanced.

Think About It

There are many gaseous oxides of nitrogen because nitrogen can occur in compounds with several different charges. In these reactions we see nitrogen's oxidation states as 0 (N_2), 1+ (N_2O), 2+ (NO), and 4+ (NO_2), and 6+ (NO_3).

3.53. **Collect and Organize**

In this question we are asked to write balanced chemical equations for the reactions described. Because we are provided with only the names, not the chemical formulas for the reactants and products, we have to be sure to correctly write the formulas to balance the equations. To balance these chemical reactions we use the three steps described in the textbook.

Analyze

To balance each equation we first write the unbalanced equation using the chemical formulas of the reactants and products. Next, we balance an element that is present in only one reactant and product. Finally, we balance the other elements present by placing coefficients in front of the species in the reaction so that the number of the atoms for each element is equal on both sides of the equation. If there are any fractional coefficients, we multiply the entire equation through to eliminate all fractions.

Solve

(a) 1. The unbalanced reaction is

$$N_2O_5(g) + Na(s) \rightarrow NaNO_3(s) + NO_2(g)$$

Element	Left Side	Right Side
N	2	$1 + 1 = 2$
O	5	$3 + 2 = 5$
Na	1	1

The numbers of Na, N, and O atoms on the reactants and products side are all equal. This reaction is already balanced!

(b) 1. The unbalanced reaction is

$$N_2O_4(g) + H_2O(\ell) \rightarrow HNO_3(aq) + HNO_2(aq)$$

Element	Left Side	Right Side
N	2	$1 + 1 = 2$
O	$4 + 1 = 5$	$3 + 2 = 5$
H	2	$1 + 1 = 2$

The numbers of N, O, and H atoms on the reactants and products side are all equal. This reaction is already balanced!

(c) 1. The unbalanced reaction is

$$NO(g) \rightarrow N_2O(g) + NO_2(g)$$

Element	Left Side	Right Side
N	1	$2 + 1 = 3$
O	1	$1 + 2 = 3$

This reaction is not balanced as the number of N and O atoms differs from the reactants to the products side.

2. We can start by balancing the N atoms. We can do this by placing a 3 in front of NO on the left side.

$$3 NO(g) \rightarrow N_2O(g) + NO_2(g)$$

Element	Left Side	Right Side
N	$3(1) = 3$	$2 + 1 = 3$
O	$3(1) = 3$	$1 + 2 = 3$

This also resulted in the O atoms being balanced. The equation is now balanced.

(d) 1. The unbalanced reaction is

$$C_2H_2(g) + O_2(g) \rightarrow CO_2(g) + H_2O(g)$$

Element	Left Side	Right Side
C	2	1
H	2	2
O	2	2 + 1 = 3

This reaction is not balanced as the number of C and O atoms differs from the reactants to the products side.
2. We can start by balancing the C atoms. We can do this by placing a 2 in front of CO_2 on the products side.

$$C_2H_2(g) + O_2(g) \rightarrow 2\,CO_2(g) + H_2O(g)$$

Element	Left Side	Right Side
C	2	2
H	2	2
O	2	2(2) + 1 = 5

3. The only thing left to balance are the O atoms. We can do this by placing a 5/2 in front of O_2 on the reactant side.

$$C_2H_2(g) + 5/2\,O_2(g) \rightarrow 2\,CO_2(g) + H_2O(g)$$

Element	Left Side	Right Side
C	2	2
H	2	2
O	5	2(2) + 1 = 5

4. Multiplying everything by 2 will remove the fractional coefficient.

$$2\,C_2H_2(g) + 5\,O_2(g) \rightarrow 4\,CO_2(g) + 2\,H_2O(g)$$

Element	Left Side	Right Side
C	4	4
H	4	4
O	10	8 + 2 = 10

The equation is now balanced.

Think About It
The first two chemical reactions were balanced as written, and there was no need to change the coefficients. When writing chemical equations, however, it is always best to be sure that the equation is balanced.

3.55. **Collect and Organize**
We are asked to consider whether having mass balance for a chemical reaction (mass of products = mass of reactants) means that the reaction is balanced.

Analyze
A balanced chemical reaction follows the law of conservation of mass. This means that we balance the numbers of a particular kind of atom (element) in reactants and products. If we balance the numbers of the kinds of atoms, the mass of the reactants will be equal to the mass of the products.

Solve
Yes, if the masses were unequal, then the equation would be missing either some reactants or products.

Think About It
When balancing equations, we most often look for atom balance; we do not often calculate a mass balance to check whether a reaction is balanced.

3.57. Collect and Organize
We need to convert the given mass of carbon by which emissions would be reduced first to moles of carbon, then to the mass of carbon dioxide.

Analyze
The mass of carbon can be converted to moles by dividing by the average molar mass of carbon after converting the mass of carbon to grams from kilograms (1000 g = 1 kg). Because 1 mole of carbon is contained in 1 mole of carbon dioxide, the moles of carbon are equal to the moles of carbon dioxide. To find the mass of CO_2 in grams, we need only to multiply the moles of CO_2 by the molar mass of CO_2, then convert that mass (which will be in grams) to kilograms of CO_2.

Solve

(a) $5.4 \times 10^9 \text{ kg C} \times \dfrac{1000 \text{ g}}{1 \text{ kg}} \times \dfrac{1 \text{ mol C}}{12.01 \text{ g}} \times \dfrac{1 \text{ mol CO}_2}{1 \text{ mol C}} = 4.5 \times 10^{11} \text{ mol CO}_2$

(b) $4.5 \times 10^{11} \text{ mol CO}_2 \times \dfrac{44.01 \text{ g}}{1 \text{ mol CO}_2} \times \dfrac{1 \text{ kg}}{1000 \text{ g}} = 2.0 \times 10^{10} \text{ kg CO}_2$

Think About It
The mass amount of CO_2 emissions that would be reduced is greater than the mass of carbon burned. This is because the molar mass of carbon dioxide is greater, so the mass of a certain molar amount of carbon dioxide is greater than that of the same molar amount of carbon.

3.59. Collect and Organize
To calculate the amount of CO_2 produced from the decomposition of 25.0 g of $NaHCO_3$, we need the balanced chemical equation to calculate the molar ratio of the CO_2 produced from a given mass of $NaHCO_3$.

Analyze
We know that the reactant for the balanced equation is $NaHCO_3$ and that the products are Na_2CO_3, H_2O, and CO_2. After we balance the equation, we can use the ratio of $NaHCO_3$ to CO_2 to find the moles of CO_2 from the moles of $NaHCO_3$ (found from the mass by dividing by the molar mass, 84.01 g/mol). From the moles of CO_2, we can find the mass of CO_2 produced using 44.01 g/mol for the molar mass of CO_2.

Solve
(a) 1. The unbalanced reaction is
$$NaHCO_3(s) \rightarrow CO_2(g) + H_2O(g) + Na_2CO_3(s)$$

Element	Left Side	Right Side
Na	1	2
H	1	2
C	1	2
O	3	6

None of the atoms are balanced in this reaction. We do notice that the number of atoms in the products is always twice that in the reactants.
2. We begin to balance the atoms by placing a 2 in front of $NaHCO_3$ on the left side.
$$2\,NaHCO_3(s) \rightarrow CO_2(g) + H_2O(g) + Na_2CO_3(s)$$

Element	Left Side	Right Side
Na	2 (1) = 2	2
H	2 (1) = 2	2
C	2 (1) = 2	2
O	2 (3) = 6	6

The equation is now balanced.

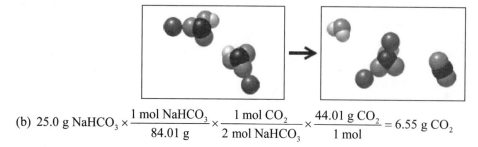

(b) $25.0 \text{ g NaHCO}_3 \times \dfrac{1 \text{ mol NaHCO}_3}{84.01 \text{ g}} \times \dfrac{1 \text{ mol CO}_2}{2 \text{ mol NaHCO}_3} \times \dfrac{44.01 \text{ g CO}_2}{1 \text{ mol}} = 6.55 \text{ g CO}_2$

Think About It
The mass of the CO_2 produced is quite a bit less than the 25 g of $NaHCO_3$ decomposed not only because the molar mass of CO_2 is lower than that of $NaHCO_3$, but also because for every mole of $NaHCO_3$ decomposed, only one-half mole of CO_2 is produced.

3.61. Collect and Organize
We use stoichiometric relationships to calculate the amount of a reactant, $NaAlO_2$, required to produce a given amount of cryolite, Na_3AlF_6.

Analyze
First we need to calculate the moles of Na_3AlF_6 that are present in 1.00 kg, using 1000 g = 1 kg and the molar mass of Na_3AlF_6 (209.94 g/mol). From that result and the 3:1 ratio of $NaAlO_2$ to Na_3AlF_6 in the balanced equation, we can calculate the moles of $NaAlO_2$ required. Finally, the molar mass of $NaAlO_2$ (81.97 g/mol) is used to convert the moles into mass.

Solve

$$1.00 \text{ kg Na}_3\text{AlF}_6 \times \dfrac{1000 \text{ g}}{1 \text{ kg}} \times \dfrac{1 \text{ mol Na}_3\text{AlF}_6}{209.94 \text{ g}} \times \dfrac{3 \text{ mol NaAlO}_2}{1 \text{ mol Na}_3\text{AlF}_6}$$

$$\times \dfrac{81.97 \text{ g NaAlO}_2}{1 \text{ mol}} = 1170 \text{ g or } 1.17 \text{ kg}$$

Think About It
In this reaction we need three moles of the reactant $NaAlO_2$ to yield 1 mole of product, Na_3AlF_6. Be careful here. The mass of $NaAlO_2$ required is not three times the mass of the product; it is the molar relationship that is important.

3.63. Collect and Organize
We have to consider that only 3.0% of the coal is sulfur and that the sulfur is then converted, upon burning, into SO_2. We are to find the number of tons of SO_2 produced from 25 metric tons of coal.

Analyze
To find the mass of sulfur in the coal, we multiply the mass of the coal by 0.030 (for 3.0%). We then need to convert the tons into grams [1 t (metric ton) = 1000 kg and 1000 g = 1 kg] and then into moles using the molar mass of sulfur (32.065 g/mol). From the balanced equation for the conversion of sulfur with oxygen to sulfur dioxide, we use the molar ratio of S to SO_2 to find the moles of SO_2 produced. Finally, we convert the molar mass of SO_2 to grams, which we can then convert back into metric tons.

Solve

The unbalanced reaction is

$$S(s) + O_2(g) \rightarrow SO_2(g)$$

Element	Left Side	Right Side
S	1	1
O	2	2

The equation is already balanced.
The moles of sulfur in the coal are given by

$$25 \text{ t coal} \times 0.030 = 0.75 \text{ t S in the coal}$$

$$0.75 \text{ t S} \times \frac{1000 \text{ kg}}{1 \text{ t}} \times \frac{1000 \text{ g}}{1 \text{ kg}} \times \frac{1 \text{ mol S}}{32.06 \text{ g}} = 2.3 \times 10^4 \text{ mol S}$$

and the mass of sulfur dioxide can then be found by

$$2.3 \times 10^4 \text{ mol S} \times \frac{1 \text{ mol SO}_2}{1 \text{ mol S}} \times \frac{64.06 \text{ g SO}_2}{1 \text{ mol SO}_2} \times \frac{1 \text{ kg}}{1000 \text{ g}} \times \frac{1 \text{ t}}{1000 \text{ kg}} = 1.5 \text{ t SO}_2$$

Think About It

This problem may have looked complex, but it could be broken down into several steps: writing a balanced equation, determining the amount of S in the coal using percents, finding the moles of S, using stoichiometric ratios to find the moles of SO_2 produced, and converting that answer into mass. If we remember that the key to the problem is the stoichiometric relationship of one quantity in a chemical reaction to another, we start to notice that the other parts of the problem usually involve converting quantities to the proper units.

3.65. Collect and Organize

We are asked to write balanced chemical equations for the combustion of ethanol and octane (as a model for gasoline). Using these balanced equations, we are then asked which fuel produces more CO_2 on a per gram basis.

Analyze

As a first step, we should write and balance chemical equations for the combustion of each fuel. Using the molar masses of ethanol (46.08 g/mol) and octane (114.26 g/mol), as well as the stoichiometric coefficients from the balanced chemical equations, we may determine how many moles of carbon dioxide are formed as a result of the combustion reactions. Multiplying by the molar mass of carbon dioxide (44.01 g/mol) allows us to determine the mass of CO_2 produced for each gram of fuel.

Solve

Octane

1. The unbalanced reaction is

$$C_8H_{18}(\ell) + O_2(g) \rightarrow CO_2(g) + H_2O(g)$$

Element	Left Side	Right Side
C	8	1
H	18	2
O	2	2 + 1 = 3

This reaction is not balanced, as the number of C, H, and O atoms differs from the reactants side to the products side.

2. We start by balancing the C atoms. We can do this by placing an 8 in front of CO_2 on the products side.

$$C_8H_{18}(\ell) + O_2(g) \rightarrow 8\,CO_2(g) + H_2O(g)$$

Element	Left Side	Right Side
C	8	8
H	18	2
O	2	16 + 1 = 17

3. We can balance H atoms by placing a 9 in front of H_2O on the product side.

$$C_8H_{18}(\ell) + O_2(g) \rightarrow 8\,CO_2(g) + 9\,H_2O(g)$$

Element	Left Side	Right Side
C	8	8
H	18	18
O	2	16 + 9 = 25

4. The number of O atoms on the right side is 12.5 times the number of that on the left side. To balance the O atoms, therefore, we place 25/2 as the coefficient before O_2 on the left side of the equation.

$$C_8H_{18}(\ell) + 25/2\,O_2(g) \rightarrow 8\,CO_2(g) + 9\,H_2O(g)$$

Element	Left Side	Right Side
C	8	8
H	18	18
O	25	16 + 9 = 25

5. Multiplying everything by 2 will remove the fractional coefficient.

$$2\,C_8H_{18}(\ell) + 25\,O_2(g) \rightarrow 16\,CO_2(g) + 18\,H_2O(g)$$

Element	Left Side	Right Side
C	16	16
H	36	36
O	50	32 + 18 = 50

The equation is now balanced.

Ethanol
1. The unbalanced reaction is

$$C_2H_6O(\ell) + O_2(g) \rightarrow CO_2(g) + H_2O(g)$$

Element	Left Side	Right Side
C	2	1
H	6	2
O	1 + 2 = 3	2 + 1 = 3

This reaction is not balanced as the number of C and H atoms differs from the reactants to the products side.
2. We can start by balancing the C atoms. We can do this by placing a 2 in front of CO_2 on the products side.

$$C_2H_6O(\ell) + O_2(g) \rightarrow 2\,CO_2(g) + H_2O(g)$$

Element	Left Side	Right Side
C	2	2
H	6	2
O	1 + 2 = 3	4 + 1 = 5

3. We have unbalanced the O atoms! (Don't worry, we will come back to them presently.) There are three times as many H atoms on the left side as on the right side, so we can balance H atoms by placing a 3 in front of H_2O on the product side.

$$C_2H_6O(\ell) + O_2(g) \rightarrow 2\,CO_2(g) + 3\,H_2O(g)$$

Element	Left Side	Right Side
C	2	2
H	6	6
O	1 + 2 = 3	4 + 3 = 7

4. The number of O atoms on the right side is greater than the number of O atoms on the left side. If we balance by adjusting the coefficient on ethanol, we will unbalance C and H atoms. We need an additional 7 − 3 = 4 O atoms on the left side. To balance the O atoms, therefore, we place 3 as the coefficient before O_2 on the left side of the equation.

$$C_2H_6O(\ell) + 3\,O_2(g) \rightarrow 2\,CO_2(g) + 3\,H_2O(g)$$

Element	Left Side	Right Side
C	2	2
H	6	6
O	1 + 6 = 7	4 + 3 = 7

The equation is now balanced.
The mass of CO_2 produced by 1 g of octane is

$$1.00\text{ g C}_8\text{H}_{18} \times \frac{1\text{ mol C}_8\text{H}_{18}}{114.26\text{ g C}_8\text{H}_{18}} \times \frac{16\text{ mol CO}_2}{2\text{ mol C}_8\text{H}_{18}} \times \frac{44.01\text{ g CO}_2}{1\text{ mol CO}_2} = 3.08\text{ g CO}_2$$

The mass of CO_2 produced by 1 g of ethanol is

$$1.00\text{ g C}_2\text{H}_6\text{O} \times \frac{1\text{ mol C}_2\text{H}_6\text{O}}{46.08\text{ g C}_2\text{H}_6\text{O}} \times \frac{2\text{ mol CO}_2}{1\text{ mol C}_2\text{H}_6\text{O}} \times \frac{44.01\text{ g CO}_2}{1\text{ mol CO}_2} = 1.91\text{ g CO}_2$$

Ethanol produces less CO_2 per gram of fuel.

Think About It
In order to solve this problem, it is critical to correctly balance the equations for the combustion reactions. Without the correct stoichiometric ratio of fuel to CO_2, we cannot determine the correct mass ratio. You may be wondering why we use gasoline and not ethanol, if ethanol emits less CO_2 per gram of fuel. For a closer look, see Problem 5.99 in Chapter 5 of this text!

3.67. Collect and Organize
In converting chalcopyrite ($CuFeS_2$) to copper, we have to take into account that there is 1 mole of copper atoms in the formula for this mineral.

Analyze
The problem asks how much copper could be produced from 1.00 kg of the mineral and looks like many other stoichiometry problems. The molar mass of the mineral (183.52 g/mol) will have to be used to find the moles of the mineral. From there, knowing that there is 1 mole of copper atoms in 1 mole of the mineral, we can use the molar mass of copper (63.55 g/mol) to calculate the copper that would be produced.

Solve

$$1.00\text{ kg CuFeS}_2 \times \frac{1000\text{ g}}{1\text{ kg}} \times \frac{1\text{ mol CuFeS}_2}{183.52\text{ g CuFeS}_2} \times \frac{1\text{ mol Cu}}{1\text{ mol CuFeS}_2} \times \frac{63.55\text{ g Cu}}{1\text{ mol Cu}} = 346\text{ g Cu}$$

Think About It

Our calculation tells us that the ore is 34.6% Cu by mass.

3.69. Collect and Organize

We are asked to distinguish between empirical formula and molecular formula.

Analyze

An empirical formula gives the simplest whole-number ratio of atoms of the elements in a molecule, whereas a molecular formula gives the actual number of the atoms in a molecule.

Solve

An empirical formula is concerned with the lowest whole-number ratios of atoms in a substance. A molecular formula is concerned with the actual numbers of each kind of atom that compose one molecular unit of the substance.

Think About It

In some cases the molecular formula is equivalent to the empirical formula—the atoms in the molecular formula are in their lowest whole-number ratios.

3.71. Collect and Organize

We are asked whether the atom in a molecular formula with the largest atomic mass is *always* the element present in the highest percentage by mass.

Analyze

The percent composition of an element is the total mass of the element in the compound divided by the atomic mass of the compound. In calculating the percent mass for each element, we need to take into account how many atoms of that element are present in the molecular formula.

Solve

No, lighter elements may be present in sufficient quantities to be of a greater percent mass than a heavier element.

Think About It

A good example is SiO_2. Silicon is the heavier element (28 g/mol), but the presence of two oxygen atoms (16 g/mol × 2) makes O 53% by mass and Si only 47% by mass.

3.73. Collect and Organize

We are asked if three compounds with the same molar mass and percent composition have the same empirical and molecular formula.

Analyze

The percent composition and empirical formula of a sample are related to the mass ratio of component atoms. The molar mass of a compound allows us to link the empirical and molecular formulas.

Solve

Since all three compounds have the same percent composition, they must also have the same empirical formula. Given that they also have the same molar mass, all three compounds must also have the same relationship between molecular and empirical formulas. Both empirical and molecular formulas are the same for all three.

Think About It

The molecular formula reflects both the number of atoms of each kind in one molecule and the moles of those kinds of atoms that make up 1 mole of the substance.

3.75. Collect and Organize

We are asked if the four compounds listed have the same empirical formula.

Analyze

Recall that the empirical formula of a compound is the smallest whole number ratio of atoms in the compound. We can compare these directly for the given compounds.

Solve

The empirical formulas for the given compounds are

Molecular Formula	C_6H_{14}	C_7H_{16}	C_8H_{18}	C_9H_{20}
Empirical Formula	C_3H_7	C_7H_{16}	C_4H_9	C_9H_{20}

None of these compounds have the same empirical formula.

Think About It

Though these compounds all have different empirical and molecular formulas, they have similar boiling points.

3.77. **Collect and Organize**

To calculate the percent composition for the elements in each compound we divide the molar mass of each element from the periodic table by the molar mass for the compound and convert to a percentage.

Analyze

All of the chemical formulas are given for the compounds. Assume we have 1 mole of each compound. We first compute the molar mass. Then, to find the percentage of each element, divide the mass of each element present in the compound, taking into account the presence of multiple atoms of the element if appropriate, by the molar mass of the compound and multiply by 100.

Solve

(a) Molar mass of Na_2O = 61.98 g/mol.

$$\% \text{ Na} = \frac{(22.99 \text{ g} \times 2) \text{ Na}}{61.98 \text{ g}} \times 100 = 74.19\% \text{ Na}$$

$$\% \text{ O} = \frac{16.00 \text{ g O}}{61.98 \text{ g}} \times 100 = 25.81\% \text{ O}$$

(b) Molar mass of NaOH = 40.00 g/mol.

$$\% \text{ Na} = \frac{22.99 \text{ g Na}}{40.00 \text{ g}} \times 100 = 57.48\% \text{ Na}$$

$$\% \text{ O} = \frac{16.00 \text{ g O}}{40.00 \text{ g}} \times 100 = 40.00\% \text{ O}$$

$$\% \text{ H} = \frac{1.008 \text{ g H}}{40.00 \text{ g}} \times 100 = 2.52\% \text{ H}$$

(c) Molar mass of $NaHCO_3$ = 84.01 g/mol.

$$\% \text{ Na} = \frac{22.99 \text{ g Na}}{84.01 \text{ g}} \times 100 = 27.37\% \text{ Na}$$

$$\% \text{ H} = \frac{1.01 \text{ g H}}{84.01 \text{ g}} \times 100 = 1.20\% \text{ H}$$

$$\% \text{ C} = \frac{12.01 \text{ g C}}{84.01 \text{ g}} \times 100 = 14.30\% \text{ C}$$

$$\% \text{ O} = \frac{(16.00 \text{ g} \times 3) \text{ O}}{84.01 \text{ g}} \times 100 = 57.14\% \text{ O}$$

(d) Molar mass of Na_2CO_3 = 105.99 g/mol.

$$\% \text{ Na} = \frac{(22.99 \text{ g} \times 2) \text{ Na}}{106.0 \text{ g}} \times 100 = 43.38\% \text{ Na}$$

$$\% \text{ C} = \frac{12.01 \text{ g C}}{106.0 \text{ g}} \times 100 = 11.33\% \text{ C}$$

$$\% \text{ O} = \frac{(16.00 \text{ g} \times 3) \text{ O}}{106.0 \text{ g}} \times 100 = 45.28\% \text{ O}$$

Think About It

For all of these common salts of sodium, notice that the percentage of sodium is different in each. This is due not only to the different atom ratios of sodium present in the compounds, but also to the different molar masses of the compounds.

3.79. **Collect and Organize**

We are not able to tell simply by looking at the chemical formula which compound has the greatest percentage of carbon by mass. We have to find the percent composition of hydrogen and carbon in each and then compare the compounds' percent carbon (%C). We are also asked to determine if any of the given compounds have identical empirical formulas.

Analyze

For each compound, assume we have 1 mole of the substance and then compute the molar mass. For the percentage of carbon, divide the mass of all the carbon present in 1 mole by the molar mass and multiply by 100 to find the percentage.

Solve
(a) Naphthalene, $C_{10}H_8$:

$$\% \text{ C} = \frac{(12.01 \times 10) \text{ g C}}{128.2 \text{ g}} \times 100 = 93.68\% \text{ C}$$

(b) Chrysene, $C_{18}H_{12}$:

$$\% \text{ C} = \frac{(12.01 \times 18) \text{ g C}}{228.3 \text{ g}} \times 100 = 94.69\% \text{ C}$$

(c) Pentacene, $C_{22}H_{14}$:

$$\% \text{ C} = \frac{(12.01 \times 22) \text{ g C}}{278.4 \text{ g}} \times 100 = 94.91\% \text{ C}$$

(d) Pyrene, $C_{16}H_{10}$:

$$\% \text{ C} = \frac{(12.01 \times 16) \text{ g C}}{202.3 \text{ g}} \times 100 = 94.99\% \text{ C}$$

Pyrene, $C_{16}H_{10}$, has the highest %C by mass of all these hydrocarbons.
The empirical formulas for the given compounds are

Molecular Formula	$C_{10}H_8$	$C_{18}H_{12}$	$C_{22}H_{14}$	$C_{16}H_{10}$
Empirical Formula	C_5H_4	C_3H_2	$C_{11}H_7$	C_8H_5

None of these compounds has the same empirical formula.

Think About It

These compounds all have relatively close percent compositions, which is not at all obvious by looking only at their chemical formulas.

3.81. **Collect and Organize**

We are asked to compare the percent carbon by mass in CH_4 and CF_4.

Analyze

To determine the percentage of C by mass in each compound, we assume that we have 1 mole of the substance. We then have to divide the mass of carbon present by the molar mass of either CH_4 (16.05 g/mol) or CF_4 (88.01 g/mol). Since the molar masses are significantly different, we expect a large difference in the percentage C by mass.

Solve

The %C by mass in CH_4 is

$$\% \, C = \frac{12.01 \text{ g C}}{16.05 \text{ g CH}_4} \times 100\% = 74.83\% \text{ C}$$

The %C by mass in CF_4 is

$$\% \, C = \frac{12.01 \text{ g C}}{88.01 \text{ g CF}_4} \times 100\% = 13.65\% \text{ C}$$

CH_4 has a greater mass percentage carbon.

Think About It

The large difference in percent C by mass for these two compounds reminds us that mass percentage and mole percentage may be significantly different from one another.

3.83. **Collect and Organize**

We are to determine the empirical formula for surgical-grade titanium, an alloy, given the percentages of titanium, aluminum, and vanadium it contains.

Analyze

The empirical formula is the lowest whole-number ratio of atoms in a compound. If we assume 100 grams of surgical-grade titanium, we have 64.39 g Ti, 24.19 g Al, and 11.42 g V in the sample. After we calculate the moles of each of these elements by dividing the mass by the molar mass of each element, we divide the molar amounts obtained by the lowest molar amount to find the lowest whole-number ratio of the elements.

Solve

$$\text{mol Ti} = \frac{64.39 \text{ g}}{47.87 \text{ g/mol}} = 1.345 \text{ mol Ti}$$

$$\text{mol Al} = \frac{24.19 \text{ g}}{26.98 \text{ g/mol}} = 0.8966 \text{ mol Al}$$

$$\text{mol V} = \frac{11.42 \text{ g}}{50.94 \text{ g/mol}} = 0.2242 \text{ mol V}$$

Dividing by the lowest molar amount we get a titanium–aluminum–vanadium ratio of 6:4:1. Therefore, the empirical formula for surgical-grade titanium is Ti_6Al_4V.

Think About It

This titanium alloy is useful as a surgical alloy because it resists corrosion, is lightweight and strong, and is biocompatible.

3.85. **Collect and Organize**

We are given the masses of the reactants and products and are asked to determine the empirical formula for the product of the reaction of magnesium with oxygen. Then we need to write a balanced chemical reaction.

Analyze

The masses of the reactants add up to give the mass of the products; so no other products are formed in this reaction. Therefore, the product contains 2.43 g Mg and 1.60 g O. The mass composition of the product, then, is the mass of each element divided by the total mass of the product, multiplied by 100. We can then assume that there are 100 g of product. In that case, the percent composition gives the mass of each element in the product, which we can then convert to moles using the molar mass of the elements. We can get the mole ratio for the two elements in the products by determining the lowest whole-number ratio of the moles. From that we can write the empirical formula and the chemical equation for balancing.

Solve

(a) The mass percentage of each element in the product is

$$\% \text{ mass of Mg} = \frac{2.43 \text{ g}}{4.03 \text{ g}} \times 100 = 60.3\%$$

$$\% \text{ mass of O} = \frac{1.60 \text{ g}}{4.03 \text{ g}} \times 100 = 39.7\%$$

Assuming 100 g of product,

$$60.3 \text{ g Mg} \times \frac{1 \text{ mol}}{24.31 \text{ g}} = 2.48 \text{ mol Mg}$$

$$39.7 \text{ g O} \times \frac{1 \text{ mol}}{16.00 \text{ g}} = 2.48 \text{ mol O}$$

This is a 1:1 molar ratio of Mg to O; so the empirical formula is MgO.

(b) The balanced equation is $2 \text{ Mg}(s) + \text{O}_2(g) \longrightarrow 2 \text{ MgO}(s)$.

Think About It

This question reflects how we would experimentally determine the formula for a new compound.

3.87. **Collect and Organize**

We use the percent composition of the asbestos mineral chrysotile to determine the empirical formula.

Analyze

If we assume 100 g of the chrysotile, the percent composition (26.31% Mg, 20.27% Si, 1.45% H, and the rest O) gives us the grams of each element. These can be converted to moles of each element via the molar masses of the elements from the periodic table. The empirical formula will be the lowest whole-number ratio of the moles of the elements in the chrysotile.

Solve

Oxygen is the only element that is not specified with a mass percentage. Therefore, oxygen's percentage can be determined by

$$100 - (26.31 + 20.27 + 1.45) = 51.97\%$$

The moles of each element present are given by

$$26.31 \text{ g Mg} \times \frac{1 \text{ mol}}{24.31 \text{ g}} = 1.082 \text{ mol Mg}$$

$$20.27 \text{ g Si} \times \frac{1 \text{ mol}}{28.09 \text{ g}} = 0.7216 \text{ mol Si}$$

$$1.45 \text{ g H} \times \frac{1 \text{ mol}}{1.01 \text{ g}} = 1.44 \text{ mol H}$$

$$51.97 \text{ g O} \times \frac{1 \text{ mol}}{16.00} = 3.248 \text{ mol O}$$

Dividing these by the lowest molar amount (0.7216) gives a magnesium–silicon–hydrogen–oxygen ratio of 1.5:1:2:4.5. Multiplying these by 2 gives a whole-number ratio of 3:2:4:9 for an empirical formula of $Mg_3Si_2H_4O_9$.

Think About It
The trick here is to recognize that the mass percentage of oxygen was not given in the original statement of the problem. Be sure to determine the moles for each of the elements present in the compound.

3.89. Collect and Organize
From the percent composition of a compound containing only carbon, hydrogen, and nitrogen, we are to determine the empirical and molecular formula. We have to convert the mass to moles, then find the lowest whole-number molar ratio for the elements in the compound. Using the molar mass, we must then determine the molecular formula for adenine.

Analyze
If we assume 100 g of the compound, the percentage of each of the elements (44.44% C, 3.73% H, 51.84% N) gives us the mass of each element in grams. Those masses can be converted into moles using the molar masses of the elements from the periodic table. Then we compare the moles to find the molar ratio (empirical formula). Dividing the molar mass for adenine by the molar mass of the formula unit will tell us how many formula units are in this molecule.

Solve

$$44.44 \text{ g C} \times \frac{1 \text{ mol C}}{12.01 \text{ g C}} = 3.70 \text{ mol C}$$

$$3.73 \text{ g H} \times \frac{1 \text{ mol H}}{1.01 \text{ g H}} = 3.69 \text{ mol H}$$

$$51.84 \text{ g N} \times \frac{1 \text{ mol N}}{14.01 \text{ g N}} = 3.70 \text{ mol N}$$

Dividing each of these mole amounts by the lowest mole amount (3.69) gives a C:H:N ratio of 1:1:1. The empirical formula for the compound therefore is CHN. Using the molar mass and the formula mass, we may determine the number of formula units in the molecule

$$\frac{\mathcal{M}_{molecular}}{\mathcal{M}_{empirical}} = \frac{135.14 \text{ g/mol}}{27.03 \text{ g/mol}} = 5$$

This means that the molecular formula is 5(CHN) or $C_5H_5N_5$.

Think About It
We know that adenine contains only carbon, hydrogen, and nitrogen because the percent abundances for these three elements total 100%.

3.91. **Collect and Organize**

We consider why combustion analysis must be carried out in excess amounts of oxygen.

Analyze

In combustion analysis, compounds (usually organic) are burned in oxygen and the masses of recovered CO_2 and H_2O produced in the reaction are related to the percentages of C and H in the original compound.

Solve

The excess of oxygen is required in combustion analysis to ensure the complete reaction of the hydrogen and carbon to form water and carbon dioxide.

Think About It

Combustion in an atmosphere deficient in oxygen gives CO instead of CO_2 as the main gaseous carbon product.

3.93. **Collect and Organize**

We are asked whether combustion analysis can ever give the true molecular formula for a compound.

Analyze

Combustion analysis gives us the percent mass of C, H, and, by calculation of the missing mass for some compounds, O in the compound, from which we can derive the empirical formula.

Solve

Yes, the combustion analysis can give the true molecular formula for a compound, but only if the empirical formula is the same as the molecular formula and only if the compound contains only C, H, and O.

Think About It

We can confirm the molar mass of a compound by other methods such as boiling point elevation, freezing point depression, or osmotic pressure.

3.95. **Collect and Organize**

We are asked to predict the identity of the nitrogen-containing species of combustion in excess oxygen.

Analyze

Complete combustion in excess oxygen will generate a nitrogen-containing compound with a high percentage of oxygen atoms. We should consider commonly-observed nitrogen oxides such as NO, NO_2, and N_2O. We can determine percentage oxygen in a compound by dividing the mass of all oxygen atoms in the molecule by the total mass of the molecule.

Solve

$$\% \, O = \frac{16.00 \text{ g O}}{30.01 \text{ g NO}} \times 100\% = 53.32\% \, O$$

$$\% \, O = \frac{16.00 \text{ g O}}{44.02 \text{ g N}_2\text{O}} \times 100\% = 36.35\% \, O$$

$$\% \, O = \frac{32.00 \text{ g O}}{46.01 \text{ g NO}_2} \times 100\% = 69.55\% \, O$$

NO_2 has the highest percentage oxygen, so it is the most likely product.

Think About It

Nitrogen oxides contribute to smog and air pollution, as we will discuss later in this text.

3.97. **Collect and Organize**

From the combustion data of a given mass of a compound containing only hydrogen and carbon and the molar mass of the compound, we are to determine the empirical and molecular formulas.

Analyze

This compound contains only hydrogen and carbon. The water (135.0 mg) resulted from the combustion of the hydrogen, and the carbon dioxide (440.0 mg) resulted from the combustion of the carbon. First, we determine the mass of hydrogen and oxygen present in the water and the carbon dioxide. From those results, we determine the mass percentage of the hydrogen and carbon in the compound (we know the original mass of the sample used in the analysis, 135.0 mg). From the mass percentage, we can find moles and the mole ratio of carbon and hydrogen in the compound and from there determine the empirical and molecular formulas (knowing the molar mass of the compound is 270 g/mol).

Solve

The mass and percentage of carbon and hydrogen in the compound are

$$440.0 \text{ mg CO}_2 \times \frac{1 \text{ g}}{1000 \text{ mg}} \times \frac{1 \text{ mol CO}_2}{44.01 \text{ g}} \times \frac{1 \text{ mol C}}{1 \text{ mol CO}_2} = 0.009998 \text{ mol C}$$

$$135.0 \text{ mg H}_2\text{O} \times \frac{1 \text{ g}}{1000 \text{ mg}} \times \frac{1 \text{ mol H}_2\text{O}}{18.02 \text{ g}} \times \frac{2 \text{ mol H}}{1 \text{ mol H}_2\text{O}} = 0.01498 \text{ mol H}$$

Dividing 0.01498 mol H by 0.009998 mol C gives a C:H ratio of 1:1.499. Multiplying by 2 to obtain a whole-number ratio gives an empirical formula of C_2H_3. The molar mass of this empirical formula is 27.05 g/mol. The molar mass of the compound is 270 g/mol, which is 10 times the molar mass of the empirical formula. Therefore, the molecular formula is $C_{20}H_{30}$.

Think About It

The empirical formula is derived from the molar ratio of the elements in the compound. Because combustion analysis gives us the amount of carbon and hydrogen in the compound, we need only relate the moles of CO_2 and H_2O to the C and H present in the compound.

3.99. **Collect and Organize**

From the data obtained from the combustion analysis of 175 mg of geraniol, we are to determine its empirical formula.

Analyze

The moles and masses of C and H can be calculated directly from the combustion analysis results. The oxygen content is the difference between the mass of the carbon plus hydrogen in the compound and the mass of the 175 mg sample. The moles of oxygen can then be determined using the molar mass of oxygen from the periodic table, and the ratio of carbon to hydrogen to oxygen can be calculated and used to determine the empirical formula.

Solve

$$0.499 \text{ g CO}_2 \times \frac{1 \text{ mol CO}_2}{44.01 \text{ g}} \times \frac{1 \text{ mol C}}{1 \text{ mol CO}_2} = 1.13 \times 10^{-2} \text{ mol C}$$

$$1.13 \times 10^{-2} \text{ mol C} \times \frac{12.01 \text{ g C}}{1 \text{ mol}} = 0.136 \text{ g C}$$

$$0.184 \text{ g H}_2\text{O} \times \frac{1 \text{ mol H}_2\text{O}}{18.02 \text{ g}} \times \frac{2 \text{ mol H}}{1 \text{ mol H}_2\text{O}} = 2.04 \times 10^{-2} \text{ mol H}$$

$$2.04 \times 10^{-2} \text{ mol H} \times \frac{1.008 \text{ g H}}{1 \text{ mol}} = 0.0206 \text{ g H}$$

$$\text{Total mass of C and H} = 0.136 \text{ g} + 0.0206 \text{ g} = 0.157 \text{ g}$$

$$\text{Mass of O present} = 0.175 \text{ g} - 0.157 \text{ g} = 0.018 \text{ g O}$$

$$\text{Moles of O in compound} = 0.018 \text{ g O} \times \frac{1 \text{ mol}}{15.999 \text{ g}} = 1.1 \times 10^{-3} \text{ mol O}$$

Dividing the moles of C, H, and O by the lowest molar amount (1.1×10^{-3} mol) gives a ratio of 10:18:1. This indicates an empirical formula of $C_{10}H_{18}O$.

Think About It
This problem involves an additional step to determine the mass of carbon and hydrogen present so that we can calculate the mass (and therefore the moles) of oxygen in the compound.

3.101. **Collect and Organize**

The compound from the bark contains oxygen as well as carbon and hydrogen. In this problem, therefore, we must determine the mass of carbon and hydrogen in the given mass of the compound to find the mass and moles of oxygen. From there we are able to get the molar ratios of C, H, and O.

Analyze
The moles and masses of C and H can be calculated directly from the combustion analysis results. The oxygen content is the difference between the mass of the carbon plus hydrogen in the compound and the mass of the 100 mg sample. The moles of oxygen can then be determined using the molar mass of oxygen from the periodic table, and the ratio of carbon to hydrogen to oxygen found can be used to determine the empirical formula.

Solve

$$220 \text{ mg CO}_2 \times \frac{1 \text{ g}}{1000 \text{ mg}} \times \frac{1 \text{ mol CO}_2}{44.01 \text{ g CO}_2} \times \frac{1 \text{ mol C}}{1 \text{ mol CO}_2} = 4.999 \times 10^{-3} \text{ mol C}$$

$$4.999 \times 10^{-3} \text{ mol C} \times \frac{12.01 \text{ g C}}{1 \text{ mol C}} = 0.06004 \text{ g C}$$

$$40.3 \text{ mg H}_2\text{O} \times \frac{1 \text{ g}}{1000 \text{ mg}} \times \frac{1 \text{ mol H}_2\text{O}}{18.02 \text{ g H}_2\text{O}} \times \frac{2 \text{ mol H}}{1 \text{ mol H}_2\text{O}} = 4.473 \times 10^{-3} \text{ mol H}$$

$$4.473 \times 10^{-3} \text{ mol H} \times \frac{1.01 \text{ g H}}{1 \text{ mol H}} = 0.004518 \text{ g H}$$

$$\text{Total mass of C and H} = 0.06004 \text{ g} + 0.004518 \text{ g} = 0.06456 \text{ g}$$

$$\text{Mass of O present} = 0.100 \text{ g} - 0.06456 \text{ g} = 0.035 \text{ g O}$$

$$\text{Moles of O} = 0.035 \text{ g O} \times \frac{1 \text{ mol O}}{16.00 \text{ g O}} = 2.188 \times 10^{-3} \text{ mol O}$$

Dividing the moles of C, H, and O by the lowest molar amount (2.188×10^{-3} mol) gives a C:H:O ratio of 2.28:2:1. Multiplying this by a factor of four gives a whole number ratio of 9:8:4, corresponding to an empirical formula of $C_9H_8O_4$.

Think About It
This problem involves an additional step to determine the mass of carbon and hydrogen present so that we can calculate the mass (and therefore the moles) of oxygen in the compound.

3.103. **Collect and Organize**

Given a reaction that starts with equal masses of iron and sulfur, we consider the mass of iron(II) sulfide that can be produced in the reaction.

Analyze

The elements react to give a molar Fe:S ratio of 1:1. Because there is a difference in the molar mass of these elements (Fe = 55.845 and S = 32.065 g/mol), a mass of S contains more moles of sulfur than the same mass of iron contains moles of iron.

Solve

There is excess sulfur at the end of the reaction, so the mass of FeS produced is (c) less than the sum of the masses of Fe and S to start.

Think About It

The limiting reactant in this case is iron.

3.105. **Collect and Organize**

We are to provide two reasons why the actual yield for a reaction is usually less than the theoretical yield.

Analyze

The actual yield is determined experimentally, and the theoretical yield assumes that all of the limiting reactant is chemically transformed into product.

Solve

The observed, or actual, yield for a reaction is usually less than the theoretical yield because reactions do not always go to completion; the reaction may be slow or may have, for a portion of the reaction, created different products than expected.

Think About It

Chemists try hard to maximize yields for important products by changing the chemical reaction path of the synthesis or by changing the conditions (temperature, pressure, solvent, etc.) under which the reaction is run.

3.107. **Collect and Organize**

We need to determine the maximum amount of hollandaise sauce that can be made with the ingredients on hand.

Analyze

The ingredient that would produce the least amount of the sauce will be the limiting ingredient; therefore, that amount is the highest possible amount of sauce that can be made.

Solve

Because one cup of the sauce requires $\frac{1}{2}$ c (cup) butter, $\frac{1}{2}$ c water, 4 egg yolks, and the juice of one lemon, we can determine how many cups of sauce could be made from the ingredients on hand.

Two cups of butter would be enough to prepare 4 c of sauce.
Unlimited amounts of hot water are enough to prepare an unlimited amount of sauce.
Twelve eggs would be enough to prepare 3 c of sauce.
Four lemons would be enough to prepare 4 c of sauce.

The limiting ingredient is eggs; all the other ingredients are in sufficient supply to make 4 cups of hollandaise sauce. Based on the limited number of eggs, 3 cups of sauce can be made.

Think About It

The maximum amount of sauce that we can make is limited by the ingredient most limited in supply. This is true for chemical reactions as well, although we are thinking in moles, not cups of butter or number of eggs.

3.109. Collect and Organize

To solve this problem we first write a balanced chemical equation. To find the amount of excess reactant left over at the end of the reaction, we have to determine the limiting reactant. Knowing that, we can compute the number of moles (and then mass) of the excess reactant used in the reaction to subtract from what was initially present before the reaction.

Analyze

The molecular formulas for the compounds are needed to write the balanced chemical equation. Ammonia is NH_3, hydrogen chloride is HCl, and ammonium chloride is NH_4Cl. We need the molar masses of the reactants to compute the moles of each present. The molar mass of NH_3 is 17.04 g/mol, and for HCl the molar mass is 36.46 g/mol.

Solve

The balanced chemical equation is

$$NH_3(g) + HCl(g) \longrightarrow NH_4Cl(s)$$

The moles of NH_3 present at the start of the reaction:

$$3.0 \text{ g NH}_3 \times \frac{1 \text{ mol}}{17.04 \text{ g}} = 0.18 \text{ mol NH}_3$$

The moles of HCl present at the start of the reaction:

$$5.0 \text{ g HCl} \times \frac{1 \text{ mol}}{36.46 \text{ g}} = 0.14 \text{ mol HCl}$$

Comparing these two mole amounts and considering that 1 mole of NH_3 reacts with 1 mole of HCl in the balanced equation, we see that HCl is the limiting reactant. Also, we know that, in the reaction 0.14 mol of NH_3 is used up in the reaction and therefore

$$0.18 \text{ mol} - 0.14 \text{ mol} = 0.04 \text{ mol of NH}_3$$

will be leftover. We can convert this excess of NH_3 in moles to grams with the molar mass of NH_3.

$$0.04 \text{ mol NH}_3 \times \frac{17.04 \text{ g}}{1 \text{ mol NH}_3} = 0.7 \text{ g NH}_3$$

Thus, 0.7 g of NH_3 remains as excess at the end of the reaction.

Think About It

Although there was a higher gram amount of HCl in the reaction, it proved to be the limiting reactant. Notice, too, for this problem we did not even need to consider the theoretical yield of the product. We focused only on the amounts and molar relationships of the reactants.

3.111. Collect and Organize

We first need a balanced chemical equation. We then consider how much phosgenite can be produced from 10.0 g of PbO and NaCl; we have to determine which of the two reactants is limiting. Finally, we are asked to calculate a percent yield of phosgenite based on our limiting reagent calculations in (b).

Analyze

To determine the limiting reactant and theoretical yield, we use the balanced equation in (a) and the molar masses of PbO (223.2 g/mol), NaCl (58.44 g/mol), and phosgenite ($Pb_2Cl_2CO_3$, 545.3 g/mol).

Solve

(a) The balanced equation is

$$2 \text{ PbO}(s) + 2 \text{ NaCl}(aq) + H_2O(\ell) + CO_2(g) \longrightarrow Pb_2Cl_2CO_3(s) + 2 \text{ NaOH}(aq)$$

(b) From 10.0 g of PbO, we can theoretically obtain

$$10.0 \text{ g PbO} \times \frac{1 \text{ mol PbO}}{223.2 \text{ g PbO}} \times \frac{1 \text{ mol Pb}_2\text{Cl}_2\text{CO}_3}{2 \text{ mol PbO}} \times \frac{545.3 \text{ g Pb}_2\text{Cl}_2\text{CO}_3}{1 \text{ mol Pb}_2\text{Cl}_2\text{CO}_3} = 12.2 \text{ g Pb}_2\text{Cl}_2\text{CO}_3$$

From 10.0 g of NaCl, we can theoretically obtain

$$10.0 \text{ g NaCl} \times \frac{1 \text{ mol NaCl}}{58.44 \text{ g NaCl}} \times \frac{1 \text{ mol Pb}_2\text{Cl}_2\text{CO}_3}{2 \text{ mol NaCl}} \times \frac{545.3 \text{ g Pb}_2\text{Cl}_2\text{CO}_3}{1 \text{ mol Pb}_2\text{Cl}_2\text{CO}_3} = 46.7 \text{ g Pb}_2\text{Cl}_2\text{CO}_3$$

PbO is the limiting reactant, and 12.2 g of the phosgenite will be produced.
(c) The percent yield for this reaction is

$$\% \text{ yield} = \frac{\text{actual yield}}{\text{theoretical yield}} \times 100\% \qquad \frac{2.72 \text{ g}}{12.2 \text{ g}} \times 100\% = 22.3\%$$

Think About It
Because the molar mass of sodium chloride is low, the number of moles of NaCl in 10 g is much greater than the moles of PbO present in 10.0 g. It is not surprising, therefore, that PbO is the limiting reactant in this problem.

3.113. Collect and Organize
In this problem, since oxygen is the reactant in excess, we focus only on the amount of carbon present before the reaction. We still should write a balanced chemical reaction for the process. We have to convert the mass of carbon into moles, relate those moles to moles of CO_2 product in the balanced equation, and compute the theoretical yield.

Analyze
We need the molar masses of C (12.011 g/mol) and CO_2 (44.01 g/mol). The percent yield for the reaction is given by

$$\% \text{ yield} = \frac{\text{observed experimental yield}}{\text{theoretical yield}} \times 100\%$$

Solve
The balanced chemical equation is

$$C(s) + O_2(g) \longrightarrow CO_2(g)$$

The theoretical yield of carbon dioxide is

$$3.0 \text{ g C} \times \frac{1 \text{ mol C}}{12.011 \text{ g C}} \times \frac{1 \text{ mol CO}_2}{1 \text{ mol C}} \times \frac{44.01 \text{ g CO}_2}{1 \text{ mol CO}_2} = 11 \text{ g CO}_2$$

The percent yield for this reaction is

$$\% \text{ yield} = \frac{6.5 \text{ g}}{11 \text{ g}} \times 100\% = 59\%$$

Think About It
This reaction, like many reactions, did not give 100% of the desired product. The loss may be due to a side reaction. For example, incomplete combustion might give CO as a side product.

3.115. Collect and Organize
We first have to balance the equation for the conversion of glucose into ethanol. We need the molar ratio of the reactant to the ethanol product to determine the theoretical yield of ethanol for the fermentation of 100.0 g of glucose that produces 50.0 mL of ethanol.

Analyze

We must write the balanced equation for the process for (a). For (b), we need the molar masses of $C_6H_{12}O_6$ (180.16 g/mol) and C_2H_5OH (46.07 g/mol). The density of the ethanol (0.789 g/mL) is needed to convert the grams of ethanol produced theoretically in the reaction to milliliters to compute the percent yield for the reaction, which is given by

$$\% \text{ yield} = \frac{\text{observed experimental yield}}{\text{theoretical yield}} \times 100\%$$

Solve

(a) The balanced equation is

$$C_6H_{12}O_6(aq) \longrightarrow 2\,C_2H_5OH(\ell) + 2\,CO_2(g)$$

(b) The theoretical yield of C_2H_5OH is

$$100.0 \text{ g } C_6H_{12}O_6 \times \frac{1 \text{ mol } C_6H_{12}O_6}{180.16 \text{ g } C_6H_{12}O_6} \times \frac{2 \text{ mol } C_2H_5OH}{1 \text{ mol } C_6H_{12}O_6} \times \frac{46.07 \text{ g } C_2H_5OH}{1 \text{ mol } C_2H_5OH}$$

$$\times \frac{1 \text{ mL } C_2H_5OH}{0.789 \text{ g } C_2H_5OH} = 64.82 \text{ mL } C_2H_5OH$$

The percent yield for this reaction is

$$\% \text{ yield} = \frac{50.0 \text{ mL}}{64.82 \text{ mL}} \times 100\% = 77.1\%$$

Think About It

The conversion of glucose by fermentation into ethanol is fairly efficient.

3.117. Collect and Organize

We are to determine the mass percentage of water in the hydrated salt of copper(II) sulfate and to calculate the percentage of mass lost when the water is driven away by heating the compound.

Analyze

The percentage of water in the compound can be determined by assuming 1 mole of the salt and then dividing the mass of water (taking into account that 5 molecules of water are present in the crystal structure) by the total mass of the salt. To find out how much mass is lost for (b), we divide the difference in the masses of the dehydrated copper sulfate salt and hydrated salt by the mass of the hydrated salt (again, assuming 1 mole of each salt). We need the molar masses of H_2O (18.02 g/mol), of $CuSO_4 \cdot 5H_2O$ (249.68 g/mol), and of $CuSO_4$ (159.61 g/mol).

Solve

(a) The percentage by mass of water in the hydrated copper sulfate salt is

$$\frac{(5 \times 18.02 \text{ g})}{249.68 \text{ g}} \times 100\% = 36.09\% \text{ H}_2\text{O}$$

(b) Finding the mass percentage of the compound driven off as steam by heating should give us the same answer, but let's examine it closely.

$$\frac{(249.68 \text{ g} - 159.61 \text{ g})}{249.68 \text{ g}} \times 100\% = 36.07\%$$

Think About It

The two answers are the same (slight difference is due to rounding) because they are asking the same question: What is the mass percentage of water in the hydrated compound? You can determine the amount of water in hydrated salts by determining the mass loss of water upon dehydration, as shown in (b).

3.119. Collect and Organize

We are asked to derive the amount of copper ore necessary to produce the copper used to make pennies. We have to work our way back through a series of reactions for which we are given the balanced chemical equations.

Analyze

We need the molar masses of copper (63.55 g/mol) and chalcopyrite ($CuFeS_2$, 183.52 g/mol). We have to take into account that the reactions do not go to completion to answer (b) and (c).

Solve

(a) One dollar's worth of pennies is 100 pennies. If each penny weighs 3.0 g, then there are 300 g of copper in a dollar's worth of pennies. Using this, the amount of chalcopyrite needed to make the 300 g of pennies is

$$300 \text{ g Cu} \times \frac{1 \text{ mol Cu}}{63.55 \text{ g Cu}} \times \frac{1 \text{ mol Cu}_2\text{S}}{2 \text{ mol Cu}} \times \frac{2 \text{ mol CuS}}{1 \text{ mol Cu}_2\text{S}} \times \frac{2 \text{ mol CuFeS}_2}{2 \text{ mol CuS}}$$

$$\times \frac{183.52 \text{ g CuFeS}_2}{1 \text{ mol}} = 8.7 \times 10^2 \text{ g CuFeS}_2$$

(b) If the first step's yield is 85%, we have to account for this in the molar amounts required.

$$300 \text{ g Cu} \times \frac{1 \text{ mol Cu}}{63.55 \text{ g Cu}} \times \frac{1 \text{ mol Cu}_2\text{S}}{2 \text{ mol Cu}} \times \frac{2 \text{ mol CuS}}{1 \text{ mol Cu}_2\text{S}} \times \frac{(2 \times 100) \text{ mol CuFeS}_2}{(2 \times 85) \text{ mol CuS}}$$

$$\times \frac{183.52 \text{ g CuFeS}_2}{1 \text{ mol}} = 1.0 \times 10^2 \text{ g CuFeS}_2$$

(c) If all of the steps are 85% yield, the amount of the ore would be

$$300 \text{ g Cu} \times \frac{1 \text{ mol Cu}}{63.55 \text{ g Cu}} \times \frac{(1 \times 100) \text{ mol Cu}_2\text{S}}{(2 \times 85) \text{ mol Cu}} \times \frac{(2 \times 100) \text{ mol CuS}}{(1 \times 85) \text{ mol Cu}_2\text{S}}$$

$$\times \frac{(2 \times 100) \text{ mol CuFeS}_2}{(2 \times 85) \text{ mol CuS}} \times \frac{183.52 \text{ g CuFeS}_2}{1 \text{ mol}} = 1.4 \times 10^2 \text{ g CuFeS}_2$$

Think About It

As the yields of the reactions go down, we need more of the ore to produce the same amount of copper to make the pennies.

3.121. Collect and Organize

For this problem we must use percent compositions of the compounds that form upon heating $UO_x(NO_3)_y(H_2O)_z$ to ultimately determine the values of x, y, and z.

Analyze

We need the molar masses of uranium (238.03 g/mol) and oxygen (16.00 g/mol) to determine the molar ratios of U to O in the oxides. We also are given that the charge on uranium in the oxides may range from 3+ to 6+.

Solve

(a) If U_aO_b is 83.22% U, then it is 16.78% O. Assuming 100 g of the compound, and using the molar masses of these elements we obtain their molar ratios:

$$83.22 \text{ g U} \times \frac{1 \text{ mol U}}{238.03 \text{ g}} = 0.3496 \text{ mol U}$$

$$16.78 \text{ g O} \times \frac{1 \text{ mol O}}{16.00 \text{ g}} = 1.049 \text{ mol O}$$

Multiplying these molar amounts by 3 gives a whole-number ratio for an empirical formula of UO_3 for this oxide. In this formula $a = 1$ and $b = 3$ with a charge on the U of 6+ (since O is 2−).

(b) If U_cO_d is 84.8% U, then it is 15.2% O. Assuming 100 g of the compounds, and using the molar masses of these elements we obtain their molar ratios:

$$84.8 \text{ g U} \times \frac{1 \text{ mol U}}{238.03 \text{ g}} = 0.356 \text{ mol U}$$

$$15.2 \text{ g O} \times \frac{1 \text{ mol O}}{16.00 \text{ g}} = 0.950 \text{ mol O}$$

Dividing these molar amounts by the lowest number of moles (0.356) gives a ratio of 1 U to 2.67 O. To reach a whole-number ratio, we multiply by 3 to get U_3O_8, in which $c = 3$ and $d = 8$. The average charge on the U atoms is 16/3+.

(c) Upon gentle heating, $UO_x(NO_3)_y(H_2O)_z$ loses the water according to the following equation:

$$UO_x(NO_3)_y(H_2O)_z \longrightarrow UO_x(NO_3)_y + z\,H_2O$$

More heating of $UO_x(NO_3)_y$ gives the reaction

$$UO_x(NO_3)_y \longrightarrow U_nO_m + \text{nitrogen oxides}$$

Putting these equations together

$$UO_x(NO_3)_y(H_2O)_z \longrightarrow UO_x(NO_3)_y + z\,H_2O \rightarrow U_nO_m + \text{nitrogen oxides}$$

The continued heating of the compound will yield U_3O_8 (part b). The balanced reaction then is

$$3\,UO_x(NO_3)_y(H_2O)_z \longrightarrow 3\,UO_x(NO_3)_y + 3z\,H_2O \rightarrow U_3O_8 + \text{nitrogen oxides}$$

The amount in moles of H_2O present based on 0.742 g of U_3O_8 is

$$0.742 \text{ g } U_3O_8 \times \frac{1 \text{ mol } U_3O_8}{842 \text{ g } U_3O_8} \times \frac{3z \text{ mol } H_2O}{1 \text{ mol } U_3O_8} \times \frac{18.02 \text{ g } H_2O}{1 \text{ mol } H_2O} = 0.0476z \text{ g } H_2O$$

The mass of water lost in this process is $1.328 - 1.042 = 0.286$ g H_2O. The moles of water lost are therefore

$$0.0476z = 0.286$$
$$z = 6$$

The amount in moles of $UO_x(NO_3)_y$ from the 0.742 g of U_3O_8 is

$$0.742 \text{ g } U_3O_8 \times \frac{1 \text{ mol } U_3O_8}{842 \text{ g } U_3O_8} \times \frac{3 \text{ mol } UO_x(NO_3)_y}{1 \text{ mol } U_3O_8} = 0.00264 \text{ mol } UO_x(NO_3)_y$$

Because we know the mass of $UO_x(NO_3)_y$, for this number of moles the molar mass of $UO_x(NO_3)_y$ is

$$\frac{1.042 \text{ g}}{0.00264 \text{ mol}} = 395 \text{ g/mol}$$

This molar mass is expressed by

$$395 \text{ g/mol} = 238 \text{ g/mol} + [(x + 3y)(16.00 \text{ g/mol})] + y(14.00 \text{ g/mol})$$

$$157 \text{ g/mol} = 16x + 62y$$

We know that the total charge on $UO_x(NO_3)_y$ is 0 (it is a neutral compound). This can be expressed as

$$-2(x) + -1(y) + c = 0$$

where x = number of O^{2-} ions, y = number of NO_3^- ions, and c = charge on U in the molecular formula. We are given that c could be 3+, 4+, 5+, or 6+. When $c = 6$, $-2x - y = -6$ or $2x + y = 6$. Rearranging this gives $y = 6 - 2x$, which when substituted into 157 g/mol = $16x + 62y$, gives

$$157 \text{ g/mol} = 16x + 62(6 - 2x)$$
$$157 \text{ g/mol} = -108x + 372$$
$$-215 \text{ g/mol} = -108x$$
$$2 = x$$

and then

$$y = 6 - 2x = 2$$

If c = 5+, 4+, or 3+, the solution does not work. The formula for $UO_x(NO_3)_y(H_2O)_z$ is $UO_2(NO_3)_2(H_2O)_6$.

Think About It
The value of the charge on U in (b) must mean that it is a mixed oxidation state compound. In this case one-third of the U atoms have a 6+ charge and two-thirds of the U atoms have a 5+ charge.

3.123. Collect and Organize
To determine the molecules of active compound in each medication, we convert the mass of each to moles using the molar masses and then multiply by Avogadro's number.

Analyze
The molar mass of ibuprofen ($C_{13}H_{18}O_2$) is 206.31 g/mol; for calcium carbonate ($CaCO_3$), the molar mass is 100.09 g/mol; and for chlorpheniramine ($C_{16}H_{19}N_2Cl$), the molar mass is 274.82 g/mol. The value of Avogadro's number is 6.022×10^{23} molecules/mol.

Solve
(a) The number of molecules of ibuprofen in 200.0 mg (0.2000 g) is

$$0.2000 \text{ g} \times \frac{1 \text{ mol}}{206.31 \text{ g}} \times \frac{6.022 \times 10^{23} \text{ molecules}}{\text{mol}} = 5.838 \times 10^{20} \text{ molecules } C_{13}H_{18}O_2$$

(b) The number of molecules of calcium carbonate in 500.0 mg (0.5000 g) is

$$0.5000 \text{ g} \times \frac{1 \text{ mol}}{100.09 \text{ g}} \times \frac{6.022 \times 10^{23} \text{ molecules}}{\text{mol}} = 3.008 \times 10^{21} \text{ molecules } CaCO_3$$

(c) The number of molecules of chlorpheniramine in 4 mg (0.004 g) is

$$0.004 \text{ g} \times \frac{1 \text{ mol}}{274.82 \text{ g}} \times \frac{6.022 \times 10^{23} \text{ molecules}}{\text{mol}} = 9 \times 10^{18} \text{ molecules } C_{16}H_{19}N_2Cl$$

Think About It
Notice that, even though we started with more than twice the mass of calcium carbonate compared to the mass of the ibuprofen sample, the number of molecules of $CaCO_3$ is more than four times higher because of the difference in molar masses of these two medications.

3.125. Collect and Organize
In examining two compounds of manganese and oxygen, we are asked to name them, calculate the percent Mn by mass in each, and explain how the law of multiple proportions applies to the compounds.

Analyze
We can use the rules set forth in Chapter 2 to name the compounds, keeping in mind that, because Mn may have different charges in different compounds, we have to indicate the charge on the Mn atom in the name.

We can then calculate the percentage of Mn in each compound by assuming 1 mole of the substance and dividing the mass of the Mn present by the molar mass to get percentage of Mn by mass.

Solve
(a) Mn_2O_3 is manganese(III) oxide and MnO_2 is manganese(IV) oxide.
(b) Assuming 1 mole of each substance,

$$\% \text{ Mn in } Mn_2O_3 = \frac{(2 \times 54.938 \text{ g) Mn}}{157.87 \text{ g } Mn_2O_3} \times 100 = 69.599\%$$

$$\% \text{ Mn in } MnO_2 = \frac{54.938 \text{ g Mn}}{86.94 \text{ g } MnO_2} \times 100 = 63.19\%$$

(c) These compounds are consistent with the law of multiple proportions in that they contain the same elements, but in different atom ratios.

Think About It
Manganese takes on a wide range of oxidation numbers in compounds, including 2+, 3+, 4+, 6+, and 7+. For example, the oxide compound in which manganese has a 7+ oxidation number is Mn_2O_7.

3.127. **Collect and Organize**
This problem asks for the mass of FeO product for the given balanced equation. We have to determine the limiting reactant (FeS or H_2CO_3) to determine the theoretical yield and then compute the actual yield knowing that the reaction proceeds with 78.5% yield. In (b) we are asked to determine the chemical formula of $(CH_2O)_n$ given a molar mass of 300 g/mol.

Analyze
To determine the theoretical yield of FeO, we need the molar mass of FeS (87.91 g/mol) and the molar mass of FeO (71.84 g/mol). The empirical formula for $(CH_2O)_n$ is CH_2O whose molar mass is 30.03 g/mol.

Solve
(a) If FeS is the limiting reactant, the theoretical yield would be

$$1.50 \text{ g FeS} \times \frac{1 \text{ mol FeS}}{87.91 \text{ g FeS}} \times \frac{2 \text{ mol FeO}}{2 \text{ mol FeS}} \times \frac{71.84 \text{ g FeO}}{1 \text{ mol FeO}} = 1.23 \text{ g FeO}$$

If H_2CO_3 is the limiting reactant, the theoretical yield would be

$$0.525 \text{ mol } H_2CO_3 \times \frac{2 \text{ mol FeO}}{1 \text{ mol } H_2CO_3} \times \frac{71.84 \text{ g FeO}}{1 \text{ mol FeO}} = 75.4 \text{ g FeO}$$

We see that the limiting reactant is FeS and that the theoretical yield is 1.23 g.

$$\text{Actual yield} = \% \text{ yield} \times \text{theoretical yield} = 0.785 \times 1.23 \text{ g} = 0.966 \text{ g FeO}$$

(b) If the molar mass of the compound is 300 g/mol and the empirical formula's mass is 30 g/mol, then the chemical formula is $(CH_2O)_{10}$, or $C_{10}H_{20}O_{10}$.

Think About It
Very few reactions proceed to 100% of the desired products, and the percent yield for this reaction, at 78.5%, is actually quite high.

3.129. **Collect and Organize**
Given that there are 6×10^{15} solar-wind ions on average in every cubic kilometer of near-Earth space, we are asked to convert this into moles.

Analyze
All we need for this problem is Avogadro's number, 6.022×10^{23}.

Solve

$$\frac{6 \times 10^{15} \text{ ions}}{1 \text{ km}^3} \times \frac{1 \text{ mol}}{6.022 \times 10^{23} \text{ ions}} = 1 \times 10^{-8} \text{ mol/km}^3$$

Think About It

There are very few solar-wind ions spread out over a large volume in interplanetary space.

3.131. **Collect and Organize**

The alternative fuel E-85 is 85% (by volume) ethanol, and we are to determine how many moles of ethanol are in 1 gal of the fuel.

Analyze

If we first convert 1 gal to milliliters (1 gal = 3785 mL), we can find the milliliters of ethanol in the fuel by multiplying by 0.85 (the percent ethanol by volume in the fuel). From the milliliters of ethanol, we can calculate the mass of ethanol by multiplying by the density (0.79 g/mL). From the mass and the molar mass of ethanol (C_2H_6O, 46.07 g/mol), we can determine the moles of ethanol in 1 gal of the fuel.

Solve

$$(3785 \text{ mL} \times 0.85) \times \frac{0.79 \text{ g}}{\text{mL}} \times \frac{1 \text{ mol ethanol}}{46.07 \text{ g ethanol}} = 55 \text{ mol ethanol}$$

Think About It

Notice that we could do this calculation all in one step, without separately calculating the milliliters of ethanol, the mass, and then the moles.

3.133. **Collect and Organize**

From the mass of the salt before and after dehydration (0.6240 and 0.5471 g, respectively) and the known ratio of M to Cl to H_2O in the compound [as given by the molecular formula, $MCl_2(H_2O)_2$], we are to determine the identity (through calculation of the molar mass) of M.

Analyze

From the difference of the masses, we can obtain the mass of water lost and from that value calculate the moles of water lost. The moles of water lost are equal to the moles of Cl in the compound and are twice the moles of M in the compound. From the moles of Cl, we can find the mass of Cl in the sample. The mass of M in the sample is the total mass of the sample minus the combined masses of the water and the chlorine. Once we know the mass of M, we can divide by the moles of M in the sample, found earlier, to determine the molar mass of M, which identifies the metal.

Solve

$$(0.6240 \text{ g} - 0.5471 \text{ g}) \times \frac{1 \text{ mol H}_2\text{O}}{18.02 \text{ g H}_2\text{O}} = 4.27 \times 10^{-3} \text{ mol H}_2\text{O}$$

From the formula we also know that $Cl = 4.27 \times 10^{-3}$ mol and $M = 2.14 \times 10^{-3}$ mol.

The molar mass of M is

$$4.27\times10^{-3} \text{ mol Cl}\times\frac{35.453 \text{ g Cl}}{1 \text{ mol Cl}}=0.151 \text{ g Cl}$$

$$\text{Total mass of } H_2O \text{ and } Cl = 0.0769 \text{ g} + 0.151 \text{ g} = 0.228 \text{ g}$$

$$0.6240 \text{ g} - 0.228 \text{ g} = 0.396 \text{ g M}$$

$$\frac{0.396 \text{ g M}}{2.14\times10^{-3} \text{ mol M}}=185 \text{ g/mol}$$

The identity of M is Re.

Think About It
This problem relies heavily on our being able to relate the moles of atoms in a compound's formula to each other.

3.135. **Collect and Organize**
To determine the percent yield of ammonia in the reaction described, we need a balanced chemical equation for the reaction of 6.04 kg of H_2 with N_2 to give 28.0 kg of NH_3. We have excess N_2 in the reaction, so the theoretical yield is based solely on the moles of H_2 present at the beginning of the reaction.

Analyze
The balanced equation for the reaction is

$$N_2(g) + 3 H_2(g) \rightarrow 2 NH_3(g)$$

Solve
The theoretical yield of ammonia in this reaction is

$$6.04 \text{ kg } H_2 \times \frac{1000 \text{ g}}{1 \text{ kg}}\times\frac{1 \text{ mol } H_2}{2.016 \text{ g } H_2}\times\frac{2 \text{ mol } NH_3}{3 \text{ mol } H_2}$$
$$\times\frac{17.03 \text{ g } NH_3}{1 \text{ mol } NH_3}=34,015 \text{ g or } 34.0 \text{ kg } NH_3$$

The percent yield is

$$\frac{28.0 \text{ kg}}{34.0 \text{ kg}}\times100\%=82.4\%$$

Think About It
The clue that this is not a limiting reactant problem is that there is a stated excess of nitrogen in the problem.

3.137. **Collect and Organize**
Knowing the percentage by mass of S in the coal burned by the power plant, we calculate the mass of SO_2 produced by burning the coal and then consider the reaction of SO_2 with water and oxygen in the atmosphere to produce acid rain, ultimately calculating the mass of sulfuric acid produced by the 100 metric tons (t) of coal burned.

Analyze
(a) Using the percent S in the coal, we can calculate the mass of S present in 100 metric tons. From that we can use the equation that converts S into SO_2 upon burning in oxygen to calculate the mass of SO_2 produced.

$$S(s) + O_2(g) \rightarrow SO_2(g)$$

(b) We have to balance the following reaction:

$$SO_2(g) + H_2O(g) + O_2(g) \rightarrow H_2SO_4(\ell)$$

(c) From our answer in (a) and the equation in (b), we use stoichiometric relationships to calculate the mass of H_2SO_4 produced.

Solve

(a) $100 \text{ t C} \times \dfrac{1000 \text{ kg}}{1 \text{ t}} \times \dfrac{1000 \text{ g}}{1 \text{ kg}} \times 0.030 = 3.0 \times 10^6 \text{ g S in the coal}$

$3.0 \times 10^6 \text{ g S} \times \dfrac{1 \text{ mol S}}{32.07 \text{ g S}} \times \dfrac{1 \text{ mol SO}_2}{1 \text{ mol S}} \times \dfrac{64.06 \text{ g SO}_2}{1 \text{ mol SO}_2} \times \dfrac{1 \text{ kg}}{1000 \text{ g}}$

$\times \dfrac{1 \text{ t}}{1000 \text{ kg}} = 6.0 \text{ metric tons SO}_2$

(b) $2 SO_2(g) + 2 H_2O(g) + O_2(g) \rightarrow 2 H_2SO_4(\ell)$

(c) $6.0 \text{ t SO}_2 \times \dfrac{1000 \text{ kg}}{1 \text{ t}} \times \dfrac{1000 \text{ g}}{1 \text{ kg}} \times \dfrac{1 \text{ mol SO}_2}{64.06 \text{ g SO}_2} \times \dfrac{2 \text{ mol H}_2\text{SO}_4}{2 \text{ mol SO}_2}$

$\times \dfrac{98.08 \text{ g H}_2\text{SO}_4}{1 \text{ mol H}_2\text{SO}_4} \times \dfrac{1 \text{ kg}}{1000 \text{ g}} \times \dfrac{1 \text{ t}}{1000 \text{ kg}} = 9.2 \text{ metric tons H}_2\text{SO}_4$

Think About It

Notice that we would not have to strictly convert from metric tons to grams in these calculations because we converted back to metric tons for our answers. The conversion factors, therefore, cancelled each other out.

3.139. **Collect and Organize**

We need the balanced equation through which sulfuric acid is produced from burning sulfur to form SO_2, which when mixed with water in the atmosphere forms H_2SO_4. Then we use stoichiometry to calculate how many grams of sulfuric acid are formed from 1 g of sulfur through this process.

Analyze

The two reactions (unbalanced) are

$$S(s) + O_2(g) \rightarrow SO_2(g)$$

$$SO_2(g) + H_2O(g) + O_2(g) \rightarrow H_2SO_4(\ell)$$

Solve

Combining the unbalanced reactions into one overall reaction gives the balanced equation:

$$2 S(s) + 2 H_2O(g) + 3 O_2(g) \rightarrow 2 H_2SO_4(\ell)$$

The amount of sulfuric acid is

$$1.00 \text{ g S} \times \dfrac{1 \text{ mol S}}{32.07 \text{ g}} \times \dfrac{2 \text{ mol H}_2\text{SO}_4}{2 \text{ mol S}} \times \dfrac{98.08 \text{ g}}{1 \text{ mol H}_2\text{SO}_4} = 3.06 \text{ g H}_2\text{SO}_4$$

Think About It

In terms of mass, the amount of sulfuric acid is greater than the original mass of sulfur burned. This is due to the greater molar mass of H_2SO_4 compared to that of S.

3.141. **Collect and Organize**

Using the percent composition of a mineral, we are to determine the formula.

Analyze

The percent composition is given in percentage by mass. If we assume 100 g of the mineral, 34.55 g of it is Mg, 19.96 g of it is Si, and 45.49 g of it is O. Using the molar masses of these elements, we can calculate the moles that these masses represent for each element and then find the whole-number ratio of the elements in the compound to give the formula of the mineral.

Solve

$$34.55 \text{ g Mg} \times \frac{1 \text{ mol Mg}}{24.31 \text{ g Mg}} = 1.421 \text{ mol Mg}$$

$$19.96 \text{ g Si} \times \frac{1 \text{ mol Si}}{28.09 \text{ g Si}} = 0.7106 \text{ mol Si}$$

$$45.49 \text{ g O} \times \frac{1 \text{ mol O}}{16.00 \text{ g O}} = 2.843 \text{ mol O}$$

This gives a molar ratio of Mg:Si:O of 2:1:4; the formula for the mineral is Mg_2SiO_4.

Think About It

The SiO_4^{4-} ion is the silicate ion, and this mineral's name is therefore magnesium silicate.

3.143. **Collect and Organize**

Several questions about underwater breathing devices based on KO_2 and Na_2O_2 are posed. First, we are asked to determine the mass of KO_2 needed to supply "air" for a two hour underwater excursion. We are then asked to compare the mass of Na_2O_2 required for the same task. Lastly, we are asked to compare the volume of each compound required for the task in (a). In addition to the quantities and densities given in the problem, we will need to calculate the molar masses for KO_2 and Na_2O_2. Note that we must ensure both equations are properly balanced before proceeding.

Analyze

(a) This is best thought of as a dimensional analysis problem; our starting point will be the two hours of underwater breathing required. From this point, we may convert time to the number of breaths taken, and from there, to the volume of oxygen gas consumed over the two hour period. This volume may be converted to a mass of O_2 using the density provided (1.429 g/L), and then to the number of moles of O_2 using the molar mass of O_2 (32.00 g/mol). Using the ratio of stoichiometric coefficients from the balanced chemical equation and the molar mass of KO_2 (71.097 g/mol), we arrive at the mass of KO_2 needed. When correctly balanced, the equation for the reaction of KO_2 is

$$2 \text{ KO}_2(s) + \text{CO}_2(g) \rightarrow \text{K}_2\text{CO}_3(s) + \tfrac{3}{2}\text{O}_2(g)$$

(b) Using the same mass of O_2 calculated in (a) and the molar mass of Na_2O_2 (79.978 g/mol), we may calculate the mass of Na_2O_2 required. Since the molar mass of Na_2O_2 is greater than that of KO_2, we might expect that more mass is required to produce the same quantity of oxygen.

(c) Dividing the masses obtained in (a) and (b) by their respective densities will tell us the volume of KO_2 and Na_2O_2 required. Since Na_2O_2 has a larger density than KO_2, an equal mass of both will mean a smaller volume for Na_2O_2.

Solve

(a) $2 \text{ hours} \times \dfrac{60 \text{ min}}{1 \text{ hr}} \times \dfrac{12 \text{ breaths}}{1 \text{ min}} \times \dfrac{0.500 \text{ L O}_2}{1 \text{ breath}} \times \dfrac{1.429 \text{ g O}_2}{1 \text{ L O}_2} = 1028.88 \text{ g O}_2$

$1028.88 \text{ g O}_2 \times \dfrac{1 \text{ mol O}_2}{32.00 \text{ g O}_2} \times \dfrac{2 \text{ mol KO}_2}{\frac{3}{2} \text{ mol O}_2} \times \dfrac{71.097 \text{ g KO}_2}{1 \text{ mol KO}_2} = 3.05 \times 10^3 \text{ g KO}_2$

(b) $1028.88 \text{ g O}_2 \times \dfrac{1 \text{ mol O}_2}{32.00 \text{ g O}_2} \times \dfrac{2 \text{ mol Na}_2\text{O}_2}{1 \text{ mol O}_2} \times \dfrac{79.978 \text{ g Na}_2\text{O}_2}{1 \text{ mol Na}_2\text{O}_2} = 5.14 \times 10^3 \text{ g Na}_2\text{O}_2$

We would need more Na_2O_2 than KO_2 to supply an equivalent quantity of oxygen gas.

(c) $3.05 \times 10^3 \text{ g KO}_2 \times \dfrac{1 \text{ mL KO}_2}{2.14 \text{ g KO}_2} \times \dfrac{1 \text{ L}}{1000 \text{ mL}} = 1.43 \text{ L for KO}_2$

$5.01 \times 10^3 \text{ g Na}_2\text{O}_2 \times \dfrac{1 \text{ mL Na}_2\text{O}_2}{2.805 \text{ g Na}_2\text{O}_2} \times \dfrac{1 \text{ L}}{1000 \text{ mL}} = 1.79 \text{ L for Na}_2\text{O}_2$

KO_2 will occupy a smaller volume.

Think About It

Although Na_2O_2 is more dense, the lower mass of KO_2 required for two hours of underwater breathing is enough to compensate for the lower density of KO_2.

CHAPTER 4 | Solution Chemistry: The Hydrosphere

4.1. Collect and Organize

This question asks us to differentiate between a strong binary acid and a weak binary acid. A binary acid is an acid containing hydrogen and another element, such as HCl. A strong acid completely dissociates, meaning that all binary molecules of HX are present in solution as H^+ and X^-. A weak acid does not completely dissociate, meaning that some of the HX molecules are in the form of H^+ and X^- and some are present in the form HX.

Analyze

In Figure P4.1, all of the gray spheres have a + charge, so these must represent the H^+ in the binary acids. The green (G), yellow (Y), and magenta (M) spheres carrying a – charge must represent the Xs in the binary acids HX. If the X^- are free in solution and not combined with H^+, then that HX must be a strong acid. If, however, HX is found in the solution as a molecule along with H^+ and X^-, then that HX must be a weak acid.

Solve

All the green and magenta spheres are present in solution as ions, not in combination with H^+ as HX. Therefore, HG and HM must both be strong acids. HY (yellow), however, is represented as two HY molecules and three Y^- ions. Because only some of the HY has dissociated, HY is the weak acid.

Think About It

The extent to which a weak acid may be dissociated can vary. If this were a strong weak acid, perhaps only one HY (yellow) molecule would be represented in Figure P4.1 along with four Y^- anions.

4.3. Collect and Organize

We have to consider the possible oxidation states for the highlighted elements in each of the substances. All of the elements are nonmetals and may have positive or negative oxidation states, depending on what other elements they are combined with.

Analyze

In each of the substances, oxygen adopts an oxidation state of 2– and hydrogen adopts an oxidation state of 1+. Knowing that, we can determine the oxidation state of the X atom in all the compounds. The elements that are shown in Figure P4.3 are green = nitrogen, blue = phosphorus, orange = sulfur, purple = chlorine, and yellow = argon. Nitrogen and phosphorus could have oxidation states of 3+, 5+, and 3–; sulfur may have an oxidation state of 4+, 6+, and 2–; chlorine may have oxidation states of 1–, 5+, and 7+; and argon is a noble gas that does not lose or gain electrons to become charged and so it does not combine with other elements, even ones like fluorine and oxygen.

Solve

(a) In HX, X has an oxidation state of 1–, so HX is HCl (purple).
(b) In H_2XO_4, X has an oxidation state of 6+, so H_2XO_4 is H_2SO_4 (orange).
(c) In HXO_3, X has an oxidation state of 5+, which may be either HNO_3 or HPO_3; nitric acid is a common acid and HPO_3 is not very common, so HXO_3 is HNO_3 (green).
(d) In H_3XO_4, X has an oxidation state of 5+, which may be either H_3NO_4 or H_3PO_4; phosphoric acid is a common acid and H_3NO_4 is not very common, so H_3XO_4 is H_3PO_4 (blue).

Think About It

We will find out later, based on electron count, Lewis structures, and formal charge, why H_3NO_4 is not common but H_3PO_4 is.

4.5. Collect and Organize

We are to classify each of the depictions in Figure P4.5 as either a strong electrolyte, a weak acid, a weak electrolyte, a strong acid, or a nonelectrolyte.

Analyze

When considering the depictions in Figure P4.5, we should pay attention to the extent of dissociation, especially with respect to hydronium ions. Strong acids and other strong electrolytes will dissociate completely, while weak acids and other weak electrolytes will dissociate only partially. Nonelectrolytes will not dissociate. It may be helpful to re-draw the diagrams in Figure P4.5 with charges in place in order to assess the extent of dissociation.

Solve

(a) All ions are dissociated; this is a strong electrolyte.

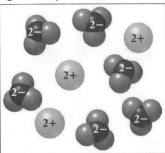

(b) Some of the ions depicted are dissociated, but not all. The cation in this case is hydronium, making this an acid. This is both a weak electrolyte and a weak acid.

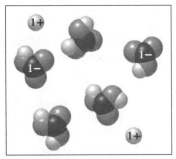

(c) All ions are dissociated, and the cation is hydronium; this is a strong acid and a strong electrolyte.

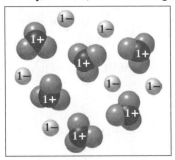

(d) None of the molecules depicted have dissociated. This is a nonelectrolyte.

Think About It

All strong acids (e.g., HCl) are strong electrolytes, but not all strong electrolytes (e.g., NaCl) are strong acids.

4.7. **Collect and Organize**

We are asked which of the ions depicted in Figure P4.7 will remain in solution. We will need to consult the solubility rules in Table 4.5.

Analyze

We must consider all combinations of cations and anions; if any of these combinations is insoluble, it will precipitate out. Anything that does not precipitate will remain in solution. The four species that may be formed from the ions listed in Figure P4.7 are HBr, HNO_3, $PbBr_2$, and $Pb(NO_3)_2$.

Solve

Of the species listed, $PbBr_2$ is insoluble and will precipitate out, as depicted below. The hydronium and nitrate ions will remain in solution.

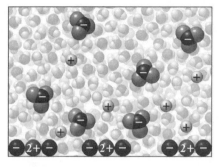

Think About It

The ions that remain in solution in this reaction are spectator ions and would not appear in the net ionic equation describing this reaction.

4.9. **Collect and Organize**

A solution consists of a solute and a solvent. To decide which is the solvent we have to interpret the definitions of these terms.

Analyze

A solvent is something (often a liquid, but not always) in which the solute dissolves. The solute is a substance that dissolves into a solvent. The solvent is the species present in greater proportion in the solution.

Solve

When the solute is a salt and the solvent is water, we could define the solvent as the liquid component of the solution. When the solvent and solute are both liquids or both solids, however, that definition is muddied. The solvent though is usually in greater quantity (volume) than the solute, so defining the solvent as the component that is present in the greatest amount may be more broadly applied.

Think About It

A solid–solid solution that involves two metals is called an alloy. An example of an alloy is bronze.

4.11. **Collect and Organize**

Molarity is expressed in moles per liter (mol/L). We need to convert from 1.00 mmol in 1 mL to mol/L for this problem.

Analyze

To convert from mmol to mol, we need to divide the mmol by 1000. To convert mL to liters, divide mL by 1000.

Solve

Because 1.00 mmol is 1.00×10^{-3} mol and 1.00 mL is 1.00×10^{-3} L,

$$\frac{1.00 \text{ mmol}}{1.00 \text{ mL}} = \frac{1.00 \times 10^{-3} \text{ mol}}{1.00 \times 10^{-3} \text{ L}} = 1.00 \text{ mol/L} = 1.00 \; M$$

Think About It

Because 1 mmol/mL = 1 mol/L, we can use this relationship to express molarity comfortably in either units.

4.13. **Collect and Organize**

The molarity of a solution is the moles of solute in one liter of solution. For each part of this problem, we are given the moles of solute in a volume (in mL) of solution. Molarity is abbreviated as M (e.g., 2.00 M).

Analyze

To find the molarity, we need only divide the moles of solute by the volume of solution in liters. To get volume in liters, we simply divide the milliliters of solution by 1000 or, even more simply, move the decimal three places to the left (e.g., 100.0 mL = 0.1000 L).

Solve

(a) $\dfrac{0.56 \text{ mol}}{0.1000 \text{ L}} = 5.6 \ M \ BaCl_2$

(b) $\dfrac{0.200 \text{ mol}}{0.2000 \text{ L}} = 1.00 \ M \ Na_2CO_3$

(c) $\dfrac{0.325 \text{ mol}}{0.2500 \text{ L}} = 1.30 \ M \ C_6H_{12}O_6$

(d) $\dfrac{1.48 \text{ mol}}{0.2500 \text{ L}} = 5.92 \ M \ KNO_3$

Think About It

Notice that these calculations can be done quickly if you recognize, for example, that 100 mL is one-tenth of a liter. Therefore the molarity of the solution will be 10 times the number of moles.

4.15. **Collect and Organize**

The molarity of ions in a solution is found similarly to the molarity of a solute: divide the moles of the ion present by the volume of the solution.

Analyze

In this problem, all of the volumes are given in milliliters. To convert these volumes to liters we move the decimal three places to the left (e.g., 100.0 mL = 0.1000 L). The quantity of ions is given in grams. To find moles of solute we use the molar mass of each ion (recall that the mass of missing or added electrons is negligible).

Solve

(a) $\dfrac{0.33 \text{ g} \times \left(\dfrac{1 \text{ mol Na}^+}{29.99 \text{ g Na}^+} \right)}{0.1000 \text{ L}} = 0.11 \ M \ Na^+$

(b) $\dfrac{0.38 \text{ g} \times \left(\dfrac{1 \text{ mol Cl}^-}{35.45 \text{ g Cl}^-} \right)}{0.1000 \text{ L}} = 0.11 \ M \ Cl^-$

(c) $\dfrac{0.46 \text{ g} \times \left(\dfrac{1 \text{ mol SO}_4^{2-}}{96.06 \text{ g SO}_4^{2-}} \right)}{0.0500 \text{ L}} = 0.096 \ M \ SO_4^{2-}$

(d) $\dfrac{0.40 \text{ g} \times \left(\dfrac{1 \text{ mol Ca}^{2+}}{40.08 \text{ g Ca}^{2+}} \right)}{0.0500 \text{ L}} = 0.20 \ M \ Ca^{2+}$

Think About It

Calculating molarity given a mass of a solute involves first calculating moles of solute using the molar mass.

4.17. **Collect and Organize**
We are asked to calculate the grams of a solute needed to prepare a solution of a specific concentration.

Analyze
First we can find the moles of solute needed for the solution by multiplying the molarity by the volume (in liters!). Once we have moles, we can then use the molar mass of the solute to calculate the mass of solute needed.

Solve

(a) $1.000 \text{ L} \times \dfrac{0.200 \text{ mol NaCl}}{1 \text{ L}} \times \dfrac{58.44 \text{ g NaCl}}{1 \text{ mol NaCl}} = 11.7 \text{ g NaCl}$

(b) $0.2500 \text{ L} \times \dfrac{0.125 \text{ mol CuSO}_4}{1 \text{ L}} \times \dfrac{159.61 \text{ g CuSO}_4}{1 \text{ mol CuSO}_4} = 4.99 \text{ g CuSO}_4$

(c) $0.5000 \text{ L} \times \dfrac{0.400 \text{ mol CH}_3\text{OH}}{1 \text{ L}} \times \dfrac{32.04 \text{ g CH}_3\text{OH}}{1 \text{ mol CH}_3\text{OH}} = 6.41 \text{ g CH}_3\text{OH}$

Think About It
This is a practical calculation for preparing solutions when we know that we want to have a solution with a particular concentration.

4.19. **Collect and Organize**
From the concentration of ions, we are asked to calculate the total mass of the ions in 2.75 L of river water.

Analyze
We first have to convert the millimolar (mM) concentrations of each ion to molar ($1000 \ mM = 1 \ M$) by moving the decimal three places to the left (e.g., $0.100 \ mM = 1.00 \times 10^{-4} \ M$). From the molar concentration of each ion we can find the moles (and subsequently the mass) of each through multiplication of the molarity by the volume (2.75 L). Finally, we need to add all the masses of the ions together.

Solve

Mass of Ca^{2+}: $2.75 \text{ L} \times \left(\dfrac{8.20 \times 10^{-4} \text{ mol Ca}^{2+}}{\text{L}} \right) \times \left(\dfrac{40.08 \text{ g Ca}^{2+}}{1 \text{ mol Ca}^{2+}} \right) = 0.0904 \text{ g Ca}^{2+}$

Mass of Mg^{2+}: $2.75 \text{ L} \times \left(\dfrac{4.30 \times 10^{-4} \text{ mol Mg}^{2+}}{\text{L}} \right) \times \left(\dfrac{24.31 \text{ g Mg}^{2+}}{1 \text{ mol Mg}^{2+}} \right) = 0.0287 \text{ g Mg}^{2+}$

Mass of Na^{+}: $2.75 \text{ L} \times \left(\dfrac{3.00 \times 10^{-4} \text{ mol Na}^{+}}{\text{L}} \right) \times \left(\dfrac{22.99 \text{ g Na}^{+}}{1 \text{ mol Na}^{+}} \right) = 0.0190 \text{ g Na}^{+}$

Mass of K^{+}: $2.75 \text{ L} \times \left(\dfrac{2.00 \times 10^{-2} \text{ mol K}^{+}}{\text{L}} \right) \times \left(\dfrac{39.10 \text{ g K}^{+}}{1 \text{ mol K}^{+}} \right) = 2.15 \text{ g K}^{+}$

Mass of Cl^{-}: $2.75 \text{ L} \times \left(\dfrac{2.50 \times 10^{-4} \text{ mol Cl}^{-}}{\text{L}} \right) \times \left(\dfrac{35.45 \text{ g Cl}^{-}}{1 \text{ mol Cl}^{-}} \right) = 0.0244 \text{ g Cl}^{-}$

Mass of SO_4^{2-}: $2.75 \text{ L} \times \left(\dfrac{3.80 \times 10^{-4} \text{ mol SO}_4{}^{2-}}{\text{L}} \right) \times \left(\dfrac{96.06 \text{ g SO}_4{}^{2-}}{1 \text{ mol SO}_4{}^{2-}} \right) = 0.100 \text{ g SO}_4{}^{2-}$

$$\text{Mass of } HCO_3^- : 2.75 \text{ L} \times \left(\frac{1.82 \times 10^{-3} \text{ mol } HCO_3^-}{\text{L}} \right) \times \left(\frac{61.02 \text{ g } HCO_3^-}{1 \text{ mol } HCO_3^-} \right) = 0.305 \text{ g } HCO_3^-$$

Total mass of ions: $0.0904 \text{ g} + 0.0287 \text{ g} + 0.0190 \text{ g} + 2.15 \text{ g} + 0.0244 \text{ g} + 0.100 \text{ g} + 0.305 \text{ g} = 2.72 \text{ g}$

Think About It
The mass of the ions dissolved in natural water is sometimes called "total dissolved solids."

4.21. **Collect and Organize**
For each of the pesticides, we are given the volume and concentration of the solution. From this information we can find the moles of pesticide by multiplying the volume by the concentration.

Analyze
We have to watch our units here for concentration and volume. We need to use the fact that 1000 mL = 1 L and 1000 mmol = 1 mol.

Solve

(a) $0.400 \text{ L} \times \dfrac{0.024 \text{ mol}}{1 \text{ L}} = 9.6 \times 10^{-3} \text{ mol or } 9.6 \text{ mmol lindane}$

(b) $1.65 \text{ L} \times \dfrac{4.73 \times 10^{-4} \text{ mol}}{1 \text{ L}} = 7.80 \times 10^{-4} \text{ mol or } 0.780 \text{ mmol dieldrin}$

(c) $25.8 \text{ L} \times \dfrac{3.4 \times 10^{-3} \text{ mol}}{1 \text{ L}} = 8.8 \times 10^{-2} \text{ mol or } 88 \text{ mmol DDT}$

(d) $154 \text{ L} \times \dfrac{2.74 \times 10^{-2} \text{ mol}}{1 \text{ L}} = 4.22 \text{ mol aldrin}$

Think About It
Converting mL to L, L to mL, mmol to mol, and mol to mmol by moving the decimal is convenient but be sure that you do not mistakenly move it in the wrong direction (e.g., it would be wrong to say 2.56 mmol = 2560 mol).

4.23. **Collect and Organize**
The table gives information on both sample size and the mass of DDT in each groundwater sample. In order to compare the DDT amounts among samples, we are asked to calculate the DDT in millimoles per liter and ppm.

Analyze
We have to first compute the millimoles of DDT in each sample by dividing the mass (in mg) by the molar mass of DDT ($C_{14}H_9Cl_5$, 354.49 mg/mmol). To find the concentration in each sample, we divide this result by the volume of the sample in liters. The concentration in ppm is equal to 1 mg per kilogram of solution.

Solve

$$\text{Sample from the orchard: } \frac{0.030 \text{ mg DDT} \times \left(\dfrac{1 \text{ mmol DDT}}{354.49 \text{ mg DDT}} \right)}{0.2500 \text{ L}} = 3.4 \times 10^{-4} \text{ mmol/L}$$

$$\frac{0.030 \text{ mg DDT}}{0.2500 \text{ L}} \times \frac{1 \text{ L}}{1000 \text{ mL}} \times \frac{1 \text{ mL}}{1 \text{ g}} \times \frac{1 \text{ g}}{1000 \text{ mg}} = 1.2 \times 10^{-7} \text{ ppm}$$

$$\text{Sample from the residential area: } \frac{0.035 \text{ mg DDT} \times \left(\dfrac{1 \text{ mmol DDT}}{354.49 \text{ mg DDT}} \right)}{1.750 \text{ L}} = 5.6 \times 10^{-5} \text{ mmol/L}$$

$$\frac{0.035 \text{ mg DDT}}{1.750 \text{ L}} \times \frac{1 \text{ L}}{1000 \text{ mL}} \times \frac{1 \text{ mL}}{1 \text{ g}} \times \frac{1 \text{ g}}{1000 \text{ mg}} = 2.0 \times 10^{-8} \text{ ppm}$$

$$\text{Sample from the residential area after storm: } \frac{0.57 \text{ mg DDT} \times \left(\dfrac{1 \text{ mmol DDT}}{354.49 \text{ mg DDT}} \right)}{0.0500 \text{ L}} = 3.2 \times 10^{-2} \text{ mmol/L}$$

$$\frac{0.57 \text{ mg DDT}}{0.0500 \text{ L}} \times \frac{1 \text{ L}}{1000 \text{ mL}} \times \frac{1 \text{ mL}}{1 \text{ g}} \times \frac{1 \text{ g}}{1000 \text{ mg}} = 1.1 \times 10^{-5} \text{ ppm}$$

Think About It

With all the concentrations of DDT in the samples now expressed in millimoles per liter (mM), we can make comparisons. The orchard is a little less contaminated than the residential area. The big surprise is that after a storm the groundwater contains nearly 600 times more DDT than before the storm.

4.25. Collect and Organize

We are given the concentration of NF_3 in air in parts-per-trillion and asked to express this in milligrams NF_3 per kg of air.

Analyze

Since we are given parts-per-trillion NF_3 in air, we first have to convert to a mass-based expression (kg NF_3 per kg air), then kg NF_3 to milligrams NF_3.

Solve

$$1 \text{ kg air} \times \frac{0.454 \text{ kg } NF_3}{10^{12} \text{ kg air}} \times \frac{1000 \text{ g}}{1 \text{ kg}} \times \frac{10^3 \text{ mg}}{1 \text{ g}} = 0.454 \text{ mg } NF_3 \text{ per kg air}$$

Think About It

This problem is made even shorter when you are comfortable immediately stating that 0.454 kg = 454 g NF_3.

4.27. Collect and Organize

The concentration of Cu^{2+} in mol/L is to be calculated knowing the mass percent of $CuSO_4$, the sample size, and the volume of the solution.

Analyze

We can find the mass of $CuSO_4$ in the sample by multiplying the percent (0.07%) by the sample size. Next, we can find the moles of $CuSO_4$ in the sample using the molar mass of $CuSO_4$ (159.61 g/mol). Because 1 mole of Cu^{2+} is in 1 mole of $CuSO_4$, this value is also the amount of moles of Cu^{2+}. The concentration of Cu^{2+}, then, is the moles of Cu^{2+} divided by the volume of the solution (2.0 L).

Solve

Mass of $CuSO_4$ in fertilizer sample: $20 \text{ g} \times 0.0007 = 1.4 \times 10^{-2} \text{ g}$

Moles of Cu^{2+} in sample: $1.4 \times 10^{-2} \text{ g } CuSO_4 \times \dfrac{1 \text{ mol } CuSO_4}{159.61 \text{ g } CuSO_4} \times \dfrac{1 \text{ mol } Cu^{2+}}{1 \text{ mol } CuSO_4} = 8.8 \times 10^{-5} \text{ mol } Cu^{2+}$

Concentration of Cu^{2+}: $\dfrac{8.8 \times 10^{-5} \text{ mol } Cu^{2+}}{2.0 \text{ L}} = 4.4 \times 10^{-5} \ M$

Think About It
Because very little of the fertilizer's mass is copper, we expect a low concentration of Cu^{2+} in the solution.

4.29. Collect and Organize
In diluting a solution, the final concentration is less than the original concentration. Each solution is diluted to 25.0 mL.

Analyze
For each dilution, we can use Equation 4.3:

$$V_{initial} \times M_{initial} = V_{final} \times M_{final}$$

Since we are calculating the final concentration C_{final}, the equation can be rearranged to

$$M_{final} = \frac{V_{initial} \times M_{initial}}{V_{final}}$$

Solve
(a) When 1.00 mL of 0.452 M Na^+ is diluted to 25.0 mL, we have $V_{initial} = 1.00$ mL, $M_{initial} = 0.452$ M, and $V_{final} = 25.0$ mL. The final concentration after diluting will be

$$M_{final} = \frac{1.00 \text{ mL} \times 0.452 \text{ } M}{25.0 \text{ mL}} = 1.81 \times 10^{-2} \text{ } M \text{ } Na^+$$

(b) When 2.00 mL of 3.4 mM LiCl is diluted to 25.0 mL, the final concentration will be

$$M_{final} = \frac{2.00 \text{ mL} \times 3.4 \text{ m}M}{25.0 \text{ mL}} = 2.7 \times 10^{-1} \text{ m}M \text{ LiCl}$$

(c) When 5.00 mL of 6.42×10^{-2} mM Zn^{2+} is diluted to 25.0 mL, the final concentration after diluting will be

$$M_{final} = \frac{5.00 \text{ mL} \times 6.42 \times 10^{-2} \text{ m}M}{25.0 \text{ mL}} = 1.28 \times 10^{-2} \text{ m}M \text{ } Zn^{2+}$$

Think About It
Notice that the milliliter volume units for these calculations do not need to be converted to another unit. As long as $V_{initial}$ and V_{final} are in the same units, the units will cancel in the calculations.

4.31. Collect and Organize
We must determine how much water needs to be added to a seawater solution to make the Na^+ concentration equivalent (isotonic) to that of human cytosol. This will require a dilution, as seawater is a more concentrated solution.

Analyze
Using the volume of seawater (1.50 mL), we multiply by the molar concentration, and multiply by 1000 to obtain the number of millimoles of Na^+ contained in the sample. Dividing by the desired concentration of cytosol (12 mM) tells us the final volume of the solution we will need to prepare. The volume of water required is the difference between the solution, and the seawater we started with. Seawater is significantly more concentrated than cytosol, so the volume of water required should be significant.

Solve

The total volume of the prepared solution is

$$1.50 \text{ mL seawater} \times \frac{0.481 \text{ mol Na}^+}{1000 \text{ mL seawater}} \times \frac{1000 \text{ mmol}}{1 \text{ mol}}$$

$$\times \frac{1000 \text{ mL solution}}{12 \text{ mmol Na}^+} = 60.125 \text{ mL}$$

To prepare the solution, the volume of water needed is

$$60.125 \text{ mL} - 1.50 \text{ mL} = 58.63 \text{ mL H}_2\text{O must be added}$$

Think About It

This answer makes sense because the seawater is approximately 40 times more concentrated than cytosol.

4.33. Collect and Organize

This "dilution" question is reversed: the concentration of Na^+ increases as the water in the puddle evaporates during the summer day. Given the initial concentration of Na^+ (0.449 M) and the percent evaporation, we are asked to find the final concentration of Na^+.

Analyze

We first have to compute volumes from the percent volume of the puddle after evaporation. If we assume a 1000 mL puddle, a reduction of the puddle volume to 23% would mean that 230 mL of the water in the puddle remains. Rearranging the dilution equation for the final concentration gives

$$M_{final} = \frac{V_{initial} \times M_{initial}}{V_{final}}$$

Solve

Here, $V_{initial} = 1000$ mL, $M_{initial} = 0.449$ M, and $V_{final} = 230$ mL. The final concentration of Na^+ in the puddle after evaporation is

$$M_{final} = \frac{1000 \text{ mL} \times 0.449 \text{ } M}{230 \text{ mL}} = 1.95 \text{ } M$$

Think About It

The concentration of the Na^+ in the puddle increased, as we would expect.

4.35. Collect and Organize

We are to use the change in absorbance of a solution of Cu^{2+} to determine the change in concentration of the solution. We are to use this change in concentration to determine the volume of water added to the original solution.

Analyze

Using the ratio of absorbances, we may determine the ratio of concentrations by applying Beer's Law (Equation 4.4). Knowing the volume and concentration of the original solution, we can determine the number of moles of Cu^{2+} present in the sample. Since this quantity will not change with dilution, and we know the final concentration of the solution, we can solve for the volume of the final solution. The dilution is depicted in Figure SM4.35.

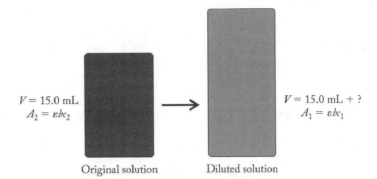

Original solution   Diluted solution

Solve

The concentration of the diluted solution is

$$\frac{A_1}{A_2} = \frac{c_1}{c_2} = \frac{0.55}{1.00} = \frac{c_1}{1.00\ M}$$

$$c_1 = 0.55\ M$$

The number of moles of Cu^{2+} in the original solution is

$$15.0\ mL \times \frac{1.00\ mol}{1000\ mL} = 0.015\ mol$$

The volume of the new solution is

$$0.015\ mol \times \frac{1000\ mL}{0.55\ mol} = 27.3\ mL$$

Subtracting the original volume of solution, we find the volume of water added

$$27.3\ mL - 15.0\ mL = 12.3\ mL\ added$$

Think About It

The absorption of light by the new solution is slightly less than half of the original solution, so it makes sense that we added approximately 50% of the original volume when diluting.

4.37. Collect and Organize

A platinum–tin complex is prepared by mixing a platinum complex and a tin complex. Given the molar absorptivity ($1.3 \times 10^4\ M^{-1}cm^{-1}$) and path length (1.0 cm), we are asked to calculate the absorbance of the solution.

Analyze

Initially, this is a limiting reagent problem. We must use the given masses and molar masses of $(NH_4)_2PtCl_6$ (5.2 mg, 443.88 g/mol) and $SnCl_2$ (2.2 mg, 189.61 g/mol) to determine which reagent is limiting. Using this limiting reagent and dividing by the final volume of solution will yield the molarity of the Pt/Sn complex. Substituting this value into Beer's Law ($A = \varepsilon bc$) will yield the absorbance of the final solution.

Solve

$$5.2 \text{ mg } (NH_4)_2PtCl_6 \times \frac{1 \text{ g}}{1000 \text{ mg}} \times \frac{1 \text{ mol } (NH_4)_2PtCl_6}{443.88 \text{ g } (NH_4)_2PtCl_6} \times \frac{1 \text{ mol Pt}}{1 \text{ mol } (NH_4)_2PtCl_6} = 1.1715 \times 10^{-5} \text{ mol Pt}$$

$$2.2 \text{ mg } SnCl_2 \times \frac{1 \text{ g}}{1000 \text{ mg}} \times \frac{1 \text{ mol } SnCl_2}{189.61 \text{ g } SnCl_2} \times \frac{1 \text{ mol Sn}}{1 \text{ mol } SnCl_2} = 1.1603 \times 10^{-5} \text{ mol Sn}$$

The limiting reagent is tin. The concentration of the Pt/Sn complex is

$$\frac{1.1603 \times 10^{-5} \text{ mol Sn}}{0.100 \text{ L} + 0.100 \text{ L}} = 5.801 \times 10^{-5} \ M$$

The absorbance of the final solution is

$$A = 1.3 \times 10^4 \ M^{-1}\text{cm}^{-1} \left(1.00 \text{ cm}\right)\left(5.801 \times 10^{-5} \ M\right) = 0.75$$

Think About It

Our calculations assume that the excess platinum complex does not absorb any light at the wavelength we are observing. Small differences in observed absorbance may be due to the extremely dilute platinum complex.

4.39. Collect and Organize

Electricity can conduct through a solution if it contains mobile ions.

Analyze

Table salt produces Na^+ and Cl^- ions in solution when it dissolves. Sugar does not dissociate into ions because it is not a salt. Ions are required to conduct electricity.

Solve

Sugar is not a good conductor of electricity in solution.

Think About It

A solution that conducts electricity is called an electrolyte and one that does not conduct electricity is a nonelectrolyte.

4.41. Collect and Organize

Electricity can be conducted through a liquid if there are mobile ions.

Analyze

Liquid methanol does not contain any ions because it does not dissociate. Molten sodium hydroxide, on the other hand, contains mobile Na^+ and OH^- ions.

Solve

The lack of ions in methanol means that the liquid is nonconductive. Molten NaOH, however, has freely moving Na^+ and OH^- ions, which can conduct electricity.

Think About It

Solid NaOH, however, is a poor conductor because the ions are locked into position in the structure of the solid.

4.43. Collect and Organize

The ability to conduct electricity rises with the presence of more ions in solution.

Analyze

All of the solutions contain salts, but they have different numbers of ions in their formula units and different concentrations. For each salt, we have to determine the number of ions (in terms of molarity).

Solve

(a) $1.0\ M\ NaCl$ contains $1.0\ M\ Na^+ + 1.0\ M\ Cl^- = 2.0\ M$ ions
(b) $1.2\ M\ KCl$ contains $1.2\ M\ K^+ + 1.2\ M\ Cl^- = 2.4\ M$ ions
(c) $1.0\ M\ Na_2SO_4$ contains $2.0\ M\ Na^+ + 1.0\ M\ SO_4^{2-} = 3.0\ M$ ions
(d) $0.75\ M\ LiCl$ contains $0.75\ M\ Li^+ + 0.75\ M\ Cl^- = 1.5\ M$ ions
Therefore the order of the solutions in decreasing ability to conduct electricity is (c) > (b) > (a) > (d).

Think About It

Notice that although the Na_2SO_4 and NaCl solutions are both $1.0\ M$ in salt concentration, the solution of Na_2SO_4 is more conductive because it contains $3.0\ M$ ions versus NaCl's $2.0\ M$ ions.

4.45. **Collect and Organize**

The concentration of Na^+ ions in each solution depends on the concentration of the solution as well as the number of sodium ions in the chemical formula of the salt. The concentration of all these salt solutions is 0.025 *M*.

Analyze

If a substance has one Na^+ in its formula, then the concentration of Na^+ ions in a solution of that salt is equal to the concentration of the substance in solution. If, however, there are, for example, two Na^+ ions in the salt's formula, then the concentration of Na^+ ions in the solution is twice the concentration of the salt.

Solve

(a) $0.025\ M\ NaBr$ is $0.025\ M\ Na^+$
(b) $0.025\ M\ Na_2SO_4$ is $2 \times 0.025\ M = 0.050\ M\ Na^+$
(c) $0.025\ M\ Na_3PO_4$ is $3 \times 0.025\ M = 0.075\ M\ Na^+$

Think About It

Salts dissociate into their constituent ions and so a solution may become more concentrated in a particular ion than the original concentration of salt.

4.47. **Collect and Organize**

In this question, we are comparing the number of particles in solution for various soluble species. We have to consider whether the species breaks into ions upon dissolution and, if so, how many ions are produced in the dissolution.

Analyze

All solutions are $1\ M$ so we need only consider the number of ions produced in solution (if any).

Solve

(a) NaCl forms Na^+ and Cl^- in solution. Two particles for every one of NaCl dissolved are released into the solution, so 1 L of a $1\ M$ NaCl solution contains a total of 2 moles of particles per liter.
(b) $CaCl_2$ forms Ca^{2+} and $2\ Cl^-$ in solution. Three particles for every one of $CaCl_2$ dissolved are released into the solution, so a $1\ M\ CaCl_2$ solution contains a total of 3 moles of particles per liter.
(c) Ethanol dissolves in water but is not a salt and therefore does not dissociate. A $1\ M$ solution of ethanol contains 1 mole of particles per liter.
(d) Acetic acid is a weak acid and partially dissociates in solution. A $1\ M$ solution of acetic acid contains between 1 and 2 moles of particles per liter.
Therefore, (b) contains the largest number of particles in solution.

Think About It

The number of particles that dissolve for $CaCl_2$ versus NaCl is a determinant in evaluating each salt's ability to melt ice on roads and sidewalks in winter.

4.49. Collect and Organize / Analyze

In the context of acid–base reactions, we are asked to give a name for the proton-donating species.

Solve

A proton donor is a Brønsted–Lowry acid.

Think About It

A Brønsted–Lowry base, therefore, is a proton acceptor by this acid–base definition.

4.51. Collect and Organize

In the context of acid–base reactions, we are asked to name two strong acids and two weak acids.

Analyze

A strong acid is one that completely dissociates into $H^+ + A^-$ in solution. A weak acid only partially dissociates and gives a mixture of H^+, A^-, and HA in solution.

Solve

Strong acids include HCl, HNO_3, $HClO_4$, H_2SO_4, HI, HBr; weak acids include CH_3COOH, HCOOH, HF, H_3PO_4.

Think About It

Strong bases include NaOH and KOH; weak bases include NH_3 and anions of weak acids such as the acetate ion, CH_3COO^-.

4.53. Collect and Organize / Analyze

In the context of acid–base reactions, we are asked to give a name for the proton-accepting species.

Solve

A proton acceptor is a Brønsted–Lowry base.

Think About It

A Brønsted–Lowry acid, therefore, is a proton donor by this acid–base definition.

4.55. Collect and Organize / Analyze

In the context of acid–base reactions, we are asked to name two strong bases and two weak bases. A strong base is one that produces 100% OH^- in solution, either by releasing OH^- upon dissolution, or by reacting completely with water to generate OH^- ions. A weak base only partially reacts with water to give OH^- ions, generating a mixture of BH^+, OH^-, and B in solution.

Solve

Strong bases include NaOH, KOH, CsOH, LiOH, RbOH, $Ba(OH)_2$, $Sr(OH)_2$, $Ca(OH)_2$; weak bases include NH_3, CH_3NH_2, C_5H_5N.

Think About It

Strong acids include HBr and HNO_3; weak acids include acetic acid and formic acid.

4.57. Collect and Organize

In each part of this problem, we identify the acid (proton donor) and base (proton acceptor). To write the net ionic equations we must identify the spectator ions and remove them from the ionic equation.

Analyze

For each reaction, write the species present in aqueous solution (showing dissociation). From these species you can identify the acid and base. Then eliminate any spectator ions in the ionic equation to give the net ionic equation.

Solve

(a) The acid is H_2SO_4; the base is $Ca(OH)_2$. Ionic and net ionic equation:

$$2\,H^+(aq) + SO_4^{2-}(aq) + Ca^{2+}(aq) + 2\,OH^-(aq) \rightarrow CaSO_4(s) + 2\,H_2O\,(\ell)$$

(b) $PbCO_3$ is the base; sulfuric acid is the acid. Ionic and net ionic equation:

$$PbCO_3(s) + 2\,H^+(aq) + SO_4^{2-}(aq) \rightarrow PbSO_4(s) + CO_2(g) + H_2O\,(\ell)$$

(c) Calcium is a spectator ion. $Ca(OH)_2$ is the base; CH_3COOH is the acid. Ionic equation:

$$Ca^{2+}(aq) + 2\,OH^-(aq) + 2\,CH_3COOH(aq) \rightarrow$$

$$Ca^{2+}(aq) + 2\,CH_3COO^-(aq) + 2\,H_2O\,(\ell)$$

Net ionic equation:

$$OH^-(aq) + CH_3COOH(aq) \rightarrow CH_3COO^-(aq) + H_2O\,(\ell)$$

Think About It

Reactions (a) and (c) are neutralization reactions whereas reaction (b) is an acid–base reaction that also forms a precipitate ($PbSO_4$) and a gas (CO_2).

4.59. **Collect and Organize**

We need to write molecular formulas for the reactants from the chemical names, determine the formulas for the products, write a balanced molecular equation, and then write the net ionic equation.

Analyze

All of the reactions involve an acid–base reaction in which a proton is transferred from the acid to the base. We can use the rules from Chapter 2 to write the formulas from the chemical names. In the net ionic equation, we have to be sure to eliminate all spectator ions. For (d), sulfur trioxide gas is first dissolved in water, then reacts with an aqueous solution of sodium hydroxide. Summing these two equations will yield the overall molecular equation.

Solve

(a) Molecular equation:

$$Mg(OH)_2(s) + H_2SO_4(aq) \rightarrow MgSO_4(aq) + 2\,H_2O\,(\ell)$$

Ionic and net ionic equations:

$$Mg(OH)_2(s) + 2\,H^+(aq) + SO_4^{2-}(aq) \rightarrow Mg^{2+}(aq) + SO_4^{2-}(aq) + 2\,H_2O\,(\ell)$$

$$Mg(OH)_2(s) + 2\,H^+(aq) \rightarrow Mg^{2+}(aq) + 2\,H_2O\,(\ell)$$

(b) Molecular equation:

$$MgCO_3(s) + 2\,HCl(aq) \rightarrow MgCl_2(aq) + H_2CO_3(aq)$$

The carbonic acid reacts in solution to give CO_2 and H_2O so the ionic and net ionic equations are as follows:

$$MgCO_3(s) + 2\,H^+(aq) + 2\,Cl^-(aq) \rightarrow Mg^{2+}(aq) + 2\,Cl^-(aq) + H_2O\,(\ell) + CO_2(g)$$

$$MgCO_3(s) + 2\,H^+(aq) \rightarrow Mg^{2+}(aq) + H_2O\,(\ell) + CO_2(g)$$

(c) Molecular equation:

$$NH_3(g) + HCl(g) \rightarrow NH_4Cl(s)$$

This is also the ionic and net ionic equations because these species are not in aqueous solution and are unable to form ions.

(d) Balanced molecular equations:

$$SO_3(g) + H_2O\,(\ell) \rightarrow H_2SO_4(aq)$$

$$H_2SO_4(aq) + 2\,NaOH(aq) \rightarrow Na_2SO_4(aq) + 2\,H_2O\,(\ell)$$

Overall molecular equation:

$$SO_3(g) + 2\,NaOH(aq) \rightarrow Na_2SO_4(aq) + H_2O\,(\ell)$$

Ionic and net ionic equations:

$$SO_3(g) + 2\,Na^+(aq) + 2\,OH^-(aq) \rightarrow 2\,Na^+(aq) + SO_4^{2-}(aq) + H_2O\,(\ell)$$

$$SO_3(g) + 2\,OH^-(aq) \rightarrow SO_4^{2-}(aq) + H_2O\,(\ell)$$

Think About It

Notice that species that are solids or gases do not appear as ionic species. Only those species that are soluble and are dissolved in water appear with the designation *aq* in the ionic equations.

4.61. Collect and Organize

We are given that lead(II) carbonate ($PbCO_3$) and lead(II) hydroxide [$Pb(OH)_2$] dissolve in acidic solutions (containing H_3O^+).

Analyze

To write the net ionic equations, we need to determine the acid–base reaction that might be occurring. Here the acid is in solution as H_3O^+, which we can write as $H^+(aq)$. This species must react with the anions of the solid salts (CO_3^{2-} and OH^-).

Solve

Lead(II) carbonate:

$$PbCO_3(s) + 2\,H^+(aq) \rightarrow Pb^{2+}(aq) + H_2CO_3(aq)$$

Carbonic acid reacts in solution to give $H_2O\,(\ell)$ and $CO_2(g)$:

$$PbCO_3(s) + 2\,H^+(aq) \rightarrow Pb^{2+}(aq) + CO_2(g) + H_2O\,(\ell)$$

Lead (II) hydroxide:

$$Pb(OH)_2(s) + 2\,H^+(aq) \rightarrow Pb^{2+}(aq) + 2\,H_2O\,(\ell)$$

Think About It

Both of the solids, by reacting with acid to form either CO_2 with water or just water, dissolve the solid, releasing toxic Pb^{2+} ions into the water.

4.63. Collect and Organize

All of the titrations involve a neutralization reaction. The moles of base (OH^-) required must be equal to the moles of acid (H^+) in the sample.

Analyze

First, we need to calculate the number of moles of acid from the volume and concentration of acid in the samples. Because the stoichiometry of the neutralization reaction is 1 mole OH^- to 1 mole H^+, the moles of base required is equal to the moles of H^+ in the sample. We can find the volume of base needed by dividing moles of OH^- required by the concentration of base used.

Solve

(a) $10.0 \text{ mL} \times \dfrac{0.0500 \text{ mol HCl}}{1000 \text{ mL}} \times \dfrac{1 \text{ mol H}^+}{1 \text{ mol HCl}} \times \dfrac{1 \text{ mol OH}^-}{1 \text{ mol H}^+} \times \dfrac{1000 \text{ mL}}{0.100 \text{ mol NaOH}} = 5.00 \text{ mL}$

(b) $25.0 \text{ mL} \times \dfrac{0.126 \text{ mol HNO}_3}{1000 \text{ mL}} \times \dfrac{1 \text{ mol H}^+}{1 \text{ mol HNO}_3} \times \dfrac{1 \text{ mol OH}^-}{1 \text{mol H}^+} \times \dfrac{1000 \text{ mL}}{0.100 \text{ mol NaOH}} = 31.5 \text{ mL}$

(c) $50.0 \text{ mL} \times \dfrac{0.215 \text{ mol H}_2\text{SO}_4}{1000 \text{ mL}} \times \dfrac{2 \text{ mol H}^+}{1 \text{ mol H}_2\text{SO}_4} \times \dfrac{1 \text{ mol OH}^-}{1 \text{ mol H}^+} \times \dfrac{1000 \text{ mL}}{0.100 \text{ mol NaOH}} = 215 \text{ mL}$

Think About It
Because sulfuric acid has two H^+ ions (it is a diprotic acid), we need twice as many moles of OH^- to neutralize it compared to the same concentration of a monoprotic acid such as HNO_3.

4.65. Collect and Organize
Using the solubility of Ca(OH)_2 we first must calculate the moles of Ca(OH)_2 in the solution; then we can find the volume of the $\text{HCl}(aq)$ solution to neutralize the Ca(OH)_2 solution.

Analyze
To find the moles of Ca(OH)_2 in the saturated solution, we multiply the volume of the solution by the solubility of Ca(OH)_2. This gives the grams of Ca(OH)_2 in the solution, which can then be converted into moles by dividing the grams of Ca(OH)_2 by the molar mass of Ca(OH)_2. Because there are two moles of OH^- in Ca(OH)_2, the moles of OH^- to neutralize must be twice the moles of Ca(OH)_2. We can then use the 1:1 molar ratio of OH^- to H^+ in the neutralization reaction and the concentration of the HCl solution to find the volume of HCl required to neutralize the Ca(OH)_2 solution.

Solve
Moles of Ca(OH)_2 in the saturated solution:

$$10.0 \text{ mL} \times \dfrac{0.185 \text{ g}}{100.0 \text{ mL}} \times \dfrac{1 \text{ mol Ca(OH)}_2}{74.09 \text{ g Ca(OH)}_2} = 2.50 \times 10^{-4} \text{ mol Ca(OH)}_2$$

Volume (mL) of HCl required to neutralize:

$$2.50 \times 10^{-4} \text{ mol Ca(OH)}_2 \times \dfrac{2 \text{ mol OH}^-}{1 \text{ mol Ca(OH)}_2} \times \dfrac{1 \text{ mol H}^+}{1 \text{ mol OH}^-}$$

$$\times \dfrac{1 \text{ mol HCl}}{1 \text{ mol H}^+} \times \dfrac{1000 \text{ mL}}{0.00100 \text{ mol HCl}} = 500 \text{ mL}$$

Think About It
Calcium hydroxide is not very soluble in water, but this neutralization requires a large volume of HCl solution because the HCl solution is fairly dilute.

4.67. Collect and Organize
This problem asks us to calculate how much $0.10 \, M$ acid one dose of antacid could neutralize.

Analyze
First, we need to calculate the moles of Mg(OH)_2 in the dose using the molar mass of Mg(OH)_2. Next, we have to use the balanced chemical equation for the neutralization:

$$\text{Mg(OH)}_2(s) + 2 \text{ HCl}(aq) \rightarrow \text{MgCl}_2(aq) + 2 \text{ H}_2\text{O}(\ell)$$

which shows that 2 moles of HCl are required for 1 moles of Mg(OH)_2 to determine the moles of HCl that would be neutralized. The volume of HCl can then be found from the moles of HCl neutralized and the concentration of the HCl in stomach acid.

Solve

$$0.830 \text{ g Mg(OH)}_2 \times \dfrac{1 \text{ mol Mg(OH)}_2}{58.32 \text{ g}} \times \dfrac{2 \text{ mol HCl}}{1 \text{ mol Mg(OH)}_2} \times \dfrac{1 \text{ L}}{0.10 \text{ mol HCl}} = 0.28 \text{ L or } 280 \text{ mL}$$

Think About It
One 10 mL dose can neutralize nearly 300 mL of stomach acid.

4.69. **Collect and Organize**

Solutions that contain a solute may be classified as either unsaturated, saturated, or supersaturated. We are to distinguish saturated from supersaturated solutions.

Analyze

More of the solute can dissolve in a unsaturated solution, but a saturated solution contains all of the solute it can hold in the solution.

Solve

A saturated solution contains the maximum concentration of a solute. A supersaturated solution *temporarily* contains *more* than the maximum concentration of a solute at a given temperature.

Think About It

A supersaturated solution eventually precipitates out some solute (until it reaches the saturation point).

4.71. **Collect and Organize**

The compound that precipitates first from an evaporating solution will be the one that is the least soluble.

Analyze

The potential salts that could form are all salts of Ca^{2+}: $CaCl_2$, $CaCO_3$, and $Ca(NO_3)_2$. The solubility rules state that nitrate and chloride salts are soluble for Ca^{2+} but imply that the carbonate salt of calcium is insoluble.

Solve

The most insoluble salt, $CaCO_3$, precipitates first from the evaporating solution.

Think About It

Calcium carbonate may not have precipitated from the original, more dilute solution because $CaCO_3$, being somewhat soluble, had not yet become concentrated enough to be a saturated solution. Once the saturation point is reached through evaporation, the salt will precipitate.

4.73. **Collect and Organize**

We are asked to compare a saturated solution with a concentrated solution.

Analyze

A saturated solution is one in which no more solute can dissolve. A concentrated solution is one that contains a high amount of solute.

Solve

A saturated solution may not be a concentrated solution if the solute is only sparingly or slightly soluble in the solution. In that case, the solution is a saturated dilute solution.

Think About It

Be careful when using the terms unsaturated/saturated and dilute/concentrated. These terms have precise meanings in chemistry.

4.75. **Collect and Organize**

Solubility can be predicted using the rules in Table 4.5. All of the compounds listed in the problem are ionic salts and are being dissolved in water.

Analyze

Soluble salts include those of the alkali metals and the ammonium cation and those with the acetate or nitrate anion. There are exceptions to the general solubility of halide salts (Ag^+, Cu^+, Hg_2^{2+}, and Pb^{2+} halides are insoluble) and sulfates (Ba^{2+}, Ca^{2+}, Hg_2^{2+}, Pb^{2+}, and Sr^{2+} sulfates are insoluble). All other salts are insoluble except the hydroxides of Ba^{2+}, Ca^{2+}, and Sr^{2+}.

Solve

(a) Barium sulfate is insoluble.

(b) Barium hydroxide is soluble.

(c) Lanthanum nitrate is soluble.

(d) Sodium acetate is soluble.

(e) Lead hydroxide is insoluble.

(f) Calcium phosphate is insoluble.

Think About It

Knowing well the few simple rules of solubility can help us easily predict which salts dissolve in water and which do not.

4.77. Collect and Organize

We are to write balanced molecular and net ionic equations for any precipitation reactions. For each reaction, we must determine if the mix of cations and anions present in solution will result in an insoluble salt.

Analyze

We use the solubility rules in Table 4.5 to determine which, if any, species precipitate when the two solutions are mixed. The net ionic equation can be written from the ionic equation by eliminating any of the spectator ions, ions that are not involved in forming the insoluble precipitate.

Solve

(a) The reactants in aqueous solution are Pb^{2+}, NO_3^-, Na^+, and SO_4^{2-}. If $Pb(NO_3)_2$ and Na_2SO_4 switched anionic partners, $PbSO_4$ and $NaNO_3$ would form. Of these two salts, $PbSO_4$ is insoluble. The ionic equation describing this reaction is

$$Pb^{2+}(aq) + 2\,NO_3^-(aq) + 2\,Na^+(aq) + SO_4^{2-}(aq) \rightarrow PbSO_4(s) + 2\,Na^+(aq) + 2\,NO_3^-(aq)$$

The balanced reaction is

$$Pb(NO_3)_2(aq) + Na_2SO_4(aq) \rightarrow PbSO_4(s) + 2\,NaNO_3(aq)$$

The net ionic equation is

$$Pb^{2+}(aq) + SO_4^{2-}(aq) \rightarrow PbSO_4(s)$$

(b) The reactants in aqueous solution are Ni^{2+}, Cl^-, NH_4^+, and NO_3^-. If $NiCl_2$ and NH_4CO_3 switched anionic partners $Ni(NO_3)_2$ and NH_4Cl would form. Both of those salts are soluble; therefore no precipitation reaction occurs.

(c) The reactants in aqueous solution are Fe^{2+}, Cl^-, Na^+, and S^{2-}. If $FeCl_2$ and Na_2S switched anionic partners FeS and NaCl would form. Of these two salts, FeS is insoluble. The ionic equation describing this reaction is

$$Fe^{2+}(aq) + 2\,Cl^-(aq) + 2\,Na^+(aq) + S^{2-}(aq) \rightarrow FeS\,(s) + 2\,Na^+(aq) + 2\,Cl^-(aq)$$

The balanced reaction is

$$FeCl_2(aq) + Na_2S(aq) \rightarrow FeS(s) + 2\,NaCl(aq)$$

The net ionic equation is

$$Fe^{2+}(aq) + S^{2-}(aq) \rightarrow FeS(s)$$

(d) The reactants in aqueous solution are Mg^{2+}, SO_4^{2-}, Ba^{2+}, and Cl^-. If $MgSO_4$ and $BaCl_2$ switched anionic partners $MgCl_2$ and $BaSO_4$ would form. Of these two salts, $BaSO_4$ is insoluble. The ionic equation describing this reaction is

$$Mg^{2+}(aq) + SO_4^{2-}(aq) + Ba^{2+}(aq) + 2\,Cl^-(aq) \rightarrow Mg^{2+}(aq) + 2\,Cl^-(aq) + BaSO_4(s)$$

The balanced reaction is

$$MgSO_4(aq) + BaCl_2(aq) \rightarrow MgCl_2(aq) + BaSO_4(s)$$

The net ionic equation is

$$Ba^{2+}(aq) + SO_4^{2-}(aq) \rightarrow BaSO_4(s)$$

Think About It

The net ionic equation for a precipitation reaction describes the formation of the insoluble salt from the aqueous cations and anions.

4.79. **Collect and Organize**

To determine how much $MgCO_3$ precipitates in this reaction, we have to determine whether either Na_2CO_3 or $Mg(NO_3)_2$ is the limiting reactant. The net ionic equation for the reaction is

$$Mg^{2+}(aq) + CO_3^{2-}(aq) \rightarrow MgCO_3(s)$$

Analyze

From the given volume of each reactant and its concentration, we first need to calculate the moles of Mg^{2+} and CO_3^{2-} present in the mixed solution. They react in a 1:1 molar ratio to form $MgCO_3$. By comparison of the moles of Mg^{2+} and CO_3^{2-}, we are able to determine the limiting reactant. Because 1 mol of $MgCO_3$ will form from 1 mol of either Mg^{2+} or CO_3^{2-}, the moles of the limiting reactant must equal the moles of $MgCO_3$ formed. Which ever ion is present in the smaller amount is the limiting reactant. From the moles of $MgCO_3$ formed, we can calculate the mass formed using the molar mass of $MgCO_3$ (84.31 g/mol).

Solve

$$mol\ CO_3^{2-} = 10.0\ mL\ Na_2CO_3 \times \frac{0.200\ mol}{1000\ mL} \times \frac{1\ mol\ CO_3^{2-}}{1\ mol\ Na_2CO_3} = 2.00 \times 10^{-3}\ mol\ CO_3^{2-}$$

$$mol\ Mg^{2+} = 5.00\ mL\ Mg(NO_3)_2 \times \frac{0.0500\ mol}{1000\ mL} \times \frac{1\ mol\ Mg^{2+}}{1\ mol\ Mg(NO_3)_2} = 2.50 \times 10^{-4}\ mol\ Mg^{2+}$$

The limiting reactant is $Mg(NO_3)_2$ and 2.50×10^{-4} $MgCO_3$ will form.
The mass of $MgCO_3$ produced is

$$2.50 \times 10^{-4}\ mol \times \frac{84.31\ g\ MgCO_3}{1\ mol\ MgCO_3} = 2.11 \times 10^{-2}\ g\ MgCO_3$$

Think About It

In every stoichiometric equation, the moles of the reactants are important. For species in solution, the moles can be found by multiplying the volume of the solution by the concentration, just as finding moles from a mass of substance involves dividing the mass of substance by the molar mass.

4.81. **Collect and Organize**

From the balanced equation, 1 mole of O_2 is required to react with 4 moles of $Fe(OH)^+$, the Fe(II) species. Knowing the volume and concentration of Fe(II), we are asked to find the grams of O_2 needed to form the insoluble $Fe(OH)_3$ product.

Analyze

The moles of $Fe(OH)^+$ in solution can be found by multiplying the volume of the Fe(II) solution (75 mL) by its concentration (0.090 M). The number of moles of O_2 required in the reaction is one-fourth of the moles of $Fe(OH)^+$ present. From moles of O_2 we can use the molar mass of O_2 (32.00 g/mol) to calculate the grams of O_2 needed.

Solve

$$mol\ Fe(OH)^+ = 75\ mL \times \frac{0.090\ mol\ Fe(OH)^+}{1000\ mL} = 6.75 \times 10^{-3}\ mol\ Fe(OH)^+$$

$$Mass\ of\ O_2 = 6.75 \times 10^{-3}\ mol\ Fe(OH)^+ \times \frac{1\ mol\ O_2}{4\ mol\ Fe(OH)^+} \times \frac{32.00\ g\ O_2}{1\ mol} = 5.4 \times 10^{-2}\ g\ O_2$$

Think About It

Because the molar ratio of $Fe(OH)^+$ to O_2 is 1:4, we require fewer moles of O_2 than the Fe(II) species.

4.83. **Collect and Organize**

To remove 90% of the phosphate from the 4.5×10^6 L of drinking water, we first have to determine the moles of PO_4^{3-} present, then multiply by 0.90 to obtain the mass of PO_4^{3-} that needs to be removed. The balanced equation tells us that 5 moles of $Ca(OH)_2$ are required to react with 3 moles of PO_4^{3-}. From that information, we can calculate the amount of $Ca(OH)_2$ needed.

Analyze

The moles of PO_4^{3-} in the water can be found from the volume of the water and the concentration of the PO_4^{3-}. Because that concentration is given as 25 mg/L, we need to use the molar mass of PO_4^{3-} (94.97 g/mol) to calculate mol/L. This amount then has to be multiplied by 0.90 (so that 90% of the phosphates are removed). Because the molar ratio for the reaction between PO_4^{3-} and $Ca(OH)_2$ is 3:5, we see that the moles of $Ca(OH)_2$ needed are $\frac{5}{3}$ times the moles of phosphate in the water. Once we know moles of $Ca(OH)_2$, we can use the molar mass (74.09 g/mol) to calculate the mass of $Ca(OH)_2$ required.

Solve

$$4.5 \times 10^6 \text{ L} \times \frac{25 \text{ mg}}{\text{L}} \times \frac{1 \text{ g}}{1000 \text{ mg}} \times \frac{1 \text{ mol PO}_4^{3-}}{94.97 \text{ g PO}_4^{3-}} = 1185 \text{ mol PO}_4^{3-}$$

$$90\% \text{ of the phosphate} = 1184.6 \text{ mol PO}_4^{3-} \times 0.90 = 1066 \text{ mol PO}_4^{3-}$$

$$1066 \text{ mol PO}_4^{3-} \times \frac{5 \text{ mol Ca(OH)}_2}{3 \text{ mol PO}_4^{3-}} \times \frac{74.09 \text{ g Ca(OH)}_2}{1 \text{ mol Ca(OH)}_2} = 1.3 \times 10^5 \text{ g or } 130 \text{ kg Ca(OH)}_2$$

Think About It

When doing any stoichiometry problem, convert the mass or volume given to moles.

4.85. **Collect and Organize**

We are asked to determine which ions from those listed will experience a decrease in concentration and which will stay the same.

Analyze

If a precipitate or pure liquid forms, the concentration of the ionic components of that precipitate or pure liquid will decrease, while the concentration of any spectator ions will remain the same. We must determine if a precipitate or a pure liquid will form in each case.

Solve

(a) AgCl will precipitate out of solution, so $[Ag^+]$ and $[Cl^-]$ will decrease, while $[Na^+]$ and $[NO_3^-]$ remain the same.

(b) Water will form, so $[H^+]$ and $[OH^-]$ will decrease, while $[Na^+]$ and $[Cl^-]$ remain the same.

(c) There is no net reaction, so all ionic concentrations will remain the same.

Think About It

Although we do not encounter reactions that form a gas in this problem, they would be similar to reactions in which a pure liquid or solid is formed.

4.87. **Collect and Organize**

Anion and cation exchangers swap unwanted cations and anions in water with other ions. We are asked to explain how they can deionize water.

Analyze

To deionize water (i.e., remove all the cations and anions) we would have to swap the cations and anions with ones that, when combined, produce H_2O.

Solve

To deionize water, cations such as Na^+ and Ca^{2+} are exchanged for H^+ at cation-exchange sites. Anions such as Cl^- and SO_4^{2-} are exchanged for OH^- at the anion-exchange sites. The released ions (H^+ and OH^-) at these sites combine to form H_2O.

Think About It

The electrical neutrality of the water is preserved. If we swap every Cl^- and Na^+ for OH^- and H^+ ions, we have kept the number of ions in the solution the same. However, because OH^- and H^+ combine to form neutral H_2O, we no longer have ions in the water—it is deionized.

4.89. Collect and Organize

In deionizing water all of the cations and anions are removed from the water.

Analyze

To completely remove the anions and cations, we have to have a cation exchange combined with an anion exchange that together produce H_2O.

Solve

To deionize water, the cation at the cation-exchange site must be H^+ and the anion at the anion-exchange site must be OH^-. When these combine, they form H_2O.

Think About It

The electrical neutrality of the water is preserved. If we swap every Cl^- and Na^+ for OH^- and H^+ ions, we have kept the number of ions in the solution the same. However, because OH^- and H^+ combine to form neutral H_2O, we no longer have ions in the water—it is deionized.

4.91. Collect and Organize

Here we are asked to define the connection between losses or gains of electrons and changes in oxidation numbers.

Analyze

Oxidation numbers are assigned to atoms in compounds based on the number of electrons they formally bring to the species. The loss of electrons (oxidation) means that the oxidation number becomes more positive. The gain of electrons (reduction) means that the oxidation number becomes more negative.

Solve

The number of electrons gained or lost is directly related to the change in oxidation number of a species. If a species loses two electrons, the oxidation number of one of the atoms in the species will increase by two (e.g., from +1 to +3).

Think About It

Oxidation numbers can help us decide which species is oxidized and which is reduced in a redox reaction.

4.93. Collect and Organize

The charges of all the ions are shown as superscripts for the species. We are to determine the sum of oxidation numbers for each species.

Analyze

Because the sum of the oxidation numbers of the atoms must equal the total charge on the polyatomic ion, the sum of the oxidation numbers for each species is simply the charge on the species.

Solve

(a) –1 for OH^-
(b) +1 for NH_4^+
(c) –2 for SO_4^{2-}
(d) –3 for PO_4^{3-}

Think About It
Recall that we can use the charge on a species to determine the oxidation state of an atom in the species that might have a variable oxidation state. For example, the oxidation state for S in SO_4^{2-} is +6 because the sum of the oxidation states of the 4 oxygen atoms is –8 (4 O^{2-}) and the overall charge is 2– on the anion.

4.95. Collect and Organize
Both silver and gold are placed into sulfuric acid, but only silver dissolves. We are asked which metal is the better reducing agent.

Analyze
In order for the metals to dissolve they must be oxidized from their metallic state to a soluble cation (Au^{3+} or Ag^+). When a metal is oxidized, it is acting as a reducing agent. In this case Au does not reduce sulfuric acid, but Ag does.

Solve
Since silver dissolves (is oxidized), but gold does not, in sulfuric acid, silver is more easily oxidized and is therefore the stronger reducing agent of the two metals.

Think About It
Oxidation and reduction reactions always occur in pairs. If a substance is reduced, it acts as an oxidizing agent; if a substance is oxidized, it acts as a reducing agent.

4.97. Collect and Organize
We are asked to describe what is occurring at the ionic level when electricity is passed through molten sodium chloride by writing the half-reactions.

Analyze
Electrolysis is the reverse of a spontaneous redox reaction. If a species is present as a cation, it will likely gain an electron (be reduced). Anionic species, on the other hand, will likely lose electrons (be oxidized).

Solve
Electrolysis of molten sodium chloride will reduce the Na^+ ions to Na(s) and oxidize the Cl^- ions to $Cl_2(g)$. The half-reactions are

$$Na^+ + e^- \rightarrow Na(s)$$

$$2\ Cl^- \rightarrow Cl_2(g) + 2\ e^-$$

Think About It
From the electrolysis of NaCl, we can produce elemental sodium, a highly reactive metal not found as a free element in nature.

4.99. Collect and Organize
The oxidation number for chlorine in these species varies depending on the number of oxygens to which the Cl atom is bound and the overall charge on the species.

Analyze
The oxidation number (O.N.) for oxygen in these species is –2. For hydrogen it is +1. The O.N. for chlorine in each species, therefore, must be positive and can be determined by

O.N. on Cl =
(charge on species) – [(number of O atoms) × (–2) + (number of H atoms) × (+1)]

Solve
(a) HClO: oxidation number on Cl = 0 – [1(–2) + 1(+1)] = +1
(b) $HClO_3$: oxidation number on Cl = 0 – [3(–2) + 1(+1)] = +5
(c) ClO_4^-: oxidation number on Cl = –1 – [4(–2)] = +7

Think About It
You may also determine the oxidation number by considering the oxidation number of each ion and then adding them together to get the overall oxidation number for the species. For example, in $HClO_3$ we have H^+, O^{2-}, O^{2-}, O^{2-}, and Cl^{n+} with the sum of +1, –2, –2, –2, and $n+$ equaling zero, so $n+$ must be 5+.

4.101. **Collect and Organize**
Each of the reactions is either an oxidation or reduction half-reaction. We need to balance each by adding the correct number of electrons to either the reactant side or the product side of the equation to balance the charge.

Analyze
A reduction reaction has electrons added to a reactant so the electrons, after balancing, appear on the left side of the equation. An oxidation reaction loses electrons so the electrons, after balancing, appear on the right side of the equation.

Solve
(a) As written, the reactant side has a charge of 0 and the product side has a charge of 2–. We need to add 2 electrons to the reactant side to balance the charge.

$$2\,e^- + Br_2(\ell) \rightarrow 2\,Br^-(aq)$$

This reaction is a reduction.
(b) As written, the reactant side has a charge of 2– and the product side has a charge of 0. We need to add 2 electrons to the product side to balance the charge.

$$Pb(s) + 2\,Cl^-(aq) \rightarrow PbCl_2(s) + 2\,e^-$$

This reaction is an oxidation.
(c) As written, the reactant side has a charge of 2+ and the product side has a charge of 0. We need to add 2 electrons to the reactant side to balance the charge.

$$2\,e^- + O_3(g) + 2\,H^+(aq) \rightarrow O_2(g) + H_2O(\ell)$$

This reaction is a reduction.
(d) As written, the reactant side has a charge of 1+ and the product side has a charge of 2–. We need to add 2 electrons to the reactant side to balance the charge.

$$2\,H_2SO_3(aq) + H^+(aq) + 2\,e^- \rightarrow 2\,H_2O(\ell) + HS_2O_4^-(aq)$$

This reaction is a reduction.

Think About It
Remember to compute the total charge on each side of the half-reaction. The total charge of 2 moles of H^+ is 2+ not 1+.

4.103. **Collect and Organize**
For every species we are asked to find the oxidation number for each atom. From those oxidation numbers, we can see which species is oxidized and which species is reduced.

Analyze
All of these reactions involve species of iron, silicon, oxygen, and hydrogen. Oxygen typically has an oxidation number of –2; hydrogen typically has an oxidation number of +1. Oxidation numbers of pure elements are zero. Iron is the atom most likely to have a variable oxidation number. Silicon's oxidation number is usually +4, consistent with its position in group 14 of the periodic table. Because all the compounds are neutral, the sum of the oxidation numbers for the atoms must be zero.

Solve
(a) Reactants   Products
SiO_2: Si = +4, O = –2 Fe_2SiO_4: Fe = +2, Si = +4, O = –2
Fe_3O_4: Fe = +8/3, O = –2 O_2: O = 0
Notice that we compute an oxidation state for Fe in Fe_3O_4 as +8/3. Actually this compound consists of FeO (Fe^{2+}) and Fe_2O_3 (Fe^{3+}).

Oxygen is oxidized (O^{2-} to O_2) and iron is reduced (Fe^{3+} to Fe^{2+}).
(b) Reactants   Products
SiO_2: Si = +4, O = –2 Fe_2SiO_4: Fe = +2, Si = +4, O = –2
Fe: Fe = 0
O_2: O = 0
Iron is oxidized (Fe^0 to Fe^{2+}) and oxygen is reduced (O_2 to O^{2-}).
(c) Reactants   Products
FeO: Fe = +2, O = –2 $Fe(OH)_3$: Fe = +3, O = –2, H = +1
O_2: O = 0
H_2O: H = +1, O = –2
Iron is oxidized (Fe^{2+} to Fe^{3+}) and oxygen is reduced (O_2 to O^{2-}).

Think About It
Molecular elemental oxygen is reduced in equations b and c and, therefore, acts as an oxidizing agent.

4.105. Collect and Organize
The reduction half-reaction of O_2 is to be combined with oxidation half-reactions and balanced.

Analyze
We are provided with the balanced half-reactions for both the oxidation and reduction. To combine them, we must balance the electrons produced in the oxidation with the electrons gained in the reduction reaction. To do so we may have to multiply each half-reaction by a factor to give the same number of electrons in both.

Solve
(a) The reduction of oxygen involves 4 electrons and the oxidation of $FeCO_3$ involves 2 e$^-$. To obtain a balanced overall equation, we must multiply the oxidation half-reaction by two.

$$O_2 + 4\,H^+ + 4\,e^- \rightarrow 2\,H_2O$$
$$\underline{\left(2\,FeCO_3 + H_2O \rightarrow Fe_2CO_3 + 2\,CO_2 + 2\,H^+ + 2\,e^-\right) \times 2}$$
$$O_2 + 4\,H^+ + 4\,FeCO_3 + 2\,H_2O \rightarrow 2\,H_2O + 2\,Fe_2O_3 + 4\,CO_2 + 4\,H^+$$

After canceling species that appear on both sides of the equation, this simplifies to

$$O_2(g) + 4\,FeCO_3(s) \rightarrow 2\,Fe_2O_3(s) + 4\,CO_2(g)$$

(b) The reduction of oxygen involves 4 electrons and the oxidation of $FeCO_3$ involves 2 e$^-$. To obtain a balanced overall equation, we must multiply the oxidation half-reaction by two.

$$O_2 + 4\,H^+ + 4\,e^- \rightarrow 2\,H_2O$$
$$\underline{\left(3\,FeCO_3 + H_2O \rightarrow Fe_3O_4 + 3\,CO_2 + 2\,H^+ + 2\,e^-\right) \times 2}$$
$$O_2 + 4\,H^+ + 6\,FeCO_3 + 2\,H_2O \rightarrow 2\,H_2O + 2\,Fe_3O_4 + 6\,CO_2 + 4\,H^+$$

After canceling species that appear on both sides of the equation, this simplifies to

$$O_2(g) + 6\,FeCO_3(s) \rightarrow 2\,Fe_3O_4(s) + 6\,CO_2(g)$$

(c) The reduction of oxygen involves 4 electrons and the oxidation of Fe_3O_4 involves 2 e$^-$. To obtain a balanced overall equation, we must multiply the oxidation half-reaction by two.

$$O_2 + 4\,H^+ + 4\,e^- \rightarrow 2\,H_2O$$
$$\underline{\left(2\,Fe_3O_4 + H_2O \rightarrow 3\,Fe_2O_3 + 2\,H^+ + 2\,e^-\right) \times 2}$$
$$O_2 + 4\,H^+ + 4\,Fe_3O_4 + 2\,H_2O \rightarrow 2\,H_2O + 6\,Fe_2O_3 + 4\,H^+$$

After canceling species that appear on both sides of the equation, this simplifies to

$$O_2(g) + 4\,Fe_3O_4(s) \rightarrow 6\,Fe_2O_3(s)$$

Think About It

All of these redox reactions simplify to overall reactions that do not need acid as a reactant, even though the half-reactions do depend on having H^+ available for the reaction.

4.107. Collect and Organize

Ammonium ions (NH_4^+) are oxidized by oxygen gas to give nitrate ions (NO_3^-). We are asked to balance the reaction in acid solution.

Analyze

To balance the reaction, first write the unbalanced half-reactions. In each, balance all atoms except hydrogen and oxygen, then use H_2O to balance oxygen and H^+ to balance hydrogen (in that order). Finally, balance charge with electrons. If the reactions have unequal amounts of electrons produced and consumed, multiply each half-reaction by a factor to give the same number of electrons in each. Then combine the half-reactions.

Solve

The oxidation of NH_4^+ involves 8 electrons and the reduction of O_2 requires 4 electrons. To obtain a balanced overall equation, we must multiply the reduction half-reaction by two.

$$3\,H_2O + NH_4^+ \rightarrow NO_3^- + 10\,H^+ + 8\,e^-$$

$$\underline{\left(4\,e^- + 4\,H^+ + O_2 \rightarrow 2\,H_2O\right)\times 2}$$

$$3\,H_2O + NH_4^+ + 8\,H^+ + 2\,O_2 \rightarrow NO_3^- + 10\,H^+ + 4\,H_2O$$

After canceling species that appear on both sides of the equation, this simplifies to

$$NH_4^+(aq) + 2\,O_2(g) \rightarrow NO_3^-(aq) + 2\,H^+(aq) + H_2O(\ell)$$

Think About It

This is an 8-electron oxidation in which the oxidation number of the nitrogen atom changes from –3 in NH_4^+ to +5 in NO_3^-.

4.109. Collect and Organize

We must consider a redox reaction between $HCrO_4^-$ and H_2S. We are asked to assign oxidation numbers to all species in the reaction and to balance the reaction.

Analyze

The oxidation number for sulfur in SO_4^{2-} and H_2S and for chromium in $HCrO_4^-$ and Cr_2O_3 can be determined considering that the typical oxidation numbers for oxygen and hydrogen are –2 and +1, respectively. We can balance the half-reactions for oxidation and reduction using H_2O and H^+ to balance for oxygen and hydrogen as needed, since the reaction is run under acidic conditions. Then, if needed, we multiply by factors to compensate for unequal numbers of electrons and add the half-reactions together.

Solve

(a) For SO_4^{2-}, each oxygen has a 2– charge and the overall species charge is 2–. This means that sulfur has an oxidation number of +6. For H_2S, each hydrogen is 1+. Since the species is neutral overall, sulfur has an oxidation number of –2. The change in oxidation state for S is –2 to +6. For Cr_2O_3, each oxygen has a 2– charge, and the overall species charge is zero. This means that the chromium has an oxidation number of +3. For $HCrO_4^-$, each oxygen has a 2– charge, and each hydrogen is 1+, and the overall charge is 1–. This means that the oxidation number of chromium is 6+. The change in oxidation state for Cr is +6 to +3.

(b) The oxidation of H_2S involves 8 electrons while the reduction of $HCrO_4^-$ involves 6 electrons. To obtain a balanced overall equation, we must multiply the reduction half-reaction by four, and the oxidation half-reaction by three.

$$\left(2\,HCrO_4^- + 8\,H^+ + 6\,e^- \rightarrow Cr_2O_3 + 5\,H_2O\right)\times 4$$

$$\underline{\left(H_2S + 4\,H_2O \rightarrow SO_4^{2-} + 10\,H^+ + 8\,e^-\right)\times 3}$$

$$8\,HCrO_4^- + 32\,H^+ + 24\,e^- + 3\,H_2S + 12\,H_2O \rightarrow 4\,Cr_2O_3 + 20\,H_2O + 3\,SO_4^{2-} + 30\,H^+ + 24\,e^-$$

After canceling species that appear on both sides of the equation, this simplifies to

$$8\ HCrO_4^- + 2\ H^+ + 3\ H_2S \rightarrow 4\ Cr_2O_3 + 8\ H_2O + 3\ SO_4^{2-}$$

(c) Six electrons are transferred in the reduction of $HCrO_4^-$, though two Cr atoms are involved in the half-reaction. This means that three electrons are transferred for each atom of Cr that reacts.

Think About It
It is important to check both charge and atom balance after balancing a redox reaction. In this case, the left side has a charge of $(-8 + 2 = -6)$, and the right side has a charge of -6. The atoms are balanced as well, as seen in the table

	Left Side	Right Side
H	$8 + 2 + 3(2) = 16$	16
Cr	8	$4(2) = 8$
O	$8(4) = 32$	$4(3) + 8(1) + 3(4) = 32$
S	3	3

4.111. Collect and Organize
We can balance the reaction by breaking it up into the half-reactions. The problem does not specify whether the freshwater stream is acidic or basic; we assume acidic conditions here.

Analyze
The half-reactions involve the oxidation of manganese (+2 to +4) and the reduction of iron (+3 to +2). We use the steps for balancing atoms and electrons for acidic conditions.

Solve
The half-reactions are

$$Fe(OH)_2^{\ +} \rightarrow Fe^{2+}$$
$$Mn^{2+} \rightarrow MnO_2$$

Balancing for oxygen and hydrogen using H_2O and H^+ gives us

$$2\ H^+ + Fe(OH)_2^{\ +} \rightarrow Fe^{2+} + 2\ H_2O$$
$$2\ H_2O + Mn^{2+} \rightarrow MnO_2 + 4\ H^+$$

Balancing for charge with electrons:

$$e^- + 2\ H^+ + Fe(OH)_2^{\ +} \rightarrow Fe^{2+} + 2\ H_2O$$
$$2\ H_2O + Mn^{2+} \rightarrow MnO_2 + 4\ H^+ + 2\ e^-$$

We have to multiply the reduction reaction by two to balance the electrons:

$$\left(e^- + 2\ H^+ + Fe(OH)_2^{\ +} \rightarrow Fe^{2+} + 2\ H_2O\right) \times 2$$

$$\underline{2\ H_2O + Mn^{2+} \rightarrow MnO_2 + 4\ H^+ + 2\ e^-}$$

$$4\ H^+ + 2\ Fe(OH)_2^{\ +} + 2\ H_2O + Mn^{2+} \rightarrow 2\ Fe^{2+} + 4\ H_2O + MnO_2 + 4\ H^+$$

This simplifies to

$$2\ Fe(OH)_2^{\ +}(aq) + Mn^{2+}(aq) \rightarrow 2\ Fe^{2+}(aq) + 2\ H_2O(\ell) + MnO_2(s)$$

Think About It
The assumption that this reaction can be balanced under acidic conditions is a good one, since dissolved CO_2 in natural waters makes them slightly acidic due to the presence of carbonic acid, H_2CO_3.

4.113. Collect and Organize
Use the half-reaction method to balance the equations.

Analyze
The reactions occur in basic solution. We first balance them for acidic conditions and then add OH⁻ to both sides of the equations to give a basic solution.

Solve
(a) The half-reactions are

$$MnO_4^- \rightarrow MnS$$
$$S^{2-} \rightarrow S$$

Because we need S for balancing atoms in the first equation, we have to add S^{2-} to MnO_4^-. This will preserve the reduction reaction of MnO_4^- because the S^{2-} we are adding has the same oxidation number as the S atom in MnS.

$$MnO_4^- + S^{2-} \rightarrow MnS$$
$$S^{2-} \rightarrow S$$

Balancing for O and H (in acidic solution first):

$$8\,H^+ + MnO_4^- + S^{2-} \rightarrow MnS + 4\,H_2O$$
$$S^{2-} \rightarrow S$$

Balancing for charge:

$$5\,e^- + 8\,H^+ + MnO_4^- + S^{2-} \rightarrow MnS + 4\,H_2O$$
$$S^{2-} \rightarrow S + 2\,e^-$$

Adding the two half-reactions together after multiplying each so the electrons balance:

$$\left(5\,e^- + 8\,H^+ + MnO_4^- + S^{2-} \rightarrow MnS + 4\,H_2O\right) \times 2$$
$$\left(S^{2-} \rightarrow S + 2\,e^-\right) \times 5$$

$$\overline{16\,H^+(aq) + 2\,MnO_4^-(aq) + 2\,S^{2-}(aq) + 5\,S^{2-}(aq) \rightarrow 2\,MnS(s) + 8\,H_2O(\ell) + 5\,S(s)}$$

Adding 16 OH⁻ to both sides to account for the basic solution:

$$16\,OH^- + 16\,H^+ + 2\,MnO_4^- + 2\,S^{2-} + 5\,S^{2-} \rightarrow 2\,MnS + 8\,H_2O + 5\,S + 16\,OH^-$$
$$16\,H_2O + 2\,MnO_4^- + 2\,S^{2-} + 5\,S^{2-} \rightarrow 2\,MnS + 8\,H_2O + 5\,S + 16\,OH^-$$
$$8\,H_2O(\ell) + 2\,MnO_4^-(aq) + 7\,S^{2-}(aq) \rightarrow 2\,MnS(s) + 5\,S(s) + 16\,OH^-(aq)$$

(b) The half-reactions are

$$MnO_4^- \rightarrow MnO_2$$
$$CN^- \rightarrow CNO^-$$

Balancing for O and H (in acidic solution first):

$$4\,H^+ + MnO_4^- \rightarrow MnO_2 + 2\,H_2O$$
$$H_2O + CN^- \rightarrow CNO^- + 2\,H^+$$

Balancing for charge:

$$3\text{ e}^- + 4\text{ H}^+ + \text{MnO}_4^- \rightarrow \text{MnO}_2 + 2\text{ H}_2\text{O}$$

$$\text{H}_2\text{O} + \text{CN}^- \rightarrow \text{CNO}^- + 2\text{ H}^+ + 2\text{ e}^-$$

Adding the half-reactions:

$$\left(3\text{ e}^- + 4\text{ H}^+ + \text{MnO}_4^- \rightarrow \text{MnO}_2 + 2\text{ H}_2\text{O}\right) \times 2$$

$$(\text{H}_2\text{O} + \text{CN}^- \rightarrow \text{CNO}^- + 2\text{ H}^+ + 2\text{ e}^-) \times 3$$

$$\overline{8\text{ H}^+(aq) + 2\text{ MnO}_4^-(aq) + 3\text{ H}_2\text{O}(\ell) + 3\text{ CN}^-(aq) \rightarrow 2\text{ MnO}_2(s) + 4\text{ H}_2\text{O}(\ell) + 3\text{ CNO}^-(aq) + 6\text{ H}^+(aq)}$$

Simplifying and adding OH⁻ to account for the basic solution:

$$2\text{ H}^+ + 2\text{ MnO}_4^- + 3\text{ CN}^- \rightarrow 2\text{ MnO}_2 + \text{H}_2\text{O} + 3\text{ CNO}^-$$

$$2\text{ OH}^- + 2\text{ H}^+ + 2\text{ MnO}_4^- + 3\text{ CN}^- \rightarrow 2\text{ MnO}_2 + \text{H}_2\text{O} + 3\text{ CNO}^- + 2\text{ OH}^-$$

$$2\text{ H}_2\text{O} + 2\text{ MnO}_4^- + 3\text{ CN}^- \rightarrow 2\text{ MnO}_2 + \text{H}_2\text{O} + 3\text{ CNO}^- + 2\text{ OH}^-$$

$$\text{H}_2\text{O}(\ell) + 2\text{ MnO}_4^-(aq) + 3\text{ CN}^-(aq) \rightarrow 2\text{ MnO}_2(s) + 3\text{ CNO}^-(aq) + 2\text{ OH}^-(aq)$$

(c) The half-reactions are

$$\text{MnO}_4^- \rightarrow \text{MnO}_2$$

$$\text{SO}_3^{2-} \rightarrow \text{SO}_4^{2-}$$

Balancing for O and H (in acidic solution first):

$$4\text{ H}^+ + \text{MnO}_4^- \rightarrow \text{MnO}_2 + 2\text{ H}_2\text{O}$$

$$\text{H}_2\text{O} + \text{SO}_3^{2-} \rightarrow \text{SO}_4^{2-} + 2\text{ H}^+$$

Balancing for charge:

$$3\text{ e}^- + 4\text{ H}^+ + \text{MnO}_4^- \rightarrow \text{MnO}_2 + 2\text{ H}_2\text{O}$$

$$\text{H}_2\text{O} + \text{SO}_3^{2-} \rightarrow \text{SO}_4^{2-} + 2\text{ H}^+ + 2\text{ e}^-$$

Adding the half-reactions:

$$\left(3\text{ e}^- + 4\text{ H}^+ + \text{MnO}_4^- \rightarrow \text{MnO}_2 + 2\text{ H}_2\text{O}\right) \times 2$$

$$(\text{H}_2\text{O} + \text{SO}_3^{2-} \rightarrow \text{SO}_4^{2-} + 2\text{ H}^+ + 2\text{ e}^-) \times 3$$

$$\overline{8\text{ H}^+(aq) + 2\text{ MnO}_4^-(aq) + 3\text{ H}_2\text{O}(\ell) + 3\text{ SO}_3^{2-}(aq) \rightarrow 2\text{ MnO}_2(s) + 4\text{ H}_2\text{O}(\ell) + 3\text{ SO}_4^{2-}(aq) + 6\text{ H}^+(aq)}$$

Simplifying and adding OH⁻ to account for the basic solution:

$$2\text{ H}^+ + 2\text{ MnO}_4^- + 3\text{ SO}_3^{2-} \rightarrow 2\text{ MnO}_2 + \text{H}_2\text{O} + 3\text{ SO}_4^{2-}$$

$$2\text{ OH}^- + 2\text{ H}^+ + 2\text{ MnO}_4^- + 3\text{ SO}_3^{2-} \rightarrow 2\text{ MnO}_2 + \text{H}_2\text{O} + 3\text{ SO}_4^{2-} + 2\text{ OH}^-$$

$$2\text{ H}_2\text{O} + 2\text{ MnO}_4^- + 3\text{ SO}_3^{2-} \rightarrow 2\text{ MnO}_2 + \text{H}_2\text{O} + 3\text{ SO}_4^{2-} + 2\text{ OH}^-$$

$$\text{H}_2\text{O}(\ell) + 2\text{ MnO}_4^-(aq) + 3\text{ SO}_3^{2-}(aq) \rightarrow 2\text{ MnO}_2(s) + 3\text{ SO}_4^{2-}(aq) + 2\text{ OH}^-(aq)$$

(d) The half-reactions are

$$\text{CN}^- + \text{Ag} \rightarrow \text{Ag(CN)}_2^-$$

$$\text{O}_2 \rightarrow \text{H}_2\text{O}$$

Balancing for heavy atoms, O, and H (in acidic solution first):

$$2\ CN^- + Ag \rightarrow Ag(CN)_2^-$$

$$4\ H^+ + O_2 \rightarrow 2\ H_2O$$

Balancing for charge:

$$2\ CN^- + Ag \rightarrow Ag(CN)_2^- + 1\ e^-$$

$$4e^- + 4\ H^+ + O_2 \rightarrow 2\ H_2O$$

Adding the half-reactions:

$$\left(2\ CN^- + Ag \rightarrow Ag(CN)_2^- + 1\ e^- \right) \times 4$$

$$\underline{4\ e^- + 4\ H^+ + O_2 \rightarrow 2\ H_2O}$$

$$8\ CN^- + 4\ Ag + 4\ e^- + 4\ H^+ + O_2 \rightarrow 4\ Ag(CN)_2^- + 4\ e^- + 2\ H_2O$$

Simplifying and adding OH^- to account for the basic solution:

$$8\ CN^- + 4\ Ag + 4\ H^+ + 4\ OH^- + O_2 \rightarrow 4\ Ag(CN)_2^- + 2\ H_2O + 4\ OH^-$$

$$8\ CN^-(aq) + 4\ Ag(s) + 2\ H_2O(\ell) + O_2(g) \rightarrow 4\ Ag(CN)_2^-(aq) + 4\ OH^-(aq)$$

Think About It
The half-reaction method of balancing redox equations, once learned, gives a reliable way to systematically balance any redox reaction.

4.115. Collect and Organize
We are asked which of the listed metals will reduce an aqueous solution of Fe^{2+}.

Analyze
The activity series lists oxidation half-reactions in order. From the text, "A metal cation on the list in Table 4.6 will oxidize any metal above it in the activity series." This also means that metals above a cation in the activity series will reduce that cation.

Solve
From the list, the metals above Fe^{2+} are zinc and aluminum. Both zinc and aluminum metal will reduce Fe^{2+} to iron metal.

Think About It
We explore the activity series further in Chapter 19 of this text.

4.117. Collect and Organize
On the basis of the experiments described, we are to determine the relative position of aluminum, vanadium, and scandium in the activity series. We are also asked to suggest a metal that could be used to confirm the position of scandium in the activity series.

Analyze
A metal will be oxidized by any cation below it on the activity series. Each experiment tells us if the metal cation is above or below aluminum metal.

Solve
Aluminum is oxidized by V^{3+}, so vanadium must lie below aluminum on the activity series. Aluminum is not oxidized by Sc^{3+}, so scandium must lie above aluminum on the activity series. Based on these experiments, the order of the activity series should be

$$Sc(s) \rightarrow Sc^{3+}(aq) + 3\ e^-$$

$$Al(s) \rightarrow Al^{3+}(aq) + 3\ e^-$$

$$V(s) \rightarrow V^{3+}(aq) + 3\ e^-$$

We could use magnesium metal to test the proposed position of scandium. If magnesium metal is oxidized by Sc^{3+}, scandium must lie between aluminum and magnesium.

Think About It

Any metal above aluminum could be used to test the position of scandium in the activity series, but magnesium is closest to aluminum, and thus would give us the best confirmation.

4.119. **Collect and Organize**

We will need to use the half-reaction method to balance the reaction in acidic solution. We are also asked to determine the concentration of the Fe^{2+} solution, given the volumes of solution and concentration of titrant used in a redox titration.

Analyze

Balancing in acidic solution involves writing the half-reactions, balancing the atoms, balancing the charge, and adding the half-reactions to obtain the overall redox equation. We can determine the concentration of the Fe^{2+} solution by multiplying the volume (15.2 mL) and concentration (0.135 *M*) of $Cr_2O_7^{2-}$ solution employed, and multiplying by the stoichiometric ratio from the balanced redox equation. Dividing by the volume (100.0 mL) of Fe^{2+} solution will yield the concentration of the Fe^{2+} solution.

Solve

(a) The half-reactions are

$$Cr_2O_7^{2-} \rightarrow Cr^{3+}$$

$$Fe^{2+} \rightarrow Fe^{3+}$$

Balancing for O and H:

$$14\ H^+ + Cr_2O_7^{2-} \rightarrow 2\ Cr^{3+} + 7\ H_2O$$

$$Fe^{2+} \rightarrow Fe^{3+}$$

Balancing for charge:

$$14\ H^+ + Cr_2O_7^{2-} + 6\ e^- \rightarrow 2\ Cr^{3+} + 7\ H_2O$$

$$Fe^{2+} \rightarrow Fe^{3+} + 1\ e^-$$

Adding the half-reactions and simplifying:

$$\left(Fe^{2+} \rightarrow Fe^{3+} + 1\ e^- \right) \times 6$$

$$\underline{14\ H^+ + Cr_2O_7^{2-} + 6\ e^- \rightarrow 2\ Cr^{3+} + 7\ H_2O}$$

$$6\ Fe^{2+} + Cr_2O_7^{2-} + 6\ e^- + 14\ H^+ \rightarrow 6\ Fe^{3+} + 6\ e^- + 2\ Cr^{3+} + 7\ H_2O$$

$$6\ Fe^{2+}(aq) + Cr_2O_7^{2-}(aq) + 14\ H^+(aq) \rightarrow 6\ Fe^{3+}(aq) + 2\ Cr^{3+}(aq) + 7\ H_2O(\ell)$$

(b) $15.2 \text{ mL Cr}_2\text{O}_7^{2-} \times \dfrac{0.135 \text{ mol Cr}_2\text{O}_7^{2-}}{1000 \text{ mL Cr}_2\text{O}_7^{2-}} \times \dfrac{6 \text{ mol Fe}^{2+}}{1 \text{ mol Cr}_2\text{O}_7^{2-}} = 0.012312 \text{ mol Fe}^{2+}$

$$\left[\text{Fe}^{2+}\right] = \dfrac{0.012312 \text{ mol Fe}^{2+}}{0.100 \text{ L}} = 0.123 \text{ } M$$

Think About It
Just like an acid–base titration, a redox titration may be used to determine the concentration of a solution.

4.121. Collect and Organize
In this precipitation titration, a known volume of barium nitrate is titrated into a solution of unknown sulfate concentration. We are asked to calculate how much sulfate is in the solution and express the concentration of sulfate in moles per liter (M).

Analyze
Using the concentration of the $Ba(NO_3)_2$ titrant and the volume of titrant used, we first find the moles of $Ba(NO_3)_2$ used in the titration. From the balanced equation

$$Ba(NO_3)_2 + SO_4^{2-} \rightarrow BaSO_4 + 2 NO_3^-$$

we see that 1 mole $Ba(NO_3)_2$ reacts with 1 mole SO_4^{2-}. This 1:1 ratio means that the moles of titrant used equals the moles of SO_4^{2-} in the sample. To calculate concentration of SO_4^{2-} we divide the moles of SO_4^{2-} by the volume of the sample in liters.

Solve

$$3.19 \text{ mL Ba(NO}_3)_2 \times \dfrac{0.0250 \text{ mol Ba(NO}_3)_2}{1000 \text{ mL}} \times \dfrac{1 \text{ mol SO}_4^{2-}}{1 \text{ mol Ba(NO}_3)_2} \times \dfrac{1}{0.1000 \text{ L}} = 7.98 \times 10^{-4} \text{ } M \text{ SO}_4^{2-}$$

Think About It
Precipitation titrations give us the concentration of a species in solution and are only accurate when the salt formed has very low solubility. If we form a salt of marginal solubility, then we leave some of the species in solution and our calculation underestimates the concentration of the species in solution.

4.123. Collect and Organize
From the mass percent and density of concentrated HCl we are asked to determine the molarity of HCl. Then we calculate the amount needed to prepare a dilute solution from the concentrated HCl and figure out how much sodium bicarbonate is needed to neutralize a spill of concentrated HCl.

Analyze
Parts b and c depend on the answer to part a. If the solution is 36.0% by mass, then 100 g of concentrated HCl contains 36.0 g HCl. Using the molar mass of HCl (36.46 g/mol) we can convert the grams into moles. To obtain the volume of acid, we divide the 100 g of acid by the density (1.18 g/mL). We now have moles of HCl and volume of HCl to compute the molarity. To determine the volume of concentrated HCl needed to prepare a more dilute solution of HCl in part b, we use Equation 4.3:

$$V_{initial} M_{initial} = V_{final} M_{final}$$

For part c we need the chemical equation that describes the neutralization:

$$NaHCO_3 + HCl \rightarrow NaCl + H_2O + CO_2$$

Solve

(a) Molarity of concentrated HCl solution:

$$36.0 \text{ g HCl solution} \times \frac{1 \text{ mol HCl}}{36.46 \text{ g}} = 0.987 \text{ mol HCl}$$

$$100 \text{ g solution} \times \frac{1 \text{ mL}}{1.18 \text{ g}} = 84.7 \text{ mL or } 0.0847 \text{ L}$$

$$\text{Molarity HCl(conc)} = \frac{0.987 \text{ mol}}{0.0847 \text{ L}} = 11.7 \ M$$

(b) Volume of HCl(conc) required to make 0.250 L of 2.00 M solution:

$$V_{\text{initial}} M_{\text{initial}} = V_{\text{final}} M_{\text{final}}$$

$$V_{\text{initial}} \times 11.7 \ M = 250 \text{ mL} \times 2.00 \ M$$

$$V_{\text{initial}} = 42.7 \text{ mL}$$

(c) Mass of $NaHCO_3$ required to neutralize the spill:

$$1.75 \text{ L HCl(conc)} \times \frac{11.7 \text{ mol}}{1 \text{ L}} \times \frac{1 \text{ mol } NaHCO_3}{1 \text{ mol HCl}} \times \frac{84.01 \ NaHCO_3}{1 \text{ mol}} = 1720 \text{ g or } 1.72 \text{ kg}$$

Think About It

This problem puts together many of the chapter's topics, including the definition of molarity, the preparation of diluted solutions, and neutralization. It also involves concepts from previous chapters including density and reaction stoichiometry.

4.125. Collect and Organize

This is a redox reaction so we have to use the half-reaction method to first balance the equation. From the species involved we can then identify the oxidizing and reducing agents. Finally, we are asked to calculate the amount of $Na_2S_2O_4$ needed to remove a certain amount of CrO_4^{2-}.

Analyze

The redox reaction occurs in base. We can first balance the half-reactions as if in acid, and then add sufficient OH^- to both sides of the equation to neutralize H^+. From the half-reactions and oxidation numbers, we can identify the species that are oxidized and reduced and assign species as oxidizing or reducing agents. We need the balanced equations for the stoichiometry of reactants to calculate how much $Na_2S_2O_4$ is required to remove the CrO_4^{2-} from the wastewater. We also need the molar mass of $Na_2S_2O_4$.

Solve

(a) The reaction can be described in the following unbalanced equation:

$$S_2O_4^{2-}(aq) + CrO_4^{2-}(aq) \rightarrow SO_3^{2-}(aq) + Cr(OH)_3(s)$$

Balancing this by the half-reaction method in acidic solution first gives

$$\left(2 \ H_2O + S_2O_4^{2-} \rightarrow 2 \ SO_3^{2-} + 4 \ H^+ + 2 \ e^-\right) \times 3$$

$$\left(3 \ e^- + 5 \ H^+ + CrO_4^{2-} \rightarrow Cr(OH)_3 + H_2O\right) \times 2$$

$$\overline{6 \ H_2O + 3 \ S_2O_4^{2-} + 10 \ H^+ + 2 \ CrO_4^{2-} \rightarrow 6 \ SO_3^{2-} + 12 \ H^+ + 2 \ Cr(OH)_3 + 2 \ H_2O}$$

This simplifies to

$$4 \ H_2O + 3 \ S_2O_4^{2-} + 2 \ CrO_4^{2-} \rightarrow 6 \ SO_3^{2-} + 2 \ H^+ + 2 \ Cr(OH)_3$$

Placing in base, we add OH⁻:

$$2\,OH^- + 4\,H_2O + 3\,S_2O_4^{2-} + 2\,CrO_4^{2-} \rightarrow 6\,SO_3^{2-} + 2\,H^+ + 2\,Cr(OH)_3 + 2\,OH^-$$

Simplifying:

$$2\,OH^-(aq) \;+\; 2\,H_2O(\ell) \;+\; 3\,S_2O_4^{2-}(aq) \;+\; 2\,CrO_4^{2-}(aq) \;\rightarrow\; 6\,SO_3^{2-}(aq) \;+\; 2\,Cr(OH)_3(s)$$

(b) Because $S_2O_4^{2-}$ is the reactant in the oxidation half-reaction, we must look at $S_2O_4^{2-} \rightarrow SO_3^{2-}$. Here sulfur is oxidized from +3 to +4.

Because CrO_4^{2-} is the reactant in the reduction half-reaction, we must look at $CrO_4^{2-} \rightarrow Cr(OH)_3$. Here chromium is reduced from +6 to +3.

(c) The oxidizing agent is the species that is itself reduced: CrO_4^{2-}. The reducing agent is itself oxidized: $S_2O_4^{2-}$.

(d) The amount of $Na_2S_2O_4$ required is

$$100.0\text{ L wastewater} \times \frac{0.00148\text{ mol }CrO_4^{2-}}{1\text{ L}} \times \frac{3\text{ mol }S_2O_4^{2-}}{2\text{ mol }CrO_4^{2-}} \times \frac{1\text{ mol }Na_2S_2O_4}{1\text{ mol }S_2O_4^{2-}} \times \frac{174.11\text{ g }Na_2S_2O_4}{1\text{ mol }Na_2S_2O_4} = 38.7\text{ g}$$

Think About It

Be sure to always start with a balanced chemical equation before calculating the amount of reactants required or the amount of products in a reaction.

4.127. **Collect and Organize**

In this problem we examine the tarnishing of silver to Ag_2S and the conversion of Ag_2S back to Ag in the presence of aluminum metal. Both of these are redox reactions.

Analyze

We are provided with the chemical formulas for all the reactants and products. We can use the half-reaction method to write and balance the equations described and use the oxidation number assignment rules in Section 4.9 to assign oxidation numbers. Both Ag and H_2 are reduction products from Ag_2S and H_2O. We can place these in the same reduction half-reaction.

Solve

(a) The tarnishing of Ag occurs in the presence of H_2S to form Ag_2S.

$$2\,Ag + H_2S \rightarrow Ag_2S + H_2$$

Oxidation numbers:

Reactants	Products
$Ag = 0$	$Ag_2S : Ag = +1, S = -2$
$H_2S: H = +1, S = -2$	$H_2: H = 0$

Because 2 moles of electrons are lost for every 2 moles of Ag tarnished, 1 mole of electrons are transferred per mole of Ag.

(b) The reaction of Ag_2S with Al and water is described as

$$\left(4\,e^- + 4\,H^+ + Ag_2S \rightarrow 2\,Ag + H_2S + H_2\right) \times 3$$

$$\underline{\left(3\,H_2O + Al \rightarrow Al(OH)_3 + 3\,H^+ + 3\,e^-\right) \times 4}$$

$$12\,H^+ + 3\,Ag_2S + 12\,H_2O + 4\,Al \rightarrow 6\,Ag + 3\,H_2S + 3\,H_2 + 4\,Al(OH)_3 + 12\,H^+$$

Simplifying:

$$3\,Ag_2S(s) + 12\,H_2O(\ell) + 4\,Al(s) \rightarrow 6\,Ag(s) + 3\,H_2S(g) + 3\,H_2(g) + 4\,Al(OH)_3(s)$$

Think About It

Baking soda does not become involved in the reaction but makes the solution basic.

4.129. Collect and Organize

We are asked to write overall and net ionic equations for an acid–base titration, a possible precipitation reaction, and the reaction of a metal oxide with water.

Analyze

For part a we must remember that an acid–base titration forms a salt and water as products. Acetic acid has the formula $HC_2H_3O_2$, so upon reaction with KOH, potassium acetate ($KC_2H_3O_2$) and water form. In part b we have to consider whether the possible products (NaCl and $CaCO_3$) are insoluble. In part c CaO reacts with water to give a basic solution because CaO is a base anhydride. For all of these, the net ionic equation must not display any spectator ions.

Solve

(a) Overall equation:

$$HC_2H_3O_2(aq) + KOH(aq) \rightarrow H_2O(\ell) + KC_2H_3O_2(aq)$$

Net ionic (all potassium salts are soluble):

$$HC_2H_3O_2(aq) + K^+(aq) + OH^-(aq) \rightarrow H_2O(\ell) + K^+(aq) + C_2H_3O_2^-(aq)$$
$$HC_2H_3O_2(aq) + OH^-(aq) \rightarrow H_2O(\ell) + C_2H_3O_2^-(aq)$$

(b) Overall equation ($CaCO_3$ is an insoluble product):

$$Na_2CO_3(aq) + CaCl_2(aq) \rightarrow CaCO_3(s) + 2\,NaCl(aq)$$

Net ionic (all sodium salts are soluble as is the chloride salt of Ca^{2+}):

$$2\,Na^+(aq) + CO_3^{2-}(aq) + Ca^{2+}(aq) + 2\,Cl^-(aq) \rightarrow CaCO_3(s) + 2\,Na^+(aq) + 2\,Cl^-(aq)$$
$$CO_3^{2-}(aq) + Ca^{2+}(aq) \rightarrow CaCO_3(s)$$

(c) Overall reaction:

$$CaO(s) + H_2O(\ell) \rightarrow Ca(OH)_2(aq)$$

Net ionic equation (hydroxide salt of Ca^{2+} is soluble):

$$CaO(s) + H_2O(\ell) \rightarrow Ca^{2+}(aq) + 2\,OH^-(aq)$$

Think About It

Acetic acid is not a strong acid so we do not write it as $H^+(aq) + C_2H_3O_2^-(aq)$ in water. It remains in water mostly as a molecular species.

4.131. Collect and Organize

Considering the values we are given for the stream contamination by perchlorate, the flow rate of the stream, and the advisory range for perchlorate in drinking water, we are asked to determine the amount of perchlorate that flows into Lake Mead and the volume of perchlorate-free water that would be needed to reduce concentrations from 700.0 to 4 µg/L. We also compare results from three labs on replicate samples of water for perchlorate and determine which lab provided the most precise results.

Analyze

(a) To write the formulas for sodium perchlorate and ammonium perchlorate, we must write neutral chemical formulas using Na^+ with ClO_4^- and NH_4^+ with ClO_4^-. The cation and anion each have a charge of one, so the cation and anion in each salt are present in a 1:1 ratio.

(b) To calculate how much ClO_4^- enters the lake from the stream we need to multiply the flow rate (after converting to liters per day) by the concentration of ClO_4^- in the stream (and convert μg into kg).

(c) In this part we can use Equation 4.3

$$V_{initial} \times M_{initial} = V_{final} \times M_{final}$$

where $M_{initial}$ = 700 μg/L, and M_{final} = 4 μg/L, and $V_{initial}$ is 1.61×10^8 gal (volume that flows in the stream each day). V_{final} is the final total volume to reduce the concentration of ClO_4^- to 4 μg/L. The volume of lake water that must be mixed with the stream water will be $V_{final} - 1.61 \times 10^8$ gal.

(d) In comparing the sample data for MD, MA, and NM, the most precise data for the replicate samples must have the narrowest range of values.

Solve

(a) Sodium perchlorate, $NaClO_4$
Ammonium perchlorate, NH_4ClO_4

(b) Perchlorate flow into Lake Mead each day:

$$\frac{1.61 \times 10^8 \text{ gal}}{1 \text{ day}} \times \frac{3.785 \text{ L}}{1 \text{ gal}} \times \frac{700 \text{ μg}}{1 \text{ L}} \times \frac{1 \text{ g}}{1 \times 10^6 \text{ μg}} \times \frac{1 \text{ kg}}{1000 \text{ g}} = 427 \text{ kg}$$

(c) Using the dilution equation:

$$1.61 \times 10^8 \text{ gal} \times 700 \text{ μg/L} = 4 \text{ μg/L} \times V_{final}$$

$$V_{final} = 2.82 \times 10^{10} \text{ gal}$$

Volume of lake water required: 2.82×10^{10} gal $- 1.61 \times 10^8$ gal $= 2.80 \times 10^{10}$ gal

(d) The data for the samples from MD range from 0.9 to 1.4 μg/L or a range of 0.5 μg/L; the data from MA range from 0.90 to 0.95 μg/L or a range of 0.05 μg/L; and the data from NM range from 1.1 to 1.3 μg/L or a range of 0.2 μg/L. Because the range of data is lowest for the MA sample, this lab produced the most precise results.

Think About It

Looking closely at the dilution in part c, we see that the stream water must be diluted by

$$\frac{2.82 \times 10^{10} \text{ gal}}{161 \times 10^6 \text{ gal}} = 175$$

or nearly 200 times!

4.133. Collect and Organize

From the balanced equations that we write for the two fermentation steps for the conversion of sugar to acetic acid, we are to calculate how much acetic acid could be produced (with 100% theoretical yield) from 100 g of sugar.

Analyze

The first step in the fermentation of apple juice is an anaerobic process, so only the sugars are converted to ethanol and carbon dioxide. The acid fermentation of ethanol to give acetic acid and water, however, requires oxygen as a reactant. Oxidation states for the carbon atoms in both the reactants and products of these two fermentation reactions can be deduced using the typical oxidation states of H (+1) and O (−2). Because all the carbon species are neutral, the sum of the oxidation numbers must be zero. Therefore,

Oxidation number on C = 0 − [number H atoms × (+1) + number of oxygen atoms × (−2)]

To calculate the maximum acetic acid produced from the fermentation of 100 g sugar, we first need to calculate the moles of sugar in 100 g using the molar mass of CH_2O (30.03 g/mol). Then we use the molar ratios in the balanced equations to find the moles of acetic acid that can be produced (3:1 ratio for

$CH_2O:C_2H_5OH$ and 1:1 ratio for $C_2H_3OH:HC_2H_3O_2$). Using the molar mass of acetic acid (60.05 g/mol) we can calculate the maximum mass of acetic acid that could be produced.

Solve
(a) The fermentation of the sugars is described by

$$3\ CH_2O \rightarrow CO_2 + C_2H_5OH$$

(b) The fermentation of ethanol is described by

$$C_2H_5OH + O_2 \rightarrow HC_2H_3O_2 + H_2O$$

(c) Oxidation states for carbon in reactants and products

CH_2O: $C = 0 - [2(+1) + 1(-2)] = 0$

CO_2: $C = 0 - [2(-2)] = +4$

C_2H_5OH: $C = 0 - [6(+1) + 1(-2)] = -4$ over two carbon atoms, so oxidation number on each carbon $= -2$

$HC_2H_3O_2$: $C = 0 - [1(+1) + 3(+1) + 2(-2)] = 0$

(d) The maximum amount of acetic acid that could be produced assumes that both fermentation reactions give 100% yield.

$$100\ g\ CH_2O \times \frac{1\ mol\ CH_2O}{30.03\ g} \times \frac{1\ mol\ C_2H_5OH}{3\ mol\ CH_2O} \times \frac{1\ mol\ HC_2H_3O_2}{1\ mol\ C_2H_5OH} \times \frac{60.05\ g\ HC_2H_3O_2}{1\ mol} = 66.7\ g\ acetic\ acid$$

Think About It
In the first step of the fermentation process the carbon of the sugar is being oxidized and reduced to CO_2 and C_2H_5OH, respectively. In the second step, ethanol is being oxidized and oxygen is being reduced.

4.135. **Collect and Organize**
For the titration of sulfuric acid with barium hydroxide, we are asked to write the overall equation and then consider how the conductivity of the solution would change over the course of the titration.

Analyze
The reaction in the titration is an acid–base neutralization reaction. Therefore, the products are water and a salt ($BaSO_4$, which is insoluble). To think about how the conductivity changes, we need the ionic equation. Conductivity will be high for the solution when there are a lot of ions present in solution. When no ions are present, the conductivity is that of pure water (zero on our graph).

Solve
(a) Overall equation for the titration is

$$H_2SO_4(aq) + Ba(OH)_2(aq) \rightarrow BaSO_4(s) + 2\ H_2O(\ell)$$

(b) The ionic equation is

$$2\ H^+(aq) + SO_4^{2-}(aq) + Ba^{2+}(aq) + 2\ OH^-(aq) \rightarrow BaSO_4(s) + 2\ H_2O(\ell)$$

Before the titration begins, we have only a solution of H_2SO_4, which is ionized in solution to $2\ H^+(aq)$ and $SO_4^{2-}(aq)$. The conductivity of this solution is high (and above that of pure water). As the titration proceeds, the conductivity will decrease because of the formation of low solubility $BaSO_4$ and nonionized H_2O. At the equivalence point all the H_2SO_4 has reacted, only $BaSO_4$ and H_2O are present, and therefore the conductivity will be 0 (that of pure water). As we titrate with $Ba(OH)_2$ beyond the equivalence point, the conductivity will increase because of the presence of excess Ba^{2+} and OH^- ions. The graph that describes these conductivity changes is c.

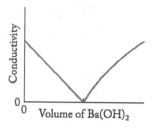

Think About It

Before the equivalence point, conductivity is due to unreacted H_2SO_4, but after the equivalence point the conductivity is due to excess $Ba(OH)_2$.

4.137. Collect and Organize

We are asked whether the addition of NaOH to a solution of acetic acid will cause a lightbulb connected through the solution to glow more brightly, the same, or less brightly. We are also asked to write a net ionic equation for the reaction that occurs.

Analyze

The brightness of the lightbulb depends on the conductivity of the solution. If the number of ions in the solution increases, the conductivity will increase and the lightbulb will get brighter. If the number of ions in the solution decreases, the conductivity will decrease and the lightbulb will dim. If the number of ions in solution remains the same, the lightbulb will remain the same brightness. We must determine the extent of ionization in solution before and after the addition of NaOH. Net ionic equations are written from ionic equations by canceling any spectator ions that are present as both reactants and products.

Solve

Acetic acid is a weak acid, and so it is a weak electrolyte. Upon addition of one equivalent of NaOH, the following reaction occurs:

Balanced: $CH_3COOH(aq) + NaOH(aq) \rightarrow CH_3COONa(aq) + H_2O(\ell)$

Net ionic: $CH_3COOH(aq) + OH^-(aq) \rightarrow CH_3COO^-(aq) + H_2O(\ell)$

After the addition, all of the acetic acid has been converted to the acetate anion, CH_3COO^-. The acetate is fully ionized, so the conductivity will increase, and the bulb will become brighter.

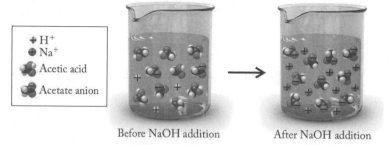

Before NaOH addition   After NaOH addition

Think About It

Despite being a weak electrolyte, acetic acid is written as an undissociated compound in a net ionic equation.

4.139. Collect and Organize

We are asked to identify the oxidation and reduction half-reactions, then balance the given equation.

Analyze

The first step in identifying an oxidation and reduction reaction is to assign oxidation numbers to all species in the unbalanced equation. We are then to balance the equation in acidic solution using the half-reaction method, which involves balancing the atoms in each half-reaction, balancing the charge, and adding the half-reactions to obtain the overall redox equation.

Solve

(a) The oxidation numbers for the unbalanced reaction are

	Left Side	Right Side
H	+1	+1
O	−1/2	−1 (H_2O_2) and 0 (O_2)

The half-reactions are

$$\text{Reduction} \qquad O_2^-(aq) \rightarrow H_2O_2(aq)$$

$$\text{Oxidation} \qquad O_2^-(aq) \rightarrow O_2(aq)$$

(b) Balancing for O and H:

$$2\,H^+(aq) + O_2^-(aq) \rightarrow H_2O_2(aq)$$

$$O_2^-(aq) \rightarrow O_2(aq)$$

Balancing for charge:

$$1\,e^- + 2\,H^+(aq) + O_2^-(aq) \rightarrow H_2O_2(aq)$$

$$O_2^-(aq) \rightarrow O_2(aq) + 1\,e^-$$

Adding the half-reactions and simplifying:

$$1\,e^- + 2\,H^+(aq) + O_2^-(aq) \rightarrow H_2O_2(aq)$$

$$\underline{O_2^-(aq) \rightarrow O_2(aq) + 1\,e^-}$$

$$1\,e^- + 2\,H^+(aq) + O_2^-(aq) + O_2^-(aq) \rightarrow H_2O_2(aq) + O_2(aq) + 1\,e^-$$

$$2\,H^+(aq) + 2\,O_2^-(aq) \rightarrow H_2O_2(aq) + O_2(aq)$$

Think About It

Notice that the superoxide ion is both oxidized and reduced in this reaction.

4.141. Collect and Organize

For the two reactions that describe the formation of $CaSO_4$ (gypsum), we are asked to identify whether either is a redox reaction and to write a net ionic equation for the reaction of sulfuric acid with calcium carbonate for an alternate form of the equation.

Analyze

A redox reaction is indicated by a change in oxidation state in one of the elements in going from reactant to product in an equation. Net ionic equations are written from ionic equations by canceling any spectator ions that are present as both reactants and products.

Solve

(a) In the first reaction, the oxidation number of hydrogen stays the same (+1), but the oxidation number of S in H_2S (−2) increases to +6 in H_2SO_4. Coupled with this oxidation is the reduction of oxygen in O_2 (oxidation number = 0) to oxidation number −2 in H_2SO_4. Eight electrons are transferred in this reaction to oxidize S^{2-} to S^{6+}. In the second reaction, no oxidation numbers change (Ca = +2, C = +4, O = −2, H = +1, S = +6). This reaction is not a redox reaction and would be classified as an acid–base reaction instead.

(b) Ionic equation:

$$2\,H^+(aq) + SO_4^{2-}(aq) + CaCO_3(s) \rightarrow CaSO_4(s) + H_2O(\ell) + CO_2(g)$$

This is also the net ionic equation.

(c) The ionic equation would change slightly on the product side:

$$2\,H^+(aq) + SO_4^{2-}(aq) + CaCO_3(s) \rightarrow CaSO_4(s) + 2\,H^+(aq) + CO_3^{2-}(aq)$$

The net ionic equation would be

$$SO_4^{2-}(aq) + CaCO_3(s) \rightarrow CaSO_4(s) + CO_3^{2-}(aq)$$

Think About It

Notice that the decomposition of H_2CO_3 to CO_2 and H_2O is not a redox reaction, since no atoms change oxidation number.

4.143. Collect and Organize

From four reactions involving the element calcium or its compounds, we are to find the redox reactions.

Analyze

In redox reactions, there must be a change in oxidation number in an element. By assigning oxidation numbers to the elements in each compound and then comparing products versus reactants, we can find the redox reactions.

Solve

(a) Reactant oxidation numbers: Ca = +2, C = +4, O = –2
Product oxidation numbers: Ca = +2, C = +4, O = –2
This is not a redox reaction.
(b) Reactant oxidation numbers: Ca = +2, O = –2, S = +4
Product oxidation numbers: Ca = +2, O = –2, S = +4
This is not a redox reaction.
(c) Reactant oxidation numbers: Ca = +2, Cl = –1
Product oxidation numbers: Ca = 0, Cl = 0
This is a redox reaction.
(d) Reactant oxidation numbers: Ca = 0, N = 0
Product oxidation numbers: Ca = +2, N = –3
This is a redox reaction.

Think About It

In redox reactions, both reduction and oxidation must be present. For example, in (c), calcium is reduced and chlorine is oxidized.

CHAPTER 5 | Thermochemistry: Energy Changes in Reactions

5.1. Collect and Organize

We are asked what the brick's kinetic energy is at two different distances above street level. The energy of the brick at the top of the building (Figure P5.1) is all potential energy. As the brick falls, the potential energy is converted into kinetic energy.

Analyze

At any point in the fall, the sum of the potential energy and the kinetic energy is equal to the brick's starting potential energy. If the brick falls halfway to the street, half of the potential energy will have been converted to kinetic energy. Therefore, we can simply determine the ratio of how far the brick has fallen from its original height and multiply by the initial potential energy.

Solve

The brick's initial position is 50 feet and the point to which it falls is 35 feet. The potential energy at 35 feet is:

$$\frac{35 \text{ ft}}{50 \text{ ft}} \times 500 \text{ J} = 350 \text{ J}$$

The kinetic energy then is 500 J – 350 J = 150 J.

When the brick hits the street surface, all the potential energy has been converted into kinetic energy. Therefore, PE = 0 kJ and KE = 500 J.

Think About It

This problem may have been solved by a more "scenic route" by first using PE = mgh to determine the mass of the brick (using h = 50 ft). Knowing that $PE_{50 \text{ ft}} = PE_{35 \text{ ft}} - KE$, we could compute the potential energy at 35 ft, then solve for kinetic energy (KE). The result is the same as given.

5.3. Collect and Organize

A gas enclosed in a container with a movable piston (Figure P5.3) has different volumes at different temperatures. As gases are heated, the volume increases. Likewise, as gases are cooled, the volume decreases.

Analyze

The cylinder is heated, so the gas expands. The piston, therefore, moves up to increase the volume inside the cylinder for the heated gas.

Solve

(a)

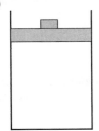

(b) The piston is higher in the cylinder.
(c) Energy has been added to the gas (the system) inside the piston.
(d) Because the gas expanded, the system (the gas) pushed up the piston and $V_f > V_i$.

$$w = P\Delta V = P(V_{fi} - V)$$

Because the gas expansion did work against the pressure of the atmosphere, the system did work on the surroundings.

Think About It

In this case, $P\Delta V$ is positive, so the work is positive. When a gas is compressed, however, $V_f < V_i$ and the sign of work and $P\Delta V$ is negative.

5.5. **Collect and Organize**

The box on the table (Figure P5.5) is the system under consideration. Everything else, including the table, is the surroundings.

Analyze

The box is made of metal, so it is not insulated and is affected by temperature changes. The box is closed, so no material can get into or out of the box.

Solve

(a) This is a closed system because energy may be transferred but no new material can be added.
(b) The internal energy of the system will increase because the system would absorb energy from the burning table.
(c) Since the box is rigid and closed, no work can be done because the volume cannot change in order for $P\Delta V$ work to be done by the system.

Think About It

If the box were a cylinder with a movable piston, the piston would have moved up, allowing the volume of the cylinder to increase, and the system would have done work on the surroundings during the fire. If the fire was then put out, and the system cooled, the piston would go down in the cylinder, and the surroundings would be doing work on the system.

5.7. **Collect and Organize**

The enthalpy diagram in Figure P5.7 shows the enthalpy of formation (ΔH_f°, kJ/mol) for the elements C, H_2, and O_2 along with those of C_2H_2, CO, H_2O, and CO_2.

Analyze

All the elements have $\Delta H_f^\circ = 0$. The ΔH_f° is positive for C_2H_2 and negative for CO, H_2O, and CO_2. For reactions, the enthalpy is calculated by subtracting the enthalpy of the reactants from the enthalpy of the products.

Solve

(a) The elements are all put on the same horizontal line ($\Delta H_f^\circ = 0$) because the enthalpy of formation of an element in its standard state is defined as zero.
(b) C_2H_2 is an endothermic compound because the enthalpy of its formation is positive. This means that energy would have to be added to the elements in order for C_2H_2 to form.
(c) The energy of reaction from the balanced equation

$$2\,C_2H_2(g) + 3\,O_2(g) \rightarrow 4\,CO(g) + 2\,H_2O(g)$$

is calculated by subtracting the sum of the enthalpies of formation of the reactants (multiplied by the number of moles in the balanced equation for each product) from the sum of enthalpies of formation of the products (again multiplied by their molar amounts from the balanced equation).
For this particular equation:

$$\Delta H_{rxn}^\circ = \left[4\text{ mol }(\Delta H_f^\circ)CO(g) + 2\text{ mol}(\Delta H_f^\circ)H_2O(g)\right] - \left[2\text{ mol}(\Delta H_f^\circ)C_2H_2(g) + 3\text{ mol}(\Delta H_f^\circ)O_2(g)\right]$$

Think About It

The reverse reaction of the formation of C_2H_2 is exothermic and gives off energy to the surroundings.

5.9. Collect and Organize
Energy and work must be related, since from our everyday experience we know that it takes energy to do work.

Analyze
In this context, energy is defined as the capacity to do work. Work is defined as moving an object with a force over some distance. Energy is also thought to be a fundamental component of the universe. The Big Bang theory postulates that all matter originated from a burst of energy, and Albert Einstein proposed that $m = E/c^2$ (mass equals energy divided by the speed of light squared).

Solve
Energy is needed to do work, and doing work uses energy.

Think About It
A system with high energy has the potential to do a lot of work.

5.11. Collect and Organize / Analyze
The question asks us to explain what is meant by a state function.

Solve
The value of a state function is independent of the path taken in reaching a particular state; only the initial and final values are important.

Think About It
Examples of state functions are internal energy and enthalpy, but not work and heat.

5.13. Collect and Organize
Potential energy can take several forms.

Analyze
The basic definition of potential energy is the energy of position from PE = *mgh*. But potential energy is also present at the molecular level in the form of stored energy in the chemical bonds of a substance or the ability to transfer electrons.

Solve
(a) The potential energy in a battery consists of the energy stored in the molecules, which is released when they react via a redox reaction.
(b) The potential energy in a gallon of gasoline consists of the energy stored in the chemical bonds in the fuel molecules, which is released when the fuel is combusted.
(c) The potential energy of the crest of a wave is due to its position above the ground. This energy is released as kinetic energy when the wave crashes on the shore.

Think About It
Energy that we use for transportation and for electrical devices in the form of batteries derives from stored chemical potential energy.

5.15. Collect and Organize / Analyze
We are to define the terms *system* and *surroundings*.

Solve
The system is that part of the universe that we are interested in. The surroundings are everything else, extending to the entire universe.

Think About It

Systems may be isolated (no matter or energy can be exchanged with the surroundings), closed (only energy can be exchanged with the surroundings), or open (both matter and energy can be exchanged with the surroundings).

5.17. Collect and Organize

ΔE is the sum of the heat (q) and work (w) for the process.

Analyze

In an endothermic process, the system gains energy from the surroundings (the reaction vessel feels cold), so $q > 0$. If work is done on the surroundings for a given reaction, $w < 0$. By considering cases where heat and work are very large and very small relative to one another, we can determine if the sign of ΔE is variable.

Solve

Since the heat and work have opposite signs, the sign of ΔE is depends on the magnitude of q and w.

Think About It

The same chemical reaction may have a different value for ΔE if it does a different quantity of work.

5.19. Collect and Organize

For the processes listed, we are asked to identify which are exothermic and which are endothermic.

Analyze

An exothermic process transfers energy from the system to the surroundings. Thus, something feels warm or hot in an exothermic process. An endothermic process transfers energy from the surroundings to the system. Thus, something feels cool or cold in an endothermic process.

Solve

(a) When a candle burns, energy is generated; this process is exothermic.

(b) When alcohol evaporates, energy is absorbed by the system (alcohol) from the surroundings (the skin of the hand); this process is endothermic.

(c) When a supersaturated solution crystallizes, the temperature of the solution rises, telling us that the crystallization is releasing heat in the reaction. This process must be exothermic.

(a)

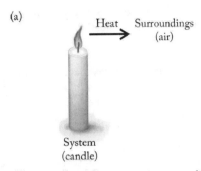

Heat transferred from system to surroundings

(b)

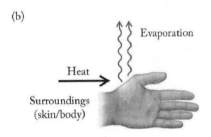

Heat transferred from surroundings to system

(c)

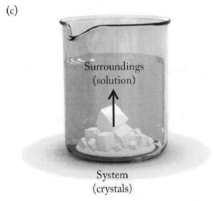

Heat transferred from system to surroundings

Think About It

When a supersaturated solution crystallizes, releasing heat, the energy comes from the attractive interaction between ions forming a lattice.

5.21. **Collect and Organize**

Internal energy is defined as

$$\Delta E = q + w$$

where q = energy transferred and w = work done on a system ($-P\Delta V$). Commonly, the energy transfer is caused by a temperature difference, so q is called "heat".

Analyze
When a process releases energy, the internal energy decreases and q is negative. The reverse is true (ΔE increases and q is positive) if the surroundings transfer energy to the system. If the volume of a system increases (ΔV is positive) then work is negative and the internal energy of the system decreases. The reverse is true (ΔE increases and w is positive) when the volume of the system decreases (e.g., a gas is compressed).

Solve
When a liquid vaporizes at its boiling point, energy is absorbed from the surroundings. Thus, q is positive. The volume increases ($\Delta V > 0$) so work is done by the system on the surroundings and the sign of w is negative. More energy is transferred to the liquid than work done, so the sign of $q + w$ is positive; therefore, $\Delta E > 0$.

Think About It
Remember to focus on work and energy transfer from the system's point of view. This will help you be clear about the sign conventions as you learn more about thermochemistry.

5.23. Collect and Organize
The work being done is due to an expansion of the gas from 250.0 mL to 750.0 mL.

Analyze
Work is expressed as $-P\Delta V$. In this case, P is constant at 1.00 atm and the volume change is 500.0 mL or 0.5000 L. We are to express work in both L · atm and in joules. We can convert from one unit to another using 101.32 J/L · atm.

Solve
$$w = -P\Delta V = (1.00 \text{ atm})(0.5000 \text{ L}) = -0.500 \text{ L} \cdot \text{atm}$$

In joules, this work is

$$-0.500 \text{ L} \cdot \text{atm} \times \frac{101.32 \text{ J}}{\text{L} \cdot \text{atm}} = -50.7 \text{ J}$$

Think About It
Because work was done by the system on the surroundings, the sign of work is negative.

5.25. Collect and Organize
For each part, we are to calculate internal energy (ΔE) from energy and work values.

Analyze
The formula for internal energy from energy transferred as heat and work is

$$\Delta E = q + w$$

We need only add the values of q and w given.

Solve
(a) $E = 120.0 \text{ J} + (-40.0 \text{ J}) = 80.0 \text{ J}$
(b) $E = 9.2 \times 10^3 \text{ J} + 0.70 \text{ J} = 9200.70 \text{ J}$, or 9.20070 kJ, which rounds to 9.2 kJ
(c) $\Delta E = -625 \text{ J} + (-315 \text{ J}) = -940 \text{ J}$

Think About It
When adding q and w, be sure to add values with consistent units.

5.27. Collect and Organize
Work will be done by a system on the surroundings when the volume of the system shown in Figure P5.29 increases.

Analyze

The volume of gas is proportional to the number of moles of gas (n) at constant temperature and pressure. Because both temperature and pressure are constant for each reaction, if n increases in going from reactants to products, the volume of the system increases, and work is done by the system on the surroundings. Throughout this process the temperature is constant at 110°C.

Solve

(a) In this reaction, 3 moles of gaseous reactants form 3 moles of gaseous products. The Δn for the reaction is 0, so this reaction does not do work on the surroundings.

(b) In this reaction, 6 moles of gaseous reactants form 7 moles of gaseous products. The Δn for the reaction is +1, so this reaction does work on the surroundings.

(c) In this reaction, 3 moles of gaseous reactants form 2 moles of gaseous products. The Δn for the reaction is –1, so this reaction does not do work on the surroundings.

Think About It

Reaction c has $+w$ (from $-P\Delta V$ where $\Delta V = V_f - V_i$ and $V_i > V_f$, so ΔV is negative). This means that the surroundings did work on the system. The system was compressed.

5.29. Collect and Organize

We are asked to determine the amount of pressure–volume work done by a chemical reaction on the surroundings, given the mass quantity of reactant. Using this value for the pressure–volume work, we are asked to determine ΔE for the same system. The pressure volume work for a system is $-P\Delta V$, expressed in Joules (101.325 J = 1 L · atm). Internal energy is defined as

$$\Delta E = q + w = q - P\Delta V$$

Analyze

Using the mass of sodium azide provided, we can determine the number of moles of nitrogen gas generated using the molar mass of sodium azide (65.02 g/mol) and the stoichiometric coefficients from the balanced chemical equation. We may obtain the change in the volume of N_2 by multiplying the molar quantity of nitrogen gas by the molar mass of N_2 (28.02 g/mol) and dividing by the density (1.165 g/L). We should assume that the initial volume is zero.

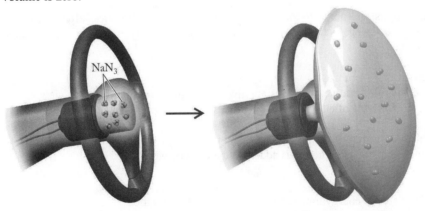

Solve

(a) Determining the change in volume of nitrogen gas

$$2.25 \text{ g NaN}_3 \times \frac{1 \text{ mol NaN}_3}{65.02 \text{ g NaN}_3} \times \frac{3 \text{ mol N}_2}{2 \text{ mol NaN}_3} \times \frac{28.02 \text{ g N}_2}{1 \text{ mol N}_2} \times \frac{1 \text{ L}}{1.165 \text{ g N}_2} = 1.248 \text{ L}$$

$$\Delta V = 1.248 \text{ L} - 0 \text{ L} = 1.248 \text{ L}$$

The pressure–volume work for the system is

$$-P\Delta V = 1.00 \text{ atm}\left(1.248 \text{ L}\right) = -1.25 \text{ L}\cdot\text{atm} \times \frac{101.325 \text{ J}}{1 \text{ L}\cdot\text{atm}} = -127 \text{ J}$$

(b) To determine ΔE

$$\Delta E = q - P\Delta V = \left(-2340 \text{ J}\right) + \left(-127 \text{ J}\right) = -2467 \text{ J} = -2.47 \text{ kJ}$$

Think About It
Only gaseous reactants and products affect the volume in the reaction.

5.31. **Collect and Organize**
We are asked to define a change in enthalpy.

Analyze
We can refer to the mathematical expression for the change in enthalpy (ΔH) to answer this question.

$$\Delta H = \Delta E + P\Delta V$$

Solve
Interpreting the equation above, a change in enthalpy is the sum of the change of internal energy and the product of the system's pressure and change in volume.

Think About It
Because $\Delta E = q + w$, internal energy changes may involve changes in energy or work or both.

5.33. **Collect and Organize**
We are asked why we assign a negative sign to the ΔH for an exothermic process.

Analyze
In an exothermic process, the system transfers energy to the surroundings.

Solve
If the system transfers energy to the surroundings, its energy will be less after the process than at the start of the process, so q is negative. If the pressure is constant and only PV work is done, then this q is ΔH, so ΔH must also be negative.

Think About It
The signs of thermodynamic quantities are always assigned from the point of view of the system. This can be confusing, since we are often observing the process from the point of view of the surroundings.

5.35. **Collect and Organize**
A drain pipe gets hot when Drano® is added. We are asked what is the sign for ΔH for this process.

Analyze
Enthalpy is related to energy transferred as heat (q) by the equation

$$\Delta H = q_P$$

Solve
Because energy is released in the reaction between the Drano® and the clog in the pipe, q is released from the system to the surroundings so q is negative and therefore ΔH is also negative.

Think About It
All reactions that are exothermic have a negative ΔH.

5.37. **Collect and Organize**

If O_2 is the stable form of oxygen under standard conditions, it must have lower energy than other forms. The reaction describes breaking the oxygen–oxygen bond in O_2. We are asked to identify the sign of ΔH in this process.

Analyze

If a species is at low energy, then it must require an input of energy to take it out of its most stable form. Breaking the oxygen–oxygen bond, therefore, must require an input of energy.

Solve

The process of breaking the oxygen–oxygen bond is endothermic, and the sign of ΔH will be positive.

Think About It

Breaking chemical bonds is always an endothermic process.

5.39. **Collect and Organize**

Compression of H_2 gas will give solid H_2. To predict the sign of the enthalpy change of this transformation, we have to consider the enthalpy of the phase change from gas to solid.

Analyze

In order to solidify substances, we must cool them down to reduce their molecular motion.

Solve

To solidify hydrogen gas, we would remove energy from the gas, so the sign of q from the point of view of the system would be negative. Because $\Delta H = q_P$, the sign of ΔH is also negative.

Think About It

The reverse reaction in which $H_2(s)$ sublimes into $H_2(g)$ is endothermic.

5.41. **Collect and Organize**

We look at the definitions and units of heat capacity and specific heat to differentiate these two terms.

Analyze

Heat capacity is the amount of energy needed to raise the temperature of an object 1°C. Specific heat is the amount of energy needed to raise the temperature of 1 g of a substance 1°C.

Solve

The difference in the terms lies in the specificity. Heat capacity does not take into account how much of a substance there is; it is defined for a given object. Specific heat is specified for a gram of the substance.

Think About It

Specific heat for a substance is characteristic of that substance.

5.43. **Collect and Organize**

To compare the heat of fusion of a substance to the heat of vaporization of that same substance we need to consider the processes that are occurring for each of the phase changes.

Analyze

The enthalpy of fusion is the heat required to melt a solid substance. The heat of vaporization is the energy required to vaporize the substance from a liquid to gaseous state.

Solve

Melting and vaporizing are very different processes so we would not expect the heat of both to be the same.

Think About It
Usually the heat of vaporization is more endothermic than the heat of fusion because in the phase change from liquid to gas, individual molecules must be completely separated from each other.

5.45. Collect and Organize
To compare the advantage of water-cooled engines over air-cooled engines, we have to compare the relative heat capacities of water and air.

Analyze
The specific heat capacity of water (4.18 J/g · °C) is higher than that of air (1.01 J/g · °C) at typical room conditions.

Solve
Water's high heat capacity relative to air means that water carries away more energy from the engine for every degree Celsius rise in temperature, making water a better choice for cooling automobile engines.

Think About It
On a volume basis, water has an even higher heat capacity than air.

5.47. Collect and Organize
The amount of energy needed to raise the temperature of a substance depends on the substance's heat capacity and the change in temperature of the substance.

Analyze
The energy required to raise the temperature of a substance is related to the molar heat capacity by the equation
$$q = nc_P \Delta T$$
where q = energy, n = moles of the substance, c_P = molar heat capacity of the substance, and ΔT = difference in temperature, $T_f - T_i$.

Solve

$$q \text{ required} = \left(100.0 \text{ g} \times \frac{1 \text{ mol}}{18.02 \text{ g}} \right) \times \left(\frac{75.3 \text{ J}}{\text{mol} \cdot {}^\circ\text{C}} \right) \times \left(100.0\,{}^\circ\text{C} - 30.0\,{}^\circ\text{C} \right) = 29,300 \text{ J or } 29.3 \text{ kJ}$$

Think About It
If the water were cooled from 100°C to 30°C the sign of ΔT would be negative and, therefore, q would be negative, showing that the system was cooled.

5.49. Collect and Organize
A heating curve plots temperature as a function of energy added to a substance.

Analyze
Methanol at –100°C is a solid. As energy is added, the solid methanol increases in temperature until it reaches its melting point, at which point the added energy does not change the temperature of the methanol until all the solid has melted. Only when methanol is entirely liquid will added energy increase the temperature of the liquid until the boiling point of methanol is reached. During boiling, the temperature of the methanol does not change. Once the methanol is converted to gaseous form, added energy increases the temperature of the gaseous methanol.
Relevant equations for finding the energy required for each step are as follows:
  $q = nc_P \Delta T$ for energy added to the solid, liquid, and gas phases where n = 1 mol, c_P = heat capacity for that phase, and ΔT = temperature change for that phase
  $q = n\Delta H_{fus}$ energy for melting
  $q = n\Delta H_{vap}$ energy for vaporization

Solve

Step in Heating Curve	T_i, °C	T_f, °C	q, kJ	Total q, kJ
Heating methanol ice	−100	−94	q = 1 mol × 48.7 J/(mol · °C) × 6°C = 0.292 kJ	0.29
Melting methanol ice	−94	−94	q_{fus} = 1 mol × 3.18 kJ/mol = 3.18 kJ	3.47
Warming liquid methanol	−94	65	q = 1 mol × 81.1 J/(mol · °C) × 159°C = 12.9 kJ	16.37
Boiling methanol	65	65	q_{vap} = 1 mol × 37 kJ/mol = 37 kJ	53
Heating methanol gas	65	100	q = 1 mol × 43.9 J/(mol · °C) × 35°C = 1.54 kJ	55

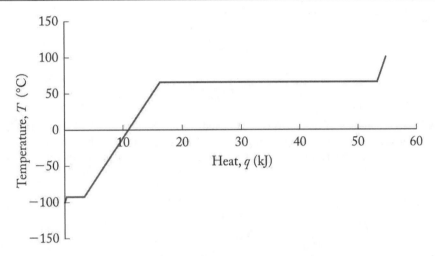

Think About It

Methanol has a wide temperature range as a liquid and that makes it useful as a solvent.

5.51. **Collect and Organize**

Sweating helps to cool athletes. During a workout, an athlete generates 2000.0 kJ of energy. The question asks us to calculate how much water the athlete would lose if all the energy is used to evaporate water.

Analyze

The energy generated by the athlete must vaporize the water so the pertinent equation is

$$q = n\Delta H_{vap}$$

where ΔH_{vap} for water is 40.67 kJ/mol.

Solve

The amount of water in moles that 2000.0 kJ of energy would vaporize is

$$2000.0 \text{ kJ} = n \times 40.67 \text{ kJ/mol}$$

$$n = 49.18 \text{ mol H}_2\text{O}$$

Converting this to mass

$$49.18 \text{ mol} \times \frac{18.02 \text{ g H}_2\text{O}}{1 \text{ mol}} = 886.2 \text{ g H}_2\text{O}$$

Think About It
Since the density of water is 1.00 g/mL, the athlete would use 886.2 mL of water to dissipate the heat. To rehydrate, the athlete should drink about a liter of water.

5.53. Collect and Organize
The three metals have very similar molar heat capacities. Because the final temperature is determined by the number of moles of each metal present, the moles (n), not the heat capacity, determine the final temperature of the metals when the same amount of energy is added to each.

Analyze
The equation for energy and molar heat capacity is

$$q = nc_P\Delta T$$

Rearranging this to find ΔT and T_f, we get

$$\Delta T = \frac{q}{nc_p} = T_f - T_i$$

$$T_f = \frac{q}{nc_p} + T_i$$

where $T_i = 25°C$, n = moles of metal (to be calculated from the molar mass of the metal), and c_P is the molar heat capacity for the metal. Since we are not given a value for q, only that the same amount of energy is added to each metal, we can assume any number. Let's assume $q = 100$ J for this calculation.

Solve
For gold

$$T_f = \frac{100 \text{ J}}{\left(10.00 \text{ g Au} \times \dfrac{1 \text{ mol Au}}{196.967 \text{ g}} \times \dfrac{25.41 \text{ J}}{\text{mol} \cdot °C}\right)} + 25°C = 103°C$$

For magnesium

$$T_f = \frac{100 \text{ J}}{\left(10.00 \text{ g Mg} \times \dfrac{1 \text{ mol Mg}}{24.305 \text{ g}} \times \dfrac{24.79 \text{ J}}{\text{mol} \cdot °C}\right)} + 25°C = 35°C$$

For platinum

$$T_f = \frac{100 \text{ J}}{\left(10.00 \text{ g Pt} \times \dfrac{1 \text{ mol Pt}}{195.084 \text{ g}} \times \dfrac{25.95 \text{ J}}{\text{mol} \cdot °C}\right)} + 25°C = 100°C$$

Therefore, gold has the highest temperature.

Think About It
Because the molar heat capacities of the metals are all very close in value, the metal with the least number of moles in 10 g is the metal with the highest final temperature. The metal with the highest molar mass, in this case gold, has the lowest number of moles in 10 g and, therefore, will reach the highest temperature of all the metals.

5.55. Collect and Organize
When the water is converted into steam, energy must be lost by the skillet in order to heat and boil the water completely away.

Analyze
The equation describing the energy exchange between the skillet and the water is

$$q_{\text{water gained}} = -q_{\text{skillet lost}}$$

where

$$q_{\text{water gained}} = nc_{\text{p}} \Delta T + n\Delta H_{\text{vap}}$$

$$q_{\text{skillet lost}} = nc_{\text{p}} \Delta T$$

Solve

$$q_{\text{water gained}} = -q_{\text{skillet lost}}$$

$$\left[10.0 \text{ g} \times \frac{1 \text{ mol}}{18.02 \text{ g}} \times \frac{75.3 \text{ J}}{\text{mol} \cdot {}^\circ\text{C}} \times (100.0{}^\circ\text{C} - 25.0{}^\circ\text{C}) \right] + \left[10.0 \text{ g} \times \frac{1 \text{ mol}}{18.02 \text{ g}} \times \frac{40{,}670 \text{ J}}{\text{mol}} \right] =$$

$$- \left(1200 \text{ g} \times \frac{1 \text{ mol Fe}}{55.845 \text{ g Fe}} \right) \times \frac{25.19 \text{ J}}{\text{mol} \cdot {}^\circ\text{C}} \times \Delta T$$

$$\Delta T = -47.5{}^\circ\text{C}$$

Think About It
The temperature of a stovetop burner may range from 100°C to over 200°C. Because the iron skillet has a high molar heat capacity and contains much iron (in terms of moles) the temperature does not change much, relative to its initial temperature.

5.57. Collect and Organize
To explain why the calorimeter constant (i.e., the heat capacity of the calorimeter) is important, we need to define the system and surroundings for the calorimetry experiment.

Analyze
The system in a calorimetry experiment is defined as the substance for which, for example, the heat capacity, is being measured. The calorimeter is everything but the system (i.e., the calorimeter is the surroundings).

Solve
Because energy is transferred between the system and the surroundings, the heat capacity of the calorimeter (the surroundings) is important because we need to know how much energy (generated or absorbed by the system) is required to change the temperature of the surroundings (the calorimeter) in order to calculate the heat capacity or final temperature of the system in an experiment.

Think About It
Calorimeter constants vary from calorimeter to calorimeter.

5.59. Collect and Organize / Analyze
The calorimeter constant is the heat capacity for the surroundings. By replacing water in a calorimeter with another liquid, we are changing the surroundings.

Solve
The heat capacity of the new liquid is different from that of water. The liquid is part of the calorimeter and therefore part of the surroundings. Yes, the calorimeter constant must be redetermined.

Think About It
If the system is expected to transfer a lot of energy to the calorimeter, then it might be necessary to use a liquid with a higher heat capacity than water.

5.61. Collect and Organize
We are asked to consider which of the three salts would provide the greatest drop in temperature when dissolved in water. We are given the molar enthalpy of solution for each salt.

Analyze

The drop in temperature is related to the quantity of energy absorbed by the salt as it dissolves to form the solution. The salt with the largest quantity of energy absorbed per gram will have the greatest drop in temperature. The energy absorbed by each salt may be obtained by dividing the molar enthalpy of solution by the molar mass of each salt. Since it has the largest molar enthalpy of solution, NH_4NO_3 is a reasonable guess for the greatest temperature drop per gram.

Solve

For NH_4Cl

$$\frac{14.6\,\text{kJ}}{1\,\text{mol}} \times \frac{1\,\text{mol}}{53.50\,\text{g}} = 0.273\,\text{kJ/g}$$

For NH_4NO_3

$$\frac{25.7\,\text{kJ}}{1\,\text{mol}} \times \frac{1\,\text{mol}}{80.06\,\text{g}} = 0.321\,\text{kJ/g}$$

For $NaNO_3$

$$\frac{20.4\,\text{kJ}}{1\,\text{mol}} \times \frac{1\,\text{mol}}{85.00\,\text{g}} = 0.240\,\text{kJ/g}$$

NH_4NO_3 absorbs the most energy per gram, so it is the best choice for a cold pack.

Think About It

Even with the relatively high molar mass, NH_4NO_3 still absorbs the most energy per gram.

5.63. Collect and Organize

Benzoic acid is often used to determine calorimeter constants. As mentioned in the text, when 1.0 g of it combusts, 26.38 kJ are released to the surroundings.

Analyze

The calorimeter constant is defined as

$$C_{\text{calorimeter}} = \frac{q}{\Delta T}$$

In the combustion of benzoic acid for this calorimeter, we use 5.000 g of benzoic acid and get a temperature change of 16.397°C.

Solve

$$C_{\text{calorimeter}} = \frac{\left(\dfrac{26.38\,\text{kJ}}{\text{g benzoic acid}}\right) \times 5.000\,\text{g}}{16.397\,°\text{C}} = 8.044\,\text{kJ/}°\text{C}$$

Think About It

Be sure to account for how many grams of benzoic acid are used in measuring the calorimeter constant.

5.65. Collect and Organize

In a bomb calorimeter $q_{\text{system}} = \Delta E_{\text{comb}}$, but since the PV work is usually small, $\Delta E_{\text{comb}} \approx \Delta H_{\text{comb}}$. So we may assume that $\Delta H_{\text{comb}} = -q_{\text{calorimeter}}$ since $q_{\text{rxn}} = -q_{\text{calorimeter}}$.

Analyze

We can find ΔH_{comb} through

$$\Delta H_{\text{comb}} = -q_{\text{calorimeter}} = -C_{\text{calorimeter}}\,\Delta T$$

The ΔH_{comb} we find is for the combustion of 1.200 g of cinnamaldehyde. To find ΔH_{comb} in terms of kilojoules per mole we need to divide the calculated ΔH_{comb} by the moles of cinnamaldehyde (C_9H_8O).

Solve

$$\Delta H_{comb} = -3.640 \text{ kJ/}^\circ C \times 12.79\,^\circ C = -46.56 \text{ kJ}$$

$$\text{molar } \Delta H_{comb} = \frac{-46.56 \text{ kJ}}{\left(1.200 \text{ g} \times \dfrac{1 \text{ mol}}{132.2 \text{ g}}\right)} = -5129 \text{ kJ/mol}$$

Think About It

Expressing the enthalpies of reactions in terms of molar enthalpies allows us to compare reactions on a per mole basis.

5.67. **Collect and Organize**

In this problem, we are asked to work backward from the molar enthalpy of combustion of dimethylphthalate to the final temperature of the calorimeter.

Analyze

First, we have to calculate ΔH_{comb} from the molar heat of combustion and the grams (which we convert into moles) of dimethylphthalate ($C_{10}H_{10}O_4$). We can then find T_f from the rearrangement of the equation for ΔH_{comb}.

$$\Delta H_{comb} = -C_{calorimter}\,\Delta T$$

$$\Delta T = \frac{\Delta H_{comb}}{C_{cal}} = T_f - T_i$$

Solve

The ΔH_{comb} for 1.00 g of dimethylphthalate is

$$1.00 \text{ g} \times \frac{1 \text{ mol}}{194.19 \text{ g}} \times \frac{4685 \text{ kJ}}{1 \text{ mol}} = 24.13 \text{ kJ}$$

The final temperature of the calorimeter is

$$T_f - 20.215\,^\circ C = \frac{24.13 \text{ kJ}}{7.854 \text{ kJ/}^\circ C} = 3.072\,^\circ C$$

$$T_f = 23.29\,^\circ C$$

Think About It

Although the molar ΔH_{comb} is large for dimethylphthalate, this experiment combusts so little dimethylphthalate that the change in temperature is small.

5.69. **Collect and Organize**

Consider the reaction between magnesium metal and hydrochloric acid. We are asked to write a net ionic equation that describes this reaction and determine the enthalpy of reaction.

Analyze

Consulting the activity series, we see that magnesium is oxidized by strong acids to generate H_2 gas and Mg^{2+} ions. Using the volume and density of the solution and dividing by the molar mass of water (18.02 g/mol), we may determine the number of moles of solution present. To determine the heat produced during the reaction, we can multiply the number of moles of solution, the molar heat capacity for water, and the temperature change. Dividing the mass of magnesium by the molar mass of magnesium (24.305 g/mol) will yield the number of moles of magnesium. The molar enthalpy of reaction may be obtained by dividing the heat produced by the number of moles of magnesium present.

Solve

(a) The balanced chemical equation for the reaction that occurs is

$$Mg(s) + 2\,HCl(aq) \rightarrow H_2(g) + MgCl_2(aq)$$

Removing chloride spectator ions, we obtain the following net ionic reaction:

$$Mg(s) + 2 H^+(aq) \rightarrow H_2(g) + Mg^{2+}(aq)$$

(b) The molar quantities of reactants and solution are

$$95.0 \text{ mL solution} \times \frac{1.00 \text{ g}}{1 \text{ mL}} \times \frac{1 \text{ mol H}_2\text{O}}{18.02 \text{ g H}_2\text{O}} = 5.272 \text{ mol H}_2\text{O}$$

$$2.00 \text{ g Mg} \times \frac{1 \text{ mol Mg}}{24.305 \text{ g Mg}} = 0.0823 \text{ mol Mg}$$

The heat evolved during the reaction is

$$q_{\text{solution}} = nc_P\Delta T = 5.272 \text{ mol}\left(\frac{75.3 \text{ J}}{\text{mol} \cdot {}^\circ\text{C}} \times \frac{1 \text{ kJ}}{1000 \text{ J}}\right)(9.2\,{}^\circ\text{C}) = 3.652 \text{ kJ}$$

$$q_{\text{rxn}} = -q_{\text{solution}} = -3.652 \text{ kJ}$$

Calculating the enthalpy of reaction

$$\Delta H_{\text{rxn}} = \frac{q_{\text{rxn}}}{n} = \frac{-3.652 \text{ kJ}}{0.0823 \text{ mol Mg}} = -44.4 \text{ kJ/mol}$$

Think About It
The enthalpy of reaction is negative because heat is released to the surroundings.

5.71. Collect and Organize
We are to calculate the quantity of glucose that would provide sufficient energy to evaporate 1.00 g of water at 37°C.

Analyze
We will assume that the enthalpy of vaporization of water (40.67 kJ/mol) is relatively constant throughout the liquid phase. Knowing the mass of water, we may determine the number of moles of water present, and, multiplying by ΔH_{vap}, determine the heat required to vaporize this number of moles. This heat comes from the combustion of glucose, so dividing by $\Delta H_{\text{comb,glucose}}$ and multiplying by the molar mass of glucose (180.16 g/mol) will reveal how many grams of glucose must have been metabolized. One gram of water is a modest quantity, so we might guess that the mass of glucose will be similarly small, likely less than a gram.

Solve
The heat required to vaporize the water is

$$n = 1.00 \text{ g H}_2\text{O} \times \frac{1 \text{ mol H}_2\text{O}}{18.02 \text{ g H}_2\text{O}} = 0.05549 \text{ mol H}_2\text{O}$$

$$q_{\text{solution}} = n\Delta H_{\text{vap}} = 0.05549 \text{ mol} \times \frac{40.67 \text{ kJ}}{\text{mol}} = 2.2568 \text{ kJ}$$

The mass of glucose required to generate this quantity of heat is

$$q_{\text{glucose}} = -q_{\text{solution}} = -2.2568 \text{ kJ}$$

$$-2.2568 \text{ kJ} \times \frac{1 \text{ mol glucose}}{-2803 \text{ kJ}} \times \frac{180.16 \text{ g glucose}}{1 \text{ mol glucose}} = 0.145 \text{ g glucose}$$

Think About It
As expected, only a small sample of glucose is required to vaporize 1.00 grams of water.

5.73. **Collect and Organize**
Compare Hess's law with the law of conservation of energy.

Analyze
The law of conservation of energy states that energy cannot be created or destroyed; it can be converted from one form into another. Hess's law states that the enthalpy change for a reaction can be obtained by summing the enthalpies of constituent reactions.

Solve
When we apply Hess's law, all the energy is accounted for in the reaction; energy is neither created nor destroyed when using Hess's law.

Think About It
Hess's law makes it easy for us to calculate energy changes for chemical reactions from those for other chemical reactions.

5.75. **Collect and Organize**
We combine the first two chemical reactions to give the overall reaction of methane with ammonia to give hydrogen cyanide and dihydrogen.

Analyze
The first equation is alright as is. The second equation must be flipped so that methane is a reactant and dihydrogen is a product. Adding these reactions will now generate the third reaction.

Solve

$$CO(g) + NH_3(g) \rightarrow HCN(g) + H_2O(g)$$

$$\underline{CH_4(g) + H_2O(g) \rightarrow CO(g) + 3\,H_2(g)}$$

$$CH_4(g) + NH_3(g) \rightarrow HCN(g) + 3\,H_2(g)$$

Think About It
In deciding how to sum chemical reactions, you should focus on species that are the reactants and products in the overall reaction. Reverse reactions as necessary to place species on the correct side of the chemical equation.

5.77. **Collect and Organize**
To calculate the enthalpy change for a reaction, we use Hess's law to add chemical reactions and their associated enthalpy changes.

Analyze
If we reverse the first reaction, $ClO(g)$ will be a product as it is in the overall desired equation. This exothermic reaction will now be endothermic. Adding the second equation to this equation gives us the overall equation.

Solve

$$Cl(g) + 2\,O_2(g) \rightarrow ClO(g) + O_3(g) \qquad \Delta H^{\circ}_{rxn} = 29.90 \text{ kJ}$$

$$\underline{2\,O_3(g) \rightarrow 3\,O_2(g) \qquad\qquad \Delta H^{\circ}_{rxn} = 24.18 \text{ kJ}}$$

$$Cl(g) + O_3(g) \rightarrow ClO(g) + O_2(g) \qquad \Delta H^{\circ}_{rxn} = 54.08 \text{ kJ}$$

Think About It
The overall reaction is endothermic meaning that energy must be supplied by the surroundings to form $ClO(g)$.

5.79. **Collect and Organize**
When we sum together ΔH°_{f} values for compounds in a reaction we are using Hess's law.

Analyze

The ΔH_f° for a compound is defined as the enthalpy of the reaction to form the compound from its constituent elements in their standard states. If we write out the chemical equations for the ΔH_f° for the reactants and products in reactions, they will add up using Hess's law to the overall reaction.

Solve

Let's look at this by example. For the overall reaction

$$NH_3(g) + HCl(g) \rightarrow NH_4Cl(s)$$

the ΔH_f° chemical equations are

$$\tfrac{1}{2} N_2(g) + \tfrac{3}{2} H_2(g) \rightarrow NH_3(g) \qquad\qquad \Delta H_{f,\,NH_3}^{\circ}$$

$$\tfrac{1}{2} H_2(g) + \tfrac{1}{2} Cl_2(g) \rightarrow HCl(g) \qquad\qquad \Delta H_{f,\,HCl}^{\circ}$$

$$\tfrac{1}{2} N_2(g) + 2 H_2(g) + \tfrac{1}{2} Cl_2(g) \rightarrow NH_4Cl(s) \qquad\qquad \Delta H_{f,\,NH_4Cl}^{\circ}$$

For the sum of these equations to add up to the overall equation for the reaction of ammonia with hydrogen chloride, we need to reverse the first two reactions and use the negatives of $\Delta H_{f,\,NH_3}^{\circ}$ and $\Delta H_{f,\,HCl}^{\circ}$

$$NH_3(g) \rightarrow \tfrac{1}{2} N_2(g) + \tfrac{3}{2} H_2(g) \qquad -\Delta H_{f,\,NH_3}^{\circ}$$

$$HCl(g) \rightarrow \tfrac{1}{2} H_2(g) + \tfrac{1}{2} Cl_2(g) \qquad -\Delta H_{f,\,HCl}^{\circ}$$

$$\underline{\tfrac{1}{2} N_2(g) + 2 H_2(g) + \tfrac{1}{2} Cl_2(g) \rightarrow NH_4Cl(s) \qquad\qquad \Delta H_{f,\,NH_4Cl}^{\circ}}$$

$$NH_3(g) + HCl(g) \rightarrow NH_4Cl(s)$$

Thus, $\Delta H_{rxn}^{\circ} = \Delta H_{f,\,NH_4Cl}^{\circ} - \Delta H_{f,\,NH_3}^{\circ} - \Delta H_{f,\,HCl}^{\circ}$

which is equivalent to the expression

$$\Delta H_{rxn}^{\circ} = \sum n\Delta H_{f,\,NH_4Cl}^{\circ} - \sum m\Delta H_{f,\,NH_3 \text{ and } HCl}^{\circ}$$

Think About It

By Hess's law, a chemical reaction can be thought of as the reactants decomposing into the elements then reforming into the products.

5.81. Collect and Organize

The standard enthalpy of formation is the energy that is absorbed or produced when 1 mole of a substance is formed from the elements, all in their standard states.

Analyze

Standard conditions are 1 atm and the specified temperature for the reaction. Both ozone (O_3) and elemental oxygen (O_2) exist under these conditions.

Solve

Because ozone and elemental oxygen are different forms of oxygen, their standard enthalpies of formation are different. From Appendix 4, ΔH_f° for O_2 is 0 kJ/mol (because it is an element in its most stable form under standard conditions) and ΔH_f° for O_3 is 142.7 kJ/mol.

Think About It

Because ΔH_f° for O_3 is more positive than that of O_2, ozone is less stable than oxygen.

5.83. Collect and Organize

The enthalpy of formation is reflected in a reaction when (1) one mole of the substance is produced, (2) the substance is produced under standard state conditions, and (3) it is produced from the substance's constituent elements in their standard state.

Analyze

Each reaction must meet all the criteria for ΔH°_{rxn} to be classified as an enthalpy of formation.

Solve

(a) One mole of CO_2 is produced from elemental carbon and oxygen so ΔH°_{rxn} for this reaction represents ΔH°_f.

(b) Because two moles of CO are produced from CO_2 (which is not an element) and C, this reaction does not represent ΔH°_f.

(c) Because two substances are produced and one of the reactants (CO_2) is not an element, this reaction does not represent ΔH°_f.

(d) One mole of CH_4 is produced from elemental carbon and hydrogen; therefore this reaction represents ΔH°_f.

Think About It

Enthalpy of formation must involve the reaction of the *elements* to form compounds. Remember, though, that some elements such as O_2, H_2, and N_2 are diatomic in their elemental state.

5.85. Collect and Organize

The enthalpy of a reaction can be computed by finding the difference between the sum of the enthalpies of formation of the products and the sum of the enthalpies of formation of the reactants.

Analyze

We have to take into account the moles of products formed and the moles of reactants used as well, since enthalpy is a stoichiometric quantity.

$$\Delta H^\circ_{rxn} = \sum n\Delta H^\circ_{f,\,products} - \sum m\Delta H^\circ_{f,\,reactants}$$

Values for ΔH°_f for the reactants and products are found in Appendix 4.

Solve

$$\Delta H^\circ_{rxn} = \left[(1 \text{ mol } CH_4)(-74.8 \text{ kJ/mol}) + (2 \text{ mol } H_2O)(-241.8 \text{ kJ/mol})\right]$$
$$- \left[(4 \text{ mol } H_2)(0 \text{ kJ/mol}) + (1 \text{ mol } CO_2)(-393.5 \text{ kJ/mol})\right]$$
$$\Delta H^\circ_{rxn} = -164.9 \text{ kJ}$$

Think About It

Be careful to note and find the appropriate ΔH°_f for a compounds, which may exist in different phases. For example, ΔH°_f of $H_2O(g) = -241.8$ kJ/mol but ΔH°_f of $H_2O(\ell) = -285.8$ kJ/mol.

5.87. Collect and Organize

In order to calculate ΔH°_{rxn} for the decomposition of NH_4NO_3 to N_2O and H_2O vapor, we need the balanced equation because the enthalpy of the reaction depends on the moles of reactants consumed and moles of products formed in the reaction.

Analyze

From the balanced chemical equation and the values of ΔH°_f of the reactants and products, we use

$$\Delta H^\circ_{rxn} = \sum n\Delta H^\circ_{f,\,products} - \sum m\Delta H^\circ_{f,\,reactants}$$

Because the reaction is run at 250–300°C, the water product is in the gaseous phase.

Solve

The balanced equation for this reaction is

$$NH_4NO_3(s) \rightarrow N_2O(g) + 2\,H_2O(g)$$

We use the coefficients in the equation for ΔH°_{rxn} :

$$\Delta H^\circ_{rxn} = \left[(1 \text{ mol } N_2O)(82.1 \text{ kJ/mol}) + (2 \text{ mol } H_2O)(-241.8 \text{ kJ/mol})\right]$$
$$- \left[(1 \text{ mol } NH_4NO_3)(-365.6 \text{ kJ/mol})\right]$$
$$\Delta H^\circ_{rxn} = -35.9 \text{ kJ}$$

Think About It

The reaction is exothermic and 36 kJ are released for every mole of NH_4NO_3 decomposed.

5.89. Collect and Organize

We are given the balanced chemical equation for the explosive reaction of fuel oil with ammonium nitrate in the presence of oxygen.

Analyze

To calculate ΔH°_{rxn} we use

$$\Delta H^\circ_{rxn} = \sum n\Delta H^\circ_{f, \text{products}} - \sum m\Delta H^\circ_{f, \text{reactants}}$$

Solve

$$\Delta H^\circ_{rxn} = \left[(3 \text{ mol } N_2)(0.0 \text{ kJ/mol}) + (17 \text{ mol } H_2O)(-241.8 \text{ kJ/mol}) + (10 \text{ mol } CO_2)(-393.5 \text{ kJ/mol})\right]$$
$$- \left[(3 \text{ mol } NH_4NO_3)(-365.6 \text{ kJ/mol}) + (1 \text{ mol } C_{10}H_{22})(249.7 \text{ kJ/mol}) + (14 \text{ mol } O_2)(0.0 \text{ kJ/mol})\right]$$
$$\Delta H^\circ_{rxn} = -7198 \text{ kJ}$$

Think About It

This is a very exothermic reaction that occurs very fast and is therefore explosive.

5.91. Collect and Organize

We are asked to determine the standard enthalpy of formation for CO from the enthalpy of formation for CO_2 and the enthalpy of combustion of CO.

Analyze

The enthalpy may be obtained by manipulating equations for the processes described in the question and applying Hess's law. Recall that when an equation is flipped, the sign of ΔH is also flipped. The equations describing the formation of CO_2 and combustion of CO, respectively, are

$$CO(g) + \tfrac{1}{2}O_2(g) \rightarrow CO_2(g) \qquad\qquad \Delta H^\circ_f = -283.0 \text{ kJ}$$
$$C(s) + O_2(g) \rightarrow CO_2(g) \qquad\qquad \Delta H^\circ_{comb} = -393.5 \text{ kJ}$$

After flipping the equation for the formation of CO_2, we may simply add the two reactions and their respective enthalpies to arrive at the equation and enthalpy for the formation of CO.

Solve

$$C(s) + O_2(g) \rightarrow CO_2(g) \qquad\qquad \Delta H^\circ_{comb} = -393.5 \text{ kJ}$$
$$\underline{CO_2(g) \rightarrow \tfrac{1}{2}O_2(g) + CO(g) \qquad\qquad \Delta H^\circ_{rxn} = +283.0 \text{ kJ}}$$
$$C(s) + \tfrac{1}{2}O_2(g) \rightarrow CO(g) \qquad\qquad \Delta H^\circ_f = -110.5 \text{ kJ}$$

Think About It

Hess's law makes it easy for us to calculate energy changes for chemical reactions from those for other chemical reactions, even those that would be difficult to measure directly.

5.93. **Collect and Organize**

In this problem, we combine two chemical equations to calculate the enthalpy of formation of NO_2Cl.

Analyze

The first equation must be reversed so that NO_2Cl is a product in the overall equation. This results in an endothermic reaction.

Solve

$$NO_2(g) + \tfrac{1}{2} Cl_2(g) \rightarrow NO_2Cl(g) \qquad \Delta H^{\circ}_{rxn} = -20.6 \text{ kJ}$$

$$\underline{\tfrac{1}{2} N_2(g) + O_2(g) \rightarrow NO_2(g) \qquad \Delta H^{\circ}_{rxn} = 33.2 \text{ kJ}}$$

$$\tfrac{1}{2} N_2(g) + O_2(g) + \tfrac{1}{2} Cl_2(g) \rightarrow NO_2Cl(g) \qquad \Delta H^{\circ}_{f} = 12.6 \text{ kJ}$$

Think About It

The enthalpy of formation of NO_2Cl is endothermic. This means that the product, NO_2Cl, is unstable (higher in energy) with respect to the elements N_2, O_2, and Cl_2.

5.95. **Collect and Organize / Analyze**

We are asked to define and describe the term *fuel value*.

Solve

Because we often are concerned about the energy a fuel provides per mass, fuel values give the energy per gram a fuel releases upon burning.

Think About It

For some fuels, like gasoline, it might be more convenient to think of energy per volume (liter or gallon).

5.97. **Collect and Organize / Analyze**

The units for molar enthalpies of combustion are kilojoules per mole (kJ/mol). We are asked how these can be converted into kilojoules per gram (kJ/g), the units for fuel values.

Solve

Because the moles of a substance are related to the grams of a substance by molar mass, to convert kJ/mol to kJ/g, we need only divide by the molar mass.

$$\frac{kJ}{mol} \times \frac{mol}{g} = \frac{kJ}{g}$$

Think About It

The fuel value could be converted into the molar enthalpy of combustion by multiplying the fuel value by the molar mass of the fuel.

5.99. **Collect and Organize**

We are asked to calculate the fuel values for gasoline (C_9H_{20}) and ethanol (C_2H_5OH). The enthalpy of combustion for gasoline is provided (–6160 kJ/mol), and the enthalpy of formation for ethanol may be looked up in Appendix 4 (–235.1 kJ/mol). The fuel value is a measure of the energy stored in one gram of a fuel.

Analyze

The fuel value may be obtained by dividing the enthalpy of combustion or formation by the molar mass of gasoline (128.29 g/mol) or ethanol (46.069 g/mol).

Solve

The fuel value for gasoline is

$$\frac{6160 \text{ kJ}}{1 \text{ mol}} \times \frac{1 \text{ mol}}{128.29 \text{ g}} = 48.02 \text{ kJ/g}$$

The fuel value for ethanol is

$$\frac{235.1\,kJ}{1\,mol} \times \frac{1\,mol}{46.069\,g} = 5.103\,kJ/g$$

Gasoline has a higher fuel value.

Think About It
Gasoline's relatively high fuel value makes it an efficient fuel.

5.101. Collect and Organize
Once we compute the fuel value of C_5H_{12}, we use it to calculate the energy released when 1.00 kg of C_5H_{12} is burned and how much C_5H_{12} is needed to increase the temperature of 1.00 kg of water by 70.0°C.

Analyze
(a) The fuel value can be calculated by dividing the absolute value of the given ΔH°_{comb} (–3535 kJ/mol) by the molar mass of C_5H_{12} (72.15 g/mol).
(b) The energy released when 1.00 kg of C_5H_{12} is burned can be found by multiplying the mass in grams by the fuel value.
(c) The molar heat equation is

$$q = nc_P\Delta T$$

where moles (n) can be determined from m = 1.00 kg (1000 g) water, c_P = 75.3 J/(mol · °C), and ΔT = 70.0°C. This gives us the energy (in J) required to heat the water. The amount of C_5H_{12} needed to heat the water can then be calculated by dividing the q value by the fuel value for C_5H_{12}.

Solve
(a) Fuel value of $C_5H_{12} = \dfrac{3535\,kJ}{mol} \times \dfrac{1\,mol}{72.15\,g} = 48.99\,kJ/g$

(b) Heat released by 1.00 kg $C_5H_{12} = 1000\,g \times \dfrac{48.99\,kJ}{g} = 4.90 \times 10^4\,kJ$

(c) Energy needed to raise 1.00 kg water from 20.0°C to 90.0°C

$$q = 1000\,g \times \frac{1\,mol}{18.02\,g} \times \frac{75.3\,J}{mol \cdot °C} \times 70.0°C = 2.925 \times 10^5\,J\;or\;293\,kJ$$

Mass of C_5H_{12} needed to generate this energy $= 292.5\,kJ \times \dfrac{1\,g}{48.99\,kJ} = 5.97\,g$

Think About It
Only about 6 grams of fuel is required from the camper's stove to heat up the water. The white gas has a relatively high fuel value.

5.103. Collect and Organize
We are to calculate the fuel value and fuel density for diethyl ether, given the enthalpy of combustion and density of diethyl ether, and compare these to the values for diesel. The fuel value is the energy contained in one gram of diethyl ether, and the fuel density is the energy contained in one liter of diethyl ether.

Analyze
The fuel value may be calculated by dividing the enthalpy of combustion by the molar mass of diethyl ether (74.122 g/mol). The fuel density may be calculated by multiplying the fuel value by the density of diethyl ether.

Solve

The fuel value is

$$\frac{2726.3\,\text{kJ}}{1\,\text{mol}} \times \frac{1\,\text{mol}}{74.122\,\text{g}} = 36.781\,\text{kJ/g}$$

The fuel density is

$$\frac{36.781\,\text{kJ}}{1\,\text{g}} \times \frac{0.7134\,\text{g}}{1\,\text{mL}} \times \frac{1000\,\text{mL}}{1\,\text{L}} = 2.624 \times 10^4\,\text{kJ/L}$$

The fuel value and fuel density for diesel are 40.0 kJ/g and 3.05×10^4 kJ/L. Both of these values are higher for diesel than for diethyl ether.

Think About It

Although diethyl ether is volatile and extremely flammable, its fuel value and fuel density are lower than those for diesel.

5.105. **Collect and Organize**

Given the molar heat capacity of CCl_4, we are to calculate the energy required to boil 275 g CCl_4 (bp 77°C) from room temperature (22°C).

Analyze

The energy required to raise the temperature of a substance is related to the molar heat capacity by the equation

$$q = nc_P\Delta T$$

where q = energy, n = moles of the substance, c_P = molar heat capacity of the substance, and ΔT = difference in temperature, $T_f - T_i$.

Solve

$$q = \left(275\,\text{g} \times \frac{1\,\text{mol}}{153.81\,\text{g}}\right) \times \frac{131.3\,\text{J}}{\text{mol}\cdot\text{°C}} \times \left(77\,\text{°C} - 22\,\text{°C}\right) = 13\,\text{kJ}$$

Think About It

Once the CCl_4 boils, the added energy is used to vaporize CCl_4, and the temperature of the liquid CCl_4 does not change.

5.107. **Collect and Organize**

Given the specific heat capacity for sodium metal, we are to calculate the mass of sodium that will absorb 1.00×10^3 kJ with a temperature rise of only 1K.

Analyze

The energy required to raise the temperature of a substance is related to the specific heat capacity by the equation

$$q = mc_s\Delta T$$

where for this problem $q = 1.00 \times 10^3$ kJ, n = grams of sodium, $c_s = 1.23$ J/g · °C, and $\Delta T = 1$K. Once we calculate the mass of sodium needed, we can convert it to moles by diving by the molar mass of sodium metal, 22.99 g/mol.

Solve

$$1.00 \times 10^3\,\text{kJ} \times \frac{1000\,\text{J}}{\text{kJ}} = m \times \frac{1.23\,\text{J}}{\text{g}\cdot\text{°C}} \times 1\text{K}$$

$$m = 8.13 \times 10^5\,\text{g}$$

$$n = 8.13 \times 10^5\,\text{g} \times \frac{1\,\text{mol}}{22.99\,\text{g}} = 3.54 \times 10^4\,\text{mol}$$

Think About It
To solve this problem, we could also convert the specific heat capacity of sodium into the molar heat capacity using

$$c_s \text{ (in J/g} \cdot \text{K)} \times \mathcal{M} \text{ (in g/mol)} = c_P \text{ (in J/mol} \cdot \text{K)}$$

and then use $q = nc_P\Delta T$ to solve for the moles of sodium needed.

5.109. Collect and Organize
Given the value of ΔH_f° for NH_3 we are to determine the ΔH° for the reaction of N_2 with H_2 to produce 2 moles of NH_3 and for the reaction in which a mole of NH_3 decomposes to N_2 and H_2.

Analyze
The ΔH_f° value is for the formation of one mole of NH_3 according to the equation

$$\tfrac{1}{2} N_2(g) + \tfrac{3}{2} H_2(g) \rightarrow NH_3(g)$$

When the reaction is multiplied to produce more than one mole, the enthalpy of formation must also be multiplied by the same factor. When a reaction is reversed, the sign of the enthalpy of formation is multiplied by -1.

Solve
(a) $\Delta H_{rxn}^\circ = 2 \times -46.1 \text{ kJ/mol} = -92.2 \text{ kJ}$
(b) $\Delta H_{rxn}^\circ = -(-46.1 \text{ kJ/mol}) = 46.1 \text{ kJ}$

Think About It
The reverse reaction, the decomposition reaction of ammonia, is endothermic.

5.111. Collect and Organize
In this problem, CF_2Cl_2 is evaporated to cool 200.0 g of water 10.0°C.

Analyze
The energy lost by the water is given as

$$q_{water} = nc_P\Delta T$$

The energy gained by the CF_2Cl_2 to evaporate is

$$q_{vap} = n\Delta H_{vap}^\circ$$

The energy lost by the water will be due to the vaporization of n moles of CF_2Cl_2:

$$q_{vap} = -q_{water}$$
$$n\Delta H_{vap}^\circ = -nc_P\Delta T$$

Solve

$$n \times (1.74 \times 10^4 \text{ J/mol}) = -\left(200.0 \text{ g} \times \frac{1 \text{ mol}}{18.02 \text{ g}}\right) \times \left(75.3 \text{ J/mol} \cdot {}^\circ\text{C}\right) \times \left(-10.0 {}^\circ\text{C}\right)$$
$$n = 0.480 \text{ mol } CF_2Cl_2$$

The mass of CF_2Cl_2 needed would be

$$0.480 \text{ mol} \times \frac{121 \text{ g}}{1 \text{ mol}} = 58.1 \text{ g } CF_2Cl_2$$

Think About It
We did not need to use ΔH_{vap}° of water to solve this problem. However, you should notice that more energy (in kJ) will be needed (per mole) to vaporize water compared to 1 mole CF_2Cl_2 because CF_2Cl_2 is more volatile than water.

5.113. Collect and Organize

When NaOH and H_2SO_4 solutions are mixed, an exothermic reaction occurs, which heats up the solution. In this problem, we are asked to determine the final temperature of the solution, T_f.

Analyze

From Problem 5.112, we know the molar ratio of H_2SO_4 to NaOH in the balanced equation is 1:2. We must first find out how much H_2SO_4 reacts by computing the moles of H_2SO_4 and NaOH. The q for the reaction then may be found by multiplying ΔH_{rxn} (–114 kJ/mol H_2SO_4) by the moles of H_2SO_4 used in the reaction. Then ΔT can be found from

$$q = mc_s\Delta T$$

where m = total mass of the solution (165 g) and c_s = 4.184 J/(g · °C).

Solve

Molar amounts of the reactants:

$$\text{mol } H_2SO_4 = 65.0 \text{ mL} \times \frac{1.0 \text{ mol}}{1000 \text{ mL}} = 0.065 \text{ mol}$$

$$\text{mol NaOH} = 100.0 \text{ mL} \times \frac{1.0 \text{ mol}}{1000 \text{ mL}} = 0.10 \text{ mol}$$

The limiting reactant here is NaOH, which would react with 0.050 mol H_2SO_4. Because $\Delta H_{rxn} = -114$ kJ/mol H_2SO_4, for this reaction

$$q = \frac{-114 \text{ kJ}}{\text{mol } H_2SO_4} \times 0.050 \text{ mol} = -5.7 \text{ kJ or } -5700 \text{ J}$$

$$q = 5700 \text{ J} = 165 \text{ g} \times \frac{4.184 \text{ J}}{\text{g} \cdot °C} \times \Delta T$$

$$\Delta T = 8.3°C$$

The reaction is exothermic so $T_f > T_i$ and $T_f = T_i + 8.3°C$:

$$T_f = 25.0°C + 8.3°C = 33.3°C$$

Think About It

The ΔH_{rxn} in Problem 5.112 was defined on a per mole of H_2SO_4 basis, so we needed to focus on that quantity in computing q.

5.115. Collect and Organize

We are asked to relate the enthalpy of reaction to the balanced chemical equation.

Analyze

We should be able to balance the reaction between FeO [iron(II) oxide] and O_2 by inspection. Once balanced, we can use the stoichiometric relationships to determine ΔH_{rxn} based on the value of ΔH_{rxn} for 1 mole of Fe_3O_4, which is –318 kJ.

Solve

(a) $3 \text{ FeO}(s) + \frac{1}{2} O_2(g) \rightarrow Fe_3O_4(s)$

Eliminating fractional coefficients gives

$$6 \text{ FeO}(s) + O_2(g) \rightarrow 2 \text{ Fe}_3O_4(s)$$

(b) If $\Delta H_{rxn} = -318$ kJ for 1 mole of Fe_3O_4 then for 2 moles of Fe_3O_4 in the balanced equation, $\Delta H_{rxn} = -636$ kJ.

Think About It

Remember that enthalpy is stoichiometric. The more magnetite you produce in this reaction, the greater the energy released.

5.117. Collect and Organize

We are to consider ΔH_f° of water and compare it to ΔH_{vap}°.

Analyze

Exothermic reactions have a negative ΔH; endothermic reactions have a positive ΔH. The chemical equation describing ΔH_f° is

$$H_2(g) + \tfrac{1}{2}O_2(g) \rightarrow H_2O(\ell)$$

The equation for vaporization is

$$H_2O(\ell) \rightarrow H_2O(g)$$

If we know the moles of water produced (from the mass) we can compute the energy given off by multiplying the moles of water by the ΔH_f° value.

Solve

(a) The negative sign means that the reaction to form H_2O from H_2 and O_2 is exothermic—energy is transferred from the system to the surroundings.

(b) The ΔH_f° of water has a large magnitude because in that reaction, hydrogen–hydrogen and oxygen–oxygen bonds break and hydrogen–oxygen bonds form. This is more energy intensive than the breaking of the weaker attractive forces between water molecules when they go into the gas phase.

(c) $\text{mol } H_2O = 50.0 \text{ mL} \times \dfrac{1.00 \text{ g}}{1 \text{ mL}} \times \dfrac{1 \text{ mol}}{18.02 \text{ g}} = 2.775 \text{ mol}$

Energy required to form this amount of water:

$$q = 2.775 \text{ mol} \times \dfrac{-285.8 \text{ kJ}}{\text{mol}} = -793 \text{ kJ}$$

Think About It

During vaporization of a substance, no strong intramolecular bonds are broken.

5.119. Collect and Organize

We need the balanced chemical equation to compute the enthalpy for a reaction.

Analyze

The chemical equation can be balanced by inspection. The ΔH_{rxn}° can be computed from

$$\Delta H_{rxn}^\circ = \sum n\Delta H_{f,products}^\circ - \sum m\Delta H_{f,reactants}^\circ$$

Solve

The balanced chemical equation for the reaction of iron(II) oxide with oxygen to form iron(III) oxide is

$$2\,FeO(s) + \tfrac{1}{2}O_2(g) \rightarrow Fe_2O_3(s)$$
$$4\,FeO(s) + O_2(g) \rightarrow 2\,Fe_2O_3(s)$$

We use the coefficients in the equation for ΔH_{rxn}°:

$$\Delta H_{rxn}^\circ = \left[2 \text{ mol } Fe_2O_3 \times -824.2 \text{ kJ/mol}\right]$$
$$- \left[(4 \text{ mol } FeO \times -271.9 \text{ kJ/mol}) + (1 \text{ mol } O_2 \times 0.0 \text{ kJ/mol})\right]$$
$$\Delta H_{rxn}^\circ = -560.8 \text{ kJ}$$

Think About It

This enthalpy change is for the production of 2 moles of Fe_2O_3. For 1 mole, $\Delta H_{rxn}^\circ = -280.4 \text{ kJ}$.

5.121. Collect and Organize

From the amount of energy required to convert the given mass of liquid water (0.90 g) into steam, we can find the enthalpy of vaporization (ΔH_{vap}) for one mole of water.

Analyze

We first need to determine the number of moles of water in 0.90 g. The enthalpy of the vaporization of water is simply the energy divided by the moles of water.

Solve

Moles of water converted into steam:

$$0.90 \text{ g} \times \frac{1 \text{ mol}}{18.02 \text{ g}} = 0.050 \text{ mol}$$

If 0.050 mol of water requires 2.0 kJ, then one mole would require

$$\frac{2.0 \text{ kJ}}{0.050 \text{ mol}} = 40 \text{ kJ}$$

Think About It

The ΔH_{vap} of water under standard conditions is 40.67 kJ/mol. Because the value calculated here is very close to the "known" ΔH°_{vap}, our system is probably at or near standard conditions.

5.123. **Collect and Organize**

Given the balanced overall thermochemical equation in which methanol is oxidized to formic acid, we are to determine if the reaction is exothermic or endothermic by calculating the enthalpy of the reaction. We also are asked to calculate the energy change when 60.0 g of methanol is oxidized, and then make a prediction about the magnitude of the enthalpy change for the conversion of methanol to formaldehyde, the first step in the reaction.

Analyze

If energy (in kJ) is a reactant in the thermochemical equation, the reaction is endothermic; if energy is a product in the thermochemical equation, the reaction is exothermic. ΔH°_{rxn} is the enthalpy change under standard conditions to form one mole of the product (HCOOH).

$$\Delta H^{\circ}_{rxn} = \sum n \Delta H^{\circ}_{f,products} - \sum m \Delta H^{\circ}_{f,reactants}$$

The energy change in the thermochemical equation is for 1 mole methanol consumed. We multiply this relationship by the moles of CH_3OH in 60.0 g. When converting methanol to formic acid in a two-step process, the intermediate, formaldehyde, has only one oxygen atom, whereas formic acid has two.

Solve

(a) This reaction is exothermic because energy is produced (as a "product") in the thermochemical equation.

(b) $\Delta H^{\circ}_{rxn} = \left[\left(1 \text{ mol HCOOH} \times \frac{-424.7 \text{ kJ}}{\text{mol}} \right) + \left(1 \text{ mol H}_2\text{O} \times \frac{-285.8 \text{ kJ}}{\text{mol}} \right) \right] -$

$\left[\left(1 \text{ mol O}_2 \times \frac{0.0 \text{ kJ}}{\text{mol}} \right) + \left(1 \text{ mol CH}_3\text{OH} \times \frac{-238.7 \text{ kJ}}{\text{mol}} \right) \right] = -471.8 \text{ kJ}$

(c) The number of moles of methanol in 60.0 g is

$$60.0 \text{ g} \times \frac{1 \text{ mol}}{32.04 \text{ g}} = 1.87 \text{ mol}$$

For this number of moles, the enthalpy is

$$1.87 \text{ mol} \times \frac{-471.8 \text{ kJ}}{\text{mol}} = -882.3 \text{ kJ}$$

(d) We can expect that the first step, that converts methanol to formaldehyde where only one oxygen atom is present, would be less exothermic than the overall conversion of methanol to formic acid, so the enthalpy of reaction for the formation of formaldehyde from methanol is expected to be smaller (less exothermic) than the overall enthalpy of the reaction to form formic acid.

Think About It

This reaction, because it is exothermic, is favored by enthalpy.

5.125. Collect and Organize

We use Hess's law to add two equations to get a third and write an expression for calculating the overall ΔH_{rxn}.

Analyze

Because species D and E are reactants in the overall reaction, we have to reverse equation 1. The same is true for equation 2 because F(s) is a product in the overall reaction. Reversing the reactions changes the sign of the ΔH for each reaction.

Solve

$$
\begin{array}{ll}
D(s) + E(g) \rightarrow F(g) & -\Delta H_1 \\
\underline{F(g) \rightarrow F(s)} & \underline{-\Delta H_2} \\
D(s) + E(g) \rightarrow F(s) & \Delta H_3 = -\Delta H_1 + -\Delta H_2
\end{array}
$$

Think About It

Remember that reversing a reaction does not change the magnitude of the enthalpy; it only reverses the sign.

5.127. Collect and Organize

We combine three reactions to give the overall reaction and calculate the enthalpy for the formation of $NiSO_3$ from nickel, sulfur, and oxygen.

Analyze

Because $NiSO_3$ is a product in the overall reaction, the first reaction will have to be reversed. Because S, O_2, and Ni are reactants, neither reaction 2 nor reaction 3 needs to be changed.

Solve

$$
\begin{array}{ll}
NiO(s) + SO_2(g) \rightarrow NiSO_3(s) & \Delta H^{\circ}_{rxn} = -156 \text{ kJ} \\
\tfrac{1}{8}S_8(s) + O_2(g) \rightarrow SO_2(g) & \Delta H^{\circ}_{rxn} = -297 \text{ kJ} \\
\underline{Ni(s) + \tfrac{1}{2}O_2(g) \rightarrow NiO(s)} & \underline{\Delta H^{\circ}_{rxn} = -241 \text{ kJ}} \\
Ni(s) + \tfrac{1}{8}S_8(s) + \tfrac{3}{2}O_2(g) \rightarrow NiSO_3(s) & \Delta H^{\circ}_{rxn} = -694 \text{ kJ}
\end{array}
$$

Multiplying by 2 gives the desired equations:
$$
2\,Ni(s) + \tfrac{1}{4}S_8(s) + 3\,O_2(g) \rightarrow 2\,NiSO_3(s) \qquad \Delta H^{\circ}_{rxn} = -1390 \text{ kJ}
$$

Think About It

All the reactions that were added together are exothermic. The overall reaction, therefore, must also be exothermic. Notice, too, that the overall enthalpy of reaction is also the enthalpy of formation of $NiSO_3$.

5.129. Collect and Organize

Use the ΔH°_f for the reactants and products to calculate the enthalpy of reaction for the combustion of 1 mole of ethanol.

Analyze

The balanced equation for the combustion of ethanol is
$$
C_2H_5OH(\ell) + 3\,O_2(g) \rightarrow 2\,CO_2(g) + 3\,H_2O(g)
$$
To calculate the enthalpy of the combustion, use the equation
$$
\Delta H^{\circ}_{comb} = \sum n\Delta H^{\circ}_{f,products} - \sum n\Delta H^{\circ}_{f,reactants}
$$

Solve

$$\Delta H^{\circ}_{comb} = \left[(2 \text{ mol } CO_2 \times -393.5 \text{ kJ/mol}) + (3 \text{ mol } H_2O \times -241.8 \text{ kJ/mol}) \right]$$
$$- \left[(1 \text{ mol } C_2H_5OH \times -277.7 \text{ kJ/mol}) + (3 \text{ mol } O_2 \times 0 \text{ kJ/mol}) \right]$$
$$\Delta H^{\circ}_{comb} = -1234.7 \text{ kJ}$$

Think About It

Notice that the value calculated is for one mole of ethanol that is burned.

5.131. **Collect and Organize**

The dissolution of two solids (NaOH and KOH) in water raises the temperature of the solution from 23°C. We are asked to calculate the final temperature of the solution.

Analyze

We can calculate the heat released upon dissolution for NaOH and KOH by dividing the mass of each solid by the correct molar mass and multiplying the result by the enthalpy of solution. The sum of these heats is absorbed by the water, so we can calculate the change in temperature using the equation

$$\Delta T = \frac{q_{solution}}{nc_P}$$

Solve

$$5.00 \text{ g NaOH} \times \frac{1 \text{ mol NaOH}}{40.00 \text{ g NaOH}} \times \frac{-44.3 \text{ kJ}}{1 \text{ mol}} = -5.54 \text{ kJ}$$

$$4.20 \text{ g KOH} \times \frac{1 \text{ mol KOH}}{56.108 \text{ g KOH}} \times \frac{-56.0 \text{ kJ}}{1 \text{ mol}} = -4.19 \text{ kJ}$$

$$q_{total} = (-5.54 \text{ kJ}) + (-4.19 \text{ kJ}) = -9.73 \text{ kJ}$$

$$q_{solution} = -q_{total} = 9.73 \text{ kJ}$$

$$n = 150.0 \text{ mL} \times \frac{1.00 \text{ g}}{1 \text{ mL}} \times \frac{1 \text{ mol}}{18.02 \text{ g}} = 8.324 \text{ mol}$$

$$\Delta T = \frac{9.73 \text{ kJ}}{8.324 \text{ mol} \left(\frac{75.3 \text{ J}}{1 \text{ mol}} \times \frac{1 \text{ kJ}}{1000 \text{ J}} \right)} = 15.5°C$$

$$T_f = T_i + \Delta T = 23.0°C + 15.5°C = 38.5°C$$

Think About It

Roughly half of the energy for the temperature change comes from each solid.

5.133. **Collect and Organize**

By multiplying the molar mass of various elements by their corresponding specific heats, we investigate the validity of the law of Dulong and Petit.

Analyze

When the molar mass of the element and specific heat capacity are given in the table, we compute the value for the fourth column in the table. Using the average of these values, we then compute the specific heat capacity or the molar mass for the other elements in the table.

Solve

All results are shown in the table below.

(a) The specific heat capacity values and molar masses are given for Au, Sn, Zn, Cu, and Fe. These can be used to calculate the values in the fourth column. Using the calculation for Au to determine the units gives:

$$\frac{197.0 \text{ g}}{1 \text{ mol}} \times \frac{0.125 \text{ J}}{\text{g} \cdot {}^{\circ}\text{C}} = 24.6 \text{ J/mol} \cdot {}^{\circ}\text{C}$$

(b) The average value for the fourth column values using the results for Pb, Au, Sn, Zn, Cu, and Fe is

$$\frac{25.5 + 24.6 + 25.5 + 25.4 + 25.2 + 25.7}{6} = 25.3$$

(c) Missing atomic masses for Bi and the two unknown elements in the table are found by dividing 25.3 by the specific heat capacity. The result for Bi is 210.8 g/mol, which is close to the known molar mass in the periodic table, but that mass could also be for At or Po. For the two unknown elements, the molar masses are calculated as 108.6 and 58.4 g/mol. This would identify these elements as Ag (107.9 g/mol) and either Co (58.9 g/mol) or Ni (58.7 g/mol). There is obviously some uncertainty in assigning elements based on this application of the law of Dulong and Petit.

(d) The specific heat capacity for Pt and S are calculated by dividing 25.3 by the molar mass. See table for results.

Element	$\mathcal{M}$ (g/mol)	$c_s[\text{J/(g} \cdot {}^{\circ}\text{C})]$	$\mathcal{M} \, c_s$
Bismuth	*210.8*	0.120	*25.3*
Lead	207.2	0.123	25.5
Gold	197.0	0.125	*24.6*
Platinum	195.1	*0.130*	*25.3*
Tin	118.7	0.215	25.5
Silver	*108.6*	0.233	*25.3*
Zinc	65.38	0.388	25.4
Copper	63.5	0.397	25.2
Cobalt/nickel	*58.4*	0.433	*25.3*
Iron	55.8	0.460	25.7
Sulfur	32.1	*0.788*	*25.3*
Average value in column 4			*25.3*

Think About It

The law of Dulong and Petit is of some, but limited, use.

5.135. Collect and Organize / Analyze

Energy and specific heat for both the metal and the water are related by the equation $q = mc_s\Delta T$.

Solve

First, heat the piece of metal by immersing a dry test tube containing the metal into a boiling water bath long enough for the metal to come to thermal equilibrium. The temperature of the boiling water will be the T_i of the metal. After recording T_i of the water in the Styrofoam cup, quickly transfer the metal from the test tube to the water in the cup. Record the highest temperature of the water as it heats up from the transfer of energy from the hot metal to the water. This temperature will be T_f for both the metal and the water. Because all the energy lost by the metal is gained by the water:

$$-q_{\text{lost,metal}} = q_{\text{gained,water}}$$
$$-mc_{s,\text{metal}}(T_f - T_{i,\text{metal}}) = nc_{P,\text{water}}(T_f - T_{i,\text{water}})$$

where the mass (m) of the metal, T_f, $T_{i,\text{water}}$, $T_{i,\text{metal}}$, moles (n) of the water, and the specific heat of water [75.3 J/(mol · °C)] are all known. This leaves $c_{s,\text{metal}}$ as the only unknown in the equation.

Think About It

If a chemist prepares a new alloy (a solid solution of a mixture of metals) one of the ways to characterize that new alloy is to determine its specific heat capacity in this manner.

5.137. Collect and Organize

Given the combustion reactions in an automobile engine that produce NO and NO_2, we are to add the equations and use Hess's law to calculate ΔH°_{comb} for the overall reaction and determine whether that reaction is exothermic or endothermic.

Analyze

Because 2 moles NO are produced in the first reaction and are consumed in the second reaction, NO cancels, and the reactions can be directly added.

Solve

Adding the reactions gives

$$N_2(g) + O_2(g) \rightarrow 2\ NO(g) \qquad \Delta H^\circ_{comb} = +180\ \text{kJ}$$
$$\underline{2\ NO(g) + O_2(g) \rightarrow 2\ NO_2(g) \qquad \Delta H^\circ_{comb} = -112\ \text{kJ}}$$
$$N_2(g) + O_2(g) \rightarrow 2\ NO_2(g) \qquad \Delta H^\circ_{comb} = 68\ \text{kJ}$$

This reaction is endothermic under standard conditions.

Think About It

Automobile engines, of course, run hot, so there is energy available to produce NO_2 as a product of combustion.

5.139. Collect and Organize

We must calculate the fuel density of hydrogen gas and liquid, given the densities of each. We also are asked to calculate the enthalpy change for the listed reactions, given values for ΔH_f for the reactants and products. Finally, we are asked to calculate the mass of ammonia-borane that will decompose to form 10 kg of hydrogen gas.

Analyze

Fuel density is the quantity of energy contained in one liter of fuel. For H_2, the fuel density may be calculated by dividing the enthalpy of formation (–241.8 kJ/mol) by the molar mass of H_2 (2.01588 g/mol), and multiplying by the density provided for gaseous or liquid H_2. We may use Hess's Law to calculate the enthalpy for each of the reactions provided.

$$\Delta H^\circ_r = \sum n \Delta H^\circ_{f,products} - \sum m \Delta H^\circ_{f,reactants}$$

Two of the reactions listed in (b) produce hydrogen gas, so summing these reactions, we find

$$H_3NBH_3(g) \rightarrow H_2(g) + H_2NBH_2(g)$$
$$\underline{H_2NBH_2(g) \rightarrow H_2(g) + HNBH\ (g)}$$
$$H_3NBH_3(g) \rightarrow 2\ H_2(g) + HNBH(g)$$

The overall reaction generates 2 moles of H_2 for every mole of ammonia-borane consumed. Starting from the desired quantity of H_2 gas, we may divide by the molar mass of H_2, multiply by the stoichiometric ratio from the balanced chemical equation, and multiply by the molar mass of ammonia-borane (30.8653 g/mol) to determine the mass of H_3NBH_3 required.

Solve

(a) For $H_2(g)$

$$\frac{241.8\ \text{kJ}}{1\ \text{mol}} \times \frac{1\ \text{mol}}{2.01588\ \text{g}} \times \frac{0.0899\ \text{g}}{1\ \text{L}} = 10.78\ \text{kJ/L}$$

For $H_2(\ell)$

$$\frac{241.8\ \text{kJ}}{1\ \text{mol}} \times \frac{1\ \text{mol}}{2.01588\ \text{g}} \times \frac{70.8\ \text{g}}{1\ \text{L}} = 8492\ \text{kJ/L}$$

(b) The enthalpies for each of the reactions given are
$$H_3NBH_3(g) \rightarrow NH_3(g) + BH_3(g)$$
$$\Delta H^\circ_r = \left[1\ \text{mol}\left(110.2\ \text{kJ/mol}\right) + 1\ \text{mol}\left(-46.1\ \text{kJ/mol}\right)\right] - \left[1\ \text{mol}\left(-38.1\ \text{kJ/mol}\right)\right] = 102.2\ \text{kJ}$$

$$H_3NBH_3(g) \rightarrow H_2(g) + H_2NBH_2(g)$$
$$\Delta H_r^\circ = \left[1\,mol\left(0\,kJ/mol\right) + 1\,mol\left(-66.5\,kJ/mol\right)\right] - \left[1\,mol\left(-38.1\,kJ/mol\right)\right] = -28.40\,kJ$$
$$H_2NBH_2(g) \rightarrow H_2(g) + HNBH\,(g)$$
$$\Delta H_r^\circ = \left[1\,mol\left(0\,kJ/mol\right) + 1\,mol\left(+56.9\,kJ/mol\right)\right] - \left[1\,mol\left(-66.5\,kJ/mol\right)\right] = 123.4\,kJ$$

(c) $10\,kg\,H_2 \times \dfrac{1000\,g}{1\,kg} \times \dfrac{1\,mol\,H_2}{2.01588\,g\,H_2} \times \dfrac{1\,mol\,H_3NBH_3}{2\,mol\,H_2} \times \dfrac{30.8653\,g}{1\,mol} \times \dfrac{1\,kg}{1000\,g} = 76.6\,kg\,H_3NBH_3$

Think About It
The fuel density for liquid hydrogen is significantly higher than for gaseous hydrogen.

5.141. Collect and Organize
Use the ΔH_f° for the reactants and products to calculate the enthalpy of reaction for the conversion of diamond into graphite.

Analyze
The first three equations may be added to generate the fourth. ΔH° can be obtained for the fourth reaction by summing ΔH° for the first three reactions.

Solve
$$\Delta H^\circ = -395.4\,kJ + 566.0\,kJ + -172.5\,kJ = -1.9\,kJ$$
The reaction is slightly exothermic.

Think About It
Although ΔH° is small, it is slightly negative, so the reaction is exothermic.

5.143. Collect and Organize
A ring at 100°C is placed in a volume of water at 15°C. If the final temperature of both the ring and the water is 40°C, what is the volume of water?

Analyze
We are given the masses and heat capacities for the diamond [0.20 g and 6.23 J/(mol · °C)] and gold [23 g and 25.41 J/(mol·°C)] components of the ring. We may calculate the number of moles of each material by dividing the mass by the molar mass of diamond (12.0107 g/mol) and gold (196.96657 g/mol).

We also know that the heat lost by the ring is absorbed by the water, so
$$q_{H_2O} = -q_{ring}$$

The equations describing the heat gained by the water and lost by the ring are

$$q_{H_2O} = mc_s\Delta T$$

$$q_{ring} = n_{diamond}c_{P,diamond}\Delta T_{diamond} + n_{gold}c_{P,gold}\Delta T_{gold}$$

Substituting the values given in the problem, we may solve for the mass of water, and convert this to a volume using the density of water.

Solve

$$m_{H_2O} \times \left(\frac{4.184\ J}{g\cdot°C}\right)(40°C - 15°C) =$$

$$-\left[\left(0.0167\ mol \times \frac{6.23\ J}{mol\cdot°C} \times -60°C\right) + \left(0.117\ mol \times \frac{25.41\ J}{mol\cdot°C} \times -60°C\right)\right]$$

$$104.6\ J/g\left(m_{H_2O}\right) = -\left[\left(-6.24\ J\right) + \left(-178\ J\right)\right]$$

$$m_{H_2O} = 1.8\ g$$

$$1.8\ g \times \frac{1\ mL}{1.00\ g} = 1.8\ mL\ water$$

Think About It
This is quite a small volume of water.

CHAPTER 6 | Properties of Gases: The Air We Breathe

6.1. Collect and Organize

In Figure P6.1, the barometer in the middle (b), which is at sea level, has an intermediate mercury level, between (a), a lower mercury level, and (c), a higher mercury level.

Analyze

The altitude of Denver, Colorado, is significantly higher than sea level, and the atmospheric pressure there is lower than at sea level. The height of the mercury in a barometer depends on the atmospheric pressure: the higher the pressure, the higher the column of mercury.

Solve

Because Denver has lower atmospheric pressure due to its altitude, the barometer on the left (a) reflects the pressure in Denver, CO.

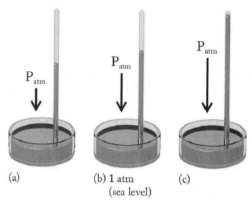

(a)   (b) 1 atm   (c)
                 (sea level)

Think About It

Altitude above or below sea level is not the only factor that changes atmospheric pressure. Weather (cold fronts, storms, etc.) also changes the atmospheric pressure at a given location.

6.3. Collect and Organize

When the atmospheric pressure outside of the balloon in Figure P6.3 increases, does the balloon shrink while keeping the molecules randomly distributed in the balloon, expand, or shrink and condense the molecules of gas into liquid?

Analyze

Boyle's law states that pressure and volume are inversely proportional to each other.

Solve

If we increase pressure, we decrease volume, according to Boyle's law. This eliminates drawing (b). In (c), the molecules of gas are shown to "stick" together and have short distances between them, indicating that they have condensed. Gases will not condense inside of a balloon simply by changing the outside pressure. Drawing (a) is the correct choice because the molecules are still in the gaseous state and the balloon has shrunk (decreased) in volume.

Think About It

If the atmospheric pressure is lowered, then the balloon expands.

6.5. Collect and Organize

When the amount of gas is increased in a balloon, and temperature and pressure remain constant (Figure P6.5), does the volume of the balloon stay the same, decrease, or increase?

Analyze

Avogadro's law states that volume and the number of moles of gas are directly proportional.

Solve

As we increase the amount of gas in a balloon, we increase the moles of gas in the balloon. Applying Avogadro's law, drawing (c) represents the addition of gas to the balloon.

Think About It

If the container is rigid (not expandable like a balloon), then adding more gas would increase the pressure but not the volume.

6.7. **Collect and Organize**

The plot in Figure P6.7 shows the volume (*V*) as a function of temperature (*T*). Line 2 is higher on the *y*-axis(*V*) than line 1. We are asked which line represents the higher pressure and whether an absolute temperature scale is shown.

Analyze

According to Charles's law, volume and temperature are directly proportional. The higher the temperature of the gas sample, the higher the volume. Pressure and volume, on the other hand, are inversely proportional according to Boyle's law. The higher the pressure, the lower the volume.

Solve

Since line 1 is always below line 2 on the plot (volume at any given temperature is smaller for the line 1 gas), line 1 represents the gas that is at a higher pressure. The *x*-axis is not on the absolute temperature scale. If it were, line 1 and line 2 would meet at $T = 0$ K.

Think About It

A plot of pressure versus temperature would also show that at 0 K the pressure is also zero.

6.9. **Collect and Organize**

The plot of volume versus temperature in Figure P6.9 contains two lines: one with a positive slope as temperature increases, the other with a negative slope as temperature increases. Which line is inconsistent with the ideal gas law?

Analyze

The ideal gas law shows the direct proportionality between *V* and *T* as

$$PV = nRT$$

$$V = \frac{nRT}{P} \text{ or } V = cT$$

where $c = nR/P$.

Solve

According to the ideal gas law, as temperature increases volume also increases. This is shown by line 1 in Figure P6.9. Line 2, therefore, is inconsistent with the ideal gas law.

Think About It

The same plot is valid for pressure as a function of temperature since

$$P = \frac{nRT}{V} = cT$$

where $c = nR/V$.

6.11. **Collect and Organize**

For Problem 6.10, the lines in Figure P6.10 were assigned as 1 = methane and 2 = nitrogen based on the molar mass of the gases. We are asked to add lines for He and NO to the plot.

Analyze

Density is related to molar mass ($\mathcal{M}$) through the equation

$$d = \frac{\mathcal{M}P}{RT}$$

As the molar mass increases at a particular P, the density increases.

Solve

The molar mass of helium is 4 g/mol. The line on the graph for this gas goes below line 1. The molar mass of NO is 30 g/mol so the line on the graph for this gas goes above line 2. All lines will converge at $P = 0$ and $d = 0$.

Think About It

The d versus P line for helium will lie well below the line for methane. Its molar mass is significantly less than methane's, and the slope of the line of d versus P will be small.

6.13. Collect and Organize

A mixture of gases is a homogeneous system of one or more gases. We are asked to choose the drawing in Figure P6.13 that best represents this definition.

Analyze

In order for the mixture to be homogeneous, the two gases, helium and neon for example, must be unreactive with each other and must be uniformly and randomly dispersed throughout the container.

Solve

The correct answer is drawing (b) because it shows He and Ne randomly dispersed as a mixture of gases. Drawing (a) shows one component in the gas phase and the other in a condensed phase (liquid or solid). Drawing (c) shows the two components as gases, but they are not uniformly dispersed in the container. Drawing (d) shows the product of a reaction (HeNe) in which the atoms have bonded to each other. This is not a mixture—it is a sample of pure HeNe gas.

Think About It

Gases form uniform mixtures because the gas particles are in constant, random motion, leading to uniform distribution of the atoms or molecules of each gas throughout the sample.

6.15. Collect and Organize

For CO_2 and SO_2 the number of molecules versus molecular speed is plotted in Figure P6.15. We are asked to identify which curve belongs to CO_2 and which belongs to SO_2 and then to determine which curve would also describe the distribution of molecular speeds for propane (C_3H_8).

Analyze

The formula that describes molecular speed (u_{rms}) is

$$u_{rms} = \sqrt{\frac{3RT}{\mathcal{M}}}$$

where R is the universal gas constant, T is the absolute temperature in Kelvin, and $\mathcal{M}$ is the molar mass of the gaseous compound. The larger the molar mass, the slower the u_{rms}. Curve 1 shows more molecules at lower speeds; curve 2 shows more molecules at higher speeds.

Solve

The molar masses of CO_2, SO_2, and C_3H_8 are 44, 64, and 44 g/mol, respectively. Because SO_2 has a higher molar mass than either CO_2 or C_3H_8, it is represented by curve 1. Curve 2 represents both CO_2 and C_3H_8 because they have the same molar mass.

Think About It

Notice that for heavier molecules, the distribution of speeds is narrower than for lighter molecules.

6.17. **Collect and Organize**

The highlighted elements in Figure P6.17 are typically found as gaseous elements. We are to determine which would leak slowest from a container with a pinhole.

Analyze

The rate at which a gas diffuses compared to another gas is given by the equation

$$\frac{r_x}{r_y} = \sqrt{\frac{\mathcal{M}_y}{\mathcal{M}_x}}$$

where r is the rate at which the gas diffuses and $\mathcal{M}$ is the molar mass of the gas. From this, we see that rate and the molar mass of the gas can be treated qualitatively as inversely proportional.

Solve

The larger the molar mass, the slower the rate of effusion. Therefore, the element with the largest molar mass diffuses the slowest. Our choices are He, Ne, Ar, F_2, Cl_2, Br_2, and O_2. The species with the highest molar mass is Br_2 (orange), so it will leak from the container the slowest.

Think About It

The gas that leaks the fastest would be the one with the lowest molar mass, namely, He (lavender).

6.19. **Collect and Organize**

Which of the graphs in Figure P6.19 best describes the relationship between the ratio of rates of effusion and the temperature?

Analyze

The equation describing the ratio of rates of effusion is

$$\frac{r_x}{r_y} = \sqrt{\frac{\mathcal{M}_y}{\mathcal{M}_x}}$$

Since both gases are at the same temperature for a given ratio, the temperature does not appear in this equation.

Solve

The temperature does not appear in the equation, so the ratio of rates of effusion is independent of temperature. The graph that best represents this is (c).

Think About It

(a) and (b) represent an increase and decrease in the rate of effusion of x with respect to y as temperature increases, respectively.

6.21. **Collect and Organize**

Effusion is the leaking of gas through a small hole in a container. We are asked which change in Figure P6.21 shows effusion of helium.

Analyze

The balloon contains fewer He atoms after effusion has taken place and, because of the presence of fewer He atoms in the balloon, the pressure inside the balloon decreases, and the balloon shrinks.

Solve

Both (a) and (b) show fewer atoms of He after effusion; however, outcome (a) is the correct choice, because the balloon in (a) shrinks, but the balloon in (b) expands.

Think About It

We witness this happening to helium balloons in our everyday experience: a helium balloon made from latex usually shows noticeable shrinkage the next day.

6.23. **Collect and Organize**

Force and pressure are often used interchangeably in physics describing speed, but what is their difference in meaning in the context of gases?

Analyze

The mathematical equations describing force and pressure are

$$F = ma$$
$$P = F/A$$

Solve

Force is the product of the mass of an object and the acceleration due to gravity. Pressure uses force in its definition: it is the force an object exerts over a given area.

Think About It

A large force over a small area results in high pressure, but that same large force distributed over a very large area results in low pressure.

6.25. **Collect and Organize / Analyze**

We are to define and relate two units of pressure, torr and atmospheres.

Solve

An atmosphere is defined as the pressure that supports a column of 760 mm of mercury in a barometer. The pressure in 1 mmHg is also a torr, so 760 torr = 1 atmosphere.

Think About It

An atmosphere would be a good unit to express fairly large pressures, but mmHg or torr is a better unit to use for low pressures.

6.27. **Collect and Organize**

Differences in density of the liquid determine how high the liquid climbs in the tube of a Torricelli barometer. We are asked which liquid (water, ethanol, or mercury) would have the tallest column in the barometer.

Analyze

Because the force of gravity on the liquid in the column of a barometer opposes the atmospheric pressure forcing the liquid up the column, the less dense the liquid, the less the liquid is affected by gravity, so atmospheric pressure raises the liquid higher in the column.

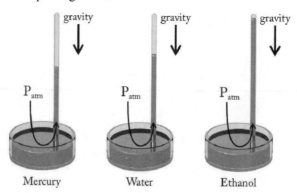

Solve

Between ethanol, water, and mercury liquids, ethanol has the lowest density, so it has the highest column of liquid in a barometer.

Think About It

Mercury, having the highest density of the liquids, has the shortest column in a barometer.

6.29. **Collect and Organize**

We are asked to explain the difference in pressure that an ice skater places on ice with a dull versus a sharp skate blade.

Analyze

Pressure is force over a unit area. The force of the ice skater on the ice is due to the skater's mass. The mass stays constant for this comparison. What must be different is the area between the skate blade and the ice.

Solve

A sharpened blade has a smaller area over which the force is distributed compared to a dull blade. Because $P = F/A$, as area (A) decreases, pressure must increase since the force (skater's mass) is constant.

Think About It

The pressure also increases if we increase the mass of the ice skater.

6.31. **Collect and Organize**

We are asked to consider why atmospheric pressure changes with altitude.

Analyze

The pressure we feel from the atmosphere is due to the overlying mass of the gases above us.

Solve

As we go up in altitude, the overlying mass of the atmosphere above us decreases, so the pressure also decreases.

Think About It

Airplanes must pressurize their cabins for passenger comfort because they fly at very high altitude where the air is "thin."

6.33. **Collect and Organize**

The pressure due to gravity for a 1.00 kg cube of iron, 5.00 cm on a side, is given by

$$P = \frac{F}{A} = \frac{ma}{A}$$

where m = mass of the iron cube, a = acceleration due to gravity, and A = area over which the force is distributed.

Analyze

We have to be sure to use consistent units in this calculation. The acceleration due to gravity is 9.8 m/s^2, so we want to express the area of the cube face in square meters.

Solve

$$P = \frac{1.00 \text{ kg} \times 9.8 \text{ m/s}^2}{(5.00 \text{ cm})^2 \times \left(\dfrac{1 \text{ m}}{100 \text{ cm}}\right)^2} = 3.9 \times 10^3 \ \frac{\text{kg}}{\text{m} \cdot \text{s}^2} = 3.9 \times 10^3 \text{ Pa}$$

Think About It

Be sure to take the square of the conversion factor for area in the denominator in problems like these.

6.35. **Collect and Organize**

We are to convert pressures expressed in units of kilopascals and millimeters of mercury to atmospheres.

Analyze

We need the following conversion factors:

$$1 \text{ atm} = 101.325 \text{ kPa} = 101,325 \text{ Pa}$$

$$1 \text{ atm} = 760 \text{ mmHg}$$

Solve

(a) $2.0 \text{ kPa} \times \dfrac{1 \text{ atm}}{101.325 \text{ kPa}} = 0.020 \text{ atm}$

(b) $562 \text{ mmHg} \times \dfrac{1 \text{ atm}}{760 \text{ mmHg}} = 0.739 \text{ atm}$

Think About It

A pascal also is equivalent to 1 N/m^2. This makes sense because the newton (N) is a unit of force. That force divided by area (in m^2) gives the units of pressure.

6.37. **Collect and Organize**

We are to express the highest recorded atmospheric pressure of 108.6 kPa in millimeters of mercury, atmospheres, and millibars.

Analyze

We need the following conversion factors:

$$1 \text{ atm} = 101.325 \text{ kPa} = 101,325 \text{ Pa}$$

$$1 \text{ atm} = 760 \text{ mmHg}$$

$$1 \text{ kPa} = 10 \text{ mbar}$$

Solve

(a) $108.6 \text{ kPa} \times \dfrac{1 \text{ atm}}{101.325 \text{ kPa}} \times \dfrac{760 \text{ mmHg}}{1 \text{ atm}} = 814.6 \text{ mmHg}$

(b) $108.6 \text{ kPa} \times \dfrac{1 \text{ atm}}{101.325 \text{ kPa}} = 1.072 \text{ atm}$

(c) $108.6 \text{ kPa} \times \dfrac{10 \text{ mbar}}{1 \text{ kPa}} = 1086 \text{ mbar}$

Think About It

The highest recorded atmospheric pressure is just a little above 1 atm at 1.072 atm.

6.39. **Collect and Organize**

We are asked to calculate the atmospheric pressure on Venus, given the mass of the atmosphere, the acceleratiuon due to gravity, and the surface area of Venus. The atmospheric pressure may be calculated using the equation

$$P = \frac{F}{A} = \frac{ma}{A}$$

where m is the mass of the atmosphere, a is the acceleration due to gravity, and A is the surface area of the planet. We are also asked to explain the difference between the atmospheres of Venus and Earth, given that the planets are roughly the same size and distance from the Sun.

Analyze

Substituting the data provided into the equation above will yield the atmospheric pressure in units of $\text{Kg} \cdot \text{m}^{-1} \text{s}^{-2}$, or Pa. It is more convenient to express this in atmospheres using the conversion factor

$$1 \text{ atm} = 101.325 \text{ kPa} = 101,325 \text{ Pa}$$

Solve

(a)

$$P = \frac{ma}{A} = \frac{4.8 \times 10^{20} \text{ kg} \left(8.87 \text{ m/s}^2\right)}{4.6 \times 10^{14} \text{ m}^2} = 9.3 \times 10^6 \text{ Pa}$$

$$9.3 \times 10^6 \text{ Pa} \times \frac{1 \text{ atm}}{101,325 \text{ Pa}} = 91 \text{ atm}$$

(b) The atmosphere on Venus has a much higher proportion of CO_2 ($\mathcal{M}$ = 44.01 g/mol), and a much lower portion of N_2 ($\mathcal{M}$ = 24.02 g/mol) than Earth. The Venusian atmosphere is much heavier than that of Earth, accounting for the higher atmospheric pressure.

Think About It

The mass of the atmosphere may have a large impact on atmospheric pressure, as we see on Earth at different elevations.

6.41. **Collect and Organize**

Amontons's law states that pressure and temperature are directly related: as temperature increases, so does pressure. We are asked to interpret this from the molecular perspective.

Analyze

Pressure originates from the collision of gas molecules with the walls of a container. The more collisions between gas molecules and the container, the greater the pressure. Temperature is a measure of how fast molecules are moving in the gas.

Solve

The higher the temperature, the faster the gas molecules move. The faster they move, the more often they collide with the walls of the container and the greater the force with which gas molecules hit the walls. Both of these result in increased pressure as temperature is raised.

Think About It

At absolute zero, there would be no molecular motion and thus no pressure.

6.43. **Collect and Organize**

A balloonist needs to decrease her rate of ascent. How should she change the temperature of the gas in the balloon?

Analyze

A balloon rises because the air inside the balloon is less dense than the air outside. This means that there are fewer air molecules inside the balloon than in an identical volume outside the balloon. Warmer gas molecules move faster and with more force, which results in a higher pressure allowing them to escape from the bottom of the balloon.

Solve

To allow more air molecules into the balloon so as to increase the mass and therefore the density of air inside the balloon, the balloonist should decrease the temperature. Fewer "hot" gas molecules will escape, and the open design of the balloon will allow an increase in the amount of cooler gas molecules that enter. This will slow down the ascent of the balloon.

Think About It

You see this effect if you ever have a chance to ride in a hot-air balloon. To avoid going too high in altitude, the balloonist will stop firing the burners.

6.45. **Collect and Organize**

A sample of gas is simultaneously cooled and compressed. We are asked to predict how the pressure changes for this sample of gas.

Analyze

Pressure is directly related to the temperature so decreasing the temperature decreases the pressure. When the volume is decreased, however, the pressure increases because pressure and volume are inversely proportional.

Solve

These two changes are in opposition to each other so we must look at their relative magnitude. A temperature drop of 10°C is 10 K on the absolute temperature scale. Starting at 20°C (293 K), this represents only a 10 K/293 K × 100 = 3.4% decrease in both temperature and pressure. Decreasing the volume by a factor of 2 means that the pressure increases by a factor of 2. In terms of percent change, pressure increases 100%. The pressure change due to the volume decrease more than offsets the effect of the decrease in temperature. Therefore, the pressure increases.

Think About It

Only a very large change in temperature would offset the effect due to the compression of the gas.

6.47. **Collect and Organize**

Using Boyle's law, find the pressure of the ammonia gas when 1.00 mol at 1.00 atm is subjected to the changes in volume in each part. The equation for Boyle's law for the compression (or expansion) of a gas is
$$P_1V_1 = P_2V_2$$

Analyze

(a) and (c) represent a decrease in the volume of the gas, so the pressure should increase, whereas (b) is an increase in the volume, so a decrease in pressure is expected. Assuming an initial volume of 1.00 L for (c), a 40% decrease results in a final volume of 0.60 L. The number of moles and temperature of ammonia gas are constant, so they can be ignored.

Solve

(a) $P_2 = \dfrac{P_1V_1}{V_2} = \dfrac{1.00\ \text{atm}\,(78.0\ \text{mL})}{39.0\ \text{mL}} = 2.00\ \text{atm}$

(b) $P_2 = \dfrac{P_1V_1}{V_2} = \dfrac{1.00\ \text{atm}\,(43.5\ \text{mL})}{65.5\ \text{mL}} = 0.664\ \text{atm}$

(c) $P_2 = \dfrac{P_1V_1}{V_2} = \dfrac{1.00\ \text{atm}\,(1.00\ \text{L})}{0.60\ \text{L}} = 1.67\ \text{atm}$

Think About It

Because the volume was halved in (a), the pressure increased by a factor of two.

6.49. **Collect and Organize**

We are asked to find the pressure of an underwater site, knowing how much expansion a balloon undergoes when rising from the site to the surface. From that pressure and the fact that for every 10 m change in depth the pressure increases by 1.0 atm, we can calculate the diver's depth.

Analyze

We can use Boyle's law ($P_1V_1 = P_2V_2$) to find P_1. Once we know the pressure, we can multiply it by 10 m/1 atm to arrive at the depth of the diver.

Underwater
$V_1 = 153$ L
$P_1 = ?$

At the surface
$V_2 = 352$ L
$P_2 = 1.00$ atm

Solve

The pressure and depth at the site of the diver's work are as follows:

$$P_1 \times 153 \text{ L} = 1.00 \text{ atm} \times 352 \text{ L}$$

$$P_1 = 2.30 \text{ atm}$$

The pressure includes atmospheric pressure (1.00 atm) so

$$P_{additional} = 1.30 \text{ atm}$$

$$1.30 \text{ atm} \times \frac{10 \text{ m}}{1.0 \text{ atm}} = 13.0 \text{ m}$$

Think About It

If the diver had been working deeper, the balloon would have undergone a greater expansion.

6.51. **Collect and Organize**

From a plot of V (y-axis) versus $1/P$ (x-axis) for 1 mole of $H_2(g)$ at 298 K, that we are to construct from the P and V data given, we are asked whether the same plot would be applicable to the same number of moles of Ar(g).

Analyze

Use Excel to plot the data for $H_2(g)$. The relationship describes Boyle's law in the form

$$P^{-1} \propto \frac{1}{V} \text{ or } PV = \text{constant}$$

Solve

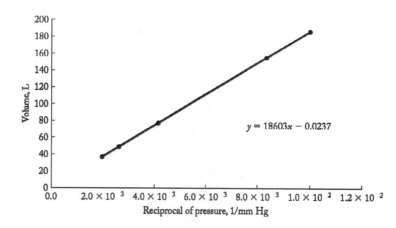

The identity of the gas does not matter in the behavior associated with Boyle's law. Therefore, the graph is exactly the same for the same number of moles of argon gas.

Think About It

The data gave a straight line because PV = constant, or, expressed another way, V = constant/P.

6.53. **Collect and Organize**

We are to plot volume (y-axis) as a function of temperature (x-axis) for 1.0 mol of He(g) at 1.00 atm. We are also asked how the graph would change if the amount of gas were halved.

Analyze

Use Excel to plot the given data. This relationship describes Charles's law in the form

$$V \propto T \text{ or } \frac{V}{T} = \text{constant or } V = T \times \text{constant}$$

Solve

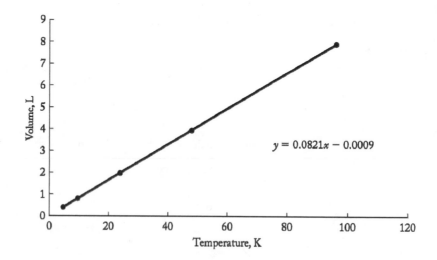

If the amount of gas were halved, the graph would look similar—the plot would still be linear. The plot would be different for a lower number of moles of gas in that at a given temperature the volume would be lower (at constant pressure)—the slope of the line would be halved.

Think About It

The graph does indeed show the direct linear relationship between the temperature and volume. We see clearly on the graph that as temperature increases the volume increases.

6.55. **Collect and Organize**

Using Charles's law, find the temperature at which the volume of a cylinder (with a moveable piston; Figure P6.55) initially at 25°C is doubled.

Analyze

Charles's law for a change in volume with a change in temperature at constant n and P is

$$\frac{V_1}{T_1} = \frac{V_2}{T_2}$$

We are not given the volume of the cylinder, but because the V–T relationship is linear, we can choose any convenient initial volume. We also have to use temperatures on the Kelvin scale (25°C = 298 K).

Solve

Assuming $V_1 = 1.00$ L, Charles's law would give the final temperature of the gas in which the volume (V_2) is 2.00 L as follows:

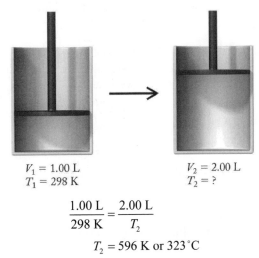

$$V_1 = 1.00 \text{ L} \qquad V_2 = 2.00 \text{ L}$$
$$T_1 = 298 \text{ K} \qquad T_2 = ?$$

$$\frac{1.00 \text{ L}}{298 \text{ K}} = \frac{2.00 \text{ L}}{T_2}$$

$$T_2 = 596 \text{ K or } 323\,°C$$

Think About It

In order for the volume to double, the absolute temperature would also have to double.

6.57. **Collect and Organize**

Determine the change in volume for a gas initially occupying 2.68 L when subjected to the changes described.

Analyze

(a) is a change in temperature, so Charles's Law applies:

$$\frac{V_1}{T_1} = \frac{V_2}{T_2}$$

Since the temperature is increasing, we expect the volume to increase as well.

(b) is a change in pressure, so Boyle's Law applies:

$$P_1 V_1 = P_2 V_2$$

Assuming an initial pressure of 1.00 atm, an increase of 33% corresponds to a final pressure of 1.33 atm. Since the pressure is increasing, we expect the volume to decrease.

(c) is a change in the number of moles of gas, so Avogadro's Law applies:

$$\frac{V_1}{n_1} = \frac{V_2}{n_2}$$

Assuming 1.00 moles of gas initially, a leak of 10% corresponds to 0.9 moles after the leak. Since the number of moles of gas is decreasing, we expect the volume to decrease.

Solve

(a) $V_2 = \dfrac{V_1 T_2}{T_1} = \dfrac{2.68 \text{ L} (398 \text{ K})}{250 \text{ K}} = 4.27 \text{ L}$

(b) $V_2 = \dfrac{P_1 V_1}{P_2} = \dfrac{1.00 \text{ atm} (2.68 \text{ L})}{1.33 \text{ atm}} = 2.02 \text{ L}$

(c) $V_2 = \dfrac{V_1 n_2}{n_1} = \dfrac{2.68 \text{ L} (0.90 \text{ mol})}{1.00 \text{ mol}} = 2.41 \text{ L}$

Think About It

Our predictions were correct: the gas expanded in (a), and contracted in (b) and (c).

6.59. Collect and Organize

We are asked to calculate the amount of work done to expand a balloon containing 1.75 moles of He at 17 °C and a pressure of 1.00 atm. The equation describing pressure–volume work is

$$w = -P\Delta V$$

Analyze

In order to calculate the work done, we must determine the change in volume using the ideal gas law. Assuming an initial volume of zero, the change in volume is

$$V = \frac{nRT}{P}$$

We must be careful to express the temperature in K:

$$T = 17 \text{ °C} + 273.15 \text{ K} = 290 \text{ K}$$

The conversion from L·atm to J is 101.325 J/(L·atm).

Solve

The volume change is

$$V = \frac{(1.75 \text{ mol})[0.08206 \text{ L} \cdot \text{atm} / (\text{mol} \cdot \text{K})](290 \text{ K})}{1.00 \text{ atm}} = 41.65 \text{ L}$$

$$\Delta V = 41.65 \text{ L} - 0 \text{ L} = 41.65 \text{ L}$$

The work done to expand the balloon is

$$w = -(1.00 \text{ atm})(41.65 \text{ L}) = -41.65 \text{ L} \cdot \text{atm} \times \frac{101.325 \text{ J}}{1 \text{ L} \cdot \text{atm}} \times \frac{1 \text{ kJ}}{1000 \text{ J}} = -4.22 \text{ kJ}$$

Think About It

Work must be done on the system to inflate the balloon, as denoted by the negative sign for the work.

6.61. Collect and Organize

We compare the effect of decreasing pressure on a gas sample (from 760 to 720 mmHg) to the effect of raising the temperature (from 10°C to 40°C).

Analyze

Boyle's law describes the effect of decreasing pressure on the volume where P/V = constant. If we decrease the pressure by 50%, the volume must increase by 50%. Charles's law shows the direct relationship between temperature and volume. Here $\%\Delta V = \%\Delta T$.

Solve

(a) Lowering the pressure from 760 mmHg to 720 mmHg gives a percent change in P of

$$\frac{720 \text{ mmHg} - 760 \text{ mmHg}}{760 \text{ mmHg}} \times 100 = -5.3\% \text{ change in pressure}$$

This will give a –5.3% change in volume.

(b) The percent change in temperature is

$$\frac{313 \text{ K} - 283 \text{ K}}{283 \text{ K}} \times 100 = +10.6\%$$

This will give a +10.6% increase in volume.

Therefore, (b), an increase in the temperature from 10°C to 40°C, gives the greatest increase in volume.

Think About It

Remember to convert temperatures to the Kelvin scale. If we had not converted to the Kelvin scale, it would have appeared that there was a 75% increase in volume in (b).

6.63. **Collect and Organize**

For various changes in temperature and external pressure, we are to predict how the volume of a gas sample changes.

Analyze

We must keep in mind the direct proportionality between V and T as shown by Charles's law and the inverse proportionality between V and P as shown by Boyle's law.

Solve

(a) When the absolute temperature doubles, the volume doubles. When the pressure is doubled, the volume is halved. Combining these yields zero change in the volume of the gas.

(b) When the absolute temperature is halved, the volume is halved. When the pressure is doubled, the volume is halved. Combining these results in a decrease in the volume of the gas sample to $\frac{1}{4}$ the original volume.

(c) Combining these effects is best looked at mathematically:

$$V_2 = \frac{P_1 V_1 \times 1.75 T_1}{T_1 \times 1.50 P_1} = 1.17 V_1 \text{ or an increase of } 17\%$$

Think About It

Notice the way to simplify this problem is to consider each change separately and then "add" the effects.

6.65. **Collect and Organize**

We are to calculate the volume of a weather balloon after 24 hours, during which helium leaks out.

Analyze

Because volume and the amount of gas are directly proportional, the appropriate form of Avogadro's law is

$$\frac{V_1}{n_1} = \frac{V_2}{n_2}$$

We are given $V_1 = 150.0$ L and $n_1 = 6.1$ mol of He and told that the helium leaks from the balloon at the rate of 10 mmol/hr for 24 hr.

Solve

The amount of gas leaked from the balloon is

$$24 \text{ hr} \times \frac{10 \text{ mmol}}{1 \text{ hr}} = 240 \text{ mmol} = 0.24 \text{ mol}$$

The amount of He in the balloon after 24 hr is

$$6.1 \text{ mol} - 0.24 \text{ mol} = 5.86 \text{ mol}$$

The volume (V_2) of the balloon is

$$\frac{150.0 \text{ L}}{6.1 \text{ mol}} = \frac{V_2}{5.86 \text{ mol}}$$
$$V_2 = 144 \text{ L}$$

Think About It

As expected, fewer moles of gas in the balloon take up less volume.

6.67. Collect and Organize

When the bicycle tire cools, we would expect the pressure to decrease according to Amontons's law.

Analyze

Amontons's law states that pressure and temperature are directly proportional:

$$\frac{P_1}{T_1} = \frac{P_2}{T_2}$$

We are given $P_1 = 7.1$ atm. We have to be sure to express $T_1 = 27°C$ and $T_2 = 5.0°C$ as absolute temperatures.

Solve

$$\frac{7.1 \text{ atm}}{300 \text{ K}} = \frac{P_2}{278.2 \text{ K}}$$
$$P_2 = 6.6 \text{ atm}$$

Think About It

As expected, the pressure of air in the tire did decrease when the temperature decreased.

6.69. Collect and Organize

Define standard temperature and pressure conditions and define the volume of 1 mole of an ideal gas at those conditions.

Analyze

We can use the ideal gas law $PV = nRT$ to calculate the volume (V) of 1 mole of gas (n) at standard temperature (T) and pressure (P).

Solve

STP is defined as 1 atm and 0°C (273 K).
The volume of 1 mole of gas at STP is

$$1 \text{ atm} \times V = 1.00 \text{ mol} \times 0.08206 \, \frac{\text{L} \cdot \text{atm}}{\text{mol} \cdot \text{K}} \times 273 \text{ K}$$
$$V = 22.4 \text{ L}$$

Think About It

The molar volume of gas at STP does not vary much depending on the identity of the gas. For this reason, we may assume that the molar volume of an ideal gas at STP is a constant, 22.41 L/mol.

6.71. Collect and Organize

In a graph of P versus $1/V$, what is the meaning of the slope?

Analyze

Let's look at the ideal gas equation and rearrange it into the form of $y = mx + b$ (a linear plot).

$$PV = nRT$$

$$P = \frac{nRT}{V} = nRT \times \frac{1}{V}$$

Solve

In the equation

$$P = nRT \times \frac{1}{V}$$

$y = P$, $x = 1/V$, and $m = nRT$. Therefore, the slope of the line represents the product of the number of moles of gas in the sample, the temperature, and the gas constant.

Think About It

If we know T and R for a given experiment and plot P versus $1/V$, we can determine the moles of gas in the sample by

$$\text{slope} = nRT \quad \text{so } n = \frac{\text{slope}}{RT}$$

6.73. Collect and Organize

Use the ideal gas law to determine the moles of air present in a bicycle tire with a volume of 2.36 L at 6.8 atm and 17.0°C.

Analyze

The ideal gas law is

$$PV = nRT$$

Rearranging to solve for n, number of moles of air:

$$n = \frac{PV}{RT}$$

With R in units of L·atm/(mol·K), V must be in liters, pressure in atmospheres, and temperature in Kelvin.

Solve

$$n = \frac{6.8 \text{ atm} \times 2.36 \text{ L}}{[0.08206 \text{ L} \cdot \text{atm}/(\text{mol} \cdot \text{K})](290 \text{ K})} = 0.67 \text{ mol}$$

Think About It

Notice that if we were to pump more moles of gas into the bicycle tire, the pressure would increase.

6.75. Collect and Organize

Given the volume of a hyperbaric chamber, the mass of oxygen in the chamber, and the temperature in the chamber, we can use the ideal gas law to calculate the pressure inside the chamber.

Analyze

Rearranging the ideal gas equation to solve for pressure, we get

$$P = \frac{nRT}{V}$$

We first have to determine the moles (n) of O_2 present in 4635 g of O_2.

Solve

$$\text{Moles } (n) \text{ } O_2 \text{ present} = 4635 \text{ g } O_2 \times \frac{1 \text{ mol } O_2}{32.00 \text{ g}} = 144.8 \text{ mol}$$

The pressure in the chamber is

$$P = \frac{(144.8 \text{ mol})[0.08206 \text{ L} \cdot \text{atm}/(\text{mol} \cdot \text{K})](298 \text{ K})}{2.36 \times 10^3 \text{ L}} = 1.50 \text{ atm}$$

Think About It

Hyper comes from the Greek meaning *over*. Thus, a hyperbaric chamber will have an *overpressure* of oxygen. Our answer therefore makes sense.

6.77. **Collect and Organize**

A weather balloon is filled and rises into the atmosphere. As it rises, temperature decreases (contracting the balloon) and pressure decreases (inflating the balloon). We are asked to calculate the volume of the balloon at 20,000 m and 50,000 m, and whether the balloon is expected to rupture at either altitude.

Analyze

We can use the combined ideal gas law:

$$\frac{P_1 V_1}{n_1 T_1} = \frac{P_2 V_2}{n_2 T_2}$$

Because $n_1 = n_2$, the equation reduces, and may be rearranged to

$$V_2 = \frac{P_1 V_1 T_2}{T_1 P_2}$$

We must be consistent in our units of P and T to use this equation. Here $P_1 = 1.00$ atm (760 mmHg), $T_1 = 20°C$ (293 K), $V_1 = 200.0$ L. For each part
(a) $P_2 = 63$ mmHg, $T_2 = 210$ K.
(b) $P_2 = 0.80$ mmHg, $T_2 = 270$ K.
If the volume at either elevation exceeds the maximum of 400.0 L, the balloon will rupture.

Solve

(a) $V_2 = \dfrac{760 \text{ mmHg} (200.0 \text{ L})(210 \text{ K})}{293 \text{ K} (63 \text{ mmHg})} = 1.8 \times 10^3 \text{ L}$

(b) $V_2 = \dfrac{760 \text{ mmHg} (200.0 \text{ L})(270 \text{ K})}{293 \text{ K} (0.80 \text{ mmHg})} = 1.8 \times 10^5 \text{ L}$

(c) The balloon will rupture at both altitudes.

Think About It

The effect of lowered pressure inflating the balloon was greater than the effect of lowered temperature contracting the balloon.

6.79. **Collect and Organize**

We are given the balanced chemical equation for the conversion of CaC_2 to C_2H_2. We are asked to calculate the moles of C_2H_2 required to light the lamp for 1 hour, then calculate the mass of CaC_2 to provide 4 hours of light.

Analyze

The ideal gas law can be used to calculate $n_{C_2H_2}$ where $P = 1.00$ atm, $V = 1.00$ L/hr, $T = 18°C$ (291 K). Once we know how much C_2H_2 is required for 1 hr, we can multiply it by 4 to arrive at the moles of C_2H_2 required for 4 hr. Because 1 mole of C_2H_2 is produced from 1 mole of CaC_2, this is also the moles of CaC_2 required. The mass of CaC_2 can then be found by multiplying n_{CaC_2} by the molar mass of CaC_2.

Solve

(a) Moles of C_2H_2 used per hour:

$$n = \frac{1.00 \text{ atm}(1.00 \text{ L/hr})}{[0.08206 \text{ L} \cdot \text{atm}/(\text{mol} \cdot \text{K})](291 \text{ K})} = 0.0419 \text{ mol/hr}$$

(b) For a 4-hour shift

$$4 \text{ hr} \times \frac{0.0419 \text{ mol C}_2\text{H}_2}{\text{hr}} \times \frac{1 \text{ mol CaC}_2}{1 \text{ mol C}_2\text{H}_2} \times \frac{64.10 \text{ g CaC}_2}{1 \text{ mol}} = 10.7 \text{ g CaC}_2 \text{ needed}$$

Think About It

Because the ideal gas law allows us to compute the moles of gas, we can use the result in subsequent stoichiometry calculations.

6.81. **Collect and Organize**

200.0 L of "pure air" (a mixture of 80% nitrogen and 20% oxygen by volume) is at a temperature of 273 K and a pressure of 0.85 atm. Assuming the nitrogen is prepared by the decomposition of ammonium dichromate, and the oxygen is prepared by the decomposition of potassium chlorate, what mass of each reactant is required to prepare this mixture?

Analyze

We may calculate the volume of N_2 and O_2 using the percentages given. In each case, we now have a volume of gas, as well as the temperature and pressure of the sample. Substituting these values and the ideal gas constant into the ideal gas law will provide the molar quantity of O_2 and N_2 required for the mixture

$$n = \frac{PV}{RT}$$

Multiplying by the stoichiometric ratios from the balanced chemical equations then multiplying by the molar mass of ammonium dichromate (252.09 g/mol) or potassium chlorate (122.55 g/mol) yields the requisite masses.

Solve

The volume of each gas is

$$V_{N_2} = 0.80(200.0 \text{ L}) = 160.0 \text{ L}$$
$$V_{O_2} = 0.20(200.0 \text{ L}) = 40.0 \text{ L}$$

The mass of ammonium dichromate is

$$n_{N_2} = \frac{0.85 \text{ atm}(160.0 \text{ L})}{[0.08206 \text{ L}\cdot\text{atm}/(\text{mol}\cdot\text{K})](273 \text{ K})} = 6.07 \text{ mol N}_2$$

$$6.07 \text{ mol N}_2 \times \frac{1 \text{ mol (NH}_4)_2\text{Cr}_2\text{O}_7}{1 \text{ mol N}_2} \times \frac{252.09 \text{ g}}{1 \text{ mol}} = 1.5\times10^3 \text{ g (NH}_4)_2\text{Cr}_2\text{O}_7$$

The mass of potassium chlorate is

$$n_{O_2} = \frac{0.85 \text{ atm}(40.0 \text{ L})}{[0.08206 \text{ L}\cdot\text{atm}/(\text{mol}\cdot\text{K})](273 \text{ K})} = 1.52 \text{ mol O}_2$$

$$1.52 \text{ mol O}_2 \times \frac{2 \text{ mol KClO}_3}{3 \text{ mol O}_2} \times \frac{122.55 \text{ g}}{1 \text{ mol}} = 1.2\times10^2 \text{ g KClO}_3$$

Think About It

Over ten times as much ammonium dichromate as potassium chlorate is required to generate this artificial air, as expected from the percentage of each gas present by volume.

6.83. **Collect and Organize**

We are given the balanced chemical reaction for changing CO_2 into O_2 using Na_2O_2. We are asked to calculate the amount of Na_2O_2 needed on the submarine per sailor in 24 hours, given the rate of CO_2 exhaled by each sailor.

Analyze

Use the ideal gas equation to calculate moles of CO_2 exhaled by a sailor in one minute, where $P = 1.02$ atm, $V = 150.0$ mL (0.1500 L), and $T = 20°C$ (293 K). Then, convert this into n_{CO_2} exhaled in 1 day using 60 min = 1 hr and 24 hr = 1 d. Finally, use the stoichiometric equation to calculate the mass of Na_2O_2 required for each sailor (molar mass of Na_2O_2 = 77.98 g/mol).

Solve

Moles of CO_2 exhaled in 1 minute:

$$n_{CO_2} = \frac{1.02 \text{ atm} \times 0.1500 \text{ L}}{[0.08206 \text{ L} \cdot \text{atm/(mol} \cdot \text{K)}](293 \text{ K})} = 6.36 \times 10^{-3} \text{ mol } CO_2$$

Moles of CO_2 exhaled in 1 day:

$$\frac{6.36 \times 10^{-3} \text{ mol } CO_2}{1 \text{ min}} \times \frac{60 \text{ min}}{1 \text{ hr}} \times \frac{24 \text{ hr}}{1 \text{ d}} = 9.16 \text{ mol } CO_2 \text{ / d}$$

Mass of Na_2O_2 required:

$$9.16 \text{ mol } CO_2 \times \frac{2 \text{ mol } Na_2O_2}{2 \text{ mol } CO_2} \times \frac{77.98 \text{ g } Na_2O_2}{1 \text{ mol}} = 715 \text{ g } Na_2O_2$$

Think About It

Because the ideal gas law allows us to compute the moles of gas, we can use the result in subsequent stoichiometry calculations.

6.85. Collect and Organize

The density of a gas is defined as its mass per unit volume. If gases are at the same T and P, will they have the same density?

Analyze

The density of gas is derived from the ideal gas equation,

$$PV = nRT$$

where "molar density" would be given by

$$\frac{n}{V} = \frac{P}{RT}$$

The moles of gas, n, are related to its mass by

$$n = \frac{m}{\mathcal{M}}$$

where $\mathcal{M}$ = molar mass.

Solve

Substituting $n = m / \mathcal{M}$ into the molar density equation and arranging to give $m/V = d$ gives

$$\frac{m/\mathcal{M}}{V} = \frac{P}{RT}$$

$$\frac{m}{V} = \frac{P\mathcal{M}}{RT} = d$$

Because each gas has a different molar mass, the densities of different gases are not necessarily the same for a particular temperature and pressure.

Think About It

From the equation above for density, we can see that as the molar mass of the gas increases, the density increases.

6.87. Collect and Organize

Density changes as a function of pressure and temperature. We are to predict how in this question.

Analyze

The equation describing the density of gas is

$$d = \frac{\mathcal{M}P}{RT}$$

Solve

From the equation we see that density is directly proportional to pressure and inversely proportional to temperature. This means that density (a) increases with increasing pressure and (b) increases with decreasing temperature.

Think About It

Notice from the equation that density also increases as the molar mass of the gas increases.

6.89. **Collect and Organize**

We use the equation for the density of a gas to calculate the density of radon. Will radon concentrations be higher in the basement or on the top floor of a building?

Analyze

The density of the radon can be calculated from

$$d = \frac{\mathcal{M}P}{RT}$$

where $\mathcal{M} = 222$ g/mol, $P = 1$ atm, and $T = 298$ K. We can compare the radon density to air's density (1.2 g/L).

Solve

(a) $d = \dfrac{(222 \text{ g/mol})(1 \text{ atm})}{[0.08206 \text{ L} \cdot \text{atm} / (\text{mol} \cdot \text{K})](298 \text{ K})} = 9.08$ g/L

(b) The density of radon is greater than the density of air (1.2 g/L) so radon is more likely to be concentrated in the basement.

Think About It

Radon testing kits all measure radon levels in basements of homes and buildings because that is where radon is most concentrated.

6.91. **Collect and Organize**

Given the mass of an unknown gas (0.391 g) in a known volume (150.0 mL) measured at a particular temperature (22°C) and pressure (750 mmHg), we are to determine the identity of the gas by calculating the gas's molar mass.

Analyze

Use the gas density equation to calculate $\mathcal{M}$:

$$d = \frac{m}{V} = \frac{\mathcal{M}P}{RT}$$

$$\mathcal{M} = \frac{mRT}{VP}$$

Be sure to use units of volume in liters, pressure in atmospheres, and temperature in Kelvin.

Solve

$$\mathcal{M} = \frac{(0.391 \text{ g})[0.08206 \text{ L} \cdot \text{atm} / (\text{mol} \cdot \text{K})](295 \text{ K})}{0.1500 \text{ L} \times \left(750 \text{ mmHg} \times \dfrac{1 \text{ atm}}{760 \text{ mmHg}} \right)} = 63.9 \text{ g/mol}$$

The molar mass of SO_2 is 64.1 g/mol and the molar mass of SO_3 is 80.1 g/mol. The gas in the flask is SO_2.

Think About It

Remember that consistent units are very important when using any equation.

6.93. Collect and Organize

For a gas we are given the density (1.107 g/L) at a particular temperature (300 K) and pressure (740 mmHg). We are to determine if the gas could be CO or CO_2.

Analyze

Use the gas density equation to calculate $\mathcal{M}$:

$$d = \frac{m}{V} = \frac{\mathcal{M}P}{RT}$$

$$\mathcal{M} = \frac{mRT}{VP} = \frac{dRT}{P}$$

Solve

$$\mathcal{M} = \frac{(1.107 \text{ g/L})[0.08206 \text{ L} \cdot \text{atm} / (\text{mol} \cdot \text{K})](300 \text{ K})}{\left(740 \text{ mmHg} \times \dfrac{1 \text{ atm}}{760 \text{ mmHg}}\right)} = 28.0 \text{ g/mol}$$

The molar masses of CO and CO_2 are 28.01 and 44.01 g/mol, respectively. The unknown gas could be CO.

Think About It

Be careful in this problem to convert mmHg to atm for the pressure.

6.95. Collect and Organize / Analyze

We are asked to define the partial pressure of a gas.

Solve

The partial pressure of a gas is the pressure that a particular gas individually contributes to the total pressure.

Think About It

The sum of the partial pressures of all the gases in a mixture adds up to the total pressure.

6.97. Collect and Organize

Of three gas samples, we are to choose the one that would have the greatest volume at 25°C and 1 atm pressure.

Analyze

From the ideal gas law, $PV = nRT$, we can see that at the same temperature and pressure, the volume of the gas depends only on n, the number of moles of gas particles present.

Solve

(a) 0.500 mol of dry H_2 has 0.500 mol of gas particles.
(b) 0.500 mol of dry N_2 has 0.500 mol of gas particles.
(c) 0.500 mol H_2 collected over water contains more than 0.500 mol of gas because the water exerts some vapor pressure. This means that in addition to 0.500 mol H_2 there is also some $H_2O(g)$ in the sample.
Thus (c) has the largest volume.

Think About It

More water particles will be in the gas phase as the temperature increases. Therefore, the volume of the sample (at constant pressure) will increase with temperature.

6.99. Collect and Organize

We are asked to figure the mole fraction of H_2 in a mixture. The mole fraction of a gas is the ratio of moles of a gas divided by the total number of moles in the gas mixture.

Analyze

To get the mole fraction for each gas, we first sum the moles of all the gaseous components (0.70 mol N_2 + 0.20 mol H_2 + 0.10 mol CH_4) to calculate the total moles of gas in the mixture. The individual mole fraction for H_2 is

$$\chi_{H_2} = \frac{n_{H_2}}{n_{total}}$$

where n_{H_2} is the number of moles of hydrogen and n_{total} is the sum of the moles of all components in the mixture.

The partial pressure of each gas is related to the total pressure: $P_x = \chi_x P_{total}$. We can calculate P_{total} from the ideal gas law where $V = 10.0$ L, n = total moles of gas in the mixture, and $T = 300$ K.

Solve

(a) Total moles of gas in the mixture equal $(0.70 + 0.20 + 0.10) = 1.00$ mol.

$$\chi_{H_2} = \frac{0.20 \text{ mol}}{1.00 \text{ mol}} = 0.20$$

(b)

$$P_{total} = \frac{(1.00 \text{ mol})[0.08206 \text{ L} \cdot \text{atm/(mol} \cdot \text{K)}](300 \text{ K})}{10.0 \text{ L}} = 2.46 \text{ atm}$$

$$P_{N_2} = 0.70 \times 2.46 \text{ atm} = 1.7 \text{ atm}$$

$$P_{H_2} = 0.20 \times 2.46 \text{ atm} = 0.49 \text{ atm}$$

$$P_{CH_4} = 0.10 \times 2.46 \text{ atm} = 0.25 \text{ atm}$$

Think About It

The mole fraction of N_2 would be 0.70, and the mole fraction of CH_4 would be 0.10. Notice that the sum of the mole fractions for all the components in the mixture adds up to 1.

6.101. Collect and Organize

Write a balanced chemical equation for the combustion of ammonia and oxygen to yield nitrogen dioxide, and determine the ratio of partial pressures of each reactant required for the reaction.

Analyze

The partial pressure of each gas is related to the number of moles, and so to the stoichiometric coefficients of the balanced equation.

Solve

(a) The unbalanced equation is

$$NH_3(g) + O_2(g) \rightarrow NO_2(g) + H_2O(g)$$

Element	Left Side	Right Side
N	1	1
H	3	2
O	2	2 + 1 = 3

The species are all present, but there are too few H atoms on the product side. We can change the coefficients on H_2O to 3 and on NH_3 to 2 so that there are 6 H atoms on each side.

$$2 NH_3(g) + O_2(g) \rightarrow NO_2(g) + 3 H_2O(g)$$

Element	Left Side	Right Side
N	2	1
H	6	6
O	2	2 + 3 = 5

Hydrogen atoms are balanced, but we need more N atoms on the product side. We can change the coefficient on NO_2 to 2.

$$2\,NH_3(g) + O_2(g) \rightarrow 2\,NO_2(g) + 3\,H_2O(g)$$

Element	Left Side	Right Side
N	2	2
H	6	6
O	2	4 + 3 = 7

There are not enough O atoms on the reactant side. We can change the coefficient on O_2 to 7/2 to fix this
$$2\,NH_3(g) + 7/2\,O_2(g) \rightarrow 2\,NO_2(g) + 3\,H_2O(g)$$

Element	Left Side	Right Side
N	2	2
H	6	6
O	7	4 + 3 = 7

Multiplying by 2 will remove fractional coefficients
$$4\,NH_3(g) + 7\,O_2(g) \rightarrow 4\,NO_2(g) + 6\,H_2O(g)$$

Element	Left Side	Right Side
N	4	4
H	12	12
O	14	14

(b) The ratio of NH_3 to O_2 is 4:7, as seen in the balanced equation.

Think About It
The ratio of partial pressures could also be calculated using the mole fractions, as determined using the stoichiometric coefficients from the balanced equation.

6.103. **Collect and Organize**
Because the oxygen is collected over water or ethanol, a portion of the volume includes gaseous water or ethanol. We are given the volume of the $O_2(g)$ and $H_2O(g)$ mixture, the temperature, and the pressure. We are asked to calculate how many moles of O_2 are present in the mixture.

Analyze
At 25°C (298 K), the vapor pressure due to water is 23.8 mmHg (Table 6.4). The pressure of O_2 in the sample is $P_{O_2} = P_{total} - P_{H_2O} = 760 - 23.8$ mmHg $= 736$ mmHg. Then, we can use the ideal gas equation ($PV = nRT$) to calculate the number of moles of oxygen gas where $P = 736$ mmHg (we have to convert this to atm), $V = 0.480$ L, and $T = 298$ K. At 25°C (298 K), the vapor pressure due to ethanol is 50 mmHg (Appendix 4). The pressure of O_2 in the sample is
$$P_{O_2} = P_{total} - P_{H_2O} = 760 - 50 \text{ mmHg} = 710 \text{ mmHg}$$
Then, we can use the ideal gas equation ($PV = nRT$) to calculate the number of moles of oxygen gas where $P = 710$ mmHg (we have to convert this to atm), $V = 0.480$ L, and $T = 298$ K.

Solve
(a) $n_{O_2} = \dfrac{\left(736 \text{ mm} \times \dfrac{1 \text{ atm}}{760 \text{ mm}}\right) \times 0.480 \text{ L}}{[0.08206 \text{ L} \cdot \text{atm/(mol} \cdot \text{K)}](298 \text{ K})} = 0.0190 \text{ mol } O_2$

(b) $n_{O_2} = \dfrac{\left(710 \text{ mm} \times \dfrac{1 \text{ atm}}{760 \text{ mm}}\right) \times 0.480 \text{ L}}{[0.08206 \text{ L} \cdot \text{atm/(mol} \cdot \text{K)}](298 \text{ K})} = 0.0183 \text{ mol } O_2$

There are fewer moles of oxygen gas in the sample collected over ethanol than in the sample over water.

Think About It
Notice that water vapor contributes little to the total pressure, so oxygen is the major gaseous component in the gas mixture.

6.105. **Collect and Organize**
We are given the balanced chemical equations for four reactions in which reactants and products are all gases. Based on the moles of gas consumed and the moles of gas produced, we are to determine how the pressure changes for each reaction.

Analyze
The greater the moles of gas in a sealed rigid container, the higher the pressure. If the reaction produces more moles of gas than it consumes, then the pressure after the reaction is complete must be greater than the pressure before the reaction took place. On the other hand, if a reaction produces fewer moles of gas than it consumes, the pressure must be lower. If there is no change in the moles of gas between reactants and products, the pressure does not change.

Solve
(a) Two moles of gas are consumed and five moles of gas are produced; $\Delta n = n_{products} - n_{reactants} = 3$; the pressure is greater at the end of the reaction.
(b) Three moles of gas are consumed and two moles of gas are produced; $\Delta n = -1$; the pressure is lower at the end of the reaction.
(c) Six moles of gas are consumed and seven moles of gas are produced; $\Delta n = 1$; the pressure is greater at the end of the reaction.
(d) Nine moles of gas are consumed and ten moles of gas are produced; $\Delta n = 1$; the pressure is greater at the end of the reaction.

Think About It
Mathematically, if $\Delta n = n_{products} - n_{reactants}$ is positive, the pressure increases; if $\Delta n = 0$, the pressure stays the same; if Δn is negative, the pressure decreases.

6.107. **Collect and Organize**
Using $PV = nRT$ to calculate the moles of O_2 present in a human lung at 8000 m with pure O_2 and one at sea level with air, we can determine which lung will have the most O_2.

Analyze
Rearranging the ideal gas equation to solve for n gives

$$n = \frac{PV}{RT}$$

At sea level, $P = 1.00$ atm and the mole fraction of oxygen in air is 0.2095. Therefore, the partial pressure of O_2 at sea level is $P_{O_2} = 1.00 \text{ atm} \times 0.2095 = 0.2095$ atm. At 8000 m, the climber is breathing 100% O_2 with $P = 0.35$ atm. We are not given the values of V and T, so we can treat these as constants. As we are comparing n_{O_2} values in a ratio of $n_{O_2,8000m} / n_{O_2,\text{sea level}}$, we see that we do not need these values. In fact, we can even leave R, the gas constant, out of our final calculation.

Solve

$$n_{O_2,8000m} = \frac{0.35 \text{ atm} \times V}{RT} \qquad n_{O_2,\text{sea level}} = \frac{0.2095 \text{ atm} \times V}{RT}$$

The ratio $n_{O_2,8000m} / n_{O_2,\text{sea level}}$ is

$$\frac{\left(\dfrac{0.35 \text{ atm} \times V}{RT}\right)}{\left(\dfrac{0.2095 \text{ atm} \times V}{RT}\right)} = \frac{0.35 \text{ atm}}{0.2095 \text{ atm}} = 1.7$$

$$\frac{1.7 - 1}{1} \times 100\% = 70\% \text{ more O}_2 \text{ at 8000 m}$$

The lung at 8000 m will contain 70% more O_2 than the lung at sea level.

Think About It

Breathing pure O_2 at that great altitude more than compensates for the pressure difference. The climber actually has more oxygen in her lungs at 8000 m than at sea level.

6.109. Collect and Organize

The reaction of carbon monoxide with oxygen to form CO_2 is described by the equation

$$2 \text{ CO}(g) + \text{O}_2(g) \rightarrow 2 \text{ CO}_2(g)$$

From the information given (680 mmHg of CO and 340 mmHg of O_2 at the beginning of the reaction), we are to calculate the final pressure of the gases in the sealed vessel at the end of the reaction.

Analyze

We need to determine the moles of CO_2 produced from the CO and O_2. We can express the moles of CO and O_2 using the ideal gas equation:

$$n_{CO} = \frac{680 \text{ mmHg} \times V}{RT} \quad \text{and} \quad n_{CO} = \frac{340 \text{ mmHg} \times V}{RT}$$

From the balanced equation we see that 2 moles of CO are required to react with 1 mole of O_2.

Solve

Notice that n_{CO} is twice that of n_{O_2}. This agrees exactly with the stoichiometric ratio of $CO:O_2$ in the balanced equation. Assuming 100% yield, both reactants will be completely consumed in the reaction to give CO_2 as the only product. There will be the same number of moles of CO_2 as there were moles of CO at the start of the reaction. Therefore, the pressure of CO_2 at the end of the reaction will be the same as the pressure that the reactant CO exerted at the beginning of the reaction, namely, 680 mmHg.

Think About It

In answering this question, we didn't have to perform any calculations; rather we just had to relate the stoichiometry to the pressures of the gases.

6.111. Collect and Organize

The balanced chemical equation for the production of ammonia from nitrogen and hydrogen is

$$N_2(g) + 3 \text{ H}_2(g) \rightarrow 2 \text{ NH}_3(g)$$

We are given the initial amounts of N_2 and H_2 (1.20×10^3 and 3.60×10^3 mol, respectively), and we are to calculate the pressure when half of the N_2 is used to produce NH_3 and compare it to the pressure at the beginning of the reaction.

Analyze

The initial pressure inside the vessel can be calculated using the ideal gas equation, where n = total moles of H_2 and N_2. To find the pressure when half of the N_2 is consumed, we first have to determine the moles of H_2, N_2, and NH_3 present in the vessel using the reaction stoichiometry. Then we calculate the pressure from the total moles of gas in the vessel. Because we are looking for a percentage decrease, we can express pressures with the unspecified "variables" R, T, and V.

Solve

Initial pressure before reaction:

$$P_i = \frac{(3600 \text{ mol} + 1200 \text{ mol})RT}{V} = 4800RT/V$$

Moles of N_2 after the reaction goes halfway: $\frac{1}{2} \times (1.20 \times 10^3) = 600$ moles

Moles of H_2 consumed in the reaction:

$$600 \text{ mol N}_2 \text{ reacted} \times \frac{3 \text{ mol H}_2}{1 \text{ mol N}_2} = 1800 \text{ mol H}_2$$

Moles of H_2 remaining after the reaction: $3600 - 1800 = 1800$ mol
Moles of NH_3 produced in the reaction:

$$600 \text{ mol N}_2 \text{ reacted} \times \frac{2 \text{ mol NH}_3}{1 \text{ mol N}_2} = 1200 \text{ mol NH}_3$$

Total moles of gases present at the end of the reaction:

$$600 \text{ mol N}_2 + 1800 \text{ mol H}_2 + 1200 \text{ mol NH}_3 = 3600 \text{ mol}$$

Pressure in vessel after the reaction:

$$P_f = 3600RT/V$$

Difference in pressure: $3600RT/V - 4800RT/V = -1200RT/V$

$$\% \text{ decrease} = \frac{1200RT/V}{4800RT/V} \times 100 = 25\%$$

Think About It
We observe a decrease in pressure as the reaction proceeds, which we expect since 4 moles of reactants produce 2 moles of products in the balanced chemical equation.

6.113. Collect and Organize / Analyze
We are asked to define the root-mean-square speed of gas molecules.

Solve
The root-mean-square speed, u_{rms}, is the speed of a molecule in a gas that has the average kinetic energy of all the molecules of the sample.

Think About It
The root-mean-square speed is not exactly the same as the most probable speed, which is the speed corresponding to the peak in the distribution diagram. Nor is it the same as the average speed, which is a simple average of all the molecular speeds.

6.115. Collect and Organize
We are asked how temperature and molar mass affect the root-mean-square speed.

Analyze
The root-mean-square speed is the speed of a molecule that has the average kinetic energy of the molecules in the sample. The equation is

$$u_{rms} = \sqrt{\frac{3RT}{\mathcal{M}}}$$

Solve
(a) As the molar mass increases, u_{rms} decreases because of the inverse relationship between u_{rms} and $\mathcal{M}$.
(b) As temperature increases, the u_{rms} increases because of the direct relationship between u_{rms} and T.

Think About It
Because u_{rms} is related to the average kinetic energy, we can also say that the average kinetic energy decreases as molar mass increases, and increases as temperature increases.

6.117. Collect and Organize
Graham's law of effusion describes the rate of effusion of gases as a function of the gases' masses. We are asked to describe how we can use Graham's law to determine the molar mass of a gas.

Analyze

The mathematical form of Graham's law is

$$\frac{r_x}{r_y} = \frac{u_{rms,x}}{u_{rms,y}} = \sqrt{\frac{\mathcal{M}_y}{\mathcal{M}_x}}$$

Solve

To determine the molar mass of an unknown gas, measure the rate of its effusion (r_x) relative to the rate of effusion of a known gas (r_y). Since we know the molar mass of the known gas, $\mathcal{M}_y$, we can use the equation to solve for the unknown $\mathcal{M}_x$.

Think About It

Alternatively, if we have the means to measure the root-mean-square speed of the known and unknown gases, we can use those values in Graham's law to calculate the molar mass of the unknown gas.

6.119. **Collect and Organize / Analyze**

By defining diffusion and effusion, we can describe the difference between them.

Solve

Diffusion is the spread of one substance into another, for example, the mixing of gasoline vapors and air inside a car's cylinders. Effusion, on the other hand, is the escape of a gas from its container through a tiny hole. The gas is escaping from a region of higher pressure to one of lower pressure. In diffusion, there is no difference in pressure, and two gases are involved in the process.

Think About It

Graham's law applies to both effusion and diffusion.

6.121. **Collect and Organize**

We are to rank the various gases according to the root-mean-square speeds.

Analyze

The equation for root-mean-square speed shows an inverse dependence of u_{rms} on the molar mass of the gas:

$$u_{rms} = \sqrt{\frac{3RT}{\mathcal{M}}}$$

Solve

According to the u_{rms} equation, the lower the molar mass, the greater the root-mean-square speed. The molar masses of SO_2, CO_2, and NO_2 are 64.06, 44.01, and 46.01 g/mol, respectively. Therefore, SO_2 has the lowest u_{rms}, and CO_2 has the highest u_{rms}. The rank order in terms of increasing root-mean-square speed is $SO_2 < NO_2 < CO_2$.

Think About It

There was not a specific need to compute the molar masses here. Because each gas is XO_2 all we need to do is to consider the molar mass of X. The largest molar mass is S, next is N, and smallest is C.

6.123. **Collect and Organize**

We are given u_{rms} values for three gases at 286 K and asked to determine which gas is oxygen.

Analyze

Use the root-mean-square speed equation to calculate each gas's molar mass and compare to oxygen's computed molar mass of 32.0 g/mol. The molar mass for each gas may be solved by

$$u_{rms} = \sqrt{\frac{3RT}{\mathcal{M}}} \quad \text{rearranged to} \quad \mathcal{M} = \frac{3RT}{u_{rms}^2}$$

We want to pause here to consider which units of the gas constant, R, to use. The units for u_{rms}^2 are m^2/s^2. In order to cancel those units to obtain molar mass (kg or g/mol) we have to use $R = 8.314$ kg $\cdot$ $m^2/s^2(mol \cdot K)$.

Solve

Gas A: $\mathcal{M} = \dfrac{3[8.314 \text{ kg} \cdot m^2/s^2(mol \cdot K)](286 \text{ K})}{(360 \text{ m/s})^2} = 0.0550$ kg/mol or 55.0 g/mol

Gas B: $\mathcal{M} = \dfrac{3[8.314 \text{ kg} \cdot m^2/s^2(mol \cdot K)](286 \text{ K})}{(441 \text{ m/s})^2} = 0.0367$ kg/mol or 36.7 g/mol

Gas C: $\mathcal{M} = \dfrac{3[8.314 \text{ kg} \cdot m^2/s^2(mol \cdot K)](286 \text{ K})}{(472 \text{ m/s})^2} = 0.0320$ kg/mol or 32.0 g/mol

The molar mass from gas C corresponds to the molar mass of oxygen, so gas C is O_2.

Think About It

Notice that the gas with the highest molar mass has the slowest u_{rms}.

6.125. **Collect and Organize**

We are asked to sketch a plot of $u_{rms,He}$ versus T for the data provided, and determine if the graph is linear.

Analyze

The relationship between the root-mean-square speed and temperature is

$$u_{rms} = \sqrt{\frac{3RT}{\mathcal{M}}}$$

Based on this equation, we expect that u_{rms} will be proportional to the square root of T, rather than T.

Solve

Plotting u_{rms} versus T and u_{rms}^2 versus T

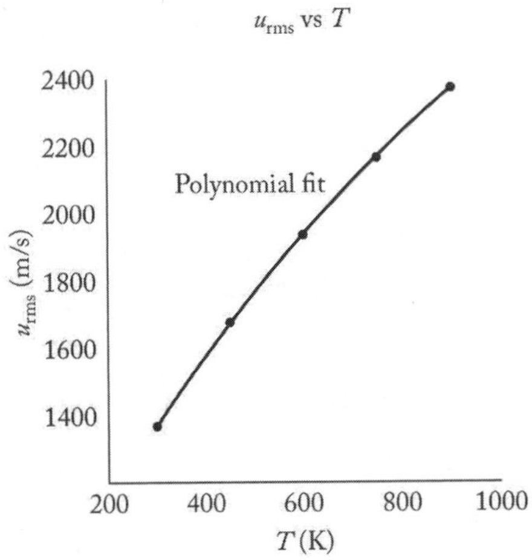

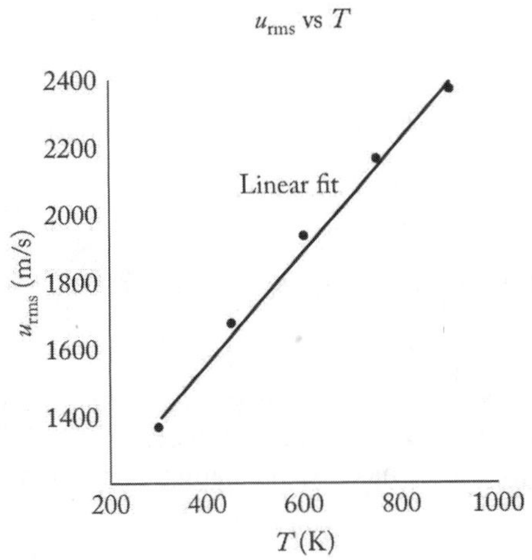

No, the graph is not linear. Although at first glance it appears that the points of the plot of u_{rms} versus T fit the trendline, we see that there are both positive and negative deviations from the line of best fit. The dependence is that of an exponential curve rather than a straight line. The plot of u_{rms}^2 versus T would display a perfect linear fit to the data points.

Think About It

The plot of u_{rms}^2 versus T matches the equation relating u_{rms} and T.

6.127. Collect and Organize

We are to calculate the ratio of the u_{rms} of deuterium, D_2, versus the u_{rms} of hydrogen, H_2. We expect that the lighter H_2 has a larger u_{rms} so we expect the ratio to be less than 1.

Analyze

We are not provided with temperature information, but because we are determining a ratio of the root-mean-square speeds, that information is not necessary. The expressions for u_{rms} for each gas are

$$u_{rms,D_2} = \sqrt{\frac{3RT}{\mathcal{M}_{D_2}}} \qquad u_{rms,H_2} = \sqrt{\frac{3RT}{\mathcal{M}_{H_2}}}$$

Squaring both sides of these equations we get

$$\left(u_{rms,D_2}\right)^2 = \frac{3RT}{\mathcal{M}_{D_2}} \qquad \left(u_{rms,H_2}\right)^2 = \frac{3RT}{\mathcal{M}_{H_2}}$$

The ratio of the root-mean-square speeds would be

$$\frac{\left(u_{rms,D_2}\right)^2}{\left(u_{rms,H_2}\right)^2} = \frac{\left(\frac{3RT}{\mathcal{M}_{D_2}}\right)}{\left(\frac{3RT}{\mathcal{M}_{H_2}}\right)} = \frac{\mathcal{M}_{H_2}}{\mathcal{M}_{D_2}} \quad \text{or} \quad \frac{u_{rms,D_2}}{u_{rms,H_2}} = \sqrt{\frac{\mathcal{M}_{H_2}}{\mathcal{M}_{D_2}}}$$

Solve

$$\frac{u_{rms,D_2}}{u_{rms,H_2}} = \sqrt{\frac{2.02 \text{ g/mol}}{4.00 \text{ g/mol}}} = 0.711$$

Think About It

Our prediction that the ratio would be less than 1, since the heavier D_2 molecule travels more slowly, was correct.

6.129. Collect and Organize

We are asked if a student accurately measured the rates of effusion for carbon dioxide and propane.

Analyze

The equation describing the ratio of rates of effusion is

$$\frac{r_{CO_2}}{r_{C_3H_8}} = \sqrt{\frac{\mathcal{M}_{C_3H_8}}{\mathcal{M}_{CO_2}}}$$

The molar masses of CO_2 and C_3H_8 are 44.01 g/mol and 44.11 g/mol, respectively.

Solve

Since the molar masses for the two molecules are so similar, it is reasonable to expect that they would effuse at the same rate, within experimental error.

This may be confirmed by calculating the ratio of rates of effusion

$$\frac{r_{CO_2}}{r_{C_3H_8}} = \sqrt{\frac{44.11 \text{ g/mol}}{44.01 \text{ g/mol}}} = 1.001$$

Since the ratio of the rates of effusion is nearly one, we would expect the two gasses to effuse at the same rate.

Think About It

The molar masses of the two compounds differ by only 0.23%, so their similarity is expected.

6.131. Collect and Organize

We can compare u_{rms} for an unknown gas to H_2 to determine the compound's molar mass.

Analyze

The ratio of $u_{rms,X}/u_{rms,H_2}$ is given as 1/3. The relevant equation is

$$\frac{u_{rms,X}}{u_{rms,H_2}} = \sqrt{\frac{\mathcal{M}_{H_2}}{\mathcal{M}_X}} = \frac{1}{3}$$

Rearranging to solve for $\mathcal{M}_X$ gives

$$\mathcal{M}_X = \mathcal{M}_{H_2} \times \left(\frac{u_{rms,H_2}}{u_{rms,X}}\right)^2$$

Solve

$$\mathcal{M}_X = \frac{2.02 \text{ g}}{\text{mol}} \times (3)^2 = 18.2 \text{ g/mol}$$

Think About It

If the gas diffuses slower than H_2, its molar mass must be greater than that of H_2. A common molecule with a molar mass of ~18 g/mol is water, H_2O, so this gas might be water vapor.

6.133. Collect and Organize

From the equation for Graham's law we can calculate the relative rates of diffusion of $^{13}CO_2$ and $^{12}CO_2$. Because $^{12}CO_2$ is lighter, we expect it to diffuse faster.

Analyze

The relevant equation is

$$\frac{r_{^{12}CO_2}}{r_{^{13}CO_2}} = \sqrt{\frac{\mathcal{M}_{^{13}CO_2}}{\mathcal{M}_{^{12}CO_2}}}$$

Solve

(a) $\dfrac{r_{^{12}CO_2}}{r_{^{13}CO_2}} = \sqrt{\dfrac{45.0 \text{ g/mol}}{44.0 \text{ g/mol}}} = 1.01$

(b) Because this ratio is greater than 1 (although it is just a little bit greater than 1!), $^{12}CO_2$ diffuses faster than $^{13}CO_2$.

Think About It

Our prediction was correct!

6.135. Collect and Organize

Two balloons, one with He and the other with H_2, have different sizes (both smaller, though) after 24 hours. Which one is which?

Analyze

The lower the molar mass of the gas in the balloon, the more rapidly that gas will effuse out of the balloon. The molar mass of H_2 is 2.0 g/mol; for He the molar mass is 4.0 g/mol.

Solve

Hydrogen effuses out of a balloon faster than He, so the smaller balloon contains hydrogen.

Think About It

Both balloons are smaller after 24 hours because both gases diffuse out of the balloons faster than atmospheric gases effuse into them.

6.137. **Collect and Organize / Analyze**

We are to explain why real gases deviate from ideal behavior at low temperatures and high pressures.

Solve

At low temperatures the gas particles move more slowly, and their collisions become inelastic; they stick together due to the weak attractive forces between them. The particles, therefore, do not act separately to contribute to the pressure in the container, and the pressure is lower than would be expected by the ideal gas law. Also, the gas particles take up real volume in the container and as the pressure increases, the volume of the particles takes up a greater volume of the free space in the container. This has the effect of raising the pressure–volume above what we would expect from the ideal gas law (in a plot of PV/RT versus P).

Think About It

Gases at low pressures and not-too-cold temperatures behave most ideally.

6.139. **Collect and Organize**

The van der Waals constant b is associated with the volume of the gas particles. We are asked why the value of b increases as the atomic number of noble gas elements increases.

Analyze

As atomic number increases, the number of electrons and the size of the atoms increase.

Solve

Since b is a measure of the volume that the gas particles occupy, b increases as the sizes of the particles increase.

Think About It

Notice in Table 6.5 that the units of b are in liters per mole (L/mol). This is the volume that one mole of the gas particles occupy in the volume of the gas container. As the size increases from He to Ar, b does indeed increase.

6.141. **Collect and Organize**

Given that the plot in Figure 6.38 of PV/RT versus P for CH_4 is different than the plot for H_2 in describing how they deviate from ideal behavior, we are to consider for which gas the volume of the gas particles is more important than the attractive interactions between them.

Analyze

In Figure 6.38 we see that the curve for H_2 always deviates above the ideal gas line while the curve for CH_4 first deviates below the ideal gas line at lower pressures, then deviates above the line at higher pressures. A line diverging above the ideal gas line indicates that the volume of the real molecules in the gas sample is no longer approximately equal to zero, so that the free volume no longer is a good estimate of the total volume. A line that diverges below the ideal gas line indicates attractive forces between the molecules that cause them to stick together, which decreases the force of collisions with the wall of the container; therefore, the pressure decreases compared to ideal behavior.

Solve

Figure 6.38 shows that CH_4 has attractive forces that cause deviation from ideal behavior at lower pressures and that for H_2, the only deviation is caused by the real volume occupied by the molecules. From this behavior we can say that for H_2, the effect of the volume occupied by the gas molecules is more important than the attractive forces between them.

Think About It

In a plot of PV/RT versus P, we can see that CO_2 has an even larger attractive forces effect than CH_4 because its negative deviation from ideal behavior (at pressures of 0–100 atm) is even greater than that of CH_4.

6.143. Collect and Organize

We are to calculate the pressure of 40.0 g of H_2 at 20°C (293 K) in a 1.00 L vessel using the van der Waals equation for real gases and the ideal gas equation.

Analyze

The van der Waals equation is

$$\left(P + \frac{n^2 a}{V^2}\right)(V - nb) = nRT$$

where $a = 0.244$ $L^2 \cdot$ atm/mol^2 and $b = 0.0266$ L/mol for H_2. We first have to determine the number of moles of H_2 in 40.0 g using the molar mass of H_2 (2.02 g/mol).

Solve

(a) Moles of H_2: 40.0 g × 1 mol/2.02 g = 19.8 mol H_2

Pressure calculation using the van der Waals equation:

$$\left(P + \frac{(19.8 \text{ mol})^2 \times 0.244 \text{ L}^2 \cdot \text{atm/mol}^2}{(1.00 \text{ L})^2}\right)(1.00 \text{ L} - 19.8 \text{ mol} \times 0.0266 \text{ L/mol}) =$$

$$(19.8 \text{ mol})[0.08206 \text{ L} \cdot \text{atm/(mol} \cdot \text{K)}](293 \text{ K})$$

$$(P + 95.66 \text{ atm}) = 1005.8 \text{ atm}$$

$$P = 910 \text{ atm}$$

(b) Pressure calculation using the ideal gas equation:

$$P = \frac{nRT}{V} = \frac{(19.8 \text{ mol})[0.08206 \text{ L} \cdot \text{atm/(mol} \cdot \text{K)}](293 \text{ K})}{1.00 \text{ L}} = 476 \text{ atm}$$

Think About It

At these high pressures, there is a very large deviation from ideal behavior for hydrogen as shown by the very different pressures calculated through the ideal gas equation and the van der Waals equation.

6.145. Collect and Organize

For a given volume of propane gas at a certain pressure and occupying a given volume, we are to calculate the volume when the pressure is reduced if the temperature and number of moles of propane remain constant.

Analyze

We can use the combined ideal gas law:

$$\frac{P_1 V_1}{n_1 T_1} = \frac{P_2 V_2}{n_2 T_2}$$

Because $n_1 = n_2$ and $T_1 = T_2$, the equation reduces to

$$P_1 V_1 = P_2 V_2$$

Solve

$$12.5 \text{ atm} \times 10.6 \text{ L} = 1.05 \text{ atm} \times V_2$$

$$V_2 = 126 \text{ L}$$

Think About It

The pressure on this sample of propane gas decreased by a factor of about 12 so we expect that the volume will increase also by about a factor of 12, so our answer makes sense.

6.147. Collect and Organize

We are asked to calculate the volume of 22.4 L of HCl gas after it has been heated from 15°C to 78°C. The number of moles of gas and the pressure on the sample remain constant through the temperature change.

Analyze
We can use the combined ideal gas law:

$$\frac{P_1V_1}{n_1T_1} = \frac{P_2V_2}{n_2T_2}$$

Because $n_1 = n_2$ and $P_1 = P_2$, the equation reduces to

$$\frac{V_1}{T_1} = \frac{V_2}{T_2}$$

Solve

$$\frac{22.4 \text{ L}}{288 \text{ K}} = \frac{V_2}{351 \text{ K}}$$

$$V_2 = 27.3 \text{ L}$$

Think About It
We must convert temperature from Celsius to Kelvin.

6.149. **Collect and Organize**

We are asked to determine the pressure on a gas cylinder that is driven from the Rocky Mountains of Colorado (12°F) to Death Valley, CA (143°F). The number of moles of gas and the volume of the cylinder remain constant through the temperature change.

Analyze
The original pressure of the gas cylinder is 2200 psi. In atmospheres, this pressure corresponds to

$$2200 \text{ psi} \times \frac{1 \text{ atm}}{14.7 \text{ psi}} = 149.7 \text{ atm}$$

The temperature of the gas cylinder goes from 12°F to 143°F. These temperatures correspond to

$$K = \frac{5}{9}(°F - 32) + 273.15$$

$$T_1 = \frac{5}{9}(12°F - 32) + 273.15 = 260 \text{ K}$$

$$T_2 = \frac{5}{9}(143°F - 32) + 273.15 = 335 \text{ K}$$

We can use the combined ideal gas law:

$$\frac{P_1V_1}{n_1T_1} = \frac{P_2V_2}{n_2T_2}$$

Because $n_1 = n_2$ and $V_1 = V_2$, the equation reduces to

$$\frac{P_1}{T_1} = \frac{P_2}{T_2}$$

Solve

$$\frac{149.7 \text{ atm}}{260 \text{ K}} = \frac{P_2}{335 \text{ K}}$$

$$P_2 = 192.8 \text{ atm} \times \frac{14.7 \text{ psi}}{1 \text{ atm}} = 2835 \text{ psi}$$

The gauge will read 2835 psi in Death Valley.

Think About It
We must convert psi to atm and temperature from Farenheit to Celsius to Kelvin to use the combined gas law.

6.151. **Collect and Organize**

For a partially deflated soccer ball with a volume of 1.034 L at a pressure of 0.947 atm at 27°C (300 K), we are to calculate the volume of the ball during flight in an airplane baggage hold where the pressure is 0.235 atm and the temperature is −35°C.

Analyze

We can use the combined ideal gas law:

$$\frac{P_1 V_1}{n_1 T_1} = \frac{P_2 V_2}{n_2 T_2}$$

Because $n_1 = n_2$, the equation reduces to

$$\frac{P_1 V_1}{T_1} = \frac{P_2 V_2}{T_2}$$

Solve

$$\frac{0.947\ \text{atm} \times 1.034\ \text{L}}{300\text{K}} = \frac{0.235\ \text{atm} \times V_2}{238\ \text{K}}$$

$$V_2 = 3.31\ \text{L}$$

Think About It

The dominant effect on the ball's volume is the reduced pressure in the baggage hold of the plane, considering the ball increases in volume with the reduced pressure. We would expect the volume to decrease if the reduced temperature in the airplane were the dominant effect.

6.153. **Collect and Organize**

We are to calculate u_{rms} for argon atoms at 0.00010 K.

Analyze

The u_{rms} is calculated through

$$u_{\text{rms}} = \sqrt{\frac{3RT}{\mathcal{M}}}$$

Solve

$$u_{\text{rms}} = \sqrt{\frac{3 \times 8.314\ \text{kg} \cdot \text{m}^2/\text{s}^2 (\text{mol} \cdot \text{K}) \times 0.00010\ \text{K}}{39.948\ \text{g/mol} \times \dfrac{1\ \text{kg}}{1000\ \text{g}}}} = 0.25\ \text{m/s}$$

Think About It

Helium atoms at the same temperature would have a greater root-mean-square speed (0.79 m/s) since they are lighter gas particles.

6.155. **Collect and Organize**

We can use the ideal gas equation to calculate the moles of air (and then the mass of air) that must be compressed into a scuba tank so that it will deliver 80 ft³ of air at 72°F and 1.00 atm pressure.

Analyze

To find the moles of air in 80 ft³ we use $PV = nRT$ rearranged to $n = PV/RT$, where $V = 80$ ft³ (which must be converted to liters using 1 ft = 0.3048 m and 1 m³ = 1000 L), $P = 1.00$ atm, and $T = 72°$F (which must be converted to Kelvin using °C = 5/9(°F − 32) and K = °C + 273.15). Using the molar mass of air (which we can calculate from the mole fractions and the molar mass of each component), we can convert the moles of air into grams then add them to the weight of the 15 kg scuba tank.

Solve

Moles of air required:

$$n = \frac{1 \text{ atm} \times \left(80 \text{ ft}^3 \times \frac{(0.3048 \text{ m})^3}{1 \text{ ft}^3} \times \frac{1000 \text{ L}}{1 \text{ m}^3} \right)}{[0.08206 \text{ L} \cdot \text{atm/(mol} \cdot \text{K)}] \left(\frac{5}{9}(72-32) + 273.15 \text{ K} \right)} = 93.5 \text{ mol}$$

Molar mass of dry air (data from Table 6.1):

$$\mathcal{M}_{\text{air}} = (0.7808 \times 28.01 \text{ g/mol}) + (0.2095 \times 32.00 \text{ g/mol}) + (0.00934 \times 39.948 \text{ g/mol})$$
$$+ (0.00033 \times 44.01 \text{ g/mol}) + (0.000002 \times 16.04 \text{ g/mol}) + (0.0000005 \times 2.02 \text{ g/mol}) = 28.96 \text{ g/mol}$$

Mass of air in the tank:

$$93.5 \text{ mol} \times \frac{28.96 \text{ g}}{\text{mol}} = 2.71 \times 10^3 \text{ g or } 2.71 \text{ kg}$$

Total mass of the tank: 2.71 kg + 15 kg = 18 kg.

Think About It

The air inside the tank does not add much to the mass. The tank, though, must be thick and heavy to withstand the 3000 psi of pressure needed for that amount of air.

6.157. Collect and Organize

We can use Graham's law to calculate the molar mass of the noble gas that effuses at half the rate that O_2 effuses. We would predict that the gas has a greater molar mass than O_2 because heavier gases effuse more slowly.

Analyze

Graham's law for this case is

$$\frac{r_X}{r_{O_2}} = \sqrt{\frac{\mathcal{M}_{O_2}}{\mathcal{M}_X}} = \frac{1}{2}$$

Solve

$$\frac{1}{2} = \sqrt{\frac{32.00 \text{ g/mol}}{\mathcal{M}_X}}$$

$$\frac{1}{4} = \frac{32.00 \text{ g/mol}}{\mathcal{M}_X}$$

$$\mathcal{M}_X = 128 \text{ g/mol}$$

The noble gas with a molar mass closest to 128 g/mol is xenon, Xe.

Think About It

Our prediction was correct. Notice, too, that the molar mass of a gas that effuses half as quickly as another has a molar mass that is 4 times that of the other.

6.159. Collect and Organize

We are asked to write (and balance) the equation for the reaction between NH_3 and HCl to form NH_4Cl, and then predict the position (and calculate it!) of the NH_4Cl ring that forms when NH_3 and HCl are placed at opposite ends of a 1.00 m glass tube (refer to Figure P6.159).

Analyze

The reaction to form NH_4Cl takes place when the NH_3 (17.03 g/mol) vapors meet the HCl (36.46 g/mol) vapors in the tube. The diffusion of the vapors can be calculated using Graham's law:

$$\frac{r_{NH_3}}{r_{HCl}} = \sqrt{\frac{\mathcal{M}_{HCl}}{\mathcal{M}_{NH_3}}}$$

Solve

(a) The chemical equation for this reaction is

$$NH_3(g) + HCl(g) \rightarrow NH_4Cl(s)$$

(b) The rate of diffusion of a gas is inversely related to its molar mass. The gas, then, with the highest molar mass does not diffuse as far along the tube before it reacts. Therefore, the ring of NH_4Cl should appear closer to the end with HCl.

(c) $\dfrac{r_{NH_3}}{r_{HCl}} = \sqrt{\dfrac{36.46 \text{ g/mol}}{17.03 \text{ g/mol}}} = 1.463$

This means that the NH_3 diffuses 1.463 times further along the 1.00 m tube compared to the HCl. If d = distance traveled along the tube:

$$d_{NH_3} + d_{HCl} = 1.00 \text{ m}$$

$$d_{NH_3} = 1.463 d_{HCl}, \text{ so}$$

$$1.463 d_{HCl} + d_{HCl} = 1.00 \text{ m}$$

$$d_{HCl} = 0.406 \text{ m}$$

The HCl diffuses 0.406 m, and the NH_3 diffuses 0.594 m. The ring therefore appears 0.594 m from the ammonia end of the tube.

Think About It

If HBr had been used instead of HCl, the NH_4Br ring would be even further from the ammonia end of the tube because of HBr's larger molar mass.

6.161. **Collect and Organize**

A car was built so as to be propelled by the expansion of liquid nitrogen into gaseous nitrogen. Given the amount of liquid nitrogen in a tank, the temperature, and the atmospheric pressure, we are asked to calculate the volume of the gaseous nitrogen when released.

Analyze

We use the ideal gas law to find the volume of $N_2(g)$ released, but we must first determine the moles of N_2 in a 182 L tank of liquid N_2 by multiplying the volume of liquid N_2 (182 L) by the density of the liquid (0.808 g/mL) and dividing by the molar mass. The temperature is 25°C and the pressure is 0.927 atm.

Solve

The amount, in moles, of N_2 in the tank is

$$182 \text{ L} \times \frac{0.808 \text{ g}}{1 \text{ mL}} \times \frac{1000 \text{ mL}}{1 \text{ L}} \times \frac{1 \text{ mol}}{28.01 \text{ g}} = 5.25 \times 10^3 \text{ mol}$$

The volume of N_2 gas released is

$$V = \frac{(5.25 \times 10^3 \text{ mol})[0.08206 \text{ L} \cdot \text{atm}/(\text{mol} \cdot \text{K})](298 \text{ K})}{0.927 \text{ atm}} = 1.38 \times 10^5 \text{ L}$$

Think About It

Liquid nitrogen, being a condensed gas, takes up much less volume than gaseous N_2. In this case the volume increased by a factor of

$$\frac{1.38 \times 10^5 \text{ L}}{182 \text{ L}} = 758 \text{ times}$$

6.163. **Collect and Organize**

We are to determine the pressure in a 400 L cylinder containing 10 kg of hydrogen gas at 300 K, and decide if it conforms to the maximum safe pressure of 10,000 psi.

Analyze

The mass of H_2 may be converted to a molar quantity by dividing by the molar mass of H_2 (2.01588 g/mol), and converting kilograms to grams. We may then calculate the pressure in the cylinder using the ideal gas law, as depicted

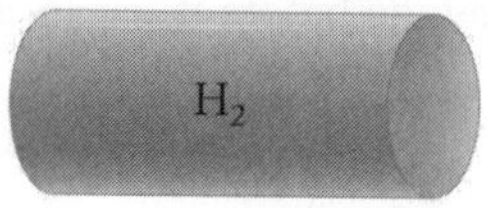

$V = 400$ L
$T = 300$ K
$m = 10$ kg

$$P = \frac{nRT}{V}$$

Solve

$$10 \text{ kg H}_2 \times \frac{1000 \text{ g}}{1 \text{ kg}} \times \frac{1 \text{ mol H}_2}{2.01588 \text{ g H}_2} = 4960.6 \text{ mol H}_2$$

$$P = \frac{4960.6 \text{ mol}[0.08206 \text{ L} \cdot \text{atm/(mol} \cdot \text{K)}](300 \text{ K})}{400 \text{ L}} = 305.3 \text{ atm}$$

$$305.3 \text{ atm} \times \frac{14.7 \text{ psi}}{1 \text{ atm}} = 4.5 \times 10^3 \text{ psi}$$

The pressure in the cylinder is lower than the highest safe pressure, so the vehicle should be safe.

Think About It

This mass of hydrogen gas should not exceed the maximum safe pressure in a 400 L tank until it reaches a temperature of 668 K!

6.165. Collect and Organize

We are asked to calculate the mass of argon in 15.0 L at standard temperature and pressure (STP).

Analyze

STP is 0°C (273 K) and 1.00 atm. Using the ideal gas equation, we can solve for *n*, the number of moles of argon, then convert those moles into mass using the molar mass of Ar (39.948 g/mol).

Solve

$$n = \frac{PV}{RT} = \frac{1 \text{ atm} \times 15.0 \text{ L}}{[0.08206 \text{ L} \cdot \text{atm/(mol} \cdot \text{K)}](273 \text{ K})} = 0.6696 \text{ mol Ar}$$

$$\text{Mass of Ar} = 0.6696 \text{ mol} \times \frac{39.948 \text{ g}}{\text{mol}} = 26.7 \text{ g}$$

Think About It

It might be confusing that "standard temperature" for the ideal gas law is 0°C whereas the "standard temperature" for the enthalpy of formation is 25°C, so try to keep this difference in mind.

6.167. Collect and Organize

We are asked to calculate the pressure of a nitrogen and hydrogen mixture at 273°C in 10.0 L.

Analyze

The total pressure depends on the total moles of gas particles present in the 10.0 L vessel. We can calculate the moles of N_2 and H_2 separately from the masses given 7.00 g N_2 and 2.00 g H_2 and their molar masses. Then, using the sum of the moles of gas, we can calculate the pressure in the vessel using $V = 10.0$ L and $T = 273°C$ (546 K) and the ideal gas equation:

$$P = \frac{nRT}{V}$$

Solve

Moles of H_2: $2.00 \text{ g} \times \dfrac{1 \text{ mol}}{2.016 \text{ g}} = 0.992 \text{ mol}$

Moles of N_2: $7.00 \text{ g} \times \dfrac{1 \text{ mol}}{28.01 \text{ g}} = 0.250 \text{ mol}$

Total moles of gas: $0.992 + 0.250 = 1.242 \text{ mol}$

$$P = \frac{(1.242 \text{ mol})[0.08206 \text{ L} \cdot \text{atm/(mol} \cdot \text{K)}](546 \text{ K})}{10.0 \text{ L}} = 5.56 \text{ atm}$$

Think About It

Remember to convert 273°C to Kelvin in this problem. This temperature looks at first glance as if it is already in Kelvin.

6.169. ## Collect and Organize

We are given the time that an equal number of moles of two gases (one is N_2, 28.01 g/mol) take to diffuse through a porous plug and are asked to calculate the molar mass of the second gas.

Analyze

$$\frac{r_{N_2}}{r_X} = \sqrt{\frac{\mathcal{M}_X}{\mathcal{M}_{N_2}}}$$

The rate at which N_2 and the unknown gas X travel can be expressed as $r_{N_2} = d_{N_2}/t_{N_2}$ and $r_X = d_X/t_X$, where $d =$ distance the gas travels over time, t. Therefore,

$$\frac{d_{N_2}/t_{N_2}}{d_X/t_X} = \sqrt{\frac{\mathcal{M}_X}{\mathcal{M}_{N_2}}} = \frac{t_X}{t_{N_2}}$$

because $d_X = d_{N_2}$ (the gases travel the same distance). The times are given as 240 s for N_2 and 530 s for X.

Solve

$$\frac{530 \text{ s}}{240 \text{ s}} = \sqrt{\frac{\mathcal{M}_X}{28.01 \text{ g/mol}}}$$

$$\mathcal{M}_X = \left(\frac{530 \text{ s}}{240 \text{ s}}\right)^2 \times 28.01 \text{ g/mol} = 137 \text{ g/mol}$$

Think About It

We can use time of travel of the gases to find the molar mass, but we have to be careful because the rate is distance/time.

6.171. **Collect and Organize**
We are asked to derive an equation relating u_{rms} and the density of a gas.

Analyze
The equations describing u_{rms} and the density of a gas are

$$\mathcal{M} = \frac{dRT}{P} \quad \text{and} \quad u_{rms} = \sqrt{\frac{3RT}{\mathcal{M}}}$$

Solve
Rearranging the density equation for a gas:

$$\frac{1}{\mathcal{M}} = \frac{P}{dRT}$$

Substituting into the equation describing u_{rms} and cancelling species common to both:

$$u_{rms} = \sqrt{3RT \times \frac{1}{\mathcal{M}}} = \sqrt{3RT \times \frac{P}{dRT}} = \sqrt{\frac{3P}{d}}$$

Think About It
Alternately, we could express the equation derived above as

$$\left(u_{rms}\right)^2 = \frac{3P}{d} \quad \text{or} \quad d = \frac{3P}{\left(u_{rms}\right)^2}$$

6.173. **Collect and Organize**
We are given the balanced equation for ammonium nitrite decomposing to nitrogen and liquid water. We are asked to calculate the change in pressure in a 10.0 L vessel, due to N_2 formation, for 1.00 L of a 1.0 M NH_4NO_2 solution decomposing at 25°C.

Analyze
Because N_2 is the only gas involved in the reaction, all of the pressure increase will be due to the production of N_2 from the reaction. From the stoichiometric equation we can calculate the moles of N_2 that will be produced (assuming 100% decomposition) and then we can use the ideal gas equation ($PV = nRT$) to calculate the pressure change. The volume occupied by the gas will be the volume of the vessel, minus the volume occupied by the solution.

Solve

$$\text{Moles of } N_2 \text{ produced: } 1.00\text{ L} \times \frac{1.0\text{ mol }NH_4NO_2}{1\text{ L}} \times \frac{1\text{ mol }N_2}{1\text{ mol }NH_4NO_2} = 1.0\text{ mol}$$

$$P_{N_2} = \frac{(1.0\text{ mol})[0.08206\text{ L}\cdot\text{atm/(mol}\cdot\text{K)}](298\text{ K})}{9.0\text{ L}} = 2.7\text{ atm}$$

Think About It
Because the molar ratio of N_2 to NH_4NO_2 is 1:1, the moles of N_2 produced in the reaction equals the moles of NH_4NO_2 consumed.

6.175. **Collect and Organize**
From the balanced chemical equation and given the amount of nitrate that enters the swamp in one day (200.0 g), we are asked first to calculate the volume of CO_2 and N_2 that would form at 17°C (290 K) and 1.00 atm then to calculate the density of the gas mixture at that temperature and pressure.

Analyze
Once we calculate the moles of NO_3^- (from the molar mass of $NO_3^- = 62.00$ g/mol), we can calculate the moles of N_2 and CO_2 produced using the stoichiometric ratios from the balanced equation (2 NO_3^-:1 N_2 and 2 NO_3^-:5 CO_2). Using the ideal gas equation, we can calculate the volume of each gas that is produced. We can

then calculate the density of the gas mixture by determining the mass of CO_2 and N_2 produced (from the molar amounts already calculated) and dividing the total mass by the total volume the gases occupy.

Solve

Volume of gases produced in the swamp:

$$\text{mol NO}_3^- = 200.0 \text{ g} \times \frac{1 \text{ mol}}{62.00 \text{ g}} = 3.226 \text{ mol}$$

$$\text{mol N}_2 \text{ produced} = 3.226 \text{ mol NO}_3^- \times \frac{1 \text{ mol N}_2}{2 \text{ mol NO}_3^-} = 1.613 \text{ mol N}_2$$

$$\text{Volume of N}_2 = \frac{nRT}{P} = \frac{(1.613 \text{ mol})[0.08206 \text{ L} \cdot \text{atm/(mol} \cdot \text{K)}](290 \text{ K})}{1.00 \text{ atm}} = 38.4 \text{ L}$$

$$\text{mol CO}_2 \text{ produced} = 3.226 \text{ mol NO}_3^- \times \frac{5 \text{ mol CO}_2}{2 \text{ mol NO}_3^-} = 8.065 \text{ mol CO}_2$$

$$\text{Volume of CO}_2 = \frac{nRT}{P} = \frac{(8.065 \text{ mol})[0.08206 \text{ L} \cdot \text{atm/(mol} \cdot \text{K)}](290 \text{ K})}{1.00 \text{ atm}} = 192 \text{ L}$$

Total volume of gases at 1.00 atm and 290 K: 38.4 L + 192 L = 230 L

The density of this gas is as follows:

$$\text{Mass of N}_2 \text{ produced} = 1.613 \text{ mol} \times \frac{28.01 \text{ g}}{1 \text{ mol}} = 45.18 \text{ g}$$

$$\text{Mass of CO}_2 \text{ produced} = 8.065 \text{ mol} \times \frac{44.01 \text{ g}}{1 \text{ mol}} = 354.9 \text{ g}$$

$$\text{Total mass of gases} = 45.18 \text{ g} + 354.9 \text{ g} = 400.1 \text{ g}$$

$$\text{Density of gas mixture} = \frac{400.1 \text{ g}}{230 \text{ L}} = 1.74 \text{ g/L or } 0.00174 \text{ g/mL}$$

Think About It

The calculation of density here drew on the simple definition of density as mass divided by volume.

6.177. Collect and Organize

Given the mass of glucose, we are asked to calculate the volume of H_2 gas that could be produced at STP by the reaction described in the question.

Analyze

Recall that STP describes a temperature of 0 °C (273 K) and a pressure of 1.00 atm. We may obtain the number of moles of H_2 gas produced by dividing the mass of glucose by the molar mass of glucose (180.158 g/mol) and multiplying by the stoichiometric coefficients from the balanced chemical equation (12 H_2:1 glucose). Substituting this number of moles of H_2 gas into the ideal gas law and solving for V as below, we may determine the volume the gas would occupy.

$$V = \frac{nRT}{P}$$

Solve

$$256 \text{ g C}_6\text{H}_{12}\text{O}_6 \times \frac{1 \text{ mol C}_6\text{H}_{12}\text{O}_6}{180.158 \text{ g C}_6\text{H}_{12}\text{O}_6} \times \frac{12 \text{ mol H}_2}{1 \text{ mol C}_6\text{H}_{12}\text{O}_6} = 17.05 \text{ mol H}_2$$

$$V = \frac{(17.05 \text{ mol})[0.08206 \text{ L} \cdot \text{atm/(mol} \cdot \text{K)}](273 \text{ K})}{1.00 \text{ atm}} = 382 \text{ L}$$

Think About It

Each gram of glucose produces nearly 1.5 liters of hydrogen gas!

6.179. **Collect and Organize**

We are to calculate the volume of a weather balloon (filled at 1.0 atm with a volume of 50.0 L) deployed into the center of a hurricane with a pressure that is 90 mbar below atmospheric pressure.

Analyze

The pressure at the center of the hurricane, in atmospheres, is

$$1013.25 \text{ mbar} - 90 \text{ mbar} = 923 \text{ mbar}$$

$$923 \text{ mbar} \times \frac{1 \text{ atm}}{1013.25 \text{ mbar}} = 0.911 \text{ atm}$$

We may use Boyle's Law ($P_1V_1 = P_2V_2$) to solve for the final volume of the balloon.

Solve

Using $P_1V_1 = P_2V_2$ and solving for V_2:

$$1.0 \text{ atm} \times 50.0 \text{ L} = 0.911 \text{ atm} \times V_2$$

$$V_2 = 54.9 \text{ L}$$

Think About It

Our result that the balloon increased in volume when the pressure decreased makes sense according to Boyle's law that volume and pressure are inversely related.

6.181. **Collect and Organize**

We are to predict the products and balance the equations when ammonia reacts with oxygen and when ammonia reacts with nitric acid.

Analyze

Ammonia with oxygen produces NO and H_2O. Ammonia with nitric acid produces NH_4NO_3.

Solve

(a) $4 \text{ NH}_3(g) + 5 \text{ O}_2(g) \rightarrow 4 \text{ NO}(g) + 6 \text{ H}_2\text{O}(g)$
(b) $\text{NH}_3(g) + \text{HNO}_3(\ell) \rightarrow \text{NH}_4\text{NO}_3(s)$

Think About It

Notice that the reaction of ammonia (a gas) with liquid nitric acid produces a solid salt, ammonium nitrate.

6.183. **Collect and Organize**

We are asked to write a balanced equation for the decomposition of NH_4NO_2 and determine whether the reaction is a redox reaction.

Analyze

The products of the decomposition are given as N_2 and H_2O gas. We have to balance for all atoms in the reaction. We can use oxidation numbers to determine whether this reaction is a redox reaction. If a species increases (more positive) the oxidation number on one of its atoms, it is oxidized; if a species decreases (more negative) the oxidation number on one of its atoms, it is reduced.

Solve

(a) The balanced chemical equation is

$$\text{NH}_4\text{NO}_2(s) \rightarrow \text{N}_2(g) + 2 \text{ H}_2\text{O}(g)$$

(b) Yes, this is a redox reaction. The oxidation state of nitrogen in NH_4^+ (the cation of the salt) changes from -3 to 0 in N_2; it is oxidized. The oxidation state of nitrogen in NO_2^- (the anion of the salt) changes from $+3$ to 0 in N_2; it is reduced.

Think About It

In this reaction the salt is both oxidized and reduced.

6.185. **Collect and Organize**

The production of N_2 to fill up automobile air bags produces liquid Na as a by-product. We are to consider why this needs to be removed in the reaction.

Analyze

This must have to do with some toxicity or flammability of the sodium metal or of its products.

Solve

Sodium is a highly reactive element. It reacts quickly with moist air to form NaOH and H_2. NaOH is caustic, since it is a strong base, and the H_2 produced could form an explosive mixture with oxygen in the air.

Think About It

Sodium azide is used because it quickly releases the desired nitrogen gas to fill the air bag upon impact. To deal with the sodium that is produced, KNO_3 and SiO_2 are added, as stated in the problem.

6.187. **Collect and Organize**

We are to calculate the enthalpy for the reaction of hydrazine as a rocket fuel in its reaction with hydrogen peroxide.

Analyze

To calculate the ΔH_{rxn}^i we use

$$\Delta H_{rxn}^{\circ} = \sum n \Delta H_{f,products}^{\circ} - \sum m \Delta H_{f,reactants}^{\circ}$$

Solve

$$\Delta H_{rxn}^{\circ} = \left[(1 \text{ mol } N_2)(0 \text{ kJ/mol}) + (4 \text{ mol } H_2O)(-241.8 \text{ kJ/mol}) \right]$$
$$- \left[(2 \text{ mol } H_2O_2)(-187.8 \text{ kJ/mol}) + (1 \text{ mol } N_2H_4)(50.63 \text{ kJ/mol}) \right]$$
$$\Delta H_{rxn}^{\circ} = -642.2 \text{ kJ}$$

Think About It

The reaction is not only exothermic, but it also produces 5 moles of gas to provide thrust for the rocket.

CHAPTER 7 | A Quantum Model of Atoms: Waves and Particles

7.1. Collect and Organize
From the highlighted elements in Figure P7.1 we can correlate position on the periodic table with orbital (*s*, *p*, *d*) filling.

Analyze
The periodic table consists of the *s*-block elements, which are the first two columns (groups 1–2), the *d*-block elements, which are the next ten columns (groups 3–12), and the *p*-block elements, which are the rightmost six columns (groups 13–18). The *f*-block elements are in the two rows at the bottom of the table. From their positions in these blocks, we can ascertain the elements' electron configurations.

Solve
(a) Group 1 elements have an ns^1 configuration, including the element shaded purple. Because half-filled and filled *d* orbitals are predicted for transition metals, the red-shaded and orange-shaded ($3d^5 4s^1$) elements also have a single *s* electron in their outermost shells.
(b) Filled sets of *s* and *p* orbitals ($ns^2 np^6$ configuration) occur for the noble gases (group 18 elements), including the element shaded blue.
(c) Filled sets of *d* orbitals would occur for elements in period 4 and below having in their electron configuration nd^{10}. Because a filled *d* orbital is stable for transition elements, the orange-shaded element is predicted to have a $5d^{10} 6s^1$ configuration.
(d) A half-filled *d*-orbital set would be predicted for the element shaded red; its electron configuration would have $3d^5 4s^1$.
(e) Filled *s* orbitals would occur in the outermost shells of the blue-shaded element ($2s^2 2p^6$) and the green-shaded element ($3s^2 3p^5$).

Think About It
Remember that the outermost shell of an atom includes all those electrons above the previous noble gas core.

7.3. Collect and Organize
Of the highlighted elements in Figure P7.1, we are to identify the elements that form common ions (cations or anions) that are larger than the element itself.

Analyze
Anions are always larger than the parent atom, so we are looking for elements that will likely form an anion (X^{n-}). Nonmetals tend to form anions (metals tend to form cations), so we are looking for an element that is located on the right side of the periodic table.

Solve
Both the green and blue elements are nonmetals and potentially form anions. Green would form X^-, picking up an electron to fill its outermost shell. Blue, however, would not form an anion; its outermost shell is already full.

Think About It
A majority of the elements in the periodic table are classified as metals, so most elements tend to form cations, not anions.

7.5. Collect and Organize
Of the highlighted elements in Figure P7.4, we are to find the elements that form cations or anions that are smaller than the parent atom.

Analyze
Cations are always smaller than the parent atom so we look for elements that likely form cations (X^{n+}). Metals tend to form cations, so we look for elements on the left side of the periodic table.

Solve

Blue, green, and orange are all metals and potentially lose electrons to form cations that are smaller than their parent atoms.

Think About It

As an atom loses more electrons, the size continues to decrease. Therefore $X^+ > X^{2+} > X^{3+}$ in size.

7.7. Collect and Organize

All the highlighted elements in Figure P7.4 are located in the same row (period) of the periodic table. In assessing the relative sizes of the monatomic ions these form, we have to consider whether the elements tend to form cations or anions. When a cation is formed the size of the atom is reduced; when an anion is formed, the atom expands in size. Also, as the charge on a cation increases, the size decreases (X^{2+} is smaller than X^+), but as the charge on an anion increases, the size increases (X^{2-} is larger than X^-).

Analyze

Metals (left side) tend to form cations:
 Blue would form 1+ ion (blue$^+$)
 Green would form 2+ ion (green^{2+})
 Orange would form 3+ ion (orange^{3+})
Nonmetals (right side) tend to form anions:
 Red would form 2– ion (red^{2-})
 Dark gray would form 1– ion (gray$^-$)

Solve

In order of increasing size: orange (Y^{3+}) < green (Sr^{2+}) < blue (Rb^+) < gray (I^-) < red (Te^{2-})

Think About It

The cations in this series (orange^{3+}, green^{2+}, blue$^+$) are isoelectronic with each other as are the anions (gray$^-$, red^{2-}).

7.9. Collect and Organize

The second ionization energy is described by the equation $X^+(g) \rightarrow X^{2+}(g) + e^-$. To determine which highlighted element in Figure P7.4 has the highest second ionization energy (IE_2) we have to consider each element's electronic configuration in this ionization reaction.

Analyze

	X^+ Configuration	X^{2+} Configuration
Blue	$5s^0$ or [Kr]	[Ar] $3d^{10}4p^54s^2$
Green	$5s^1$	[Kr]
Orange	$4d^15s^1$	$4d^1$
Red	$4d^{10}5s^25p^3$	$4d^{10}5s^25p^2$
Gray	$4d^{10}5s^25p^4$	$4d^{10}5s^25p^3$

Solve

Once an atom has attained a noble gas configuration it is difficult to remove further electrons. Therefore, the element shaded blue has the highest IE_2.

Think About It

The second ionization energy is always higher than the first ionization energy because we are trying to remove an electron (e^-) from an already positively charged atom (X^+).

7.11. Collect and Organize

The diagram in Figure P7.11 represents electronic transitions between energy levels in an atom. We are asked to identify lines corresponding to absorption and emission.

Analyze

When an electron absorbs energy, it is promoted to a higher energy level. When an electron drops to a lower energy level, energy is emitted. The arrow points in the direction of electron travel.

Solve

Lines (d), (e), and (f) represent the movement of electrons to a higher energy level, so they are absorption events. Lines (a), (b), and (c) represent movement of electrons to a lower energy level, so they are emission events.

Think About It

Lines (a) and (e) and lines (b) and (f) represent transitions with the same energy but opposite directions (absorption versus emission).

7.13. **Collect and Organize**

The diagram in Figure P7.11 represents electronic transitions between energy levels in an atom. We are asked to identify an absorption of light starting from an excited state.

Analyze

Excited states have a higher potential energy than the ground state, which is the lowest energy state. We must determine which, if any, of the lines start from a higher energy state, gaining energy as light is absorbed.

Solve

Line (e) represents an absorption starting from an excited state.

Think About It

Lines (d) and (f) are both absorptions, but both originate from the ground state.

7.15. **Collect and Organize**

We are to identify the spheres in Figure P7.15 as Na, K, or Na^+ atoms/ions.

Analyze

We can predict which sphere is which atom/ion by considering individual size relationships. Since potassium is a heavier member of the alkali metals than sodium, potassium atoms should be larger than sodium atoms. Since cations are smaller than the neutral atoms from which they are derived, Na^+ should be smaller than a sodium atom.

Solve

The relationships determined above are

$$K > Na \quad \text{and} \quad Na > Na^+$$

Thus, the largest sphere (b) is K, the smallest sphere (c) is Na^+, and the remaining sphere (a) is Na.

Think About It

Na^+ is much smaller than Na because the removal of one electron results in a completely empty $3s$ orbital.

7.17. **Collect and Organize**

We are asked which of the orbital diagrams in Figure P7.17 represents an excited state of a nitrogen atom.

Analyze

The lowest energy arrangement of electrons is called the ground state. Lower energy orbitals fill before higher energy orbitals, and degenerate orbitals fill singly at first, then electrons pair up. The ground state of nitrogen has an electron configuration of $1s^2 2s^2 2p^3$. Any configuration with seven electrons where more than three electrons are located in the $2p$ orbital will be an excited state, and any configuration where electrons in the $2p$ orbital are paired up will be an excited state.

Solve

(a) and (b) are excited states of nitrogen. (a) has paired electrons in the $2p$ orbital, rather than all three unpaired and distributed equally. In (b), one electron has been promoted from the $2s$ orbital to the $2p$, instead of completely filling lower energy orbitals first.

Think About It

Configuration (c) corresponds to the ground state of nitrogen, and (d) corresponds to the ground state of oxygen.

7.19. **Collect and Organize**

We are to determine which of the waves shown in Figure P7.19 has the lowest energy. Recall that the energy of a wave is determined using the equation

$$E = \frac{hc}{\lambda}$$

Analyze

The lowest energy wave will have the longest wavelength (the larger denominator will result in a smaller value of E). It is often easier to estimate wavelength by tracing a complete cycle of each wave, and making a mark, as shown below.

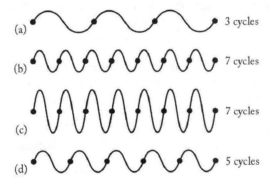

Solve

As seen above, the wave depicted in (a) completes only three cycles over the same space as the other waves. (a) has the longest wavelength, and as such the smallest energy.

Think About It

Although (c) and (d) look different, they have the same wavelength. It is easier to distinguish differences between waves using a mark to track a complete cycle of the wave.

7.21. **Collect and Organize / Analyze**

All forms of radiant energy (light) from gamma rays to low energy radio waves are called *electromagnetic radiation*. Why?

Solve

All these forms of light have perpendicular, oscillating electric and magnetic fields that travel together through space as described by Maxwell.

Think About It

All forms of electromagnetic radiation travel at the speed of light (3.00×10^8 m/s in a vacuum).

7.23. **Collect and Organize**

We are asked why a lead shield is used at the dentist's office when X-ray images are taken.

Analyze

Light interacts with matter and X-rays are high-energy light that can damage living cells.

Solve

The lead shield must be used to protect the part of our bodies that might be unnecessarily exposed to X-rays (does not need to be imaged). Lead is a very high density metal with many electrons that interact with X-rays and can absorb nearly all the X-rays before they can reach our bodies.

Think About It

Exposure to high-energy radiation (γ rays and X-rays in particular) may cause genetic damage in cells, which may lead to cancers.

7.25. Collect and Organize

Given the definition of ionizing radiation, we are asked to name other forms of radiation that might also be ionizing, such as gamma rays.

Analyze

Ionizing radiation (e.g., gamma rays) must have high energy in order to break apart molecules into ions and electrons. Using Figure 7.1 we can determine what forms of electromagnetic radiation have these high energies.

Solve

X-rays and ultraviolet radiation, having short wavelengths and high energies, could also be ionizing like gamma rays.

Think About It

We know that visible light is not ionizing because we are not harmed by exposure to it. However, the UV radiation that is not filtered out by the upper atmosphere reaches us from the Sun, damaging our cells and causing sunburn and skin cancer.

7.27. Collect and Organize

We are to calculate the frequency of light of a wavelength given in nanometers ($\lambda = 616$ nm).

Analyze

The wavelength of light is related to the frequency through the equation $\nu = c/\lambda$. Wavelength must be expressed in meters for this calculation (1 nm = 1×10^{-9} m).

Solve

$$\nu = \frac{3.00 \times 10^8 \text{ m/s}}{\left(616 \text{ nm} \times \dfrac{1 \times 10^{-9} \text{ m}}{1 \text{ nm}} \right)} = 4.87 \times 10^{14} \text{ s}^{-1}$$

Think About It

Be sure to convert nanometers to meters when using this equation.

7.29. Collect and Organize

Given the frequencies of several radio stations, we are to calculate the corresponding wavelengths.

Analyze

The equation to calculate wavelength from frequency is
$$\lambda = c/\nu$$
where λ is in meters, $c = 3.00 \times 10^8$ m/s, and ν is in hertz (per second). We need to convert megahertz to hertz for our calculation (1×10^6 Hz = 1 MHz).

Solve

(a) $\lambda_{\text{KKNB}} = \dfrac{3.00 \times 10^8 \text{ m/s}}{104.1 \times 10^6 \text{ s}^{-1}} = 2.88$ m

(b) $\lambda_{\text{WFNX}} = \dfrac{3.00 \times 10^8 \text{ m/s}}{101.7 \times 10^6 \text{ s}^{-1}} = 2.95 \text{ m}$

(c) $\lambda_{\text{KRTX}} = \dfrac{3.00 \times 10^8 \text{ m/s}}{100.7 \times 10^6 \text{ s}^{-1}} = 2.98 \text{ m}$

Think About It
Remember that the speed of electromagnetic radiation in air is approximately the same as in a vacuum (3.00×10^8 m/s).

7.31. Collect and Organize
We are to compare the frequency of 1090 kHz radio waves versus that of the green light emitted from an LED.

Analyze
We can convert the wavelength of light (550 nm) to frequency using $v = c/\lambda$. Wavelength has to be expressed in meters for this calculation (1 nm = 1×10^{-9} m).

Solve
The frequency of the LED light (b) is

$$v = \frac{3.00 \times 10^8 \text{ m/s}}{550 \times 10^{-9} \text{ m}} = 5.45 \times 10^{14} \text{ s}^{-1}$$

This frequency is much higher than 1090 kHz (1.090×10^6 Hz = 1.090×10^6 s^{-1}) from the radio station. Therefore, the radio station (a) has the lower frequency.

Think About It
The radio station has the longer wavelength since wavelength and frequency are inversely related.

7.33. Collect and Organize
Earth is 93 million miles from the Sun. We are to calculate how long it takes for the Sun's light to reach Earth.

Analyze
We can find the time it takes light to travel from the Sun to Earth by dividing the distance (93 million miles) by the speed at which the light travels (3.00×10^8 m/s). We have to convert the distance in miles to meters.

Solve

$$\text{Time} = \frac{\left(93 \times 10^6 \text{ mi} \times \dfrac{1609 \text{ m}}{1 \text{ mi}}\right)}{3.00 \times 10^8 \text{ m/s}} = 499 \text{ s or } 8.3 \text{ min}$$

Think About It
Even though light travels very fast, the large distance between Earth and the Sun means that events (e.g., solar flares) we witness on the Sun actually happened 8 minutes ago.

7.35. Collect and Organize
We are asked to compare the atomic emission and absorption spectra of hydrogen.

Analyze
Emission of light from a hydrogen atom occurs when the atoms are heated to a high temperature. Absorption of light occurs when the atoms absorb energy from an external source of energy. We are asked to compare the two spectra for atomic hydrogen.

Solve
The hydrogen absorption spectrum consists of dark lines at wavelengths specific to hydrogen. The emission spectrum has bright lines on a dark background with the lines appearing at the exact same wavelengths as the dark lines in the absorption spectrum.

Think About It

The hydrogen atom absorbs only certain amounts of energy (as seen in the absorption spectrum) and, once excited to those higher energy states, emits that same energy back (in the emission spectrum) to return to its ground (lowest) energy state.

7.37. Collect and Organize

We are asked how study of the emission spectra of the elements led to identification of the dark Fraunhofer lines in the Sun's spectrum.

Analyze

The emission lines for elements match the absorption lines exactly in energy because these processes are the exact reverse of each other.

Solve

In the study of atomic emission spectra, each element was found to have a distinctive and unique set of absorption and emission lines. Observations of the many Fraunhofer lines (dark absorbtion lines) in sunlight were found to correspond to several atomic emission spectra of elements, thus allowing us to deduce the Sun's elemental composition.

Think About It

The elemental composition of distant stars can be determined in this way, as well.

7.39. Collect and Organize / Analyze

We are to define the terms *quantum* and *photon*.

Solve

Named by Max Planck, the quantum is the smallest indivisible amount of radiant energy that an atom can absorb or emit. A photon is the smallest indivisible packet or particle of light energy.

Think About It

Planck also defined the relationship between energy of the quantum particle to its frequency ($E = h\nu$). The energy of a photon can be related to its frequency ($E = h\nu$) and wavelength ($E = hc/\lambda$).

7.41. Collect and Organize

We are asked what effect the amplitude (intensity) of incoming light has on the emission of electrons by the photoelectric effect.

Analyze

Electrons are emitted from a charged surface when the energy of the incident light is higher than the work function for the surface.

Solve

Provided the incoming light has a high enough energy to cause the emission of electrons to occur in the first place, increasing the intensity of the light causes more electrons to be emitted as a function of time.

Think About It

The energy of the incident light must be greater than the work function for the surface or no electrons will be emitted, regardless of the intensity of light.

7.43. Collect and Organize

From a list, we are to choose which have quantized values.

Analyze

Something is quantized if it is present only in discrete amounts and can only have whole-number multiples of the smallest amount.

Solve
(a) The elevation of a step on a moving escalator continuously changes—this is not a quantized value.
(b) Because the doors open only *at* the floors and not *between* the floors—this value is quantized.
(c) The speed of an automobile can change smoothly—this is not a quantized value.

Think About It
Any quantity that is quantized has to have changes occurring in discrete steps.

7.45. Collect and Organize
We are to calculate the velocity (speed) of electrons ejected from Na and K when irradiated by a 300–nm light source to determine which has the higher speed. Because the work function for potassium is lower than that of sodium, we expect that an electron ejected from potassium has a higher kinetic energy.

Analyze
For each metal, $KE_{electron} = h\nu - \Phi$ where the frequency of the light irradiating the metals would be found from $\nu = c/\lambda$ where $\lambda = 300$ nm (3.00×10^{-7} m). The speed of the electron can then be found as

$$KE = \frac{1}{2}m_e u^2 \text{ or } u = \sqrt{\frac{2KE}{m_e}}$$

where m_e is the mass of an electron, 9.11×10^{-31} kg.

Solve
Potassium:

$$KE = \left(6.626\times10^{-34}\text{ J·s}\times\frac{3.00\times10^8\text{ m/s}}{3.00\times10^{-7}\text{ m}}\right) - 3.68\times10^{-19}\text{ J} = 2.946\times10^{-19}\text{ J}$$

$$u = \sqrt{\frac{2\times2.946\times10^{-19}\text{ kg·m}^2/\text{s}^2}{9.11\times10^{-31}\text{ kg}}} = 8.04\times10^5\text{ m/s}$$

Sodium:

$$KE = \left(6.626\times10^{-34}\text{ J·s}\times\frac{3.00\times10^8\text{ m/s}}{3.00\times10^{-7}\text{ m}}\right) - 4.41\times10^{-19}\text{ J} = 2.216\times10^{-19}\text{ J}$$

$$u = \sqrt{\frac{2\times2.216\times10^{-19}\text{ kg·m}^2/\text{s}^2}{9.11\times10^{-31}\text{ kg}}} = 6.97\times10^5\text{ m/s}$$

Potassium's ejected electrons have a greater speed than sodium's.

Think About It
Notice how the units of 1 J = 1 kg·m^2/s^2 are useful in being sure that the speeds are in meters per second.

7.47. Collect and Organize
We can use the equation for the photovoltaic effect to determine if electrons could be ejected from tantalum using light with a wavelength of 500 nm.

Analyze
As long as the energy of the light that shines on the metal is greater than the work function for tantalum ($\Phi = 6.81 \times 10^{-19}$ J), the light will eject electrons, and tantalum would therefore be useful in a voltaic cell. The energy of the light can be calculated from $E = hc/\lambda$.

Solve

The energy of light of 500 nm wavelength can be found as

$$E = \frac{6.626 \times 10^{-34} \text{ J} \cdot \text{s} \times 3.00 \times 10^{8} \text{ m/s}}{5.00 \times 10^{-7} \text{ m}} = 3.98 \times 10^{-19} \text{ J}$$

This energy is less than the work function for tantalum ($\Phi = 6.81 \times 10^{-19}$ J), so tantalum could not be used to convert solar energy at 500 nm to electricity.

Think About It

The wavelength of light that would be needed to eject electrons from tantalum would be

$$\lambda = \frac{hc}{\Phi} = \frac{6.626 \times 10^{-34} \text{ J} \cdot \text{s} \times 3.00 \times 10^{8} \text{ m/s}}{6.81 \times 10^{-19} \text{ J}} = 2.92 \times 10^{-7} \text{ m or } 292 \text{ nm}$$

7.49. Collect and Organize

Using the information that a red laser's power is 1 J/s (1 Watt), we are to calculate the number of photons emitted by the laser per second.

Analyze

The energy of the photons of red laser light ($\lambda = 630$ nm) is given by

$$E = hc/\lambda$$

This corresponds to the number of photons in a second by

$$\text{Number of photons} = (1 \text{ J}) \times \left(\frac{1 \text{ photon}}{\text{energy of photon}} \right)$$

Solve

Energy of one photon with $\lambda = 630$ nm:

$$E = \frac{6.626 \times 10^{-34} \text{ J} \cdot \text{s} \times 3.00 \times 10^{8} \text{ m/s}}{6.30 \times 10^{-7} \text{ m}} = 3.155 \times 10^{-19} \text{ J/photon}$$

Number of photons per second:

$$1.00 \text{ Watt} \times \frac{1 \text{ J/s}}{\text{Watt}} \times \frac{1 \text{ photon}}{3.155 \times 10^{-19} \text{ J}} = 3.17 \times 10^{18} \text{ photons/s}$$

Think About It

Greater power (Wattage) leads to more photons being emitted each second. Also, the shorter the wavelength, the greater the energy per photon and the higher the power (Wattage) for the same number of photons emitted.

7.51. Collect and Organize

Hydrogen has the simplest atomic spectrum of all the elements, and we are asked why this might be so.

Analyze

The hydrogen atom consists of 1 proton in the nucleus and 1 electron.

Solve

The atomic spectrum of hydrogen is simple because the single electron interacts only with the proton in the nucleus, and there are no other electrons to repel it.

Think About It

The He^{+} atom, also with only 1 electron, would have a simple atomic spectrum, too.

7.53. Collect and Organize

We must consider the process of emission of light from a hydrogen atom in an excited state. Does the energy emitted depend on the values of n_1 and n_2 or only on the difference between them ($n_1 - n_2$)?

Analyze

The emission of light from an H atom in the excited state is the result of the electron dropping from a higher energy orbit to a lower-lying orbit.

Solve

Because it is the difference in the orbits' energies that is correlated to the light energy emitted by the hydrogen atom in the excited state, it is the difference between n levels that determines emission energy as the excited H atom relaxes to its ground state.

Think About It

The n levels get closer in energy as n increases, so that $n_2 - n_1 > n_3 - n_2 > n_4 - n_3$, etc.

7.55. **Collect and Organize**

In considering the absorption of energy for an electron to be promoted from a lower energy orbit to a higher energy orbit ($n_1 < n_2$), we can predict which transition requires the shortest wavelength of light.

Analyze

The transition with the shortest wavelength has the highest change in energy. The energy levels (n) in hydrogen are not evenly spaced; as n increases, the differences in energy between adjacent energy levels decreases, as shown here.

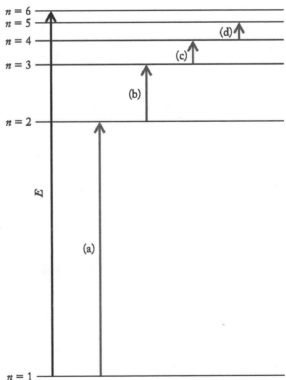

Solve

All of the transitions given involve changes between adjacent levels ($\Delta n = 1$). Because the energy levels become more closely spaced as n increases, ΔE for energy levels for lower n values are greater and therefore have shorter wavelengths associated with them. Therefore, (a), in which the electron is "promoted" from $n = 1$ to $n = 2$, has the shortest wavelength.

Think About It

The order of the transitions from longest to shortest wavelength is

(d) $n = 4$ to $n = 5$ > (c) $n = 3$ to $n = 4$ > (b) $n = 2$ to $n = 3$ > (a) $n = 1$ to $n = 2$

7.57. Collect and Organize / Analyze

The lines of the Fraunhofer series from hydrogen in the Sun's spectrum are observed in the visible region of the electromagnetic spectrum. These arise from transitions from $n = 2$ to higher energy levels ($n = 3$–6). We are asked whether any transitions from $n = 3$ to higher energy levels would be observed among the Fraunhofer lines.

Solve

From Figure 7.15 (on page 331, which shows emission not absorption, but the two are complementary), we see that the $n = 2$ to $n = 3$, 4, 5, and 6 energy states generally have larger energies associated with them than transitions from $n = 3$. It is possible, however, that transitions from $n = 3$ to much higher energy levels (e.g., $n = 10$) would have energies in the visible range. If we promote the electron to a very high energy level (let's pick $n = 1000$) then

$$\frac{1}{\lambda} = [1.097 \times 10^{-2} \, (\text{nm})^{-1}] \left(\frac{1}{3^2} - \frac{1}{1000^2} \right)$$

$$\lambda = 820.4 \text{ nm}$$

This wavelength still lies outside of the visible range, in the infrared range. There are no Fraunhofer lines for transitions from $n = 3$ in hydrogen.

Think About It

We could see $n = 3$ transitions in the Paschen series.

7.59. Collect and Organize

We are asked to explain why Balmer observed the $n = 6$ to $n = 2$ transition in hydrogen but not $n = 7$ to $n = 2$.

Analyze

As seen in Figure 7.15, the energy levels become more closely spaced as the value of n increases. We can use the equation below to calculate the wavelength of the transition.

$$\frac{1}{\lambda} = [1.097 \times 10^{-2} \, (\text{nm})^{-1}] \left(\frac{1}{n_1^2} - \frac{1}{n_2^2} \right)$$

Solve

$$\frac{1}{\lambda} = [1.097 \times 10^{-2} \, (\text{nm})^{-1}] \left(\frac{1}{2^2} - \frac{1}{7^2} \right)$$

$$\lambda = 397.0 \text{ nm}$$

At $n = 7$, the wavelength of the electron's transition ($n = 7$ to $n = 2$) has moved out of the visible region.

Think About It

The transition from the $n = 7$ energy level can be detected in the UV region of the electromagnetic spectrum.

7.61. Collect and Organize

To calculate the wavelength emitted when an electron in hydrogen undergoes a transition from $n = 4$ to $n = 3$, we can use the Rydberg equation. The region of the electromagnetic spectrum corresponding to that wavelength can be found from Figure 7.1.

Analyze

In the Rydberg equation, $n_1 = 3$ and $n_2 = 4$.

$$\frac{1}{\lambda} = [1.097 \times 10^{-2} \, (\text{nm})^{-1}] \left(\frac{1}{n_1^2} - \frac{1}{n_2^2} \right)$$

Solve

$$\frac{1}{\lambda} = [1.097 \times 10^{-2} \ (nm)^{-1}] \left(\frac{1}{3^2} - \frac{1}{4^2} \right) = 5.333 \times 10^{-4} (nm)^{-1}$$

$$\lambda = 1875 \ nm$$

This wavelength occurs in the infrared region of the electromagnetic spectrum.

Think About It

Notice that in the Rydberg equation, n_2 is a higher orbit number (n) than n_1. In this way we do not get the nonsensical result of a negative wavelength.

7.63. **Collect and Organize**

The equation given relates the photon's energy to atomic number and the energy-level transition.

Analyze

The equation shows a direct relationship between the energy and the atomic number and between the energy and the transition of the electron between n_1 and n_2 as

$$E = (2.18 \times 10^{-18} \ J)Z^2 \left(\frac{1}{n_1^2} - \frac{1}{n_2^2} \right)$$

Solve

(a) Because the energy of the photons is directly related to the atomic number in the equation, as Z increases, the energy increases. Since energy is inversely related to wavelength, the wavelength decreases as Z increases.
(b) The energy of the photon for $Z = 1$ (hydrogen) is

$$E = (2.18 \times 10^{-18} \ J)(1)^2 \left(\frac{1}{1^2} - \frac{1}{2^2} \right) = 1.635 \times 10^{-18} \ J$$

The wavelength of this photon is

$$\lambda = \frac{6.626 \times 10^{-34} \ J \cdot s \times 3.00 \times 10^8 \ m/s}{1.635 \times 10^{-18} \ J} = 1.22 \times 10^{-7} \ m \ or \ 122 \ nm$$

This wavelength is in the UV range.
As Z increases, the energy of the photon increases and the wavelength decreases, so the transition will never be observed in the visible range.

Think About It

The $n = 1$ and $n = 2$ energy levels are too far apart in energy to give an emitted photon in the visible range.

7.65. **Collect and Organize**

The difference between the hydrogen atom and the Li^{2+} ion is the charge on the nucleus. Both are one-electron species, so we can use the Bohr equation to calculate the energy (and then the wavelength) of the $n = 3$ to $n = 2$ transition in Li^{2+}.

Analyze

In the equation

$$E = \left(2.18 \times 10^{-18} \ J\right) Z^2 \left(\frac{1}{n_f^2} - \frac{1}{n_i^2} \right)$$

For Li^{2+} in this problem, $Z = 3$, $n_i = 3$ and $n_f = 2$.
To convert energy into wavelength use $\lambda = hc/E$.

Solve

$$E = \left(2.18 \times 10^{-18} \ J\right)(3)^2 \left(\frac{1}{2^2} - \frac{1}{3^2} \right) = 2.725 \times 10^{-18} \ J$$

$$\lambda = \frac{6.626 \times 10^{-34} \text{ J} \cdot \text{s} \times 3.00 \times 10^{8} \text{ m/s}}{2.725 \times 10^{-18} \text{ J}} = 7.29 \times 10^{-8} \text{ m or 72.9 nm}$$

This wavelength is much shorter than the wavelength associated with the same transition in an H atom.

Think About It
Notice that this would be the energy and wavelength of the photon *emitted* from the Li^{2+} ion.

7.67. Collect and Organize / Analyze
The de Broglie equation is

$$\lambda = \frac{h}{mv}$$

We are to define the symbols in the equation and explain how this equation shows the wavelike properties of a particle.

Solve
In the de Broglie equation, λ is the wavelength the particle, of mass m, exhibits as it travels at speed v, with h being Planck's constant. This equation states that any moving particle has wavelike properties because a wavelength can be calculated through the equation and that the wavelength of the particle is inversely related to its momentum (mass multiplied by velocity).

Think About It
From the equation, we see that as mass and/or speed increases, the wavelength of a particle decreases.

7.69. Collect and Organize / Analyze
In this question we consider whether the shape or density of an object has any effect on its de Broglie wavelength.

Solve
The de Broglie equation relates only the mass and the speed to the wavelength, so neither the shape nor the density would affect the de Broglie wavelength.

Think About It
From the de Broglie equation we see that as the mass or the speed of a particle increases, the wavelength associated with that particle decreases.

7.71. Collect and Organize
Given the mass of objects (from a muon to Earth), we are to use de Broglie's equation to calculate the wavelength of the objects moving at given speeds.

Analyze
The de Broglie equation is

$$\lambda = \frac{h}{mv}$$

where the Planck constant is $h = 6.626 \times 10^{-34}$ J·s, m is the mass of the particle in kg, and v is the speed of the particle in m/s. Recall that 1 J = 1 kg·m^2/s^2.

Solve
(a) Muon:

$$\lambda = \frac{6.626 \times 10^{-34} \text{ kg} \cdot \text{m}^2/\text{s}}{1.884 \times 10^{-28} \text{ kg} \times 325 \text{ m/s}} = 1.08 \times 10^{-8} \text{ m or 10.8 nm}$$

(b) Electron:

$$\lambda = \frac{6.626 \times 10^{-34} \text{ kg} \cdot \text{m}^2/\text{s}}{9.10939 \times 10^{-31} \text{ kg} \times 4.05 \times 10^{6} \text{ m/s}} = 1.80 \times 10^{-10} \text{ m or 0.180 nm}$$

(c) Runner:

$$\lambda = \frac{6.626 \times 10^{-34} \text{ kg} \cdot \text{m}^2/\text{s}}{80 \text{ kg} \times \left(\dfrac{1 \text{ mi}}{4 \text{ min}} \times \dfrac{1 \text{ min}}{60 \text{ s}} \times \dfrac{1609 \text{ m}}{1 \text{ mi}} \right)} = 1.24 \times 10^{-36} \text{ m or } 1.24 \times 10^{-27} \text{ nm}$$

(d) Earth:

$$\lambda = \frac{6.626 \times 10^{-34} \text{ kg} \cdot \text{m}^2/\text{s}}{6.0 \times 10^{24} \text{ kg} \times 3.0 \times 10^4 \text{ m/s}} = 3.68 \times 10^{-63} \text{ m or } 3.68 \times 10^{-54} \text{ nm}$$

Think About It
Generally, only small particles with low mass show wavelike behavior. We can only detect λ on the order of the size of atoms or particles that are 10^{-10} m. We, therefore, do not observe the athlete's or Earth's waves.

7.73. Collect and Organize
When objects of different masses and/or speeds are compared we can use the de Broglie relationship and the inverse relationship between λ and v to determine how the frequency of particles compares and whether the statements are true.

Analyze
Using $\lambda = c/v$, we can express the de Broglie relationship as

$$\lambda = \frac{c}{v} = \frac{h}{mv}$$

$$v = \frac{cmv}{h}$$

Here, frequency increases with increasing mass and speed.

Solve
(a) False. The relative frequencies of a large, fast-moving object and a lighter, faster-moving object depend upon how much larger (massive) the large object is and how much faster the light object is moving. For example, a large object with two times the mass of a light object has the same frequency as the light object moving two times faster.
(b) False. Very light (low mass) particles will have low frequencies.
(c) True. Doubling the mass while halving the speed has no effect on v:

$$v = \frac{c \times 2m \times \dfrac{1}{2}v}{h} = \frac{cmv}{h}$$

Think About It
When considering the frequency of a particle using the de Broglie equation, there is a direct relationship between the frequency and both mass and velocity of the particle.

7.75. Collect and Organize
We are asked to calculate u_{rms} and the DeBroglie wavelength for a helium atom at 500 K.

Analyze
u_{rms} may be determined using the equation

$$u_{\text{rms}} = \sqrt{\frac{3RT}{\mathcal{M}}}$$

where $\mathcal{M}$ is the molar mass of the helium atom in kg/mol (4.002602×10^{-3} kg/mol), R is 8.314 Kg·m^2/(s·k·mol), and T is the temprature in Kelvin (500 K). We can express the de Broglie relationship as

$$\lambda = \frac{c}{v} = \frac{h}{mv}$$

where the m is the mass of our particle, and v is u_{rms}. The mass of a single helium atom in Kg/atom is determined by dividing the molar mass in kg/mol by Avogadro's number.

Solve

(a) $u_{rms} = \sqrt{\dfrac{3\left(8.314 \text{ kg} \cdot \text{m}^2 / (\text{s} \cdot \text{K} \cdot \text{mol})\right)\left(500 \text{ K}\right)}{4.002602 \times 10^{-3} \text{ kg/mol}}} = 1.77 \times 10^3 \text{ m/s}$

(b)

$$m = \frac{4.002602 \times 10^{-3} \text{ kg}}{1 \text{ mol}} \times \frac{1 \text{ mol}}{6.022 \times 10^{23} \text{ atoms}} = 6.64663 \times 10^{-27} \text{ kg/atom}$$

$$\lambda = \frac{6.626 \times 10^{-34} \text{ J} \cdot \text{s}}{6.64663 \times 10^{-27} \text{ kg/atom}\left(1.77 \times 10^3 \text{ m/s}\right)} = 5.65 \times 10^{-11} \text{ m} = 56.5 \text{ pm}$$

Think About It

The frequency of this particle is much smaller than the frequency of visible light.

7.77. Collect and Organize

Heisenberg's uncertainty principle states that the uncertainty in the position (Δx) multiplied by $m\Delta v$, where m is the particle's mass and Δv is the uncertainty in the particle's speed, must be equal to or greater than $h/4\pi$, that is:

$$\Delta x \cdot m\Delta v \geq \frac{h}{4\pi}$$

Analyze

For the H_2^+ particle in the cyclotron, the 3% uncertainty in velocity would be

$$\Delta v = 4 \times 10^6 \text{ m/s} \times 0.03 = 1.2 \times 10^5 \text{ m/s}$$

The mass of the H_2^+ would be

$$\frac{2.016 \text{ g}}{1 \text{ mol}} \times \frac{1 \text{ mol}}{6.022 \times 10^{23} \text{ particles}} \times \frac{1 \text{ kg}}{1000 \text{ g}} = 3.35 \times 10^{-27} \text{ kg}$$

Solve

Rearranging Heisenberg's equation to solve for Δx:

$$\Delta x \geq \frac{h}{4\pi m \Delta v} = \frac{6.626 \times 10^{-34} \text{ kg} \cdot \text{m}^2/\text{s}}{4\pi \times 3.35 \times 10^{-27} \text{ kg} \times 1.2 \times 10^5 \text{ m/s}} = 1.3 \times 10^{-13} \text{ m}$$

Think About It

The minimum uncertainty in position is very small, smaller than what we can measure, so in principle we can accurately determine the position of the H_2^+ particle moving at this speed.

7.79. Collect and Organize / Analyze

We are to differentiate between a Bohr orbit and a quantum theory orbital.

Solve

The Bohr model orbit showed the quantized nature of the electron in the atom as a particle moving around the nucleus in concentric orbits, much like planets moving around the Sun.

In quantum theory, an orbital is a region of space where the probability of finding the electron is high. The electron is not viewed as a particle, but as a wave, and it is not confined to a clearly defined orbit; rather, we refer to the probability of the electron being at various locations around the nucleus.

Think About It

Bohr's model helped explain atomic spectra. The quantum theory of the atom helped to explain much more, including how atoms bond together and the probability of an electronic transition in an atom.

7.81. **Collect and Organize / Analyze**
To identify the orbital for an electron, we are asked how many quantum numbers we would need.

Solve
We need to describe the shell, the subshell, and the orbital's orientation to define a particular orbital. Therefore, we need three quantum numbers: n, ℓ, and m_ℓ.

Think About It
We could not use fewer quantum numbers to describe a particular orbital because there would then be confusion as to which shell or subshell an electron belonged to or what its orientation was.

7.83. **Collect and Organize**
As the principal quantum number, n, increases so too does the number of orbitals available at the n level.

Analyze
The number of orbitals at each level n is n^2.

Solve
(a) For $n = 1$, there is only 1 orbital (an s orbital).
(b) For $n = 2$, there are 4 orbitals (one s and three p orbitals).
(c) For $n = 3$, there are 9 orbitals (one s, three p, and five d orbitals).
(d) For $n = 4$, there are 16 orbitals (one s, three p, five d, and seven f orbitals).
(e) For $n = 5$, there are 25 orbitals (one s, three p, five d, seven f, and nine g orbitals).
This means there are 55 orbitals, total, in the atom.

Think About It
Notice that in each subshell, there is an odd number of orbitals, and that the number of orbitals in a particular subshell is $2\ell + 1$, where $\ell = 0$ for s orbitals, 1 for p orbitals, 2 for d orbitals, and 3 for f orbitals.

7.85. **Collect and Organize**
We are to list all the possible ℓ values when $n = 4$.

Analyze
The angular momentum quantum number is related to n as $\ell = n - 1, n - 2, n - 3, \ldots, 0$.

Solve
When $n = 4$, $\ell = 3, 2, 1, 0$.

Think About It
These ℓ values correspond to the f, d, p, and s subshells, respectively.

7.87. **Collect and Organize**
Given values for the quantum numbers n, ℓ, and m_ℓ, we are to determine the number of electrons that could occupy the orbitals described by these quantum numbers.

Analyze
The principal quantum number gives us the shell of the orbitals. This then gives the allowed values of ℓ ($n - 1$) which in turn describe the type of orbital (s, p, d, or f). The m_ℓ quantum number gives us the orientation of the orbital and its allowed values ($-\ell, -\ell + 1, \ldots, \ell - 1, \ell$), which gives us the number of orbitals available for that subshell. Each orbital can accommodate two electrons.

Solve
(a) The set of quantum numbers $n = 2$, $\ell = 0$ describes a $2s$ orbital that can be occupied by two electrons.
(b) The set of quantum numbers $n = 3$, $\ell = 1$ describes the set of $3p$ orbitals. There are three p orbitals in the subshell, so 6 electrons can occupy this orbital set.

(c) The set of quantum numbers $n = 4$, $\ell = 2$ describes the set of $4d$ orbitals. There are five d orbitals in the subshell, so 10 electrons can occupy this orbital set.

(d) The set of quantum numbers $n = 1$, $\ell = 0$ describes the $1s$ orbital that can be occupied by two electrons.

Think About It

Remember that there are one s, three p, five d, seven f, and nine g orbitals in shells for which these are allowed.

7.89. Collect and Organize

Given values for the quantum numbers n, ℓ, m_ℓ, and m_s we are to determine which combinations are allowed.

Analyze

The principal quantum number (n) can take on whole numbers starting with 1 ($n = 1, 2, 3, 4, \ldots$). The angular momentum quantum numbers (ℓ) possible for a given n value are $n - 1, n - 2, \ldots, 0$. The magnetic quantum numbers (m_ℓ) allowed for a given ℓ are $-\ell, -\ell + 1, \ldots, \ell - 1, \ell$. Allowed values for m_s are $+\frac{1}{2}$ or $-\frac{1}{2}$.

Solve

(a) For $n = 1$, the only allowed value of ℓ and m_ℓ is 0; the combination $n = 1$, $\ell = 1$, $m_\ell = 0$, $m_s = +\frac{1}{2}$ is not allowed because $\ell \neq 1$ for $n = 1$.

(b) For $n = 3$, the allowed values of ℓ are 0, 1, 2 and when $\ell = 0$ the allowed value of m_ℓ is 0; this combination of $n = 3$, $\ell = 0$, $m_\ell = 0$, $m_s = -\frac{1}{2}$ is allowed.

(c) For $n = 1$, the only allowed value for ℓ and m_ℓ is 0; the combination $n = 1$, $\ell = 0$, $m_\ell = 1$, $m_s = -\frac{1}{2}$ is not allowed because $m_\ell \neq 1$ when $\ell = 0$.

(d) For $n = 2$, the allowed values of ℓ are 0 and 1 and when $\ell = 1$ the allowed value of $m_\ell = -1, 0, 1$; this combination of $n = 2$, $\ell = 1$, $m_\ell = 2$, $m_s = +\frac{1}{2}$ is not allowed because $m_\ell \neq 2$ when $\ell = 1$.

Think About It

For the allowed combination of quantum numbers, part b describes a $3s$ orbital.

7.91. Collect and Organize / Analyze

We are asked what is meant by a degenerate orbital.

Solve

Degenerate orbitals have the same energy and are indistinguishable from each other.

Think About It

In the hydrogen atom, all the orbitals in a given n level are degenerate. This means that in hydrogen the $3s$, $3p$, and $3d$ orbitals, for example, all have the same energy. In multielectron atoms, however, these orbitals split in energy and are no longer degenerate.

7.93. Collect and Organize

In the filling of atomic orbitals, the $4s$ level fills before the $3d$. We are asked how this is evident in the periodic table.

Analyze

The two leftmost columns in the periodic table correspond to the s block and columns 3–12, starting in period 4, correspond to the d block.

Solve

As we start from an argon core of electrons, we move to potassium and calcium, which are located in the s block on the periodic table. It is not until Sc, Ti, V, etc., that we begin to put electrons into the $3d$ shell.

Think About It
Also, notice that the 6s orbitals fill (Cs and Ba) followed by the 4f orbitals (Ce–Yb) then the 5d orbitals (La–Hg).

7.95. Collect and Organize
We are asked why there are many excited-state electron configurations for an atom, but only one ground-state electron configuration.

Analyze
An electron configuration is a way of arranging electrons in the quantum mechanical orbitals m, ℓ and m_ℓ. The ground-state electron configuration is, by definition, the lowest energy arrangement of electrons in orbitals.

Solve
There are many possible arrangements of electrons in configurations that are higher in energy than the ground state because electrons may be "promoted" to many different orbitals of increasing energy. There can be only one arrangement of electrons that is the lowest in energy (the ground state).

Think About It
The ground state is really just a special term given to a specific electron configuration.

7.97. Collect and Organize
For multielectron atoms, we are to list a set of orbitals defined by their n and ℓ quantum numbers in order of increasing energy.

Analyze
The higher the energy of an orbital, the farther the electron is from the nucleus. This means that for differing n values, the order of energies is $1 < 2 < 3$, etc. For orbitals in the same n shell, the orbitals increase in energy, that is, $s < p < d < f$ for multielectron atoms.

Solve
The orbitals described are: (a) 3d, for $n = 3$, $\ell = 2$; (b) 5p, for $n = 5$, $\ell = 1$; (c) 3s, for $n = 3$, $\ell = 0$; and (d) 4p for $n = 4$, $\ell = 1$, $m_\ell = -1$.
In increasing order of energy: (c) 3s < (a) 3d < (d) 4p < (b) 5p.

Think About It
To determine the energy of an orbital, first look to the n quantum number, then to the ℓ.

7.99. Collect and Organize
We can use the periodic table and Figure 7.30 to write the electron configurations for several elemental species, including anions and cations.

Analyze
When a cation is formed, electrons are removed from the highest energy orbital. None of the species are transition metals, so we remove the electrons from the orbitals last filled in building the electron configuration of the element. To form an anion, we need to add electrons to the highest energy orbital or the next orbital up in energy. We use the previous noble gas configuration as the "core" to write the condensed form of the configurations.

Solve

Li: [He]$2s^1$
Li$^+$: [He] or $1s^2$
Ca: [Ar]$4s^2$
F$^-$: [He]$2s^22p^6$ or [Ne]

Na$^+$: [Ne] or [He]$2s^22p^6$
Mg^{2+}: [Ne] or [He]$2s^22p^6$
Al^{3+}: [Ne] or [He]$2s^22p^6$

Think About It
Because F^-, Na^+, Mg^{2+}, and Al^{3+} all have the same electron configurations and thus the same number of electrons, they are isoelectronic with each other.

7.101. **Collect and Organize**
We are to write the condensed electron configurations (using the noble gas core configuration in brackets) for several species, including cationic and anionic species. Figure 7.30 will be helpful here.

Analyze
When a cation is formed, electrons are removed from the highest energy orbital. To form an anion, we need to add electrons to the highest energy orbital or the next orbital up in energy.

Solve

K: $[Ar]4s^1$

K^+: $[Ar]$

S^{2-}: $[Ne]3s^23p^6$ or $[Ar]$

N: $[He]2s^22p^3$

Ba: $[Xe]6s^2$

Ti^{4+}: $[Ar]$ or $[Ne]3s^23p^6$

Al: $[Ne]3s^23p^1$

Think About It
Notice that K^+, S^{2-}, and Ti^{4+} are isoelectronic with each other and with Ar.

7.103. **Collect and Organize**
To determine the number of unpaired electrons in the ground-state atoms and ions, we have to first write the electron configuration for each species and then detail how the electrons are distributed among the highest energy orbitals.

Analyze
If the highest energy orbital (s, p, d, or f) is either empty or completely filled, the species has no unpaired electrons. If the highest energy orbital is partially full, Hund's rule states that electrons singly occupy the degenerate orbitals at that level before pairing up in those orbitals.

Solve

(a) N: $[He]2s^22p^3$ 3 unpaired e^-

(b) O: $[He]2s^22p^4$ 2 unpaired e^-

(c) P^{3-}: $[Ne]3s^23p^6$ 0 unpaired e^-

(d) Na^+: $[Ne]$ or $[He]2s^22p^6$ 0 unpaired e^-

Think About It
Notice that the ground-state configuration of these elements fills the s orbital completely first then places electrons into the p orbitals. This is because for a multielectron atom, $s < p$ in terms of energy for a given principal quantum level.

7.105. **Collect and Organize**
An atom with the electron configuration $[Ar]3d^24s^2$ is in the fourth period in the periodic table and is among the transition metals.

Analyze
This atom has no charge so we do not have to account for additional or lost electrons.

Solve
The $4s$ orbital is filled for the element Ca. Two additional electrons are present in the $3d$ orbitals for the second transition metal of the fourth period: titanium, Ti. The electron-filling orbital box diagram shows 2 unpaired electrons.

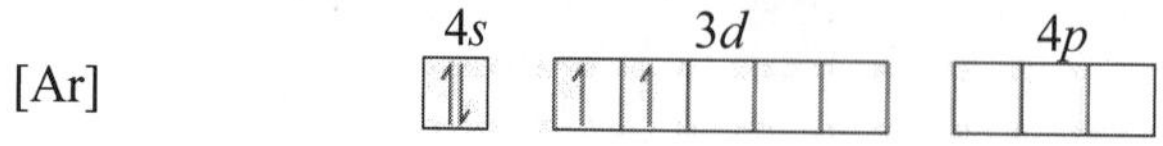

Think About It
Although we write the electron configuration so that $3d$ comes before $4s$, remember that the $4s$ orbital fills before the $3d$ in building up electron configurations.

7.107. Collect and Organize
We are to name the monatomic anion which has a filled-shell configuration of $[Ne]3s^2 3p^6$ or $[Ar]$ and determine the number of unpaired electrons in the ion in its ground state.

Analyze
Because the atom has an extra electron to form the monatomic anion, the neutral atom would have an electron configuration with one less electron.

Solve
Ion's electron configuration: $[Ne]3s^2 3p^6 = X^-$
Atom's electron configuration: $[Ne]3s^2 3p^5 = X$
This atom is chlorine and the monatomic anion is chloride, Cl^-. Because electrons fill the s and p orbitals, Cl^- has no unpaired electrons in its ground state.

Think About It
When identifying elements with the electron configurations of anions, remove the electrons associated with the anionic charge to obtain the electron configuration of the neutral atom.

7.109. Collect and Organize
We are asked if the ground-state electron configuration of Na^+ is an excited state configuration of Ne.

Analyze
To illustrate the similarities between these two species, we can write an electron configuration for each. In order to reach an excited state, an electron must be promoted to a higher energy orbital such as the $3s$ orbital, as shown follows:

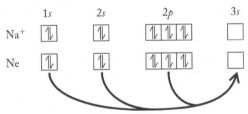

One or more must move to generate an excited state

Solve
Both Na^+ and Ne have the same electron configuration, with all ten electrons in the lowest energy configuration. Both of these are ground-state electron configurations, so no, the ground-state electron configuration of Na^+ is not an excited state configuration of Ne.

Think About It
When identifying elements with the electron configuration of cations, add or remove the electrons associated with the amount of charge to obtain the electron configuration of the neutral atom.

7.111. Collect and Organize
For Al, N, Mg, and Cs we can use the electron configurations and the positions of the elements in the periodic table to predict the charge on these elements as monatomic ions.

Analyze
The elements lose electrons to reach a noble gas configuration if they are located among the metals. If the elements are nonmetals, they gain electrons to reach a noble gas configuration.

Solve

Al loses three electrons to become Al^{3+}, with the electron configuration of Ne.
N could either gain three electrons to become N^{3-}, having the electron configuration of Ne, (more likely) or lose five electrons to become N^{5+}, with the electron configuration of He (less likely).
Mg loses two electrons to become Mg^{2+}, with the electron configuration of Ne.
Cs loses one electron to become Cs^+, with the electron configuration of Xe.

Think About It

Recall earlier in the textbook that nitrogen can have varying oxidation states from negative to positive as in NO_2, NO, and NH_3. This is reflected in nitrogen's middle position between two noble gases.

7.113. **Collect and Organize**

An electronic excited state occurs when an electron is in a higher energy orbital than would be predicted using the filling rules shown by the periodic table.

Analyze

The order of filling for the orbitals is as follows:

$$1s < 2s < 2p < 3s < 3p < 4s < 3d < 4p < 5s < 4d < 5p < 6s < 4f < 5d < 6p < 7s < 5f < 6d < 7p$$

Solve

(a) Because the $2s$ orbital is lower in energy than the $2p$ orbital, the lowest energy configuration for this atom is $[He]2s^2 2p^4$, so the configuration $[He]2s^1 2p^5$ represents an excited state.
(b) The order of filling of orbitals for atoms after krypton is $5s < 4d < 5p$. This atom has a total of 13 electrons in its outer shell: 2 fill the $5s$ orbital, 10 fill the $4d$ orbitals and one is placed in a $5p$ orbital. This configuration, $[Kr]4d^{10}5s^2 5p^1$, does not represent an excited state.
(c) The order of filling of orbitals for atoms after argon is $4s < 3d < 4p$. This atom has a total of 17 electrons in its outer shell, 2 fill the $4s$ orbital, 10 fill the $3d$ orbitals, and 5 are placed in the $4p$ orbitals. This configuration, $[Ar]3d^{10}4s^2 4p^5$, does not represent an excited state.
(d) Because the $3p$ orbital is lower in energy than the $4s$ orbital, the lowest energy configuration for this atom is $[Ne]3s^2 3p^3$, so the configuration $[Ne]3s^2 3p^2 4s^1$ represents an excited state.

Think About It

If these configurations are for neutral atoms, the elements are: (a) O, excited-state; (b) In, ground-state; (c) Br, ground-state; and (d) P, excited-state.

7.115. **Collect and Organize**

We are to write ground-state electron configurations for boron and oxygen atoms, and for the boron and oxygen ionic charges in boric acid.

Analyze

The order of filling for the orbitals is as follows:

$$1s < 2s < 2p < 3s < 3p < 4s < 3d < 4p < 5s < 4d < 5p < 6s < 4f < 5d < 6p < 7s < 5f < 6d < 7p$$

Solve

(a) Boron has five electrons and a ground-state electron configuration of $1s^2 2s^2 2p^1$. Oxygen has eight electrons and a ground-state electron configuration of $1s^2 2s^2 2p^4$.
(b) The oxidation numbers and electron configurations for the elements in boric acid are

hydrogen	+1	$1s^0$
boron	+3	$1s^2 2s^0$
oxygen	−2	$1s^2 2s^2 2p^6$

Think About It

All of the species in boric acid have a noble gas configuration, or no electrons at all in the case of hydrogen.

7.117. Collect and Organize

Iodine-131 has 53 protons, 78 neutrons, and 53 electrons as a neutral atom. We are to identify the subshell containing the highest-energy electrons and compare the electron configuration of ^{131}I to ^{127}I.

Analyze

The electron configuration for iodine is $[Kr]4d^{10}5s^25p^5$. The difference between ^{131}I and ^{127}I is that ^{131}I has four more neutrons in its nucleus.

Solve

The electron configuration of iodine shows that the highest-energy electrons are in the $5p$ subshell. Because the difference in isotopes is only the number of neutrons present, the electron configurations of ^{131}I and ^{127}I (which are based on total number of electrons in the atom) are the same.

Think About It

Electron configurations, however, do change if the atom gains or loses electrons to become either anionic or cationic, respectively.

7.119. Collect and Organize

Sodium and chlorine atoms are neutral in charge, but the sodium atom in NaCl has a charge of 1+ and the chlorine atom has a charge of 1−. These changes in charge also come with a change in size. Why?

Analyze

When we remove an electron from an atom, we reduce the repulsion for the remaining electrons in the atom. When we add electrons, we increase e^--e^- repulsion.

Solve

If electrons do not repel each other as much in Na^+ as they do in Na, they will have lower energy and be, on average, closer to the nucleus, resulting in a smaller size. When electrons are added to an atom (Cl), the e^--e^- repulsion increases, so the electrons have higher energy and they will be, on average, farther from the nucleus, thereby creating a larger size species (Cl^-).

Think About It

The change in size upon forming a cation or anion can be dramatic, as seen in Figure 7.37.

7.121. Collect and Organize

Of the group 1 elements (Li, Na, K, Rb) we are to predict the largest and explain our selection.

Analyze

The sizes of atoms increase down a group because electrons have been added to higher n levels.

Solve

Rb is the largest atom.

Think About It

The largest atoms are those situated to the lower left in the periodic table.

7.123. Collect and Organize

Ionization energy is the energy required to remove an electron from a gaseous atom.

$$X(g) \rightarrow X^-(g) + e^-$$

We are to state the trends in ionization energies down and across the periodic table.

Analyze

The ionization energy will change with effective nuclear charge (the higher the Z_{eff}, the greater the ionization energy) and with size (an electron farther away from the nucleus requires less energy to remove).

Solve

(a) As the atomic number increases down a group, electrons are added to higher n levels, leading to a decrease in ionization energy.

(b) As the atomic number increases across a period, the effective nuclear charge increases. This means that the ionization energy increases across a period of elements.

Think About It

Ionization energy trends follow atomic size trends; smaller atoms require more energy to ionize than larger atoms.

7.125. **Collect and Organize**

Fluorine and boron are located in the same period of the periodic table (period 2). Fluorine has 9 protons in its nucleus while boron has 5. We are to explain why F is more difficult to ionize than B.

Analyze

Both fluorine and boron have $2s^2 2p^n$ configurations, and the ionized electron is removed from the $2p$ orbital.

Solve

Fluorine, with a higher nuclear charge, exerts a higher Z_{eff} on the $2p$ electrons than boron, resulting in a higher ionization energy.

Think About It

The general trend across a period for ionization energies follows the trend for effective nuclear charge. As effective nuclear charge increases, so does ionization energy.

7.127. **Collect and Organize**

We have to consider the electron configurations of the cations of Br, Kr, Rb, Sr, and Y to determine which of the elements would have the smallest second ionization energy (IE_2).

Analyze

Element	Number of Protons in Nucleus	Electron Configuration	Cation (X^+) Electron Configuration
Br	35	$[Ar]3d^{10}4s^2 4p^5$	$[Ar]3d^{10}4s^2 4p^4$
Kr	36	$[Kr]$	$[Ar]3d^{10}4s^2 4p^5$
Rb	37	$[Kr]5s^1$	$[Kr]$
Sr	38	$[Kr]5s^2$	$[Kr]5s^1$
Y	39	$[Kr]4d^1 5s^2$	$[Kr]4d^1 5s^1$

Solve

Rb^+, with the noble gas configuration of Kr as Rb^+, has the highest IE_2. Both Br^+ and Kr^+ lose the second electron from a $4p$ orbital, which is lower in energy (harder to remove) than the removal of a $5s$ electron (higher in energy, easier to remove). Therefore, the IE_2 for Br^+ and Kr^+ is expected to be higher than that for Sr^+ or Y^+. Sr^+ with fewer protons in the nucleus holds onto the $5s$ electron less tightly than Y^+. Therefore, Sr is expected to have the smallest IE_2.

Think About It

In determining relative orders for second, third, etc., IEs, we have to be sure to consider the electron configuration of the cation that will lose the electron for that particular ionization step.

7.129. **Collect and Organize**

We consider an electron dropping from $n = 732$ to $n = 731$ in a hydrogen atom. We are to calculate the energy of this transition, along with its wavelength, and say what kind of telescope could detect such radiation.

Analyze
The energy difference between two n levels in the hydrogen atom is given by the equation

$$\Delta E = -2.18 \times 10^{-18} \text{ J} \left(\frac{1}{n_f^2} - \frac{1}{n_i^2} \right)$$

The wavelength of light associated with a particular energy is
$$\lambda = hc/E$$

Solve

(a) $\Delta E = -2.18 \times 10^{-18} \text{ J} \left(\frac{1}{731^2} - \frac{1}{732^2} \right) = -1.11 \times 10^{-26} \text{ J}$

Because this energy represents a loss of energy as the electron drops from a higher energy level to a lower energy level, this process is exothermic, so the sign of ΔE is negative.

(b) $\lambda = \dfrac{6.626 \times 10^{-34} \text{ J} \cdot \text{s} \times 3.00 \times 10^8 \text{ m/s}}{1.11 \times 10^{-26} \text{ J}} = 17.9 \text{ m}$

(c) This long wavelength occurs in the radio portion of the electromagnetic spectrum, so we would need a radio telescope to detect this transition.

Think About It
Our result makes sense. As n increases in the hydrogen atom, the energy levels get closer and closer together in energy and a transition between any two adjacent n levels where n is high would emit very little energy (long wavelength).

7.131. Collect and Organize
We are to write ground-state electron configurations for a copper atom, and for the copper(I) and copper(II) ions. Using these electron configurations, we are to describe the difference between copper ions.

Analyze
The order of filling for the orbitals is as follows:
$$1s < 2s < 2p < 3s < 3p < 4s < 3d < 4p < 5s < 4d < 5p < 6s < 4f < 5d < 6p < 7s < 5f < 6d < 7p$$

Solve
(a) Copper has 29 electrons, and a ground-state electron configuration of $1s^2 2s^2 2p^6 3s^2 3p^6 4s^1 3d^{10}$. The filled d orbital and half-filled s orbital in this configuration are more stable than nine electrons in the d orbital and a filled s orbital.
(b) The electron configurations for copper(I) and copper(II) are

Copper(I)	$1s^2 2s^2 2p^6 3s^2 3p^6 4s^0 3d^{10}$
Copper(II)	$1s^2 2s^2 2p^6 3s^2 3p^6 4s^0 3d^9$

The difference between copper(I) and copper(II) cations is the population of the $3d$ orbital.

Think About It
Remember that electrons in an s orbital are removed before those in a d orbital when forming a cation. This comes up when writing electron configurations for many cations of the transition metals.

7.133. Collect and Organize
Using electron configurations, we are asked to explain why silver's typical ion is Ag^+, why the heavier group 13 elements tend to form both 1+ and 3+ ions, and why the heavier group 14 and group 4 elements tend to form both 2+ and 4+ ions.

Analyze
The electron configuration for Ag is $[Kr]4d^{10}5s^1$ (completely filled d orbital). The electron configuration for a group 13 element is $[core](n-1)d^{10}ns^2np^1$. The electron configuration for a group 14 element is $[core](n-1)d^{10}ns^2np^2$. The electron configuration for a group 4 element is $[core]ns^2(n-1)d^2$.

Solve

(a) Silver forms a 1+ ion through the loss of a high-energy $5s$ electron. Palladium ($[Kr]4d^8 5s^2$) and cadmium ($[Kr]4d^{10}5s^2$) each lose two $5s$ electrons to form 2+ cations.

(b) The heavier group 13 elements may form 1+ cations through the loss of the one np electron and form 3+ cations through the loss of the np electron and the two ns electrons.

(c) The heavier group 14 elements may form 2+ cations through the loss of the two np electrons and form 4+ cations through the loss of both np electrons and the two ns electrons. The group 4 elements may lose the two ns electrons to form 2+ cations and may lose both the ns electrons and the two $(n-1)d$ electrons to form 4+ cations.

Think About It

The formation of ions that are two less than typical for the group (as in the heavier group 13 and 14 elements) is sometimes called the "inert pair effect."

7.135. Collect and Organize

We consider the effect of replacing Cl^- in photo-gray sunglasses with Br^- after defining an excited state and writing the electron configurations of Ag^+, Ag, Cl, and Cl^-.

Analyze

(a) The electron configuration of Cl^- results in a closed-shell configuration. For the Ag atom, a $5s$ electron is placed into the $4d$ shell to complete that subshell. For the Ag^+ ion, the electron in Ag is removed from the $5s$ orbital.

(b) In a ground state, all electrons are in their lowest energy orbital according to the aufbau principle. An excited state occurs when an electron absorbs light and moves to a higher energy orbital.

(c) The ionization energies of species decrease as we descend a group in the periodic table, so the IE of Br^- should be less than that of Cl^-.

(d) According to (c), less energy is needed to ionize Br^-. Energy is inversely proportional to wavelength.

Solve

(a) Cl^-: $[Ne]3s^2 3p^6$ or $[Ar]$
    Cl: $[Ne]3s^2 3p^5$
    Ag: $[Kr]4d^{10}5s^1$
    Ag^+: $[Kr]4d^{10}$

(b) An excited state occurs when an electron occupies a higher energy orbital. The electron is not in its lowest energy state.

(c) More energy is needed to remove an electron from Cl^- compared to Br^- because the electron removed from Cl^- is at a lower n (principal quantum number) level and is held more tightly by the nucleus.

(d) If AgBr were used in place of AgCl, longer wavelength (lower energy) light would remove the electron. The AgBr sunglasses would darken perhaps in the infrared region.

Think About It

Extending the periodic trend in this question, AgF would be sensitive to even shorter wavelengths of light.

7.137. Collect and Organize

We are to write condensed ground-state electron configurations for Sn^{2+}, Sn^{4+}, and Mg^{2+}, determine which neutral atoms are isoelectronic with Sn^{2+} and Mg^{2+}, and which 2+ ion is isoelectronic with Sn^{4+}.

Analyze

(a) The ground-state electron configurations for the neutral atoms are: $Sn = [Kr]4d^{10}5s^2 5p^2$ and $Mg = [Ne]3s^2$. To form Sn^{2+}, remove the two $5p$ electrons; to form Sn^{4+}, remove the two $5p$ electrons and the two $5s$ electrons. To form Mg^{2+}, remove the two $3s$ electrons.

(b) The neutral atom that has the same electron configuration as Sn^{2+} would have to have two $5s$ electrons and a filled $4d$ shell. The neutral atom that has the same electron configuration as Mg^{2+} would have to have a filled $n = 2$ shell (two $2s$ electrons and six $2p$ electrons).

(c) Isoelectronic species are those that have the same number of electrons. The 2+ cation that would be isoelectronic with Sn^{4+} would have a filled $4d$ shell but no $5s$ or $5p$ electrons.

Solve
(a) Sn^{2+}: $[Kr]4d^{10}5s^2$
   Sn^{4+}: $[Kr]4d^{10}$
   Mg^{2+}: $[Ne]$ or $[He]2s^22p^6$
(b) Cadmium has the same electron configuration as Sn^{2+} and neon has the same electron configuration as Mg^{2+}.
(c) Cd^{2+} is isoelectronic with Sn^{4+}.

Think About It
When writing electron configurations for ionic species, start with the neutral atom and then add or remove electrons to form the ions.

7.139. Collect and Organize
Using the equation $Z_{eff} = Z - \sigma$, where Z is the atomic number and σ is the shielding parameter, we are to compare the Z_{eff} (effective nuclear charge) for the outermost s electrons in neon and argon.

Analyze
In the effective nuclear charge equation given, use $Z = 10$ and $\sigma = 4.24$ for Ne and $Z = 18$ and $\sigma = 11.24$ for Ar. Shielding depends on the number of electrons that are lower in energy than the electron of interest.

Solve
(a) Ne: $Z_{eff} = 10 - 4.24 = 5.76$
   Ar: $Z_{eff} = 18 - 11.24 = 6.76$
(b) The outermost s electron in argon is a $3s$ electron, which is shielded by the electrons in the $n = 2$ level (10 electrons) and the $n = 1$ level (2 electrons), whereas the outermost s electron in neon is a $2s$ electron which is shielded only by the electrons in the $n = 1$ level (2 electrons). The greater number of electrons shielding the outermost s electrons in argon is the reason the shielding parameter (σ) for argon is greater.

Think About It
Notice that Z_{eff} is greater for the outermost electron in Ar compared to that of Ne. The ionization energy of Ar, however, is lower than the ionization energy for Ne. The effective nuclear charge equation therefore, does not seem to predict the trend in decreasing ionization energy as we descend a group in the periodic table. The effective nuclear charge equation here does not take into account the n level from which the electron is removed (ionized) to form the cation. Remember that the farther away the electron is from the nucleus, the lower the energy required to remove it.

7.141. Collect and Organize
The p orbital has two lobes of different phase with a node between the lobes. We are asked how an electron gets from one lobe to the other without going through the node between them.

Analyze
When we think of an orbital, we should think of the electron not as a particle (which in this case would have to move through the node, a region of zero probability), but as a wave.

Solve
When we think of the electron as a wave, we can envision the node between the two lobes as a wave of zero amplitude and the p orbital as a standing wave.

Think About It
Remember that an orbital describes the wave function for the electron and does not specifically locate the electron as a particle.

7.143. **Collect and Organize**

The visible part of the electromagnetic spectrum is 400–750 nm. We are to calculate the wavelength of light observed from a galaxy emitting light at 656 nm that is moving away from us at half the speed of light (1.50×10^8 m/s). We can use the Doppler equation to find the frequency of the shifted light and then convert that frequency back to the wavelength using $\lambda = c/v$.

Analyze

First, we have to determine the "unshifted" frequency of 656 nm light using $v = c/\lambda$. Then, we can use the Doppler equation given in the problem to calculate the shifted frequency and convert that frequency back to wavelength using $\lambda = c/v$.

Solve

$$v_{unshifted} = \frac{3.00 \times 10^8 \text{ m/s}}{656 \text{ nm} \times \dfrac{1 \text{ m}}{1 \times 10^9 \text{ m/s}}} = 4.57 \times 10^{14} \text{ s}^{-1}$$

Using the Doppler equation to solve for the shift in frequency, v^1,

$$\frac{\left(4.57 \times 10^{14}\text{s}^{-1} - v^1\right)}{4.57 \times 10^{14}\text{s}^{-1}} = \frac{1.50 \times 10^8 \text{ m/s}}{3.00 \times 10^8 \text{ m/s}}$$

$$v^1 = 2.29 \times 10^{14}\text{s}^{-1}$$

$$\lambda_{shifted} = \frac{3.00 \times 10^8 \text{ m/s}}{2.29 \times 10^{14}\text{s}^{-1}} = 1.31 \times 10^{-6} \text{ m or } 1310 \text{ nm}$$

This wavelength is outside of the visible range and into the infrared portion of the electromagnetic spectrum.

Think About It

We could not use an ordinary light telescope to see the hydrogen in this receding galaxy. We would have to use a telescope that detects light in the IR region.

7.145. **Collect and Organize**

The heavier noble gases can form compounds with oxygen and fluorine, but the light noble gases do not. Why?

Analyze

In order to form compounds, electrons must be either exchanged or shared between two atoms. In compounds, we can assign oxidation numbers to the atoms. The oxidation number for oxygen is typically 2–, and for fluorine, it is 1–. This means that the noble gases would need to take on a positive oxidation number to form compounds with fluorine and oxygen.

Solve

The heavier noble gases are easier to ionize (IE decreases down a group in the periodic table) and therefore can combine with oxygen and fluorine.

Think About It

We will learn later that fluorine and oxygen are the most electronegative elements in the periodic table and therefore combine with most elements to form both covalent and ionic compounds.

7.147. **Collect and Organize**

Helium's name derives from *helios*, Greek for sun, where it was first discovered. Helium, as evidenced by its use in party balloons, is lighter than air. We are asked why helium was discovered extraterrestrially before being found on Earth.

Analyze

Helium's light mass gives helium atoms high velocities at normal temperatures, as expressed by the root-mean-square speed equation.

Solve
The high velocity of helium atoms means that once helium atoms are released into the atmosphere, they can escape Earth's gravitational pull. Therefore, Earth's atmosphere contains very little helium.

Think About It
Helium on Earth is found trapped with natural gas underground and is a result of radioactive decay (α particles) of heavier elements. The United States is the world's largest supplier of helium.

CHAPTER 8 | Chemical Bonds: What Makes a Gas a Greenhouse Gas?

8.1. Collect and Organize

In the periodic table shown in Figure P8.1, groups 1, 14, 16, and 18 are highlighted. We are to determine which groups have 1, 4, and 6 valence electrons.

Analyze

The number of valence electrons in an element is equal to the number of electrons in the outermost shell. For the groups highlighted, the electron configurations for the valence electrons are

Group 1	ns^1
Group 14	ns^2np^2
Group 16	ns^2np^4
Group 18	ns^2np^6

Solve

(a) Group 1 (red) elements have 1 valence electron.
(b) Group 14 (blue) elements have 4 valence electrons.
(c) Group 16 (purple) elements have 6 valence electrons.

Think About It

The number of valence electrons for neutral atoms is related to the group number. For groups 1 and 2, the number of valence electrons equals the group number. For the other representative (main group) elements, groups 13–18, the number of valence electrons is 10 fewer than the group number ($13 - 10 = 3$ valence electrons for group 13 elements).

8.3. Collect and Organize

Magnesium is in group 2 of the periodic table and the neutral atom has an electron configuration of $[Ne]3s^2$.

Analyze

Magnesium loses its two outermost electrons to form the Mg^{2+} cation with an electron configuration of [Ne], which leaves no electrons in the valence shell of Mg^{2+}.

Solve

The Lewis symbol must correctly show both the charge and the number of valence electrons on the species. Here the charge is +2 and there are no valence electrons, so the correct Lewis structure is
$$Mg^{2+}$$

Think About It

The only other correct charge–valence electron choice in this problem for Mg is
$$[Mg\bullet]^+$$
but a charge of +1 is not the most stable for the Mg cation. The other Lewis structures either have too few or too many valence electrons for the Mg^{n+} ion shown.

8.5. Collect and Organize

Bonding capacity is the number of bonds that an atom forms with other atoms in stable compounds. For the metallic elements, ionic compounds are usually formed, and the bonding capacity depends on the charge associated with the cation (Na^+ "bonds" with 1 Cl^- to form NaCl, so Na has a bonding capacity of 1). For the nonmetallic elements, the bonding capacity depends on the number of covalent bonds the atom forms in order to fill its octet (C forms bonds with 4 other atoms).

Analyze

Applying the definition of bonding capacity gives

Group 1 (red)	= bonding capacity of 1
Group 2 (yellow)	= bonding capacity of 2

Group 13 (green) = bonding capacity of 3
Group 14 (blue) = bonding capacity of 4
Group 15 (purple) = bonding capacity of 3

Solve
The element highlighted blue in Figure P8.5, in group 14 (carbon), has the highest bonding capacity.

Think About It
Bonding capacity varies with group number in the periodic table.

8.7. **Collect and Organize**
Ionic bonds are formed between elements of very different (>2.0) values of electronegativity. All the elements highlighted in Figure P8.6 are in period 2 of the periodic table.

Analyze
As we go across a row in the periodic table, electronegativity increases.

Solve
According to the periodic trend in electronegativity, the element on the far left (Li, red) is the least electronegative and the element on the far right (Ne, peach) would be the most electronegative. However, because of neon's completely filled shell, it does not form a compound with Li. Fluorine (lilac), the element to the left of Ne, is very electronegative. Therefore, the most ionic bond (greatest electronegativity difference) will be for LiF (red and lilac).

Think About It
Bonds formed between the low electronegative metals (on the left side of the periodic table) and high electronegative nonmetals (on the right side) tend to be ionic.

8.9. **Collect and Organize**
From the three drawings given in Figure P8.9, we are to determine which best describes the electron density in lithium fluoride.

Analyze
The distribution of electron density in LiF depends on the electronegativities of the atoms in the bond. Lithium has an electronegativity of 1.0 and fluorine has an electronegativity of 4.0. The higher the negative charge, the more red the atom and the higher the positive charge, the more blue the atom as shown in Figure 8.4.

Solve
The difference in electronegativity between lithium and fluorine is very large. Because fluorine has the highest electronegativity, most of the electron density (carrying a negative charge) resides on the fluorine atom in LiF. Because the difference in electronegativity is greater than 2.0, the bond between Li and F is ionic and best written as Li^+F^-. This is best shown by drawing (b) in Figure P8.9.

Think About It
For this problem, we need not know the exact values for the electronegativities of Li and F. From the periodic trends for electronegativity (increases as we go across a period) we know that the electronegativity of F is much greater than the electronegativity of Li.

8.11. **Collect and Organize**
Given possible structures for S_2O, we are to explain why they are not all resonance forms.

Analyze
Resonance structures show more than one valid Lewis structure for a compound. They have the same arrangement of atoms but different arrangements of electrons.

Solve

The arrangement of the atoms in two of the structures is S—O—S and in the other two structures it is S—S—O. Because the arrangement of atoms differs, they, are not resonance structures. Also, for each arrangement, the structures do not show a different arrangement of electrons on the atoms, the only difference is that the bonds are drawn bent, not straight. The "bent form" and "linear form" are not resonance forms of each other if the numbers of lone pairs and bonding pairs of electrons on each atom are the same.

Think About It

Valid resonance structures of the atoms arranged as S—S—O are

$$:S=\ddot{S}-\ddot{O}: \longleftrightarrow :\ddot{S}-\ddot{S}=\ddot{O}:$$

8.13. Collect and Organize

Of the two drawings of bent triatomic molecules in Figure P8.13, we are to choose the one that represents the electron density distribution in sulfur dioxide, SO_2, and explain our choice.

Analyze

Differences in electron density within a molecule depend on the different pulling powers of the atoms in the molecule for electrons (electronegativity). Because sulfur is first in the molecular formula, we can assume that it is the central atom in the molecule. In the drawings, the higher the negative charge, the more red the atom, and the higher the positive charge, the more blue the atom, as shown in Figure 8.4.

Solve

Sulfur has a lower electronegativity than oxygen because as we descend down a group in the periodic table, electronegativity decreases. The oxygen atoms have the higher electron density so drawing (a) best represents SO_2.

Think About It

Drawing (b) shows the reverse polarity in which a higher electron density is on the sulfur atom. This implies that the electronegativity of sulfur is greater than oxygen, which we know is not the case.

8.15. Collect and Organize

The elements highlighted in Figure P8.15 are lithium (red), carbon (green), oxygen (light blue), fluorine (purple), phosphorus (dark blue), and gallium (yellow). We are asked which of these elements is reactive enough to form compounds with krypton and xenon.

Analyze

Only the most electronegative elements form compounds with Kr and Xe. Electronegativity is highest for elements at the top of a group and at the end of a period.

Solve

The most reactive (most electronegative) element highlighted is fluorine (purple). Oxygen is also quite reactive. These two elements would be expected to form compounds with Kr and Xe.

Think About It

In order to form a bond with a noble gas, the other atom must force the noble gas atom to share an electron in a covalent bond. Only the most electronegative elements are able to force the noble gas atom to share electrons.

8.17. Collect and Organize

We are asked to revise the Lewis structures depicted in Figure P8.17. This revision could include ensuring the correct number of electrons are present, shifting the position and number of multiple bonds, and adding lone pairs as required.

Analyze

Each of the molecules drawn has fewer electrons than we would predict based on the number of valence electrons for the component atoms. Adding the correct number of electrons, or removing multiple bonds where atoms cannot exceed their octet will restore the correct number of valence electrons. We may also add formal charges to the molecule as drawn to highlight areas of concern; unusually large formal charges, or formal charges that do not sum to the charge on the ion should be investigated further.

Solve

(a) As drawn, H_2NO_3S has 25 valence electrons, though we would expect it to contain 31 valence electrons. The location of the unpaired electron can be determined using formal charges.

(b) As drawn, NO_2Cl has 22 valence electrons, though we would expect it to contain 24 valence electrons. One oxygen atom is completely missing lone pairs (thus the unusually high formal charge), and nitrogen exceeds the octet as drawn.

(c) As drawn, S_2F_2 has 24 valence electrons, though we would expect it to contain 26 valence electrons. We are missing one lone pair from each of the S atoms, and we may remove the S=S double bond to reduce formal charge.

(d) As drawn, Br_3^- has 18 valence electrons, though we would expect it to contain 22 valence electrons. The central bromine atom is missing two lone pairs, and the Br=Br double bond may be removed to reduce formal charges.

The atoms in need of attention are circled, with the corrected structures at right.

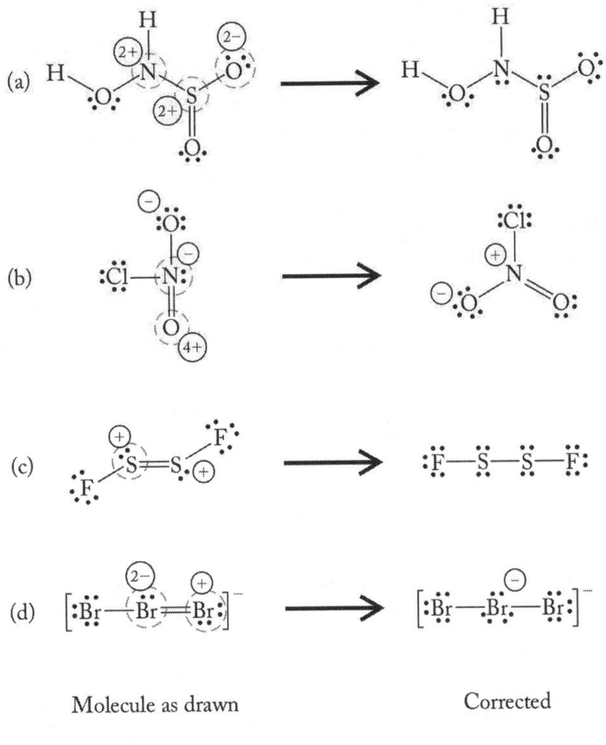

Molecule as drawn   Corrected

Think About It

When drawing Lewis structures, we are to minimize formal charges, especially when opposite signs are adjacent to one another.

8.19. Collect and Organize

We are asked to determine if the number of valence electrons is ever the same as the atomic number.

Analyze

The atomic number of an element is the number of protons in the nucleus. For a neutral atom, the number of electrons equals the number of protons. The number of valence electrons is the number of electrons in the highest n shell.

Solve
For the first and second row elements:

Element	Atomic Number	Number of Valence e⁻
H	1	1
He	2	2
Li	3	1
Be	4	2
B	5	3
C	6	4
N	7	5
O	8	6
F	9	7
Ne	10	8

Only the atomic number of hydrogen and helium equal the number of valence electrons.

Think About It
Notice that for the second row elements,

Number of valence electrons = atomic number − 2

For the third row elements,

Number of valence electrons = atomic number − 10

Try to write a similar formula for the fourth row elements (you will have to distinguish between the main group elements and the transition metals).

8.21. **Collect and Organize**
We are to determine if all the elements within a group have the same number of valence electrons.

Analyze
The number of valence electrons is determined by an atom's electron configuration. Elements in the same group have the same electron configuration.

Solve
Yes, because the elements in a group have the same valence electron configuration, they have the same number of valence electrons.

Think About It
Having the same number of valence electrons means that the elements in a group display very similar reactivities.

8.23. **Collect and Organize**
We are asked to consider how Lewis electron counting might be considered double counting.

Analyze
Lewis counts all electrons surrounding the atom in a bond including all of the electrons in shared pairs as well as those electrons in lone pairs on the atom.

Solve
In the diatomic molecule XY shown here

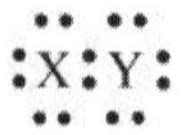

$$\overset{\bullet\bullet}{\underset{\bullet\bullet}{:}}X\overset{\bullet\bullet}{:}Y\overset{\bullet\bullet}{\underset{\bullet\bullet}{:}}$$

Lewis counts 6 e⁻ in 3 lone pairs on both X and Y. He also counts the two electrons shared between X and Y separately (2 e⁻ for X and 2 e⁻ for Y). However, there are not 4 e⁻ being shared, only 2 e⁻. It seems that the Lewis counting scheme counts the shared electrons twice.

Think About It
The octet rule uses the double counting to surround each nonhydrogen atom in a Lewis structure with 8 electrons.

8.25. **Collect and Organize**

We are to consider why water has a bonding pattern of H—O—H instead of H—H—O.

Analyze

We consider the Lewis structures of each of these compounds. For both structures, the total number of valence electrons is 1 e⁻ (H) + 1 e⁻ (H) + 6 e⁻ (O) = 8 e⁻. Each oxygen atom wants 8 e⁻ and each hydrogen atom wants 2 e⁻ for a total of 12 e⁻. The difference in the number of valence electrons and the number the molecule wants is 12 – 8 e⁻ = 4 e⁻. This means that there are two covalent bonds in water.

Solve

For the H—O—H bonding pattern, the oxygen of the central atom forms bonds to the two hydrogen atoms. This uses 4 of the 8 e⁻ leaving 4 e⁻ left over for the 2 lone pairs. Each hydrogen atom has a duet of electrons, so the lone pairs reside on oxygen and form an octet on oxygen.

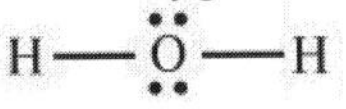

$$H—\overset{..}{\underset{..}{O}}—H$$

For H—H—O bonding, the two covalent bonds again use 4 of the 8 e⁻, leaving 4 e⁻ for 2 lone pairs. If these are placed on the oxygen atom as shown here,

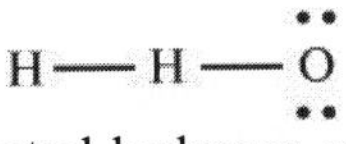

$$H—H—\overset{..}{\underset{..}{O}}$$

oxygen does not complete its octet and the central hydrogen atom has 4 e⁻, not a duet. This structure would violate the Lewis structure formalism.

Think About It

Because hydrogen does not expand its duet in covalent bonding, the H atom is always terminal and never a central atom in a Lewis structure.

8.27. **Collect and Organize**

We are to draw the Lewis symbols for the neutral atoms Li, Mg, and Al.

Analyze

Lewis symbols show the number of valence electrons as dots around the element symbol. Li has one valence electron, Mg has two valence electrons, and Al has three valence electrons.

Solve

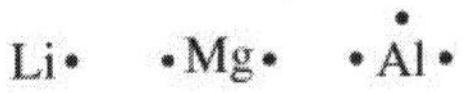

$$Li\cdot \quad \cdot Mg\cdot \quad \cdot \overset{\cdot}{Al}\cdot$$

Think About It

The particular placement of the electrons around the element symbol is not crucial to correct Lewis symbols. The electron dots are placed around the four sides of the element symbol and, generally, the electrons are not "paired" up on a side until the other sides all have an electron dot.

8.29. **Collect and Organize**

We are to find the error in each of the Lewis symbols for the ions depicted in Figure P8.29.

Analyze

Recall that the Lewis symbol for an atom depicts the arrangement of valence electrons. We should determine the number of valence electrons in each of the ions shown.

Solve

(a) Na^+ has no valence electrons, having lost a $3s$ electron to attain the same closed-shell configuration as neon.
(b) Pb^{2+} has lost two of the four valence electrons in the neutral atom to form a 2+ cation with two valence electrons.
(c) The sulfur atom gains two electrons to form S^{2-} with the same closed-shell electron configuration as argon. S^{2-} has eight valence electrons.

(d) Al^{3+} has no valence electrons, having lost all three valence electrons from the neutral aluminum atom to attain the same closed-shell configuration as neon.

The corrected Lewis symbols are

$$\left[Na\right]^{+} \qquad \left[\cdot Pb\cdot\right]^{2+} \qquad \left[:\overset{\cdot\cdot}{\underset{\cdot\cdot}{S}}:\right]^{2-} \qquad \left[Al\right]^{3+}$$

Think About It

Although the 5*d* subshell is located between the 6*s* and 6*p* subshells on the periodic table, the 10 electrons located there are not typically counted as valence electrons. It would be difficult to arrange 12 electrons around a Pb^{2+} atom!

8.31. Collect and Organize

We are to draw the Lewis symbols for the ions In^{+}, I^{-}, Ca^{2+}, and Sn^{2+}.

Analyze

To form the ions, we have to remove or add the appropriate number of electrons based on each neutral element atom.

Element/ Ion	Number of Valence e⁻ in Neutral Atom	Number of e⁻ in Ion
In/In⁺	3	2
I/I⁻	7	8
Ca/Ca²⁺	2	0
Sn/Sn²⁺	4	2

Solve

$$\left[\cdot In\cdot\right]^{+} \qquad \left[:\overset{\cdot\cdot}{\underset{\cdot\cdot}{I}}:\right]^{-} \qquad \left[Ca\right]^{2+} \qquad \left[\cdot Sn\cdot\right]^{2+}$$

I^{-} and Ca^{2+} have a complete valence shell octet.

Think About It

Calcium loses electrons to form a cation with a noble gas configuration whereas the metal tin loses two electrons to form a cation in which two electrons remain in the 5*s* orbital. The metal indium can exist in either the +3 oxidation state (to have a noble gas configuration) or as the +1 oxidation state in which two electrons remain in the 5*s* orbital.

8.33. Collect and Organize

Given the charge of an ion and the number of valence electrons, we are to write the Lewis symbol.

Analyze

In the Lewis symbol, the valence electrons are represented as dots and the charge, if any, is indicated outside brackets that contain the Lewis symbol.

Solve

(a) For a 1+ charge and 1 valence electron:

$$\left[X\cdot\right]^{+}$$

(b) For a 3+ charge and no valence electrons:

$$\left[X\right]^{3+}$$

Think About It

The Lewis symbol for the neutral species in (a) would be

$$X\!:$$

and that in (b) would be

$$\cdot \overset{\cdot\cdot}{X} \cdot$$

8.35. Collect and Organize

For the diatomic species BN, HF, OH⁻, and CN⁻, we are to determine the total number of valence electrons.

Analyze

For each species we need to add the valence electrons for each atom. If the species is charged, we need to reduce or increase the number of electrons as necessary to form cations or anions, respectively.

Solve

(a) 3 valence e⁻ (B) + 5 valence e⁻ (N) = 8 valence e⁻
(b) 1 valence e⁻ (H) + 7 valence e⁻ (F) = 8 valence e⁻
(c) 6 valence e⁻ (O) + 1 valence e⁻ (H) + 1 e⁻ (negative charge) = 8 valence e⁻
(d) 4 valence e⁻ (C) + 5 valence e⁻ (N) + 1 e⁻ (negative charge) = 10 valence e⁻

Think About It

For each of these, we can predict the number of covalent bonds between the atoms by finding the difference between what each species needs (to fill a duet for H and an octet for all other atoms) and the number of valence electrons for the molecule.

Molecule or Ion	Species Needs	Number of Valence e⁻	Number of Covalent Bonds
BN	16 e⁻	8 e⁻	4
HF	10 e⁻	8 e⁻	1
OH⁻	10 e⁻	8 e⁻	1
CN⁻	16 e⁻	10 e⁻	3

8.37. Collect and Organize

We are to draw correct Lewis structures satisfying the octet rule for all atoms in the diatomic molecules and ions CO, O_2, ClO⁻, and CN⁻.

Analyze

To draw the Lewis structures, we first must determine the number of valence electrons in each of the structures. Then, we arrange the atoms to show the bonding in the molecule by connecting the atoms with single covalent bonds. Finally, we complete the octets of the atoms bonded to the central atoms and then complete the octet of the central atom.

Solve

(a) For CO

(Step 1) The number of valence electrons in CO is

Element	C		O	
Valence electrons per atom	4	+	6	= 10

(Step 2) There are only two atoms bonded together so neither is the central atom.

$$C\!-\!O$$

(Step 3) We complete the octet on the oxygen atom by adding three lone pairs.

(Step 4) In this structure there are 8 electrons from three lone pairs and one bond pair. We need 2 more electrons (one pair) to match the valence electrons determined in step 1. We add the lone pair to the carbon atom.

$$:C\!-\!\overset{\cdot\cdot}{\underset{\cdot\cdot}{O}}:$$

(Step 5) To complete the octet on the carbon atom, we convert two lone pairs on the O atom to give a triple bond between the oxygen atom and the carbon atom.

$$: C \longrightarrow \ddot{O} :$$

$$: C \equiv O :$$

The Lewis structure is now complete.

(b) For O_2
   (Step 1) The number of valence electrons in O_2 is

Element	2O
Valence electrons per atom	$(2 \times 6) = 12$

   (Step 2) There are only two atoms bonded together, so neither is the central atom.

   $$O \longrightarrow O$$

   (Step 3) We complete the octet on one of the oxygen atoms by adding three lone pairs.

   $$O \longrightarrow \ddot{O} :$$

   (Step 4) In this structure there are 8 electrons from three lone pairs and one bond pair. We need 4 more electrons (two pairs) to match the valence electrons determined in step 1. We add the lone pairs to the other oxygen atom.

   $$: \ddot{O} \longrightarrow \ddot{O} :$$

   (Step 5) To complete the octet on the left oxygen atom in the structure we convert a lone pair on the right O atom to give a double bond between the oxygen atoms.

   $$: \ddot{O} \longrightarrow \ddot{O} :$$

   $$: \ddot{O} = \ddot{O} :$$

The Lewis structure is now complete.

(c) For ClO^-
   (Step 1) The number of valence electrons in ClO^- is

Element	Cl		O	
Valence electrons per atom	7	+	6	= 13
Gain of electron due to charge				+1
Total valence electrons				14

   (Step 2) There are only two atoms bonded together so neither is the central atom.

   $$Cl \longrightarrow O$$

   (Step 3) We complete the octet on the oxygen atoms by adding three lone pairs.

   $$Cl \longrightarrow \ddot{O} :$$

   (Step 4) In this structure there are 8 electrons from three lone pairs and one bond pair. We need 6 more electrons (three pairs) to match the valence electrons determined in step 1. We add the lone pairs to the chlorine atom.

   $$: \ddot{Cl} \longrightarrow \ddot{O} :$$

   (Step 5) This Lewis structure is complete. To indicate the charge on this ion we add brackets for the structure and the charge.

   $$\left[: \ddot{Cl} \longrightarrow \ddot{O} : \right]^-$$

(d) For CN⁻

    (Step 1) The number of valence electrons in CN⁻ is

Element	C		N	
Valence electrons per atom	4	+	5	= 9
Gain of electron due to charge				+1
Total valence electrons				10

    (Step 2) There are only two atoms bonded together so neither is the central atom.

$$C-N$$

    (Step 3) We complete the octet on the nitrogen atom by adding three lone pairs.

$$C-\ddot{\underset{..}{N}}:$$

    (Step 4) In this structure there are 8 electrons from three lone pairs and one bond pair. We need 2 more electrons (one pair) to match the valence electrons determined in step 1. We add the lone pair to the carbon atom.

$$:C-\ddot{\underset{..}{N}}:$$

    (Step 5) To complete the octet on the carbon atom we convert two lone pairs on the N atom to give a triple bond between the nitrogen atom and the carbon atom. Finally, we add brackets to the structure and indicate the charge on this anion.

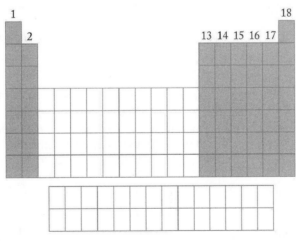

The Lewis structure is now complete.

Think About It

When writing Lewis structures for ionic species, remember to enclose the structure in brackets and indicate the charge on the ion as shown in this problem for ClO⁻ and CN⁻.

8.39. Collect and Organize / Analyze

We are asked which of the main group elements have an odd number of valence electrons. The elements of the main group are

Solve

Groups 1, 13, 15, and 17 have an odd number of electrons.

Think About It
The groups with an odd number of electrons are the odd-numbered groups.

8.41. Collect and Organize
Using the method described in the textbook, we are to draw Lewis structures for three greenhouse gases.

Analyze
To draw the Lewis structures, we first must determine the number of valence electrons in each of the structures, then arrange the atoms to show the bonding in the molecule by connecting the atoms with single covalent bonds. Finally, we complete the octets of the atoms bonded to the central atoms, then complete the octet of the central atom.

Solve
(a) CF_2Cl_2
  (Step 1) The number of valence electrons in CF_2Cl_2 is

Element	C	2F	2Cl	Total
Valence electrons per atom	4 +	(2×7) +	(2×7)	= 32

  (Step 2) Carbon has the most unpaired electrons (4) in its Lewis symbol and therefore has the highest bonding capacity and will be the central atom in the structure. The fluorine and chlorine atoms will each be bonded to the carbon.

  (Step 3) We complete the octets on the chlorine and fluorine atoms by adding three lone pairs to each.

  (Step 4) In this structure there are 32 electrons from 12 lone pairs and four bond pairs. We do not need any more valence electrons in this structure.
  (Step 5) With carbon satisfied with its octet, the Lewis structure is complete.

(b) Cl_2FCCF_2Cl
  (Step 1) The number of valence electrons in Cl_2FCCF_2Cl is

Element	2C	3F	3Cl	Total
Valence electrons per atom	(2×4) +	(3×7) +	(3×7)	= 50

  (Step 2) The carbon atoms have the most unpaired electrons (4) in their Lewis symbols and therefore have the highest bonding capacity and will be the central atoms in the structure. We are given that there is a C—C bond. The fluorine and chlorine atoms will each be bonded to the carbon atoms.

  (Step 3) We complete the octets on the chlorine and fluorine atoms by adding three lone pairs to each.

(Step 4) In this structure there are 50 electrons from 18 lone pairs and seven bond pairs. We do not need any more valence electrons in this structure.
(Step 5) With the carbon atoms satisfied with their octets, the Lewis structure is complete.

(c) C_2Cl_3F
    (Step 1) The number of valence electrons in C_2Cl_3F is

Element	2C	F	3Cl	Total
Valence electrons per atom	$(2 \times 4) +$	7 +	(3×7)	$= 36$

(Step 2) The carbon atoms have the most unpaired electrons (4) in their Lewis symbols and therefore have the highest bonding capacity and will be the central atoms in the structure. We are given that there is a C=C bond. The fluorine and chlorine atoms will each be bonded to the carbon atoms.

(Step 3) We complete the octets on the chlorine and fluorine atoms by adding three lone pairs to each.

(Step 4) In this structure there are 36 electrons from 12 lone pairs and six bond pairs. We do not need any more valence electrons in this structure.
(Step 5) With the carbon atoms satisfied with their octets, the Lewis structure is complete.

Think About It
In determining the skeletal structure for these molecules, it is helpful to know that carbon commonly bonds to 4 atoms and can double or triple bond to itself.

8.43. Collect and Organize
Using the method described in the textbook, we are to draw Lewis structures for $CH_3CH_2CH_2CH_2SH$ and H_2S.

Analyze
To draw the Lewis structures, we first must determine the number of valence electrons in each of the structures, then arrange the atoms to show the bonding in the molecule by connecting the atoms with single covalent bonds. Finally, we complete the octets of the atoms bonded to the central atoms, then complete the octet of the central atom. Considering that hydrogen is always terminal, the carbon atoms must be bonded together in a chain as indicated in the formula given in the problem.

Solve
Butanethiol
    (Step 1) The number of valence electrons in $CH_3CH_2CH_2CH_2SH$ is

Element	4C	10H	S	Total
Valence electrons per atom	$(4 \times 4) +$	$(10 \times 1) +$	6	$= 32$

(Step 2) Carbon has the most unpaired electrons (4) in its Lewis symbol and therefore has the highest bonding capacity and will be the central atom in the structure. The hydrogen and sulfur atoms will be bonded to the carbon as indicated in the chemical formula. Also, as indicated in the formula, one H atom is bonded to the sulfur atom

(Step 3) We complete the octet on the sulfur atom by adding two lone pairs to it.

$$H-\overset{\overset{\displaystyle H}{|}}{\underset{\underset{\displaystyle H}{|}}{C}}-\overset{\overset{\displaystyle H}{|}}{\underset{\underset{\displaystyle H}{|}}{C}}-\overset{\overset{\displaystyle H}{|}}{\underset{\underset{\displaystyle H}{|}}{C}}-\overset{\overset{\displaystyle H}{|}}{\underset{\underset{\displaystyle H}{|}}{C}}-\ddot{\ddot{S}}-H$$

(Step 4) In this structure there are 32 electrons from two lone pairs and 14 bond pairs. We do not need any more valence electrons in this structure.

(Step 5) With carbon satisfied with its octets and hydrogen satisfied with its duets, the Lewis structure is complete.

Hydrogen sulfide

(Step 1) The number of valence electrons in H_2S is

Element	2H	S	Total
Valence electrons per atom	(2×1) +	6	= 8

(Step 2) Sulfur has the most unpaired electrons (2) in its Lewis symbol and therefore has the highest bonding capacity and will be the central atom in the structure.

$$H-S-H$$

(Step 3) We complete the octet on the sulfur atom by adding two lone pairs to it.

$$H-\ddot{\ddot{S}}-H$$

(Step 4) In this structure there are 8 electrons from two lone pairs and two bond pairs. We do not need any more valence electrons in this structure.

(Step 5) With hydrogen satisfied with its duets, the Lewis structure is complete.

Think About It

Carbon atoms are often bonded together in chains, as seen in butanethiol. The bonding of an elemental atom to the same elemental atom to form chains is called *catenation*.

8.45. Collect and Organize

Using the method in the textbook, we are to draw Lewis structures for Cl_2O and ClO_3^-.

Analyze

To draw the Lewis structures, we first must determine the number of valence electrons, then arrange the atoms to show the bonding in the molecules by connecting the atoms with single covalent bonds. Finally, we complete the octets of the atoms bonded to the central atoms, then complete the octet of the central atom.

Solve

Cl_2O

(Step 1) The number of valence electrons in Cl_2O is

Element	2Cl	O	Total
Valence electrons per atom	(2×7) +	6	= 20

(Step 2) We are given that one of the chlorine atoms is the central atom in this structure.

$$Cl-Cl-O$$

(Step 3) We complete the octets on the oxygen and terminal chlorine atoms by adding three lone pairs to each.

$$:\ddot{Cl}-Cl-\ddot{O}:$$

(Step 4) In this structure there are 16 electrons from six lone pairs and two bond pairs. We need 4 more electrons (two pairs) to match the valence electrons determined in step 1. We add the lone pairs to the central chlorine atom.

$$:\ddot{Cl}-\ddot{Cl}-\ddot{O}:$$

(Step 5) The central chlorine atom is satisfied with its octet, so this Lewis structure is complete.

ClO_3^-

    (Step 1) The number of valence electrons in ClO_3^- is

Element	Cl	3O	
Valence electrons per atom	7	$+ (3 \times 6)$	$= 25$
Gain of electron due to charge			$+1$
Total valence electrons			26

    (Step 2) We are given that the chlorine atom is the central atom in this structure.

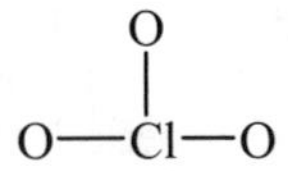

$$\begin{array}{c} O \\ | \\ O\!-\!\overset{}{Cl}\!-\!O \end{array}$$

    (Step 3) We complete the octets on the oxygen atoms by adding three lone pairs to each.

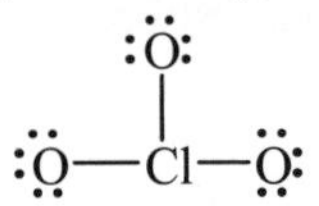

$$\begin{array}{c} :\ddot{O}: \\ | \\ :\ddot{O}\!-\!Cl\!-\!\ddot{O}: \end{array}$$

    (Step 4) In this structure there are 24 electrons from nine lone pairs and three bond pairs. We need 2 more electrons (one pair) to match the valence electrons determined in step 1. We add the lone pair to the central chlorine atom.

$$\begin{array}{c} :\ddot{O}: \\ | \\ :\ddot{O}\!-\!\ddot{Cl}\!-\!\ddot{O}: \end{array}$$

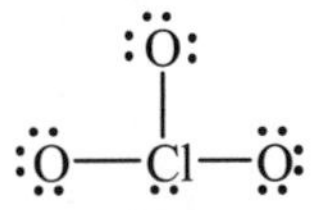

    (Step 5) The central chlorine atom is satisfied with its octet so this Lewis structure is complete. We add brackets and the charge to indicate the ion.

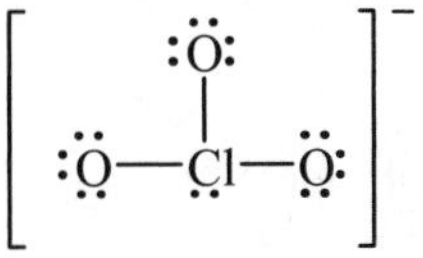

$$\left[\begin{array}{c} :\ddot{O}: \\ | \\ :\ddot{O}\!-\!\ddot{Cl}\!-\!\ddot{O}: \end{array}\right]^-$$

Think About It

In ClO_3^-, Cl is the central atom as we would guess from the formula, but in Cl_2O, oxygen could be the central atom. As shown below, the arrangement of the molecule with Cl as the central atom gives nonzero formal charges for the atoms (see Section 8.6). In Section 8.6, we will also learn that to reduce the formal charge on Cl in ClO_3^-, we can form a double bond between Cl and one of the O atoms.

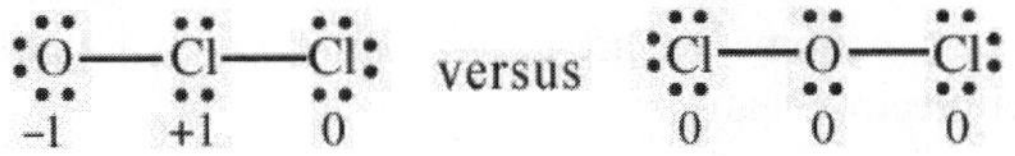

$$\underset{-1}{:\ddot{O}}\!-\!\underset{+1}{\ddot{Cl}}\!-\!\underset{0}{\ddot{Cl}:} \quad \text{versus} \quad \underset{0}{:\ddot{Cl}}\!-\!\underset{0}{\ddot{O}}\!-\!\underset{0}{\ddot{Cl}:}$$

8.47. Collect and Organize

We are asked how we can use electronegativity to define whether a bond is ionic or covalent.

Analyze

When there is a large difference in electronegativity, the transfer of an electron from one atom to another is likely, and an ionic bond will form. When the electronegativities of two atoms are similar, the electrons will be shared in a covalent bond.

Solve

The general rule is that if there is an electronegativity difference of 2.0 or greater, the bond between the atoms is ionic. Below 2.0, the bond is covalent.

Think About It

Large differences in electronegativity, and thus the occurrence of ionic bonding, are likely between a metallic atom (low electronegativity) and a nonmetallic atom (high electronegativity).

8.49. Collect and Organize
We are asked to explain why electronegativity is related to atomic size.

Analyze
Small atoms such as oxygen and fluorine have high electronegativities; large atoms such as cesium have low electronegativities.

Solve
The size of the atom is the result of the number of electrons, and the pull of the nucleus on those electrons. The higher the nuclear charge, the stronger the pull on the electrons within a given valence shell. This is why the size of the atoms generally decreases across a period. A small atom will form a shorter bond with another atom, and the electrons in the bond will feel a strong pull from the nucleus of a smaller atom since the bonding electrons will be "closer" to the nucleus. This stronger pull is the reason for the higher electronegativity for smaller atoms.

Think About It
Both size and electronegativity trends are fundamentally a reflection of the trends in effective nuclear charge and the *n* level of the valence electrons of an atom.

8.51. Collect and Organize
We are to define *polar covalent bond.*

Analyze
A covalent bond forms between atoms of close electronegativities and involves the sharing, not the transfer, of electrons between the atoms. When a bond is polar, there is an uneven distribution of electron density.

Solve
A polar covalent bond is one in which the electrons are shared, but not equally, by the atoms.

Think About It
The more electronegative atom in the bond pulls more of the electron density toward itself, making that atom slightly rich in electron density (δ^-) leaving the other atom slightly deficient in electron density (δ^+).

8.53. Collect and Organize
Of the bonds listed between two atoms, we are to determine which are polar and, in those bonds that are polar, which atom has the greater electronegativity.

Analyze
Polar bonds form between any two dissimilar atoms that have different electronegativity values. To determine the more electronegative atom, we can use knowledge of the periodic trends or electronegativity values in Figure 8.5 in the textbook.

Solve
The polar bonds and the atoms with the greater electronegativity (underlined) are C—Se, C—O, N—H, and C—H.

Think About It
The Cl—Cl and O=O bonds are not polar because the bonded atoms are identical.

8.55. Collect and Organize
From the list of compounds, we are to determine which has the least ionic character.

Analyze
When the difference in electronegativity between the atoms is zero, the bond is nonpolar. If the electronegativity difference is below 2.0, the bond is polar covalent. If the electronegativity difference is 2.0 or

greater, the bond is ionic. The bond with the smallest difference in electronegativities will be the least ionic. We can use the electronegativity values from Figure 8.5 in the textbook.

Solve

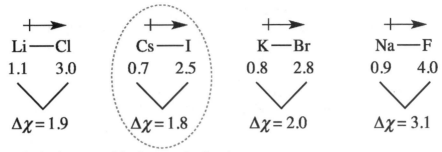

CsI has the least ionic character of the four species listed.

Think About It

Although one might assume that all alkali metal/halide compounds would be ionic, we see here that both CsI and LiCl may be classified as polar covalent.

8.57. **Collect and Organize**

We are asked to compare the atmospheric greenhouse gases to the glass of a greenhouse.

Analyze

The atmospheric gases that are considered to be greenhouse gases include CO and CH_4. By acting as greenhouse gases they trap radiation, acting as a blanket to warm the earth.

Solve

Like the panes of glass in a greenhouse, the greenhouse gases in the atmosphere are transparent to visible light. The visible light warms the surface of the earth and is reemitted as infrared (lower energy) light. The greenhouse gases absorb the infrared light, accumulating and holding heat in the atmosphere just as the panes of greenhouse glass hold heat inside a greenhouse.

Think About It

In order for a molecule to absorb infrared radiation and act as a greenhouse gas, it must have a molecular vibration that has a change in dipole moment upon stretching or bending. Nitrogen (N_2) and oxygen (O_2) with no dipole moment change upon stretching and are not greenhouse gases.

8.59. **Collect and Organize**

We are asked to consider which bond stretching in N_2O is responsible for the absorption of infrared radiation.

Analyze

Infrared radiation is absorbed by molecules with polar bonds when the fluctuating electric fields in the molecule do not cancel each other out (asymmetric stretching). The Lewis structure of N_2O shows that this molecule is linear.

$$:N\equiv N-\ddot{\underset{..}{O}}: \quad \text{or} \quad :\ddot{N}=N=\ddot{\underset{..}{O}}:$$

Solve

The N–N bond is not polar and so the stretching of that bond would not absorb IR radiation. However, the N–O bond is polar and would be expected to absorb IR radiation upon stretching.

Think About It

N_2O is estimated to be about three hundred times more potent a greenhouse gas than CO_2. It is, however, present in much lower concentrations (320 ppb versus 385 ppm) than CO_2.

8.61. Collect and Organize

We are asked to consider whether carbon monoxide is infrared active (can absorb IR radiation).

Analyze

Infrared radiation is absorbed by a molecular vibration (bond stretch or bend) when the bond is polar and when the fluctuating electric fields do not cancel each other out. The Lewis structure of carbon monoxide is

$$:C\equiv O:$$

Solve

The C—O bond is polar due to the difference in electronegativity of carbon and oxygen. Stretching the linear C—O bond in carbon monoxide would give a fluctuating electric field, and therefore CO does absorb IR radiation.

Think About It

Carbon monoxide is a weak greenhouse gas, but it reacts with hydroxyl (OH) radicals, which then cannot react with other greenhouse gases like methane to "neutralize" them and reduce their effects.

8.63. Collect and Organize

For this question we consider why infrared radiation causes vibrations, but not breakage, of chemical bonds.

Analyze

Infrared radiation has wavelengths in the range of 10^{-6} to 10^{-4} m, while ultraviolet radiation has wavelengths in the range of 10^{-6} to 10^{-8} m.

Solve

The shorter the wavelength, the higher the energy of the radiation. Thus, infrared radiation with its longer wavelengths and lower energy causes chemical bonds only to stretch and bend. Higher energy ultraviolet radiation can cause chemical bonds to break.

Think About It

Ultraviolet radiation, along with X-ray and gamma radiation, can cause bonds to break and are classified as ionizing radiation.

8.65. Collect and Organize

For this question we compare the bonds in CO_2 to CO to determine if the energy required to vibrate the C–O bond in CO is greater or less than the energy to vibrate the C–O bond in CO_2.

Analyze

The Lewis structure of CO shows that the C–O bond is a triple bond, whereas the C–O bond in CO_2 is a double bond.

$$:C\equiv O: \qquad :\ddot{O}=C=\ddot{O}:$$

Solve

Because the triple bond in CO is stronger than the double bond in CO_2, the energy required to vibrate the triple bond in CO is higher.

Think About It

As we see in Section 8.8, the bond strength is also related to bond length. The stronger the bond, the shorter the bond distance. The triple bond in CO is 113 pm; the double bond in CO_2 is 123 pm.

8.67. Collect and Organize

We are to explain the concept of resonance.

Analyze

Resonance structures are equivalent Lewis structures that differ only in the placement of electrons.

Solve

Resonance occurs when two or more valid Lewis structures may be drawn for a molecular species. The true structure of the species is a hybrid of the structures drawn.

Think About It

When drawing resonance structures for a molecule, there are two important rules to keep in mind: (1) each Lewis structure must be valid (i.e., the atoms must have complete octets, or duets if hydrogen atoms), and (2) the positions of the atoms must not change—only the distribution of the electrons in bonding pairs and lone pairs will differ between resonance structures.

8.69. Collect and Organize

We are asked to describe what factors determine resonance in molecules or ions.

Analyze

We find out if a molecule has resonance structures by drawing out their Lewis structures.

Solve

A molecule or ion shows resonance when there is more than one correct Lewis structure; that is, when the electrons in the correct Lewis structure may be distributed in more than one way.

Think About It

Remember that resonance structures differ from each other only in the arrangement of the electrons, not in the atoms of the structure. Often, when the central atom has both a single and a double bond, resonance is possible.

8.71. Collect and Organize

We are to explain why NO_2 is more likely to exhibit resonance than CO_2. Drawing the Lewis structures of these two molecules will be useful.

Analyze

The Lewis structures of NO_2 and CO_2 are:

$$:\ddot{O}\!-\!\dot{N}\!=\!\ddot{O}: \qquad :\ddot{O}\!=\!C\!=\!\ddot{O}:$$

Solve

Either N–O bond in the NO_2 structure could be double-bonded, and the formal charges for each structure are identical, so there is more than one correct Lewis structure, and NO_2 will exhibit resonance.

$$\overset{-1}{:\ddot{O}}\!-\!\overset{+1}{\dot{N}}\!=\!\overset{+0}{\ddot{O}}: \longleftrightarrow \overset{+0}{:\ddot{O}}\!=\!\overset{+1}{N}\!-\!\overset{-1}{\ddot{O}}:$$

The resonance forms of CO_2 show that one is dominant (the one in which all formal charges are zero) and so the other forms contribute little to the true structure of CO_2.

$$\overset{+0}{:\ddot{O}}\!=\!\overset{+0}{C}\!=\!\overset{+0}{\ddot{O}}: \longleftrightarrow \overset{+1}{:O}\!\equiv\!\overset{+0}{C}\!-\!\overset{-1}{\ddot{O}}: \longleftrightarrow \overset{-1}{:\ddot{O}}\!-\!\overset{+0}{C}\!\equiv\!\overset{+1}{O}:$$

Think About It

The following is *not* a correct Lewis structure for CO_2.

$$:O\!\equiv\!C\!=\!\ddot{O}:$$

8.73. Collect and Organize

Fulminic acid has a linear structure with atom connectivity as described by the molecular formula given. From this framework we are to draw valid resonance structures for HCNO.

Analyze

We first draw one of the valid Lewis structures by the method described in the textbook, then redistribute the bonding pairs and lone pairs in the structure to draw resonance forms.

Solve

HCNO, fulminic acid:

(Step 1) The number of valence electrons is

Element	H		C		N		O		Total
Valence electrons per atom	1	+	4	+	5	+	6	=	16

(Step 2) We are given that fulminic acid is a linear molecule with the connectivity of the atoms as

$$\text{H—C—N—O}$$

(Step 3) We complete the octets on the oxygen atom by adding three lone pairs. The duet on the terminal H atom is already satisfied.

$$\text{H—C—N—}\ddot{\underset{\displaystyle ..}{\text{O}}}:$$

(Step 4) In this structure there are 12 electrons from three lone pairs and three bond pairs. We need 4 more electrons (two pairs) to match the valence electrons determined in step 1. The nitrogen and carbon atoms do not have octets yet so we will add lone pairs.

$$\text{H—}\underset{\displaystyle ..}{\text{C}}\text{—}\overset{\displaystyle ..}{\text{N}}\text{—}\ddot{\underset{\displaystyle ..}{\text{O}}}:$$

(Step 5) We can complete the octet for the carbon and nitrogen atoms by forming a triple bond between them.

$$\text{H—}\underset{\displaystyle ..}{\text{C}}\overset{\curvearrowright}{}\overset{\displaystyle ..}{\text{N}}\text{—}\ddot{\underset{\displaystyle ..}{\text{O}}}: \qquad \text{H—C}\equiv\text{N—}\ddot{\underset{\displaystyle ..}{\text{O}}}:$$

The electrons could also be distributed in two additional resonance forms that also complete the octet on all the atoms:

$$\text{H—C}\equiv\text{N—}\ddot{\underset{\displaystyle ..}{\text{O}}}: \longleftrightarrow \text{H—}\overset{\displaystyle ..}{\text{C}}\text{=N=}\ddot{\underset{\displaystyle ..}{\text{O}}}: \longleftrightarrow \text{H—}\underset{\displaystyle ..}{\overset{\displaystyle ..}{\text{C}}}\text{—N}\equiv\text{O}:$$

Think About It

The following resonance form is not valid because it has more than a duet for the H atom and less than an octet for the C atom.

$$\text{H=C—}\overset{\displaystyle ..}{\text{N}}\text{=}\ddot{\underset{\displaystyle ..}{\text{O}}}:$$

8.75. Collect and Organize

For N_2O_2 and N_2O_3 we are to draw Lewis structures and show all possible resonance forms.

Analyze

To draw the Lewis structures we must first determine the number of valence electrons, then the number of covalent bonds in each structure, then complete the octets (duets for hydrogen) as necessary, and check the structure with electron bookkeeping. Once one Lewis structure is drawn, we can then consider alternate structures in resonance with the first.

Solve

N_2O_2:

(Step 1) The number of valence electrons is

Element	2N		2O		Total
Valence electrons per atom	(2×5)	+	(2×6)	=	22

(Step 2) Nitrogen has more unpaired electrons (3) than oxygen and is less electronegative, so the two nitrogen atoms are the central atoms in the structure.

$$\text{O—N—N—O}$$

(Step 3) We complete the octets on the oxygen atoms by adding three lone pairs to each.

$$:\ddot{\underset{\displaystyle ..}{\text{O}}}\text{—N—N—}\ddot{\underset{\displaystyle ..}{\text{O}}}:$$

(Step 4) In this structure there are 18 electrons from six lone pairs and three bond pairs. We need 4 more electrons (two pairs) to match the valence electrons determined in step 1. The nitrogen atoms do not have octets yet so we will add one lone pair to each N atom.

$$:\ddot{O}—\ddot{N}—\ddot{N}—\ddot{O}:$$

(Step 5) We can complete the octet for each nitrogen by forming double bonds between the oxygen and nitrogen atoms.

$$:\ddot{O}—\ddot{N}—\ddot{N}—\ddot{O}: \qquad :\ddot{O}=\ddot{N}—\ddot{N}=\ddot{O}:$$

The electrons could also be distributed in five more resonance forms that also complete the octet on all the atoms:

$$\ddot{O}=\ddot{N}—\ddot{N}=\ddot{O} \longleftrightarrow :O\equiv N—\ddot{N}—\ddot{O}: \longleftrightarrow \ddot{O}=N=\ddot{N}—\ddot{O}: \longleftrightarrow$$

$$:\ddot{O}—N\equiv N—\ddot{O}: \longleftrightarrow :\ddot{O}—\ddot{N}—N\equiv O: \longleftrightarrow :\ddot{O}—\ddot{N}=N=\ddot{O}$$

N_2O_3:

(Step 1) The number of valence electrons is

Element	2N	3O	Total
Valence electrons per atom	(2×5) +	(3×6) =	28

(Step 2) Nitrogen has more unpaired electrons (3) than oxygen and is less electronegative, so the two nitrogen atoms are the central atoms in the structure.

$$\begin{array}{c} O \\ | \\ O—N—N—O \end{array}$$

(Step 3) We complete the octets on the oxygen atoms by adding three lone pairs to each.

$$\begin{array}{c} :\ddot{O}: \\ | \\ :\ddot{O}—N—N—\ddot{O}: \end{array}$$

(Step 4) In this structure there are 26 electrons from nine lone pairs and four bond pairs. We need 2 more electrons (one pair) to match the valence electrons determined in step 1. The nitrogen atoms do not have octets yet so we will add one lone pair to a N atom.

$$\begin{array}{c} :\ddot{O}: \\ | \\ :\ddot{O}—N—\ddot{N}—\ddot{O}: \end{array}$$

(Step 5) We can complete the octet for each nitrogen by forming double bonds between the oxygen and nitrogen atoms.

$$\begin{array}{c} :\ddot{O}: \\ | \\ :\ddot{O}—N—\ddot{N}—\ddot{O}: \end{array} \qquad \begin{array}{c} \ddot{O}: \\ || \\ :\ddot{O}=N—\ddot{N}—\ddot{O}: \end{array}$$

The electrons could also be distributed in three more resonance forms that also complete the octet on all the atoms:

$$:\ddot{O}=\ddot{N}—N\diagup^{\ddot{O}:}_{\diagdown\ddot{O}:} \longleftrightarrow :\ddot{O}=\ddot{N}—N\diagup^{\ddot{O}:}_{\diagdown O} \longleftrightarrow$$

$$:\ddot{O}=N=N\diagup^{\ddot{O}:}_{\diagdown\ddot{O}:} \longleftrightarrow :O\equiv N—\ddot{N}\diagup^{\ddot{O}:}_{\diagdown\ddot{O}:}$$

Think About It

In N_2O_3 a resonance structure that has a triple bond between the N atoms would violate the octet rule for the N bound to two O atoms and for one of the O atoms.

$$:\ddot{O}-N\equiv N\overset{\displaystyle \ddot{O}:}{\underset{\displaystyle \ddot{O}:}{}}$$

8.77. Collect and Organize

We are to account for the short O–O bond length in F_2O_2 by drawing Lewis structures and comparing these to the Lewis structure of H_2O_2.

Analyze

We first draw valid Lewis structures for H_2O_2 and F_2O_2 using the method described in the textbook, and then consider the ionic form of F_2O_2, redistributing the bonding pairs and lone pairs in the structure to draw resonance forms as appropriate.

Solve

(a) H_2O_2:

(Step 1) The number of valence electrons is

Element	2 H	2 O	Total
Valence electrons per atom	(2×1) +	(2×6)	= 14

(Step 2) The oxygen atoms have the highest bonding capacity and they are the central atoms in the structure, including an O–O bond.

$$H-O-O-H$$

(Step 3) We complete the octets on the oxygen atoms by adding two lone pairs to each.

$$H-\ddot{O}-\ddot{O}-H$$

(Step 4) In this structure there are 14 electrons from four lone pairs and three bond pairs. No more electrons are needed for this Lewis structure.

F_2O_2:

(Step 1) The number of valence electrons is

Element	2 F	2 O	Total
Valence electrons per atom	(2×7) +	(2×6)	= 26

(Step 2) The oxygen atoms have the highest bonding capacity, and they are the central atoms in the structure, including an O–O bond.

$$F-O-O-F$$

(Step 3) We complete the octets on the fluorine atoms by adding three lone pairs to each.

$$:\ddot{F}-O-O-\ddot{F}:$$

(Step 4) In this structure there are 18 electrons from six lone pairs and three bond pairs. Eight more electrons are needed for this Lewis structure. We may complete the octets on the oxygen atoms by adding two lone pairs to each.

$$:\ddot{F}-\ddot{O}-\ddot{O}-\ddot{F}:$$

(Step 5) In this structure there are 26 electrons from ten lone pairs and three bond pairs. No more electrons are needed for this Lewis structure.

(b) We may draw the ionic form of F_2O_2 by considering the structure as $[FO_2]^+[F]^-$. Starting from the Lewis structure for F_2O_2 as determined in (a), the removal of F^- results in an oxygen atom with an incomplete octet.

$$:\ddot{F}-\ddot{O}-\ddot{O}\curvearrowleft\ddot{F}: \longrightarrow :\ddot{F}-\ddot{O}-\ddot{O}^{\oplus} \quad \overset{\displaystyle :\ddot{F}:^{\ominus}}{} \longrightarrow :\ddot{F}-\ddot{O}=\ddot{O}: \quad \overset{\displaystyle :\ddot{F}:^{\ominus}}{}$$

incomplete octet
on oxygen

The completion of the octet on oxygen may be accomplished by forming a double bond. The presence of this double bond results in an overall decrease in bond distance.

Think About It
That the bond length is 20% shorter in F_2O_2 than in H_2O_2 suggests that the ionic form of F_2O_2 accounts for at least 20% of the overall bonding description.

8.79. **Collect and Organize**
We are asked to draw Lewis structures for all of the valid resonance forms of ClSeNSO.

Analyze
We first draw one of the valid Lewis structures by the method described in the textbook, and then redistribute the bonding pairs and lone pairs in the structure to draw resonance forms. Sulfur, selenium, and chlorine may expand their octets if doing so reduces formal charges.

Solve
(Step 1) The number of valence electrons is

Element	Cl	Se	N	S	O	Total
Valence electrons per atom	7 +	6 +	5 +	6 +	6	= 30

(Step 2) The atoms are connected in the order written.

$$Cl—Se—N—S—O$$

(Step 3) We complete the octets on the chlorine and oxygen atoms by adding three lone pairs to each.

$$:\ddot{C}l—Se—N—S—\ddot{O}:$$

(Step 4) In this structure there are 20 electrons from six lone pairs and four bond pairs. Ten more electrons are needed for this Lewis structure. We may complete the octets on sulfur and selenium by adding two lone pairs to each. The remaining two electrons may be distributed by adding one lone pair to nitrogen.

$$:\ddot{C}l—\ddot{S}e—\ddot{N}—\ddot{S}—\ddot{O}:$$

(Step 5) We can complete the octet on nitrogen by forming a double bond between nitrogen and either selenium or sulfur. The electrons could also be distributed in three additional resonance forms, including one in which sulfur exceeds its octet to reduce formal charges.

Think About It
Because the resonance structure in which the octet on sulfur is expanded has no formal charge, it is likely to be the most significant contributor to the overall bonding description.

8.81. **Collect and Organize**
We are to explain how we can use formal charges to choose the best molecular structure for a given chemical formula.

Analyze

Formal charge is not a real charge on the atoms but, rather, a method to assign the apparent charges on atoms in covalently bonded compounds. Formal charge (FC) is determined by

$$FC = (\text{number of valence } e^- \text{ for the atom}) - [(\text{number of } e^- \text{ in lone pairs}) + (\tfrac{1}{2} \times \text{number of } e^- \text{ in bonding pairs})]$$

Solve

The best possible structure for a molecule judging by formal charges is the structure in which the formal charges are minimized and the negative formal charges are on the most electronegative atoms in the structure.

Think About It

The sum of the formal charges on the atoms in a structure must equal the charge on the molecule.

8.83. Collect and Organize

In a sulfur–oxygen bond, we can predict which atom would carry the negative formal charge for the structure most likely to contribute to the bonding using the electronegativity values of S and O. We are asked if a structure with a negative formal charge on S rather than O is more likely to contribute to bonding in a molecule containing S and O atoms.

Analyze

The more electronegative atom is more likely to carry the negative formal charge.

Solve

No. The electronegativity of oxygen (3.5) is higher than that of sulfur (2.5), so the negative formal charge must be on the O atom in the structure that contributes most to the bonding.

Think About It

The most electronegative elements are F, O, N, and Cl, and these are the elements most likely to carry negative formal charges, if they must, in Lewis structures that contribute significantly to the bonding.

8.85. Collect and Organize

After drawing the Lewis structures for HNC and HCN and assigning the formal charges to the atoms, we are asked to analyze the differences in their formal charges (and choose the best, most stable, arrangement for the atoms).

Analyze

After drawing the Lewis structures for both HNC and HCN, we assign the formal charge (FC) for each atom based on the formula

$$FC = (\text{number of valence } e^- \text{ for the atom}) - [(\text{number of } e^- \text{ in lone pairs}) + (\tfrac{1}{2} \times \text{number of } e^- \text{ in bonding pairs})]$$

Solve

For both HNC and HCN there are 10 valence electrons, and the Lewis structures with formal charges are

$$\overset{0}{\text{H}}-\overset{+1}{\text{N}}\equiv\overset{-1}{\text{C}}\!:\qquad\qquad \overset{0}{\text{H}}-\overset{0}{\text{C}}\equiv\overset{0}{\text{N}}\!:$$

The formal charges are zero for all the atoms in HCN, whereas in HNC the carbon atom, with a lower electronegativity than N, has a –1 formal charge.

Think About It

The HCN arrangement, being more stable, is the more significant contributor to this.

8.87. Collect and Organize

We are to draw Lewis structures for cyanamide, H_2NCN, and assign formal charges to each atom.

Analyze

Because we are asked to draw *structures* for the compound, we suspect that the compound may show resonance. After drawing the possible resonance structures, we assign formal charges to all atoms in each structure using

$$FC = (\text{number of valence } e^- \text{ for the atom}) - [(\text{number of } e^- \text{ in lone pairs}) +$$
$$(\tfrac{1}{2} \times \text{number of } e^- \text{ in bonding pairs})]$$

Solve

For cyanamide, H_2NCN, there are 16 valence electrons, and the possible structures with formal charges assigned for the atoms are

The preferred structure is the one with the C triple bonded to N because all the formal charges on the structure are zero.

Think About It

Be careful in drawing the resonance structures. The structure below is not valid because the octet rule for both nitrogen atoms is violated.

8.89. Collect and Organize

We are to draw Lewis structures for all the valid resonance forms of $[N_2S_2As]$.

Analyze

Using the connectivity shown in Figure P8.89, we first draw one of the valid Lewis structures by the method described in the textbook, then redistribute the bonding pairs and lone pairs in the structure to draw resonance forms.

Solve

(Step 1) The number of valence electrons is

Element	As		2 S		2 N	Total
Valence electrons per atom	5	+	(2×6) +		(2×5)	= 27

(Step 2) Using the framework provided on Figure P8.89, we may satisfy the octet on sulfur by adding two lone pairs to each, and on arsenic by adding one lone pair and one unpaired electron to arsenic.

(Step 3) In this structure there are 21 electrons from five lone pairs, one unpaired electron, and five bond pairs. The remaining six electrons may be distributed by placing a total of three lone pairs on the nitrogen atoms.

(Step 4) In this structure there are 27 electrons from eight lone pairs, one unpaired electron, and five bond pairs. This matches the number of electrons required for the structure.

(Step 5) We can complete the octet on nitrogen by forming a double bond between nitrogen and an adjacent sulfur atom. The electrons could be distributed in four additional resonance forms:

Think About It

The resonance structure featuring an As=S double bond is the least significant contributor to the overall bonding description.

8.91. Collect and Organize

For the arrangement of atoms in nitrous oxide (N_2O) in which oxygen is the central atom, we are to assign formal charges and suggest why this structure is not stable.

Analyze

After drawing the possible resonance structures, we assign formal charges to all atoms in each structure using

$$FC = \text{(number of valence e}^- \text{ for the atom)} - [\text{(number of e}^- \text{ in lone pairs)} +$$
$$(\tfrac{1}{2} \times \text{number of e}^- \text{ in bonding pairs)}]$$

Solve

For nitrous oxide, N_2O, there are 16 valence electrons, and the Lewis structures with formal charges assigned to the atoms are

None of these structures is likely to be stable because oxygen is more electronegative than nitrogen and would be predicted by electronegativity to be negative, but the formal charge on O in all of these is positive.

Think About It

The Lewis structure for the arrangement N—N—O is far better by formal charge, particularly with the resonance structure in which there is a N to N triple bond.

8.93. Collect and Organize

We are asked to consider how odd-electron molecules are inconsistent with the octet rule.

Analyze

When the octet rule is applied to the drawing of any Lewis structure, the number of electrons needed will be a multiple of 8 (ignoring any duets required for H).

Solve

The number of electrons (multiple of 8) needed to follow the octet rule is always even; therefore, yes, odd-electron molecules are always exceptions to the octet rule.

Think About It

When an odd-electron molecule (radical) either gains or loses an electron, it may then satisfy the octet rule for the atoms in the molecule.

8.95. Collect and Organize

We are asked why C, N, O, and F always obey the octet rule in Lewis structures.

Analyze

C, N, O, and F are all second period elements with electron configurations of $[He]2s^22p^x$ where $x = 2, 3, 4, 5$. Once the $2p$ shell is filled with 6 e$^-$, a closed-shell configuration is formed.

Solve

In order for the atom to accommodate more than 8 e^- in covalently bonded molecules, it would require the use of orbitals beyond *s* and *p*. The *d* orbitals are not available to the small elements in the second period but do become available for the third period (and subsequent periods!) elements such as P, S, and Cl.

Think About It

The octet rule strictly applies for only second period elements but remains a starting place for drawing Lewis structures for compounds where larger elements are the central atoms in the structure.

8.97. **Collect and Organize**

To determine which of the sulfur–fluorine molecules have an expanded octet, we need to consider the number of electrons around the central atom required to form the compound.

Analyze

In each of these compounds sulfur is the central atom, as it is the least electronegative and has the highest bonding capacity. If the number of electrons in bonding pairs and lone pairs on the sulfur atom in the Lewis structure of each compound is greater than eight, then sulfur in that compound has an expanded octet.

Solve

Molecule	Lewis structure	Number of electrons around S
(a) SF_6		12
(b) SF_5		11
(c) SF_4		10
(d) SF_2		8

SF_6, SF_5, and SF_4 (a–c) require sulfur to expand its octet.

Think About It

Notice that SF_5 is an odd-electron (radical) species.

8.99. **Collect and Organize**

To determine the number of electrons in the covalent bonds around each central atom in the molecules, we first draw the Lewis structures for each.

Analyze

For these Lewis structures, we might have to expand octets for the sulfur atom (sulfur bonded to >2 atoms has to have greater than 8 e^- to form the compound), or the compound may contain a central atom with fewer than eight valence electrons. We also have to consider whether or not the expansion of the octet on sulfur through double bonding, for example, reduces the formal charges on the atoms in the structure. Some elements of Group 13 (B and Al) are capable of forming stable compounds with an incomplete octet.

Solve

(a) Al(CH$_3$)$_3$ has 24 valence electrons, and its Lewis structure is

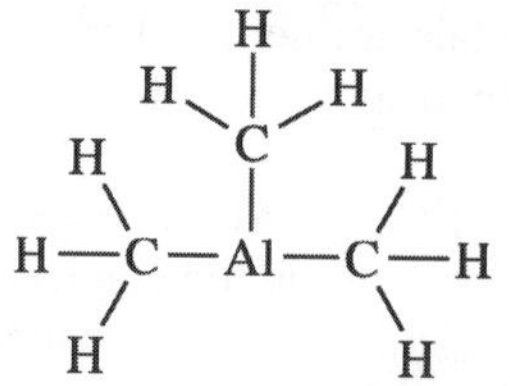

Aluminum may form stable compounds with an incomplete octet, so this structure is best represented as above, with no formal charges. There are 6 e$^-$ in three covalent bonds around aluminum in Al(CH$_3$)$_3$.

(b) B$_2$Cl$_4$ has 34 valence electrons, and its Lewis structure with formal charges assigned to its atoms is

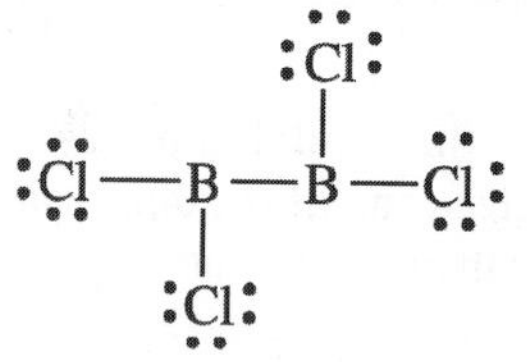

Boron may form stable compounds with an incomplete octet, so this structure is best represented as above, with no formal charges This gives 6 e$^-$ in three covalent bonds on each boron in B$_2$Cl$_4$.

(c) SO$_3$ has 24 valence electrons, and its Lewis structure with formal charges assigned to its atoms is

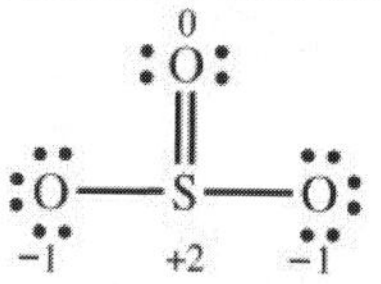

To reduce the formal charges on S and O, we could add double bonds between the other oxygen atoms and sulfur.

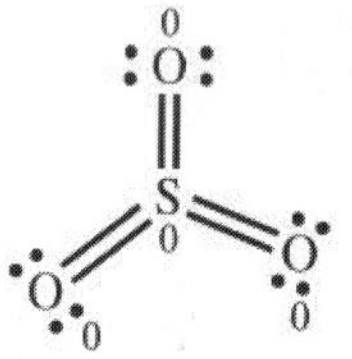

This gives 12 e$^-$ in six covalent bonds on sulfur in SO$_3$.

(d) SF$_5^-$ has 42 valence electrons, and its Lewis structure with formal charges assigned to its atoms is

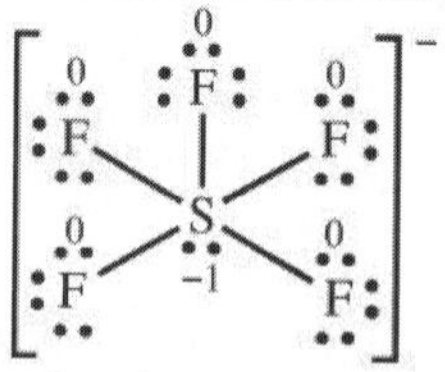

If we were to add a double bond between a fluorine atom and the sulfur atom, the formal charge build up would *not* be preferred over the previous structure.

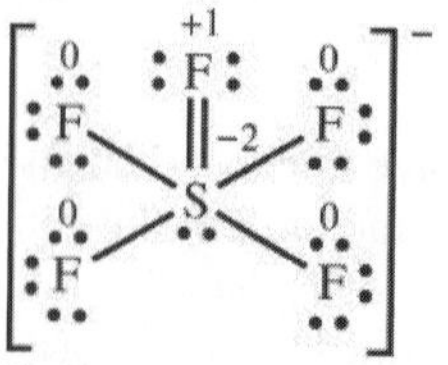

There are 10 e$^-$ in five covalent bonds in SF$_5^-$.

Think About It

Double bonding of an atom to fluorine as in the second structure in (d) gives a positive formal charge on F. This is never preferred since fluorine is the most electronegative element!

8.101. **Collect and Organize**

By drawing the Lewis structures of NOF_3 and POF_3, we are to describe the differences in bonding between these molecules.

Analyze

Nitrogen is a second period element that is not able to expand its octet, but phosphorus, as a third period element, can expand its octet.

Solve

Both molecules have 32 valence electrons. The Lewis structures with formal charges are as follows:

Because oxygen is more electronegative than N, these structures with a −1 formal charge on the O atom seem reasonable. However, the formal charges on P and O in POF_3 can be reduced to zero because P can expand its octet to form a double bond with O.

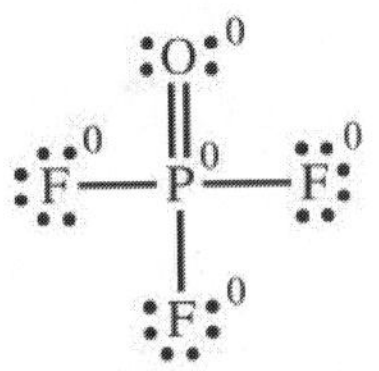

In POF_3 there is a double bond and no formal charges; in NOF_3 there are only single bonds and formal charges are present on the N and O atoms.

Think About It

In Lewis structures, always try to minimize formal charges. For elements in the third or higher periods, formal charge reduction can be accomplished by expanding the octets to form double bonds.

8.103. **Collect and Organize**

By drawing the Lewis structures of SeF_4 and SeF_5^- we can determine in which structure the Se atom has expanded its octet.

Analyze

Selenium, in the fourth period, expands its octet by making use of its $4d$ orbitals. For each Lewis structure, we use the method in the textbook. If the number of electrons around the central Se atom in the structures is greater than eight, then selenium expands its octet to form the compound.

Solve

SeF_4 has 34 valence electrons, and its Lewis structure shows that there are 10 electrons around the central selenium atom.

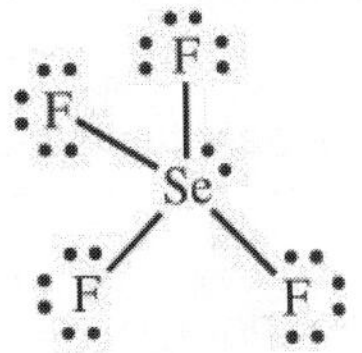

SeF_5^- has 42 valence electrons, and its Lewis structure shows that there are 12 electrons around the central selenium atom.

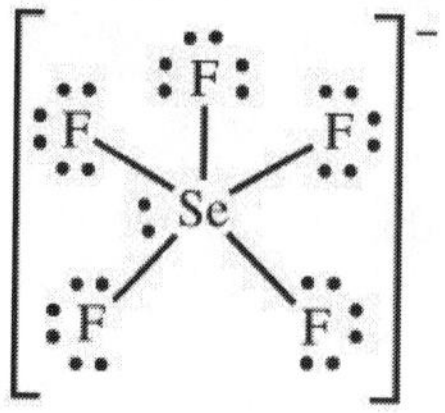

In both SeF_4 and SeF_5^-, Se has more than 8 valence electrons.

Think About It
Notice that in both SeF_4 and SeF_5^-, a lone pair of electrons is present on the selenium atom to give 10 and 12 electrons around Se, respectively.

8.105. ### Collect and Organize
Based on the arrangement of atoms given in Figure P8.105 for Cl_2O_2, we are to draw the Lewis structure and determine if either of the Cl atoms needs to expand its octet in order to form the molecule.

Analyze
The arrangement of atoms in Cl_2O_2 requires 26 valence electrons.

Solve
The Lewis structure for Cl_2O_2 is

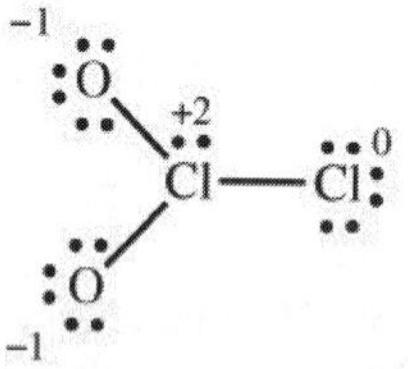

In this structure, neither Cl atom needs to expand its octet. However, the formal charges on the atoms of this structure are fairly high. To reduce this we could form double bonds between the Cl and O atoms.

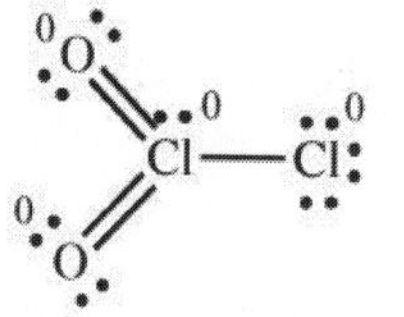

In this structure, all of the formal charges of all the atoms are zero, and the central chlorine atom has an expanded octet.

Think About It
Use formal charges to determine when an atom will expand its octet. At first glance Cl_2O_2 did not appear to have any more than 8 e$^-$ surrounding the central Cl atom.

8.107. ### Collect and Organize
For each molecule combining Cl with O, we are to determine which are odd-electron molecules.

Analyze
To answer this we need only add up the valence electrons for each molecule.

Solve
(a) Cl_2O_7 has $(2Cl \times 7 \text{ e}^-) + (7O \times 6 \text{ e}^-) = 56 \text{ e}^-$
(b) Cl_2O_6 has $(2Cl \times 7 \text{ e}^-) + (6O \times 6 \text{ e}^-) = 50 \text{ e}^-$
(c) ClO_4 has $(1Cl \times 7 \text{ e}^-) + (4O \times 6 \text{ e}^-) = 31 \text{ e}^-$
(d) ClO_3 has $(1Cl \times 7 \text{ e}^-) + (3O \times 6 \text{ e}^-) = 25 \text{ e}^-$

(e) ClO_2 has $(1Cl \times 7 \text{ e}^-) + (2O \times 6 \text{ e}^-) = 19 \text{ e}^-$
The odd-electron molecules are (c) ClO_4, (d) ClO_3, and (e) ClO_2.

Think About It
For these chlorine–oxygen molecules, notice that the odd-electron species have an odd number of Cl atoms in their formulas.

8.109. **Collect and Organize**
From the Lewis structures given we can use formal charge arguments to determine which structure contributes most to the bonding in CNO.

Analyze
The resonance structure that contributes the most to the bonding has the lowest possible formal charges on the atoms and, if formal charges are present, then the negative formal charges should be on the most electronegative atoms in the structure.

Solve
The formal charge assignments on each of the structures are as follows:

(a) $\overset{-2}{\cdot\ddot{C}}-\overset{+1}{N}\equiv\overset{+1}{O}\colon$ (b) $\overset{-1}{\colon C}=\overset{+1}{N}=\overset{0}{\ddot{O}}\colon$

(c) $\overset{-1}{\colon C}\equiv\overset{+1}{N}-\overset{0}{\ddot{O}}\cdot$ (d) $\overset{0}{\cdot C}\equiv\overset{+1}{N}-\overset{-1}{\ddot{O}}\colon$

The structure that contributes the most to the bonding in CNO is (d).

Think About It
Since all of these structures are in resonance, there is some contribution to the bonding from each resonance form. However, because structure (d) is favored, with the –1 formal charge placed on the oxygen atom, it contributes the most to the bonding.

8.111. **Collect and Organize**
Compare the Lewis structures for $(CH_3)_4Al_2Cl_2$ and $(CH_3)_2AlCl$, and determine why $(CH_3)_4Al_2Cl_2$ is preferred.

Analyze
Aluminum may be found in an electron deficient state, but may also satisfy its octet if possible. The most stable molecules will feature a complete octet.

Solve

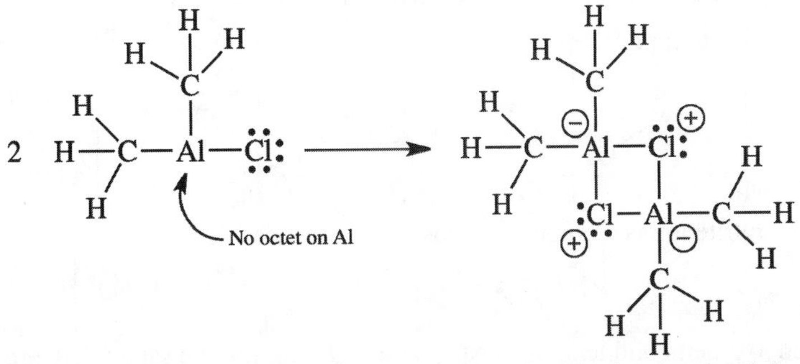

The lone pairs on chlorine may form a coordinate covalent bond with aluminum, giving rise to the dimer at right. Because the dimer features a complete octet for all atoms (except hydrogen, which has a duet), it is more stable than the monomer at left.

Think About It
Notice that the structure at right has formal charges, whereas the monomer at left does not. It is often difficult to predict which of two competing guidelines (i.e., satisfying an octet and minimizing formal charge) will be more important, so we rely on experiments to guide us!

8.113. Collect and Organize / Analyze

Compare resonance structures for SF_6 with an expanded octet on sulfur, and an ionic form $[SF_4]^{2+}[F]_2^-$ without an expanded octet on sulfur. We must determine which resonance structure is preferred based on formal charges.

Solve

SF_6 contains 48 valence electrons. If we expand the octet on sulfur

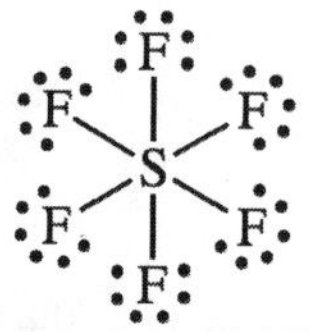

If we remove two fluoride anions from the structure above, we arrive at a structure with no expanded octet, but a formal charge of –2 on sulfur.

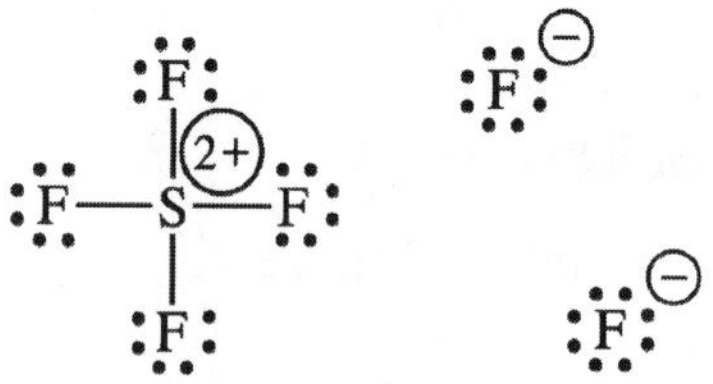

The presence of formal charges makes the ionic form of SF_6 less preferable to the resonance structure with an expanded octet on sulfur.

Think About It

The formal charge of 2+ on sulfur is especially unfavorable because sulfur has a relatively high electronegativity value of 2.5.

8.115. Collect and Organize

Using the Lewis structures for the nitrate ion (NO_3^-) and the nitrite ion (NO_2^-), we can determine if we expect their N–O bond lengths to be the same.

Analyze

In Lewis structures, single bonds are longer than double bonds, which are longer than triple bonds. We must also consider any resonance forms that these molecules might have.

Solve

Each bond in the nitrate ion is 1.33 bonds because of resonance.

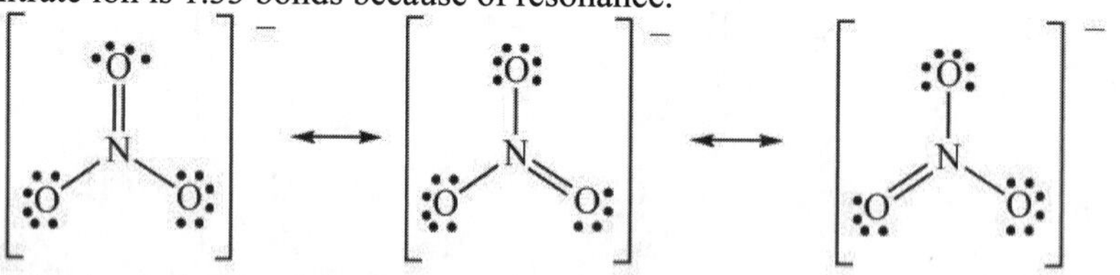

Each bond in the nitrite ion is 1.5 bonds because of resonance.

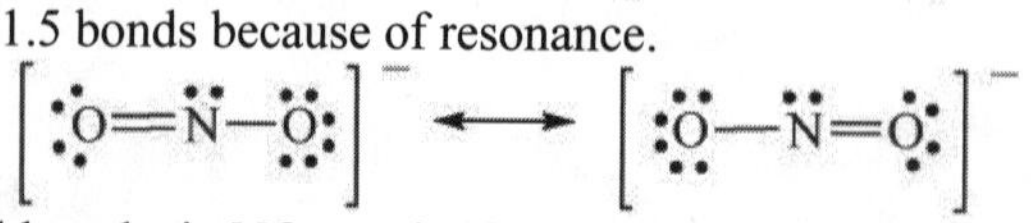

No, the nitrogen–oxygen bond lengths in NO_3^- and NO_2^- are not the same; they are different.

Think About It

We would expect the nitrogen–oxygen bond length in NO_2^- to be shorter than the bond length in NO_3^-.

8.117. Collect and Organize

Using the Lewis structures (with resonance forms if necessary), we can explain why the nitrogen–oxygen bonds in N_2O_4 and N_2O are nearly identical in length.

Analyze

In Lewis structures single bonds are longer than double bonds, which are longer than triple bonds. We must also consider any resonance forms that these molecules might have.

Solve

The nitrogen–oxygen bond in N_2O_4 has a bond order of 1.5 due to four equivalent resonance forms.

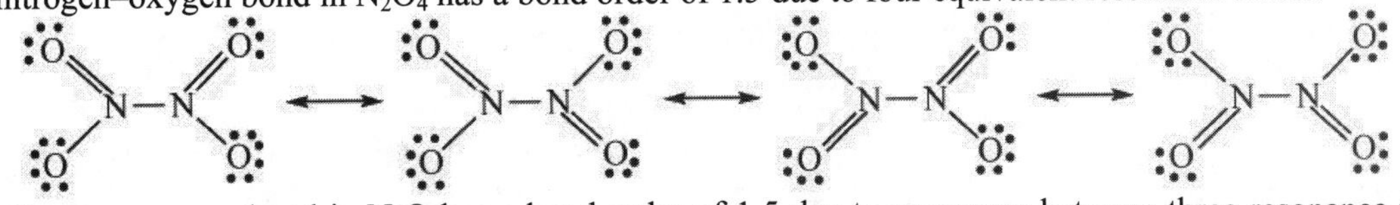

The nitrogen–oxygen bond in N_2O has a bond order of 1.5 due to resonance between three resonance forms (although, the last resonance structure shown does not significantly contribute to the structure of the molecule because of the high formal charges).

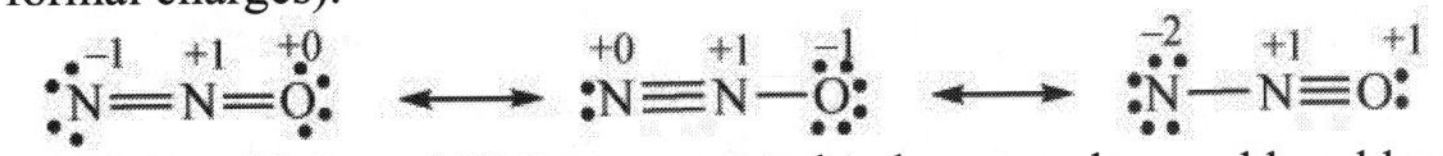

Therefore, owing to resonance, N_2O_4 and N_2O are expected to have nearly equal bond lengths.

Think About It

Remember that all resonance forms, even though they may not contribute equally, do contribute some to the structure.

8.119. Collect and Organize

In order to rank the bond lengths and bond energies in NO_2^-, NO^+, and NO_3^-, we need to draw the Lewis structures, with resonance forms if necessary.

Analyze

In Lewis structures single bonds are longer than double bonds, which are longer than triple bonds. We must also consider any resonance forms that these molecules might have. In Lewis structures, single bonds have the lowest bond energy. Double bonds are stronger (have higher bond energies) than single bonds and triple bonds are stronger than double bonds. The higher the bond order, the higher the bond energy.

Solve

For NO_2^-, the bond order for the N–O bond is 1.5 due to resonance.

For NO^+, the bond order for the N–O bond is 3.0.

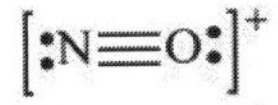

For NO_3^-, the bond order for the N–O bond is 1.33 due to resonance.

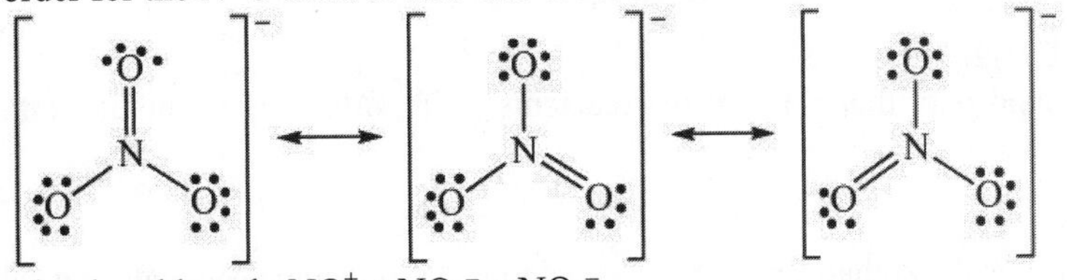

(a) In order of increasing bond length: $NO^+ < NO_2^- < NO_3^-$.
(b) In order of increasing bond energy: $NO_3^- < NO_2^- < NO^+$.

Think About It

Resonance has quite an effect on the bond order and length.

8.121. Collect and Organize

We are to determine if the B–F bond energy is the same in BF_3 and F_3BNH_3.

Analyze

The chemical environment of a bond affects the energy required to break it. Since BF_3 and F_3BNH_3 have different bonding environments, it is reasonable to expect that the bond energies will also be different. We should draw valid Lewis structures for each molecule using the method described in the text, and compare the environment of the B–F bonds for each.

Solve

The Lewis strucutres for BF_3 and F_3BNH_3 are

The electronegative nitrogen atom in F_3BNH_3 will pull some electron density away from the boron atom, weakening the B–F bond. We should not expect the B–F bond energies of the molecules to be the same.

Think About It

Despite the B–F bonds looking similar in both molecules, the presence of an electronegative atom like nitrogen can affect the B–F bond strength.

8.123. **Collect and Organize**

We are to explain why we need to know the stoichiometry of the reaction to estimate the enthalpy change for the reaction using bond energies.

Analyze

Bond breaking is endothermic, and bond formation is exothermic. The stoichiometry of the reaction tells us how many bonds break and how many form.

Solve

We must account for all the bonds that break and all the bonds that form in the reaction. To do this, we must start with a balanced chemical reaction. If we miss a bond that breaks, our calculated enthalpy of reaction would be too negative. If we miss a bond that forms, our calculated enthalpy of reaction would be too positive.

Think About It

Before using an equation to calculate enthalpy change, whether from enthalpies of formation or from bond energies, always start with a balanced chemical equation.

8.125. **Collect and Organize**

We are to explain why the phase of the reactants is important in calculating enthalpy changes using bond energies.

Analyze

The bond energy is the enthalpy change required to break one mole of bonds in a substance in the *gas phase*.

Solve

If the compounds are in the solid or liquid phase, interactions between molecules may slightly change the bond energy for a given bond.

Think About It

Having bond energy data tabulated only for the gaseous phase ensures that the measured energies are only for the bonds breaking and do not include any intermolecular interactions. Later, you will learn more about intermolecular forces. Some can be very strong (ion–ion forces) and others quite weak (van der Waals forces).

8.127. **Collect and Organize**
We can use bond energy data in Table 8.3 to estimate the enthalpy of these reactions.

Analyze
The enthalpy of a reaction as estimated by bond energies is given as

$$\Delta H_{rxn} = \sum \Delta H_{bondbreaking} + \sum \Delta H_{bondforming}$$

where $\Delta H_{bondbreaking}$ and $\Delta H_{bondforming}$ are average bond energies for the bonds in the reactants and products, respectively. In computing $\Delta H_{bondbreaking}$ and $\Delta H_{bondforming}$ we have to take into account the number (or moles) of a particular type of bond that breaks or forms. We must also keep in mind that bond breaking requires energy ($+\Delta H$), whereas bond formation releases energy ($-\Delta H$).

Solve
(a) $2\ {:}N{\equiv}N{:}\ +3\ H{-}H\ \longrightarrow\ 2\ H{-}\overset{\displaystyle ..}{N}{-}H$ with H below

$$\Delta H_{rxn} = [(2 \times 941\ \text{kJ/mol}) + (3 \times 436\ \text{kJ/mol})] + [-(6 \times 388\ \text{kJ/mol})] = 862\ \text{kJ}$$

(b) ${:}N{\equiv}N{:}\ +\ 2\ H{-}H\ \longrightarrow\ $ (H₂N–NH₂ structure)

$$\Delta H_{rxn} = [(1 \times 941\ \text{kJ/mol}) + (2 \times 436\ \text{kJ/mol})] + [-(1 \times 163\ \text{kJ/mol}) - (4 \times 388\ \text{kJ/mol})] = 98\ \text{kJ}$$

(c) $2\ {:}N{\equiv}N{:}\ +\ {:}O{=}O{:}\ \longrightarrow\ 2\ {:}N{\equiv}N{-}O{:}$

$$\Delta H_{rxn} = [(2 \times 941\ \text{kJ/mol}) + (1 \times 495\ \text{kJ/mol})] + [-(2 \times 941\ \text{kJ/mol}) - (2 \times 201\ \text{kJ/mol})] = 93\ \text{kJ}$$

Think About It
We have to use the "best" Lewis structure for these calculations of reaction enthalpy. In (c), an alternate form (but contributing less to the bonding by formal charge arguments) for N_2O would be

$2\ {:}N{\equiv}N{:}\ +\ {:}O{=}O{:}\ \longrightarrow\ 2\ {:}\overset{..}{N}{=}N{=}O{:}$

$$\Delta H_{rxn} = [(2 \times 941\ \text{kJ/mol}) + (1 \times 495\ \text{kJ/mol})] + [-(2 \times 418\ \text{kJ/mol}) - (2 \times 607\ \text{kJ/mol})] = 327\ \text{kJ}$$

8.129. **Collect and Organize**
Given the overall enthalpy for the reaction of carbon monoxide with oxygen to give carbon dioxide, we are to calculate the energy of the $C{\equiv}O$ bond in carbon monoxide.

Analyze
The enthalpy of a reaction as estimated by bond energies is given as

$$\Delta H_{rxn} = \sum \Delta H_{bondbreaking} + \sum \Delta H_{bondforming}$$

where $\Delta H_{bondbreaking}$ and $\Delta H_{bondforming}$ are average bond energies for the bonds in the reactants and products, respectively. Notice that we are given the enthalpy of combustion of 1 mole of CO. If we use the balanced equation that gives whole-number ratios of the reactants and products, we have to multiply the enthalpy of reaction by 2. To solve this we will need the bond energies of the $O{=}O$ bond in oxygen (495 kJ/mol) and the $C{=}O$ bond in carbon dioxide (799 kJ/mol) from Table 8.3.

Solve

$2\ {:}C{\equiv}O{:}\ +\ {:}O{=}O{:}\ \longrightarrow\ 2\ {:}O{=}C{=}O{:}$

$$\Delta H_{rxn} = (2 \times -283\ \text{kJ/mol}) = [(1 \times 495\ \text{kJ/mol}) + (2x)] + [-(4 \times 799\ \text{kJ/mol})]$$

$$x = 1068\ \text{kJ/mol}$$

Think About It
This compares well to the average $C{\equiv}O$ bond energy given in Table 8.3.

8.131. **Collect and Organize**

We can use bond energies to compare the reaction enthalpies of the two reactions

$$CH_4 + \tfrac{3}{2} O_2 \rightarrow CO + 2 H_2O$$

$$CH_4 + 2 O_2 \rightarrow CO_2 + 2 H_2O$$

Analyze

From the Lewis structures and the average bond energy values in Table 8.3, we can calculate the reaction enthalpies. The enthalpy of a reaction as estimated by bond energies is given as

$$\Delta H_{rxn} = \sum \Delta H_{bondbreaking} + \sum \Delta H_{bondforming}$$

where $\Delta H_{bondbreaking}$ and $\Delta H_{bondforming}$ are average bond energies for the bonds in the reactants and products, respectively.

Solve

For the incomplete combustion of CH_4 to CO:

$$\Delta H_{rxn} = [(4 \times 413 \text{ kJ/mol}) + (\tfrac{3}{2} \times 495 \text{ kJ/mol})] + [-(1 \times 1072 \text{ kJ/mol}) - (4 \times 463 \text{ kJ/mol})]$$

$$\Delta H_{rxn} = -530 \text{ kJ/mol}$$

For the complete combustion of CH_4 to CO_2:

$$\Delta H_{rxn} = [(4 \times 413 \text{ kJ/mol}) + (2 \times 495 \text{ kJ/mol})] + [-(2 \times 799 \text{ kJ/mol}) - (4 \times 463 \text{ kJ/mol})]$$

$$\Delta H_{rxn} = -808 \text{ kJ/mol}$$

The complete combustion reaction releases -808 kJ, which is 278 kJ more than the incomplete combustion reaction.

Think About It

Although weaker C$=$O bonds are formed in the complete combustion reaction, there are two such bonds formed in CO_2, which outweighs the strong C$\equiv$O bond in carbon monoxide in the incomplete combustion reaction.

8.133. **Collect and Organize**

For the reaction of ammonia with oxygen to give water and nitrogen dioxide, we can use the bond energies of the N—H (388 kJ/mol), O$=$O (495 kJ/mol), N$=$O (607 kJ/mol), N—O (201 kJ/mol), and O—H (463 kJ/mol) bonds to estimate the enthalpy of the reaction.

Analyze

To be sure which bonds are breaking and which bonds are being formed, it is helpful to draw the Lewis structures of each of the products and reactants. The enthalpy of a reaction as estimated by bond energies is given as

$$\Delta H_{rxn} = \sum \Delta H_{bondbreaking} + \sum \Delta H_{bondforming}$$

where $\Delta H_{bondbreaking}$ and $\Delta H_{bondforming}$ are average bond energies for the bonds in the reactants and products, respectively.

Solve

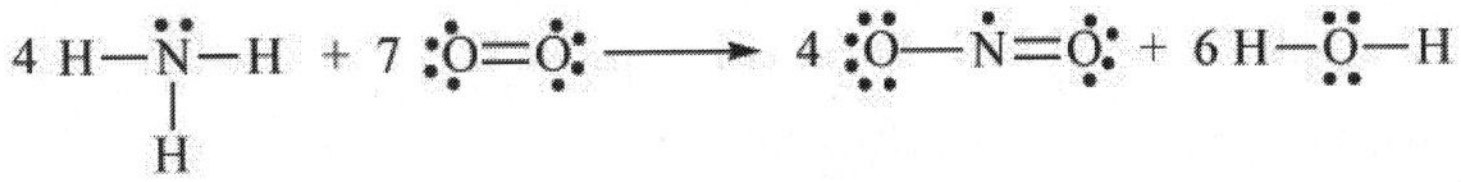

$$4 \; H-\overset{\overset{\displaystyle H}{|}}{\underset{|}{N}}-H \; + \; 7 \; \ddot{\ddot{O}}=\ddot{\ddot{O}} \longrightarrow 4 \; \ddot{\ddot{O}}-N=\ddot{\ddot{O}} \; + \; 6 \; H-\ddot{\ddot{O}}-H$$

$$\Delta H_{rxn} = [(12 \times 388 \text{ kJ/mol}) + (7 \times 495 \text{ kJ/mol})] + [-(4 \times 201 \text{ kJ/mol}) - (4 \times 607 \text{ kJ/mol}) - (12 \times 463 \text{ kJ/mol})]$$

$$\Delta H_{rxn} = -667 \text{ kJ}$$

Think About It

The bonding in NO_2 is not strictly one single bond and one double bond, as shown by the resonance structures:

$$\ddot{\ddot{O}}-\ddot{N}=\ddot{O}: \longleftrightarrow :\ddot{O}=\ddot{N}-\ddot{\ddot{O}}$$

For calculations using bond energies, however, we can use one of the "frozen" resonance structures to assign bond energy values.

8.135. Collect and Organize

We are given that the enthalpy change for the reaction of 1 mol of CS_2 with 3 moles of O_2 is -1102 kJ/mol. From this information and the data in Table 8.3, we can calculate the carbon–sulfur bond energy in a single molecule of CS_2.

Analyze

Figure P8.135 shows the following reaction:

$$3 \; O_2 + CS_2 \rightarrow CO_2 + 2 \; SO_2$$

To calculate the bond energy of the carbon–sulfur bond in a molecule of CS_2, we can use the $O=O$ (495 kJ/mol), $C=O$ (799 kJ/mol), and $S=O$ (523 kJ/mol) bond energies. From the Lewis structures and the average bond energy values in Table 8.3, we can calculate the reaction enthalpy. The enthalpy of a reaction as estimated by bond energies is given as

$$\Delta H_{rxn} = \sum \Delta H_{bondbreaking} + \sum \Delta H_{bondforming}$$

where $\Delta H_{bondbreaking}$ and $\Delta H_{bondforming}$ are average bond energies for the bonds in the reactants and products, respectively.

Solve

$$3 \; \ddot{\ddot{O}}=\ddot{\ddot{O}} + :\ddot{S}=C=\ddot{S}: \longrightarrow :\ddot{O}=C=\ddot{O}: + 2 \; :\ddot{O}=\ddot{S}=\ddot{O}:$$

$$\Delta H_{rxn} = -1102 \text{ kJ/mol} = [(3 \times 495 \text{ kJ/mol}) + (2x)] + [-(2 \times 799 \text{ kJ/mol}) - (4 \times 523 \text{ kJ/mol})]$$

$$x = 552 \text{ kJ/mol of } CS_2$$

The energy of a single carbon–sulfur bond in CS_2 is

$$C=S \text{ bond energy} = \frac{552 \text{ kJ}}{\text{mol C–S bonds}} \times \frac{1 \text{ mol C–S bonds}}{6.022 \times 10^{23} \text{ bonds}} = 9.17 \times 10^{-22} \text{ kJ or } 9.17 \times 10^{-19} \text{ J}$$

Think About It

Notice that to reduce formal charges on the atoms in SO_2, the sulfur atom has expanded its octet so as to double bond with both oxygen atoms.

8.137. Collect and Organize

From the Lewis structure of C_3O_2 we are to predict whether or not the carbon–oxygen bond lengths are equal.

Analyze

Both C and O are second period elements and strictly adhere to the octet rule. The molecule has 24 valence electrons.

Solve

$$:\!O\!=\!C\!=\!C\!=\!C\!=\!O\!:$$

All of the formal charges on the atoms of this Lewis structure are zero, so it is the preferred resonance structure. Both carbon–oxygen bonds are equal.

Think About It

The other two resonance structures for C_3O_2 are

$$:O\!\equiv\!C\!-\!\overset{..}{C}\!=\!C\!=\!\overset{..}{O}: \quad\longleftrightarrow\quad :\overset{..}{O}\!=\!C\!=\!\overset{..}{C}\!-\!C\!\equiv\!O:$$

We would predict from these structures that the carbon–oxygen bonds would be the same, with a bond order of 2.5.

8.139. Collect and Organize

We are to select the preferred symbol in each pair listed.

Analyze

In order to reflect the valence electrons available for bonding, the electron dot placement around the element symbol in a Lewis symbol should show unpaired lone electrons that may form covalent bonds with other atoms. The valence electrons available for bonding in the atoms are Be = 2, Al = 3, C = 4, He = 0.

Solve

(a) $\cdot$Be$\cdot$ (b) $\cdot\dot{Al}\cdot$ (c) $\cdot\dot{\underset{\cdot}{C}}\cdot$ (d) He$\!:$

Think About It

This can be extended to other atoms. For example, oxygen, which tends to form two bonds to other atoms, has the Lewis dot symbol

$$\cdot\overset{..}{\underset{..}{O}}\cdot$$

8.141. Collect and Organize

Carbon disulfide could have either carbon or sulfur as the central atom in its structure. Using the formal charge assignments in the two possible Lewis structures, we are to determine which is the preferred structure.

Analyze

A structure is preferred when the formal charges on the atoms are minimized and when any negative formal charges are located on the most electronegative elements. The electronegativities of carbon and sulfur are the same, namely, 2.5.

Solve

Carbon disulfide has 16 valence electrons.

$$:\overset{0}{\underset{..}{S}}\!=\!\overset{0}{C}\!=\!\overset{0}{\underset{..}{S}}: \qquad :\overset{0}{\underset{..}{S}}\!=\!\overset{+2}{S}\!=\!\overset{-2}{\underset{..}{C}}:$$

When C is the central atom, the atoms all carry zero formal charge; this structure is the preferred structure for carbon disulfide.

Think About It

Because sulfur can expand its octet, the formal charges on CSS may be reduced:

$$:\overset{0}{\underset{..}{S}}\!=\!\overset{+1}{S}\!\equiv\!\overset{-1}{C}:$$

This still leaves formal charges on the atoms. For reasons you will see when you study valence bond theory, carbon is not able to quadruple bond to reduce the formal charges on the atoms in this structure down to zero.

8.143. Collect and Organize

For the poisonous substance phosgene, $COCl_2$, we are to draw its Lewis structure and write and balance the equation describing its reaction with water.

Analyze

$COCl_2$ has 24 valence electrons. Carbon has the largest bonding capacity and is the least electronegative of the atoms; we are also told that it is the central atom in the structure. From the statement of the problem, we know that when water reacts with phosgene, CO_2 and HCl are produced.

Solve

(a)

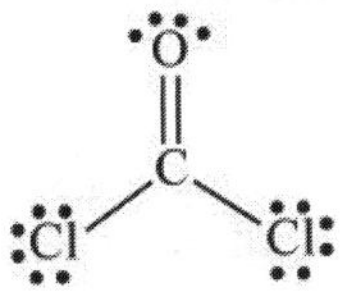

(b)

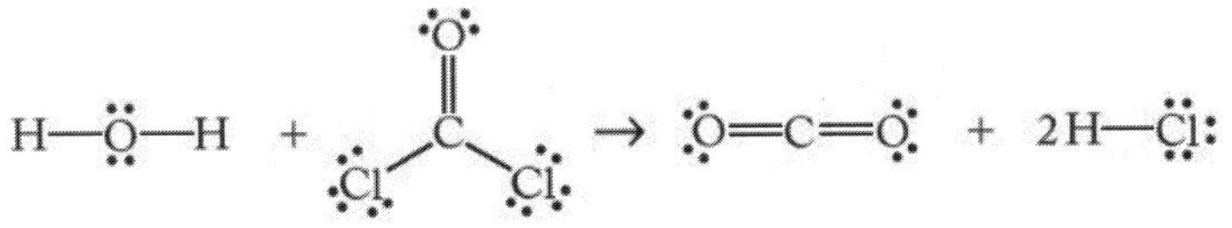

Think About It

Symptoms of human phosgene inhalation include choking, painful breathing, severe eye irritation, and skin burns. Death may result from lack of oxygen.

8.145. **Collect and Organize**

Neutral OCN reacts with itself to form OCNNCO. OCN⁻ reacts with BrNO to give OCNNO, and with Br_2 and NO_2 to give OCN(CO)NCO. For these three products we are to draw Lewis structures with any appropriate resonance forms.

Analyze

(a) OCNNCO has 30 valence electrons.
(b) BrNO has 18 valence electrons.
(c) OCN(CO)NCO has 40 valence electrons.

Solve

(a)

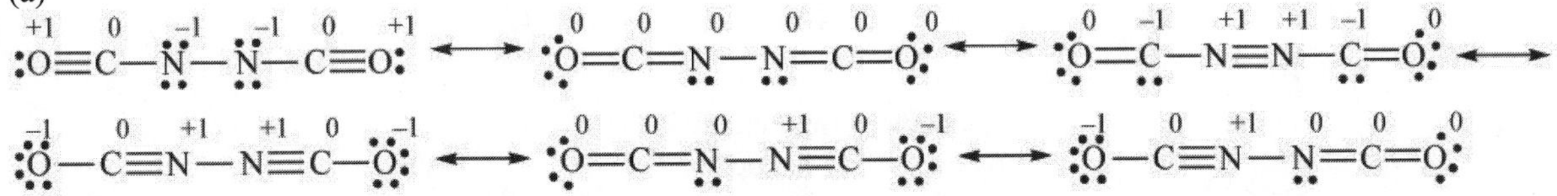

(b)

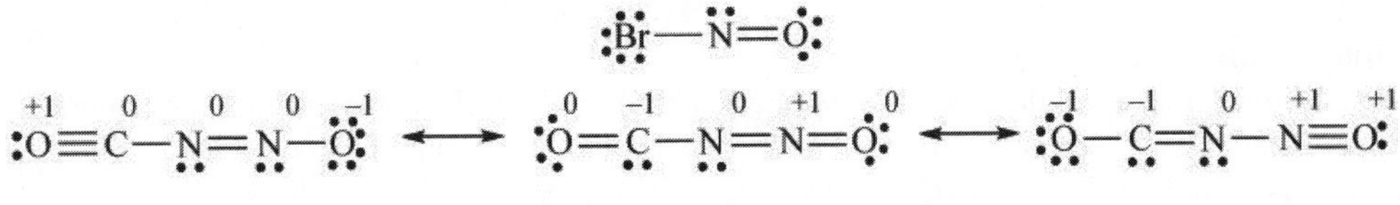

(c)

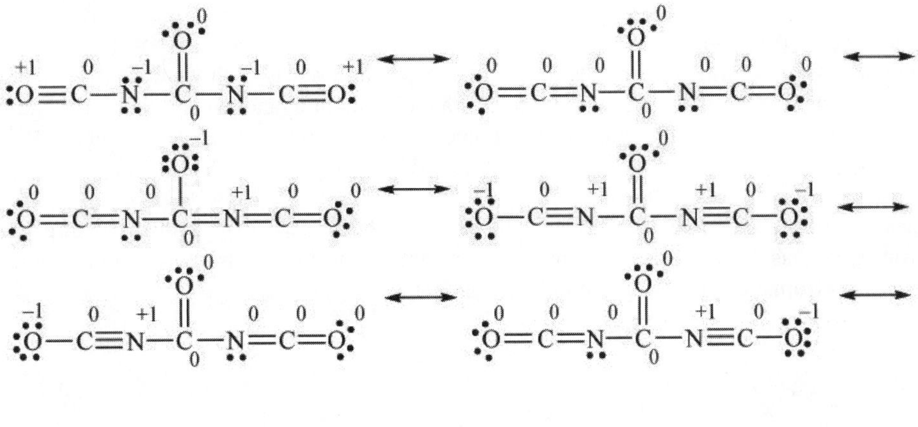

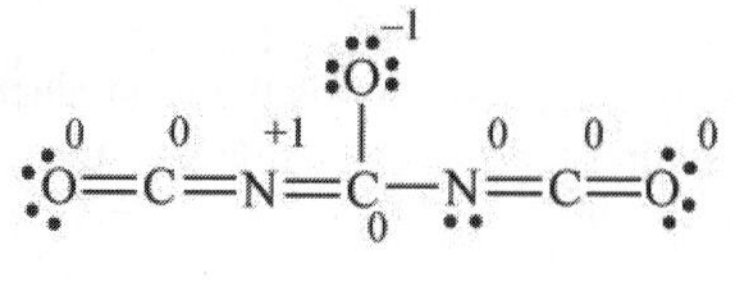

$$\overset{-1}{:\!\ddot{O}\!:} \\ \overset{0}{:\!\ddot{O}}=\overset{0}{C}=\overset{+1}{N}=\overset{+1}{\underset{0}{C}}-\overset{0}{N}=\overset{0}{C}=\overset{0}{\ddot{O}\!:}$$

Think About It

The resonance structure that contributes the most to the bonding in each of these compounds is the one that minimizes formal charges on the atoms.

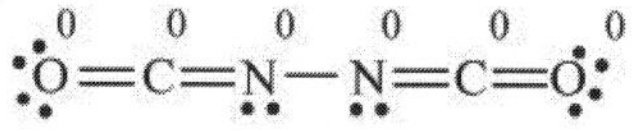

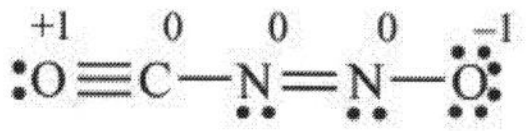

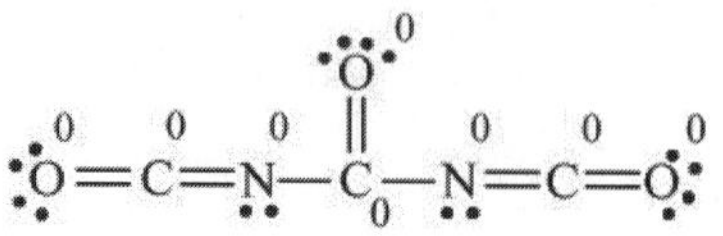

$$\overset{0}{:\!\ddot{O}}=\overset{0}{C}=\overset{0}{\ddot{N}}-\overset{0}{\ddot{N}}=\overset{0}{C}=\overset{0}{\ddot{O}\!:} \qquad \overset{+1}{:\!O}\equiv\overset{0}{C}-\overset{0}{\ddot{N}}=\overset{0}{N}-\overset{-1}{\ddot{O}\!:} \qquad \overset{0}{:\!\ddot{O}}=\overset{0}{C}=\overset{0}{N}-\overset{0}{\underset{0}{C}}-\overset{0}{N}=\overset{0}{C}=\overset{0}{\ddot{O}\!:}$$

8.147. Collect and Organize

We are to draw Lewis structures of both ClO_2 and Cl_2O_6. For Cl_2O_6, we are to consider one structure with a chlorine–chlorine bond and one with a Cl—O—Cl arrangement of atoms.

Analyze

Cl_2O_6 has 50 valence electrons. ClO_2 has 19 valence electrons, and as an odd-electron species we expect that the structure will have 1 unpaired electron.

Solve

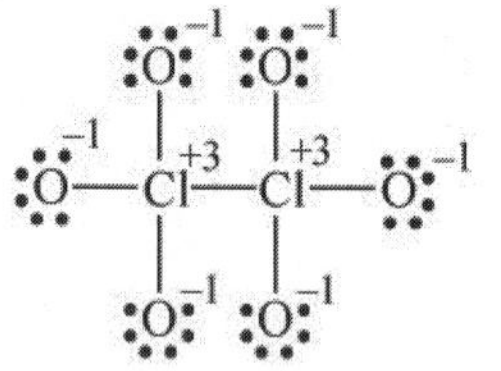

(Lewis structures for Cl_2O_6 with Cl—Cl bond, showing formal charges.)

Reducing formal charges by expanding the octets on Cl atoms gives:

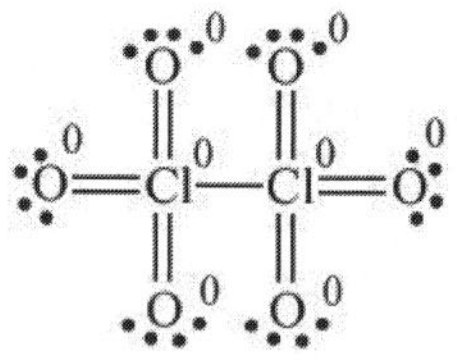

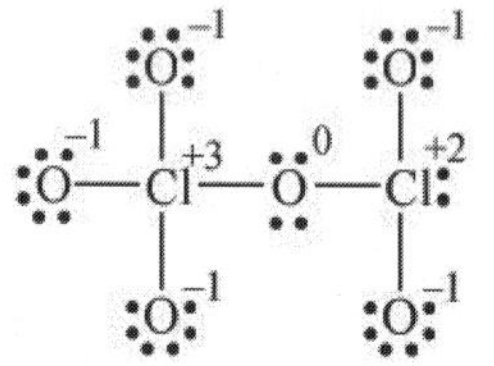

(Lewis structures for Cl_2O_6 with Cl—O—Cl arrangement, showing formal charges.)

Reducing formal charges by expanding the octets on Cl atoms gives:

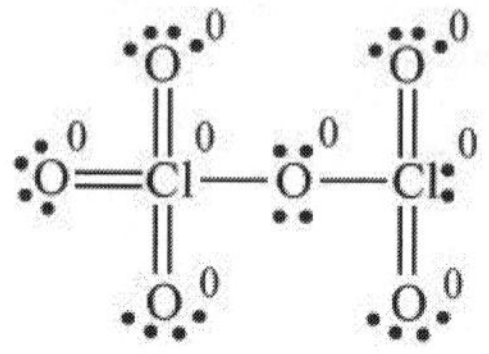

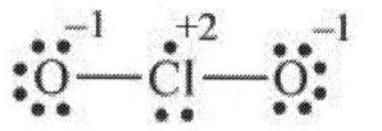

(Lewis structures for ClO_2, showing formal charges.)

Reducing formal charges by expanding the octets on Cl atoms gives:

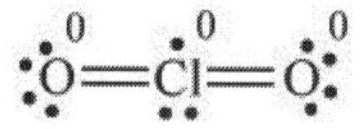

Think About It

Based on formal charges alone, we would not be able to predict the actual atom arrangement in Cl_2O_6 because we were able to draw structures for both arrangements in which all atoms have zero formal charge. The actual structure is thought to be as a perchlorate salt, $ClO_2^+ClO_4^-$.

8.149. Collect and Organize

Cyanogen (C_2N_2) is formed from CN, a radical species, and reacts with water to give oxalic acid. We are to draw the Lewis structures for the two possible arrangements of cyanogen and then, through the comparison of the structures with oxalic acid, rationalize a choice for the actual structure of cyanogen.

Analyze

CN has 9 valence electrons, and as an odd-electron species we expect the structure to have 1 unpaired electron. C_2N_2 has 18 valence electrons. We see that oxalic acid's structure contains a C—C bond, which would derive from a carbon–carbon bond in cyanogen.

Solve

(a) CN has the Lewis structure

$$\cdot C\equiv N\colon$$

The more likely structure for cyanogen is the one with no formal charges on the atoms. This is the one that contains the C—C bond:

$$\overset{0}{:}N\overset{0}{\equiv}C\overset{0}{-}C\overset{0}{\equiv}N: \qquad \overset{-1}{:}C\overset{+1}{\equiv}N\overset{+1}{-}N\overset{-1}{\equiv}C:$$

(b) Because oxalic acid has a C—C bond, we would expected that oxalic acid retained this C—C bond from the cyanogen from which it formed in reaction with water. This supports the predicted structure for cyanogen from formal charge analysis.

Think About It

Formal charge analysis is a method that works often, but not always, in predicting the atom connectivity in a molecule or the major contributing resonance form in bonding. The prediction, however, must stand up to experimental scrutiny.

8.151. **Collect and Organize**

The structure of F_3SCN shows bond lengths of 116 pm (C—N), 174 pm (S—C), and 160 pm (F—S). We can compare these bond lengths to those in Table 8.3 to determine the type of bond each might be (single, double, or triple). This information can help us to draw the Lewis structure for the molecule and assign formal charges.

Analyze

From Table 8.3, the only bond length listed that corresponds to any of the bond lengths in F_3SCN is $C\equiv N$ (116 pm). To draw the Lewis structure, 36 valence electrons are needed.

Solve

Formal charges are nonzero and the carbon–nitrogen bond is not triple.

All formal charges are zero and the carbon–nitrogen bond is triple.

Think About It

In the preferred structure, sulfur has expanded its octet.

8.153. **Collect and Organize**

Tellurium will expand its octet to form $TeOF_6^{2-}$. Using this information we can draw the Lewis structure for this ion.

Analyze

$TeOF_6^{2-}$ has 56 valence electrons, and tellurium is the central atom because it is the least electronegative atom in the structure. Using formal charges, we can draw the most likely Lewis structure for the ion.

Solve

This structure has the lowest formal charges. Oxygen is highly electronegative so it will carry a negative formal charge.

Think About It

Another resonance structure of $TeOF_6^{2-}$ would be

but this places all the negative formal charge on the least electronegative atom in the structure, so it is not the preferred structure.

8.155. **Collect and Organize**

To determine the number of pairs of shared electrons that the xenon atom has in each of the listed substances, we can draw the Lewis diagram for each molecule or ion.

Analyze

The number of electron pairs around xenon may exceed four because xenon is in the fifth period and, therefore, can expand its octet.

Solve

(a)

Two electron pairs

(b)

Four electron pairs

(c)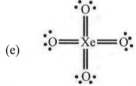

One electron pair

(d)

Five electron pairs

(e)

Eight electron pairs

Think About It

Xenon has expanded its octet in every one of these compounds except for (c) XeF^+.

8.157. **Collect and Organize**

Electron diffraction characterizes bond lengths in molecules, and if there is a difference in the number of electrons around two atoms, it can also tell us that one atom has a higher electron density compared to the other atom. We must consider the value of electron diffraction in distinguishing between the bond order between bonded atoms and bond lengths.

Analyze
We assume that the number of electrons, and thus the electron density, for atom X is different from atom A.

Solve
(a) If we can distinguish atoms by electron density using this technique, we would be able to distinguish X—A—A from A—X—A.

(b) Electron diffraction cannot distinguish resonance forms. Remember that resonance forms are not "real"—the molecule does not fluctuate between the resonance forms, but is a hybrid.

Think About It
If the resonance forms for A—X—A shown are all equally weighted (none is more preferred than another), we would expect the average X—A bond to be a double bond as in $A{=}X{=}A$. If one resonance form contributes more to the bonding in a molecule, we would see that resonance form reflected in the bond distances being shortened or lengthened.

8.159. Collect and Organize
By drawing the Lewis structures for the ionic compound NH_4SH, we can see why the compound cannot have a covalent bond between nitrogen and sulfur.

Analyze
The ions in this compound are NH_4^+ and SH^-. Both NH_4^+ and SH^- have 8 valence electrons.

Solve

$$\left[\begin{array}{c} H \\ | \\ H{-}N{-}H \\ | \\ H \end{array}\right]^{+} \qquad \left[\ddot{\underset{..}{S}}{-}H\right]^{-}$$

There cannot be a nitrogen–sulfur covalent bond because the nitrogen atom in NH_4^+ has a complete octet through its bonding with hydrogen and it cannot expand its octet because it is a second period element.

Think About It
Only elements of the third or greater periods in the periodic table have available *d* orbitals through which they may expand their octets.

8.161. Collect and Organize
We are asked to draw the Lewis structures for all resonance forms of linear N_4, assign formal charges to predict which structure best describes N_4, and draw a Lewis structure for cyclic N_4.

Analyze
N_4 has 20 valence electrons.

Solve
(a, b)

$$\overset{0}{:}N{\equiv}\overset{+1}{N}{-}\overset{0}{N}{=}\overset{-1}{\ddot{N}}: \quad\longleftrightarrow\quad \overset{-1}{:}\ddot{N}{=}\overset{+1}{N}{=}\overset{+1}{N}{=}\overset{-1}{\ddot{N}}: \quad\longleftrightarrow\quad \overset{-1}{:}\ddot{N}{=}\overset{0}{N}{-}\overset{+1}{N}{\equiv}\overset{0}{N}:$$

All of these resonance forms have atoms with nonzero formal charges. The middle structure has the most nonzero formal charges across three bond lengths, so this one is least preferred. The first and last resonance structures are preferred and are indistinguishable from each other.

(c) The resonance structures for cyclic N_4 have no formal charges on any of the nitrogen atoms.

Think About It
We would predict the structure of N_4 to be cyclic, based on formal charges.

8.163. Collect and Organize
We are asked to draw the Lewis structure for $AlFCl_2$.

Analyze
$AlFCl_2$ has 24 valence electrons. Aluminum has a higher bonding capacity, so it is the best choice as a central atom; there are three bonds to the two chlorine atoms, and one bond to the fluorine atom.

Solve

This structure, however, places formal charges on Cl and Al with the negative formal charge not on Cl (the more electronegative atom) but on Al.
A structure that minimizes formal charges is

Here all formal charges are zero, but this leaves Al without a complete octet.

Think About It
Group 13 elements commonly form "electron-deficient" compounds that have less than 8 electrons around the central atom in the Lewis structure.

8.165. Collect and Organize
By drawing the Lewis structures for each of the molecules with the aim of minimizing the formal charges on the atoms, we can determine which molecules contain an atom with an expanded octet.

Analyze
All of these molecules contain Cl, which may expand its octet since it is a third period element.

Solve

(a)   All formal charges are 0; no expanded octet

(b)   All formal charges are 0; expanded octet on Cl

(c)   All formal charges are 0; expanded octet on Cl

(d)   Formal charges are minimized and negative formal charge is on the more electronegative element; no expanded octet

Both (b) ClF$_3$ and (c) ClI$_3$ have an atom with an expanded octet.

Think About It
Remember to not expand octets on atoms so that you obtain less preferred formal charges on the atoms. For example,

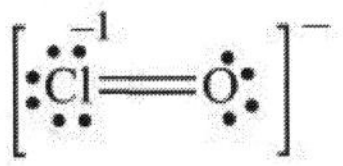

$$\left[:\overset{-1}{\underset{..}{\ddot{Cl}}}=\overset{..}{\underset{..}{\ddot{O}}}: \right]^{-}$$

is not preferred over the structure in (d) above.

8.167. **Collect and Organize**
After drawing the Lewis structure (with resonance forms) for N$_5^-$, we can determine, using formal charges, which resonance forms contribute the most to the bonding in the molecule and compare the bonding of N$_5^-$ to that of N$_3^-$ in terms of average bond order.

Analyze
N$_5^-$ has 26 valence electrons.

Solve
(a, b) All of the possible resonance structures for N$_5^-$ with formal charge assignment are as follows:

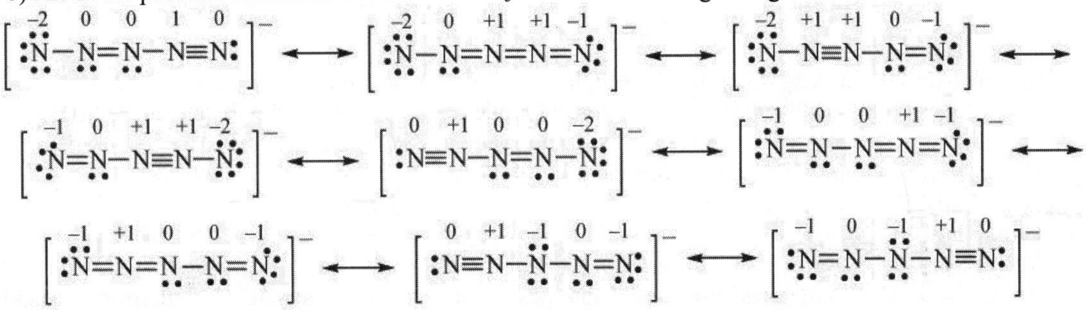

The structures that contribute most have the lowest formal charges (last four structures shown). This means that the terminal nitrogen–nitrogen bonds will be close to the length of a double bond and the middle nitrogen–nitrogen bonds in the structure will be close to a bond order of 1.5 (between a single and double bond).

(c) N$_3^-$ has the Lewis structures

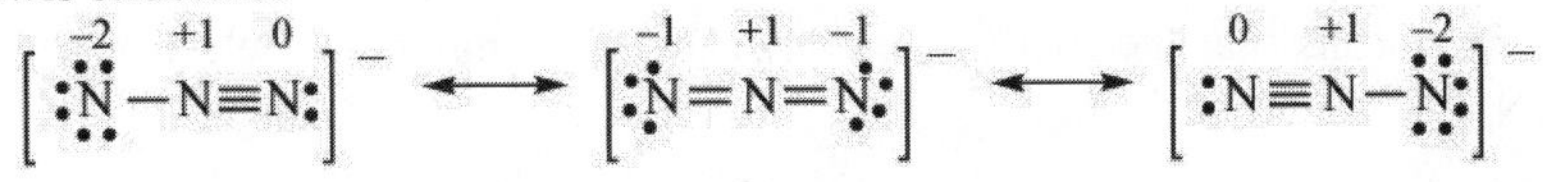

From these resonance structures we see that on average, each bond is predicted to be of double bond character in N$_3^-$. Therefore, in N$_5^-$ there are two longer N—N bonds (single) than in N$_3^-$. Thus, N$_3^-$ has the higher average bond order.

Think About It
The longer (and weaker) N—N bonds in N$_5^-$ are likely to be the ones broken in a chemical reaction.

8.169. **Collect and Organize**
By plotting the electronegativity (*y*-axis) versus the ionization energy (*x*-axis) for the elements in the second period (*Z* = 3–9), we can determine if the trend is linear and then estimate the electronegativity for neon, knowing that its first ionization energy is 2080 kJ.

Analyze
The electronegativities of the second period main group elements are shown in Figure 8.5, and the ionization energies are listed in Appendix 3. Using graphing software we can plot the values for Li, Be, B, C, N, O, and F. We can estimate the electronegativity of Ne using the equation for the line for the trend seen for the other second period elements.

Solve

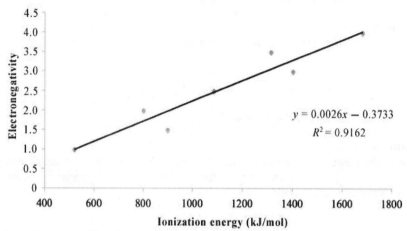

The plot shows that a linear trendline does not accurately fit the data.

Using the equation for the best-fit line where x = the ionization energy of neon (2080 kJ/mol) gives a value of y (electronegativity) of neon:

$$y = 0.0026(2080) - 0.3733 = 5.0$$

Think About It

The calculated electronegativity for neon is higher than that of fluorine, as we would expect from periodic trends. Electronegativity values, however, for the noble gases (especially the lighter ones like He and Ne) are not particularly meaningful since He and Ne form no known compounds. Remember that electronegativity is the power of an atom *in a bond* to attract electrons to itself.

8.171. Collect and Organize

We are asked what it means to be isoelectronic and must consider the Lewis structure of N_2F^+, including its possible resonance structures, the formal charges on its atoms in each resonance structure, and whether fluorine could be the central atom in the molecule.

Analyze

N_2F^+ has 16 valence electrons.

Solve

(a) Isoelectronic means that the two species have the same number of electrons. N_2O also has 16 valence electrons $[(2N \times 5e^-) + (1O \times 6e^-)]$.

(b–d)

$$\left[:N\equiv\overset{+1}{N}-\overset{0}{\ddot{\underset{..}{F}}}: \right]^+ \longleftrightarrow \left[\overset{-1}{:\ddot{N}}=\overset{+1}{N}=\overset{+1}{\ddot{F}}: \right]^+ \longleftrightarrow \left[\overset{-2}{:\ddot{\underset{..}{N}}}-\overset{+1}{N}\equiv\overset{+2}{F}: \right]^+$$

The central nitrogen atom in all the resonance structures always carries a +1 formal charge. The first resonance form shown is preferred because it has the minimal formal charges on the atoms.

(e) Yes, fluorine could be the central atom in the molecule, but this would place significant positive formal charge on the fluorine atom (the most electronegative element). These structures are unlikely:

$$\left[:N\equiv\overset{+3}{F}-\overset{-2}{\ddot{\underset{..}{N}}}: \right]^+ \longleftrightarrow \left[\overset{-1}{:\ddot{N}}=\overset{+3}{F}=\overset{-1}{\ddot{N}}: \right]^+ \longleftrightarrow \left[\overset{-2}{:\ddot{\underset{..}{N}}}-\overset{+3}{F}\equiv\overset{0}{N}: \right]^+$$

Think About It

We could try to reduce the formal charge on fluorine in (e) by drawing

$$\left[:N\equiv\overset{+1}{F}\equiv\overset{0}{N}: \right]^+$$

but this is not a viable structure because fluorine, as a second period element, is not able to expand its octet.

8.173. Collect and Organize
We are to write a balanced equation for the reaction of fluorine with water, identify the oxidizing and reducing agents, and draw the Lewis structures for all the species in the reaction, using arrows to show bond polarity for bonds between dissimilar atoms.

Analyze
The textbook shows the reaction products for the reaction as HF, O_2, H_2O_2, and OF_2. By looking at the oxidation states for the atoms in the reactants and products, we can determine which species (F_2 or H_2O) is the reducing agent (is itself oxidized) and which is the oxidizing agent (is itself reduced). Using the textbook method, we can the draw all the Lewis structures for the reactants and products in this reaction.

Solve
The balanced chemical reaction is
$$5 F_2(g) + 5 H_2O(\ell) \rightarrow 8 HF(g) + O_2(g) + H_2O_2(\ell) + OF_2(g)$$
(a) In this reaction, F_2 is reduced to HF and OF_2 and so is an oxidizing agent. The oxygen in H_2O is oxidized to O_2 and to H_2O_2 (where the oxidation number of O is −1) so H_2O is the reducing agent.
(b)

Think About It
Any bond between two similar atoms (with the exact same electronegativity value) is nonpolar.

8.175. Collect and Organize
We are asked to draw arrows on each bond in the molecule, showing the direction of polarity, and suggest why C_2F_4, which has four polar covalent bonds, is a nonpolar molecule.

Analyze
Individual bond polarities in a molecule are determined by the electronegativity difference in the bonded atoms. Fluorine is more electronegative than carbon, and therefore each C—F bond is polarized so that the partial negative charge is on the fluorine atoms. The bond polarities can be shown by drawing an arrow on each bond where the start of the arrow represents the location in the bond of partial positive charge and the arrowhead represents the location of partial negative charge.

Solve

This molecule is nonpolar overall because the individual bond dipoles are equal in magnitude and, as vectors, they cancel each other out.

Think About It
We will see in Chapter 9 that symmetric molecules will be nonpolar and that the geometry of molecules can become quite important for determining the polarity of a molecule.

CHAPTER 9 | Molecular Geometry: Shape Determines Function

9.1. Collect and Organize

Dipole moments are a result of permanent partial charge separation in a molecule due to differences in atom electronegativities and molecular geometry. We can use the given structures for S_2F_2 to determine if we can distinguish them by their dipole moments.

Analyze

The electronegativity of S is 2.5, and for F the electronegativity is 4.0. The polarity of the S—F bond, therefore, is

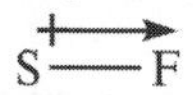

$$S \xrightarrow{\quad\quad} F$$

where the S atom carries a partial positive charge (δ^+) and the F atom carries a partial negative charge (δ^+).

Solve

The bond dipoles represented as vectors for the two isomers of S_2F_2 are

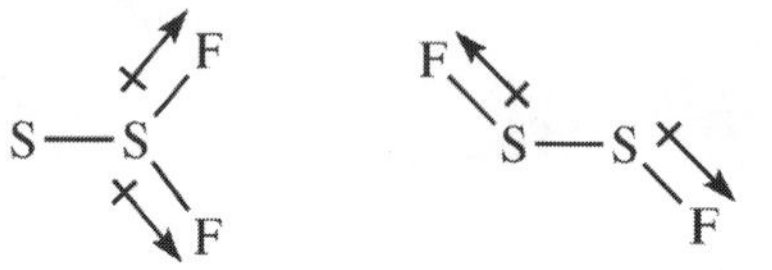

No dipole moment
Vectors cancel
(equal in magnitude and opposite in direction)

Yes, we can differentiate between these structures by their dipole moments. The arrangement of atoms on the left has a dipole moment, whereas the one on the right does not.

Think About It

Not only the difference in electronegativities of bonded elements, but also the molecular geometry is important in establishing molecular polarity.

9.3. Collect and Organize

For the molecules shown, N_2F_2, H_2NNH_2, and NCCN, we are to determine whether the molecules are planar and if any of the molecules have delocalized π electrons through resonance structures.

Analyze

To determine the planarity of the molecules, we need to draw the Lewis structures for each and determine the molecular geometry.

Solve

Electron-pair geometry on N = trigonal planar Molecular geometry = bent	Electron-pair geometry on N = tetrahedral Molecular geometry = trigonal pyramidal	Electron-pair geometry on C = linear Molecular geometry = linear

Planar No delocalized π electrons	Not planar No delocalized π electrons	Planar No delocalized π electrons No resonance forms

N_2F_2 and NCCN are planar molecules. There are no delocalized π electrons in any of these molecules.

Think About It

In order for a molecule with two central atoms to be planar, its molecular geometry must be linear, trigonal planar, or square planar.

9.5. **Collect and Organize**

For each species, O_2^+ and O_2^{2+}, we are to fill the MO diagram for homonuclear diatomic oxygen. From the filled diagram, we can determine if O_2^+ has more or fewer electrons in antibonding molecular orbitals than O_2^{2+}.

Analyze

The number of valence electrons in O_2^+ is $(2\,O \times 6\,e^-) - 1\,e^- = 11\,e^-$ and in O_2^{2+}, the number of electrons is $(2\,O \times 6\,e^-) - 2\,e^- = 10\,e^-$. Antibonding orbitals in the MO diagram are designated with an asterisk, "*".

Solve

For O_2^+ filling of the MO diagram gives
$$(\sigma_{2s})^2(\sigma_{2s}^*)^2(\sigma_{2p})^2(\pi_{2p})^4(\pi_{2p}^*)^1$$
3 electrons in antibonding orbitals

For O_2^{2+} filling of the MO diagram gives
$$(\sigma_{2s})^2(\sigma_{2s}^*)^2(\sigma_{2p})^2(\pi_{2p})^4$$
2 electrons in antibonding orbitals

Therefore O_2^+ has more electrons populating antibonding orbitals than O_2^{2+}.

Think About It

For these species, the bond orders are as follows:
$$O_2^+ \text{ bond order} = \tfrac{1}{2}(8-3) = 2.5$$
$$O_2^{2+} \text{ bond order} = \tfrac{1}{2}(8-2) = 3$$

9.7. Collect and Organize

ReF$_7$ has a geometry in which a pentagon is "capped" on the top and bottom by atoms.

We are asked to calculate the F—Re—F bond angles in this molecule.

Analyze

There are 3 F—Re—F angles present in the molecule: axial-F—Re—axial-F, axial-F—Re—equatorial-F, and equatorial-F—Re—equatorial-F.

Solve

Looking at the diagram, we see that the axial-F—Re—axial-F bond is linear, so the angle is 180°. The axial-F—Re—equatorial-F angle is 90°. Finally, because the sum of the internal angles of the regular pentagon must add up to 360°, the equatorial-F—Re—equatorial-F bonds are all equal: 360°/5 = 72°.

Think About It

The equatorial fluorine atoms are quite crowded together around the Re metal center with small bond angles.

9.9. Collect and Organize

The shape of a molecule depends on the repulsions around the central atom(s) between electrons in lone pairs and bonding pairs. We are asked why it is the repulsions between electron pairs, not the nuclei, that determine molecular geometry.

Analyze

Atoms contain very small, positively charged nuclei surrounded by relatively large electron clouds.

Solve

Because the electrons take up most of the space in the atom and because the nucleus is located in the center of the electron cloud, the electron clouds repel each other before the nuclei get close enough to interact.

Think About It

All of chemistry, in essence, is due to the behavior of the electrons (are they lost, gained, or shared?); that behavior, of course, is influenced by the attraction of the electrons to the nucleus.

9.11. Collect and Organize

To determine why SO$_3$ and BF$_3$ have the same trigonal planar structure despite having different numbers of bonds, we must draw their Lewis structures and then deduce their molecular structures by VSEPR (valence-shell electron-pair repulsion) theory.

Analyze

After drawing the Lewis structure for each molecule, we can determine the steric number (SN) for the central atoms, then locate the other atoms about the central atom to see the bond angles, and finally determine the molecular shape based on the location of the atoms.

Solve

SN = 3
Electron-pair geometry = trigonal planar
There are 3 atoms bound to the central atom with no lone pairs.
Molecular geometry = trigonal planar

SN = 3
Electron-pair geometry = trigonal planar
There are 3 atoms bound to the central atom with no lone pairs.
Molecular geometry = trigonal planar

Think About It

Notice that multiple bonds count as one steric number. Also, note that BF_3 is an electron-deficient compound. Even if we were to consider the possibility of the resonance form where an F atom is double bonded to the boron atom, the steric number would still be 3.

9.13. Collect and Organize

Using ammonia as an example, we consider why lone pair–bonding pair interactions are more repulsive than bonding-pair–bonding-pair interactions.

Analyze

The electrons in a bonding pair are located between two nuclei and therefore "feel" the attractive force from both nuclei. Lone pairs of electrons are only attracted by one nucleus.

Solve

Because the lone pair feels attraction from only one nucleus, it is less confined than bonding pairs and therefore occupies more space around the central N atom in ammonia. It will therefore "bump into" and repel a neighboring bonding pair that takes up less space.

Think About It

The presence of the lone pair in ammonia means that the H—N—H bond angles are less than the ideal 109.5° for a steric number of 4.

9.15. Collect and Organize

In considering the two structures for AB_4, seesaw and trigonal pyramidal, we are to use VSEPR theory to explain why the seesaw geometry has the lower energy.

Analyze

In VSEPR theory, the repulsive interactions between bonding pairs (bp) and lone pairs (lp) at 90° decrease in order as follows:

$$lp—lp \gg lp—bp \gg bp—bp$$

The lowest energy geometry has the lowest number of lp—lp interactions. If there are no lp—lp interactions, or if they are equal for two geometries, then the geometry with the lowest number of lp—bp interactions has the lowest energy.

Solve

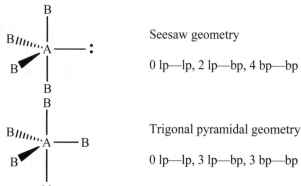

Seesaw geometry

0 lp—lp, 2 lp—bp, 4 bp—bp

Trigonal pyramidal geometry

0 lp—lp, 3 lp—bp, 3 bp—bp

The seesaw geometry has only two lp—bp interactions at 90° (compared to trigonal pyramidal's three), so it has the lower energy.

Think About It

Remember that in comparing possible geometries by VSEPR theory, we need only look at interactions at 90° or less.

9.17. **Collect and Organize**

We are to rank the trigonal planar, octahedral, and tetrahedral molecular geometries in order of increasing bond angle for the smallest bond angle in each geometry.

Analyze

In trigonal planar molecules, all bond angles are 120°. In tetrahedral molecules, all the bond angles are 109.5°. In octahedral molecules, the bond angles are 90° (and 180°).

Solve

In order of increasing bond angle, (b) octahedral < (c) tetrahedral < (a) trigonal planar.

Think About It

In general, as steric number increases, the smallest bond angle for the geometry decreases.

9.19. **Collect and Organize**

Of all the molecular geometries, we are to list those that have more than one B—A—B bond angle.

Analyze

The molecular geometries based on steric number (SN) and electron-pair geometry are as follows:

SN	Electron-Pair Geometry	Molecular Geometry	Ideal Bond Angles
2	Linear	Linear	180°
3	Trigonal planar	Trigonal planar	120°
		Bent	120°
4	Tetrahedral	Tetrahedral	109.5°
		Trigonal pyramidal	109.5°
		Bent	109.5°
5	Trigonal bipyramidal	Trigonal bipyramidal	90°, 120°, 180°
		Seesaw	90°, 120°, 180°
		T-Shaped	90°, 180°
		Linear	180°
6	Octahedral	Octahedral	90°, 180°
		Square pyramidal	90°, 180°
		Square planar	90°, 180°

Solve

Molecular geometries with more than one characteristic bond angle are trigonal bipyramidal, seesaw, T-shaped, octahedral, square pyramidal, and square planar.

Think About It

For SN = 2, 3, and 4 only one bond angle is characteristic. High steric number molecules have more than one characteristic bond angle with the exception of AB_2E_3 (trigonal bipyramidal geometry with 2 atoms bonded to the central atom and three lone pairs), which has a linear molecular geometry.

9.21. **Collect and Organize**

We are to determine if we can obtain linear triatomic molecules when we remove atoms from molecules with tetrahedral, octahedral, and T-shaped molecular geometries.

Analyze

For each geometry we will take atoms away to obtain the VSEPR geometry until we obtain a triatomic molecule.

Solve

(a)

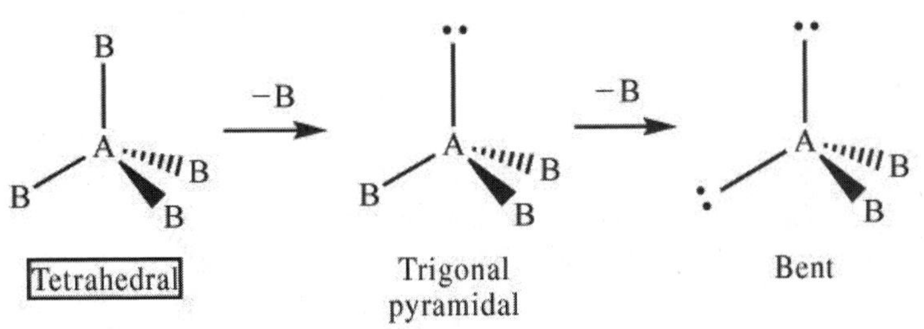

(b)

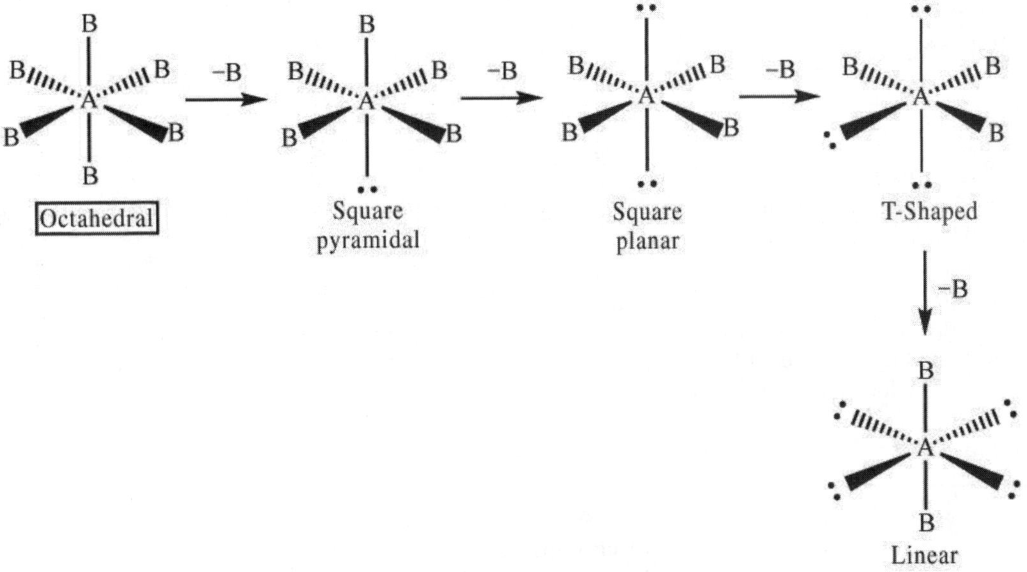

(c)

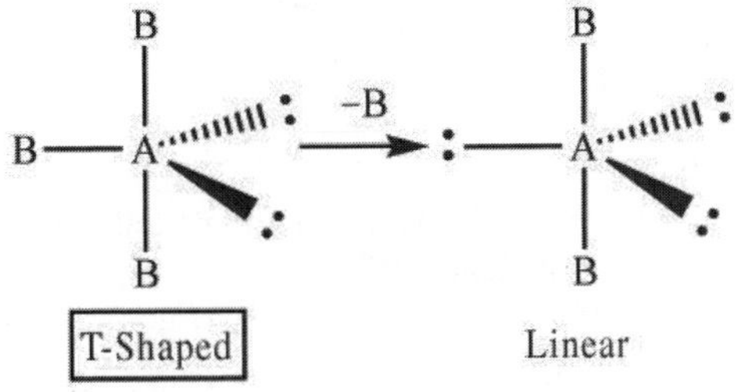

Tetrahedral (a) does not lead to a linear molecule if the molecule is triatomic. The T-shaped and octahedral geometries (b, c) lead to a linear geometry for their triatomic species.

Think About It
Linear molecules are, of course, also achieved when the central atom has a steric number equal to two.

9.23. Collect and Organize
We consider what molecular geometries might result if we replace one atom from AB_7 with a lone pair of electrons.

Analyze
The pentagonal bipyramidal AB_7 molecule has two different types of B atoms, namely, those in the axial positions and those lying in the equatorial plane.

Solve

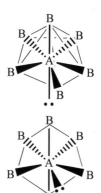

Removing an atom from one of the axial positions gives a pentagonal pyramidal geometry.

Removing an atom from one of the equatorial positions gives a distorted octahedral geometry.

Think About It
By VSEPR theory, the lone pair would have the lowest energy when placed in an equatorial position (4 lp—bp) rather than at an axial position (5 lp—bp).

9.25. Collect and Organize
Using Lewis structures and VSEPR theory, we can determine the molecular geometries of GeH_4, PH_3, H_2S, and $CHCl_3$.

Analyze
After drawing the Lewis structure for each molecule, we can determine the steric number for the central atom, then locate the atoms about the central atom to see the bond angles, and finally determine the molecular shape based on the location of the atoms.

Solve
(a)

H—Ge—H (with H above and below)

SN = 4
Electron-pair geometry = tetrahedral
No lone pairs
Molecular geometry = tetrahedral

(b)

H—P—H (with H above and lone pair below)

SN = 4
Electron-pair geometry = tetrahedral
One lone pair
Molecular geometry = trigonal pyramidal

(c)

SN = 4
Electron-pair geometry = tetrahedral
Two lone pairs
Molecular geometry = bent

(d)

SN = 4
Electron-pair geometry = tetrahedral
No lone pairs
Molecular geometry = tetrahedral

Think About It
Depending on the number of lone pairs on the central atom, molecules with SN = 4 may have molecular geometries of either tetrahedral, trigonal pyramidal, bent, or linear. HF is an example of a linear molecule with an SN of 4:

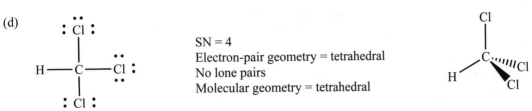

9.27. Collect and Organize
Using Lewis structures and VSEPR theory, we can determine the bond angles from the molecular geometries of NH_4^+, SO_3^{2-}, NO_2^-, and XeF_5^+.

Analyze
After drawing the Lewis structure for each molecule, we can determine the steric number for the central atom, then locate the atoms about the central atom to see the bond angles, and finally determine the molecular shape based on the location of the atoms.

Solve
(a)

SN = 4
Electron-pair geometry = tetrahedral
No lone pairs
Molecular geometry = tetrahedral
Bond angles = 109.5°

(b)

SN = 3
Electron-pair geometry = tetrahedral
One lone pair
Molecular geometry = trigonal pyramidal
Bond angles = <109.5°

(c)

SN = 3
Electron-pair geometry = trigonal planar
One lone pair
Molecular geometry = bent
Bond angles = 120°

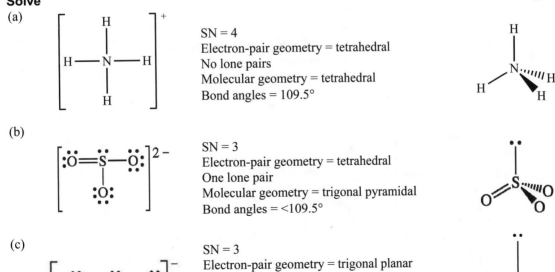

(d)

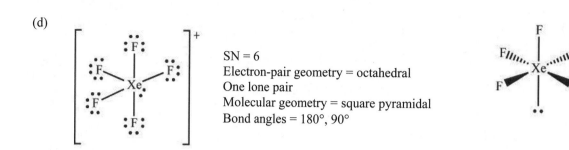

SN = 6
Electron-pair geometry = octahedral
One lone pair
Molecular geometry = square pyramidal
Bond angles = 180°, 90°

Think About It
Notice how the other resonance structure of NO_2^- also has a bent geometry and a bond angle of 120° because the steric number remains 3 at the central N atom.

9.29. Collect and Organize
Using Lewis structures and VSEPR theory, we can determine the molecular geometries of $S_2O_3^{2-}$, PO_4^{3-}, NO_3, and NCO.

Analyze
After drawing the Lewis structure for each molecule, we can determine the steric number for the central atom, then locate the atoms about the central atom to see the bond angles, and finally determine the molecular shape based on the location of the atoms.

Solve
(a)

SN = 4
Electron-pair geometry = tetrahedral
No lone pairs
Molecular geometry = tetrahedral

(b)

SN = 4
Electron-pair geometry = tetrahedral
No lone pairs
Molecular geometry = tetrahedral

(c)

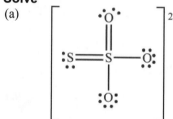

SN = 3
Electron-pair geometry = trigonal planar
No lone pairs
Molecular geometry = trigonal planar

(d)

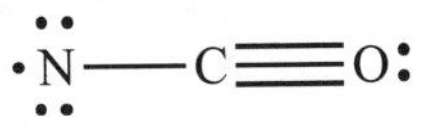

:Ṅ——C≡≡O:

SN = 2
Electron-pair geometry = linear
No lone pairs
Molecular geometry = linear

N——C≡≡O

Think About It
Remember that the presence of resonance structures for a molecule does not change its geometry.

9.31. Collect and Organize
By drawing the Lewis structures and determining the molecular geometry of O_3, SO_2, N_2O, S_2O, and CO_2 we can determine if any of these molecules have the same molecular geometry.

Analyze
After drawing the Lewis structure for each molecule, we can determine the steric number for the central atom, then locate the atoms about the central atom to see the bond angles, and finally determine the molecular shape based on the location of the atoms.

Solve

:Ö==Ö——Ö:

SN = 3
Electron-pair geometry = trigonal planar
One lone pair
Molecular geometry = bent

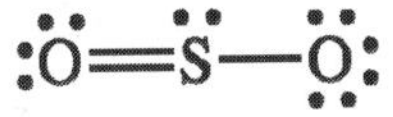

:Ö==S̈——Ö:

SN = 3
Electron-pair geometry = trigonal planar
One lone pair
Molecular geometry = bent

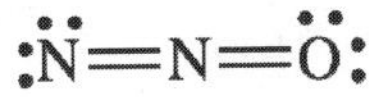

:Ṅ==N==Ö:

SN = 2
Electron-pair geometry = linear
No lone pairs
Molecular geometry = linear

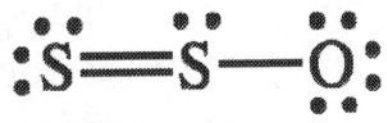

:S̈==S̈——Ö:

SN = 3
Electron-pair geometry = trigonal planar
One lone pair
Molecular geometry = bent

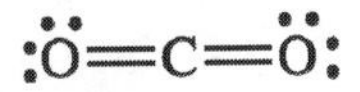

:Ö==C==Ö:

SN = 2
Electron-pair geometry = linear
No lone pairs
Molecular geometry = linear

O_3, S_2O, and SO_2 are all bent, with ~120° angles; CO_2 and N_2O are both linear, with 180° angles.

Think About It
At first these triatomic molecules may all appear to have the same geometry from their formulas. Be careful to draw correct Lewis structures because the presence of lone pairs on the central atom is important in determining the overall geometry.

9.33. Collect and Organize
We are to determine which resonance form of the $C(CN)_3^-$ anion is favored by comparing the molecular geometry of each Lewis structure to the observed trigonal planar geometry.

Analyze

After drawing the Lewis structure for each resonance form, we can determine the steric number for the central atom, then locate the atoms about the central atom to see the bond angles, and finally determine the molecular shape based on the location of the atoms. The resonance structure with a trigonal planar geometry will be preferred.

Solve

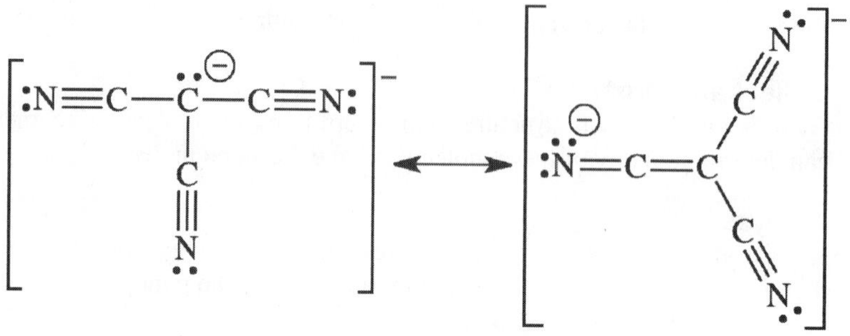

SN	4	3
Electron-pair geometry	Tetrahedral	Trigonal planar
Number of lone pairs	1	0
Molecular geometry	Trigonal pyramidal	Trigonal planar
Bond angle	<109.5°	120°

The structure at right, with a C=C double bond is preferred.

Think About It

We could also predict that the structure at right is preferred using formal charges. In the structure at right, the formal charge is placed on the more electronegative nitrogen atom.

9.35. **Collect and Organize**

We are asked to explain why trimethylamine, $N(CH_3)_3$ and trisilylamine, $N(SiH_3)_3$ have different bond angles. We can do this by drawing Lewis structures for both, including resonance forms.

Analyze

After drawing the Lewis structure for each molecule, we can determine the steric number for the central atom, then locate the atoms about the central atom to see the bond angles, and finally determine the molecular shape based on the location of the atoms. A resonance structure for $N(SiH_3)_3$ with a trigonal planar geometry (bond angle = 120°) will be preferred.

Solve

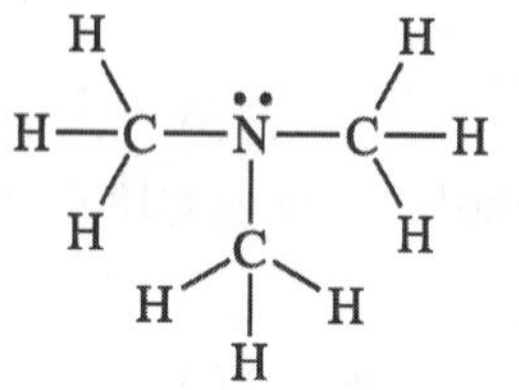

SN = 4
Electron-pair geometry = tetrahedral
One lone pair
Molecular geometry = trigonal planar
Bond angles = <109.5°

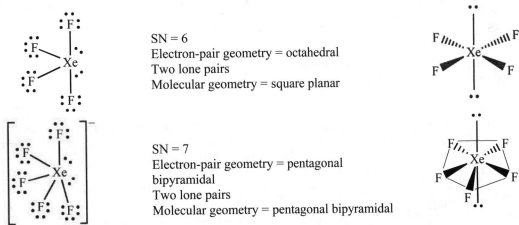

SN	4	3
Electron-pair geometry	Tetrahedral	Trigonal planar
Number of lone pairs	1	0
Molecular geometry	Trigonal pyramidal	Trigonal planar
Bond angle	<109.5°	120°

Because silicon may have an expanded octet, $N(SiH_3)_3$ may adopt the trigonal planar geometry at right.

Think About It
We could assess the extent to which the nonpolar resonance structure at right is favored by measuring the dipole moment of N–Si bond order for $N(SiH_3)_3$.

9.37. Collect and Organize
After drawing the Lewis structures for XeF_4 and XeF_5^- we can use VSEPR theory to predict the molecular structure for XeF_4. We are also asked to sketch the structure of XeF_5^- from the information on the crystal structure that identifies it as pentagonal bipyramidal for all valence electrons.

Analyze
Xenon brings 8 electrons, a closed-shell configuration, so it must expand its octet (as we saw in Chapter 8) in order to bond with F atoms in XeF_4 and XeF_5^-.

Solve

SN = 6
Electron-pair geometry = octahedral
Two lone pairs
Molecular geometry = square planar

SN = 7
Electron-pair geometry = pentagonal bipyramidal
Two lone pairs
Molecular geometry = pentagonal bipyramidal

Think About It
Placing the lone pairs in the axial positions of the pentagonal bipyramid gives the lowest energy geometry because there are no lp—lp interactions for this structure.

9.39. Collect and Organize
We are asked to draw Lewis structures for ClO_2^+, ClO_2, and ClO_2^- that are consistent with the bond order and bond angle data provided.

Analyze
ClO_2^+ has the shortest Cl–O bond length and the largest bond angle. ClO_2^- has the longest Cl–O bond length, and the shortest bond angle. ClO_2 has an intermediate value for both bond angle and bond length. The shorter

bond length for ClO_2^+ corresponds to a greater Cl–O bond order, and the large bond angle means that ClO_2^+ must have the smallest steric number. Similarly, ClO_2^- will have the smallest bond order and greatest steric number, and ClO_2 will lie in between the other two.

Solve

ClO_2^+

SN = 3
Electron-pair geometry = trigonal planar
One lone pair
Molecular geometry = bent (120°)
Bond order = 2

ClO_2^-

SN = 4
Electron-pair geometry = tetrahedral
Two lone pairs
Molecular geometry = bent (109.5°)
Bond order = 1.5

ClO_2

SN = 4
Electron-pair geometry = tetrahedral
Two lone pairs
Molecular geometry = bent (109.5°)
Bond order = 2/1.5

Think About It
While other Lewis structures are possible for these molecules, those described above fit the data given. It is important to remember that Lewis structures are an approximation, not the only way to describe bonding.

9.41. Collect and Organize
Both molecules and bonds may be polar. The definitions of the two terms allow us to differentiate between a polar bond and a polar molecule.

Analyze
A bond is polar when two bonded atoms have different electronegativities. Molecular polarity is the result of bond polarity and molecular geometry.

Solve
A polar bond occurs only between two atoms in a molecule. The more electronegative atom in the bond carries a partial negative charge and the least electronegative atom in the bond carries a partial positive charge. Molecular polarity takes into account all the individual bond polarities and the geometry of the molecule. A polar molecule has a permanent, measurable dipole moment.

Think About It
In order to determine molecular polarity we have to first determine the individual bond polarities.

9.43. Collect and Organize
We are asked whether a nonpolar molecule may contain polar covalent bonds.

Analyze
Molecular polarity is determined by adding the vectors of the individual bond polarities.

Solve
Yes. As long as the individual bond polarities are equal in magnitude and opposite in direction (as vectors), a molecule may be nonpolar overall even if the bonds themselves are polar.

Think About It
If the bond polarities (as vectors) do not cancel, then the molecule will be polar.

9.45. **Collect and Organize**
We can look at the bond polarities and the molecular structure by VSEPR theory to determine which molecules (CCl_4, $CHCl_3$, CO_2, H_2S, and SO_2) contain polar bonds, and which are polar, and which are nonpolar.

Analyze
All of the individual bonds in all of these molecules are polar, so the molecular geometry of each compound will decide the overall molecular polarity. We can represent each bond polarity with a vector with the head of the arrow pointed toward the more electronegative atom, which carries a partial negative charge. We then visually inspect the molecule to see whether the individual bond dipoles add up or cancel out.

Solve

(a)

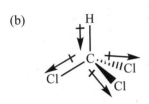

All bond polarities are equal in magnitude and cancel each other, so CCl_4 is nonpolar.

(b)

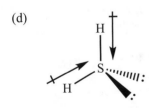

The electronegativity of the atoms in $CHCl_3$ are in order $Cl > C > H$. Because the bond polarities do not cancel, $CHCl_3$ is polar.

(c)

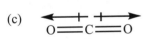

The bond polarities in CO_2 cancel, so it is nonpolar.

(d)

The molecular geometry of H_2S is bent, so it is polar.

(e)

The molecular geometry of SO_2 is bent, so it is polar.

All of the individual bonds in all of these molecules are polar. The polar molecules are (b) $CHCl_3$, (d) H_2S, and (e) SO_2. The nonpolar molecules are (a) CCl_4 and (c) CO_2.

Think About It
Molecules with polar bonds are nonpolar only for highly symmetrical geometries (linear, trigonal planar, tetrahedral, trigonal bipyramidal, and octahedral).

9.47. **Collect and Organize**
By looking at the molecular structures and individual bond polarities present in $CFCl_3$, CF_2Cl_2, and Cl_2FCCF_2Cl we can determine which of these Freons are polar and which are nonpolar.

Analyze

To determine bond polarities we need the electronegativity values for C (2.5), F (4.0), and Cl (3.0). In each of these molecules the halogens are bonded to the carbon atoms. Because the electronegativities of the halogens are higher than that of carbon, the bonds are polarized so that the halogen carries a partial negative charge.

Solve

(a)   (b)   (c)

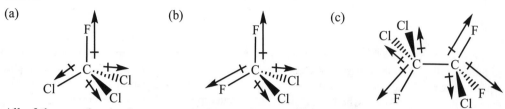

All of these molecules have a tetrahedral geometry around the carbon atoms. Because of the different bond polarities of C—F and C—Cl, however, none of the molecules have bond dipoles that cancel. All of these molecules (a–c) are polar.

Think About It

Only molecules with completely symmetric geometries with all the same atoms attached to the central atoms are nonpolar.

9.49. Collect and Organize

To determine which molecule is more polar in each pair given, we need to compare the molecules' geometries.

Analyze

In each case, we will make an assumption about the relative size of the bond dipoles in order to determine if these bond dipoles cancel. Regardless of the magnitude of the bond dipoles, if individual bond dipoles cancel, the molecule will be nonpolar.

Solve

(a) The bond dipoles in *ii* cancel, meaning *i* is more polar.

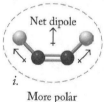

i.

More polar

Bond dipoles cancel, no net dipole

ii.

(b) The bond dipoles in *ii* cancel, meaning *i* is more polar.

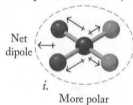

i.

More polar

Bond dipoles cancel, no net dipole

ii.

(c) The bond dipoles in *i* cancel, meaning *ii* is more polar.

Bond dipoles cancel,
no net dipole

i.

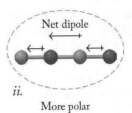

Net dipole

ii.

More polar

Think About It

If the bond dipoles pointed in the opposite directions, the same molecules would be more polar, but the direction of the net dipole would be opposite.

9.51. **Collect and Organize**

We are to determine the geometry and polarity of the bridged structure Al_2Cl_6.

Analyze

When determining the geometry around aluminum, we can determine the steric number for the aluminum atom, then locate the atoms about the central atom to see the bond angles, and finally determine the molecular shape based on the location of the atoms. To determine bond polarities we need the electronegativity values for Al (1.5) and Cl (3.0).

Solve

(a) Consider the geometry about aluminum

SN = 4
Electron-pair geometry = tetrahedral
No lone pairs
Molecular geometry = tetrahedral

(b) The bond dipoles all cancel, meaning Al_2Cl_6 is nonpolar.

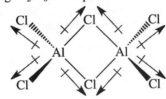

Think About It

It can be difficult to determine the polarity of molecules with more than one central atom. Choosing specific bond dipoles that cancel each other out may help with this.

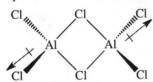 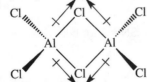 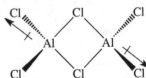

9.53. **Collect and Organize**

We consider the difference between sigma and pi bonds.

Analyze

Both sigma and pi bonds represent a two-electron interaction, frequently between two atomic centers. We should identify any differences between the shape of sigma and pi overlaps, and the type of orbitals used in each case.

Solve

A sigma bond is formed from the overlap of two atomic or hybridized atomic orbitals along a bonding axis. A pi bond is formed from unhybridized *p* or *d* orbitals above and below a bonding axis.

Think About It

We may test the statement about the difference between sigma and pi bonds by imagining the rotation of a molecule along its bond axis. We should observe no change in the distribution of electron density for a sigma bond, while a pi bond will experience a change in the distribution of electron density.

9.55. Collect and Organize

After drawing resonance structures for N_2O we can compare the hybridization of the central N atom among the structures.

Analyze

N_2O has 16 e^- and needs 24 e^- to complete the octets on all the atoms. The difference of 8 e^- corresponds to four covalent bonds. This leaves 8 e^- in four lone pairs to complete the structure.

Solve

$$ \underset{-1}{:\!N}\!=\!\!\underset{+1}{N}\!=\!\!\underset{0}{\ddot{O}:} \quad\longleftrightarrow\quad \underset{0}{:\!N}\!\equiv\!\!\underset{+1}{N}\!-\!\!\underset{-1}{\ddot{\ddot{O}}:} \quad\longleftrightarrow\quad \underset{-2}{:\!\ddot{N}}\!-\!\!\underset{+1}{N}\!\equiv\!\!\underset{+1}{O:} $$

For each resonance structure the central N atom has a steric number of 2. Yes, in all of these resonance structures, the central N atom is *sp* hybridized.

Think About It

If we had a change in hybridization then we would also have a change in structure (how the atoms are arranged in space).

9.57. Collect and Organize

From the steric number obtained from the Lewis structures of NO_2^+, NO_2^-, N_2O, N_2O_5, and N_2O_3, we can determine the hybridization of each of the nitrogen atoms.

Analyze

Hybridization is directly obtained from the steric number (SN). When SN = 2, the hybridization is *sp*; when SN = 3, the hybridization is sp^2; when SN = 4, the hybridization is sp^3; when SN = 5, the hybridization is sp^3d; when SN = 6, the hybridization is sp^3d^2.

Solve

(a) $\left[:\!\ddot{O}\!=\!N\!=\!\ddot{O}:\right]^{+}$ SN = 2   *sp* hybridized

(b) $\left[:\!\ddot{O}\!=\!\ddot{N}\!-\!\ddot{O}:\right]^{-}$ SN = 3 sp^2 hybridized

(c) $:\!N\!\equiv\!N\!-\!\!\ddot{O}:$ SN = 2 for both N atoms   *sp* hybridized

(d) [structure of N_2O_5]   SN = 3 for both N atoms sp^2 hybridized

(e) [structure of N_2O_3]   SN = 3 for both N atoms sp^2 hybridized

Think About It

The resonance forms we can draw for these molecules give the same hybridization for the central N atom.

9.59. Collect and Organize

Using Lewis structures and valence bond theory we can compare the electron distribution and hybridization of the central N atom in N_3^- and N_3F.

Analyze

N_3^- has 16 e^- and needs 24 e^-, giving a difference of 8 e^- in four covalent bonds and leaving 8 e^- in four lone pairs to complete the structure. N_3F has 22 e^- and needs 32 e^-, giving a difference of 10 e^- in five covalent bonds and leaving 12 e^- in six lone pairs to complete the structure. The hybridization of the central N atom depends on the steric number. When SN = 2, the hybridization is *sp*; when SN = 3, the hybridization is sp^2; when SN = 4, the hybridization is sp^3.

Solve

The electron distribution in N_3^- and N_3F is very similar. Both have three resonance forms: one with two N==N bonds and two with an N≡N and an N—N bond. In both molecules SN = 2 for the central N atom, which means that the hybridization at both central N atoms is *sp*.

Think About It

The *sp* hybridization on the central N atom leaves two *p* orbitals on N unhybridized so that they may form two π bonds.

9.61. Collect and Organize

Using Lewis structures and the steric number around the central S atom, we can determine how orbital hybridization of S changes in SF_2, SF_4, and SF_6.

Analyze

We draw the Lewis structures in the usual way. Hybridization is directly obtained from the steric number (SN). When SN = 2, the hybridization is *sp*; when SN = 3, the hybridization is sp^2; when SN = 4, the hybridization is sp^3; when SN = 5, the hybridization is sp^3d; when SN = 6, the hybridization is sp^3d^2.

Solve

Think About It

When S has to expand its octet in order to bond to four and six fluorine atoms in SF_4 and SF_6, it uses *d* orbitals in its hybridization.

9.63. Collect and Organize

We are to complete the Lewis structure for Minoxidil ($C_{10}H_{16}N_4O$) by adding lone pairs and formal charges, and to describe the bonding environment around the nitrogen atom in the N–O group.

Analyze

Minoxidil contains 82 valence electrons. As drawn in the problem, there are 70 electrons in 32 sigma bonds and three pi bonds. This leaves 12 electrons to be distributed as lone pairs.

Solve

The oxygen atom has an incomplete octet, as do the nitrogen atoms in the lower ring and in the amine groups. Adding the six remaining lone pairs

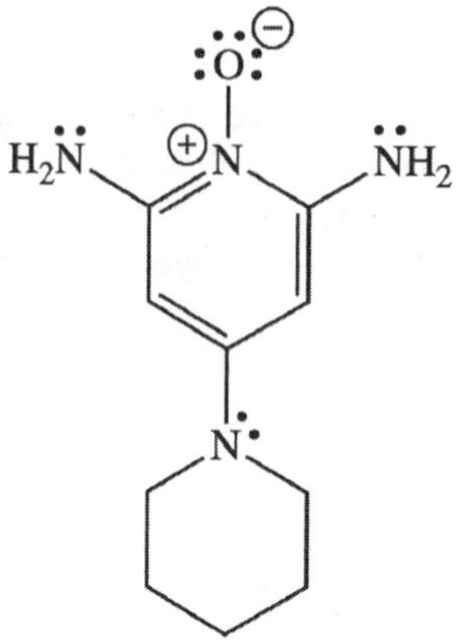

The nitrogen in the N–O group has formed a sigma bond with oxygen as an overlap of N_{sp^2}/O_{sp^3} , two sigma bonds with adjacent carbon atoms as an overlap of N_{sp^2}/C_{sp^2} , and one pi bond between carbon and nitrogen from the overlap N_{2p}/C_{2p} .

Think About It

Despite the presence of adjacent formal charges, we cannot form an N=O double bond, since nitrogen cannot exceed the octet.

9.65. **Collect and Organize**

After drawing the Lewis structures for ClO_4^- for which the formal charges are minimized, we can determine the molecular shape and the hybridization on the central Cl atom.

Analyze

We draw the Lewis structures in the usual way. The structure contributing the most to the actual geometry of the molecule is that with the lowest formal charges and with the most electronegative atom (in this case, oxygen) carrying a negative formal charge if necessary.

Solve

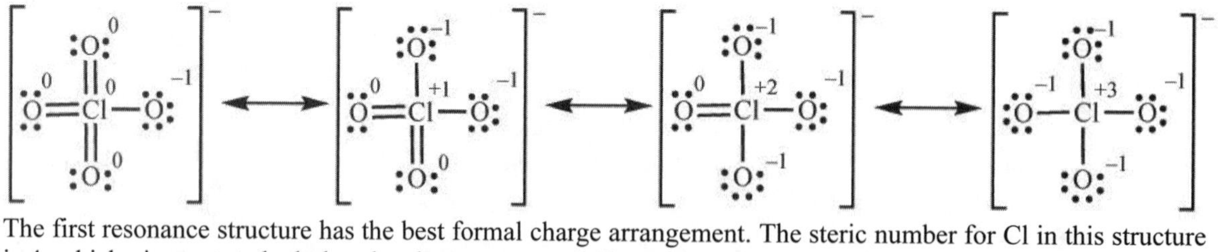

The first resonance structure has the best formal charge arrangement. The steric number for Cl in this structure is 4, which gives a tetrahedral molecular geometry. At first glance this would also mean that the hybridization would be assigned as sp^3. However, notice that the Cl forms three π bonds to three of the oxygen atoms. This requires that three of the p orbitals on Cl not be involved in the hybridization so that it can form parallel π bonds. Therefore, Cl must use low-lying d orbitals in place of the p orbitals for sd^3 hybridization to form the four σ bonds to oxygen.

Think About It

On central atoms with expanded octets we need to use d orbitals in place of p orbitals for the σ-bonded hybrid orbitals so as to leave the unhybridized p orbitals available for π bonding.

9.67. Collect and Organize

From the Lewis structure of HArF we can use the steric number to determine the molecular geometry and the hybridization of the central Ar atom.

Analyze

We draw the Lewis structure in the usual way. HArF has 16 e^- and needs 4 e^- to form the two covalent bonds from Ar to F and H. This leaves 12 e^- in six lone pairs on Ar and F to complete the structure. Since this is a compound of a noble gas, we expect Ar to have to an expanded octet. Hybridization is directly obtained from the steric number (SN). When SN = 2, the hybridization is sp; when SN = 3, the hybridization is sp^2; when SN = 4, the hybridization is sp^3; when SN = 5, the hybridization is sp^3d; when SN = 6, the hybridization is sp^3d^2.

Solve

$$H - \overset{..}{\underset{..}{Ar}} - \overset{..}{\underset{..}{F}}:$$

SN = 5
sp^3d hybridized
Electron-pair geometry = trigonal bipyramidal
Molecular geometry = linear
H—Ar—F bond angle = 180°

Think About It

This question has placed many of the components of structure and bonding theories together—Lewis structure, VSEPR theory, and valence bond theory—to fully describe the molecular structure of HArF.

9.69. Collect and Organize

We are asked to consider the Lewis structure of $[(CH_3)_4N]^+[SO_2F_3]^-$ (tetramethyammonium trifluorosulfate), and determine the geometry of both the cation and the anion, as well as to describe the bonding in the tetramethylammonium cation by valence bond theory.

Analyze

$[(CH_3)_4N]^+$ has 32 valence electrons, while $[SO_2F_3]^-$ has 40 valence electrons. The S—O single bond in sulfuric acid has a length of 157 pm, meaning that the S—O bonds in the trifluorosulfate anion are shorter, and so, have a greater bond order.

Solve

(a)

Tetrahedral C and N

SN = 4
sp^3 hybridized
Electron-pair geometry = tetrahedral
Molecular geometry = tetrahedral

All C—N bonds are an overlap of $C - sp^3$ and $N - sp^3$ hybrid orbitals.

(b) The S—O bonds are likely to be double bonds, accounting for the shorter bond length. This would also lead to a Lewis structure with only one formal charge (–1) on the sulfur atom.

(c) For the anion depicted in (b), the steric number of the sulfur atom is 5, leading to an electron-pair geometry of trigonal bipyramidal. None of these are lone pairs, so the molecular geometry is also trigonal bipyramidal.

Think About It

If we had not drawn the Lewis structure to account for all the electrons in this molecule, it would have been easy to miss the electron pair on the central S atom.

9.71. **Collect and Organize**
We consider whether it is possible for a molecule with more than one central atom to have resonance forms.

Analyze
Resonance forms are Lewis structures that show alternative (yet still valid) electron distributions in a molecule.

Solve
We have already seen several examples of molecules that have several central atoms and that have multiple resonance forms in Chapter 8. In this chapter, a good example of resonance in a molecule with more than one central atom is benzene. Yes, molecules with more than one central atom can indeed have resonance forms.

Think About It
As long as there is another way to distribute the electrons in a molecule, we have resonance.

9.73. **Collect and Organize**
We are asked to explain whether resonance structures are examples of the delocalization of electrons in a molecule.

Analyze
Resonance structures show different possible electron distributions over the atoms in the molecule. Each one contributes to the actual structure of the molecule.

Solve
To obtain the actual molecular structure, the resonance forms are mixed together. The molecule does not exist in one form at one instant and another form the next. The electron distribution is blurred across all the resonance forms, which is, essentially, the delocalization of electrons.

Think About It
Resonance forms help us see which atoms and bonds are involved in sharing the delocalized electrons.

9.75. **Collect and Organize**
For four ordinary objects we are to determine whether or not they are chiral.

Analyze
An object is chiral if it is not superimposable on its mirror image.

Solve
(a) A baseball (including the seams) is not chiral.
(b) A pair of scissors is chiral.
(c) A boot is chiral.
(d) A fork (plain) is not chiral.

Think About It
One of the simple ways to determine chirality of a molecule is to see if the molecule has an internal mirror plane; if it is symmetrical in this way, then the molecule will not be chiral.

9.77. **Collect and Organize**

For the cyclic hydrocarbons C_6H_6, C_4H_8, and C_4H_6 we can draw the Lewis structures to determine the steric number around each carbon atom so that the molecular geometry and hybridization can be identified.

Analyze

C_6H_6 has 30 e⁻ and needs 60 e⁻, giving a difference of 30 e⁻ in 15 covalent bonds and leaving no lone pairs of electrons on the structure. C_4H_8 has 24 e⁻ and needs 48 e⁻, giving a difference of 24 e⁻ in 12 covalent bonds and leaving no lone pairs of electrons on the structure. C_4H_6 has 22 e⁻ and needs 44 e⁻, giving a difference of 22 e⁻ in 11 covalent bonds and leaving no lone pairs of electrons on the structure.

Solve

Each carbon in C_6H_6 has SN = 3 and sp^2 hybridization with trigonal planar geometry.

Each carbon in C_4H_8 has SN = 4 and sp^3 hybridization with tetrahedral geometry.

The double-bonded carbon atoms in C_4H_6 have SN = 3 and sp^2 hybridization with trigonal planar geometry. The single-bonded carbon atoms in C_4H_6 have SN = 4 and sp^3 hybridization with tetrahedral geometry.

Think About It

For carbon compounds, we can make the general statement that singly bonded C atoms are usually sp^3 hybridized, doubly bonded C atoms are usually sp^2 hybridized, and triply bonded C atoms are usually sp hybridized.

9.79. **Collect and Organize**

We are asked to complete the Lewis depiction of acesulfame potassium by adding lone pairs and determining the geometry at each of the ring-member atoms. We are asked to identify which hybrid orbitals on carbon, oxygen and nitrogen overlap to form the C—O and C—N bonds and in which orbital the extra electron on N is located.

Analyze

Steric numbers depend on the number of bonded atoms and lone pairs of electrons around a particular atom in a Lewis structure. This steric number determines the electron geometry. Based on the presence of lone pairs, we may determine the molecular geometry. After completing the Lewis structure, we may describe the bonding according to the hybrid orbitals involved. Acesulfame potassium contains 56 valence electrons. As drawn in the problem, there are 36 electrons in 14 sigma bonds and four pi bonds. This leaves 20 electrons to be distributed as lone pairs.

Solve

Including lone pairs, the complete Lewis structure is

The geometry of each ring atom is

SN = 4
Electron-pair geometry = tetrahedral
No lone pairs
Molecular geometry = tetrahedral

SN = 4
Electron-pair geometry = tetrahedral
Two lone pairs
Molecular geometry = bent (<109.5°)

SN = 3
Electron-pair geometry = trigonal planar
No lone pairs
Molecular geometry = trigonal planar

SN = 3
Electron-pair geometry = trigonal planar
No lone pairs
Molecular geometry = trigonal planar

SN = 3
Electron-pair geometry = trigonal planar
No lone pairs
Molecular geometry = trigonal planar

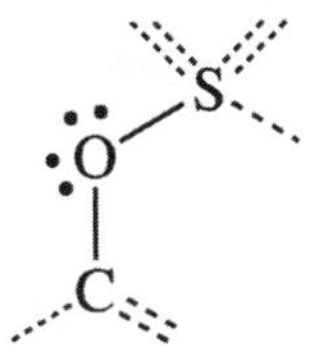

SN = 4
Electron-pair geometry = tetrahedral
Two lone pairs
Molecular geometry = bent (<109.5°)

The ring C—O bond is an overlap of C_{sp^2}/O_{sp^3} ; the C=O bond is an overlap of C_{sp^2}/O_{sp^2} and C_{2p}/O_{2p} . The C—N bond is formed by an overlap of C_{sp^2}/N_{sp^3} . The extra electron on N is located in an sp^3 hybrid orbital.

Think About It
Be careful when equating the molecular geometry and hybridization of a molecule. The lone pairs on the ring oxygen atom are located in sp^3 hybrid orbitals, as we can tell from the electron-pair geometry (tetrahedral).

9.81. **Collect and Organize**
Given four molecular structures, we are to determine whether or not each is chiral.

Analyze
For any of these molecules to be chiral it would have to contain an sp^3 hybridized carbon atom bonded to four different groups.

Solve
(a) Chiral

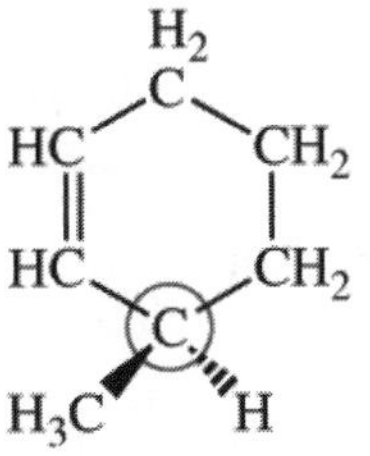

(b) Not chiral   The carbon atoms either are not sp^3 hybridized (the doubly bonded C atoms) or do not have four different substituents (the –CH₃ groups).

(c) Chiral

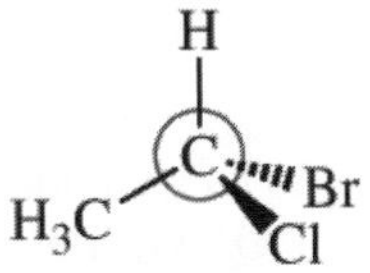

(d) Not chiral   One carbon atom is not sp^3 hybridized (the doubly bonded C=O) or it does not have four different substituents (the –CH₃ group).

Think About It
Remember to look carefully for four different substituents on the sp^3 hybridized carbons in the structures.

9.83. **Collect and Organize**
We are asked whether all σ molecular orbitals are from the overlap of one *s* orbital with another *s* orbital.

Analyze
Sigma (σ) bonds are defined as those bonds where the highest electron density is along the internuclear axis between the bonded atoms.

Solve
No. Although it is true that *s*–*s* overlap always gives σ molecular orbitals, there are other orbitals that may overlap to also give σ bonds such as the following:

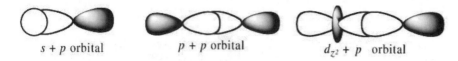

s + p orbital   *p + p* orbital $d_{z^2} + p$ orbital

Think About It

When these atomic orbitals mix to form molecular orbitals, remember that two molecular orbitals form: the sigma bonding (σ) and the sigma antibonding (σ^*) orbitals.

9.85. Collect and Organize

When an *s* orbital overlaps with another *s* orbital a sigma (σ) bond forms. We are to consider the effectiveness of the *s–s* orbital overlap between orbitals of different *n* values.

Analyze

A difference between a 1*s* orbital and a 2*s* orbital, for example, is the volume that the electrons occupy. A 2*s* orbital is larger than a 1*s* orbital.

Solve

No, because the overlap of 1*s* and 2*s* orbitals is not as efficient as 1*s*–1*s* or 2*s*–2*s* overlaps. The mismatch in size and energy is prohibitive.

Think About It

One of the guidelines for molecular orbital diagrams for homonuclear diatomic molecules states that the better mixing of orbitals from the same *n* level leads to greater bond stabilization.

9.87. Collect and Organize

We are asked to consider how molecules with an even number of valence electrons could be paramagnetic.

Analyze

Paramagnetic species contain unpaired electrons. Without violating Hund's rule or the Aufbau principle, we should consider situations in which an even number of electrons would not all be paired in the ground state.

Solve

If a molecule or ion has an even number of valence electrons distributed over an odd number of orbitals (e.g., the six valence electrons in Fe^{2+} distributed over the five 3*d* orbitals), or if the highest occupied molecular orbital is a π orbital (e.g., O_2), paramagnetic species could result.

Think About It

Many of the paramagnetic species commonly observed are transition metal ions, rather than main group species such as O_2 with a π-type highest occupied molecular orbital (HOMO). Because of the large number of *d* orbitals, it is likely that the electron configurations for many transition metal ions will contain unpaired electrons.

9.89. Collect and Organize

We are to make a sketch to show the overlap of two 1*s* orbitals to form σ_{1s} and σ_{1s}^* molecular orbitals.

Analyze

When mixing atomic orbitals (AOs) to give molecular orbitals (MOs), the number of MOs equals the number of AOs. In this case, we obtain two MOs because we are mixing two 1*s* orbitals. One MO is bonding (lower in energy) and the other is antibonding (higher in energy).

Solve

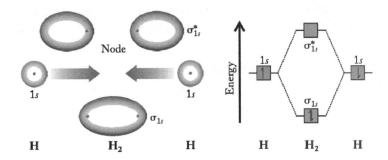

Think About It

This molecular orbital diagram is appropriate for neutral and ionic species of H_2 and He_2.

9.91. Collect and Organize

For the species N_2^+, O_2^+, C_2^+, and Br_2^{2-} we are to place electrons into the appropriate molecular orbital energy levels to predict the bond order for each diatomic molecule.

Analyze

Due to s–p orbital mixing the order of MOs for Li_2 to N_2 is

$$\sigma_{2s}\sigma_{2s}^*\pi_{2p}\sigma_{2p}\pi_{2p}^*\sigma_{2p}^*$$

For O_2 to Ne_2, which have less s–p orbital mixing, the order of the MOs is

$$\sigma_{2s}\sigma_{2s}^*\sigma_{2p}\pi_{2p}\pi_{2p}^*\sigma_{2p}^*$$

For each species we fill the MO energy levels from lowest to highest energy with the total number of electrons.

The bond order (BO) is calculated from

$$BO = \tfrac{1}{2}\,(\text{number of } e^- \text{ in bonding MOs } - \text{ number of } e^- \text{ in antibonding MOs})$$

For Br_2^{2-} we assume the same MO energies as for F_2, but the MOs involve the $4s$ and $4p$ atomic orbitals.

Solve

N_2^+ Total number of electrons = 9 e^-

$$(\sigma_{2s})^2(\sigma_{2s}^*)^2(\pi_{2p})^4(\sigma_{2p})^1$$
$$BO = \tfrac{1}{2}\,(7-2) = 2.5$$

O_2^+ Total number of electrons = 11 e^-

$$(\sigma_{2s})^2(\sigma_{2s}^*)^2(\sigma_{2p})^2(\pi_{2p})^4(\pi_{2p}^*)^1$$
$$BO = \tfrac{1}{2}\,(8-3) = 2.5$$

C_2^+ Total number of electrons = 7 e^-

$$(\sigma_{2s})^2(\sigma_{2s}^*)^2(\pi_{2p})^3$$
$$BO = \tfrac{1}{2}(5-2) = 1.5$$

Br_2^{2-} Total number of electrons = 16 e^-

$$(\sigma_{4s})^2(\sigma_{4s}^*)^2(\sigma_{4p})^2(\pi_{4p})^4(\pi_{4p}^*)^4(\sigma_{4p}^*)^2$$
$$BO = \tfrac{1}{2}\,(8-8) = 0$$

All species with nonzero bond order (N_2^+, O_2^+, and C_2^+) are expected to exist.

Think About It

Notice that N_2^+ and O_2^+ have the same bond order but very different MO filling. N_2^+ has two fewer electrons than O_2^+.

9.93. Collect and Organize

For the species N_2^+, O_2^+, C_2^+, Br_2^{2-}, O_2^-, O_2^{2-}, N_2^{2-}, and F_2^+ we can place electrons into the appropriate molecular orbital energy levels to predict which species have one or more unpaired electrons.

Analyze

Due to s–p orbital mixing the order of MOs for Li_2 to N_2 is

$$\sigma_{2s}\sigma_{2s}^*\pi_{2p}\sigma_{2p}\pi_{2p}^*\sigma_{2p}^*$$

For O_2 to Ne_2, which have less s–p orbital mixing, the order of the MOs is

$$\sigma_{2s}\sigma_{2s}^*\sigma_{2p}\pi_{2p}\pi_{2p}^*\sigma_{2p}^*$$

For Br_2^{2-}, we assume the same MO energies as for F_2, but the MOs involve the $4s$ and $4p$ atomic orbitals. The species will have unpaired electrons after filling when a σ or σ^* orbital has 1 electron in it or when a π or π^* orbital has 1, 2, or 3 electrons in it.

Solve

(a) N_2^+ Total number of electrons = 9 e⁻

$$(\sigma_{2s})^2(\sigma_{2s}^*)^2(\pi_{2p})^4(\sigma_{2p})^1$$

One unpaired electron

(b) O_2^+ Total number of electrons = 11 e⁻

$$(\sigma_{2s})^2(\sigma_{2s}^*)^2(\sigma_{2p})^2(\pi_{2p})^4(\pi_{2p}^*)^1$$

One unpaired electron

(c) C_2^{2+} Total number of electrons = 7 e⁻

$$(\sigma_{2s})^2(\sigma_{2s}^*)^2(\pi_{2p})^3$$

One unpaired electron

(d) Br_2^{2-} Total number of electrons = 16 e⁻

$$(\sigma_{4s})^2(\sigma_{4s}^*)^2(\sigma_{4p})^2(\pi_{4p})^4(\pi_{4p}^*)^4(\sigma_{4p}^*)^2$$

No unpaired electrons

(e) O_2^- Total number of electrons = 13 e⁻

$$(\sigma_{2s})^2(\sigma_{2s}^*)^2(\sigma_{2p})^2(\pi_{2p})^4(\pi_{2p}^*)^3$$

One unpaired electron

(f) O_2^{2-} Total number of electrons = 14 e⁻

$$(\sigma_{2s})^2(\sigma_{2s}^*)^2(\sigma_{2p})^2(\pi_{2p})^4(\pi_{2p}^*)^4$$

No unpaired electrons

(g) N_2^{2-} Total number of electrons = 12 e⁻

$$(\sigma_{2s})^2(\sigma_{2s}^*)^2(\pi_{2p})^4(\sigma_{2p})^2(\pi_{2p}^*)^2$$

Two unpaired electrons

(h) F_2^+ Total number of electrons = 13 e⁻

$$(\sigma_{2s})^2(\sigma_{2s}^*)^2(\sigma_{2p})^2(\pi_{2p})^4(\pi_{2p}^*)^3$$

One unpaired electron

The species with one or more unpaired electrons are (a) N_2^+, (b) O_2^+, and (c) C_2^{2+}, (e) O_2^-, (g) N_2^{2-}, and (h) F_2^+.

Think About It

The π orbital filling according to Hund's rule shows how the π orbitals have unpaired electrons when there are 1, 2, or 3 electrons occupying them, but not with four electrons.

π π π π

9.95. **Collect and Organize**

Using the MO diagrams from the text as a guide, we are asked to draw an MO diagram for the diatomic molecule ClO and determine if the unpaired electron is located in a bonding or an antibonding orbital.

Analyze

Using the MO diagram for NO (Figure 9.50) as a guide, we can create an MO diagram illustrating the overlap of the Cl $3s$ and $3p$ orbitals and the O $2s$ and $2p$ orbitals. Adding the 13 valence electrons to this diagram, following the Aufbau principle and Hund's rule, will generate the ground state configuration.

Solve

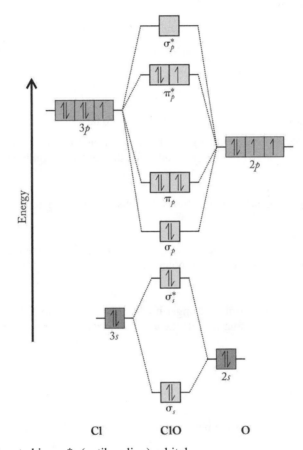

The unpaired electron is located in a π^*_p (antibonding) orbital.

Think About It

Species with an unpaired electron in an antibonding orbital such as ClO will increase in bond order by losing the unpaired electron, depopulating the antibonding orbital.

9.97. **Collect and Organize**

For B_2, C_2, N_2, and O_2 we are to determine which increases its bond order upon acquiring two electrons to become a dianion.

Analyze

Bond order increases when the two extra electrons are placed into bonding MOs. If the two extra electrons are placed into antibonding MOs, the bond order decreases.

Solve

(a) B_2 has the MO configuration of

$$(\sigma_{2s})^2(\sigma_{2s}^*)^2(\pi_{2p})^2$$

The two electrons added to form the dianion are placed into the π_{2p} orbital, so the bond order increases.

(b) C_2 has the MO configuration of

$$(\sigma_{2s})^2(\sigma_{2s}^*)^2(\pi_{2p})^4$$

The two electrons added to form the dianion are placed into the σ_{2p} orbital, so the bond order increases.

(c) N_2 has the MO configuration of

$$(\sigma_{2s})^2(\sigma_{2s}^*)^2(\pi_{2p})^4(\sigma_{2p})^2$$

The two electrons added to form the dianion are placed into the π_{2p}^* orbital, so the bond order decreases.

(d) O_2 has the MO configuration of

$$(\sigma_{2s})^2(\sigma_{2s}^*)^2(\sigma_{2p})^2(\pi_{2p})^4(\pi_{2p}^*)^2$$

The two electrons added to form the dianion are placed into the π_{2p}^* orbital, so the bond order decreases.

Because the bond order increases with a gain of two electrons for (a) B_2 and (b) C_2, these two molecules will form an anion with a 2– charge that is more stable than the neutral diatomic molecule.

Think About It

The species above that have unpaired electrons (and thus are paramagnetic) are B_2, N_2^{2-}, and O_2.

9.99. **Collect and Organize**

For the diatomic 1+ cations of Li_2, Be_2, B_2, C_2, N_2, O_2, F_2, and Ne_2, we are to consider whether the cations always have shorter bonds than the neutral molecules.

Analyze

Shorter bonds have higher bond orders. Longer bonds have lower bond orders. Bond order decreases when the electron is removed from a bonding MO. If the electron is removed from an antibonding MO, the bond order increases.

Solve

Li_2 has the MO configuration of

$$(\sigma_{2s})^2$$

Removing one electron decreases the bond order, and the bond is lengthened.
Be_2 has the MO configuration of

$$(\sigma_{2s})^2(\sigma_{2s}^*)^2$$

Removing one electron increases the bond order, and the bond is shortened.

B_2 has the MO configuration of

$$(\sigma_{2s})^2(\sigma_{2s}^*)^2(\pi_{2p})^2$$

Removing one electron decreases the bond order, and the bond is lengthened.
C_2 has the MO configuration of

$$(\sigma_{2s})^2(\sigma_{2s}^*)^2(\pi_{2p})^4$$

Removing one electron decreases the bond order, and the bond is lengthened.
N_2 has the MO configuration of

$$(\sigma_{2s})^2(\sigma_{2s}^*)^2(\pi_{2p})^4(\sigma_{2p})^2$$

Removing one electron decreases the bond order, and the bond is lengthened.
O_2 has the MO configuration of

$$(\sigma_{2s})^2(\sigma_{2s}^*)^2(\sigma_{2p})^2(\pi_{2p})^4(\pi_{2p}^*)^2$$

Removing one electron increases the bond order, and the bond is shortened.
F_2 has the MO configuration of

$$(\sigma_{2s})^2(\sigma_{2s}^*)^2(\sigma_{2p})^2(\pi_{2p})^4(\pi_{2p}^*)^4$$

Removing one electron increases the bond order, and the bond is shortened.
Ne_2 has the MO configuration of

$$(\sigma_{2s})^2(\sigma_{2s}^*)^2(\sigma_{2p})^2(\pi_{2p})^4(\pi_{2p}^*)^4(\sigma_{2p}^*)^2$$

Removing one electron increases the bond order, and the bond is shortened.
No, the cations N_2^+, C_2^+, B_2^+, and Li_2^+, which lose an electron from bonding orbitals in the corresponding neutral molecules, decrease their bond order and thus have longer bond lengths.

Think About It
All of the 1+ cations will be paramagnetic.

9.101. **Collect and Organize**
We are asked to draw the Lewis structures and determine the molecular geometries of NH_4^+ and ClO_4^-.

Analyze
First, we must draw the Lewis structures of each ion. Then, through the steric number (SN), we can determine the electron-pair geometry. If there are lone electron pairs on the central atom, we have to take that into account to translate the electron-pair geometry into the molecular geometry.

Solve

SN = 4
Electron-pair geometry = tetrahedral
Molecular geometry = tetrahedral

SN = 4
Electron-pair geometry = tetrahedral
Molecular geometry = tetrahedral

Think About It
The structure drawn for ClO_4^- is the one that has the lowest formal charges on the atoms. The expanded octet on Cl is possible because it is a third period element.

9.103. **Collect and Organize**
We are given the skeletal structure of glycine. By completing the Lewis structure and applying VSEPR theory, we can determine the N—C—C, O—C—O, and C—O—H bond angles.

Analyze
From the Lewis structure and steric number (SN) we can determine the electron-pair geometry around the carbon and oxygen atoms. If the electron-pair geometry is linear, the bond angles are 180°; if it is trigonal planar, the bond angles are 120°; if it is tetrahedral, the bond angles are 109.5°.

Solve

SN = 3
Electron-pair geometry = trigonal planar
O—C—O bond angle = 120°

SN = 4
Electron-pair geometry = tetrahedral
C—O—H bond angle = 109.5°

SN = 4
Electron-pair geometry = tetrahedral
N—C—C bond angle = 109.5°

Think About It
Remember, these are idealized bond angles. Because lone pairs take up more space than bonding pairs, the C—O—H bond angle is likely <109.5°.

9.105. **Collect and Organize**
We are given two alternate skeletal structures for Cl_2O_2. We are to complete the Lewis structures, and from those we can find the molecular geometry to determine whether either of the two isomers is linear and if they have the same dipole moment.

Analyze
Cl_2O_2 has 26 e⁻ and needs 32 e⁻, giving a difference of 6 e⁻ in three covalent bonds. This leaves 20 e⁻ in ten lone pairs to complete the octets on the atoms. Chlorine may expand its octet to minimize formal charges on the atoms in the Lewis structure.

Solve

:Cl̈—Ö—Ö—C̈l: :Cl̈—Ö—C̈l=O:

(a) All of the central atoms (O—O and O—Cl) have SN = 4, so their electron-pair geometries are tetrahedral. Each of the central atoms also has two lone pairs and two bonding pairs, which gives them a bent molecular geometry. Therefore, neither of the two molecules is linear.
(b) Only one structure (ClOOCl) is symmetrical, and would be expected to have bond dipoles that cancel one another, generating a nonpolar molecule. The other molecule (ClOClO) would instead be polar, as the O—Cl bond dipoles would not cancel. The two molecules do not have the same dipole moment.

Think About It
These compounds may be drawn three-dimensionally as

9.107. **Collect and Organize**
For the diatomic ion ClO^+, we are to draw the Lewis structure and complete the molecular orbital (MO) diagram (Figure P9.107) to determine the Cl—O bond order.

Analyze
ClO^+ has 12 e⁻. To complete the octets on the atoms for the Lewis structure, ClO^+ would need 16 e⁻, giving a difference of 4 e⁻ in two covalent bonds. This leaves 8 e⁻ in four lone pairs to complete the structure. Chlorine may expand its octet to reduce the formal charges on the atom in the Lewis structure.

Solve

(a) $\left[:\ddot{Cl}=\ddot{O}:\right]^{+}$

(b) The MO diagram would fill as

$$(\sigma_{3s})^2(\sigma_{3s}^*)^2(\sigma_{3p})^2(\pi_{3p})^4(\pi_{3p}^*)^2$$
$$BO = \tfrac{1}{2}(8-4) = 2$$

Think About It

For ClO^+ the bond order drawn in the Lewis structure matches the bond order calculated by MO theory.

9.109. **Collect and Organize**

Given the skeletal structure of phosphoric acid, we are to complete its Lewis structure and use VSEPR theory to determine the molecular geometry around the phosphorus atom.

Analyze

H_3PO_4 has 32 e^- and needs 46 e^-, giving a difference of 14 e^- in seven covalent bonds. This leaves 18 e^- in nine lone pairs to complete the octets on the atoms in the structure. Phosphorus may expand its octet to reduce the formal charges on the atoms.

Solve

All formal charges = 0
SN at phosphorus = 4
Electron-pair and molecular geometry around P = tetrahedral

Think About It

In this structure phosphorus forms three σ bonds to OH, and a σ plus a π bond to oxygen.

9.111. **Collect and Organize**

For BBCO and OCBBCO we are to draw the Lewis structures that minimize the formal charges on the atoms. Using these structures we can predict if either molecule contains electron-deficient atoms (with incomplete octets) and then use VSEPR theory to predict the molecular geometries.

Analyze

BBCO has 16 e^- and needs 32 e^-, giving a difference of 16 e^- in 8 covalent bonds. This leaves no lone pairs of electrons. OCBBCO has 26 e^- and needs 48 e^-, giving a difference of 22 e^- in 11 covalent bonds. This leaves 4 e^- in two lone pairs to complete the octets on the atoms. Boron in compounds may be electron deficient (e.g., in BF_3) so, if necessary, we can arrange the electrons to satisfy the octets on the atoms other than B, and leave B with an incomplete octet.

Solve

$:B-B=C=\ddot{O}:$

All formal charges = 0
Both B atoms have incomplete octets
On both central B and C atoms SN = 2
Molecular geometry = linear

$:\ddot{O}=C=B-B=C=\ddot{O}:$

All formal charges = 0
Both B atoms have incomplete octets
On all central B and C atoms SN = 2
Molecular geometry = linear

Think About It

These compounds might be predicted to gain some stability through the other (less favorable) resonance forms in which the formal charges do not equal zero.

9.113. Collect and Organize

We are to draw the resonance structures of methyl isothiocyanate (CH_3NCS) and use formal charges to identify the form that contributes most to the bonding. From the Lewis structure, we can determine the steric number (SN) at the carbon atoms in the molecule and predict the molecular geometry at each carbon atom.

Analyze

CH_3NCS has 22 e^- and needs 38 e^-, giving a difference of 16 e^- in eight covalent bonds. This leaves 6 e^- in three lone pairs on the molecule.

Solve

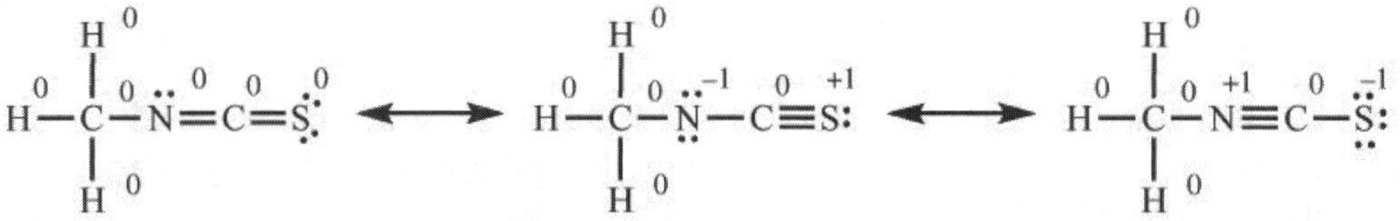

The resonance structure in which all the formal charges equal zero contributes the most to the bonding. At the methyl (CH_3) carbon, SN = 4, so this carbon is tetrahedral. At the isothiocyanate (NCS) carbon, SN = 2, so the molecular geometry at this carbon is linear.

Think About It

Notice that the molecular geometries at the carbon atoms stay the same in all of the resonance structures.

9.115. Collect and Organize

If borazine is isoelectronic with benzene, it has the same number of electrons. To determine if there are also delocalized π electrons in borazine, we must draw the Lewis structure of $B_3N_3H_6$ with resonance forms.

Analyze

$B_3N_3H_6$ has 30 e^- and needs 60 e^-, giving a difference of 30 e^- in 15 covalent bonds. This leaves no lone pairs in the molecule.

Solve

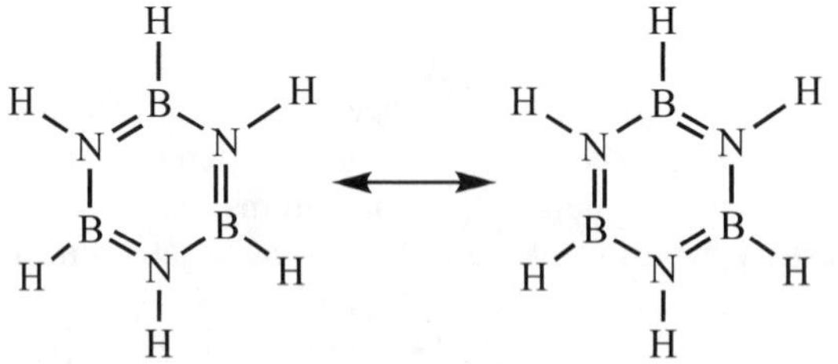

Yes, in these resonance structures, the π electrons are delocalized just as in benzene. In the resonance forms, the formal charge on B is −1 and that on N is +1.

Think About It

The resonance structure that minimizes the formal charges on all the atoms gives a structure in which the B atoms are electron deficient.

9.117. Collect and Organize

If the molecule HArF contains ArF^-, we can use molecular orbital theory to determine the bond order in ArF^-.

Analyze

ArF^- would have 16 valence electrons, and bonding would involve overlap of the $3s$ and $3p$ orbitals on Ar with the $2s$ and $2p$ orbitals on F. Although the overlap would not be as effective as the overlap of orbitals of the same n level, we can assume that the overlap gives similar molecular orbitals. The MO diagram would look similar to that for F_2.

Solve

σ_p^*

π_p^*

π_p

$$BO = \tfrac{1}{2}(8-8) = 0$$
ArF^- is not expected to be stable because its bond order is zero.

σ_p

σ_s^*

σ_s

Think About It

The argon–fluorine bond would be stable, however, as a neutral species (ArF) or as a cation (ArF^+).

9.119. Collect and Organize

In order to determine the polarity of N_2O_2, N_2O_5, and N_2O_3, we must draw Lewis structures and then consider the direction and magnitude of the individual bond dipoles.

Analyze

We draw the Lewis structures in the usual way, and then using VSEPR theory, we draw the structure's geometries based on the steric number of the central N and O atoms. The electronegativity for O is greater than that for N, thus each N—O bond is polarized so that partial negative charge is on the oxygen atom. Once we have assigned the individual bond dipoles, we can then visually see if the vectors representing those bond dipoles cancel to give a nonpolar molecule or add together to give a polar molecule.

Solve

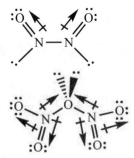

The N_2O_2 molecule is nonpolar as drawn, but we can imagine the N–N bond rotating to give a polar molecule.

Due to free rotation around the N–N bond, N_2O_2 is nonpolar.

The N_2O_5 molecule is polar.

The N_2O_3 molecule is polar.

Think About It

Be careful to consider geometry in assigning polarity to a molecule. It is the bent geometry around the central oxygen atom in N_2O_5 that makes this molecule polar.

9.121. **Collect and Organize**

From the molecular orbital diagram for S_2^{2-} we can show that the bond order is one and determine whether this species is paramagnetic or diamagnetic.

Analyze

We can assume that the MO diagram for S_2 is similar to that of O_2 except that $3s$ and $3p$ atomic orbitals are used to make the molecular orbitals. S_2^{2-} has 14 total valence electrons.

Solve

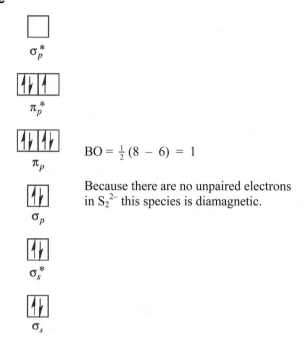

$BO = \frac{1}{2}(8 - 6) = 1$

Because there are no unpaired electrons in S_2^{2-} this species is diamagnetic.

Think About It

Like O_2, neutral S_2 is expected to have a bond order of 2, which is consistent with its Lewis diagram:

The Lewis structure, however, does not predict that we also expect S_2 to be paramagnetic like O_2.

9.123. **Collect and Organize**

To determine the orbital hybridization of the atoms in cyclic S_8, we can use the steric number (SN) of the S atoms obtained from the Lewis structure.

Analyze

S_8 has 48 e^- and needs 64 e^- for a difference of 16 e^- in eight covalent bonds. This leaves 32 e^- in 16 lone pairs to complete the octets on the S atoms.

Solve

Each S atom has SN = 4 so the electron-pair geometry is tetrahedral.
Orbital hybridization on S = sp^3

Think About It

With two lone pairs on each S atom, the molecular geometry at each S atom is predicted to be bent with an angle of approximately 109.5°. The actual bond angles (108°) are less than the ideal because the two lone pairs occupy more space and therefore the bond pair–bond pair angle is reduced.

9.125. **Collect and Organize**

We are given that ozone, even with only one kind of atom, has a permanent dipole moment. We have to look closely at the molecular geometry of O_3 to explain this molecule's polarity.

Analyze

For the Lewis structure, O_3 has 18 e⁻ and needs 24 e⁻, giving a difference of 6 e⁻ in three covalent bonds and leaving 12 e⁻ in six lone pairs to complete the octets on the oxygen atoms. The molecular geometry is obtained from the steric number (SN) around the central O atom in the structure.

Solve

SN = 3 for an electron-pair geometry of trigonal planar
The presence of one lone pair on the central O atom makes the molecular geometry at the central O atom bent.

This molecule is polar because, although the oxygen–oxygen bonds themselves are nonpolar, the lone pair has its own "pull" on the electrons in the molecule. Also, the π bonds between the oxygen atoms place slightly more electron density on the terminal O atoms, making the "nonpolar" O—O bond polar.

Think About It

Geometry around an atom, not just the difference in electronegativity, is important to the overall polarity of a molecule.

9.127. **Collect and Organize**

By completing the Lewis structures for alliin and allicin, we can determine the steric number and thus the molecular geometry around each S atom. We can then compare these geometries to those of the sulfur atom in H_2S, CH_3SH, and $(CH_3)_2S$.

Analyze

Alliin ($C_6H_{11}NO_3S$) has 64 e⁻ and needs 110 e⁻, giving a difference of 46 e⁻ in 23 covalent bonds and leaving 18 e⁻ in nine lone pairs to complete the octets. The structure given in Figure P9.127, however, shows 24 covalent bonds, which leaves 16 e⁻ in eight lone pairs. Allicin ($C_6H_{10}OS_2$) has 52 e⁻ and needs 92 e⁻, giving a difference of 40 e⁻ in 20 covalent bonds and leaving 12 e⁻ in six lone pairs to complete the octets. The structure in Figure P9.127, however, shows 21 covalent bonds, leaving 10 e⁻ in five lone pairs. The sulfur atom in both these molecules may expand its octet in order to reduce the formal charges on the atoms in the Lewis structure. The sulfur atoms in H_2S, CH_3SH, and $(CH_3)_2S$ all are predicted to have a bent geometry with sp^3 hybridization and bond angles of approximately 109.5°.

Solve

(a) Alliin

SN = 4, trigonal pyramidal molecular geometry

Allicin

SN = 4, bent molecular geometry

SN = 4, trigonal pyramidal molecular geometry

(b) Yes, the C–S–S bond angle is predicted to be the same as the S bond angles in H_2S, CH_3SH, and $(CH_3)_2S$.

Think About It

Do not take the complex skeletal structure "as is." Count up the electrons for the entire structure and make sure that all electron pairs are represented as either lone pairs or bonding pairs in the structure. It would be easy in both of these structures to miss the lone pair on the $S\!\!=\!\!O$ sulfur atom.

CHAPTER 10 | Intermolecular Forces: The Uniqueness of Water

10.1. Collect and Organize

In Figure P10.1, the two similarly sized spheres of opposite charge represent KF. In KI, the iodide ion is shown negatively charged and larger than the K^+ ion. We are to determine which substance would have a higher melting point by virtue of having stronger ion–ion forces.

Analyze

Coulomb's law describes the strength of the ion–ion interactions in these substances as

$$E \propto \frac{Q_1 \times Q_2}{d}$$

In the case of KF and KI, Q_1 and Q_2 are the same ($Q_{K^+} = +1$, Q_{F^-} and $Q_{I^-} = -1$). What does change is the distance between the ion centers.

Solve

Coulomb's law states that as the distance between ions increases, the energy of the interaction decreases. Because I^- is larger than F^-, the KI interaction is weaker. Therefore, the stronger ion–ion interaction results in a higher melting point for KF than KI.

Think About It

The strength of the ion–ion interaction is even more influenced by the charge on the ions. The melting point of CaF_2 (where $Q_1 = +2$ and $Q_2 = -1$) would be predicted to be even higher than that of KF.

10.3. Collect and Organize

Given the boiling points of two trigonal pyramidal molecules, we are to determine which substance is NH_3 and which is PH_3.

Analyze

Both molecules might be polar due to their pyramidal geometry. The more polar molecule has the stronger intermolecular forces. The dipole strength may be estimated using electronegativity values for N (3.0), P (2.1), and H (2.1).

Solve

Because the electronegativity of P is equal to that of H, the phosphine, PH_3, is predicted to be less polar. Ammonia, NH_3, on the other hand, is more polar; moreover, because H is bonded to the very electronegative N atom, it forms strong hydrogen bonds with other NH_3 molecules. Ammonia therefore has the higher boiling point ($-33°C$ YH_3 in Figure P10.3). Phosphine, with much weaker forces between its atoms, has a low ($-88°C$) boiling point and is represented by XH_3 in Figure P10.3.

Think About It

The phosphine molecule is not strictly nonpolar. The lone pair will "pull" differently than the N atoms do on the electrons on the phosphorus atom, so PH_3 is slightly polar.

10.5. Collect and Organize

Considering the graphs for this problem, we are to indicate which graph indicates a liquid with stronger intermolecular forces than the other.

Analyze

The graphs in Figure P10.5 are described by the Clausius–Clapeyron equation

$$\ln(P_{vap}) = -\frac{\Delta H_{vap}}{R}\left(\frac{1}{T}\right) + C$$

The slope of the trend line in each graph is proportional to the size of ΔH_{vap}, so the graph with the larger slope has a larger enthalpy of vaporization. A large value of ΔH_{vap} corresponds to strong intermolecular forces.

Solve

The graph on the left has a larger slope, and so greater intermolecular forces.

Think About It

Because intermolecular attractive forces must be overcome when molecules are vaporized, a molecule with strong intermolecular forces will have a high value of ΔH_{vap}.

10.7. Collect and Organize

By closely examining the phase diagram in Figure P10.7, we can determine how the freezing point changes when the pressure is increased.

Analyze

The red line separating the solid and liquid phases shows the freezing point. The solid phase is present at low temperatures (left side of the phase diagram) and the liquid phase is present at higher temperatures (middle section of the phase diagram).

Solve

The solid–liquid line on the phase diagram (red line in Figure P10.7) slopes up to the right. This means that the freezing point increases with increasing pressure.

Think About It

The phase diagram for water shows the opposite slope. The freezing point of water, therefore, decreases when pressure is applied.

10.9. Collect and Organize

From the list of salts given we are to determine which contains the largest anion.

Analyze

$BaCl_2$ contains Cl^- anions, AlF_3 contains F^- anions, KI contains the I^- anion, and $SrBr_2$ contains Br^- anions.

Solve

According to periodic trends, halide ion size increases down the group: $F^- < Cl^- < Br^- < I^-$. KI (c) contains the largest anion.

Think About It

Of these substances, the one with the largest cation is KI because K^+ is the largest cation of the series: $Al^{3+} < Sr^{2+} < Ba^{2+} < K^+$. K^+ is very close in size to Ba^{2+} however, with radii of 138 pm and 135 pm, respectively.

10.11. Collect and Organize

We are to explain why $CaSO_4$ is less soluble in water than NaCl. For this we need to keep in mind that both of these compounds are ionic.

Analyze

If an ionic bond is strong, its ions may not dissolve in water easily. In salt dissolution processes, the strength of the ion–ion bond must be balanced with the strength of the ion–solvent interactions.

Solve

The ion–ion bond in $CaSO_4$ is stronger than in NaCl because of the higher charges on the cation and anion. For $CaSO_4$, this is greater than the ion–dipole interactions that would occur when Ca^{2+} and SO_4^{2-} dissolve, so $CaSO_4$ is not very soluble in water. NaCl has a lower ion–ion bond strength and its ion–dipole interactions with water are strong, so it dissolves in water.

Think About It

Be careful in making this argument. The ion–dipole forces of Ca^{2+}–water and SO_4^{2-}–water might actually be stronger than that of Na^+–water and Cl^-–water due to the high charges on Ca^{2+} and SO_4^{2-}. However, the strengths of the ion–dipole interactions do not match (or exceed) the strength of the Ca^{2+}—SO_4^{2-} ionic bond.

10.13. Collect and Organize

We can use Coulomb's law to rank KBr, $SrBr_2$, and CsBr in order of increasing ionic attraction.

Analyze

Coulomb's law states that the attraction between oppositely charged ions is directly proportional to the product of their charges and is inversely proportional to their separation distance:

$$E \propto \frac{Q_1 \times Q_2}{d}$$

The distance between the ions is taken as the sum of the ions' radii (Figure 10.2).

Solve

For KBr $\quad E \propto \dfrac{(+1)(-1)}{(138+196)} = -0.00299$

For $SrBr_2$ $\quad E \propto \dfrac{(+2)(-1)}{(118+196)} = -0.00637$

For CsBr $\quad E \propto \dfrac{(+1)(-1)}{(170+196)} = -0.00273$

In order of increasing ionic attraction: $CsBr < KBr < SrBr_2$.

Think About It

Because the ions in CsBr are large compared to KBr, CsBr has a lower ion–ion attraction. $SrBr_2$ has the highest ion–ion attraction because of the +2 charge on Sr.

10.15. Collect and Organize

We are to describe how individual water molecules are oriented around dissolved chloride ions.

Analyze

The water molecule is polar due to its bent geometry, with a partial negative charge on the oxygen atom and a partial positive charge on the hydrogen ends.

Solve

The water molecule is oriented around Cl^- so as to point the partially positive hydrogen atoms toward the Cl^- ion. This results in attractive forces between the water molecules and the Cl^- ion.

Think About It

This interaction is an ion–dipole interaction.

10.17. Collect and Organize

We are to explain the differences in strength between dipole–dipole interactions (weaker) and ion–dipole interactions (stronger).

Analyze

The dipole–dipole interaction involves attractions between two polar molecules with slight charge separation (partial positive and negative charges) on the molecule. An ion–dipole interaction involves attractions between an ion with a full positive or negative charge and a polar molecule.

Solve

Coulomb's law states that as charge increases, the attraction of two oppositely charged species for each other increases. Because of the full positive or negative charge on the ion, the ion–dipole interaction is stronger than the dipole–dipole interaction.

Think About It

Ion–ion is the strongest of all interactions between molecules.

10.19. **Collect and Organize**

We are to explain why hydrogen bonds are considered to be a special class of dipole–dipole interactions.

Analyze

Hydrogen bonds can form when hydrogen is bonded to a very electronegative element (F, O, N). The hydrogen bond is very polar.

Solve

The charge buildup on H (partially positive) and the electronegative element (partially negative) means that the X—H bond is polar but not ionic. It is still a dipole–dipole interaction except that its strength is noticeably higher than other dipole–dipole interactions.

Think About It

Hydrogen bonds are also important in explaining why ice floats and how proteins fold.

10.21. **Collect and Organize**

We are to explain which ion, Cl^- or I^-, hydrates more strongly in water.

Analyze

Strong hydration of an ion occurs when there is a strong attraction of the ion for the surrounding water molecules. Ions with smaller size or higher charge, therefore, are more strongly hydrated.

Solve

Because Cl^- is smaller than I^-, Cl^- hydrates more strongly in solution.

Think About It

The F^- ion would be predicted to be even more strongly hydrated than the Cl^- anion.

10.23. **Collect and Organize**

We are to explain why CH_3F (mp $-142^\circ C$) has a higher melting point than CH_4 (mp $-182^\circ C$).

Analyze

The higher melting point of CH_3F is indicative of stronger intermolecular forces between CH_3F molecules compared to those between CH_4 molecules.

Solve

CH_3F is a polar molecule and therefore has stronger intermolecular forces than the nonpolar molecules of CH_4, which have only the weak dispersion forces. As it takes more energy to overcome strong intermolecular forces, CH_3F has a higher melting point than CH_4.

Think About It

Molecular polarity and the degree of charge separation are important considerations when comparing some of the physical properties of compounds (e.g., boiling point or vapor pressure).

10.25. **Collect and Organize**

We are asked to explain why CH_3F does not show hydrogen bonding but HF does.

Analyze

Hydrogen bonds can form only when hydrogen is bonded to a very electronegative element (F, O, N).

Solve

The H in CH_3F has just a single bond to the relatively low electronegativity C atom and, therefore, the carbon–hydrogen bond is not polar enough to exhibit hydrogen bonding. In HF, however, the H atom is bonded to fluorine, the most electronegative element on the Pauling scale.

Think About It

It is important to consider the connectivity of atoms in a molecule when determining the intermolecular forces in a molecule. For example, ethanol (CH_3CH_2OH) will exhibit hydrogen bonding, while dimethyl ether (CH_3OCH_3) does not, due to the absence of an O–H bond in dimethyl ether.

10.27. Collect and Organize

For the covalent molecules CF_4, CF_2Cl_2, and CCl_4 we are to determine which we would expect to have the strongest dipole–dipole interactions.

Analyze

We need to first determine whether the molecules are polar or nonpolar. Polar molecules have permanent dipoles that attract each other (δ^- to δ^+); nonpolar molecules have only weak dispersion forces between them. If all the molecules are polar, then the one with the smallest dipole moment (as determined by the differences in electronegativities between the atoms) would be the molecule with the weakest intermolecular forces.

Solve

From the Lewis structures of these molecules, we know that both CF_4 and CCl_4 are nonpolar tetrahedral molecules. These have only dispersion forces as the intermolecular force between the molecules. CF_2Cl_2 (b) is a polar tetrahedral molecule and has the strongest dipole–dipole interactions.

Think About It

Because both CF_4 and CCl_4 are nonpolar, there are no dipole–dipole attractions between them.

10.29. Collect and Organize

We are to determine which molecules are capable of forming hydrogen bonds among themselves.

Analyze

Molecules containing N–H, O–H, and F–H bonds may form hydrogen bonds with one another.

Solve

(a) Methanol and (d) acetic acid are capable of forming hydrogen bonds with themselves.

Think About It

Even if oxygen, nitrogen, or fluorine atoms are present, they must be bonded directly to a hydrogen atom to form hydrogen bonds. (c) Dimethyl ether, cannot form a hydrogen bond because the oxygen atom is bonded to two carbon atoms.

10.31. Collect and Organize

We are to name the type of intermolecular force that exists in all substances.

Analyze

All substances have electron "clouds" that can be temporarily distorted.

Solve

The temporary distortions of an electron cloud around an atom or molecule are dispersion forces (also called London forces) and all molecules have these.

Think About It
Dispersion forces, being weak, are often "trumped" by the stronger intermolecular forces (ion–ion, ion–dipole, dipole–dipole, and hydrogen bonding).

10.33. **Collect and Organize**
We are to explain the nonideal behavior of all gases at high pressures and low temperatures.

Analyze
At high pressures, gas molecules undergo many collisions with the walls of the container. This means that there are greater numbers of gas molecules in a smaller volume. At low temperatures, gas particles are moving with less kinetic energy.

Solve
In the ideal gas equation, the volume, V, assumes that all the volume of the container is "free volume," that is, the actual volume of the gas molecules is insignificant relative to the total volume of the container. The nonideal behavior at high pressures arises from the true volume of the gas molecules.
As the temperature is lowered, the kinetic energy of the gas particles is reduced enough that when particles collide, they temporarily stick together via their intermolecular forces. The P term in the ideal gas law assumes that all particles collide elastically and act independently. Cold temperatures reduce the elasticity of the collisions.

Think About It
In the real gas equation, we use the pressure correction term, a, to account for the intermolecular forces between real atoms and the volume correction term, b, to account for the volume real atoms occupy.

10.35. **Collect and Organize**
To explain why the boiling point of CH_2F_2 is lower than that of CH_2Cl_2 despite the greater dipole moment of CH_2F_2, we need to consider all the intermolecular forces that act between the molecules in each of these substances.

Analyze
Both molecules are polar so dipole–dipole interactions are present in each substance. Weak dispersion forces are also present between the molecules. These dispersion forces are greater for CH_2Cl_2 because chlorine has more electrons and is more polarizable than fluorine.

Solve
The substance with the higher boiling point is the one that has the larger sum of intermolecular forces. In this case, the greater dispersion forces of CH_2Cl_2 add to the dipole–dipole interactions to give stronger intermolecular forces between the CH_2Cl_2 molecules compared to those of CH_2F_2 molecules. Also, the molar mass of CH_2Cl_2 is higher than that of CH_2F_2, so it takes more energy to vaporize.

Think About It
Usually dispersion forces are so much weaker that they do not significantly add to the strength of the dominant intermolecular force between molecules. In this problem, we see a clear example where the consideration of dispersion forces is necessary.

10.37. **Collect and Organize**
For each pair of substances we are to determine which one has the stronger dispersion forces.

Analyze
The greater the number of atoms and electrons in a compound, the greater the dispersion forces between the molecules.

Solve
(a) CCl_4 has stronger dispersion forces than CF_4 because Cl is a larger element with more electrons than F.
(b) C_3H_8 has stronger dispersion forces than CH_4 because it has more atoms in its molecular structure than CH_4.

Think About It

The greater number of electrons and atoms in a compound gives rise to stronger dispersion forces because there are more polarizable electrons on the atoms that make up the compound.

10.39. **Collect and Organize**

We are asked to determine which of the molecules depicted in Figure P10.39 is a solid, and which is a liquid. Using the atom color code, we may identify these molecules as PCl_3 and PCl_5.

Analyze

The bond dipoles in PCl_5 cancel, meaning the molecule is nonpolar and will have only the London dispersion forces from five P–Cl bonds. PCl_3 adopts a trigonal pyramidal geometry, meaning the molecule is polar and will have dipole–dipole interactions as well as the London dispersion forces from three P–Cl bonds.

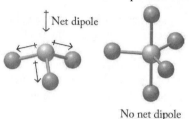

No net dipole

Solve

Although PCl_3 is polar, the summed dispersion forces for PCl_5 are greater than the sum of all dipole–dipole and dispersion forces for PCl_3. This may be predicted because of the larger surface area for PCl_5, leading to stronger dispersion interactions than for PCl_3. PCl_5 is a solid at room temperature; PCl_3 is a liquid.

Think About It

You may not have been expecting the intermolecular forces for PCl_5 to be greater than those for PCl_3. Despite the presence of dipole–dipole interactions, the dispersion interactions for PCl_3 are significantly smaller overall.

10.41. **Collect and Organize**

We can define *miscible* and *insoluble* to delineate the difference between these terms.

Analyze

Two substances are miscible when they dissolve completely (in all proportions) into each other. A substance is insoluble when it does not dissolve in a solvent; it remains phase-separated as a solid in the solvent.

Solve

Miscible and insoluble are opposites of each other in the range of possible solubilities of a solute in a solvent. At one end, miscible solutes and solvents dissolve completely in each other. At the other extreme, an insoluble solute does not dissolve at all.

Think About It

The classification between soluble and insoluble is indistinct. Generally speaking, a solute that dissolves at less than 0.1 g in 1.00 L of a solvent is considered to be insoluble.

10.43. **Collect and Organize**

We are asked whether 1,1-dichloroethane or 1,2-dichloroethane is expected to be soluble in water at 20 °C.

Analyze

Water is a polar solvent, so it is reasonable to expect that the more polar isomer will be water soluble.

Solve

In Problem 10.40, we determined that 1,1-dichloroethane is polar, and 1,2-dichloroethane is nonpolar.

1,1-Dichloroethane expected to be soluble in water.

Think About It

Although 1,2-dichloroethane was found to have greater intermolecular forces overall, the dipole–dipole interactions with water govern the solubility of 1,1-dichloroethane.

10.45. **Collect and Organize**

We are to relate the solubility of substances in water with the terms *hydrophilic* and *hydrophobic*.

Analyze

Hydrophilic means "water-loving" and *hydrophobic* means "water-hating."

Solve

Hydrophilic substances dissolve in water. Hydrophobic substances do not dissolve, or are immiscible, in water.

Think About It

Ethanol is hydrophilic because it is miscible with water to give a homogeneous solution, but olive oil is hydrophobic because it forms a heterogeneous mixture with water that separates into oil and water layers.

10.47. **Collect and Organize**

For each pair of compounds, we are to determine which is more soluble in H_2O.

Analyze

Water is a polar solvent capable of forming hydrogen bonds to dissolved substances with X—H bonds (X = F, O, N). In each pair of compounds, the more soluble compound is the more polar molecule or the one that forms hydrogen bonds. In considering whether a salt is soluble in water, we have to consider the relative strengths of the ionic bonds as well as the relative strengths of the ion–dipole interactions formed on dissolution.

Solve

(a) $CHCl_3$ is polar whereas CCl_4 is not. $CHCl_3$ is more soluble in water.
(b) CH_3OH is more polar because it has a smaller hydrocarbon chain compared to $C_6H_{11}OH$. CH_3OH is more soluble in water.
(c) NaF has a weaker ionic bond than MgO. NaF is more soluble in water.
(d) BaF_2 has a weaker ionic bond than CaF_2 because Ba^{2+} is larger than Ca^{2+}. BaF_2 is more soluble in water.

Think About It

Solubility is determined by many factors: polarity, ability to hydrogen bond, and strength of the intermolecular forces between molecules of the solute.

10.49. **Collect and Organize**

From the listed pairs of substances, we are asked to identify pairs that are expected to be miscible.

Analyze

Substances with similar polarity and intermolecular forces will be miscible.

Solve

(a) **Br——Br**   Br$_2$ is nonpolar

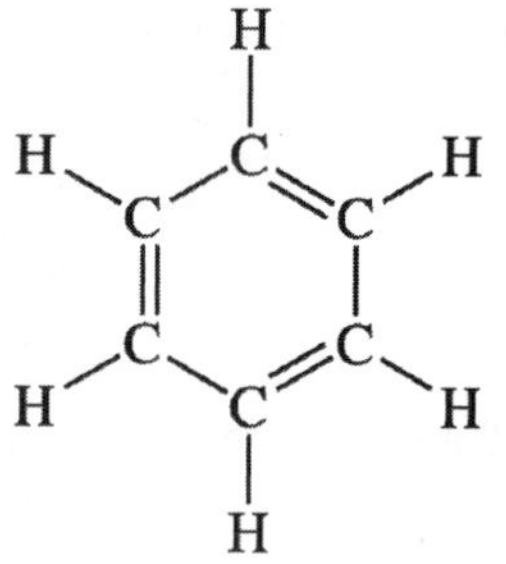

Benzene is nonpolar

(b)

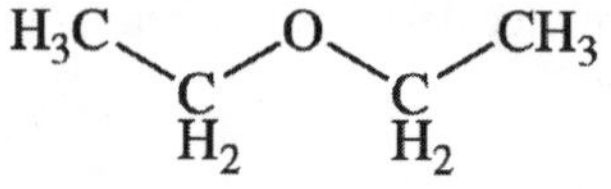

H$_3$C, C, H$_2$, O, C, H$_2$, CH$_3$ Diethyl ether is polar

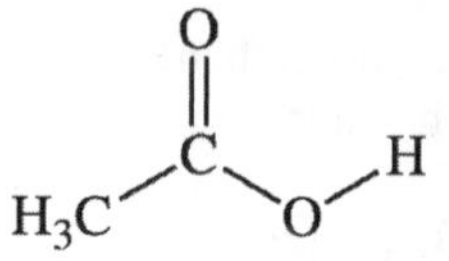

Acetic acid is polar

(c)

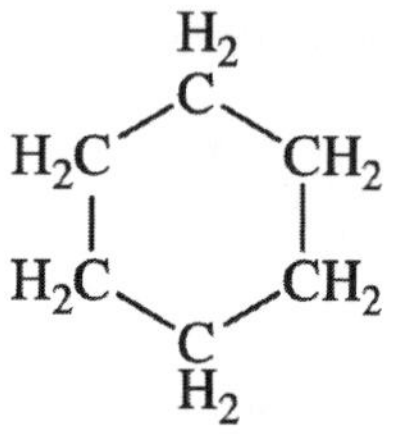

H$_2$C, C, H$_2$, CH$_2$, H$_2$C, CH$_2$, C, H$_2$ Cyclohexane is nonpolar

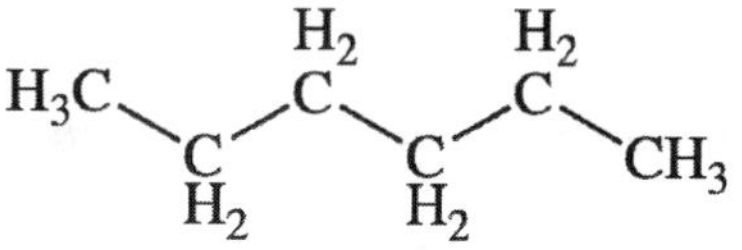

H$_3$C, C, H$_2$, C, H$_2$, CH$_3$ Hexane is nonpolar

(d) **S=C=S**   CS$_2$ is nonpolar

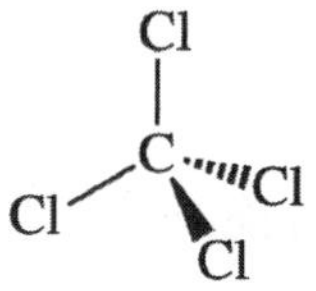

CCl$_4$ is nonpolar

All of the listed pairs are expected to be miscible. (a), (c), and (d) are all nonpolar, and thus, will only have dispersion interactions with each other. The components of (b) are both polar, and will both have dipole–dipole interactions.

Think About It
Acetic acid may also form hydrogen bonds with diethyl ether, ensuring that solute–solvent interactions will be stronger than solute–solute and solvent–solvent interaction.

10.51. Collect and Organize
Of the ionic compounds listed [NaCl, KI, Ca(OH)$_2$, and CaO], we are to determine which would be most soluble in water.

Analyze
The weaker the ionic bond, the easier the bond breaks for the cation and anion to dissolve in water. Ionic bonds are weakest for large ions of low charge.

Solve
KI (b) has the largest ions of lowest (1+ and 1–) charge, so it is the most soluble in water because it has the weakest ion–ion bond.

Think About It
CaO with a 2+ cation and 2– anion would be expected to be the least soluble in water.

10.53. Collect and Organize
Of the covalent compounds listed [CH$_3$(CH$_2$)$_2$CH$_2$OH, CH$_3$(CH$_2$)$_4$CH$_2$OH, CH$_3$(CH$_2$)$_6$CH$_2$OH, and CH$_3$(CH$_2$)$_8$CH$_2$OH] we are to predict which would be the least soluble in water.

Analyze

All of these compounds contain –OH which is capable of forming hydrogen bonds with water. These compounds differ significantly, however, in the length of their hydrocarbon chain.

Solve

Compounds with longer hydrocarbon chains are more hydrophobic and therefore are less soluble in water. $CH_3(CH_2)_8CH_2OH$ (d) is the least soluble of the compounds in water.

Think About It

The most soluble of these compounds in water is $CH_3(CH_2)_2CH_2OH$, the compound with the shortest hydrophobic hydrocarbon chain.

10.55. **Collect and Organize**

Of the two sulfur oxides, SO_2 and SO_3, we are to predict which is more soluble in nonpolar solvents.

Analyze

First we should consider the Lewis structures of SO_2 and SO_3.

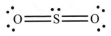 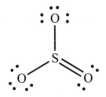

The presence of the lone pair on S gives this molecule a bent geometry.

This molecule is trigonal planar.

Solve

Because like dissolves like, a nonpolar solvent best dissolves a nonpolar molecule. From their geometries, we see that SO_2 is polar and SO_3 is nonpolar, so SO_3 dissolves best in nonpolar solvents.

Think About It

SO_2 has a permanent dipole moment and therefore would dissolve best in a polar solvent such as water.

10.57. **Collect and Organize**

We are to determine which of the listed factors affect the vapor pressure of a sample of a pure liquid.

Analyze

Section 10.5 describes factors influencing the vapor pressure of a liquid.

Solve

(a) Provided some liquid is present, the volume of liquid in the container will have no effect on the vapor pressure of the gas.
(b) When the temperature is increased, the average kinetic energy of the liquid molecules increases, more of the molecules can escape the liquid phase and enter the gas phase. More molecules in the gas phase increase the vapor pressure.
(c) Increasing the surface area of the liquid will allow more molecules on the surface of the liquid to escape the liquid and enter the gas phase. This will increase the vapor pressure.

Think About It

The type and magnitude of intermolecular forces also affect the vapor pressure of a liquid. Vapor pressure could be used to help distinguish compounds from one another.

10.59. **Collect and Organize**

We are to consider whether molecules in the vapor phase above a liquid have the same chemical formula as those in the liquid phase.

Analyze

Consider the vaporization process, and recall that this is a physical change (liquid to vapor) rather than a chemical change. Since no chemical transformation is happening, bonds are neither broken nor created.

Solve

The chemical identity of molecules in the vapor and liquid phases is identical, as bonds are neither broken nor created in the vaporization process. The molecules in both the vapor and liquid phases must have the same chemical formula.

Think About It

The energy required to vaporize a liquid is used to overcome intermolecular attractive forces, not to break bonds.

10.61. **Collect and Organize**

We are to rank the listed compounds in order of increasing vapor pressure.

Analyze

All of the compounds (CH_3CH_2OH, CH_3OCH_3, and $CH_3CH_2CH_3$) have about the same molar mass (46 g/mol or 44 g/mol). The vapor pressure differences, then, are due to differences in the strength of the intermolecular forces between the molecules of each substance.

Solve

CH_3CH_2OH is polar and may have fairly strong hydrogen bonds. CH_3OCH_3 is polar and has dipole–dipole forces between its molecules. $CH_3CH_2CH_3$ is nonpolar and has only weak dispersion forces between its molecules. The vapor pressure is expected to increase in the order (a) CH_3CH_2OH < (b) CH_3OCH_3 < (c) $CH_3CH_2CH_3$.

Think About It

This prediction is borne out by the facts that CH_3CH_2OH is a relatively high-boiling liquid, CH_3OCH_3 has a low boiling point (is volatile), and $CH_3CH_2CH_3$ is a gas at room temperature and pressure.

10.63. **Collect and Organize**

Using the Clausius–Clapeyron equation we can calculate the ΔH_{vap} of pinene given the data of vapor pressure as a function of temperature.

Analyze

We can calculate ΔH_{vap} from the relationship:

$$\ln\left(P_{vap}\right) = \frac{-\Delta H_{vap}}{R}\left(\frac{1}{T}\right) + c$$

where P_{vap} is the vapor pressure measured at temperature T (in Kelvin), R is the gas constant [8.315 J/(mol · K)], and ΔH_{vap} is the enthalpy of vaporization (in joules per mole). By plotting ln P_{vap} versus $1/T$, the slope of the line will be equal to $-\Delta H_{vap}/R$.

Solve

Plotting the data as ln P_{vap} versus $1/T$ gives a straight line with a slope of -4936.37:

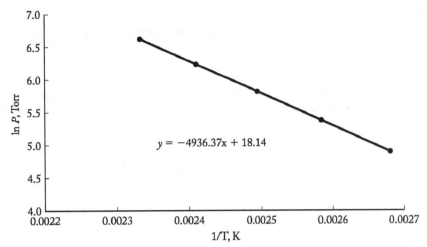

The value of ΔH_{vap} therefore is

$$\Delta H_{vap} = -\text{slope} \times R = 4936.37 \text{ K}^{-1} \times 8.314 \text{ J/mol} \cdot \text{K}$$
$$= 41041 \text{ J/mol or}$$
$$41.0 \text{ kJ/mol}$$

Think About It
We expect ΔH_{vap} to have a positive sign because the process of boiling a liquid is an endothermic process.

10.65. Collect and Organize
We are asked to differentiate between *sublimation* and *evaporation*.

Analyze
Sublimation describes the process in which a substance goes from the solid to the gas phase. Evaporation describes the process in which a substance goes from the liquid to the gas phase.

Solve
Although both processes end with the substance in the gas phase, sublimation "skips" a step in that the solid does not first liquefy before evaporating.

Think About It
A familiar substance that sublimes at room temperature and pressure is dry ice (solid CO_2).

10.67. Collect and Organize
We are asked to explain the effect of temperature and pressure on the phase of a subtance.

Analyze
We may consult a phase diagram, and consider which phase predominates as we increase the pressure (move upwards), or the temperature (move to the right).

Solve
Increasing the pressure of a system favors the phase with greater density (i.e., liquid to solid). Increasing the temperature of a system leads to greater average kinetic energy for the molecules or atoms, and thus, a greater fraction of particles capable of entering the less ordered phase (i.e., liquid to gas).

Think About It
Where the equilibrium lines meet is called the *triple point*. This is the temperature–pressure combination where all three phases (gas, liquid, and solid) are present and stable. Any change to the temperature or pressure will shift the system away from the triple point.

10.69. **Collect and Organize**

We are to predict the phase most likely to predominate at two different temperature–pressure combinations.

Analyze

At high temperatures, the atoms or molecules of a substance have high kinetic energies and can partially or fully break the intermolecular forces between them. At high pressures the atoms or molecules are in close proximity to each other and therefore are attracted to each other through intermolecular forces.

Solve

(a) At low temperatures and high pressures we would expect a solid phase to be present.
(b) At high temperatures and low pressures we would expect a gas phase to be present.

Think About It

As we decrease the pressure at low temperatures, we could melt the solid and even perhaps vaporize (sublime) it.

10.71. **Collect and Organize**

When freeze-drying food, we sublime the ice in frozen food into the gas phase. We are asked whether the pressure for this process must be below the pressure at the triple point of water.

Analyze

The triple point is where the gas–solid, solid–liquid, and liquid–gas phase boundaries meet in a phase diagram.

Solve

From the phase diagram for water we see that above the triple point, the solid phase must change to the liquid phase to enter the gaseous phase. Below the triple point, changing the temperature at a given pressure will sublime solid water into the gas phase. Yes, the pressure used for the sublimation process for freeze-drying must be below the pressure at the triple point.

Think About It

The triple point is characteristic of a particular substance. It is different for ethanol than for water.

10.73. **Collect and Organize**

By consulting Figures 10.22, 10.24, and 10.25 in the text, we are asked to estimate the normal boiling point of bromine.

Analyze

The normal boiling point is the temperature at which P_{vap} is equal to 760 mmHg.

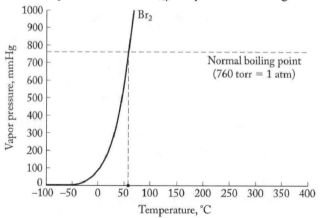

Solve

The normal boiling point of bromine is approximately 55°C to 60°C.

Think About It
Given a phase diagram for bromine, we also could have estimated the normal boiling point by drawing a horizontal line through $P = 1$ atm and finding the temperature at which this line intersects the liquid–gas boundary.

10.75. Collect and Organize
By consulting Figures 10.22, 10.24, and 10.25 in the text, we are asked to determine whether ethylene glycol or ethanol has stronger intermolecular attractive forces.

Analyze
A molecule with strong intermolecular attractive forces will require more energy to overcome the intermolecular forces and enter the vapor phase. The higher boiling species will have stronger intermolecular forces. We expect that ethylene glycol will have stronger intermolecular forces than ethanol because it has a greater number of hydrogen bonding sites.

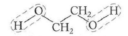

Ethanol
One hydrogen bonding site

Ethylene glycol
Two hydrogen bonding sites

Solve
Ethylene glycol has a higher boiling point, and as such, stronger intermolecular forces than ethanol.

Think About It
Our prediction, based on the number of hydrogen bonding sites, was correct.

10.77. Collect and Organize
We are to list the steps needed to convert water at room temperature and pressure (25°C, 1 atm) to its triple point.

Analyze
The triple point of water is 0.01°C and 0.006 atm.

Solve
To reach the triple point we would reduce the temperature from 25°C to 0.01°C, and reduce the pressure from 1 atm to 0.006 atm.

Think About It
At the triple point, all three phases (gas, liquid, and solid) exist in equilibrium with each other.

10.79. Collect and Organize
We consider in this question the phase changes water, initially at 5.0 atm and 100°C, undergoes when the pressure is reduced to 0.5 atm while maintaining temperature at 100°C.

Analyze
In the phase diagram for water, the phase of the water at 5.0 atm and 100°C is liquid. The phase of water at 0.5 atm and 100°C is gas.

Solve
Water at 100°C vaporizes from liquid to gas when the pressure is reduced from 5.0 atm to 0.5 atm.

Think About It
At 100°C, water boils at 1.0 atm. At pressures lower than 1.0 atm, water at 100°C is entirely in the gaseous state.

10.81. **Collect and Organize**

From the phase diagram for CO_2 (Figure 10.25) we can determine the temperature below which $CO_2(s)$ sublimes to $CO_2(g)$ simply by lowering the pressure.

Analyze

The direct solid to gaseous conversion occurs below the triple point (–57°C, 5.1 atm).

Solve

The triple point of CO_2 is at –57°C. At any temperature below the triple point, $CO_2(s)$ sublimes directly into $CO_2(g)$ by lowering the pressure.

Think About It

Because the triple point of CO_2 is at such a low temperature, we do not observe dry ice melting under ambient conditions (25°C, 1 atm).

10.83. **Collect and Organize**

We can use the phase diagram for water shown in Figure 10.24 to determine which phases of water are present at different temperature and pressure combinations.

Analyze

We use the phase diagram just like a map, locating each pressure and temperature combination and "reading" the phase at that location on the diagram. We are shown locations on the map to orient us: the normal pressure (1 atm), melting point (0°C), and boiling point (100°C) lines are indicated as well as the conditions for the triple point (0.01°C, 0.006 atm).

Solve

(a) 2 atm and 110°C: liquid
(b) 200 atm and 380°C: gas
(c) 6.0×10^{-3} atm and 0°C: gas

Think About It

At high pressures, water has a higher boiling point than at lower pressure, as seen in (a).

10.85. **Collect and Organize**

Water and methanol are both polar liquids capable of hydrogen bonding. We are asked why a needle floats on water but not on methanol.

Analyze

Surface tension is the resistance of a liquid to increase its surface area by moving the molecules of the liquid apart. The greater the intermolecular forces between the molecules in the liquid, the greater the surface tension.

Solve

A needle floats on water but not on methanol because of the high surface tension of water. This is because water can hydrogen bond through two O—H bonds with other water molecules, whereas methanol has only one O—H bond through which to form strong hydrogen bonds. The intermolecular forces between the C—H groups on the CH_3OH molecules are only weak dispersion forces.

Think About It

The high surface tension of water also allows some insects to literally walk on water.

10.87. **Collect and Organize**

We are to explain why water pipes are in danger of bursting when the temperature is below the freezing point of water.

Analyze

At temperatures below freezing, the water in the pipes freezes. Because the density of ice is less than that of liquid water, the water expands as it freezes.

Solve

The expansion of water in the pipes upon freezing may create sufficient pressure on the wall of the pipes to cause them to burst.

Think About It

To prevent pipes from freezing during the winter months, we must drain the water from the portion of the pipe that is exposed to freezing temperatures.

10.89. **Collect and Organize**

We are to explain why the meniscus of liquid mercury is convex, rather than concave as it is for most liquids.

Analyze

The shape of the meniscus is due to the competing adhesive forces (liquid to glass surface that has Si—O—H bonds) and cohesive forces (liquid to liquid).

Solve

The cohesive forces within mercury are stronger than the adhesive forces of the mercury to the glass. This results in a convex meniscus.

Think About It

The strong metallic bonding between mercury atoms is not balanced out by the adhesive Hg to Si—O—H bonds.

10.91. **Collect and Organize**

We are to describe the origin of surface tension from a molecular viewpoint.

Analyze

Surface tension is the resistance of a liquid to increase its surface area by moving the molecules of the liquid apart. The greater the intermolecular forces between the molecules in the liquid, the greater the surface tension.

Solve

Molecules in the bulk liquid are "pulled" by all the other liquid molecules surrounding them and they are, therefore, "suspended" in the bulk liquid. Molecules on the surface of a liquid, however, are only pulled by molecules under and beside them, creating a tight film of molecules on the surface that we call surface tension.

Think About It

When the surface tension is greater than the force of gravity on a small object placed on top of a liquid, the object floats.

10.93. **Collect and Organize**

As temperature increases, the viscosity and surface tension change. We are to describe and explain these changes.

Analyze

Both surface tension and viscosity are a result of the intermolecular forces between molecules. As the temperature increases, more molecules have sufficient energy to break (at least temporarily) the intermolecular attractions between themselves and their neighboring molecules.

Solve

As temperature increases, the surface tension decreases because the molecular "film" on the surface of the liquid has fewer molecules held together by tight intermolecular forces. Likewise, the viscosity decreases as the temperature increases because molecules have more energy to readily break the intermolecular forces to enable them to slide past each other more freely.

Think About It

The greater the intermolecular forces, the greater the heat (energy) needed to decrease the viscosity of a liquid.

10.95. **Collect and Organize**

We are to determine which liquid, ethanol or water, rises higher into a capillary tube.

Analyze

A liquid rises in a capillary tube until the force of gravity balances the cohesive forces between the molecules of the liquid and the adhesive forces between the liquid and the inside surface of the capillary tube.

Solve

Water is able to form more hydrogen bonds with the glass surface (composed of Si—O—H bonds) than ethanol, so water rises higher in the capillary tube.

Think About It

Another way to express this answer is that water has a greater adhesion to glass than ethanol.

10.97. **Collect and Organize**

Given the boiling points of two liquids, we are to determine which has the higher surface tension and viscosity.

Analyze

High boiling points, high surface tensions, and high viscosities of liquids are all directly related to strong intermolecular forces.

Solve

The liquid with the higher boiling point (B) is expected to have the higher surface tension and higher viscosity because that liquid has a greater magnitude of intermolecular forces between its molecules.

Think About It

Remember that many weak intermolecular forces such as London (dispersion) forces between long and large molecules may add up to very strong attractions between molecules. Therefore, long-chain polymers, such as oils, although nonpolar, have high viscosities and surface tensions.

10.99. **Collect and Organize**

We are to determine whether the sublimation point of ice increases or decreases with increasing pressure and explain why.

Analyze

By consulting Figure 10.24, we see that the equilibrium line between the solid and vapor phases of water increases upwards and to the right of the phase diagram.

Solve

As the pressure increases, the temperature at which sublimation occurs will also increase, until the ice reaches the triple point (0.0060 atm and 0.010 °C)

Think About It

At pressures greater than 0.0060 atm, solid ice will not sublime, but will instead melt, then evaporate as the temperature increases.

10.101. **Collect and Organize**

We are to explain why methanol boils at a lower temperature than water (64.7°C versus 100°C) even though methanol has a greater molar mass (32.04 g/mol versus 18.02 g/mol).

Analyze

Both the methanol and water are held together in the condensed phases (liquid and solid) by weak van der Waals forces and hydrogen bonds.

Solve

Although the dispersion forces between methanol molecules are greater than those between water molecules, because methanol has more electrons and greater molar mass, water can form two hydrogen bonds compared to methanol's one hydrogen bond. This greater number of stronger interactions between water molecules raises the boiling point of water above that of methanol.

Think About It

As we add carbons to the alcohol chain for the series R—OH, we see an increase in the boiling point due to increases in the van der Waals forces between the molecules.

$$CH_3OH < CH_3CH_2OH < CH_3CH_2CH_2OH < CH_3CH_2CH_2CH_2OH$$
bp 64.7°C   bp 78.4°C   bp 97.2°C   bp 117.7°C

10.103. **Collect and Organize**

Using a phase diagram for water, we can determine whether the sublimation point of ice increases or decreases as the pressure is increased.

Analyze

In the phase diagram, the solid–vapor phase boundary slopes up to the right.

Solve

From the slope of the solid–vapor phase boundary, we see that as pressure increases, the temperature at which ice sublimes increases.

Think About It

With increased pressure, fewer water molecules enter the gas phase from the solid phase. This is also true for the liquid–vapor transition: as pressure increases, the boiling point increases.

10.105. **Collect and Organize**

Given that the melting point of hydrogen is at a higher temperature than its triple point, we are to determine whether H_2 expands or contracts upon freezing.

Analyze

If the triple point is at a lower temperature than the melting point, then the solid–liquid phase boundary must slope up and to the right in the phase diagram. This positive slope means that the solid phase has a higher density than the liquid phase.

Solve

Hydrogen contracts as it freezes because the phase diagram tells us that the solid phase is denser than the liquid phase.

Think About It

At very high pressures, solid hydrogen forms in which the H—H bond of the diatomic molecule no longer exists and the solid hydrogen behaves like a metal!

10.107. **Collect and Organize**

We are to explain why ice floats on water.

Analyze

Water molecules in both the liquid and solid phases are held together by strong hydrogen bonds. In the liquid phase, however, the hydrogen bonds are "fluid" and do not hold the water molecules in exact positions relative to each other.

Solve

Water in the solid (ice) form has a lower density than in the liquid form due to the open lattice formed by extensive hydrogen bonding between the water molecules, so ice floats on liquid water.

Think About It

Water is very different from other liquids. Most other liquids are less dense than their solids, so their "ices" sink rather than float.

10.109. **Collect and Organize**

From among the four molecules shown we are to predict which one would be soluble in both water and octanol.

Analyze

In order for a substance to dissolve well in both water and octanol it should have both hydrophilic and hydrophobic groups. Hydrophilic groups are groups that might form hydrogen bonds with water or have strong bond dipoles. Hydrophobic groups are groups that are nonpolar.

Solve

(a) This molecule has few polar groups and is dominated by its nonpolar carbon–hydrogen regions.
(b) This molecule has a balance of polar (–COOH and –NH$_2$) groups with nonpolar carbon–hydrogen bonds.
(c) This molecule is nonpolar overall, as it is very symmetrical despite its polar C–F bonds.
(d) This molecule is dominated by polar OH groups with fewer nonpolar regions.
Because (b) has a balance of hydrophilic (polar) and hydrophobic (nonpolar) groups, we expect this molecule to have similar solubility in water and in octanol.

Think About It

Molecule (c) is similar to Teflon, which is used to coat cooking utensils to prevent foods from sticking.

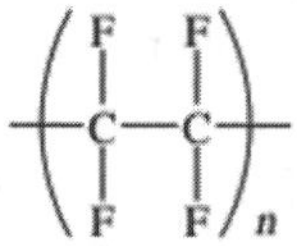

10.111. **Collect and Organize**

We are asked to describe the intermolecular forces displayed by ammonia, sulfur dioxide, and hexafluoroethane, respectively.

Analyze

The intermolecular forces depend on the polarity of the molecule. We may determine polarity by drawing a valid Lewis structure for each molecule, then determining the VSEPR shape and polarity for each.

Solve

Ammonia has $3 + 5 = 8$ valence electrons, and the Lewis structure below. The VSEPR description and structure for ammonia is

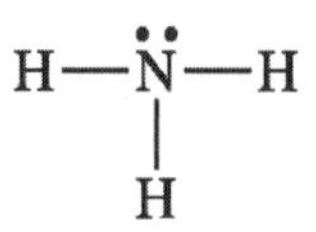

SN = 4
Electron-pair geometry = tetrahedral
One lone pair
Molecular geometry = trigonal pyramidal

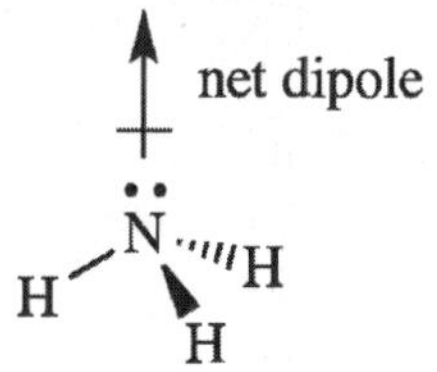

Because NH_3 has a net dipole and N–H bonds, it will form hydrogen bonds and have dipole–dipole and London dispersion interactions.

Sulfur dioxide has $6 + 12 = 18$ valence electrons, and the Lewis structure below. The VSEPR description and structure for sulfur dioxide is

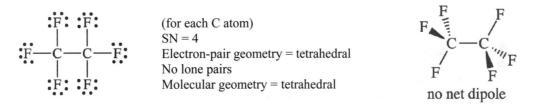

SN = 3
Electron-pair geometry = trigonal planar
One lone pair
Molecular geometry = bent (120°)

net dipole

Because SO_2 has a net dipole but no N–H, O–H, or F–H bonds, it will have dipole–dipole and London dispersion interactions.

Hexafluoroethane has $8 + 42 = 50$-valence electrons, and the Lewis structure below. The VSEPR description and structure for hexafluoroethane is

(for each C atom)
SN = 4
Electron-pair geometry = tetrahedral
No lone pairs
Molecular geometry = tetrahedral

no net dipole

Because C_2F_6 has no net dipole, it will have only London dispersion interactions.

Think About It
All three refrigerants have different intermolecular forces, based on having very different polarity and hydrogen bonding sites.

10.113. **Collect and Organize**
We are to explain why Br_2 and I_2 are so volatile at room temperature and 1 atm pressure.

Analyze
The melting points of Br_2 and I_2 are –7°C and 114°C, respectively.

Solve
Both of these diatomic elements are volatile because only weak van der Waals (dispersion) forces attract one X_2 molecule to another.

Think About It
The dispersion forces in I_2 are stronger than in Br_2, making I_2 a solid and Br_2 a liquid at room temperature and pressure. I_2 has more electrons that may be instantaneously polarized. However, I_2 is still quite volatile and has a high enough vapor pressure under normal conditions that you can see purple fumes in a closed jar of $I_2(s)$.

10.115. **Collect and Organize**
We are to name the least abundant halogen.

Analyze
We can do a quick Internet search to find the abundances of the halogens as well as use the information in the section, The Halogens: The Salt of the Earth.

Solve
Astatine is the least abundant of the halogens. Indeed, it may be the least abundant of the naturally occurring terrestrial elements. It is estimated that there is less than 28 g in Earth's crust at any given time!

Think About It
Astatine is highly radioactive with a short half-life and is produced from the decay of uranium.

10.117. Collect and Organize
We are to write the formula and draw the chemical structure for chloramine to determine its molecular shape and decide whether it is soluble in water.

Analyze
We can look up the formula for chloramine either in the textbook or using a Web search engine.

Solve
Chloramine (also called monochloramine) has the molecular formula NH_2Cl. It has a trigonal pyramidal structure:

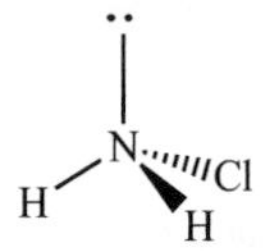

The NH_2Cl molecule is polar and it has the ability to form hydrogen bonds with water due to the presence of the N—H bonds, so it is soluble in water.

Think About It
Hydrogen bonding of chloramine to water might look like this:

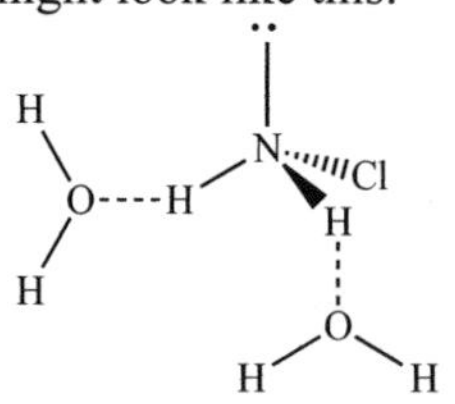

10.119. Collect and Organize
For each of the halogen species named we are to provide a formula.

Analyze
These species all are named in the textbook in the section, The Halogens: The Salt of the Earth, and in Table 2.4.

Solve
(a) hypochlorite = ClO^-
(b) chlorite = ClO_2^-
(c) chlorate = ClO_3^-
(d) perchlorate = ClO_4^-

Think About It
These all have corresponding acids: hypochlorous acid ($HClO$), chlorous acid ($HClO_2$), chloric acid ($HClO_3$), and perchloric acid ($HClO_4$).

10.121. Collect and Organize
We are to write and balance the redox reaction that occurs in basic solution between the iodate and hydrogen sulfite ions.

Analyze
The reactants are IO_3^- and HSO_3^-. The products are I^- and SO_4^{2-}. We can use the half-reaction method to balance the reaction.

Solve

First, balance the half-reactions in acidic solution

$$6\,e^- + 6\,H^+ + IO_3^- \rightarrow I^- + 3\,H_2O$$
$$\underline{(H_2O + HSO_3^- \rightarrow SO_4^{2-} + 3\,H^+ + 2e^-) \times 3}$$
$$6\,H^+ + IO_3^- + 3\,H_2O + 3\,HSO_3^- \rightarrow I^- + 3\,H_2O + 3\,SO_4^{2-} + 9\,H^+$$

This simplifies in acid solution to

$$IO_3^- + 3\,HSO_3^- \rightarrow I^- + 3\,SO_4^{2-} + 3\,H^+$$

Placing this in basic solution by adding 3 OH^- to both sides gives

$$3\,OH^-(aq) + IO_3^-(aq) + 3\,HSO_3^-(aq) \rightarrow I^-(aq) + 3\,SO_4^{2-}(aq) + 3\,H_2O(\ell)$$

Think About It

In this reaction IO_3^- is the oxidizing agent, and HSO_3^- is the reducing agent.

CHAPTER 11 | Solutions: Properties and Behavior

11.1. Collect and Organize

In considering the diagram for this problem we are to choose the one that best describes how the solubility of a gas is affected by pressure.

Analyze

In the middle of the diagram, the gas and the liquid contained in a cylinder are shown before the pressure is increased by moving the piston down in the cylinder. In the middle diagram there are 12 molecules in the gas phase and 12 molecules in the liquid phase. As the pressure is increased, we expect some of the molecules in the gas phase to become dissolved into the liquid phase to reduce the pressure of the gas in the cylinder.

Solve

In diagram a there are more molecules in the gas phase (14) than there were before the change in pressure and more than there are in the liquid phase (10), indicating that the increased pressure caused molecules to evaporate from the solution. This is not correct. In diagram b there are fewer molecules in the gas phase (8) and more molecules in the liquid phase (16). This is a reasonable expectation when the pressure of the system is increased, so choice (b) is correct.

Think About It

Cylinder (a) would mean the pressure in the cylinder increases as the volume decreases, counter to our expectations from Boyle's law.

11.3. Collect and Organize

Figure P11.3 depicts a piston before (a) and immediately after (b) the pressure of the cylinder is increased at a constant temperature and quantity of gas. We are asked to determine which cylinder in (c) best depicts the piston at dynamic equilibrium following the increase in pressure.

Analyze

Increasing the pressure in the cylinder will cause a decrease in the volume of the gas. As the pressure of the gas phase increases, we expect particles of gas to leave the gas phase and dissolve in the liquid, resulting in a slightly lower pressure in the gas phase at dynamic equilibrium than immediately following the change in volume. This is an example of Le Châtelier's principle, by which a system at equilibrium, when disturbed, will react to restore equilibrium. We expect a decrease in the number of particles in the gas phase and an increase in the number in the liquid phase.

Solve

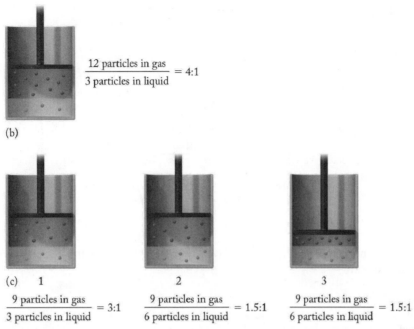

12 particles in gas / 3 particles in liquid = 4:1

(b)

(c) 1

9 particles in gas / 3 particles in liquid = 3:1

2

9 particles in gas / 6 particles in liquid = 1.5:1

3

9 particles in gas / 6 particles in liquid = 1.5:1

Although (1) has a decreased number of particles in the gas phase relative to (b), no additional gas particles are present in the liquid phase, implying a loss of particles from the sealed system. In (3), the number of particles in each phase is what we would expect, but the volume of the cylinder is lower than in (b), which is the opposite of what would be expected based on Le Châtelier's principle. (2) is the best description of the expected behavior.

Think About It

A more in depth discussion of equilibrium may be found in Chapter 15.

11.5. **Collect and Organize**

We are asked to determine the number of components in the sample described by the graph in Figure P11.5 based on the observed temperature profile from distillation. We are also asked to determine the boiling points of all components in the sample.

Analyze

The distillation depicted in Figure P11.5 is relatively simple, with all components boiling at a consistent temperature, then changing temperature as the next component boils. Each plateau should correspond to a different component of the mixture.

Solve

The sample is a mixture of three substances. The first (A) boils at approximately 80°C and has a volume of 20 mL. The second (B) boils at approximately 95°C and has a volume of 50 mL. The third (C) boils at approximately 130°C and has a volume of 30 mL. The approximate ratio of volumes is 2:5:3.

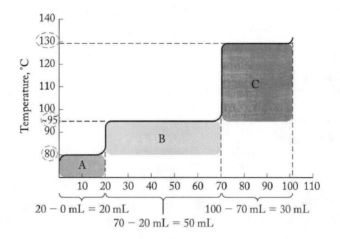

20 − 0 mL = 20 mL   100 − 70 mL = 30 mL

70 − 20 mL = 50 mL

Think About It
The ratio of volumes assumes we can obtain a perfect separation of fractions, with no carryover of one component to the next.

11.7. Collect and Organize
From the direction of solvent flow indicated in Figure P11.7, we can deduce which solution (A or B) is more concentrated.

Analyze
In osmosis, solvent passes through the semipermeable membrane from a low concentration of solute to a higher concentration of solute.

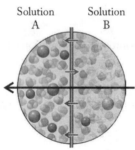

Solution A   Solution B

Solve
The flow of solvent in Figure P11.7 is from solution B to solution A. Therefore, solution A must be the more concentrated solution.

Think About It
The solvent continues to flow through the membrane until the solute concentration on both sides is equal.

11.9. Collect and Organize
We are to define the term *nonvolatile solute*.

Analyze
A substance that is volatile readily enters the gas phase.

Solve
A nonvolatile solute is a compound that dissolves into a solvent and does not enter appreciably into the gas phase, under conditions that maintain the solution.

Think About It

There can be covalent (such as sugar) as well as ionic (such as NaCl) nonvolatile solutes in aqueous solutions.

11.11. Collect and Organize

We are to explain why the vapor pressure of a liquid increases with increasing temperature.

Analyze

As energy is applied to increase the temperature, the average kinetic energy of the molecules in a liquid increases.

Solve

When the average kinetic energy of the liquid molecules increases, more of the molecules can escape the liquid phase and enter the gas phase. An increase in the number of molecules in the gas phase causes an increase in the vapor pressure.

Think About It

Warm apple pie has a stronger aroma than cold apple pie.

11.13. Collect and Organize

We have two beakers contained in a closed vessel; one contains pure ethanol to the brim and the other is half-full of a solution of sugar in ethanol. We are to explain why the beaker containing sugar in ethanol solution will overflow eventually.

Analyze

The vapor pressure of pure ethanol is greater than the vapor pressure of the solution, which contains dissolved solutes. The greater the vapor pressure, the higher the rate of evaporation. The lower the vapor pressure, the greater the rate of condensation.

Solve

The vapor pressure of pure ethanol is greater than the vapor pressure of the sugar in ethanol solution. The rate of evaporation will be higher for the pure ethanol. On the other hand, the rate of condensation in the sugar solution beaker is higher than the rate of evaporation, so eventually the sugar solution beaker will overflow, as the volume of the solution increases.

Think About It

Because ethanol evaporates from the beaker with pure ethanol and condenses into the beaker of the ethanol–sugar solution, the pure ethanol beaker will eventually become empty.

11.15. Collect and Organize

Given the quantity of water and glucose in a solution (3.5 mol and 1.5 mol, respectively) and the vapor pressure of water at 25°C (23.8 torr), we are to calculate the mole fraction of water in the solution and the vapor pressure of the solution.

Analyze

The mole fraction is defined as the number of moles of a substance divided by the total number of moles of all substances in a mixture:

$$\chi = \frac{n_x}{n_{total}}$$

The vapor pressure of a solution is given by Raoult's law:

$$P_{solution} = \chi_{solvent} P_{solvent}$$

Solve

Mole fraction of water in solution:

$$\chi = \frac{3.5 \text{ mol}}{3.5 \text{ mol} + 1.5 \text{ mol}} = 0.70$$

Vapor pressure of solution at 25°C:

$$P_{\text{solution}} = 0.70 \times 23.8 \text{ torr} = 17 \text{ torr}$$

Think About It

Vapor pressure depends on temperature. The higher the temperature, the higher the vapor pressure. Therefore, the vapor pressure is always described for a specific temperature.

11.17. Collect and Organize

We are asked to use Equation 11.1 to prove that the mole fraction of the solute is equal to $\left(P^{\circ}_{\text{solvent}} - P_{\text{soln}}\right)\big/P^{\circ}_{\text{solvent}}$.

Analyze

We will require Equation 11.1

$$P_{\text{soln}} = \chi_{\text{solvent}} \cdot P^{\circ}_{\text{solvent}}$$

as well as the following relationship for a two component solution

$$\chi_{\text{solvent}} + \chi_{\text{solute}} = 1$$

By rearranging these equations and substituting from one into the other, we arrive at the desired relationship.

Solve

Rearranging for χ_{solvent} and substituting into Equation 11.1

$$\chi_{\text{solvent}} = 1 - \chi_{\text{solute}}$$

$$P_{\text{soln}} = \left(1 - \chi_{\text{solute}}\right) \cdot P^{\circ}_{\text{solvent}}$$

Distributing the mole fraction term, we may start to isolate the mole fraction

$$P_{\text{soln}} = P^{\circ}_{\text{solvent}} - P^{\circ}_{\text{solvent}} \cdot \left(\chi_{\text{solute}}\right)$$

$$P^{\circ}_{\text{solvent}} \cdot \left(\chi_{\text{solute}}\right) = P^{\circ}_{\text{solvent}} - P_{\text{soln}}$$

Dividing by $P^{\circ}_{\text{solvent}}$ we arrive at the desired relationship.

$$\chi_{\text{solute}} = \frac{\left(P^{\circ}_{\text{solvent}} - P_{\text{soln}}\right)}{P^{\circ}_{\text{solvent}}}$$

Think About It

Since $P^{\circ}_{\text{solvent}}\big/P^{\circ}_{\text{solvent}} = 1$, our relationship may be reduced to

$$\chi_{\text{solute}} = 1 - \frac{P_{\text{soln}}}{P^{\circ}_{\text{solvent}}}$$

11.19. Collect and Organize

We can use our knowledge of the behavior of gases at low temperatures to explain why the solubility of a gas increases with decreasing temperatures.

Analyze

At low temperatures, gas molecules have low kinetic energies.

Solve

At low temperatures, the solubility of a gas increases because fewer gas molecules dissolved in the solvent have sufficient (kinetic) energy to overcome the intermolecular forces between the solvent molecules. At low temperatures the gas molecules are "trapped" in the solvent and cannot "escape" into the gas phase.

Think About It
Soda stored uncapped in the refrigerator takes longer to go flat than soda left out at room temperature.

11.21. Collect and Organize
We can look closely at Henry's law to determine if the solubility constant of air in water is the sum of $k_{\text{H,N}_2}$ and $k_{\text{H,O}_2}$.

Analyze
Henry's law is $C_{\text{air}} = k_{\text{H,air}}P_{\text{air}}$. The solubilities of O_2 and N_2 in the solvent would be expressed as
$$C_{\text{O}_2} = k_{\text{H,O}_2}P_{\text{O}_2}$$
$$C_{\text{N}_2} = k_{\text{H,N}_2}P_{\text{N}_2}$$
By the law of partial pressures (and ignoring the minor gases in the atmosphere),
$$P_{\text{O}_2} + P_{\text{N}_2} = P_{\text{air}}$$
Also, the concentration of air in the water is the sum of the concentrations of the oxygen and nitrogen:
$$C_{\text{O}_2} + C_{\text{N}_2} = C_{\text{air}}$$
Because the atmosphere is composed of approximately 78% N_2 and 22% O_2,
$$P_{\text{O}_2} = 0.22P_{\text{air}}$$
$$P_{\text{N}_2} = 0.78P_{\text{air}}$$

Solve
$$C_{\text{air}} = k_{\text{H,air}}P_{\text{air}} = C_{\text{O}_2} + C_{\text{N}_2}$$
$$k_{\text{H,air}}P_{\text{air}} = k_{\text{H,O}_2}P_{\text{O}_2} + k_{\text{H,N}_2}P_{\text{N}_2}$$
$$= k_{\text{H,O}_2}\left(0.22P_{\text{air}}\right) + k_{\text{H,N}_2}\left(0.78P_{\text{air}}\right)$$
$$= 0.22k_{\text{H,O}_2}P_{\text{air}} + 0.78k_{\text{H,N}_2}P_{\text{air}}$$
$$= \left(0.22k_{\text{H,O}_2} + 0.78k_{\text{H,N}_2}\right)P_{\text{air}}$$
$$k_{\text{H,air}} = 0.22k_{\text{H,O}_2} + 0.78k_{\text{H,N}_2}$$

This derivation shows that the solubility constant for air is not a simple sum of the solubility constants of the components of air.

Think About It
The term $k_{\text{H,air}}$ is the weighted average of the k_{H} values of the components of the air.

11.23. Collect and Organize
Given that the mole fraction of O_2 in air is 0.209 and that arterial blood has 0.25 g of O_2 per liter at 37°C and 1 atm, we can calculate k_{H} for O_2 in blood using Henry's law.

Analyze
Henry's law is defined as
$$C_{\text{gas}} = k_{\text{H}}P_{\text{gas}}$$
Where C_{gas} is the concentration of dissolved gas, k_{H} is Henry's law constant, and P_{gas} is the pressure of the gas. Rearranging the equation to solve for k_{H} gives
$$k_{\text{H}} = \frac{C_{\text{gas}}}{P_{\text{gas}}}$$
The units of k_{H} are usually expressed as moles per liter per atmosphere [mol/(L · atm)].
The concentration of O_2 in blood in moles per liter (mol/L) is
$$\frac{0.25 \text{ g O}_2}{\text{L}} \times \frac{1 \text{ mol}}{32.00 \text{ g}} = 7.81 \times 10^{-3} \text{ mol/L}$$

The partial pressure of O_2 in the air is

$$0.209 \times 1 \text{ atm} = 0.209 \text{ atm}$$

Solve

$$k_H = \frac{7.81 \times 10^{-3} \text{ mol/L}}{0.209 \text{ atm}} = 3.7 \times 10^{-2} \text{ mol/L} \cdot \text{atm}$$

Think About It

As the temperature changes, so too does the k_H. As temperature increases, less O_2 is soluble and the value of k_H decreases.

11.25. Collect and Organize

Using the k_H value of 3.7×10^{-2} mol/(L $\cdot$ atm) from Problem 11.23, we are to calculate the solubility of O_2 in the blood of a climber on Mt. Everest ($P_{atm} = 0.35$ atm) and a scuba diver ($P_{atm} = 3$ atm).

Analyze

The concentration (solubility) of O_2 in the blood can be calculated using Henry's law:

$$C_{O_2} = k_H P_{O_2}$$

For each case we need the pressure of O_2 in the atmosphere. Assuming that the mole fraction of O_2 stays at 0.209, the partial pressure of O_2 for each individual is as follows:

$$\text{For the alpine climber, } P_{O_2} = 0.209 \times 0.35 \text{ atm} = 7.32 \times 10^{-2} \text{ atm}$$

$$\text{For the scuba diver, } P_{O_2} = 0.209 \times 3 \text{ atm} = 0.627 \text{ atm}$$

Solve

(a) For the alpine climber

$$C_{O_2} = \frac{3.7 \times 10^{-2} \text{ mol}}{\text{L} \cdot \text{atm}} \times 7.32 \times 10^{-2} \text{ atm} = 2.7 \times 10^{-3} \text{ M}$$

(b) For the scuba diver

$$C_{O_2} = \frac{3.7 \times 10^{-2} \text{ mol}}{\text{L} \cdot \text{atm}} \times 0.627 \text{ atm} = 2.3 \times 10^{-2} \text{ M}$$

Think About It

The scuba diver has more than eight times the concentration of O_2 in her arterial blood than the alpine climber.

11.27. Collect and Organize

From the graph in Figure P11.27, we can determine the volume of O_2 soluble in a liter of water at different temperatures. From this we are to calculate Henry's law constant for O_2 in water at each temperature.

Analyze

Henry's law constant may be calculated using

$$k_H = \frac{C_{gas}}{P_{gas}}$$

The units of k_H are usually expressed as moles per liter per atmosphere [mol/(L $\cdot$ atm)], so we need to convert the volume of O_2 expressed in L to moles of O_2 dissolved in the water using the ideal gas equation:

$$n = \frac{PV}{RT}$$

where n = moles of dissolved O_2 in the water, P is assumed to be 1.00 atm, V = volume of oxygen in the water in liters, R = gas constant, and T = temperature in kelvins.

Solve

At 10°C (283 K), 0.038 L of O_2 is soluble in 1.00 L of water.

$$\text{Moles } O_2 \text{ in } 1.00 \text{ L } H_2O = \frac{(1.00 \text{ atm} \times 0.038 \text{ L})}{\left(0.0821 \dfrac{\text{L} \cdot \text{atm}}{\text{mol} \cdot \text{K}} \times 283 \text{ K}\right)}$$

$$= 1.64 \times 10^{-3} \text{ mol}$$

$$k_H = \frac{1.64 \times 10^{-3} \text{ mol/L}}{1.00 \text{ atm}} = 1.6 \times 10^{-3} \text{ mol/L} \cdot \text{atm}$$

At 20°C (293 K), 0.031 L of O_2 is soluble in 1.00 L of water.

$$\text{Moles } O_2 \text{ in } 1.00 \text{ L } H_2O = \frac{(1.00 \text{ atm} \times 0.031 \text{ L})}{\left(0.0821 \dfrac{\text{L} \cdot \text{atm}}{\text{mol} \cdot \text{K}} \times 293 \text{ K}\right)}$$

$$= 1.29 \times 10^{-3} \text{ mol}$$

$$k_H = \frac{1.29 \times 10^{-3} \text{ mol/L}}{1.00 \text{ atm}} = 1.3 \times 10^{-3} \text{ mol/L} \cdot \text{atm}$$

At 30°C (303 K), 0.026 L of O_2 is soluble in 1.00 L of water.

$$\text{Moles } O_2 \text{ in } 1.00 \text{ L } H_2O = \frac{(1.00 \text{ atm} \times 0.026 \text{ L})}{\left(0.0821 \dfrac{\text{L} \cdot \text{atm}}{\text{mol} \cdot \text{K}} \times 303 \text{ K}\right)}$$

$$= 1.05 \times 10^{-3} \text{ mol}$$

$$k_H = \frac{1.05 \times 10^{-3} \text{ mol/L}}{1.00 \text{ atm}} = 1.1 \times 10^{-3} \text{ mol/L} \cdot \text{atm}$$

Think About It

As temperature of a solution increases, the solubility of a gas decreases and the value of k_H decreases.

11.29. Collect and Organize

For the enthalpy of hydration we are asked why there is both an ion–dipole term and a dipole–dipole term.

Analyze

The enthalpy of hydration describes the change in enthalpy when a gas-phase ion is dissolved in water.

Solve

When a gaseous ion, M^+ or X^-, dissolves in water, new ion–dipole interactions are formed between the water molecules and the ions. In order for these interactions to form, some water–water (dipole–dipole) interactions must be broken. Both of these terms combine to give the enthalpy of hydration.

Think About It

The formation of ion–dipole interactions in the hydration of an ionic compound is exothermic, but the breaking of the dipole–dipole interactions between water molecules is endothermic.

11.31. Collect and Organize

We are asked to justify the observations made in Chapter 4 that all metal nitrates are soluble, whereas nearly all metal sulfides are insoluble.

Analyze

Water solubility depends on the strength of water–solute interactions relative to water–water and solute–solute interactions.

Solve

The sulfide anion (S^{2-}) has a greater charge density than the nitrate anion (NO_3^-). Sulfides will have a stronger ion–ion interaction than nitrates, requiring a greater offset in energy from the water–ion interactions when dissolved. If the ion–ion interactions are stronger than any offsetting water–ion interactions (as they presumably are for S^{2-}), the ionic compound will not dissolve.

Think About It

This relationship is true for other polar solvents, not just water. Since solubility depends on the strength of solvent–solute interactions, compounds that are insoluble or sparingly soluble in water may be more soluble in a different solvent.

11.33. Collect and Organize

We are asked to predict how the melting point relates to the atomic number for X in NaX (X = F, Cl, Br, I).

Analyze

Melting point increases for NaX as the ionic attraction between Na^+ and X^- increases. The differences among the ion–ion attractions for the series NaX depend on the size of X^-. As seen from Coulomb's law

$$E \propto \frac{Q_1 Q_2}{d}$$

as d increases, E decreases. Therefore, as the size of X increases, d increases and E decreases.

Solve

The anions in the NaX series increase in size as $F^- < Cl^- < Br^- < I^-$. Melting point, therefore, decreases down this series and thus decreases as the atomic number of X increases.

Think About It

The actual melting points of these compounds are:

NaF	993°C
NaCl	801°C
NaBr	755°C
NaI	660°C

This confirms our prediction.

11.35. Collect and Organize

From the list of four magnesium compounds, we are to select the one with the least negative (weakest) lattice energy.

Analyze

In each of the compounds, magnesium has a 2+ charge. The anions differ not in charge, but in size: $F^- < Cl^- < Br^- < I^-$. The lattice energy will be strongest for a small anion and weakest for a large anion, according to the increasing distance between the cation and anion (d).

$$U = \frac{k(Q_1 Q_2)}{d}$$

Solve

MgI_2 (a) will have the least negative lattice energy.

Think About It

Charge (Q_1 and Q_2) also has an effect on the lattice energy. The greater the charge, the more negative the lattice energy and the higher the strength of the ion–ion bond.

11.37. Collect and Organize

We are to construct a Born–Haber cycle to calculate the lattice energy of KCl from the data given.

Analyze

We can use the Born–Haber cycle shown for NaCl in Figure 11.6 as a guide. We have to be sure to account for the need of just one mole of Cl⁻ ions from diatomic Cl_2 within the cycle.

Solve

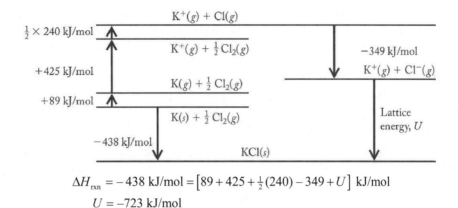

$$\Delta H_{rxn} = -438 \text{ kJ/mol} = \left[89 + 425 + \tfrac{1}{2}(240) - 349 + U \right] \text{ kJ/mol}$$

$$U = -723 \text{ kJ/mol}$$

Think About It

Lattice energies for ionic compounds are usually quite large and exothermic. This is due to the strong attraction between the anion and cation.

11.39. **Collect and Organize**

We can calculate the lattice energy of KBr from the data provided in Table 11.4 and the enthalpy of formation given as 19.9 kJ/mol.

Analyze

We may calculate the lattice energy (U_{KBr}) by rearranging Equation 11.7:

$$U_{KBr} = \Delta H_{\text{hydration,KBr}(aq)} - \Delta H_{\text{solution,KBr}(aq)}$$

where $\Delta H_{\text{hydration,KBr}(aq)}$ is determined by substituting values from Table 11.4, as in Sample Exercise 11.5.

$$\Delta H_{\text{hydration,KBr}(aq)} = \Delta H_{\text{hydration,K}^+(g)} + \Delta H_{\text{hydration,Br}^-(g)}$$

By consulting the entries for similar compounds in Table 11.2, we may estimate that the lattice energy will be somewhere between –500 and –1000 kJ/mol.

Solve

$$\Delta H_{\text{hydration,KBr}(aq)} = \left(-355 \text{ kJ}\right) + \left(-355 \text{ kJ}\right) = -670. \text{ kJ/mol}$$

$$U_{KBr} = -670. \text{ kJ/mol} - 19.9 \text{ kJ/mol} = -690. \text{ kJ/mol}$$

Think About It

The lattice energy of KBr is within the expected range for a 1+/1– ion.

11.41. **Collect and Organize**

We are to identify the physical property we can use to separate crude oil into its various components.

Analyze

Crude oil consists mainly of a mixture of hydrocarbons from C_1 to greater than C_{36}.

Solve

The components of crude oil can be separated by fractional distillation, which uses differences in boiling points of the compounds.

Think About It

In the fractional distillation process, components in crude oil of relatively low molar mass distill first. The residue left after distillation is used in asphalts.

11.43. Collect and Organize

We can consider the relative volatilities (boiling points) of C_5H_{12} and C_7H_{16} to determine which substance is present in higher concentration in the vapor phase above an equimolar mixture of the two components.

Analyze

Boiling point is an indication of volatility and thus vapor pressure of a substance. Because C_5H_{12} has fewer atoms in its structure, it has a lower boiling point than C_7H_{16}. A lower boiling point means that C_5H_{12} is more volatile.

Solve

Because C_5H_{12} is more volatile, it is present in the higher concentration in the vapor phase above a mixture of C_5H_{12} and C_7H_{16}.

Think About It

In a fractional distillation of these two compounds, C_5H_{12} is separated first from the mixture.

11.45. Collect and Organize

To calculate the vapor pressure of a mixture of 25 g methanol and 75 g ethanol, we can use the knowledge that the total vapor pressure of a mixture of two volatile components is the sum of the vapor pressures of each component multiplied by the mole fraction of that component in the mixture. We are given the vapor pressures of pure methanol (92 torr) and pure ethanol (45 torr).

Analyze

The total pressure can be calculated by

$$P_{total} = \chi_{methanol} P^{\circ}_{methanol} + \chi_{ethanol} P^{\circ}_{ethanol}$$

where P° is the vapor pressure of the pure methanol or ethanol at standard temperature and χ is the mole fraction of each component.

Solve

The moles of each component are

$$25 \text{ g} \times \frac{1 \text{ mol}}{32.04 \text{ g}} = 0.780 \text{ mol methanol}$$

$$75 \text{ g} \times \frac{1 \text{ mol}}{46.07 \text{ g}} = 1.63 \text{ mol ethanol}$$

Total moles $= 1.63 + 0.7802 = 2.41$ mol

The mole fraction of each component is

$$\chi_{methanol} = \frac{0.780 \text{ mol}}{2.41 \text{ mol}} = 0.324$$

$$\chi_{ethanol} = \frac{1.63 \text{ mol}}{2.41 \text{ mol}} = 0.676$$

The vapor pressure of the mixture at 20°C is

$$P_{total} = (0.324 \times 92 \text{ torr}) + (0.676 \times 45 \text{ torr}) = 60 \text{ torr}$$

Think About It

Because methanol is more volatile than ethanol, the proportion of methanol in the vapor state compared to that of ethanol is greater than the proportion of methanol to ethanol in the liquid state.

11.47. Collect and Organize

We are asked to determine the ratio of ethylbenzene to styrene in the vapor phase as the mixture begins to boil, and the temperature at which distillate begins to come off the column.

Analyze

The mixture of 38% styrene and 62% ethylbenzene will enter the vapor phase enriched in the more volatile component. We may obtain a rough estimate of the composition of the vapor phase using the mole fractions for ethylbenzene and styrene determined in Problem 11.46 and multiplying by the vapor pressure of the pure substances. The first distillate should come off the column at the same temperature at which the mixture first boiled (the boiling point of the lowest boiling solvent).

Solve

(a) The vapor pressures of the styrene and ethylbenzene are

$$P_{\text{styrene}} = (0.3846 \times 134 \text{ torr}) = 51.5 \text{ torr}$$

$$P_{\text{ethylbenzene}} = (0.6154 \times 183 \text{ torr}) = 113 \text{ torr}$$

The mole ratio of styrene to ethylbenzene vapors is

$$\frac{P_{\text{styrene}}}{P_{\text{ethylbenzene}}} = \frac{51.5 \text{ torr}}{113 \text{ torr}} = 0.457$$

The vapor phase will consist of approximately 46% styrene and 54% ethylbenzene, a styrene to ethylbenzene ratio of 23:27.

(b) The first distillate should come off the column at the same temperature it began to boil, (ii) approximately 90°C.

Think About It

Ethylbenzene is both the larger component of the liquid phase and the more volatile component of the mixture.

11.49. Collect and Organize

We are to differentiate between *molarity* and *molality*.

Analyze

Both terms describe the concentration of a solute in a solution.

Solve

Molarity is moles of the solute in one liter of solution. Molality is moles of solute in one kilogram of solvent.

Think About It

Molality is used when a change in temperature would change the volume of a solution.

11.51. Collect and Organize

We are asked to determine if this difference between the molarity and molality of a solution of sodium chloride will become more or less pronounced as the temperature is increased from 5°C to 90°C.

Analyze

Both molarity and molality may be used to express the concentration of a solution of sodium chloride, though the concentrations are not equal to one another because of the difference in units. The density of water will decrease from 0.99997 g/mL to 0.965 g/mL as the temperature increases from 5°C to 90°C. By considering both the molarity and molality of an example solution of NaCl(*aq*) at both temperatures, we can determine whether the difference will decrease or increase.

Solve

Let us consider a solution consisting of 0.50 mol NaCl dissolved in 1000 g H_2O. The mass of NaCl present is

$$m = 0.5 \text{ mol NaCl} \times \frac{58.44 \text{ g}}{1 \text{ mol NaCl}} = 29.22 \text{ g NaCl}$$

At 5°C, the molarity and molality of the solution are

$$\text{molarity} = \frac{0.5 \text{ mol NaCl}}{\left(29.22 \text{ g NaCl} + 1000 \text{ g H}_2\text{O}\right)} \times \frac{0.99997 \text{ g}}{1 \text{ mL}} \times \frac{1000 \text{ mL}}{1 \text{ L}} = 0.486 \text{ mol/L}$$

$$\text{molality} = \frac{0.5 \text{ mol NaCl}}{1000 \text{ g H}_2\text{O}} \times \frac{1000 \text{ g}}{1 \text{ kg}} = 0.500 \text{ mol/kg}$$

At 90°C, the molarity and molality of the solution are

$$\text{molarity} = \frac{0.5 \text{ mol NaCl}}{\left(29.22 \text{ g NaCl} + 1000 \text{ g H}_2\text{O}\right)} \times \frac{0.965 \text{ g}}{1 \text{ mL}} \times \frac{1000 \text{ mL}}{1 \text{ L}} = 0.469 \text{ mol/L}$$

$$\text{molality} = \frac{0.5 \text{ mol NaCl}}{1000 \text{ g H}_2\text{O}} \times \frac{1000 \text{ g}}{1 \text{ kg}} = 0.500 \text{ mol/kg}$$

As we can see, the mass of the solution does not change over the temperature range, though the volume, and thus the molarity of the solution, does. At 90°C, the molarity has decreased from its value at 5°C. We can conclude that the difference between a concentration expressed in terms of molarity and molality will increase with the increasing temperature.

Think About It
We are assuming that the density of the solution at 5°C and 90°C is the same as the density of water at these temperatures. We could get a more accurate (though very similar) concentration if we determine the density of the solution.

11.53. Collect and Organize
We are asked why we must know if a substance is a molecular or ionic compound in order to predict the strength of its effect on the boiling or freezing point of a solvent.

Analyze
The colligative properties (boiling point, freezing point, osmotic pressure) all depend on the number of dissolved particles in the solution. The greater the number of particles, the greater the effect.

Solve
Ionic solids that are strong electrolytes completely dissociate in the solvent. This dissociation yields two or more particles in solution from one dissolved solute particle. This results in greater changes in the melting and boiling points compared to those of a solute that does not dissociate. Since molecular compounds cannot dissociate in water, whereas ionic compounds may, we must know what type of compound we have if we are to properly predict its effects, considering the van't Hoff factor.

Think About It
Because the ΔT_f or ΔT_b is directly related to the molality of a solute in the solvent, a solute that is a strong electrolyte which, for example, gives three ions upon dissolution, will give three times the ΔT_f or ΔT_b compared to the ΔT_f or ΔT_b for the same concentration of a nonelectrolyte.

11.55. Collect and Organize
For the substances CH_3OH, NaBr, and K_2SO_4 we are to explain how i, the van't Hoff factor, can be predicted from their molecular formulas.

Analyze
The van't Hoff factor is a ratio of the measured value of a colligative property to the expected value when treating the solute as molecular. For molecular species that do not dissociate, $i = 1$. For ionic species, i is equal to the total number of ions into which the salt dissolves in a solvent.

Solve
The theoretical value of i for CH_3OH is 1 because methanol is molecular and does not dissociate in a solvent such as water. NaBr has a theoretical value of $i = 2$ because it dissociates into two particles upon dissolution

(Na$^+$ and Br$^-$). K$_2$SO$_4$ has a theoretical value of $i = 3$ because it dissociates into three particles upon dissolution (2 K$^+$ and SO$_4{}^{2-}$).

Think About It

Weak electrolytes do not completely dissociate, and have intermediate values of i (greater than it would be if the substance were molecular, but less than it would be if the substance completely dissociated).

11.57. Collect and Organize

We are to define a *semipermeable membrane*.

Analyze

We can refer to the definition to describe a semipermeable membrane.

Solve

A semipermeable membrane is a boundary between two solutions through which some molecules may pass but others cannot. Usually, small molecules may pass through, but large molecules are blocked.

Think About It

Semipermeable membranes are used in reverse osmosis processes and are present as the phospholipid bilayer in cell membranes. The thin film on the inside of an egg is also semipermeable.

11.59. Collect and Organize

We are to determine the direction of solvent flow when a semipermeable membrane separates a dilute solute from a more concentrated solution.

Analyze

Solvent flows through a semipermeable membrane from a region of more dilute solution to a region of more concentrated solution. In this way osmosis balances the concentration of solutes on both sides of the membrane.

Solve

Solvent flows across a semipermeable membrane from the more dilute solution side to the more concentrated solution side to balance the concentration of solutes on both sides of the membrane.

Think About It

The liquid level on the more dilute solution side of the membrane visibly decreases (while that on the more concentrated solution side increases) as osmosis proceeds in this experiment.

11.61. Collect and Organize / Analyze

We are to describe the process of reverse osmosis. We are to list the equipment needed to purify seawater by reverse osmosis.

Solve

Reverse osmosis transfers solvent across a semipermeable membrane from a region of higher solute concentration to a region of lower solute concentration. Because reverse osmosis goes against the natural flow of solvent across the membrane, the key component needed is a pump to apply pressure to the more concentrated side of the membrane. Other components needed include a containment system, piping to introduce and remove the solutions, and a tough semipermeable membrane that can withstand the high pressures needed.

Think About It

Seawater can be purified in large quantities in desalination plants using reverse osmosis.

11.63. Collect and Organize

We are to explain why putting a salad dressing, such as a vinaigrette, on salad will result in wilted lettuce over time.

Analyze

Osmosis transfers solvent across a semipermeable membrane from a region of lower solute concentration to a region of higher solute concentration.

Solve

The salad dressing has a higher solute concentration than the plasma in the lettuce cells. Water will move from the less-concentrated cellular fluids in the salad greens to the dressing by osmosis, resulting in wilted leaves.

Think About It

The same logic is used when making coleslaw but to the opposite end—cabbage is cut and salted, pulling water out of the cabbage so the slaw will not be as wet.

11.65. **Collect and Organize**

For each of the solutions, we are to calculate the molality.

Analyze

Molality is the number of moles of solute present in 1 kilogram of solvent.

Solve

(a) $\dfrac{0.433 \text{ mol}}{2.1 \text{ kg}} = 0.21 \ m$ glucose

(b) $\dfrac{71.5 \text{ mmol}}{125 \text{ g}} \times \dfrac{1 \text{ mol}}{1000 \text{ mmol}} \times \dfrac{1000 \text{ g}}{1 \text{ kg}} = 0.572 \ m$ acetic acid

(c) $\dfrac{0.165 \text{ mol}}{375.0 \text{ g}} \times \dfrac{1000 \text{ g}}{1 \text{ kg}} = 0.440 \ m$ $NaHCO_3$

Think About It

Be sure the mass of the solvent is expressed in kilograms when calculating molality.

11.67. **Collect and Organize**

For each of the solutions of the molality listed, we are to calculate the mass of the solution that contains 0.100 mol of the solute.

Analyze

Because molality is expressed as moles per kilogram, we can calculate the mass of solution required to give 0.100 mol by dividing 0.100 mol by the molality of the solution (or multiplying by the inverse of the molality). This gives the mass of solvent in the solution that also contains 0.100 mol of the solute. For the total solution mass, we have to add the mass of the solute to the mass of the solvent.

Solve

(a) $0.100 \text{ mol} \times \dfrac{1 \text{ kg } H_2O}{0.135 \text{ mol}} = 0.7407 \text{ kg or } 740.7 \text{ g } H_2O$

$0.100 \text{ mol} \times \dfrac{80.04 \text{ g}}{1 \text{ mol}} = 8.004 \text{ g } NH_4NO_3$

Total mass of solution $= 740.7 \text{ g} + 8.004 \text{ g} = 749 \text{ g } NH_4NO_3$ solution

(b) $0.100 \text{ mol} \times \dfrac{1 \text{ kg } H_2O}{3.92 \text{ mol}} = 0.0255 \text{ kg or } 25.5 \text{ g } H_2O$

$0.100 \text{ mol} \times \dfrac{62.07 \text{ g}}{1 \text{ mol}} = 6.207 \text{ g ethylene glycol } (HOCH_2CH_2OH)$

Total mass of solution $= 25.5 \text{ g} + 6.207 \text{ g} = 31.7 \text{ g ethylene glycol solution}$

(c) $0.100 \text{ mol} \times \dfrac{1 \text{ kg H}_2\text{O}}{1.07 \text{ mol}} = 0.09346 \text{ kg or } 93.46 \text{ g H}_2\text{O}$

$0.100 \text{ mol} \times \dfrac{110.98 \text{ g}}{1 \text{ mol}} = 11.098 \text{ g CaCl}_2$

Total mass $= 93.46 \text{ g} + 11.098 \text{ g} = 104.6 \text{ g CaCl}_2$ solution

Think About It
Remember that in molality, the mass of the solvent is used and that the mass of solute must be added to be able to determine a mass of the solution needed.

11.69. Collect and Organize
We are to express lethal concentrations of ammonia, nitrite ion, and nitrate ion (1.1 mg/L, 0.40 mg/L, and 1361 mg/L, respectively) in molality units.

Analyze
Because the aqueous solutions of these species are dilute, we can assume that the densities of the solutions are 1.00 g/mL. We have to convert each milligram amount to moles, using the molar masses of ammonia (17.03 g/mol), nitrite (NO_2^-, 46.01 g/mol), and nitrate (NO_3^-, 62.00 g/mol). The mass of the solvent (water) is the mass of the solution minus the mass of solute dissolved in the solution. From that mass of solvent (expressed in kg) and the moles of solute we can calculate the molality. However, the mass of the solutes is insignificant compared to the mass of the solvent so we can safely assume that the mass of the water is 1.00 kg.

Solve

For NH_3 $\dfrac{1.1 \text{ mg}}{1 \text{ kg}} \times \dfrac{1.00 \text{ g}}{1000 \text{ mg}} \times \dfrac{1 \text{ mol}}{17.03 \text{ g}} = 6.5 \times 10^{-5} \; m \text{ NH}_3$

For NO_2^- $\dfrac{0.40 \text{ mg}}{1 \text{ kg}} \times \dfrac{1.00 \text{ g}}{1000 \text{ mg}} \times \dfrac{1 \text{ mol}}{46.01 \text{ g}} = 8.7 \times 10^{-6} \; m \text{ NO}_2^-$

For NO_3^- $\dfrac{1361 \text{ mg}}{1 \text{ kg}} \times \dfrac{1.00 \text{ g}}{1000 \text{ mg}} \times \dfrac{1 \text{ mol}}{62.00 \text{ g}} = 2.195 \times 10^{-2} \; m \text{ NO}_3^-$

Think About It
Dissolved substances are often expressed as mass/mass or mass/volume, but the numbers of particles in a mixture are best expressed as moles/mass or moles/volume.

11.71. Collect and Organize
Using the equation for boiling point elevation, we can determine the boiling point elevation of a cinnamaldehyde solution.

Analyze
The change in boiling point of a solution is

$$\Delta T_b = K_b m$$

where K_b is the boiling point elevation constant for the solvent and m is the molality of the solute in the solution. We are given that 100 mg of cinnamaldehyde (C_9H_8O, 132.16 g/mol) are dissolved in 1.00 g (or 1.00×10^{-3} kg) of carbon tetrachloride. We have to convert the mass of the cinnamaldehyde to moles by dividing by the molar mass before calculating the molality.

Solve
Molality of cinnamaldehyde solution:

$$\dfrac{0.100 \text{ g} \times \dfrac{1 \text{ mol}}{132.16 \text{ g}}}{1.00 \times 10^{-3} \text{ kg}} = 0.757 \; m$$

Boiling point elevation:

$$\Delta T_b = \frac{5.02\,°C}{m} \times 0.757\ m = 3.80\,°C$$

Think About It
The boiling point of pure CCl_4 is 76.7°C. The boiling point of this cinnamaldehyde solution will be 80.5°C.

11.73. **Collect and Organize**
We can use the melting point depression equation to determine what molality of a nonvolatile, nonelectrolyte solute would cause the melting point of camphor to change by 1.000°C.

Analyze
We can rearrange the melting point depression equation to solve for molality:

$$\Delta T_f = K_f m$$

$$m = \frac{\Delta T_f}{K_f}$$

For camphor, $K_f = 39.7$°C/m.

Solve

$$m = \frac{1.000\,°C}{39.7\,°C/m} = 2.52 \times 10^{-2}\ m$$

Think About It
Because of camphor's large K_f value, even a little solute dissolved in it greatly depresses the freezing point.

11.75. **Collect and Organize**
We are given the concentration (in mg/mL) of an aqueous solution of saccharin and the K_f for water and asked to determine the melting point of the solution.

Analyze
The melting point depression of the solution can be calculated using

$$\Delta T_f = K_f m$$

where m is the molality of the saccharin solution, which is calculated by knowing the amount of saccharin ($C_7H_5O_3NS$, 183.19 g/mol) dissolved in 1.00 mL of solution (186 mg) and the density of the solution (1.00 g/mL). We are given water's melting point depression constant ($K_f = 1.86$°C/m). Once we know how much the melting point is depressed, we can subtract that value from 0.00°C (the melting point of pure H_2O) to arrive at the new melting point of the saccharin solution.

Solve
The molality of the solution is

$$0.186\ g \times \frac{1\ mol}{183.19\ g} = 1.015 \times 10^{-3}\ mol\ saccharin$$

The mass of water in solution = 1.00 g or 1.00×10^{-3} kg

$$\text{Molality of solution} = \frac{1.015 \times 10^{-3}\ mol}{1.00 \times 10^{-3}\ kg} = 1.015\ m$$

The freezing point depression of the solution is

$$\Delta T_f = \frac{1.86\,°C}{m} \times 1.015\ m = 1.89\,°C$$

The freezing point of this solution is 0.00°C − 1.89°C = −1.89°C

Think About It
The freezing point of this solution is noticeably lower than that of pure water.

11.77. Collect and Organize
Of the aqueous solutions named, we are to determine which has the lowest freezing point.

Analyze
Freezing point depression is a colligative property and as such depends not on the identity of the solute but on the number of particles dissolved in the solution. Molecular substances dissolve as single molecules, so their particle molality is equal to the molality of the solution. Ionic substances, on the other hand, dissolve to form two or more particles (ions) in solution and therefore the molality of particles in solution is a multiple (two times, three times, etc.) of the molality of the salt in the solution.

Solve
Glucose is a molecular solid, so its particle molality in the solution is 0.5 m. Sodium chloride forms Na^+ and Cl^- in solution, so its particle molality is 0.50 $m \times 2 = 1.0$ m. Calcium chloride forms Ca^{2+} and 2 Cl^- in solution, so its particle molality is 0.5 $m \times 3 = 1.5$ m. The greater the particle molality, the larger the freezing point depression. Therefore, the aqueous solution that has the lowest freezing point is 0.5 m $CaCl_2$.

Think About It
We have to compare not only the molality of the solutions, but also the behavior of the substance in the solvent (does it break into ions when dissolved, and how many?) in comparing colligative properties.

11.79. Collect and Organize
Of the aqueous solutions named [0.0200 m CH_3CH_2OH, 0.0125 m $LiClO_4$, and 0.0100 m $Mg(NO_3)_2$], we are to determine which one has the highest boiling point.

Analyze
Boiling point elevation is a colligative property and as such depends not on the identity of the solute but on the number of particles dissolved in the solution. Molecular substances dissolve as single molecules, so their particle molality is equal to the molality of the solution. Ionic substances, on the other hand, dissolve to form two or more particles (ions) in solution and therefore the molality of particles in solution is a multiple (two times, three times, etc.) of the molality of the salt in the solution.

Solve
Ethanol, CH_3CH_2OH, is a molecular liquid, so its particle molality is 0.0200 m. Lithium perchlorate forms Li^+ and ClO_4^- in solution, so a 0.0125 m solution of $LiClO_4$ is 0.0250 m in particles. Magnesium nitrate forms Mg^{2+} and 2 NO_3^- in solution, so a 0.0100 m solution is 0.0300 m in particles. The greater the particle molality, the higher the boiling point. Therefore, the aqueous solution with the highest boiling point is the 0.0100 m $Mg(NO_3)_2$ solution.

Think About It
We have to compare not only the molality of the solutions, but also the behavior of the substance in the solvent (does it break into ions when dissolved, and how many?) in comparing colligative properties.

11.81. Collect and Organize
For each aqueous solution, we are given the concentration of a species and the van't Hoff factor with which to calculate the boiling point elevation of the solution. We are to rank the solutions in order of increasing boiling point. We need to use the boiling point elevation constant for water ($K_b = 0.52°C/m$).

Analyze
The van't Hoff factor is a ratio that compares the measured value of a colligative property to the value expected if the dissolved substance were molecular. It gives, in essence, the number of particles in solution. The calculation for boiling point elevation using the van't Hoff factor is
$$\Delta T_b = iK_b m$$

Solve

(a) For 0.06 *m* FeCl₃ (*i* = 3.4) in water:

$$\Delta T_{\text{b}} = 3.4 \times 0.52°C/m \times 0.06 \ m = 0.11°C$$

(b) For 0.10 *m* MgCl₂ (*i* = 2.7):

$$\Delta T_{\text{b}} = 2.7 \times 0.52°C/m \times 0.10 \ m = 0.14°C$$

(c) For 0.20 *m* KCl (*i* = 1.9):

$$\Delta T_{\text{b}} = 1.9 \times 0.52°C/m \times 0.20 \ m = 0.20°C$$

Therefore, in order of increasing boiling point, 0.06 *m* FeCl₃(*aq*) < 0.10 *m* MgCl₂(*aq*) < 0.20 *m* KCl(*aq*).

Think About It

Notice that the van't Hoff factors are all less than the theoretical value. This is due to ion pairing of the solute ions in solution.

11.83. Collect and Organize

Given the composition of two solutions separated by a semipermeable membrane, we are to determine which direction the solvent will flow.

Analyze

Solvent flows through a semipermeable membrane from a region of more dilute solution to a region of more concentrated solution. In this way osmosis balances the concentration of solutes on both sides of the membrane. We need only to determine which side of the membrane (A or B) has the more concentrated solution, being sure to take into account the number of particles by calculating the theoretical value of *i*, the van't Hoff factor. Solvent flows toward that more concentrated side of the membrane.

Solve

(a) Side A has 1.25 *M* NaCl, which would be 2.50 *M* in particles (*i* = 2).
Side B has 1.50 *M* KCl, which would be 3.00 *M* in particles (*i* = 2).
Side B is more concentrated in solute. Therefore, solvent flows from side A to side B.
(b) Side A has 3.45 *M* CaCl₂, which would be 10.35 *M* in particles (*i* = 3).
Side B has 3.45 *M* NaBr, which would be 6.90 *M* in particles (*i* = 2).
Side A is more concentrated in solute. Therefore, solvent flows from side B to side A.
(c) Side A has 4.68 *M* glucose, which would be 4.68 *M* in particles (*i* = 1).
Side B has 3.00 *M* NaCl, which would be 6.00 *M* in particles (*i* = 2).
Side B is more concentrated in solute. Therefore, solvent flows from side A to side B.

Think About It

Remember to compare the molarity of the dissolved particles, not just that of the solutions, when deciding the direction of flow across a semipermeable membrane in osmosis.

11.85. Collect and Organize

For the solutions described we can calculate the osmotic pressure using the formula $\Pi = iMRT$.

Analyze

For each solution we have to calculate the molarity of the solution (if not already given) and determine the theoretical value of *i* (the van't Hoff factor) for the solute. The temperature for all solutions is (20 + 273.15) = 293 K.

Solve

(a) The value of *i* for 2.39 *M* CH₃OH is 1.
The osmotic pressure is

$$\Pi = 1 \times 2.39 \ M \times \frac{0.0821 \ \text{L} \cdot \text{atm}}{\text{mol} \cdot \text{K}} \times 293 \ \text{K} = 57.5 \ \text{atm}$$

(b) The value of i for 9.45 mM MgCl$_2$ is 3.
The osmotic pressure is

$$\Pi = 3 \times \left(\frac{9.45 \text{ mmol}}{\text{L}} \times \frac{1 \text{ mol}}{1000 \text{ mmol}} \right) \times \frac{0.0821 \text{ L} \cdot \text{atm}}{\text{mol} \cdot \text{K}} \times 293 \text{ K} = 0.682 \text{ atm}$$

(c) The molarity of the glycerol ($C_3H_8O_3$) solution is

$$\frac{\left(40.0 \text{ mL} \times \dfrac{1.265 \text{ g}}{\text{mL}} \times \dfrac{1 \text{ mol}}{92.09 \text{ g}} \right)}{0.250 \text{ L}} = 2.20 \text{ } M$$

The value of i for glycerol is 1.
The osmotic pressure is

$$\Pi = 1 \times 2.20 \text{ } M \times \frac{0.0821 \text{ L} \cdot \text{atm}}{\text{mol} \cdot \text{K}} \times 293 \text{ K} = 52.9 \text{ atm}$$

(d) The molarity of the CaCl$_2$ solution is

$$\frac{\left(25.0 \text{ g} \times \dfrac{1 \text{ mol}}{110.98 \text{ g}} \right)}{0.350 \text{ L}} = 0.644 \text{ } M$$

The value of i for CaCl$_2$ is 3.
The osmotic pressure is

$$\Pi = 3 \times 0.644 \text{ } M \times \frac{0.0821 \text{ L} \cdot \text{atm}}{\text{mol} \cdot \text{K}} \times 293 \text{ K} = 46.5 \text{ atm}$$

Think About It
From the equation for osmotic pressure we see that the pressure will increase as the van't Hoff factor, the molarity of the solution, and the temperature increase.

11.87. ### Collect and Organize
Given the osmotic pressure of solutions at 25°C, we are to determine the molarity of each.

Analyze
Rearranging the osmotic pressure equation to solve for molarity gives

$$M = \frac{\Pi}{iRT}$$

where both ethanol and aspirin have $i = 1$. We are given $i = 2.47$ for the CaCl$_2$ solution. The temperature in Kelvin is $25 + 273.15 = 298$ K.

Solve
(a) $M = \dfrac{0.674 \text{ atm}}{1 \times 0.0821 \text{ L} \cdot \text{atm} / \text{mol} \cdot \text{K} \times 298 \text{ K}} = 2.75 \times 10^{-2} \text{ } M$ ethanol

(b) $M = \dfrac{0.0271 \text{ atm}}{1 \times 0.0821 \text{ L} \cdot \text{atm} / \text{mol} \cdot \text{K} \times 298 \text{ K}} = 1.11 \times 10^{-3} \text{ } M$ aspirin

(c) $M = \dfrac{0.605 \text{ atm}}{2.47 \times 0.0821 \text{ L} \cdot \text{atm} / \text{mol} \cdot \text{K} \times 298 \text{ K}} = 1.00 \times 10^{-2} \text{ } M$ CaCl$_2$

Think About It
The higher the molarity of the (particles of the) solute, the greater the osmotic pressure.

11.89. ### Collect and Organize
We are to determine whether the molarity of a NaCl solution would always be less than that of a CaCl$_2$ solution when the reverse osmotic pressures of the two solutions are the same at the same temperature.

Analyze

Osmotic pressure is given by

$$\Pi = iMRT$$

The van't Hoff factor for NaCl is $i = 2$ and for $CaCl_2$ $i = 3$.

Solve

If the reverse osmotic pressures are equal then

$$\Pi_{NaCl} = \Pi_{CaCl_2}$$

$$iM_{NaCl}RT = iM_{CaCl_2}RT$$

$$iM_{NaCl} = iM_{CaCl_2}$$

$$2M_{NaCl} = 3M_{CaCl_2}$$

$$M_{NaCl} = \tfrac{3}{2} M_{CaCl_2}$$

This statement is false. The molarity of NaCl would be greater (1.5 times) than the molarity of $CaCl_2$.

Think About It

In order to have the same reverse osmotic pressure, the concentration of the NaCl solution would have to be greater than that of $CaCl_2$ solution because the i value for NaCl is less than the i value for $CaCl_2$.

11.91. Collect and Organize

We are to describe the effect of a solute dissolving in a solvent on the osmotic pressure, freezing point, and boiling point of the solvent.

Analyze

All of these properties are colligative and as such depend only on the number of particles in the solution, not on the identity of the solute.

Solve

(a) Dissolving a solute into a solvent increases the solvent's osmotic pressure.
(b) Dissolving a solute into a solvent decreases the solvent's freezing point.
(c) Dissolving a solute into a solvent increases the solvent's boiling point.

Think About It

All of these colligative properties may be employed to determine the molar mass of a compound.

11.93. Collect and Organize

From the osmotic pressure of a solution of 188 mg of a solid nonelectrolyte dissolved in 10.0 mL of solution at 25°C, we are to calculate the molar mass of the solute.

Analyze

We can calculate the molarity from the osmotic pressure using

$$M = \frac{\Pi}{iRT}$$

where $i = 1$, $R = 0.0821$ L · atm /(mol · K), $\Pi = 4.89$ atm, and $T = 273.15 + 25°C = 298$ K. From the molarity, we can calculate the molar mass of the solute:

$$\text{Molar mass of solute (g/mol)} = \frac{\text{mass of solute (g)}}{\text{mol/L of solute} \times \text{volume of solution (L)}}$$

where the mass of solute is 0.188 g and the volume of solute is 0.0100 L.

Solve

$$M = \frac{4.89 \text{ atm}}{1 \times 0.0821 \text{ L} \cdot \text{atm} / \text{mol} \cdot \text{K} \times 298 \text{ K}} = 0.1999 \text{ mol/L}$$

$$\text{Molar mass} = \frac{0.188 \text{ g}}{0.1999 \text{ mol/L} \times 0.0100 \text{ L}} = 94.1 \text{ g/mol}$$

Think About It

Osmotic pressure is the best colligative property to use to measure molar mass because it is the most sensitive technique.

11.95. **Collect and Organize**

From the boiling point elevation (2.45°C) of a solution prepared by dissolving 0.111 g of eugenol (a nonelectrolyte) in 1.00×10^{-3} kg chloroform (with $K_b = 3.63$°C/m), we are to calculate the molar mass of eugenol. With the elemental analysis data provided we can then determine the molecular formula.

Analyze

We can calculate the molality of the solution from the measured boiling point elevation:

$$m = \frac{\Delta T_b}{K_b}$$

From the molality, we can calculate the molar mass:

$$\frac{\text{g of eugenol}}{\text{kg of chloroform} \times \text{molality of solution (mol/kg)}} = \frac{\text{g}}{\text{mol}}$$

From the elemental analysis, we can determine the empirical formula of eugenol, which we can then relate to the molar mass to write the molecular formula.

Solve

$$m = \frac{2.45 \,^{\circ}\text{C}}{3.63 \,^{\circ}\text{C}/m} = 0.6749 \text{ mol/kg}$$

$$\text{Molar mass} = \frac{0.111 \text{ g}}{1.00 \times 10^{-3} \text{ kg} \times 0.6749 \text{ mol/kg}} = 164 \text{ g/mol}$$

From the elemental analysis, if we assume 100 g,

$$73.17 \text{ g C} \times \frac{1 \text{ mol}}{12.011 \text{ g}} = 6.092 \text{ mol C}$$

$$7.32 \text{ g H} \times \frac{1 \text{ mol}}{1.008 \text{ g}} = 7.26 \text{ mol H}$$

$$19.51 \text{ g O} \times \frac{1 \text{ mol}}{15.999} = 1.219 \text{ mol O}$$

This gives a C:H:O molar ratio of 5:6:1 for an empirical formula of C_5H_6O with a formula mass of 82.10 g/mol. This is one-half the measured molar mass, so the molecular formula of eugenol is $C_{10}H_{12}O_2$.

Think About It

The formula for eugenol is indeed $C_{10}H_{12}O_2$ and this compound has the following structure:

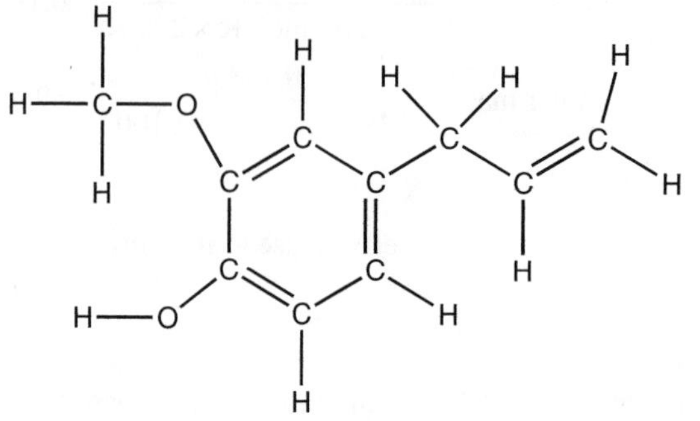

11.97. Collect and Organize

From the list of four ionic compounds we are to select the one with the least negative (weakest) lattice energy.

Analyze

In each of the compounds the cation has a 2+ charge, but they differ in size: $Mg^{2+} < Ca^{2+} < Sr^{2+}$. The anions differ not in charge, but in size: $F^- < Cl^- < Br^- < I^-$. The lattice energy will be strongest for a small anion with a small cation and weakest for a large anion with a large cation, according to the increasing distance between the cation and anion (d).

$$U = \frac{k(Q_1 Q_2)}{d}$$

Solve

SrI_2 (a) will have the least negative lattice energy.

Think About It

Charge (Q_1 and Q_2) will also have an effect on the lattice energy. The greater the charge the more negative the lattice energy and the higher the strength of the ion–ion bond.

11.99. Collect and Organize

We consider whether $CaCl_2$ would melt ice at –20°C. We are given that 70.1 g of $CaCl_2$ ($i = 2.5$) can dissolve in 100.0 g of water at this temperature.

Analyze

From the mass of $CaCl_2$ that dissolves in 100.0 g of water, we can compute the molality of a $CaCl_2$ solution. We can then use the freezing point depression equation

$$\Delta T_f = iK_f m$$

to calculate the ΔT_f. The K_f for water is 1.86°C/m. If ΔT_f is greater than 20°C, then $CaCl_2$ would melt ice at a temperature of –20°C.

Solve

$$\text{Molality of } CaCl_2 \text{ solution} = \frac{70.1 \text{ g} \times \dfrac{1 \text{ mol}}{110.98 \text{ g}}}{0.1000 \text{ kg}} = 6.316 \ m$$

Freezing point depression is $\Delta T_f = 2.5 \times 1.86°C/m \times 6.316 \ m = 29°C$.
The freezing point of the $CaCl_2$ solution would be –29°C. This is lower than –20°C. Yes, the $CaCl_2$ could melt ice at –20°C.

Think About It

$CaCl_2$ is an excellent deicer. It gives three particles, ideally, when it dissolves and is very soluble in water, so the ΔT_f for a saturated solution of $CaCl_2$ is very large.

11.101. **Collect and Organize**

Given the ΔT_f for NH_4Cl and $(NH_4)_2SO_4$ in aqueous solution, we are to calculate the value of i, the van't Hoff factor, for each solution.

Analyze

The van't Hoff factor may be calculated using

$$i = \frac{\Delta T_f}{K_f m}$$

where $\Delta T_f = 0.322°C$ for $m = 0.0935$ m NH_4Cl and $\Delta T_f = 0.173°C$ for $m = 0.0378$ m $(NH_4)_2SO_4$. K_f for water is $1.86°C/m$.

Solve

For 0.0935 m NH_4Cl

$$i = \frac{0.322°C}{1.86°C/m \times 0.0935 \; m} = 1.85$$

For 0.0378 m $(NH_4)_2SO_4$

$$i = \frac{0.173°C}{1.86°C/m \times 0.0378 \; m} = 2.46$$

Think About It

If there were no ion pairs in either of these solutions, i for NH_4Cl would be 2 and i for $(NH_4)_2SO_4$ would be 3. The greater difference between the expected i and the calculated i for the $(NH_4)_2SO_4$ solution $(3 - 2.46 = 0.54$ compared to $2 - 1.85 = 0.15$ for the NH_4Cl solution) means that the ions in the $(NH_4)_2SO_4$ solution form ion pairs to a greater extent.

11.103. **Collect and Organize**

We are to calculate the osmotic pressure of a saline solution (aqueous sodium chloride) at 37°C.

Analyze

In order to apply the osmotic pressure equation

$$\Pi = iMRT$$

we have to convert the concentration of NaCl from percent mass to molarity and account for the dilution of 100.0 mL of the solution to 350.0 mL.

Solve

We can assume that the density of the dilute saline solution is close to that of pure water (1.00 g/mL). The molarity of a 0.92% (by mass) saline solution is

$$\frac{0.92 \text{ g NaCl}}{100 \text{ g solution}} \times \frac{1.00 \text{ g solution}}{1.00 \text{ mL}} \times \frac{1000 \text{ mL}}{1 \text{ L}} \times \frac{1 \text{ mol NaCl}}{58.44 \text{ g}} = 0.157 \text{ mol/L NaCl}$$

After dilution the molarity of the saline solution is

$$M_{\text{initial}} V_{\text{initial}} = M_{\text{final}} V_{\text{final}}$$
$$0.157 \; M \times 100.0 \text{ mL} = C_{\text{final}} \times 350.0 \text{ mL}$$
$$M_{\text{final}} = 0.0449 \; M$$

The osmotic pressure of this saline solution is

$$\Pi = 2 \times 0.0449 \text{ mol/L} \times \frac{0.0821 \text{ L} \cdot \text{atm}}{\text{mol} \cdot \text{K}} \times 310 \text{ K} = 2.3 \text{ atm}$$

Think About It

Physiological saline is isotonic with blood plasma.

11.105. **Collect and Organize**

From the osmotic pressure of 6.50 torr at 23.1°C of a solution composed of 7.50 mg of a protein in 5.00 mL of aqueous solution, we are to determine the molar mass of the protein.

Analyze

Using the osmotic pressure equation, we can calculate the molarity of the protein solution:

$$M = \frac{\Pi}{iRT}$$

The protein is a nonelectrolyte molecular solid, so $i = 1$. The temperature in Kelvin is $23.1 + 273.15 = 296.25$ K. The pressure needs to be in units of atmospheres so

$$6.50 \text{ torr} \times \frac{1 \text{ atm}}{760 \text{ torr}} = 8.55 \times 10^{-3} \text{ atm}$$

Once we have calculated the molarity, we can determine the molar mass of the protein using

$$\text{molar mass of protein (g/mol)} = \frac{\text{mass of protein in sample (g)}}{\text{mol/L of protein in solution} \times \text{volume of solution (L)}}$$

Solve

$$M = \frac{8.55 \times 10^{-3} \text{ atm}}{1 \times 0.0821 \text{ L} \cdot \text{atm} / \text{mol} \cdot \text{K} \times 296.25 \text{ K}} = 3.515 \times 10^{-4} M$$

$$\text{Molar mass of protein} = \frac{7.50 \times 10^{-3} \text{ g}}{\left(3.515 \times 10^{-4} \text{ mol/L}\right)\left(0.00500 \text{ L}\right)} = 4270 \text{ g/mol}$$

Think About It

Although this molar mass may seem quite large, this is small for proteins, which may have molar masses of several hundred thousand grams per mole.

11.107. Collect and Organize

We are asked how two solutions used in hospital settings, Ringer's Lactate and 0.9% saline, compare to each other.

Analyze

The problem tells us that the total concentration of cations and anions in each solution are equal. 0.9% saline is entirely composed of Na^+ and Cl^- ions, whereas Ringer's lactate (RL) contains Na^+, K^+, and Ca^{2+} cations and Cl^- and lactate ($C_3H_5O_3^-$) anions.

Solve

The expressions describing the concentrations of cations and anions in each solution are

$$\left[Na^+\right]_{\text{saline}} = \left[Na^+\right]_{\text{RL}} + \left[K^+\right]_{\text{RL}} + \left[Ca^{2+}\right]_{\text{RL}}$$

$$\left[Cl^-\right]_{\text{saline}} = \left[Cl^-\right]_{\text{RL}} + \left[C_3H_5O_3^-\right]_{\text{RL}}$$

Think About It

Due to the other ions present in Ringer's Lactate, the concentration of Na^+ and Cl^- are lower in this solution than they are in 0.9% saline.

CHAPTER 12 | Solids: Structures and Applications

12.1. Collect and Organize

From the drawings shown in Figure P12.1, we are to choose which represents a crystalline solid and which represents an amorphous solid.

Analyze

In crystalline solids, atoms or molecules arrange themselves in regular, repeating three-dimensional patterns. In an amorphous solid, the atoms or molecules are arranged randomly, with no defined repeating pattern.

Solve

Drawings (b) and (d) are analogous to crystalline solids because they show a definite pattern while (a) and (c) are amorphous.

Think About It

In drawings (b) and (d) there are two kinds of atoms; if the drawings represent metals, these two substances would be alloys.

12.3. Collect and Organize

Using Figure P12.3, we are to determine the unit cell and write the chemical formula for the compound where element A is represented as red spheres and element B is represented as blue spheres.

Analyze

Once the unit cell is determined, the chemical formula can be deduced. Spheres on the corner are shared by eight unit cells and so are counted as $\frac{1}{8}$ in the unit cell. Likewise, spheres on the edges are shared by four unit cells and are counted as $\frac{1}{4}$ in the unit cell, and spheres on the faces of the unit cell are shared by two unit cells and are counted as $\frac{1}{2}$ in the unit cell. Any atom completely inside a unit cell belongs entirely to it and counts as 1.

Solve

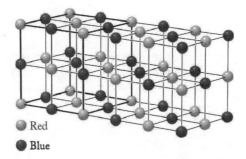

Red
Blue

There are $8\left(\frac{1}{8}\right) + 6\left(\frac{1}{2}\right) = 4$ red spheres or 4 A atoms in the unit cell. There are $12\left(\frac{1}{4}\right) + 1(1) = 4$ blue spheres or 4 B atoms in the unit cell. The chemical formula is A_4B_4 or AB.

Think About It

The empirical formula for this compound would be AB because that is the lowest whole-number ratio of elements in the substance.

12.5. Collect and Organize

We are to determine the formula of a compound in which A ions form an fcc unit cell and where half of the octahedral holes are occupied by B and one-eighth of the tetrahedral holes are occupied by C.

Analyze

In a face-centered cubic (cubic closest-packed) array of A ions, there are 8 tetrahedral holes and 4 octahedral holes.

Solve

The number of A ions in the fcc unit cell is

$$(8 \times \tfrac{1}{8}) + (6 \times \tfrac{1}{2}) = 4$$

If one-half of the octahedral holes contain B ions, then there are 2 B ions in the unit cell. If one-eighth of the tetrahedral holes contain C ions, then there is 1 C ion in the unit cell. The chemical formula for this compound is A_4B_2C.

Think About It

We may assume that the lattice is formed from an anion (the larger A spheres), and both species occupying the holes of the ccp lattice (the smaller B and C spheres) are cations.

12.7. Collect and Organize

We are to determine the formula of a lithium–sulfur compound in which the sulfide ions form an fcc arrangement and the lithium ions occupy all of the tetrahedral holes.

Analyze

In an fcc arrangement, there are 8 tetrahedral holes in the unit cell.

Solve

The number of sulfide anions in the unit cell is

$$(8 \times \tfrac{1}{8}) + (6 \times \tfrac{1}{2}) = 4$$

The number of lithium cations in the unit cell is 8. The formula for the compound is Li_2S.

Think About It

This salt is neutral because the charge on two lithium ions (1+) in the formula balances the charge on one sulfide anion (2–).

12.9. Collect and Organize

We are to identify the elements highlighted in the periodic table in Figure P12.15 that do not adopt a rock salt structure for their chloride salt.

Analyze

In the rock salt structure, the cations must fit into the octahedral holes between the closest-packed layers of Cl^-. The radius ratio of cation to anion should be between 0.41 and 0.73. The radii of the cations of the highlighted elements from Figure 10.2 are Li^+ (76 pm), Na^+ (102 pm), K^+ (138 pm), Cs^+ (170 pm), and Sr^{2+} (118 pm). The radius of Cl^- is 181 pm.

Solve

For LiCl, the radius ratio is $\dfrac{76 \text{ pm}}{181 \text{ pm}} = 0.42$

For NaCl, the radius ratio is $\dfrac{102 \text{ pm}}{181 \text{ pm}} = 0.564$

For KCl, the radius ratio is $\dfrac{138 \text{ pm}}{181 \text{ pm}} = 0.762$

For CsCl, the radius ratio is $\dfrac{170 \text{ pm}}{181 \text{ pm}} = 0.939$

For SrCl$_2$, the radius ratio is $\dfrac{118 \text{ pm}}{181 \text{ pm}} = 0.652$

By the radius ratio rule, the rock salt structure is not probable for KCl (K^+ would occupy cubic holes) nor for CsCl (Cs^+ would occupy cubic holes) or $SrCl_2$ (because it does not have a 1:1 cation-to-anion stoichiometry). LiCl is borderline. In nature, all of these adopt a rock salt structure except Cs (blue) and Sr (purple).

Think About It
In the hypothetical rock salt structure of $SrCl_2$, Sr^{2+} would occupy one-half of the octahedral holes. In nature, this ionic compound instead adopts a fluorite structure.

12.11. **Collect and Organize**
From Figure P12.11 showing the unit cell of magnesium boride, we are to determine its formula.

Analyze
We are given that one B atom is inside the trigonal prismatic unit cell. Each corner of the unit cell has an Mg atom. Because the unit cell is trigonal prismatic, each corner atom is shared with 11 other unit cells and so has one-twelfth of its volume inside the unit cell.

Solve
The number of B atoms in the unit cell is 1. The number of Mg atoms in the unit cell is $6\left(\frac{1}{12}\right) = \frac{1}{2}$. The formula from the unit cell is MgB_2 (in the lowest whole-number ratio).

Think About It
It is tricky to see that the Mg atoms here are shared between 12 unit cells. Consider the center atom in the figure on the right hand side. That atom is shared with 6 unit cells as shown, but there are also 6 more unit cells stacked on top of it that are not shown in the diagram.

12.13. **Collect and Organize**
We are asked to differentiate between cubic closest-packed (ccp) and hexagonal closest-packed (hcp) structures.

Analyze
Both structures contain layers of close-packed atoms and differ only in how the layers are stacked.

Solve
Cubic closest-packed structures have an *abcabc…* pattern, and hexagonal closest-packed structures have an *abab…* pattern.

Think About It
The unit cell for ccp is face-centered cubic (Figure 12.11), and the unit cell for hcp is hexagonal (Figure 12.5).

12.15. **Collect and Organize**
We are asked which has the greater packing efficiency, the simple cubic or the body-centered cubic structure.

Analyze
Packing efficiency is the fraction of space within a unit cell that is occupied by the atoms.

Solve
We read in Section 12.2 of the textbook that the simple cubic cell has the lowest packing efficiency of all the unit cells, so the body-centered cubic structure has a greater packing efficiency than the simple cubic structure.

Think About It
The packing efficiency of the unit cell structures can be calculated. For simple cubic, the packing efficiency is only 52%, whereas for body-centered cubic it is 68% and for face-centered and hexagonal it is 74%.

12.17. Collect and Organize

Iron can adapt either the bcc unit cell structure (at room temperature) or the fcc unit cell structure (at 1070°C). We are asked whether these two forms are allotropes.

Analyze

Allotropes are defined as different *molecular* forms of an element.

Solve

Iron is not molecular and the bcc and fcc unit cell structures describe only a difference in atom packing in the metal. Therefore, these structural forms are not allotropes.

Think About It

Elements that do have allotropes include phosphorus, sulfur, and carbon.

12.19. Collect and Organize

Knowing that $\ell = 240.6$ pm for the bcc structure of europium, we are to calculate the radius of one atom of europium.

Analyze

For the bcc structure the body diagonal contains two atoms of europium, or $4r$, and the body diagonal is equal to $\ell\sqrt{3}$ where ℓ is the unit cell edge length, 240.6 pm.

Solve

$$4r = \ell\sqrt{3}$$

$$r = \frac{\ell\sqrt{3}}{4} = \frac{240.6 \text{ pm} \times \sqrt{3}}{4} = 104.2 \text{ pm}$$

Think About It

If the structure of the unit cell were simple cubic the radius of the europium atom would simply be

$$2r = \ell$$

$$r = \frac{\ell}{2} = 120.0 \text{ pm}$$

12.21. Collect and Organize

We are to calculate the edge length of the unit cell of Ba knowing that it crystallizes in a bcc unit cell and that $r_{Ba} = 222$ pm.

Analyze

The body diagonal of a bcc unit cell has the relationship $\ell\sqrt{3} = 4r$. An edge has the length ℓ, so all we need to do is rearrange the body diagonal expression to solve for ℓ:

$$\ell = \frac{4r}{\sqrt{3}}$$

Solve

$$\ell = \frac{4r}{\sqrt{3}} = \frac{4 \times 222 \text{ pm}}{\sqrt{3}} = 513 \text{ pm}$$

Think About It

Be careful to not assume that the edge length of every unit cell is $\ell = 2r$, as in a simple cubic.

12.23. Collect and Organize

We are to determine the type of unit cell for a form of copper using the density of the crystal (8.95 g/cm^3) and the radius of the Cu atom (127.8 pm).

Analyze

For each type of unit cell (simple cubic, body-centered cubic, and face-centered cubic), we can compare the calculated density to that of the actual density given for the crystalline form of copper. Density is mass per volume. The mass of each unit cell is the mass of the copper atoms contained in each unit cell. To find this, we first have to determine the number of atoms of Cu in each unit cell based on its structure. Then, we multiply by the mass of one atom of Cu:

$$1 \text{ Cu atom} \times \frac{1 \text{ mol}}{6.022 \times 10^{23} \text{ atoms}} \times \frac{63.55 \text{ g}}{1 \text{ mol}} = 1.055 \times 10^{-22} \text{ g}$$

The volume of each unit cell is ℓ^3, which is related to r in a way that depends on the type of unit cell.

Solve

For a simple cubic unit cell where all the Cu atoms are at the corners of the cube:

$$\text{Number of Cu atoms} = 8 \times \tfrac{1}{8} = 1 \text{ Cu atom}$$
$$\ell = 2r = 2 \times 127.8 \text{ pm} = 255.6 \text{ pm}$$

Converting to centimeters for the calculation of density (g/cm³) gives

$$255.6 \text{ pm} \times \frac{1 \times 10^{-10} \text{ cm}}{1 \text{ pm}} = 2.556 \times 10^{-8} \text{ cm}$$
$$\text{Volume} = \ell^3 = (2.556 \times 10^{-8} \text{ cm})^3 = 1.670 \times 10^{-23} \text{ cm}^3$$
$$\text{Density} = \frac{1 \text{ Cu atom} \times 1.055 \times 10^{-22} \text{g/atom}}{1.670 \times 10^{-23} \text{ cm}^3} = 6.32 \text{ g/cm}^3$$

For a body-centered cubic unit cell where there is one Cu atom in the center of the unit cell and eight Cu atoms at the corners of the cube:

$$\text{Number of Cu atoms} = (8 \times \tfrac{1}{8}) + 1 = 2 \text{ Cu atoms}$$

From the body diagonal $4r = \ell\sqrt{3}$ or $\ell = 4r/\sqrt{3}$,

$$\ell = \frac{4 \times 127.8 \text{ pm}}{\sqrt{3}} = 295.1 \text{ pm}$$

Converting to centimeters for calculation of density (g/cm³) gives

$$295.1 \text{ pm} \times \frac{1 \times 10^{-10} \text{ cm}}{1 \text{ pm}} = 2.951 \times 10^{-8} \text{ cm}$$
$$\text{Volume} = \ell^3 = (2.951 \times 10^{-8} \text{ cm})^3 = 2.570 \times 10^{-23} \text{ cm}^3$$
$$\text{Density} = \frac{2 \text{ Cu atoms} \times 1.055 \times 10^{-22} \text{ g/atom}}{2.570 \times 10^{-23} \text{ cm}^3} = 8.21 \text{ g/cm}^3$$

For a face-centered cubic unit cell where there are eight Cu atoms at the corners of the unit cell and six on the faces:

$$\text{Number of Cu atoms} = (8 \times \tfrac{1}{8}) + (6 \times \tfrac{1}{2}) = 4 \text{ Cu atoms}$$

From the face diagonal $4r = \ell\sqrt{2}$ or $\ell = 4r/\sqrt{2}$,

$$\ell = \frac{4 \times 127.8 \text{ pm}}{\sqrt{2}} = 361.5 \text{ pm}$$

Converting to centimeters for calculation of density (g/cm³) gives

$$361.5 \text{ pm} \times \frac{1 \times 10^{-10} \text{ cm}}{1 \text{ pm}} = 3.615 \times 10^{-8} \text{ cm}$$
$$\text{Volume} = \ell^3 = (3.615 \times 10^{-8} \text{ cm})^3 = 4.724 \times 10^{-23} \text{ cm}^3$$
$$\text{Density} = \frac{4 \text{ Cu atoms} \times 1.055 \times 10^{-22} \text{ g/atom}}{4.724 \times 10^{-23} \text{ cm}^3} = 8.93 \text{ g/cm}^3$$

The fcc unit cell gives a density closest to the given density, so we predict that in this crystalline form of Cu the atoms pack in (c) an fcc unit cell.

Think About It
Notice that even though the fcc unit cell has the largest edge length, the unit cell contains more atoms and so it yields the densest structure.

12.25. Collect and Organize
We may use the radius and crystal structure of sodium to determine the density of Na under high pressures.

Analyze
Using the radius of a sodium atom, we may determine the edge length of the unit cell using the relationship $r = 0.4330 \ell$ for a body-centered cubic structure. We may calculate the volume by cubing the edge length, and calculate the mass of the two atoms contained in the unit cell. By dividing this mass by the calculated volume, we can determine the density of Na. Under standard conditions, the density of Na is 0.971 g/cm^3. The density should be similar under high pressure.

Solve
$$\ell = \frac{186 \text{ pm}}{0.4330} \times \frac{1 \text{ m}}{1 \times 10^{12} \text{ pm}} \times \frac{100 \text{ cm}}{1 \text{ m}} = 4.296 \times 10^{-8} \text{ cm}$$

$$\text{Volume} = \left(4.296 \times 10^{-8} \text{ cm}\right)^3 = 7.926 \times 10^{-23} \text{ cm}^3$$

$$2 \text{ Na atoms} \times \frac{1 \text{ mol Na}}{6.022 \times 10^{23} \text{ atoms Na}} \times \frac{22.9898 \text{ g Na}}{1 \text{ mol Na}} = 7.635 \times 10^{-23} \text{ g}$$

$$d = \frac{7.635 \times 10^{-23} \text{ g}}{7.926 \times 10^{-23} \text{ cm}^3} = 0.963 \text{ g/cm}^3$$

Think About It
The density is near our expected value. It is important to account for both Na atoms in the bcc unit cell.

12.27. Collect and Organize
We can use the dimensions of the hexagonal unit cell to determine the density of titanium.

Analyze
The volume of a hexagonal unit cell may be determined using the equation

$$V = \frac{3\sqrt{3}a^2 h}{2}$$

where a is the side length of the hexagon (295 pm), and h is the height of the unit cell (469 pm). Converting the cell dimensions provided in Figure P12.27 from pm to cm will yield the volume of the unit cell in cm^3. The mass of Ti in the hcp unit cell may be calculated by dividing the six Ti atoms in the unit cell by Avogadro's number and multiplying by the molar mass of titanium. The density is calculated by dividing the mass of Ti in the unit cell by the volume of the unit cell.

Solve
$$a = 295 \text{ pm} \times \frac{1 \text{ m}}{1 \times 10^{12} \text{ pm}} \times \frac{100 \text{ cm}}{1 \text{ m}} = 2.95 \times 10^{-8} \text{ cm}$$

$$h = 469 \text{ pm} \times \frac{1 \text{ m}}{1 \times 10^{12} \text{ pm}} \times \frac{100 \text{ cm}}{1 \text{ m}} = 4.69 \times 10^{-8} \text{ cm}$$

$$\text{Volume} = \frac{3\sqrt{3}\left(2.95 \times 10^{-8} \text{ cm}\right)^2 \left(4.69 \times 10^{-8} \text{ cm}\right)}{2} = 1.06 \times 10^{-22} \text{ cm}^3$$

$$6 \text{ Ti atoms} \times \frac{1 \text{ mol Ti}}{6.022 \times 10^{23} \text{ atoms Ti}} \times \frac{47.867 \text{ g Ti}}{1 \text{ mol Ti}} = 4.769 \times 10^{-22} \text{ g}$$

$$d = \frac{4.769 \times 10^{-22} \text{ g}}{1.06 \times 10^{-22} \text{ cm}^3} = 4.50 \text{ g/cm}^3$$

Think About It
Titanium is a relatively light and strong metal that is used for medical implants and other applications where a strong, light metal is required.

12.29. Collect and Organize
We are asked to describe the differences between substitutional and interstitial alloys and give an example of each.

Analyze
An alloy is a mixture of two or more metallic elements in solution with each other. The differences between substitutional and interstitial alloys involve the arrangement of atoms in the solution.

Solve
Substitutional alloys involve replacing atoms in a lattice with atoms of another element. An example of a substitutional alloy is bronze, an alloy composed of a copper lattice containing up to 30% tin. The tin atoms may be randomly distributed throughout the bronze, occupying any lattice position normally occupied by a copper atom in pure copper metal.
Interstitial alloys contain solutes in the spaces or "holes" left when the solvent lattice formed. An example of an interstitial alloy is austenite, an alloy containing carbon atoms in the octahedral holes of the iron fcc lattice.

Think About It
Both interstitial alloys and substitutional alloys have a range of compositions rather than fixed ratios of components.

12.31. Collect and Organize
We are asked what effect the substitution of one Ni atom at the center of the bcc Ti unit cell would have on the edge length of the crystal.

Analyze
From the appendix, we find that Ni has a radius of 124 pm and Ti has a radius of 147 pm. In a bcc unit cell, atoms touch along the body diagonal, but not directly along an edge.

Solve
Ni is smaller than Ti. By substituting a nickel atom for a titanium atom, the body diagonal will be shorter, and thus, the unit cell will be smaller than that for pure titanium. The edge length for NiTi will be smaller than for Ti alone.

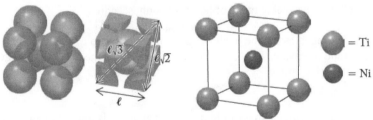

(c) Body-centered cubic:
Atoms touch along body diagonal

Think About It
Using the same logic, we could also predict that the unit cell for NiTi will be larger than that for pure Ni.

12.33. Collect and Organize
We are asked to decide if including hafnium atoms in place of magnesium atoms will generate an alloy with a greater or lesser density than pure magnesium.

Analyze

The size of the hafnium and magnesium atoms are nearly identical, so we can assume that there are no distortions to the unit cell when this substitution occurs. The atomic mass of magnesium is 24.3050 amu, and the atomic mass of hafnium is 178.49 amu.

Solve

The significant increase in mass when hafnium is substituted for magnesium will result in an alloy of greater density than that of pure magnesium.

Think About It

The density of the resulting alloy will be less than that of pure hafnium.

12.35. Collect and Organize

We are to compare the two unit cells in Figure P12.35 and decide if both are accurate depictions of NiTi.

Analyze

For a bcc unit cell, 1/8 of an atom is located in the cell at each corner, and an entire atom is located in the center. For the diagram to be accurate, the unit cells must each produce a 1:1 ratio of Ni:Ti.

Solve

For 8 X atoms at the corners and 1 Y atoms in the center of the unit cell:

(a) $\frac{1}{8}(8 \text{ Ti atoms}) + 1(1 \text{ Ni atoms}) = \text{NiTi}$

(b) $\frac{1}{8}(8 \text{ Ni atoms}) + 1(1 \text{ Ti atoms}) = \text{NiTi}$

Both unit cells produce the same ratio of Ni to Ti, so both are valid.

Think About It

This alloy structure is similar to the bcc unit cell of CsCl.

12.37. Collect and Organize

By substituting the radius of hydrogen into the radius ratio for the tetrahedral holes of an fcc unit cell, we may determine the range of metal hosts that could form an interstitial alloy with hydrogen. We are asked to determine if the interstitial alloy formed as a result of hydrogen present in the form of a hydride ion would more likely occupy octahedral or tetrahedral holes in the host metal.

Analyze

For the fcc unit cell, the acceptable range of host and nonhost radii is

$$r_{nonhost}/r_{host} = 0.22 - 0.41$$

The radius of the hydrogen atom (the nonhost) is 37 pm. The ratio of 0.41 represents the smallest difference between hydrogen and the metal. This gives the minimum radius the metal could have for hydrogen to fit into tetrahedral holes of the fcc lattice.

Solve

(a) The smallest metal host radius that would form such an alloy is

$$r_{host} = \frac{37 \text{ pm}}{0.41} = 90 \text{ pm}$$

(b) The radius of the hydride ion (146 pm) is closer to that of most metals. The hydride ion is more likely to form a substitutional alloy.

Think About It

Even if the hydride ion were to occupy the octahedral holes of a metal, the metal would have to have a radius of >200 pm!

12.39. **Collect and Organize**

By calculating the radius ratio of Sn to Ag, we can determine whether dental alloys of Sn and Ag are substitutional or interstitial alloys.

Analyze

It would be possible for an alloy to be both interstitial and substitutional as long as the voids within the lattice are about the same size as the metal atom. In particular, if the host metal has a simple cubic structure, the alloying metal may fit in a cubic void (radius ratio > 0.73) as well as substitute for the host metal in the lattice structure.

Solve

$$\frac{r_{Sn}}{r_{Ag}} = \frac{140 \text{ pm}}{144 \text{ pm}} = 0.972$$

Because these radii are within 15% of each other, an alloy of silver and tin is a substitutional alloy.

Think About It

Remember that it is also important for the metals in substitutional alloys to crystallize in the same closest-packed (hcp, ccp, simple cubic, bcc) structures.

12.41. **Collect and Organize**

We are to predict the formula of vanadium carbide if the vanadium lattice is ccp (fcc), and two octahedral holes are occupied by carbon atoms.

Analyze

The fcc arrangement of V atoms has four atoms of V in the unit cell. This fcc unit cell also contains four octahedral holes and eight tetrahedral holes.

Solve

If only two of the four octahedral holes are occupied, the ratio of atoms in this lattice is V_4C_2, or V_2C.

Think About It

If all of the octahedral holes were filled, the formula of the alloy would be VC.

12.43. **Collect and Organize**

We can use the radii of Ni and Ti to calculate the density of the nitinol alloy.

Analyze

Nitinol is an alloy composed of one Ni atom and one Ti atom per bcc unit cell. The radii of Ni (124 pm) and Ti (147 pm) may be used to calculate the body diagonal for the bcc unit cell:

$$\text{Body diagonal} = 2\left(r_{Ni}\right) + 2\left(r_{Ti}\right)$$

We may then use the relationship below to determine the edge length of the unit cell:

$$\text{Body diagonal} = \sqrt{\left(\text{edge length}\right)^2 + \left(\text{face diagonal}\right)^2} = \sqrt{\ell^2 + \left(\ell\sqrt{2}\right)^2}$$

Cubing the edge length yields the volume of the unit cell. We may determine the mass of the unit cell by multiplying the number of each atom by its molar mass and dividing by Avogadro's number. The density is obtained by dividing this mass by the volume of the unit cell determined above.

Solve

Calculating the volume of the unit cell:

$$\text{Body diagonal} = 2(124 \text{ pm}) + 2(147 \text{ pm}) = 542 \text{ pm} \times \frac{1 \text{ m}}{1 \times 10^{12} \text{ pm}} \times \frac{100 \text{ cm}}{1 \text{ m}} = 5.42 \times 10^{-8} \text{ cm}$$

$$\ell^2 + \left(\ell\sqrt{2}\right)^2 = \left(5.42 \times 10^{-8} \text{ cm}\right)^2$$

$$3\ell^2 = 2.938 \times 10^{-15} \text{ cm}^2$$

$$\ell = 3.129 \times 10^{-8} \text{ cm}$$

$$\text{Volume} = \left(3.129 \times 10^{-8} \text{ cm}\right)^3 = 3.064 \times 10^{-23} \text{ cm}^3$$

The mass of the unit cell is

$$1 \text{ Ti atom} \times \frac{1 \text{ mol Ti}}{6.022 \times 10^{23} \text{ atoms Ti}} \times \frac{47.867 \text{ g Ti}}{1 \text{ mol Ti}} = 7.949 \times 10^{-23} \text{ g Ti}$$

$$1 \text{ Ni atom} \times \frac{1 \text{ mol Ni}}{6.022 \times 10^{23} \text{ atoms Ni}} \times \frac{58.6934 \text{ g Ni}}{1 \text{ mol Ni}} = 9.746 \times 10^{-23} \text{ g Ni}$$

$$\text{Mass} = 7.949 \times 10^{-23} \text{ g Ti} + 9.746 \times 10^{-23} \text{ g Ni} = 1.770 \times 10^{-22} \text{ g}$$

The density of nitinol is

$$d = \frac{1.770 \times 10^{-22} \text{ g}}{3.064 \times 10^{-23} \text{ cm}^3} = 5.78 \text{ g/cm}^3$$

Think About It

The calculated density for nitinol is reasonable for this alloy of two first row transition metals.

12.45. Collect and Organize

We are asked to determine the type of structure depicted in Figure P12.45 and the relative proportions of each element present in the unit cell.

Analyze

Figure P12.45 depicts a body-centered cubic unit cell with opposite corners occupied by Cu atoms, and the remainder of the lattice positions occupied by Zn atoms. For a bcc unit cell, at each corner 1/8 of an atom is located in the cell, and an entire atom is located in the center.

Solve

(a) The unit cell depicted in Figure P12.45 is a bcc unit cell. The alloy is a substitutional alloy, in which some of the lattice positions of the Zn unit cell are occupied by Cu atoms.

(b) The proportions of Cu and Zn in this lattice are

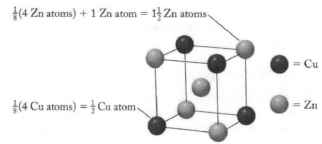

$\frac{1}{8}(4 \text{ Zn atoms}) + 1 \text{ Zn atom} = 1\frac{1}{2} \text{ Zn atoms}$

$\frac{1}{8}(4 \text{ Cu atoms}) = \frac{1}{2} \text{ Cu atom}$

$\bullet = \text{Cu}$

$\bullet = \text{Zn}$

The ratio of atoms is $Cu_{0.5}Zn_{1.5}$, or $CuZn_3$.

Think About It

Though rose gold (see Problem 12.44) and the CuZn alloy described in this problem have the same ratio of component elements, they adopt different unit cells.

12.47. Collect and Organize

We are to explain how the sea-of-electrons model accounts for the electrical conductivity of gold.

Analyze

In the sea-of-electrons model the metal atoms exist in the structure as cations with their ionized electrons not associated with particular metal ions.

Solve

Because the electrons are free to distribute themselves throughout the metal, the application of an electrical potential across the metal causes the electrons to travel freely toward the positive terminal. This movement of electrons in the structure is electrical conductivity.

Think About It

Metals have low ionization energies. This makes the electron sea somewhat easy to achieve in the lattice of metals.

12.49. **Collect and Organize**

Given that the melting and boiling points of Na are lower than those of NaCl, we are to predict the relative strengths of metallic and ionic bonds.

Analyze

Solid sodium is held together by metallic bonds, whereas sodium chloride is held together by ionic bonds. The higher the melting point or boiling point of a substance, the stronger the forces between particles of that substance.

Solve

Because the melting point and boiling point of NaCl are higher than those of Na, ionic bonds must be stronger than metallic bonds.

Think About It

Some metallic bonds are quite strong, as evidenced by the melting points of some metals. Sodium has a melting point of 97.72°C, whereas tungsten melts at 3422°C.

12.51. **Collect and Organize**

We are to consider whether band theory can explain why hydrogen at very low temperatures and high pressures might act like a metal and conduct electricity.

Analyze

Band theory is a model of bonding in which orbitals on many atoms are combined, as in molecular orbital theory, to form a fully or partially filled valence band and an empty conduction band.

Solve

Yes, by combining the $1s$ orbitals on many hydrogen atoms at very low temperatures and at high pressures, we could construct a molecular orbital diagram that would show a partially filled valence band (as for copper in Figure 12.20) in which the electrons can move into the unfilled portion of the band where they can migrate freely.

Think About It

Metallic liquid hydrogen was discovered with hydrogen at very low temperatures and very high pressures in 1996 at the Lawrence Livermore National Laboratory quite by accident.

12.53. **Collect and Organize**

We are asked to identify which groups in the periodic table contain metals with filled valence bands.

Analyze

The metals in the periodic table are in groups 1–12, with some in group 13 (Al, Ga, In, Tl), group 14 (Si, Ge, Sn, Pb), group 15 (As, Sb, Bi), group 16 (Te, Po), and group 17 (At). Filled valence bands for metals occur for electron configurations of ns^2 and nd^{10}.

Solve
Metals of groups 2 and 12 have filled valence bands.

Think About It
The orbital energy diagram for these metals will be similar to that of zinc in Figure 12.21.

12.55. Collect and Organize / Analyze
We are asked to describe the differences between n-type and p-type semiconductors. n-Type semiconductors have a small quantity of dopant with more valence electrons than the metalloid, whereas p-type semiconductors have a small quantity of dopant with fewer valence electrons than the metalloid.

Solve
Doping a metalloid generates a semiconductor with a smaller band gap than the starting material. n-Type semiconductors have "extra" electrons that are higher in energy than the valence band for the undoped material. p-Type semiconductors have "extra" holes that are lower in energy than the conduction band for the undoped material. In both types of doped semiconductor the effect is similar—the band gap is lowered, and electrons may flow through the material with greater ease than in the undoped metalloid.

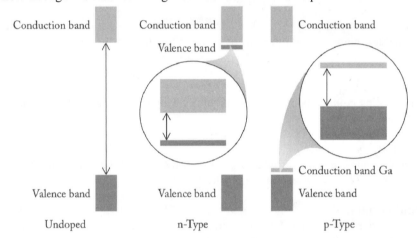

Think About It
Phosphorus, with one more electron than silicon, would form an n-type semiconductor if it were present as a dopant in silicon. Aluminum, with one fewer electron than silicon, would form a p-type semiconductor if it were present as a dopant in silicon.

12.57. Collect and Organize
We are asked to identify in which group of the periodic table we might find elements that would be useful to form a p-type semiconductor with Sb_2S_3.

Analyze
Antimony has five valence electrons, and sulfur has six valence electrons. To form a p-doped semiconductor, we would choose elements with fewer than five valence electrons.

Solve
Group 14 elements with four valence electrons would form p-type semiconductors when doped into Sb_2S_3.

Think About It
Similarly, group 13 elements, with three valence electrons, might also be used to form p-doped Sb_2S_3 semiconductors.

12.59. Collect and Organize
We consider the properties of nitrogen-doped diamond.

Analyze

(a) Nitrogen, with five valence electrons, has one more electron than carbon, with four valence electrons.

(b) In diamond, the valence band and the conduction band are separated by a large band gap, making diamond an insulator. Adding nitrogen to the diamond structure adds a partially filled band above diamond's valence band and below its conduction band.

(c) Because $E = hc/\lambda$, we can calculate the energy of the band gap when $\lambda = 4.25 \times 10^{-7}$ m.

Solve

(a) Because nitrogen has one more valence electron than carbon, nitrogen serves as an n-type dopant in diamond.

(b)

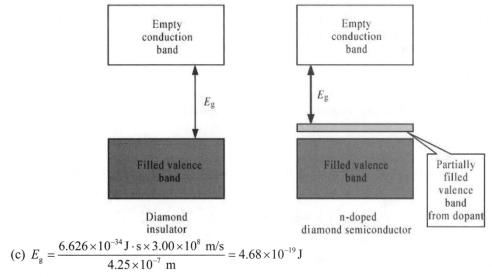

(c) $E_{g} = \dfrac{6.626 \times 10^{-34}\,\text{J} \cdot \text{s} \times 3.00 \times 10^{8}\ \text{m/s}}{4.25 \times 10^{-7}\ \text{m}} = 4.68 \times 10^{-19}\,\text{J}$

Think About It

Because nitrogen-doped diamonds absorb violet light, we observe a yellow color (the complement of violet) in the diamond.

12.61. **Collect and Organize**

We are asked to determine the effect of increased pressure on the wavelength of light emitted by ZnS. For this material, the band gap increases with increasing pressure.

Analyze

The wavelength of light absorbed or emitted by a semiconductor is inversely related to the band gap energy:

$$E_{g} = \frac{hc}{\lambda}$$

To calculate the wavelength we can rearrange this equation to

$$\lambda = \frac{hc}{E_{g}}$$

Solve

As the band gap increases with increasing pressure, the wavelength of light emitted will decrease.

Think About It

Wavelength is inversely proportional to energy. A smaller wavelength corresponds to more energetic (blue-shifted) light.

12.63. **Collect and Organize**

We are to explain why the Cl^- ions in the rock salt structure of LiCl touch in the unit cell, but those in KCl do not.

Analyze

In the fcc rock salt lattice, the alkali metal, Li^+ or K^+, is placed into the octahedral holes of the unit cell. From the radius ratio rule applied to LiCl and KCl (ionic radii are given in Figure 10.2),

$$\frac{r_{Li^+}}{r_{Cl^-}} = \frac{76 \text{ pm}}{181 \text{ pm}} = 0.42 \qquad \frac{r_{K^+}}{r_{Cl^-}} = \frac{138 \text{ pm}}{181 \text{ pm}} = 0.76$$

Solve

The radius ratios show that K^+ is large and so does not fit well into the octahedral holes (radius ratio between 0.41 and 0.73). Therefore, the rock salt structure for KCl has Cl^- ions that may not touch in order to accommodate the large K^+ ions in the fcc lattice.

Think About It

The radius ratio rule is simply a guide to predicting in which type of hole cations will fit within the anion lattice. You will find that sometimes we would predict an octahedral hole for a cation when the actual crystal places the cation into a cubic hole, for example.

12.65. **Collect and Organize**

We are to describe how CsCl could be viewed as both a simple cubic and a body-centered cubic structure.

Analyze

A simple cubic structure consists of atoms located only at the corners of a cube. A body-centered cubic structure consists of an atom inside a cube of atoms.

Solve

The radius of Cl^- is 181 pm and the radius of Cs^+ is 170 pm and so their radii are very similar. The Cs^+ ion at the center of Figure P12.65 occupies the center of the cubic cell, so CsCl could be viewed as a body-centered cubic structure when taking into account the ions' slight difference in size. However, if we look at the ions as roughly equal in size, the unit cell becomes two interpenetrating simple cubic unit cells.

Think About It

In the body-centered cubic unit cell, the cubic hole in the simple cubic unit cell is filled.

12.67. **Collect and Organize**

We can use the radius ratio rule and the formula for the salt to determine whether $CaCl_2$ could have a rock salt structure.

Analyze

From Figure 10.2, the radius of Cl^- is 181 pm and the radius of Ca^{2+} is 100 pm. The radius ratio is therefore

$$\frac{r_{Ca^{2+}}}{r_{Cl^-}} = \frac{100 \text{ pm}}{181 \text{ pm}} = 0.552$$

Solve

The radius ratio of Ca^{2+} and Cl^- shows that the Ca^{2+} ions would fit into the octahedral holes of an fcc lattice of Cl^- anions. However, this structure would not be like NaCl, where all the octahedral holes are filled in the fcc lattice. To balance charge, only half of the octahedral holes are filled in $CaCl_2$.

Think About It

$CaCl_2$ does not have a structure like CaF_2, in which the Ca^{2+} ions form an fcc array into which the F^- ions fill all the octahedral holes. Instead it adopts a rutile, TiO_2, structure.

12.69. Collect and Organize

Given that the unit cell edge length for an alloy of Cu and Sn is the same as that for pure Cu, we are to determine whether the alloy would be more dense than pure Cu.

Analyze

Density is mass per volume. If the unit cell edge length is the same for both the alloy and pure copper, then the volumes of the unit cells of both substances are the same. The density would be greater only if the mass of the unit cell alloy is greater than for the pure copper.

Solve

Because tin has a greater molar mass than copper, the mass of the unit cell in the alloy would be greater and therefore, yes, the density of the alloy would be greater than the density of pure copper.

Think About It

The unit cell volume does not change much for substitutional alloys because the alloying atoms are close in size (within 15%) to the atoms being replaced.

12.71. Collect and Organize

We are to predict the effect on density of an ionic compound of rock salt structure as the cation–anion radius ratio increases.

Analyze

When the cation–anion ratio increases, the cation is increasing, relatively, in size. If the anions remain in the rock salt structure, they might have to expand the fcc structure to accommodate larger cations.

Solve

As the cation–anion radius ratio increases, the closest-packed anions would expand so as to maintain the cations in the octahedral holes of the fcc lattice. The cell volume, therefore, increases and the calculated density would be less than the measured density.

Think About It

Another factor in density is the molar mass of the ions of the structure.

12.73. Collect and Organize

For a ccp array of O^{2-} ions with $\frac{1}{4}$ of the octahedral holes containing Fe^{3+}, $\frac{1}{8}$ of the tetrahedral holes containing Fe^{3+}, and $\frac{1}{4}$ of the octahedral holes containing Mg^{2+}, we are asked to write the formula of the ionic compound.

Analyze

In a ccp array of ions, there are four octahedral holes and eight tetrahedral holes. The ccp array has an fcc unit cell which has four closest-packed anions.

Solve

The unit cell will have the following:
   4 O^{2-} anions in the fcc unit cell
   1 Fe^{3+} in octahedral holes
   1 Fe^{3+} in tetrahedral holes
   1 Mg^{2+} in octahedral holes
The formula is $MgFe_2O_4$.

Think About It

The salt has charge balance as well: 4 O^{2-} gives a charge of 8–, which is balanced by 2 Fe^{3+} plus 1 Mg^{2+} or a charge of 8+.

12.75. Collect and Organize

We consider the structure of anatase, a form of TiO_2, found on a map of Vinland believed to date from the 1400s.

Analyze

We can predict the type of hole Ti^{4+} is likely to occupy by calculating the radius ratio $r_{Ti^{4+}}/r_{O^{2-}}$. The radius ratios for tetrahedral, octahedral, and cubic holes are 0.22–0.41, 0.41–0.73, and 0.73–1.00, respectively (Table 12.4). In the ccp structure the unit cell is fcc. This unit cell contains four octahedral and eight tetrahedral holes.

Solve

(a) From the radius ratio,

$$\frac{r_{Ti^{4+}}}{r_{O^{2-}}} = \frac{60.5 \text{ pm}}{140 \text{ pm}} = 0.432$$

Ti^{4+} is expected to occupy octahedral holes.

(b) To give charge balance there is one Ti^{4+} ion for every two O^{2-} ions in the lattice. Because the unit cell contains 4 O^{2-} ions, there must be 2 Ti^{4+} ions in the unit cell. Since there are four octahedral holes in the unit cell, this must mean that half of the octahedral holes in the unit cell are occupied.

Think About It

If the Ti^{4+} were small enough to fit into the tetrahedral holes, then one-fourth of the tetrahedral holes would be occupied.

12.77. Collect and Organize

We are asked to identify the differences between the rock salt and cesium chloride structures and explain how both structures are a suitable match for MgSe.

Analyze

In the rock salt structure, the Se^{2-} ions would be in an fcc array, with Mg^{2+} occupying all of the octahedral holes. In the cesium chloride structure the Se^{2-} ions are in a simple cubic array, with Mg^{2+} ions occupying the cubic holes.

Solve

(a) The rock salt and cesium chloride lattices have the same ratio of ions, but the relative arrangements differ. In the rock salt structure, the Se^{2-} ions would be in an fcc array, with Mg^{2+} occupying all of the octahedral holes. In the cesium chloride structure the Se^{2-} ions are in a simple cubic array, with Mg^{2+} ions occupying the cubic holes.

(b) Both structures have the same ratio of Mg to Se.

For the rock salt structure, the ratio is Mg_4Se_4, or MgSe:

$$\tfrac{1}{4}\left(12 \ Mg^{2+}\right) + 1 \ Mg^{2+} = 4 \ Mg^{2+}$$

$$\tfrac{1}{8}\left(8 \ Se^{2-}\right) + \tfrac{1}{2}\left(6 \ Se^{2-}\right) = 4 \ Se^{2-}$$

For the cesium chloride structure, the ratio is MgSe:

$$1 \ Mg^{2+} = 1 \ Mg^{2+}$$

$$\tfrac{1}{8}\left(8 \ Se^{2-}\right) = 1 \ Se^{2-}$$

Think About It

We can also imagine the rock salt structure like two interpenetrating fcc lattices, and the cesium chloride structure like two interpenetrating simple cubic lattices.

12.79. Collect and Organize

Knowing the structure of the unit cell of ReO_3 and the radii of Re and O atoms (given as 137 pm and 73 pm, respectively), we are asked to calculate the density of ReO_3.

Analyze
Given that there are Re atoms at each corner of the cubic unit cell, there is $\left(8 \times \frac{1}{8}\right)$, or 1 Re atom, in the unit cell. Likewise, given that there are 12 edge O atoms, there are $\left(12 \times \frac{1}{4}\right) = 3$ O atoms in the unit cell. We are told that the atoms touch (Re–O–Re) along an edge of the unit cell, therefore the cell edge length is
$$137 + 73 + 73 + 137 = 420 \text{ pm}$$
To calculate the density of ReO_3 we use
$$d = \frac{\text{mass of atoms in unit cell}}{\text{volume of unit cell}}$$

Solve
The mass, m, of atoms in a unit cell of ReO_3 is
$$m = \frac{1 \text{ Re atom} \times \dfrac{186.21 \text{ g}}{\text{mol}} + 3 \text{ O atoms} \times \dfrac{15.999 \text{ g}}{\text{mol}}}{6.022 \times 10^{23} \text{ atoms/mol}} = 3.889 \times 10^{-22} \text{ g}$$
The volume, V, of the unit cell (in cm^3) is
$$V = \left(420 \text{ pm} \times \frac{1 \times 10^{-10} \text{ cm}}{1 \text{ pm}}\right)^3 = 7.409 \times 10^{-23} \text{ cm}^3$$
The density, d, of the solid is then
$$d = \frac{3.889 \times 10^{-22} \text{ g}}{7.409 \times 10^{-23} \text{ cm}^3} = 5.25 \text{ g/cm}^3$$

Think About It
Remember that this is a calculated, or theoretical, density. The actual density, even if the structure is correctly described, may differ slightly from the theoretical value.

12.81. Collect and Organize
Given the structure (rock salt) and density (3.60 g/cm^3) of MgO, we are asked to calculate the length of the edge of the unit cell (ℓ).

Analyze
The volume of the unit cell is simply ℓ^3. From the density we can calculate the volume, V, as
$$V = \frac{\text{mass of unit cell}}{\text{density}}$$
From the volume of the cubic unit cell, where the lengths of all the edges are equal, we can calculate the edge length
$$\text{Edge length} = \sqrt[3]{\text{volume}}$$
In the fcc unit cell there are four Mg atoms and four O atoms to give a mass, m, of
$$m = \frac{4 \text{ Mg atoms} \times \dfrac{24.305 \text{ g}}{\text{mol}} + 4 \text{ O atoms} \times \dfrac{15.999 \text{ g}}{\text{mol}}}{6.022 \times 10^{23} \text{ atoms/mol}} = 2.677 \times 10^{-22} \text{ g}$$

Solve
$$\text{Volume} = \frac{2.677 \times 10^{-22} \text{ g}}{3.60 \text{ g/cm}^3} = 7.436 \times 10^{-23} \text{ cm}^3$$
$$\ell = \sqrt[3]{\text{volume}} = \sqrt[3]{7.436 \times 10^{-23} \text{ cm}^3} = 4.21 \times 10^{-8} \text{ cm or 421 pm}$$

Think About It

This value seems reasonable. The radius of an oxygen atom is 73 pm and that of a manganese atom is 127 pm (Appendix 3). Assuming that the O–Mn–O atoms touch along the fcc unit edge length (a good first approximation), the edge length would be

$$73 + 127 + 127 + 73 = 400 \text{ pm}$$

12.83. Collect and Organize

To answer why black phosphorus has puckered rings when graphite has planar rings, we need to consider the hybridization of the phosphorus and carbon atoms that make up these two substances.

Analyze

In both substances, the atoms are each bound to two other atoms. The phosphorus atom brings five valence electrons to its structure, but carbon brings only four to its structure.

Solve

The hybridization of the phosphorus atom in black phosphorus is sp^3 with bond angles of 102° to give a puckered ring. In graphite, the carbon atoms are sp^2 hybridized with bond angles of 120°, which gives graphite a flat geometry.

Think About It

Graphite carbon atoms also π-bond to each other since they each have an electron in the unhybridized p orbital.

12.85. Collect and Organize

We can use the edge length of the fcc lattice of C_{60} to calculate the density of buckminsterfullerene and the radius of a C_{60} molecule.

Analyze

The mass of the unit cell may be calculated by multiplying the molar mass of C_{60} (720.66 g/mol) by the four atoms in an fcc unit cell and dividing by Avogadro's number. We may determine the volume of the unit cell by cubing the edge length (1410 pm). The density is equal to the mass divided by the volume of the unit cell. Using the relationship $r = 0.3536\,\ell$, we may calculate the radius of the C_{60} molecule.

Solve

(a)

$$4\ C_{60}\ \text{molecules} \times \frac{1\ \text{mol}\ C_{60}}{6.022 \times 10^{23}\ C_{60}\ \text{molecules}} \times \frac{720.66\ \text{g}\ C_{60}}{1\ \text{mol}\ C_{60}} = 4.7868 \times 10^{-21}\ \text{g}$$

$$\text{Volume} = \left(1410\ \text{pm} \times \frac{1\ \text{m}}{1 \times 10^{12}\ \text{pm}} \times \frac{100\ \text{cm}}{1\ \text{m}}\right)^3 = 2.8032 \times 10^{-21}\ \text{cm}^3$$

$$d = \frac{4.7868 \times 10^{-21}\ \text{g}}{2.8032 \times 10^{-21}\ \text{cm}^3} = 1.708\ \text{g/cm}^3$$

(b) The radius of the C_{60} molecules is $r = 0.3536(1410 \text{ pm}) = 498.6$ pm.

Think About It

The C_{60} molecule has a radius that is roughly five to six times that of most ions or atoms, though the density is similar to that observed for the other allotropes of carbon, graphite and diamond.

12.87. Collect and Organize

Knowing the distance between the phosphorus atoms in its cubic form, we are to calculate the density.

Analyze

Density is mass per unit volume. The mass of phosphorus in one unit cell will be the number of atoms of P in one unit cell multiplied by the molar mass of P and divided by Avogadro's number. The volume of the unit cell is the cube of the length of the side of the unit cell. Density is usually expressed as grams per cubic centimeter (g/cm^3), so we have to convert picometers to centimeters.

Solve
Mass of phosphorus in one unit cell

$$8\left(\tfrac{1}{8}\right) \times \frac{30.97\ g}{1\ mol} \times \frac{1\ mol}{6.022 \times 10^{23}\ atoms} = 5.143 \times 10^{-23}\ g$$

Volume of the unit cell

$$\left(238\ pm \times \frac{1 \times 10^{-10}\ cm}{1\ pm}\right)^3 = 1.35 \times 10^{-23}\ cm^3$$

Density of cubic phosphorus

$$\frac{5.143 \times 10^{-23}\ g}{1.35 \times 10^{-23}\ cm^3} = 3.81\ g/cm^3$$

Think About It
The density of phosphorus in another form would be different. If we measure the density of a crystal of known composition, we may be able to determine what its unit cell looks like.

12.89. Collect and Organize
From a list of properties, we are to identify which describe ceramics and which are associated with metals.

Analyze
Metallic bonding in metals allows atoms to slip past each other and be easily deformed. Ceramics have either covalent or ionic bonding (or sometimes a mixture of both) that does not allow the atoms to easily slip past each other when stress is applied. Metals have either partially filled valence bands or overlapping valence and conduction bands that allow electrons to flow through the bands. Ceramics, however, have large band gaps between their valence and conduction bands that render them thermally insulative and nonconductive to the flow of electrons.

Solve
Ductility, electrical and thermal conductivity, and malleability describe metals. Ceramics can be characterized as being electrically and thermally insulative and brittle.

Think About It
The bonding differences between metals and ceramics account for their very different properties and, therefore, unique applications.

12.91. Collect and Organize
We are to write the formula for the mineral formed when Mg^{2+} replaces Al^{3+} in kaolinite [$Al_2(Si_2O_5)(OH)_4$].

Analyze
When replacing Al^{3+} with Mg^{2+}, we must be careful to put in as many Mg^{2+} as needed to balance the charge of the Al^{3+} removed. Since we replace two Al^{3+} ions (6+ charge) we need 3 Mg^{2+} to balance the charge.

Solve
$Mg_3(Si_2O_5)(OH)_4$

Think About It
Several other ions (Fe^{3+}, Be^{2+}, etc.) could replace the Al^{3+} in kaolinite.

12.93. Collect and Organize
For the reaction of $KAlSi_3O_8$ with water and carbon dioxide, we are asked to determine whether it is a redox reaction and to balance the reaction to give $Al_2(Si_2O_5)(OH)_4$, SiO_2, and K_2CO_3 as products.

Analyze

In a redox reaction the oxidation state of atoms must change. All species for this reaction contain O^{2-}, K^+, Si^{4+}, H^+, Al^{3+}, and C^{4+}.

Solve

$$2\ KAlSi_3O_8(s)\ +\ 2\ H_2O\ (\ell)\ +\ CO_2(aq)\ \rightarrow\ Al_2(Si_2O_5)(OH)_4(s)\ +\ 4\ SiO_2(s)\ +\ K_2CO_3(aq)$$

Because none of the atoms change oxidation state, this is not a redox reaction.

Think About It

In this weathering process the larger silicate anion $Si_3O_8^{4-}$ is broken down by water to $Si_2O_5^{2-}$ and SiO_2.

12.95. Collect and Organize

For the transformation of anorthite ($CaAl_2Si_2O_8$) to a mixture of grossular [$Ca_3Al_2(SiO_4)_3$], kyanite (Al_2SiO_5), and quartz (SiO_2) under high pressure, we are asked to write the balanced equation and to determine the charges on the silicate anions.

Analyze

The charge on the silicate ions is found knowing that the cations are Ca^{2+} and Al^{3+}.

Solve

(a) $3\ CaAl_2Si_2O_8(s)\ \rightarrow\ Ca_3Al_2(SiO_4)_3(s)\ +\ 2\ Al_2SiO_5(s)\ +\ SiO_2(s)$
(b) In anorthite, the silicate anion is $Si_2O_8^{8-}$.
In grossular, the silicate anion is SiO_4^{4-}.
In kyanite, the silicate anion is SiO_5^{6-}.

Think About It

There are many silicate minerals in nature.

12.97. Collect and Organize

Given the radii of Ba^{2+}, Ti^{4+}, and O^{2-} (135, 60.5, and 140 pm, respectively), we can use the radius ratio rule to determine which holes Ba^{2+} and Ti^{4+} occupy in a closest-packed arrangement of O^{2-} anions.

Analyze

The radius ratio can help us predict whether the smaller ion would fit into a cubic (radius ratio 0.73–1.00), octahedral (0.41–0.73), or tetrahedral (0.22–0.41) hole in the crystal lattice.

Solve

Cubic holes can accommodate Ba^{2+}:

$$\frac{r_{Ba^{2+}}}{r_{O^{2-}}}=\frac{135\ pm}{140\ pm}=0.964$$

Octahedral holes (maybe tetrahedral) can accommodate Ti^{4+}:

$$\frac{r_{Ti^{4+}}}{r_{O^{2-}}}=\frac{60.5\ pm}{140\ pm}=0.432$$

Think About It

Ti^{4+} has a radius ratio at the edge of the tetrahedral–octahedral ranges, so it is difficult to predict which of these holes it will occupy.

12.99. Collect and Organize

We are asked to explain why an amorphous solid does not give sharp peaks when scanned by X-ray diffraction.

Analyze
X-ray diffraction gives sharp peaks for crystalline materials that have a regular repeating array of atoms. Amorphous materials have no long-range order in their arrangement of atoms.

Solve
An amorphous solid has no regular repeating lattice to diffract the X-rays and therefore cannot give rise to the distinct constructive and destructive interference needed to produce sharp peaks in the X-ray diffraction scan.

Think About It
Amorphous materials in X-ray diffraction either give no signal or may show very broad peaks.

12.101. **Collect and Organize**
X-rays and microwaves are both electromagnetic radiation. We are to consider why X-rays, not microwaves, are used for structure determination.

Analyze
X-rays have wavelengths of the order of 10 to 10,000 pm. Microwaves have wavelengths of the order of 1 mm to 1 m.

Solve
The separation of atoms in crystal lattices is of the order of 10^{-10} m, or 100 pm. X-rays have wavelengths of the order of the separation of atoms in crystals, so constructive and destructive interference occurs. Microwaves have wavelengths too long to be diffracted by crystal lattices.

Think About It
X-ray diffraction is a useful technique, but it has difficulty "seeing" light elements, like H atoms. Complementary information on structure can be obtained from neutron-diffraction experiments.

12.103. **Collect and Organize**
We are to consider why different wavelengths of X-rays might be used to determine crystal structures.

Analyze
From the Bragg equation

$$n\lambda = 2d \sin \theta$$

we see that if λ is smaller, θ is smaller.

Solve
If a crystallographer uses a shorter λ wavelength, the data set can be collected over a smaller scanning range.

Think About It
In this way the crystallographer can see more data over a smaller scanning range so that if the camera or source can only obtain $2\theta = 30°$, for example, a shorter λ would reveal more peaks than a longer λ.

12.105. **Collect and Organize**
Given that the lattice spacing in sylvite (KCl) is larger than in halite (NaCl), we can use the Bragg equation to predict which crystal will diffract X-rays of a particular wavelength through higher 2θ values.

Analyze
Rearranging the Bragg equation to solve for d gives

$$d = \frac{n\lambda}{2\sin \theta}$$

This shows that $\sin \theta$ is inversely proportional to d.

Solve

The smaller the distance d (the lattice spacing) the larger the $\sin \theta$ and the larger the 2θ. Therefore, halite with smaller lattice spacing diffracts X-rays through larger 2θ values.

Think About It

This means that the pattern for NaCl is more spread out than that of KCl.

12.107. Collect and Organize

For galena, which shows reflections in X-ray diffraction ($\lambda = 71.2$ pm) at $2\theta = 13.98°$ and $21.25°$, we are asked to determine the value of n for these reflections and to calculate the spacing between the layers of the lattice (d).

Analyze

To determine n we must find a pattern in the angles of reflection. We notice here that for $\theta = 6.99°$ and $\theta = 10.62°$ we have a factor of 1.52. The spacing of the layers in the lattice is calculated from the Bragg equation:

$$d = \frac{n\lambda}{2\sin \theta}$$

Solve

Because the ratio of the reflection angles (θ) is $10.62°/6.99° = 1.52$, the values of n are 2 ($\theta = 6.99°$) and 3 ($\theta = 10.62°$).
The lattice spacings are

$$d = \frac{2 \times 71.2 \text{ pm}}{2\sin \left(6.99°\right)} = 585 \text{ pm}$$

$$d = \frac{3 \times 71.2 \text{ pm}}{2\sin \left(10.62°\right)} = 580 \text{ pm}$$

Average lattice spacing $(585 + 580)/2 = 582$ pm.

Think About It

The calculated lattice spacings from the reflections may not be exactly equal, but they will be close, as in the example above.

12.109. Collect and Organize

For a lattice distance of 1855 pm, we are to calculate the smallest angle of diffraction (2θ) for 154 pm wavelength X-rays diffracted in pyrophyllite.

Analyze

We can rearrange the Bragg equation to solve for $\sin \theta$:

$$\sin \theta = \frac{n\lambda}{2d}$$

where $n = 1$ for the smallest angle of diffraction. The 2θ value is twice the calculated value of θ.

Solve

$$\sin \theta = \frac{1 \times 154 \text{ pm}}{2 \times 1855 \text{ pm}} = 0.04151$$

$$\theta = 2.38°$$

$$2\theta = 4.76°$$

Think About It

For $n = 3$, the reflection would appear at

$$\sin\theta = \frac{3\times154\ \text{pm}}{2\times1855\ \text{pm}} = 0.1245$$
$$\theta = 7.15°$$
$$2\theta = 14.3°$$

12.111. Collect and Organize
For a unit cell with X at the eight corners of the cubic unit cell, Y at the center of the cube, and Z at the center of each face, we are to write the formula of the compound.

Analyze
Each atom at the corner of a cube counts as $\frac{1}{8}$ in the unit cell. Each atom on a face counts as $\frac{1}{2}$ in the unit cell. Each atom in the center of the unit cell counts as one.

Solve
$$\text{Element X} = \tfrac{1}{8}\times 8 = 1\ \text{X}$$
$$\text{Element Y} = 1\times 1 = 1\ \text{Y}$$
$$\text{Element Z} = 6\times \tfrac{1}{2} = 3\ \text{Z}$$

The formula for the compound is XYZ_3.

Think About It
An example of this kind of structure is perovskite, $CaTiO_3$.

12.113. Collect and Organize
Given the phase diagram for titanium, we can identify the predominant phase under specific conditions and track phase changes as the pressure increases at a constant temperature.

Analyze
(a) We may identify the phase at 1500 K and 6 GPa by drawing lines from these points along the x and y axes and observing the phase where the lines intersect.
(b) We may track phase changes at a temperature of 725°C (998 K) by drawing a vertical line from this point along the x-axis and observing how many phase boundary lines it crosses.

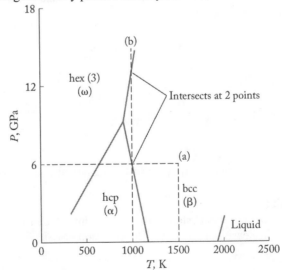

Solve
(a) At 1500 K and 6 GPa, titanium is present in the bcc (β) phase.

(b) Titanium undergoes two phase changes as we increase pressure. The transitions are hcp (α) $\rightarrow$ bcc (β) around 6 GPa, and bcc (β)$\rightarrow$ hex (3, ω) around 11 GPa.

Think About It

Don't forget to check the units on your temperature. If you had looked for phase transitions as we increase the pressure at 725 K, you would have discovered only one.

12.115. Collect and Organize

We consider cubic nanoparticles of Ag (25 nm on each side) and are asked to calculate the number of unit cells in a particle, the number of Ag atoms in a particle, and the number of silver atoms present in 1360 μg of silver nanoparticles.

Analyze

(a) We are given that the unit cells are body-centered cubic. In this arrangement the unit-cell edge length (ℓ) is $4r/\sqrt{3}$ where r is the radius of the silver atom. From the Appendix 3, the radius of a silver atom is 144 pm. Dividing the side length of a nanoparticle by the edge length of the unit cell, we may determine the number of unit cells along an edge of the nanoparticle; we may determine the number of unit cells per particle by cubing this value.

(b) The number of atoms in the nanoparticle may be found by multiplying the number of unit cells per particle, determined in (a), by four Ag particles per bcc unit cell.

(c) Converting the mass of silver to grams, dividing by the molar mass of Ag (107.8682 g/mol), multiplying by Avogadro's number, and dividing by the ratio of silver atoms per nanoparticle determined in (b) will furnish the number of nanoparticles in a sock.

Solve

(a) The edge length of a silver unit cell is

$$\ell = \frac{144 \text{ pm}}{0.3536} \times \frac{1 \text{ nm}}{1000 \text{ pm}} = 0.4072 \text{ nm}$$

The number of unit cells in a nanoparticle edge is

$$\frac{25 \text{ nm}}{0.4072 \text{ nm}} = 61.4 \text{ unit cells per edge}$$

An edge may not contain a fraction of a unit cell, so we must round down. Cubing this edge length gives $(61)^3 = 2.3 \times 10^5$ unit cells per particle.

(b) The number of Ag atoms in a nanoparticle is

$$\frac{2.3 \times 10^5 \text{ unit cells}}{\text{nanoparticle}} \times \frac{4 \text{ Ag atoms}}{1 \text{ unit cell}} = 9.1 \times 10^5 \text{ Ag atoms}$$

(c) The number of nanoparticles in one sock is

$$1360 \text{ μg Ag} \times \frac{1 \text{ g Ag}}{1 \times 10^6 \text{ μg Ag}} \times \frac{1 \text{ mol Ag}}{107.8682 \text{ g Ag}} \times \frac{6.022 \times 10^{23} \text{ atoms Ag}}{1 \text{ mol Ag}} \times \frac{1 \text{ nanoparticle}}{9.1 \times 10^5 \text{ Ag atoms}} = 8.4 \times 10^{12} \text{ nanoparticles}$$

Think About It

Even though nanoparticles are very small, they still contain thousands of Ag atoms.

12.117. Collect and Organize

We consider different possible crystal structures of Fe (radius 126 pm) and compare the densities of bcc and hcp iron and calculate the density of a crystal of 96% Fe and 4% Si.

Analyze

For each structure, the density is the mass of atoms in the unit cell divided by the volume of the cubic unit cell. For the bcc unit cell there are two atoms per unit cell and the edge length is $\ell = 4r/\sqrt{3}$. For the hcp unit cell there are two atoms per unit cell.

Solve

(a) The edge length of a bcc unit cell is

$$\ell = \frac{4r}{\sqrt{3}} = \frac{4 \times 126 \text{ pm}}{\sqrt{3}} = 291.0 \text{ pm or } 2.91 \times 10^{-8} \text{ cm}$$

The volume of the unit cell is

$$V = \left(291.0 \times 10^{-8} \text{ cm}\right)^3 = 2.464 \times 10^{-23} \text{ cm}^3$$

The mass of the unit cell is

$$2 \text{ atoms Fe} \times \frac{55.85 \text{ g}}{\text{mol}} \times \frac{1 \text{ mol}}{6.022 \times 10^{23} \text{ atoms}} = 1.855 \times 10^{-22} \text{ g}$$

The density is

$$d = \frac{1.855 \times 10^{-22} \text{ g}}{2.464 \times 10^{-23} \text{ cm}^3} = 7.53 \text{ g/cm}^3$$

(b) In the hcp unit cell there are also two atoms in the unit cell. We are given that the volume of the unit cell is 5.414×10^{-23} cm^3, so the density is

$$d = \frac{1.855 \times 10^{-22} \text{ g}}{5.414 \times 10^{-23} \text{ cm}^3} = 3.42 \text{ g/cm}^3$$

(c) If 4% of the mass is due to Si replacing Fe, the molar mass of the alloy would be

$$0.04 \times \frac{28.09 \text{ g}}{\text{mol}} + 0.96 \times \frac{55.85 \text{ g}}{\text{mol}} = 54.74 \text{ g/mol}$$

The mass of two atoms in the unit cell would be

$$2 \text{ atoms} \times \frac{54.74 \text{ g}}{\text{mol}} \times \frac{1 \text{ mol}}{6.022 \times 10^{23} \text{ atoms}} = 1.818 \times 10^{-22} \text{ g}$$

The density is

$$d = \frac{1.818 \times 10^{-22} \text{ g}}{5.414 \times 10^{-23} \text{ cm}^3} = 3.36 \text{ g/cm}^3$$

Think About It

Replacing Fe with an element of lower molar mass (silicon) gives a lower density alloy.

12.119. Collect and Organize

For the same crystalline structure of AuZn with $\ell = 319$ pm and AgZn with $\ell = 316$ pm, we can use the different molar masses of gold and silver and the length of the unit-cell edges to calculate the densities and predict which alloy is more dense.

Analyze

Because gold has a much higher molar mass than silver, we can easily predict that the density of the AuZn alloy would be greater than that of the AgZn alloy, despite the slightly smaller unit-cell edge length and therefore smaller volume.

Solve

For AuZn

$$d = \frac{\left(262.38 \text{ g/mol} \times 1 \text{ mol}/6.022 \times 10^{23}\right)}{\left(3.19 \times 10^{-8} \text{ cm}\right)^3} = 13.4 \text{ g/cm}^3$$

For AgZn

$$d = \frac{\left(173.28 \text{ g/mol} \times 1 \text{ mol}/6.022 \times 10^{23}\right)}{\left(3.16 \times 10^{-8} \text{ cm}\right)^3} = 9.12 \text{ g/cm}^3$$

The AuZn alloy is more dense than the AgZn alloy.

Think About It

Indeed, the prediction was correct and AuZn is substantially more dense than AgZn.

12.121. Collect and Organize

We are asked how Mn would incorporate into an iron lattice, either in the holes or as a substitute for iron.

Analyze

To occupy holes in the structure, Mn would have to have a distinctly smaller radius than the Fe radius. From Appendix 3, we find that the Mn radius is 127 pm and the Fe radius is 126 pm.

Solve

The radii of these metals are very similar, so Mn likely forms a substitutional alloy with iron.

Think About It

The radius of carbon (77 pm) is small enough to fit into the interstices of the iron–manganese lattice. The radius ratio of carbon to iron (77/126) gives a value of 0.61. This indicates that carbon would fit into the octahedral holes of the iron lattice.

12.123. Collect and Organize

From the formula Cu_3Al for an alloy with a bcc structure, we are to determine how the copper and aluminum atoms are distributed between the unit cells.

Analyze

The bcc unit cell contains two atoms. The radii of Al and Cu are 143 and 128, respectively. The radius ratio is 0.895.

Solve

The radius ratio indicates that Cu atoms fit into the cubic holes of the Al lattice. Another way to look at this is as a simple cubic arrangement of Cu atoms with an Al atom in the center of the unit cell. To be consistent with the formula (Cu_3Al), we would have to consider three unit cells, one of which would have one Al atom in the center.

Think About It

It is probably also possible to have an alloy of formula Al_3Cu since their atomic radii are so similar.

12.125. Collect and Organize

For each description of a reaction, we are to write a balanced chemical equation.

Analyze

We first have to write the chemical formulas of the named reactants and then complete and balance the equations.

Solve

(a) $SiO_2(s) + C(s) \rightarrow CO_2(g) + Si(s)$
(b) $Si(s) + 2\ Cl_2(g) \rightarrow SiCl_4(\ell)$
(c) $Ge(g) + 2\ Br_2(\ell) \rightarrow GeBr_4(s)$

Think About It

All of these are redox reactions in which a group 14 element is either oxidized (b and c) or reduced (a).

12.127. Collect and Organize

We can use the Bragg equation

$$n\lambda = 2d\sin\theta$$

to solve for the distance between the layers of silicon in $CaSi_2$ given 2θ values of 29.86° ($n = 2$), 45.46° ($n = 3$), and 62.00° ($n = 4$) when 154 pm X-rays are diffracted through a crystal. We are then to compute the distance between Ca^{2+} and a silicon layer, knowing that the calcium ions are located halfway between the layers. Lastly, we are to consider how $CaSi_2$ could act as a "deoxidizer."

Analyze

(a) To find the distance between the silicon layers we need to rearrange the Bragg equation to solve for d, the distance between diffraction planes or the distance between planes of silicon atoms in the structure.

$$d = \frac{n\lambda}{2\sin\theta}$$

We will compute the d value using all three 2θ values, but they should yield the same value.

(b) To find the distance between Ca^{2+} ions and a layer of silicon atoms, we need only divide the d spacing from part a by 2.

(c) From the name "deoxidizer" and the oxidation state of -1 of the Si atoms in $CaSi_2$, we can infer that the Si atoms in the calcium silicide will act as reducing agent being themselves oxidized to Si^{4+}, the most common oxidation state for silicon.

Solve

(a) For $2\theta = 29.86°$, $\theta = 14.93°$

$$d = \frac{2 \times 154 \text{ pm}}{2\sin 14.93°} = 598 \text{ pm}$$

For $2\theta = 45.46°$, $\theta = 22.73°$

$$d = \frac{3 \times 154 \text{ pm}}{2\sin 22.73°} = 598 \text{ pm}$$

For $2\theta = 62.00°$, $\theta = 31.00°$

$$d = \frac{4 \times 154 \text{ pm}}{2\sin 31.00°} = 598 \text{ pm}$$

The distance between the Si layers is 598 pm.

(b) The Ca–Si distance is $d/2 = 598$ pm/2 = 299 pm.

(c) In deoxidizing the iron, silicon is oxidized to Si^{4+} and any iron cations (such as Fe^{3+}) are reduced to Fe.

Think About It

The calcium ion, Ca^{2+}, has an ionic radius of 100 pm, so it fits well into the layers of Si atoms, which are nearly 600 pm apart.

CHAPTER 13 | Organic Chemistry: Fuels, Pharmaceuticals, Materials, and Life

13.1. Collect and Organize

Given the carbon-skeleton structures in Figure P13.1 of four hydrocarbons, we are to determine the degrees of unsaturation present in each.

Analyze

An unsaturated hydrocarbon contains one or more carbon–carbon double or carbon–carbon triple bonds. These compounds have less than the maximum amount of hydrogen for each carbon atom. A double bond has one degree of unsaturation and a triple bond has two degrees of unsaturation.

Solve

(a) This structure has one double bond. It has one degree of unsaturation.
(b) This structure has two double bonds. It has two degrees of unsaturation.
(c) This structure has neither double nor triple bonds. It is a saturated hydrocarbon with no degrees of unsaturation.
(d) This structure has one double and one triple bond. It has three degrees of unsaturation.

Think About It

Because the structure in (d) has three degrees of unsaturation, it will combine with three molecules of hydrogen (H_2) to form a saturated hydrocarbon.

13.3. Collect and Organize

From the structures of fragrant oils in Figure P13.3, we are to identify those that contain the alkene functional group.

Analyze

An alkene functional group is a carbon–carbon double bond.

Solve

Both pine oil and oil of celery contain a $C=C$ bond and so are classified as alkenes.

Think About It

Camphor is not an alkene but it does have a $C=O$ double bond. This functional group is a ketone.

13.5. Collect and Organize

We are asked to draw structures of the condensed structures provided, identifying cis and trans isomers of each.

Analyze

The terms cis and trans describe the relative position of substituents around the double bond. Cis isomers have the largest groups on the same side of the double bond, whereas trans isomers have the largest groups on opposing sides of the double bond.

Solve

(a)

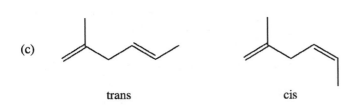

trans   cis

(b)

(both isomers equivalent)

(c)

trans   cis

Think About It
The cis and trans isomers in (b) are equivalent because there are two equivalent groups on the left side of the molecule (as drawn).

13.7. **Collect and Organize**
Of the four hydrocarbons shown in Figure P13.7, we are to identify those that are aromatic.

Analyze
Aromatic compounds have planar, hexagonal rings of six sp^2-hybridized carbon atoms with alternating single and double bonds.

Solve
(a) and (b) are aromatic.

Think About It
Aromatic compounds have a special stability due to resonance.

13.9. **Collect and Organize**
For the molecules in Problem 13.8 we are asked to identify the other functional groups, aside from the aromatic rings.

Analyze
The other typical functional groups include alkenes (C=C), alkynes (C≡C), alcohols (R—OH), ethers (R—O—R), aldehydes [R(C=O)H], ketones [R(C=O)R'], carboxylic acids [R(C=O)OH], esters [R(C=O)OR'], and amides [R(C=O)NH$_2$].

Solve
Benzyl acetate contains an ester group. Carvone contains two alkene groups and a ketone group. Cinnamaldehyde contains an alkene and an aldehyde group.

Think About It
The wide variety of functional groups possible in organic compounds means that there are many combinations of functional groups in organic chemistry. This gives rise to a seemingly endless array of different substances with potentially useful properties.

13.11. Collect and Organize

From the structure of dihydroxydimethylsilane (Figure P13.11) we are to draw the condensed structure for the repeating monomeric unit in Silly Putty.

Analyze

The condensation reaction of dihydroxydimethylsilane combines the monomers of $Si(CH_3)_2(OH)_2$ to create a larger molecule, losing water as a by-product. The –OH groups on the monomer will combine to produce the –O—Si—O– linkage for the polymer backbone.

Solve

Continuing this reaction produces a long-chain polymer with the condensed structure

Think About It

These types of siloxane polymers have found uses in soft contact lenses, oils, and greases.

13.13. Collect and Organize

We are asked to draw monomeric units of *cis*- and *trans*-polyisoprene.

Analyze

The structures of the two polyisoprenes are shown in Figure P13.13. These polymers form through addition reactions of an alkene. The monomeric unit is the smallest repeating unit in the polymer.

Solve

For *cis*-polyisoprene the monomeric unit is

For *trans*-polyisoprene the monomeric unit is

Think About It

The difference in these polymers is the orientation of how the monomeric units are joined together, which gives a different orientation of the C=C double bond in the polymer backbone.

13.15. Collect and Organize

Using the structure of Mevacor in Figure P13.15, we are to determine the number of chiral carbons this cholesterol drug contains and identify functional groups present in the molecule.

Analyze

A carbon atom is chiral when it is bonded to four different groups. Carbons that are sp^2 or sp hybridized in a structure cannot be chiral because they have only three or two groups bonded to them, respectively.

Solve

The functional groups present in Mevacor are: alcohol, ester, alkene, and alkyne.

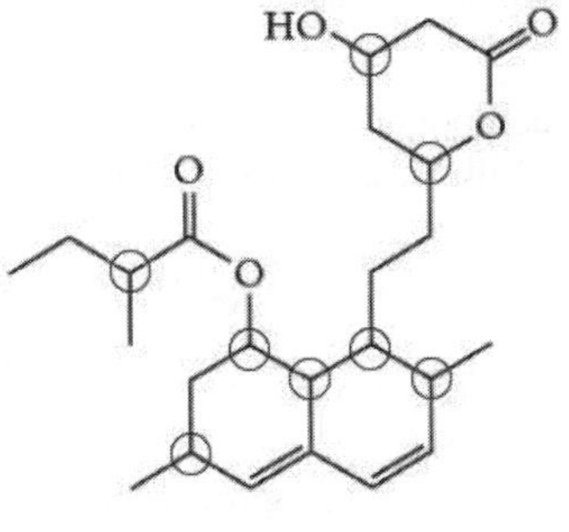

Mevacor

Mevacor has eight chiral carbon atoms.

Think About It

The more chiral carbons there are in an organic molecule, the more difficult it is, in general, to synthesize and purify it as a single enantiomer.

13.17. Collect and Organize

Carbon can have three different hybridized sets of *s* and *p* orbitals. We are asked to describe three ways in which carbon–carbon bonds form.

Analyze

There are three $2p$ orbitals on carbon that may hybridize with the one $2s$ orbital.

Solve

The sp hybrid orbital on carbon can form the σ component of a triple bond; the sp^2 hybrid orbitals can form the σ component of a double bond; the sp^3 hybrid orbitals can form single bonds.

Think About It

Carbon does not form sp^3d orbitals because there are no $2d$ orbitals (and the $3d$ orbitals are too high in energy to mix with the $2s$ and $2p$ orbitals); it does not expand its valence beyond four.

13.19. Collect and Organize

Tungsten carbide contains carbon. We are asked whether this compound is considered to be an organic compound.

Analyze

Organic compounds are compounds in which there is a carbon–element bond.

Solve

The carbon atoms in WC occupy the interstices of the closest-packed tungsten lattice and therefore are not formally bonded to the tungsten at all. No, tungsten carbide is not considered to be an organic compound.

Think About It

We generally consider organic compounds to contain carbon covalently bonded to carbon, hydrogen, oxygen, nitrogen, phosphorus, sulfur, and other atoms.

13.21. Collect and Organize
From Chapter 9 we are to identify molecules that contain more than one functional group.

Analyze
A functional group is a unit of a compound that has characteristic properties for the organic compound. The functional groups include alkanes, alkenes, alkynes, alcohols, ethers, aldehydes, ketones, carboxylic acids, esters, and amides.

Solve
Some examples from Chapter 9 are as follows:
 Acrolein (alkene + aldehyde) in Section 9.5
 Spearmint and caraway (alkene + ketone) in Figure 9.38

Think About It
The wide variety of functional groups possible in organic compounds means that there are many combinations of functional groups in organic chemistry. This gives rise to a seemingly endless array of different substances with potentially useful properties.

13.23. Collect and Organize
Given that polyethylene is composed of C_2H_4 monomeric units, we are to calculate the number of monomers of ethylene needed to give a polymer of molar mass 100,000 g/mol.

Analyze
We can find the number of monomers by dividing 100,000 g/mol by the molar mass of the C_2H_4 monomer unit. C_2H_4 has a molar mass of 28.05 g/mol.

Solve

$$\frac{100{,}000 \text{ g/mol}}{28.05 \text{ g/mol}} = 3565 \text{ monomer units}$$

Think About It
This is a relatively small polymer. Some polymers have tens of thousands or more monomeric units in their polymer chains.

13.25. Collect and Organize
We are to compare the empirical formulas for linear and branched alkanes that have the same number of carbon atoms.

Analyze
For any alkane, whether branched or linear, the empirical formula is C_nH_{2n+2}.

Solve
Yes, linear and branched alkanes with the same number of carbon atoms have the same empirical formula.

Think About It
An example of this is the following alkanes with four carbon atoms and the empirical formula C_4H_{10}:

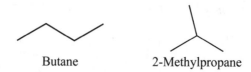

Butane   2-Methylpropane

13.27. Collect and Organize
By considering the number of covalent bonds formed by carbon in alkanes, we can determine the hybridization of the carbon atoms.

Analyze
In alkanes, carbon is singly bonded to four other atoms, either to other carbon atoms or to hydrogen atoms.

Solve
Carbon forms four single bonds when it undergoes sp^3 hybridization.

Think About It
When the carbon is sp^2 hybridized, it may form double bonds to other carbon atoms (alkenes) or to oxygen (ketones, aldehydes, carboxylic acids).

13.29. **Collect and Organize**
Given that cyclohexane (C_6H_{12}) is not planar we can draw the Lewis structure to help explain why.

Analyze
The Lewis structure for cyclohexane is

Solve
Each carbon atom in the six-membered ring is sp^3 hybridized with ideal bond angles of 109.5°. The ring, therefore, is not planar.

Think About It
The cyclohexane ring can adopt two structures; the chair structure is more stable.

Boat   Chair

13.31. **Collect and Organize**
By considering the definition of a saturated hydrocarbon, we can determine whether cycloalkanes are saturated.

Analyze
By the textbook definition (Section 13.2), a saturated hydrocarbon has the maximum ratio of hydrogen to carbon atoms in its structure and has an empirical formula of C_nH_{2n+2}.

Solve
No. A cycloalkane has a formula of C_nH_{2n}, and therefore is not a saturated hydrocarbon.

Think About It
This is a tricky definition, however. Another definition of a saturated hydrocarbon is when every carbon in the molecule is bonded to four other atoms. That definition would classify cycloalkanes as saturated hydrocarbons.

13.33. **Collect and Organize**

By considering the definition of structural isomers we can determine if they always have the same chemical properties.

Analyze

Two compounds are structural isomers if they have the same formula but different arrangements of the atoms.

Solve

No. Structural isomers have different chemical properties due to their different arrangement of atoms.

Think About It

Structural isomers also have different physical properties such as different melting points and vapor pressures.

13.35. **Collect and Organize**

For the molecular formula C_5H_{12} we are asked to draw and name all the structural isomers.

Analyze

Structural isomers all have the same molecular formula so all of our structures must contain five C atoms and 12 H atoms, just arranged differently. Naming of alkanes is described in Section 13.2 with examples in Table 13.4 for the linear alkanes.

Solve

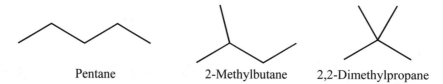

Pentane   2-Methylbutane   2,2-Dimethylpropane

Think About It

The naming convention used for these compounds uniquely describes the structure of each isomer.

13.37. **Collect and Organize**

Of the five molecules shown in Figure P13.37, we are to identify those that are structural isomers of octane. We are also asked to name the molecules.

Analyze

Any structural isomer of octane has to have the molecular formula C_8H_{18}. When naming the compound, it may help to draw the longest carbon chain, then determine which branches should be named as substituents.

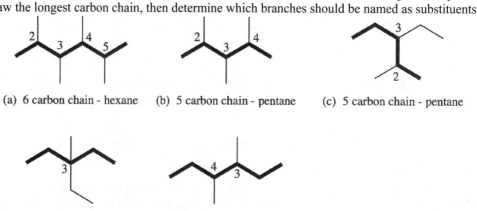

(a) 6 carbon chain - hexane   (b) 5 carbon chain - pentane   (c) 5 carbon chain - pentane

(d) 5 carbon chain - pentane   (e) 6 carbon chain - hexane

Solve

(a) This molecule is $C_{10}H_{22}$ and is not a structural isomer of octane. The name for this molecule is 2,3,4,5-tetramethylhexane.

(b) This molecule is C_8H_{18} and is therefore a structural isomer. The name for this molecule is 2,3,4-trimethylpentane.

(c) This molecule is C_8H_{18} and is therefore a structural isomer. The name for this molecule is 3-ethyl-2-methylpentane.

(d) This molecule is C_8H_{18} and is therefore a structural isomer. The name for this molecule is 3-ethyl-3-methylpentane.

(e) This molecule is C_8H_{18} and is therefore a structural isomer. The name for this molecule is 3,4-dimethylhexane.

The structural isomers of octane are (b), (c), (d), and (e).

Think About It

Even though alkanes may have the same base chain (e.g., hexane), they are not necessarily structural isomers. Here (a) and (e) are both hexane chains with various substituents, but the two are not structural isomers of one another.

13.39. **Collect and Organize**

Using the examples for hydrocarbons in Section 13.2, we can convert the line drawings of each molecule in Problem 13.37 to chemical formulas.

Analyze

In line drawings, only the carbon skeleton is shown. The end of a line is understood to be the –CH_3 group and the intersection of two lines is understood to be a –CH_2– group. If more than two lines intersect, then the number of hydrogen atoms at that carbon is for minus the number of intersecting lines. This is because carbon makes bonds to four atoms to satisfy its octet.

Solve

(a) $C_{10}H_{22}$
(b) C_8H_{18}
(c) C_8H_{18}
(d) C_8H_{18}
(e) C_8H_{18}

Think About It

Another way of representing the molecular formulas of these species is to name each group in sequence, rather than combining all carbon and hydrogen atoms. Using this system, the condensed formula for (a) is $CH_3CH(CH_3)CH(CH_3)CH(CH_3)CH(CH_3)CH_3$.

13.41. **Collect and Organize**

By adding the energy of the bonds that form ($-\Delta H$) and the energy of the bonds broken, ($+\Delta H$) in the reaction of C_2H_4 with H_2 to give C_2H_6, we can estimate $\Delta H_{hydrogenation}$.

Analyze

The relevant bond strengths from Appendix 4 for the reactants and products are as follows:

C—H   413 kJ/mol
C=C   614 kJ/mol
H—H   436 kJ/mol
C—C   348 kJ/mol

In the reactants a C=C bond and a H—H bond are broken. In forming the products, a C—C bond and two C—H bonds are formed.

Solve

$$\Delta H_{rxn} = \sum \Delta H_{\text{bond breaking}} + \Delta H_{\text{bond forming}}$$
$$= [614 + 436] + [-348 + 2(-413)] = -124 \text{ kJ}$$

Think About It

Because this reaction is exothermic, it is favored by enthalpy.

13.43. **Collect and Organize**

By examining the degree of dispersion forces in the nonpolar molecules C_3H_8, $C_{14}H_{30}$, and cyclooctane (C_8H_{16}), we can put them in order of increasing boiling point.

Analyze

The larger the dispersion forces, the higher the boiling point. Dispersion forces are greater for nonpolar molecules with more atoms and for those that are less branched or those that have longer chains.

Solve

Because $C_{14}H_{30}$ has the greatest number of atoms its boiling point is the highest. Next is cyclooctane, and C_3H_8 has the lowest boiling point. In order of increasing boiling point, $C_3H_8 < C_8H_{16} < C_{14}H_{30}$.

Think About It

The branched isomers of $C_{14}H_{30}$ have lower boiling points than the linear isomer.

13.45. **Collect and Organize**

By defining structural and stereoisomers we can differentiate between them.

Analyze

Both structural and stereoisomers have the same molecular formulas.

Solve

Structural isomers have different connectivity of the atoms; stereoisomers have the same connectivity of the atoms but a different spatial arrangement.

Think About It

As examples 1-butene, 2-butene, and 2-methylpropene are structural isomers, whereas *trans*- and *cis*-2-butene are stereoisomers.

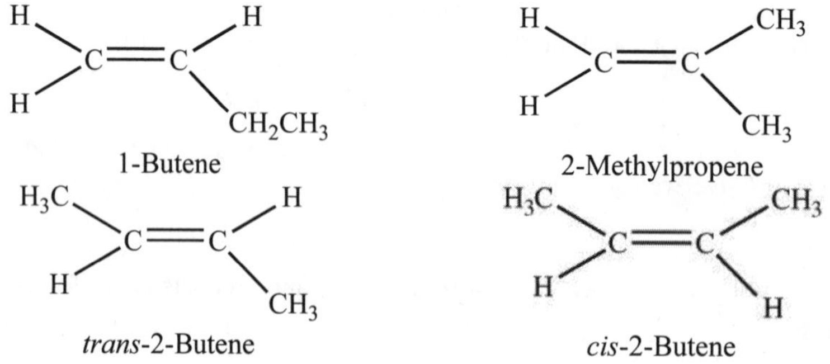

1-Butene

2-Methylpropene

trans-2-Butene

cis-2-Butene

13.47. **Collect and Organize**

By considering the empirical formula of an alkene and a cycloalkane with the same number of carbon atoms, we can determine whether combustion analysis could distinguish between the two.

Analyze

Both cycloalkanes and alkenes have the formula C_nH_{2n}.

Solve

Since there is no difference in the number of either C atoms or H atoms between the alkene and the cycloalkane, no, we cannot distinguish these by combustion analysis.

Think About It

We could, however, distinguish between an alkane (C_nH_{2n+2}) and an alkene (C_nH_{2n}) with the same numbers of C atoms.

13.49. **Collect and Organize**

We are to explain why alkenes in which the double bond involves the first (or last) carbon do not exhibit cis– and trans isomerism.

Analyze

A cis isomer has two like groups on the same side of a line drawn through the double bond. A trans isomer has two like groups on opposite sides of a line drawn through the double bond.

Solve

When the double bond is "terminal" (occurs at the end or beginning of the carbon chain) there are two like groups (H) so no cis and trans isomers are possible.

Think About It

Likewise, no cis and trans isomers would be possible for a double bond where two of the substituents on a single carbon atom are the same, as shown in the following example:

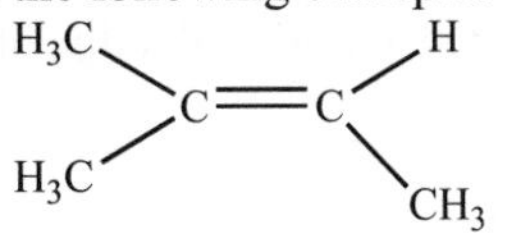

2-Methyl-2-butene

13.51. **Collect and Organize**

From the structure of carvone we are asked to explain why this molecule does not have cis and trans isomers.

Analyze

A cis isomer has two like groups on the same side of a line drawn through the double bond. A trans isomer has two like groups on opposite sides of a line drawn through the double bond.

Solve

The $C{=}C$ double bond outside of the ring does not show cis–trans isomerism because there are not two dissimilar groups on the terminal carbon atom. The $C{=}C$ double bond in the ring of carbon atoms is cis in the structure of carvone. This bond cannot be trans, or the ring of six carbon atoms would not be possible.

Think About It

A related molecule that would show trans and cis isomers is

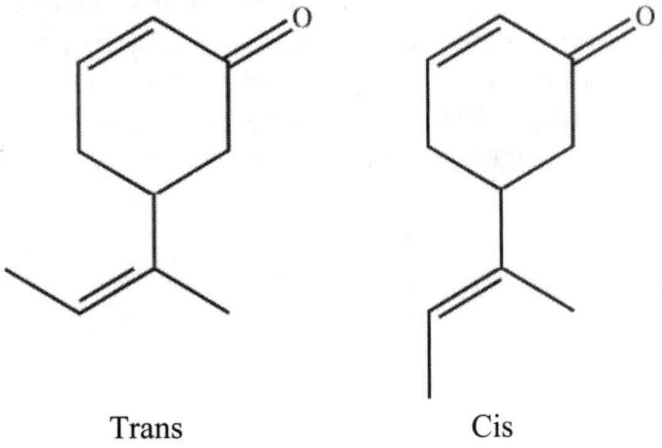

Trans   Cis

13.53. Collect and Organize

By comparing the structures of ethylene and polyethylene, we can explain why ethylene reacts with HBr but polyethylene does not.

Analyze

The structure of ethylene and polyethylene are as follows:

Ethylene   Polyethylene

Solve

Ethylene has a C=C bond with which HBr is reactive but polyethylene has only saturated C—C bonds.

Think About It

Polyethylene is produced from ethylene under high temperature and pressure.

13.55. Collect and Organize

For the two structures shown in Figure P13.55, we are asked to label the isomers as cis or trans and E or Z.

Analyze

A cis, or Z, isomer has two like groups on the same side of a line drawn through the double bond. A trans, or E, isomer has two like groups on opposite sides of a line drawn through the double bond.

Solve

(a) is trans, E, and (b) is cis, Z.

Think About It

In designating isomers, we more often encounter cis and trans than E and Z, but knowing both designations is helpful in studying organic chemistry.

13.57. Collect and Organize

Using ΔH_f° values for the reactants and products we can calculate the ΔH_{rxn}° for the controlled combustion of methane and determine whether this is an endothermic or exothermic reaction. We can predict if methane is oxidized or reduced when it is converted to CO and C_2H_2 by assigning oxidation numbers to all atoms in the equation.

Analyze

To calculate ΔH_{rxn}° we use

$$\Delta H_{rxn}^{\circ} = \sum n\, \Delta H_{f,products}^{\circ} - \sum m\, \Delta H_{f,reactants}^{\circ}$$

Oxidation numbers are assigned using the method discussed in Chapter 4. Atoms in their pure elemental state are assigned an oxidation number of 0, and the sum of the oxidation numbers for a molecule or ion must equal the charge on that molecule or ion. Fluorine is assigned an oxidation number of -1, hydrogen is assigned $+1$, and oxygen is assigned -2, except where pure elements (e.g., F_2, H_2, or O_2) are present and when superseded by a preceding rule (e.g., H is assigned $+1$ before assigning an oxidation number to O). The oxidation numbers for this reaction are:

$$6\ \overset{-4\ +1}{CH_4}(g) + \overset{0}{O_2}(g) \rightarrow 2\ \overset{-1\ +1}{C_2H_2}(g) + 2\ \overset{+2\ -2}{CO}(g) + 10\ \overset{0}{H_2}(g)$$

Solve
(a)

$$\Delta H^{\circ}_{\text{rxn}} = \big[\left(2 \text{ mol } C_2H_2 \times 226.7 \text{ kJ/mol}\right) + \left(2 \text{ mol CO} \times -110.5 \text{ kJ/mol}\right) + \left(10 \text{ mol } H_2 \times 0 \text{ kJ/mol}\right)\big]$$
$$- \big[\left(6 \text{ mol } CH_4 \times -74.8 \text{ kJ/mol}\right) + \left(1 \text{ mol } O_2 \times 0 \text{ kJ/mol}\right)\big]$$
$$= 681.2 \text{ kJ}$$

The controlled combustion of methane is endothermic.
(b) The oxidation number for C in CH_4 is –4 on the left hand side, becoming –1 (in C_2H_2) and +2 (in CO) on the right hand side. Since the oxidation number increases as the reaction proceeds, C is oxidized.

Think About It
The uncontrolled combustion of methane to give CO_2 and H_2O, however, is exothermic.

13.59. **Collect and Organize**
From the carbon-skeleton structure of vinyl acetate (Figure P13.59), we are to draw the structure of poly(vinyl acetate).

Analyze
Poly(vinyl acetate) is an addition polymer in which the $C\!=\!C$ double bonds link together to form the polymer backbone. In this polymer the acetate group is a side chain off the polymer chain.

Solve

Think About It
Because of its flexibility, poly(vinyl acetate), or PVA, is used in bookbinding and, as an emulsion in water, as an adhesive for wood, paper, and cloth.

13.61. **Collect and Organize**
By looking at the Lewis structure of benzene we can explain why it is a planar molecule.

Analyze
Benzene is a six-membered carbon ring with alternating single and double bonds.

Solve
The line structure of benzene is

In benzene, each C atom is sp^2 hybridized with bond angles of 120°. This geometry at each of the carbon atoms in the ring makes benzene a planar molecule.

Think About It
Benzene's π electrons are delocalized by resonance (Figure 13.25), lending benzene a special stability.

13.63. **Collect and Organize**

By drawing the line structures of tetramethylbenzene and pentamethylbenzene, we can determine if these molecules have any structural isomers.

Analyze

The methyl groups on these compounds are ring substitutents. That is, they take the place of H atoms in the benzene structure.

Solve

Tetramethylbenzene has three structural isomers:

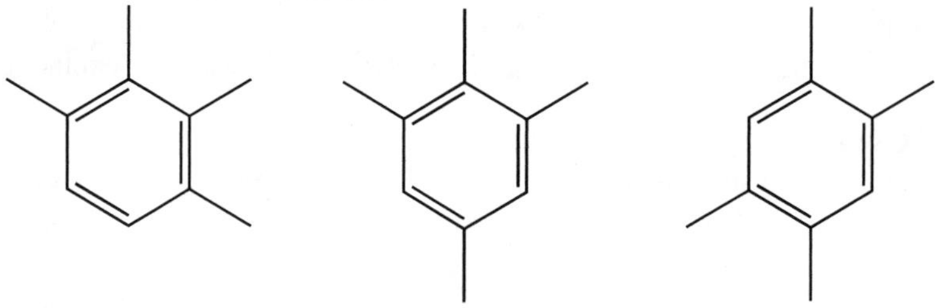

Pentamethylbenzene has no structural isomers:

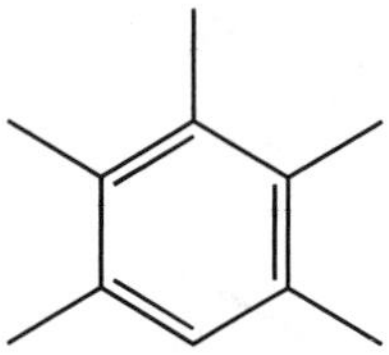

Think About It

Remember that structural isomers have distinct chemical and physical properties.

13.65. **Collect and Organize**

By examining pyridine's structure (Figure P13.65) we can determine if this compound is aromatic.

Analyze

Aromatic structures have alternating single and double bonds through which resonance delocalizes the electrons.

Solve

Yes. Pyridine is an aromatic compound because the π electrons are delocalized over all the atoms in the ring through resonance.

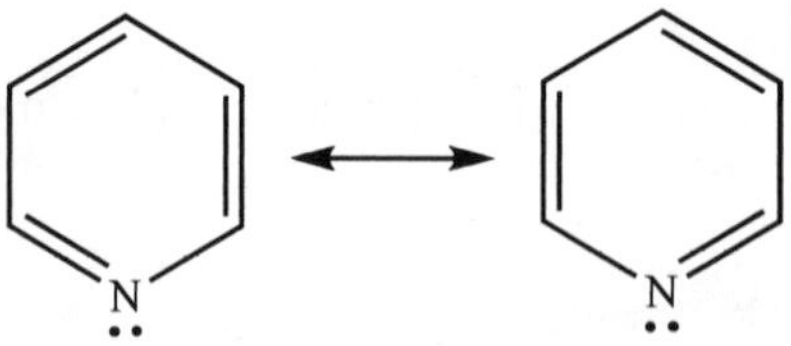

Think About It

In this aromatic compound, the alternating single and double bonds include C—N bonds.

13.67. **Collect and Organize**

We are to draw all the structural isomers of trimethylbenzene.

Analyze

The methyl groups on the benzene ring are substituents that have replaced the H atoms on benzene.

Solve

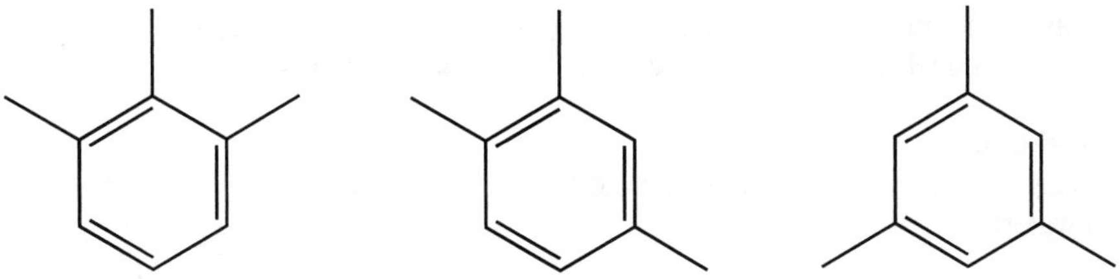

Think About It

The structures below are not additional structural isomers. They are the same as the first isomer above. To see this, we need only to rotate each of the drawings below to give the isomer above.

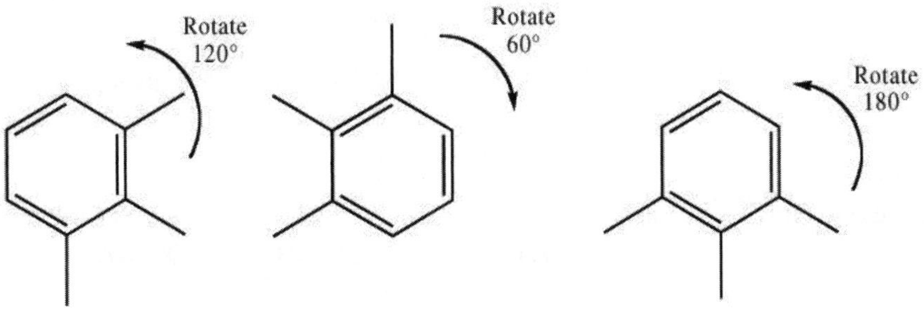

Rotate 120°

Rotate 60°

Rotate 180°

13.69. **Collect and Organize**

We are to calculate the fuel values for benzene and ethylene. We are also asked if 1 mole of benzene has a higher or lower fuel value than 3 moles of ethylene.

Analyze

The equation for calculating fuel value is ($-\Delta H_{comb}$/mass of fuel). For these two fuels the combustion reactions are

$$2\ C_6H_6\,(\ell)\ +\ 15\ O_2(g)\ \rightarrow\ 12\ CO_2(g)\ +\ 6\ H_2O\,(\ell)$$
$$C_2H_4(g)\ +\ 3\ O_2(g)\ \rightarrow\ 2\ CO_2(g)\ +\ 2\ H_2O\,(\ell)$$

Solve

The ΔH°_{comb} for 1 mole benzene:

$$\Delta H^{\circ}_{comb} = \left[\left(12 \times -393.5\right) + \left(6 \times -285.8\right)\right] - \left[\left(2 \times 49.0\right) + \left(15 \times 0\right)\right]$$
$$= -6534.8 \text{ kJ for 2 mol benzene (from balanced equation)}$$
$$= -3267.4 \text{ kJ for 1 mol benzene}$$

Fuel value:

$$\frac{3267.4 \text{ kJ/mol}}{78.11 \text{ g/mol}} = 41.83 \text{ kJ/g}$$

The ΔH°_{comb} for 3 moles ethylene:

$$\Delta H^{\circ}_{comb} = \left[\left(2 \times -393.5\right) + \left(2 \times -285.8\right)\right] - \left[\left(1 \times 52.3\right) + \left(3 \times 0\right)\right]$$
$$= -1410.9 \text{ kJ for 1 mol ethylene (from balanced equation)}$$

Fuel value:

$$= \frac{1410.9 \text{ kJ/mol}}{28.05 \text{ g/mol}} = 50.30 \text{ kJ/g}$$

Looking at this carefully for the comparison:

1 mole benzene has an energy content of 3267.4 kJ

3 moles ethylene have an energy content of 3 × 1410.9 = 4232.7 kJ

Therefore, 1 mole of benzene has a lower energy content than 3 moles of ethylene.

Think About It

Ethylene has six additional C—H in 3 moles compared to 1 mole benzene. Breaking of these bonds and the formation of additional H—O bonds in water must account for the difference in energy content.

13.71. Collect and Organize

By comparing the structure of methylamine to that of butylamine, we can explain why methylamine is more soluble in water.

Analyze

The line structures of these two amines are

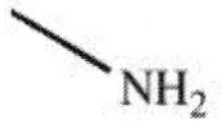

Solve

Methylamine has a smaller nonpolar hydrocarbon chain compared to butylamine, so it is more soluble in water.

Think About It

Both amines are fairly soluble in water and indeed can react with water to form basic solutions.

$$CH_3NH_2(aq) + H_2O(\ell) \rightarrow CH_3NH_3^+(aq) + OH^-(aq)$$

13.73. Collect and Organize

In the structures of amphetamine and serotonin (Figure P13.73), we are to identify the primary and secondary amine functional groups.

Analyze

Primary amines (1°) have one R group bonded to the nitrogen atom, whereas secondary amines (2°) have two R groups bonded to the nitrogen atom.

Solve

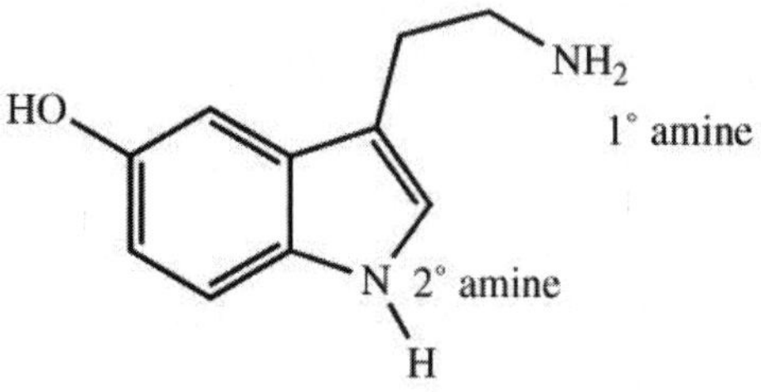

Serotonin

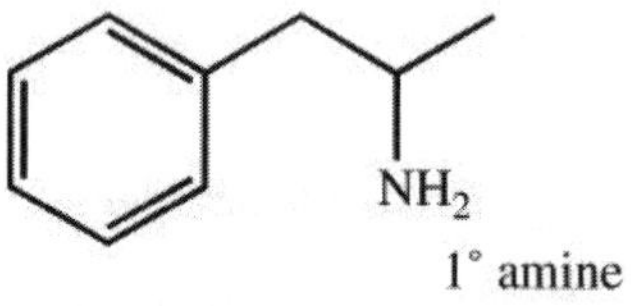

Amphetamine

Think About It

Tertiary amines (3°) have three R groups bonded to the nitrogen atom and quaternary amines (4°) have four R groups bonded to the nitrogen atom.

13.75. Collect and Organize

We can use the ΔH_f° values in Appendix 4 and the given ΔH_f° for methylamine (–23.0 kJ/mol) to calculate the enthalpy of the reaction

$$4\,CH_3NH_2(g) + 2\,H_2O(\ell) \rightarrow 3\,CH_4(g) + CO_2(g) + 4\,NH_3(g)$$

Analyze

The heat of a reaction may be calculated using ΔH_f° for the reactants and products according to the equation

$$\Delta H_{rxn}^\circ = \sum n\Delta H_{f,products}^\circ - \sum m\Delta H_{f,reactants}^\circ$$

Solve

$$\Delta H_{rxn}^{\circ} = \left[\left(3 \text{ mol CH}_4\right)\left(-74.8 \text{ kJ/mol}\right) + \left(1 \text{ mol CO}_2\right)\left(-393.5 \text{ kJ/mol}\right) + \left(4 \text{ mol NH}_3\right)\left(-46.1 \text{ kJ/mol}\right)\right]$$
$$- \left[\left(4 \text{ mol CH}_3\text{NH}_2\right)\left(-23.0 \text{ kJ/mol}\right) + \left(2 \text{ mol H}_2\text{O}\right)\left(-285.8 \text{ kJ/mol}\right)\right]$$
$$\Delta H_{rxn}^{\circ} = -138.7 \text{ kJ}$$

Think About It

Be careful to use the correct enthalpy of formation for the phases shown in the equation. In this problem ammonia and carbon dioxide are in the gas phase and water is in the liquid phase.

13.77. Collect and Organize

We are asked why the fuel values of ethanol and diethyl ether are lower than that of ethane. In comparing the fuel values we must take into account the oxygenation of the compounds.

Analyze

The more oxygenated a fuel, the lower its fuel value. All three compounds have two carbons and six hydrogens.

Solve

Because dimethyl ether and ethanol both contain oxygen in their structures and ethane does not, the fuel values of both dimethyl ether and ethanol are lower than that of ethane.

Think About It

We might expect, though, that the fuel values of dimethyl ether and ethanol will not differ much from each other, since they are isomers.

13.79. Collect and Organize

Ethers have the general structure R—O—R′, and alcohols have the general structure R—OH. We are asked to explain the difference in boiling points between the two.

Analyze

The lower the boiling point, the weaker the intermolecular forces between the molecules are. Since ethers boil at lower temperatures than isomeric alcohols, as given in the problem statement, ethers must have weaker intermolecular forces than alcohols.

Solve

Ethers have lower boiling points compared to alcohols because they have weaker dipole–dipole forces compared to alcohols, which have hydrogen bonding between their molecules.

Think About It

Both ethers and alcohols also have dispersion forces that attract their molecules to each other. These get stronger as the length of the R groups on the ether or alcohol increases.

13.81. Collect and Organize

We need to consider the evaporation of ethanol to explain why our skin feels cold after wiping with ethanol.

Analyze

When ethanol comes in contact with your skin it begins to evaporate.

Solve

Evaporation of ethanol from the skin is an endothermic process (phase change from liquid to vapor). The heat transfers from the skin to the ethanol so the skin feels cold.

Think About It

The reverse process, condensation, is an exothermic process, as discussed in Chapter 5.

13.83. **Collect and Organize**

From the structures shown in Figure P13.83, we are to identify which are ethers, which are alcohols, and place them in order of increasing boiling point.

Analyze

Ethers have the general formula R—O—R′, and alcohols have the general formula R—OH. The greater the intermolecular forces between molecules, the higher the boiling point. Because of hydrogen bonding, alcohols generally have higher boiling points than ethers. The larger the molecule (greater number of atoms) and the less branching it has, the greater the boiling point.

Solve

(a) and (d) are alcohols; (b) and (c) are ethers. In order of increasing boiling point: (b) < (c) < (a) < (d).

Think About It

Our prediction for boiling point order is nearly correct: (b) diethyl ether, 35°C < (c) isobutyl methyl ether, 59°C < (a) 3-methyl-4-heptanol, 174°C < (d) 2,5-dimethylcyclohexanol, 170°C. Notice that (a) and (d) have nearly equal boiling points.

13.85. **Collect and Organize**

We are to calculate the fuel values of butanol and diethyl ether and indicate which has the higher fuel value.

Analyze

The fuel value is equal to $-\Delta H^{\circ}_{comb}$ divided by the mass of fuel. For these fuels the combustion reactions are

$$(C_2H_5)_2O\,(\ell) + 6\,O_2(g) \rightarrow 4\,CO_2(g) + 5\,H_2O\,(\ell)$$
$$C_4H_9OH\,(\ell) + 6\,O_2(g) \rightarrow 4\,CO_2(g) + 5\,H_2O\,(\ell)$$

Solve

The fuel value for diethyl ether:

$$\Delta H^{\circ}_{comb} = \left[4(-393.5) + 5(-285.8)\right] - \left[1(-279.6) + 6(0)\right]$$
$$= -2723.4 \text{ kJ/mol}$$
$$\text{Fuel value} = \frac{2723 \text{ kJ/mol}}{74.12 \text{ g/mol}} = 36.74 \text{ kJ/g}$$

The fuel value for butanol:

$$\Delta H^{\circ}_{comb} = \left[4(-393.5) + 5(-285.8)\right] - \left[1(-327.3) + 6(0)\right]$$
$$= -2675.7 \text{ kJ}$$
$$\text{Fuel value} = \frac{2675.7 \text{ kJ/mol}}{74.12 \text{ g/mol}} = 36.10 \text{ kJ/g}$$

Diethyl ether has a slightly higher fuel value than butanol.

Think About It

We would expect these compounds to have very similar fuel values because they are isomers.

13.87. **Collect and Organize**

After calculating the fuel values of ethanol (C_2H_5OH) and methanol (CH_3OH), we can determine the validity of our prediction (Problem 13.78) that fuel value increases as carbon atoms are added to the alcohol alkyl chain.

Analyze

Fuel values are computed from ΔH_{comb}°. For methanol and ethanol the balanced combustion reactions are

$$2\ CH_3OH\,(\ell)\ +\ 3\ O_2(g) \rightarrow 2\ CO_2(g) + 4\ H_2O\,(\ell)$$

$$C_2H_5OH\,(\ell)\ +\ 3\ O_2(g) \rightarrow 2\ CO_2(g) + 3\ H_2O\,(\ell)$$

We need ΔH_f° values from the textbook (Appendix 4, Table A4.3) to compute ΔH_{comb}°.

Solve

For methanol

$$\Delta H_{comb}^{\circ} = \left[(2\ \text{mol}\ CO_2)(-393.5\ \text{kJ/mol}) + (4\ \text{mol}\ H_2O)(-285.8\ \text{kJ/mol}) \right]$$
$$- \left[(2\ \text{mol}\ CH_3OH)(-238.7\ \text{kJ/mol}) + (3\ \text{mol}\ O_2)(0.0\ \text{kJ/mol}) \right]$$

$$\Delta H_{comb}^{\circ} = -1452.8\ \text{kJ for 2 mol}\ CH_3OH\ \text{burned}$$

$$\text{Fuel value} = \frac{1452.8\ \text{kJ}}{2\ \text{mol} \times 32.04\ \text{g/mol}} = 22.67\ \text{kJ/g}$$

For ethanol

$$\Delta H_{comb}^{\circ} = \left[(2\ \text{mol}\ CO_2)(-393.5\ \text{kJ/mol}) + (3\ \text{mol}\ H_2O)(-285.8\ \text{kJ/mol}) \right]$$
$$- \left[(1\ \text{mol}\ C_2H_5OH)(-277.7\ \text{kJ/mol}) + (3\ \text{mol}\ O_2)(0.0\ \text{kJ/mol}) \right]$$

$$\Delta H_{comb}^{\circ} = -1366.7\ \text{kJ for 1 mol}\ CH_3OH\ \text{burned}$$

$$\text{Fuel value} = \frac{1366.7\ \text{kJ}}{1\ \text{mol} \times 46.07\ \text{g/mol}} = 29.67\ \text{kJ/g}$$

Yes, the answer supports the prediction made in Problem 13.78 that fuel values of alcohols increase as the number of C atoms increases.

Think About It

When comparing alcohols to alkanes, however, alcohols have lower fuel values.

13.89. **Collect and Organize**

By looking at the structures of carboxylic acids and aldehydes, we can explain why carboxylic acids are generally more soluble in water.

Analyze

Carboxylic acids have the general molecular structure

Aldehydes have the general molecular structure

Solve

Both carboxylic acids and aldehydes have polar functional groups. Carboxylic acids, however, are more soluble in water because they form strong hydrogen bonds with water.

Think About It

Remember that hydrogen bonds between a species and water are stronger than dipole–dipole interactions between a species and water.

13.91. Collect and Organize

Given the structures for butanal (an aldehye) and 2-butanone (a ketone) (Figure P13.91), we are asked if the compounds are structural isomers.

Analyze

Structural isomers have the same molecular formula but different connectivity of their atoms. The molecular formula of butanal is C_4H_8O and so is the formula of 2-butanone, C_4H_8O.

Solve

Yes, these compounds are structural isomers.

Think About It

Butanol ($C_4H_{10}O$), however, is not a structural isomer of these compounds because it has a different molecular formula.

13.93. Collect and Organize

We can compare the molecular formulas of aldehydes and ketones to determine whether we could distinguish them by combustion analysis.

Analyze

The molecular formula for aldehydes is $C_nH_{2n}O$ and for ketones it is also $C_nH_{2n}O$.

Solve

Because the empirical formulas for a ketone and for an aldehyde with the same number of carbon atoms are the same, no, we cannot distinguish between these compounds by combustion analysis.

Think About It

We would be able, however, to distinguish between a ketone or aldehyde and an alcohol, which has the molecular formula $C_nH_{2n+2}O$.

13.95. Collect and Organize

By assigning formal charges to the atoms in the two resonance structures for acetic acid (Figure P13.95), we can determine which form contributes more to bonding.

Analyze

The resonance form that has the lowest formal charges and/or the negative formal charges on the most electronegative atoms (oxygen for acetic acid) contributes the most to the structure.

Solve

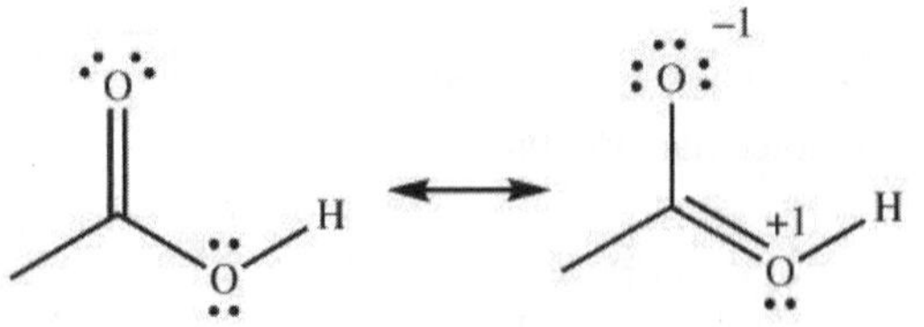

(a) All formal charges = 0   (b)

(a) contributes more to the bonding in acetic acid because all of the formal charges are zero.

Think About It
Once deprotonated, acetic acid forms the acetate anion, which has two equivalent resonance forms:

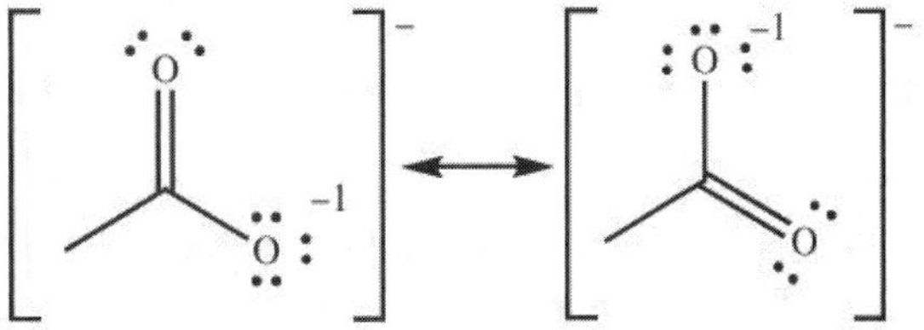

13.97. Collect and Organize
By comparing the general structures of an amine and an amide, we can distinguish between the two functional groups.

Analyze
An amine has the general structure

$$R-NH_2$$

An amide has the general structure

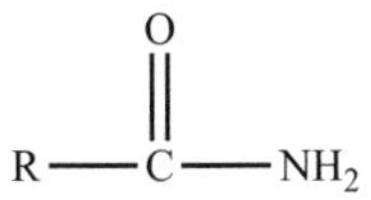

Solve
An amide includes a carbonyl (C=O) as part of its functional group in addition to the $-NH_2$ group.

Think About It
Amides are somewhat related to carboxylic acids in structure:

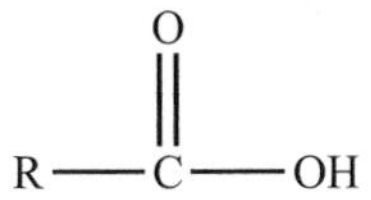

13.99. Collect and Organize
Given the molecular formula of an aldehyde ($C_5H_{10}O$), we are to choose which of four structures (Figure P13.99) are structural isomers of it.

Analyze
Structural isomers have the same molecular formula but different connectivity of their atoms.

Solve
(a) has a formula of $C_5H_{10}O$, so it is a structural isomer of the given aldehyde.
(b) has a formula of $C_5H_{10}O$, so it is a structural isomer.
(c) has a formula of C_4H_8O, so it is not a structural isomer of aldehyde.
(d) has a formula of $C_5H_{10}O$ so it is also a structural isomer of the given aldehyde.

Think About It
The names of the three structural isomers shown are (a) 3-methylbutanal, (b) 2-methylbutanal, and (d) pentanal.

13.101. Collect and Organize
From the structures shown in Figure P13.101, we are to determine which compound is a ketone.

Analyze
Ketones have the general structure

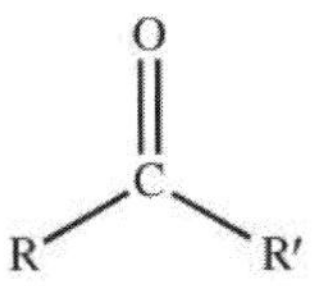

Solve
(b) is a ketone.

Think About It
(a) represents an aldehyde, (c) represents a carboxylic acid, and (d) represents an alcohol.

13.103. **Collect and Organize**
After plotting the carbon–hydrogen ratio in aldehydes as a function of the number of carbon atoms and comparing it to that of alkanes and alkenes, we can find the better correlation for aldehydes.

Analyze
Aldehydes have the general formula $C_nH_{2n}O$ so the C:H ratio is always 0.5. Alkanes have the general formula C_nH_{2n+2} so the C:H ratio changes as the number of C atoms increases. Alkenes have the general formula C_nH_{2n} so the C:H ratio is always 0.5.

Solve

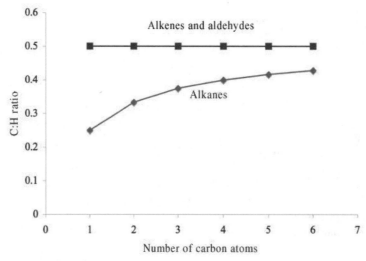

The plot of C:H ratio versus number of C atoms for aldehydes correlates exactly to that of alkenes and poorly to that of alkanes.

Think About It
Although this correlation exists, aldehydes are not structural isomers of alkenes. Aldehydes have an oxygen in their structure; alkenes do not.

13.105. **Collect and Organize**
Esters are formed in condensation reactions between carboxylic acids and alcohols. For each of the ester structures shown in Figure P13.105, we can write an equation to identify the carboxylic acid and alcohol used to synthesize them.

Analyze
The general reaction for the formation of esters is

Solve

(a) Pineapples

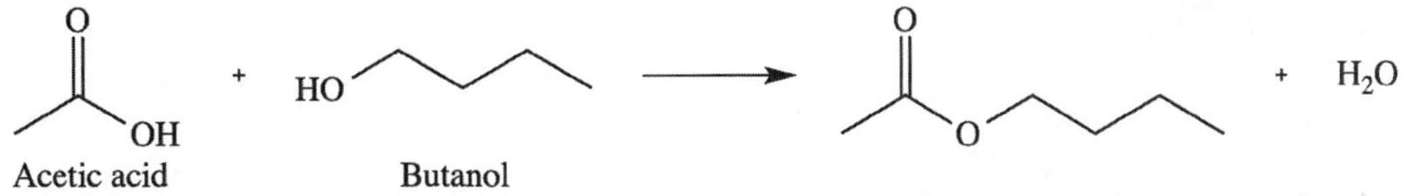
O
OH
HO
O
O
H_2O
Acetic acid
Butanol

(b) Bananas

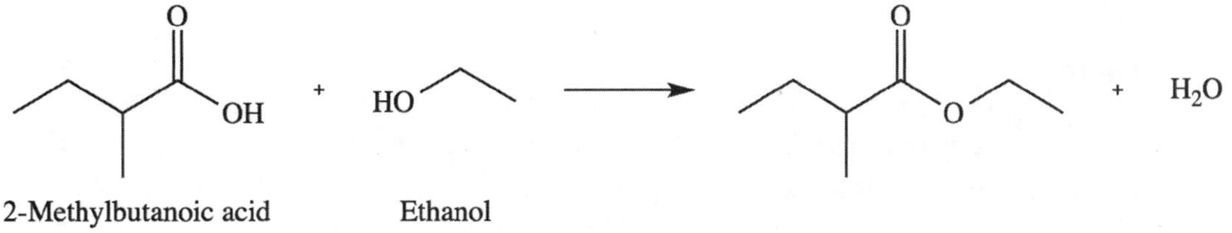
O
OH
HO
O
O
H_2O
2-Methylbutanoic acid
Ethanol

(c) Apples

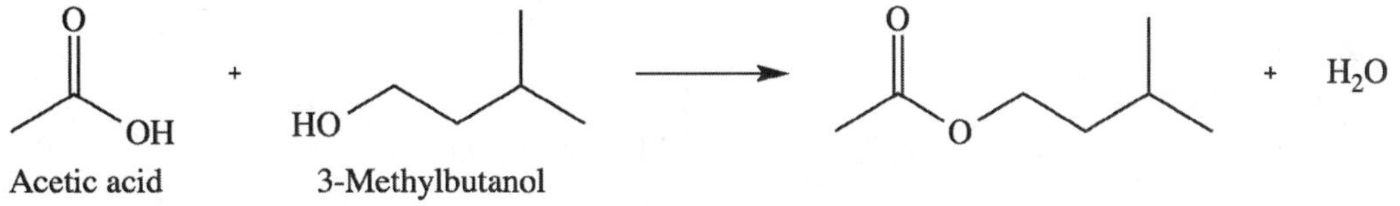
O
OH
HO
O
O
H_2O
Acetic acid
3-Methylbutanol

Think About It

These are all condensation reactions because they give a small molecule, H_2O, as the other product.

13.107. **Collect and Organize**

Given the structure of a component of the drug Truvada, we can draw the parent phosphorus acid that could be used to generate the ester linkages shown.

Analyze

Esters are formed by the reaction of a carboxylic acid and an alcohol. We can disconnect this molecule at the two ester linkages, yielding an alcohol and a carboxylic acid with the same number of carbons and other functional groups.

Solve

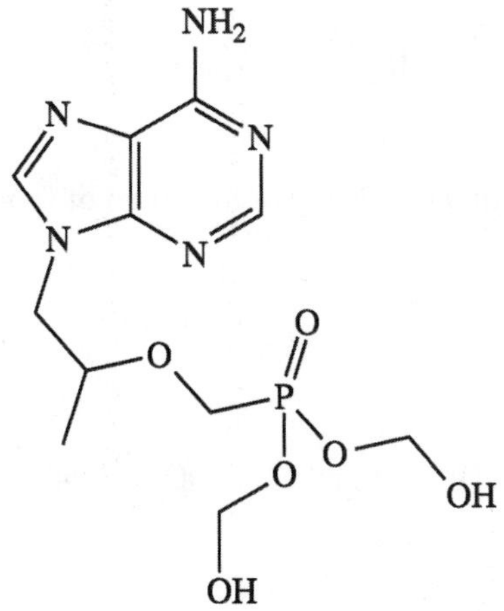
NH2
N
N
N
N
O
O
P
O
O
OH
OH

Think About It
The structure of the acid required to form the molecule depicted in Figure P13.107 is

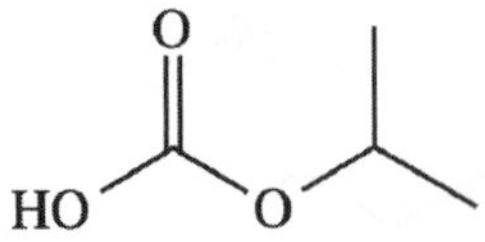

13.109. **Collect and Organize**
After calculating the fuel values of formaldehyde and formic acid (Figure P13.109), we can determine which has the higher fuel value.

Analyze
Generally the more oxygenated the fuel, the lower the fuel value. We predict that formaldehyde has the higher fuel value since it has less oxygen in its structure. The fuel value is equal to $-\Delta H^\circ_{comb}$ divided by mass of fuel. For these two fuels the combustion reactions are

$$CH_2O(g) + O_2(g) \rightarrow CO_2(g) + H_2O\,(\ell)$$
$$CH_2O_2(g) + \tfrac{1}{2}O_2(g) \rightarrow CO_2(g) + H_2O\,(\ell)$$

Solve
For formaldehyde

$$\Delta H^\circ_{comb} = \left[(-393.5)+(-285.8)\right]-\left[(-108.6)+(0)\right]$$
$$= -570.7 \text{ kJ/mol}$$
$$\text{Fuel value} = \frac{570.7 \text{ kJ/mol}}{30.03 \text{ g/mol}} = 19.00 \text{ kJ/g}$$

For formic acid

$$\Delta H^\circ_{comb} = \left[(-393.5)+(-285.8)\right]-\left[(-378.7)+(0)\right]$$
$$= -300.6 \text{ kJ}$$
$$\text{Fuel value} = \frac{300.6 \text{ kJ/mol}}{46.03 \text{ g/mol}} = 6.531 \text{ kJ/g}$$

Formaldehyde has a significantly higher fuel value than formic acid.

Think About It
Our prediction, based on the level of oxygenation of the two fuels, was correct.

13.111. **Collect and Organize**
For two reactions involving methanogenic bacteria we are to calculate ΔH°_{rxn}.

Analyze
We determine the enthalpy of each reaction using the given value of the enthalpy of formation of formic acid and the values in Appendix 4.

$$\Delta H^\circ_{rxn} = \sum n\Delta H^\circ_{f,\,products} - \sum m\Delta H^\circ_{f,\,reactants}$$

Solve
For reaction 1
$$\Delta H^\circ_{rxn} = \left[(1 \text{ mol } CH_4)(-74.8 \text{ kJ/mol})+(1 \text{ mol } CO_2)(-393.5 \text{ kJ/mol})\right]-\left[(1 \text{ mol } CH_3COOH)(-485.8 \text{ kJ/mol})\right]$$
$$\Delta H^\circ_{rxn} = 17.5 \text{ kJ}$$
For reaction 2

$$\Delta H^{\circ}_{rxn} = \left[(1 \text{ mol } CH_4)(-74.8 \text{ kJ/mol}) + (3 \text{ mol } CO_2)(-393.5 \text{ kJ/mol}) + (2 \text{ mol } H_2O)(-285.8 \text{ kJ/mol})\right]$$
$$- \left[(4 \text{ mol } HCOOH)(-378.7 \text{ kJ/mol})\right]$$

$\Delta H^{\circ}_{rxn} = -312.1 \text{ kJ}$

Think About It

The breakdown of acetic acid to methane and carbon dioxide is endothermic and therefore not favored by enthalpy, but the process that gives methanol, carbon dioxide, and water is exothermic and is favored by enthalpy.

13.113. Collect and Organize

For each of the polymers shown in Figure P13.113, which were synthesized through the condensation reaction of $H_2N(CH_2)_6NH_2$ with $HO_2C(CH_2)_nCO_2H$, we are asked to determine the number of carbon atoms in the chain (n) of the dicarboxylic acids used.

Analyze

The number of carbon atoms in the chain equals the value of n, which is the number of $-(CH_2)-$ units in the chain of the dicarboxylic acid.

Solve

(a) 4
(b) 6
(c) 8

Think About It

Because n is defined in the problem as the number of carbon atoms in the dicarboxylic acid formula $HO_2C(CH_2)_nCO_2H$, we do not count the carboxylic acid carbon atoms.

13.115. Collect and Organize

Given the reaction between dimethyl terephthalate and 1,4-di(hydroxymethyl)cyclohexane to form Kodel (Figure P13.115), we are asked to classify the reaction as either a condensation or addition reaction, give the other product of the reaction, and compare the properties of Kodel to those of Dacron, which is prepared using ethylene glycol.

Analyze

(a) In addition polymerization reactions, the atoms are joined in the monomers to form the polymeric backbone without loss of atoms. In condensation polymerization reactions, the two monomers react to form the polymeric backbone and a small molecule, like water.
(b) The structure of ethylene glycol is

Solve

(a) Kodel is a condensation polymer. Methanol (CH_3OH) is the by-product of the polymerization reaction.
(b) Kodel has a carbon six-membered ring in its backbone, not a straight chain as in Dacron. Therefore, Kodel might be better able to accept organic dyes, which are nonpolar.

Think About It

Kodel, being fairly polar because of the saturated six-membered carbon ring, is fairly resistant to water and has been used to make clothing.

13.117. Collect and Organize

Using the definitions of *enantiomer*, *achiral*, and *optically active*, we can determine whether all three terms may be applied to a single compound.

Analyze

An *enantiomer* is a molecule that is a nonsuperimposable mirror image of another molecule and is *optically active*. An *achiral* molecule is superimposable on its mirror image.

Solve

No, all three of these terms cannot describe a single compound. Whereas *enantiomer* and *optically active* describe the same chiral molecule, *achiral* does not.

Think About It

Many biomolecules are chiral.

13.119. Collect and Organize

We are to determine whether a racemic mixture is a homogeneous or heterogeneous mixture.

Analyze

For a mixture to be heterogeneous we must be able to discern by eye (or with a microscope) the different components of the mixture. A racemic mixture is a mixture of two enantiomers.

Solve

A racemic mixture is mixed at the molecular level, so it is a homogeneous mixture.

Think About It

When successfully separated, the components of a racemic mixture rotate plane-polarized light in opposite directions.

13.121. Collect and Organize

By examining the structure of glycine, we can explain why it is achiral.

Analyze

The structure of glycine from Figure P13.121 is

A carbon atom is chiral when it is bonded to four different groups. Carbons that are sp^2 or sp hybridized in a structure cannot be chiral because they have only three and two groups bonded to them, respectively.

Solve

Glycine is not chiral (achiral) because it has no chiral carbon centers. Glycine's carbons are sp^2 hybridized (on the carboxylic acid group) or have two of the same atoms or groups (H atoms in this case) bonded to the sp^3-hybridized carbon atom.

Think About It

This molecule would still be achiral if any two groups were identical, not just H atoms.

13.123. Collect and Organize

We are asked which type of orbital hybridization on a carbon atom can give rise to a chiral center.

Analyze

A carbon atom is chiral when it is bonded to four different groups.

Solve

A carbon is bonded to four groups, and thus can give rise to enantiomers, only when it has sp^3 hybridization.

Think About It
The mirror images of all the other hybridizations of carbon are superimposable.

13.125. **Collect and Organize**
Given some ordinary objects, we are to determine which are chiral.

Analyze
A chiral object is not superimposable on its mirror image.

Solve
A tennis racket (b) (if we ignore the winding of the tape on the handle) is superimposable on its mirror image; but (a) a golf club, (c) a glove, and (d) a shoe are not, so they are chiral objects.

Think About It
You might also learn later that an object is not chiral if it contains a plane of symmetry or has an inversion center.

13.127. **Collect and Organize**
From the line drawings of three carboxylic acids (Figure P13.127), we are to determine which are chiral.

Analyze
A molecule is chiral if it has at least one carbon atom bonded to four different groups.

Solve
Only (a) has a chiral carbon center and so is the only molecule shown that is chiral:

Think About It
(a) would be achiral if the –OH group were replaced by a –CH_3 group.

13.129. **Collect and Organize**
In each structure in Figure P13.129, we are to circle the chiral centers.

Analyze
There is a chiral center wherever in the molecule a carbon is bonded to four different groups.

Solve

Saccharin   Sodium cyclamate   Aspartame

Think About It

Because the ring of carbon atoms in sodium cyclamate is symmetrical, the carbon to which the $-NHSO_3^-$ group is bound is not chiral.

13.131. **Collect and Organize**

Given the structure of Crestor, we can determine if the molecule is chiral, and if so, which centers are chiral. We may also draw structural isomers showing both cis and trans disposition around the double bond.

Analyze

There is a chiral center wherever in the molecule a carbon is bonded to four different groups. Cis isomers have both substituents on the same side of the double bond, and trans isomers have substituents on opposing sides of the double bonds. The molecule drawn is the trans isomer.

Solve

(a) Yes, the anion is chiral. There are two chiral centers, both of which are circled below:

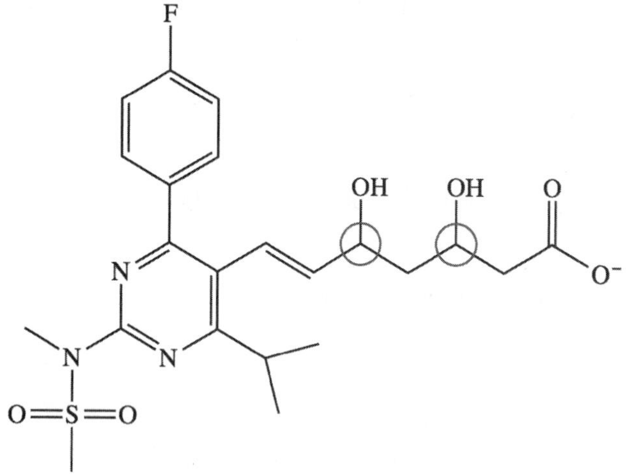

(b) The cis isomer is

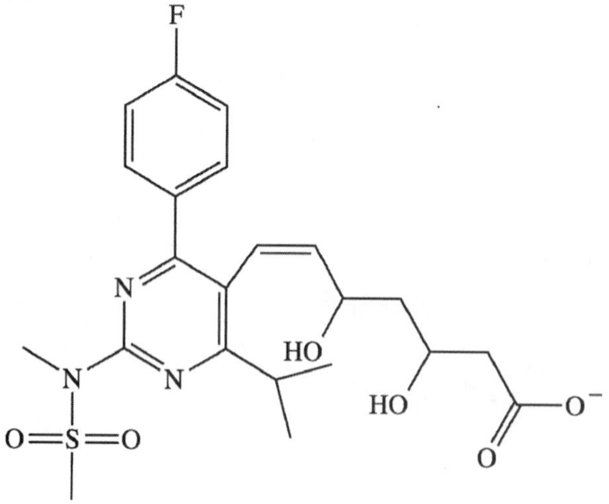

Think About It

Both cis and trans isomers have the same number and relative orientation of chiral centers.

13.133. **Collect and Organize**

For the highlighted N atoms in nicotine and Valium (Figure P13.133), we are to identify the associated functional group as either an amine or an amide.

Analyze

An amine has the following possible structures:

$$RNH_2 \qquad R_2NH \qquad R_3N$$

Primary   Secondary   Tertiary

An amide has these possible structures:

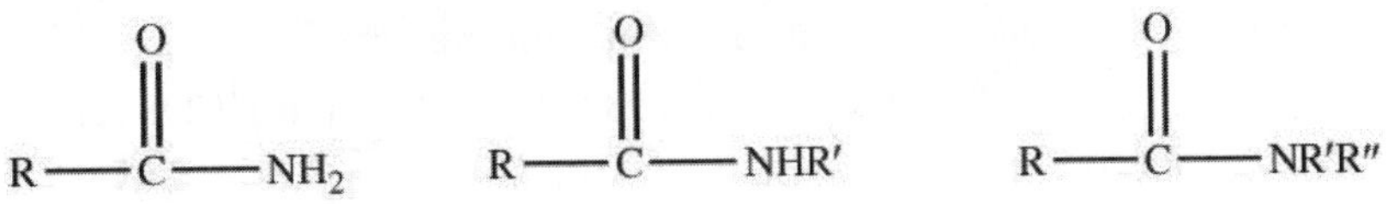

Solve

(a) Nicotine's highlighted N atom is a tertiary amine group.

(b) Valium's highlighted N atom is an amide group, as it has an adjacent $C=O$ double bond.

Think About It

Both nicotine and Valium also contain aromatic rings in their structures.

13.135. Collect and Organize

For the combustion reaction of methanol

$$CH_3OH\,(\ell) + \tfrac{3}{2}O_2(g) \rightarrow CO_2(g) + 2\,H_2O\,(\ell)$$

we are to calculate how many grams of methanol would be needed to raise the temperature of 454 g of water, from 20.0°C to 50.0°C. We are also to calculate the mass of CO_2 produced in this reaction.

Analyze

If all of the heat from burning the methanol is used to heat the water, then

$$q_{water} = -q_{comb}$$

The heat needed to raise the temperature of the water is given by

$$q_{water} = mc_s\Delta T$$

where m is the mass of the water, c_s is the specific heat capacity of water (4.184 J/g · °C), and ΔT is the change in temperature of the water (30.0°C).

We can use ΔH_f° values from Appendix 4 to calculate the enthalpy of combustion of one mole of methanol. The molar amount of methanol needed to heat the water, then, can be calculated through

$$\frac{q_{comb}}{\Delta H_{comb}} = \text{mol } CH_3OH \text{ to heat the water}$$

Once we know the moles of methanol required for the reaction, we can calculate the mass of CH_3OH needed and the mass of CO_2 produced using the balanced equation.

Solve

The heat generated by the combustion reaction is

$$q_{water} = -q_{comb} = 454\text{ g} \times 4.184\text{ J/g°C} \times 30.0°C = 57{,}000\text{ J or }57.0\text{ kJ}$$

$$\Delta H_{rxn}^\circ = \left[(1\text{ mol }CO_2)(-393.5\text{ kJ/mol}) + (2\text{ mol }H_2O)(-285.8\text{ kJ/mol})\right]$$
$$- \left[(1\text{ mol }CH_3OH)(-238.7\text{ kJ/mol}) + (\tfrac{3}{2}\text{ mol }O_2)(0.0\text{ kJ/mol})\right]$$

$$\Delta H_{rxn}^\circ = -726.4\text{ kJ for 1 mol methanol}$$

Moles of methanol to heat the water

$$\frac{-57.0\text{ kJ}}{-726.4\text{ kJ/mol}} = 7.85 \times 10^{-2}\text{ mol methanol}$$

Mass of methanol needed to heat the water

$$7.85 \times 10^{-2}\text{ mol} \times \frac{32.04\text{ g}}{\text{mol}} = 2.52\text{ g methanol}$$

Mass of CO_2 produced

$$7.85\times10^{-2}\ mol\ CH_3OH\times\frac{1\ mol\ CO_2}{1\ mol\ CH_3OH}\times\frac{44.01\ g}{mol}=3.45\ g\ CO_2$$

Think About It

This mass of CO_2 would occupy 1.92 L at 25°C and 1 atm pressure:

$$V=\frac{nRT}{P}=\frac{\left(7.85\times10^{-2}\ mol\right)\left[0.0821\ L\cdot atm/\left(mol\cdot K\right)\right]\left(298\ K\right)}{1\ atm}=1.92\ L$$

13.137. Collect and Organize

We are to use bomb calorimetry combustion reactions and the rise in temperature for two compounds to determine which is butanol (C_4H_9OH) and which is diethyl ether ($C_2H_5OC_2H_5$), both of which have a molecular formula of $C_4H_{10}O$.

Analyze

To calculate the energy released during the combustion of a compound we multiply the heat capacity of the calorimeter by the change in the temperature for the combustion of the compound:

$$\Delta H_{comb}=C_{calorimeter}\Delta T$$

This answer is in kilojoules per sample. To compare the two heats we need to convert to kilojoules per gram through

$$\frac{\Delta H_{comb}}{mass\ of\ sample}=kJ/g$$

To identify for which compounds the ΔH_{comb} values match the experimental values, we can use enthalpy of formation values (Appendix 3) to calculate ΔH_{rxn}° for the combustion reactions:

$$C_4H_9OH\,(\ell)+6\,O_2(g)\rightarrow4\,CO_2(g)+5\,H_2O\,(\ell)$$
$$C_2H_5OC_2H_5\,(\ell)+6\,O_2(g)\rightarrow4\,CO_2(g)+5\,H_2O\,(\ell)$$

Solve

From calorimetry experiments for compound A:

$$\Delta H_{comb}=\frac{3.640\ kJ}{°C}\times10.33°C=-37.60\ kJ$$

$$\Delta H_{comb}\ per\ gram=\frac{-37.60\ kJ}{0.9842\ g}=-38.20\ kJ/g$$

Compound B:

$$\Delta H_{comb}=\frac{3.640\ kJ}{°C}\times11.03°C=-40.15\ kJ$$

$$\Delta H_{comb}\ per\ gram=\frac{-40.15\ kJ}{1.110\ g}=-36.17\ kJ/g$$

For butanol the standard enthalpy of combustion is

$$\Delta H_{rxn}^{\circ}=\left[\left(4\ mol\ CO_2\right)\left(-393.5\ kJ/mol\right)+\left(5\ mol\ H_2O\right)\left(-285.8\ kJ/mol\right)\right]$$
$$-\left[\left(1\ mol\ C_4H_9OH\right)\left(-327.3\ kJ/mol\right)+\left(6\ mol\ O_2\right)\left(0.0\ kJ/mol\right)\right]$$

$$\Delta H_{rxn}^{\circ}=-2676\ kJ/mol\ of\ butanol$$

In kJ/g, this is $\Delta H_{rxn}^{\circ}=\dfrac{-2676\ kJ/mol}{74.12\ g/mol}=-36.10\ kJ/g$

For diethyl ether the standard enthalpy of combustion is

$$\Delta H^{\circ}_{rxn} = \left[\left(4 \text{ mol CO}_2\right)\left(-393.5 \text{ kJ/mol}\right) + \left(5 \text{ mol H}_2\text{O}\right)\left(-285.8 \text{ kJ/mol}\right)\right]$$
$$- \left[\left(1 \text{ mol C}_2\text{H}_5\text{OC}_2\text{H}_5\right)\left(-279.6 \text{ kJ/mol}\right) + \left(6 \text{ mol O}_2\right)\left(0.0 \text{ kJ/mol}\right)\right]$$

$\Delta H^{\circ}_{rxn} = -2723$ kJ/mol of diethyl ether

In kJ/g, this is $\Delta H^{\circ}_{rxn} = \dfrac{-2723 \text{ kJ/mol}}{74.12 \text{ g/mol}} = -36.74$ kJ/g

From the enthalpy of combustion reaction calculations, we see that ΔH_{rxn} of diethyl ether is slightly more exothermic than that of butanol. Matching this result with the calorimetry results, compound A is diethyl ether (more exothermic) and compound B is butanol (less exothermic).

Think About It
The calculated ΔH_{rxn} values for butanol and diethyl ether are close to each other because these two compounds are structural isomers.

13.139. Collect and Organize
Using the line formula for dodecenal (Figure P13.139), we are to count the carbons the molecule contains, identify its functional groups, and determine what types of isomerism are possible.

Analyze
(a) In line formulas, the end of a line represents a –CH_3 group. Carbon atoms are also present where two lines meet. Functional groups are portions of the structure that impart distinct chemical and physical properties to the compound. Isomers may be either structural or geometric.

Solve
(a) There are 12 carbon atoms in dodecenal.
(b) Dodecenal contains an alkene and an aldehyde functional group.
(c) Stereoisomers (cis and trans) are possible around the C=C bond. Structural isomers are also possible. For example, the chain may be branched, the double bond can be shifted, and the aldehyde functional group may be changed to a ketone.

Think About It
From the examples above, can you draw more structural isomers of dodecenal?

13.141. Collect and Organize
By writing the chemical formulas of naphthalene and anthracene (Figure P13.141), we can determine whether these compounds can be distinguished by combustion analysis.

Analyze
If the ratio of carbon to hydrogen atoms is different between naphthalene and anthracene, then the compounds are distinguishable by combustion analysis.

Solve

Naphthalene's chemical formula is $C_{10}H_8$ or C_nH_{n-2}. Anthracene's chemical formula is $C_{14}H_{10}$ or C_nH_{n-4}. The C:H ratio is different; therefore, yes, these compounds are distinguishable by combustion analysis.

Think About It

The balanced equation for the combustion reactions also show that we can distinguish between these compounds.

$$C_{10}H_8(s) + 12\ O_2(g) \rightarrow 10\ CO_2(g) + 4\ H_2O(g)$$
$$C_{14}H_{10}(s) + \tfrac{33}{2}\ O_2(g) \rightarrow 14\ CO_2(g) + 5\ H_2O(g)$$

13.143. Collect and Organize

Given the structure of Soman, we may identify the phosphoryl group and the parent acid from which it may be derived.

Analyze

The phosphoryl group (P=O) is structurally related to the carbonyl group. Soman (shown in Figure P13.143) contains a P–O–C linkage that may arise from a phosphorus acid (P–O–H) and an appropriate alcohol.

Solve

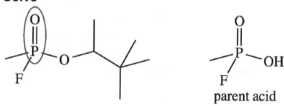

parent acid

Think About It

Many biologically relevant molecules such as ATP, DNA, and RNA contain the phosphoryl group.

13.145. Collect and Organize

From the structure of nylon we can identify functional groups that make the polymer hydrophilic.

Analyze

Hydrophilic means "water loving." The polar functional groups on nylon are hydrophilic. The structure of nylon is shown in Figure 13.48.

Solve

The amide groups [–C(=O)NH–] in nylon are hydrophilic.

Think About It

The alkyl chain –$(CH_2)_6$– on the polymer backbone of nylon is hydrophobic, "water avoiding."

13.147. Collect and Organize

We are to draw all the possible trimers possible from the condensation reaction of putrescine [$H_2N(CH_2)_4NH_2$] with adipic acid and terephthalic acid (Figure P13.147). In addition we are asked what ratio of monomers would be necessary to prepare a polymer with a 1:1 ratio of the two carboxylic acids in the polymer chain.

Analyze

In the condensation reaction, the difunctional amine can react with two carboxylic acid molecules to form a trimer.

Solve

(a)

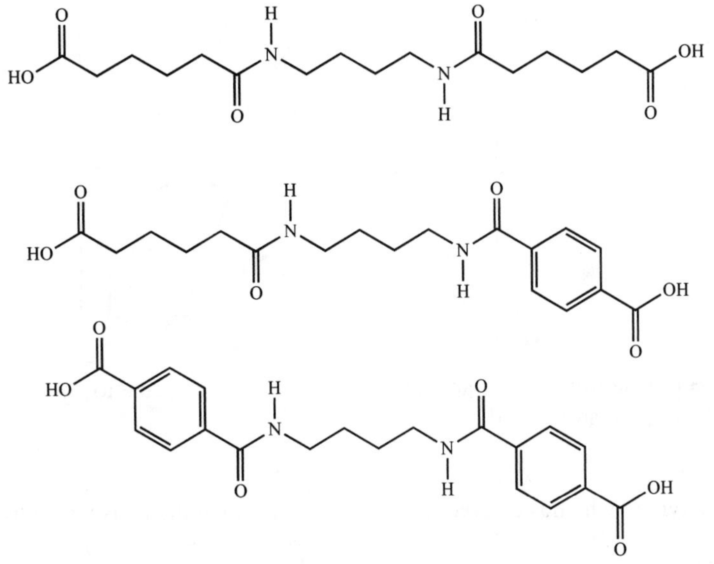

(b) 1 mole adipic acid to 1 mole terephthalic acid to 2 moles putrescine.

Think About It

The polymer from part b would have the backbone structure

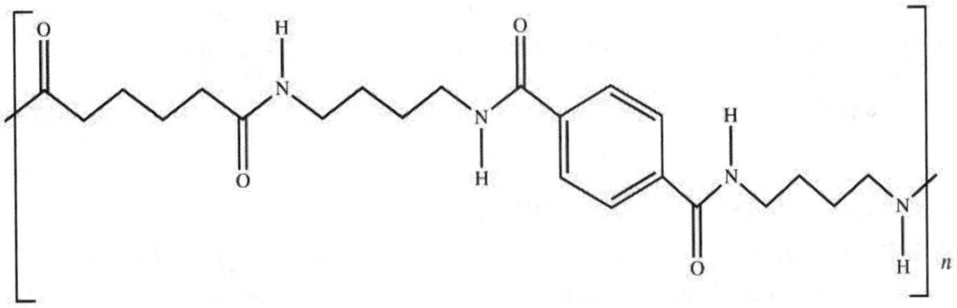

13.149. **Collect and Organize**

Using the given structures of maleic anhydride and styrene (Figure P13.149), we are to draw two structural repeating units of the polymer formed from these two monomers. By comparing the structure to that of polystyrene we can predict how the two polymers' properties might differ.

Analyze

Both polystyrene and the polymer formed from maleic anhydride and styrene are addition polymers. The structure of polystyrene is

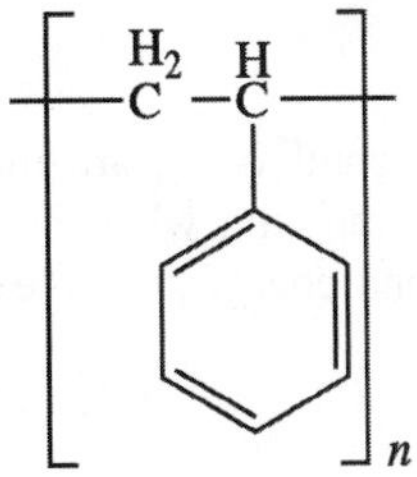

Solve

(a)

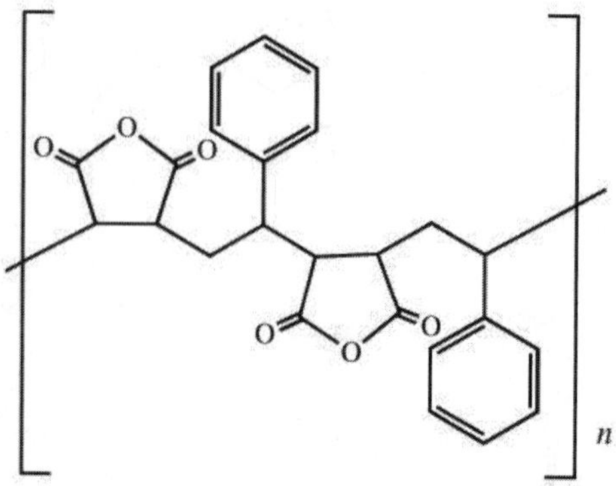

(b) The polymer made from maleic anhydride and styrene is expected to be more hydrophilic, because of the presence of the oxygen atoms, and less rigid.

Think About It

This type of polymer, which is derived from two different monomers, is a *copolymer*.

13.151. **Collect and Organize**

Considering the synthesis and properties of silicones $[R_2SiO]_n$ (Figure P13.151), we are to write two balanced equations for the reactions and explain why silicones are hydrophobic.

Analyze

(a) The two equations are (1) the reaction of R_2SiCl_2 with water to give a new monomer, whose formula is $R_2SiCl(OH)$, and HCl and (2) the reaction of two $R_2SiCl(OH)$ molecules eliminating water to give $ClR_2Si—O—SiR_2Cl$.

(b) Water-repellent polymers are nonpolar.

Solve

(a) $R_2SiCl_2(aq) + H_2O(\ell) \rightarrow R_2SiCl(OH)(aq) + HCl(aq)$

$R_2SiCl(OH)(aq) + (HO)SiClR_2(aq) \rightarrow R_2ClSi—O—SiClR_2(aq) + H_2O(\ell)$

(b) Silicones repel water because the side chains (R groups) are nonpolar.

Think About It

Silicones have many applications including cookware (because they are non-stick and heat resistant), contact lenses (because of their oxygen permeability), and lubricants (because of their clean application).

13.153. **Collect and Organize**

We are asked to explain why the enthalpy of combustion using bond energies for butane and 2-methylpropane (both with molecular formulas of C_4H_{10}) are the same, but, when the enthalpy of combustion for these compounds is determined experimentally, their values are different.

Analyze

Experimentally determined heats of combustion are measured using bomb calorimetry, which accurately measures the heat released by the compound when burned in oxygen. The heat of combustion calculated using bond energies uses average bond energies (Table 13.4). The average bond energies are derived from many examples of that type of bond.

Solve

The heat of combustion determined using experimental means is different from that calculated from average bond energies because the bond energy of a particular bond depends on the structure of the rest of the molecule.

Think About It

When using average bond energies to estimate the enthalpy of combustion for butane and 2-methylpropane, the calculated values will be equal because in both combustion reactions ten C—H bonds and four C—C bonds are broken to yield eight C=O bonds (in CO_2) and ten H—O bonds (in H_2O).

13.155. **Collect and Organize / Analyze**

We are asked to identify the chiral centers and functional groups present in the molecules depicted in Figure P13.155.

Solve

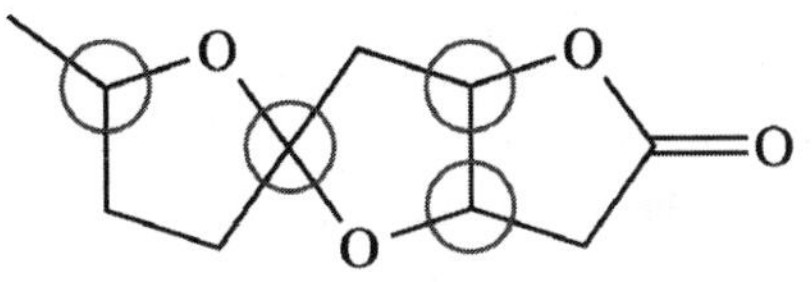

(a)

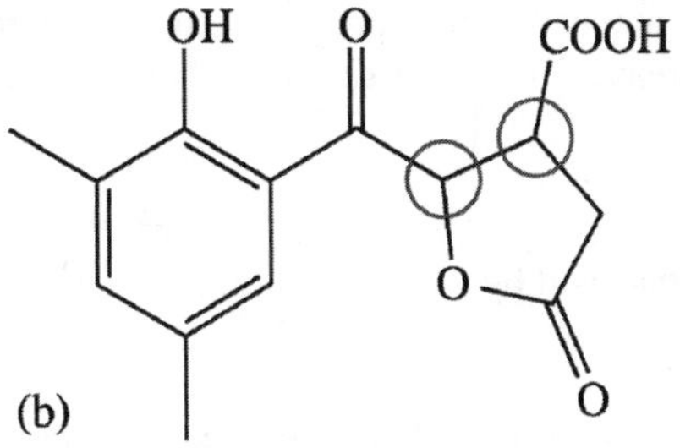

(b)

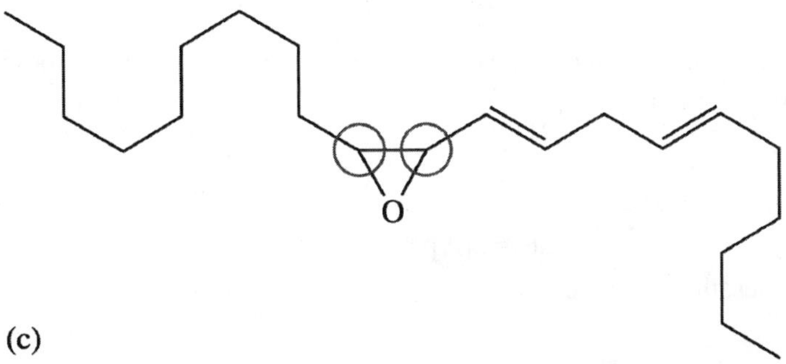

(c)

(a) The functional groups present in this molecule are: ether, ester, and alkane.
(b) The functional groups present in this molecule are: alcohol, ketone, carboxylic acid, ester, and alkane.
(c) The functional groups present in this molecule are: ether, alkene, and alkane.

Think About It

Both of the alkenes depicted in (c) are trans.

CHAPTER 14 | Chemical Kinetics: Reactions in the Air We Breathe

14.1. Collect and Organize
For the plot of Figure P14.1, we are to identify which curves represent $[N_2O]$ and $[O_2]$ over time for the conversion of N_2O to N_2 and O_2 according to the equation

$$2 N_2O(g) \rightarrow 2 N_2(g) + O_2(g)$$

Analyze
As the reaction proceeds, the concentration of the reactant, N_2O, decreases and the concentration of the product, O_2, increases. The rate at which N_2O is used up in the reaction is twice the rate at which O_2 is produced.

Solve
The $[N_2O]$ is represented by the green line and $[O_2]$ is represented by the red line.

Think About It
Notice that $[N_2]$, represented by the blue line, increases twice as fast as $[O_2]$ because there are 2 N_2 molecules produced for every O_2 molecule in the reaction.

14.3. Collect and Organize
For three different initial concentrations of reactant A shown in Figure P14.3, we are to choose which would have the fastest rate for the conversion 2 A→B.

Analyze
We are given that the reaction is second order in A. The rate law is written as follows:

$$\text{Rate} = k[A]^2$$

As the concentration of A increases, the rate increases.

Solve
(b) has the fastest reaction rate because it has the highest concentration of A.

Think About It
The higher the concentration of reactant molecules, the more often they collide, which increases the rate of the reaction.

14.5. Collect and Organize
For the three reaction profiles in Figure P14.5, we are to indicate which one has the slowest and which has the fastest reaction rate.

Analyze
The activation energy is the energy barrier that the reactants must overcome in order to form products; the lower the activation energy barrier, the faster the reaction. The rate of a reaction is determined by the activation energy (E_a) of the slowest step. All of the reactions shown consist of a single step. The E_a is the energy difference between the reactants (on the left side of the graph) and the transition state (the highest point on the reaction profile curve).

Solve
(a) Reaction profile (a) has the largest E_a and therefore is the slowest reaction.
(b) Reaction profile (b) has the lowest E_a and therefore is the fastest reaction.

Think About It
Be careful not to assume that reaction (c) is slow because it is nonspontaneous. The rate of a reaction does not depend on the thermodynamics of the reactants and the products.

14.7. Collect and Organize

We are to match the reaction profile in Figure P14.7 with the one of the reactions given.

Analyze

The reaction profile shows a two-step reaction that has a slightly larger activation energy for its second step than its first step.

Solve

(c) is correct because the graph describes a two-step reaction with the first step faster than the second.

Think About It

Reaction (b), which occurs in a single step, would show only one transition state and one activation energy in its reaction profile.

14.9. Collect and Organize

Given the reaction profile of an uncatalyzed reaction (Figure P14.9), we are to choose the reaction profile corresponding to the catalyzed reaction.

Analyze

A catalyst increases the rate of reaction by decreasing the activation energy of the reaction through an alternate pathway to the products. This alternate pathway usually involves more steps.

Solve

(b) correctly shows the catalyzed reaction.

Think About It

Reaction profiles a and c cannot be correct because the initial, uncatalyzed reaction is nonspontaneous and (a) and (c) represent spontaneous reactions. A catalyst cannot change a nonspontaneous reaction into a spontaneous reaction.

14.11. Collect and Organize

We are asked to identify the transition state and the activation energy for the forward and reverse reactions depicted in Figure P14.11.

Analyze

The transition state corresponds to the highest energy arrangement of atoms, and so to the highest point on the reaction profile. The activation energy (E_a) for a reaction is the difference in energy between the starting materials (reactants for the forward reaction, products for the reverse reaction) and the transition state.

Solve

(a) The transition state is 3*
(b) E_a in the forward direction is (a)
(c) E_a in the reverse direction is (b)

Think About It

The reaction depicted in Figure P14.11 is endothermic, since the products are higher in energy than the reactants.

14.13. Collect and Organize

Of the highlighted elements in Figure P14.13 we are to choose which forms gaseous oxides associated with photochemical smog.

Analyze

Photochemical smog is the result of sunlight interacting with NO_x produced by automobile emissions and volatile organic compounds (VOCs) released into the atmosphere.

Solve

Nitrogen (light blue) forms the volatile oxides that are components of photochemical smog.

Think About It

Sunlight causes a reaction of NO_x and VOCs to produce peroxyacyl nitrates that are very irritating to the lungs.

14.15. **Collect and Organize**

Of the highlighted elements in Figure P14.15 we are to identify which are widely used as heterogeneous catalysts.

Analyze

Heterogeneous catalysts have a different phase than the reactants. We read in Section 14.6 about the specific metals that are used in catalytic converters.

Solve

Both of the transition metals, palladium (blue) and platinum (orange), can serve as heterogeneous catalysts, and were specifically identified in the chapter as catalysts.

Think About It

Because catalytic converters contain precious metals such as rhodium, platinum, and palladium, there is great interest in recycling the metals from catalytic converters.

14.17. **Collect and Organize**

By considering the levels of O_3 during the day as seen in Figure 14.2, we are to explain why $[O_3]_{max}$ occurs later in the day than $[NO]_{max}$ and $[NO_2]_{max}$.

Analyze

Ozone in the troposphere (the lowest portion of Earth's atmosphere) is due to the reaction of O_2 with O generated from the interaction of UV light with NO_2.

Solve

The presence of NO_2 in the atmosphere and ample sunlight allows the O atoms to react with O_2 to generate O_3. The reactant NO_2 is present in the atmosphere due to automobile exhausts, building up during the day. The buildup of O_3 lags behind until later in the day, when $[NO_2]$ increases and the sunlight becomes stronger as midday approaches.

Think About It

Ozone is a very reactive gas and is irritating to lung tissues.

14.19. **Collect and Organize**

We are to explain why there is not an increase in NO concentration after the evening rush hour.

Analyze

The reaction in the troposphere (lower atmosphere) that produces NO is

$$NO_2(g) \xrightarrow{\text{sunlight}} NO(g) + O(g)$$

Solve

In the evening the sunlight (and UV radiation) is less intense, so the photochemical breakdown of NO_2 cannot occur to as great an extent as after the morning rush hour.

Think About It

The use of catalytic converters to reduce the NO_x to N_2 and O_2 in automobile exhaust has greatly helped to reduce photochemical smog in large urban and suburban centers.

14.21. Collect and Organize
We are asked to explain whether O atoms or O_2 molecules are expected to be more reactive.

Analyze
O atoms have six valence electrons, two of which are unpaired (radicals). While O_2 also has unpaired electrons, the covalent bonds present in this molecule mean that each O atom in O_2 has a complete octet.

Solve
O atoms are expected to be more reactive because of the unpaired electrons and incomplete octet.

Think About It
Oxygen radicals are powerful oxidants, leading to cell damage and the formation of other reactive species such as ozone, O_3.

14.23. Collect and Organize / Analyze
For the reaction of N_2 with O_2 to produce N_2O and N_2O_5, we are to write balanced chemical equations.

Solve
(a) $N_2(g) + \frac{1}{2} O_2(g) \rightarrow N_2O(g)$

or $2 N_2(g) + O_2(g) \rightarrow 2 N_2O(g)$

(b) $N_2(g) + \frac{5}{2} O_2(g) \rightarrow N_2O_5(g)$

or $2 N_2(g) + 5 O_2(g) \rightarrow 2 N_2O_5(g)$

Think About It
Balanced chemical equations usually are written with whole-number coefficients.

14.25. Collect and Organize
We are to explain the difference between the average rate and the instantaneous rate of a reaction.

Analyze
The rate of reaction is measured by the change in concentration of a reactant or product over time. The difference between the average and instantaneous rates is the length of the period of time over which the change in concentration is measured.

Solve
The average rate is the rate averaged over a fairly long period of time, whereas the instantaneous rate is the rate at a specific moment in time (or over a very, very short period of time).

Think About It
The rate of a reaction is always positive.

14.27. Collect and Organize
We are asked if the rate of reaction with respect to the concentration of the product will increase or decrease as the reactant concentration decreases.

Analyze

Graphically, this question is asking how the rate of reaction with respect to the concentration of product changes as the reaction proceeds (i.e., from a to c):

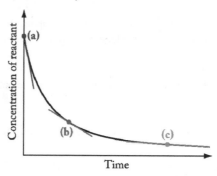

Solve

As the reaction proceeds, the concentration of product increases, so the rate, $\dfrac{\Delta[\text{Product}]}{\Delta t}$, will decrease.

Think About It

As more product is formed, the rate of change in the concentration of product will decrease. In general, the instantaneous rate of reaction at the outset of the reaction is fast, after which the reaction begins to slow down.

14.29. Collect and Organize

We may determine the relationship between the rate of formation of N_2, the rate of formation of O_2, and the rate of consumption of NO in an automotive catalytic converter by examining the reaction.

Analyze

The reaction that occurs in the catalytic converter is

$$2\,NO(g) \rightarrow N_2(g) + O_2(g)$$

From the balanced equation, the reaction may be expressed in terms of rates as

$$\text{Rate} = -\frac{1}{2}\frac{\Delta[\text{NO}]}{\Delta t} = +\frac{\Delta[\text{N}_2]}{\Delta t} = +\frac{\Delta[\text{O}_2]}{\Delta t}$$

Solve

(a) The relationship between $[O_2]$ and $[N_2]$ is

$$\frac{\Delta[\text{N}_2]}{\Delta t} = \frac{\Delta[\text{O}_2]}{\Delta t}$$

(b) The relationship between [NO] and $[N_2]$ is

$$\frac{\Delta[\text{N}_2]}{\Delta t} = -\frac{1}{2}\frac{\Delta[\text{NO}]}{\Delta t}$$

Think About It

The rate of change in concentrations of reactants and products are related to the stoichiometry of the reaction being considered.

14.31. Collect and Organize

Using the balanced equations provided, we are to write expressions to compare the rates of formation of products and the rates of consumption of reactants.

Analyze

For the reaction a A + b B → c C + d D the rate of reaction may be expressed as

$$\text{Rate} = -\frac{1}{a}\frac{\Delta[A]}{\Delta t} = -\frac{1}{b}\frac{\Delta[B]}{\Delta t} = +\frac{1}{c}\frac{\Delta[C]}{\Delta t} = +\frac{1}{d}\frac{\Delta[D]}{\Delta t}$$

Solve

(a) $\text{Rate} = -\dfrac{\Delta[F_2]}{\Delta t} = +\dfrac{\Delta[HOF]}{\Delta t} = +\dfrac{\Delta[HF]}{\Delta t}$

(b) $\text{Rate} = -\dfrac{1}{3}\dfrac{\Delta[HCl]}{\Delta t} = +\dfrac{\Delta[H_2]}{\Delta t}$

(c) $\text{Rate} = -\dfrac{1}{4}\dfrac{\Delta[NH_3]}{\Delta t} = -\dfrac{1}{4}\dfrac{\Delta[O_2]}{\Delta t} = +\dfrac{1}{2}\dfrac{\Delta[N_2]}{\Delta t} = +\dfrac{1}{6}\dfrac{\Delta[H_2O]}{\Delta t}$

Think About It

As for equilibrium expressions, pure liquids and solids are excluded from the rate law.

14.33. Collect and Organize

Given the concentration of NO(*g*) at two times along a reaction profile, we may determine the rate of the reaction.

Analyze

The reaction under consideration is

$$N_2(g) + O_2(g) \rightarrow 2\,NO(g)$$

From the balanced equation, the reaction may be expressed in terms of rates as

$$\text{Rate} = -\frac{\Delta[N_2]}{\Delta t} = -\frac{\Delta[O_2]}{\Delta t} = +\frac{1}{2}\frac{\Delta[NO]}{\Delta t}$$

where $\Delta[N_2]$ is the difference of the two provided concentrations, and Δt is the difference of the corresponding times. We may substitute the provided values for [NO] and time into this expression and solve for the rate.

Solve

$$\text{Rate} = +\frac{1}{2}\frac{\Delta[NO]}{\Delta t} = +\frac{1}{2}\left(\frac{0.407\,M - 0.375\,M}{65.7\,s - 50.0\,s}\right) = 1.0\times10^{-3}\,M/s$$

Think About It

The rate of reaction will always be positive. In this case, the +1/2 coefficient and increasing concentration of NO lead to a positive rate, but the same value could be obtained from data for the disappearance of N_2 or O_2. In these cases, the change in concentration would be negative and the coefficients are both –1, leading to a positive rate of reaction.

14.35. Collect and Organize

Using the balanced equation describing the reaction of SO_2 with CO, we are to write expressions to compare the rates of formation of products and the rates of consumption of reactants.

Analyze

From the balanced equation we see that the reaction may be expressed as

$$\text{Rate} = -\frac{\Delta[SO_2]}{\Delta t} = -\frac{1}{3}\frac{\Delta[CO]}{\Delta t} = \frac{1}{2}\frac{\Delta[CO_2]}{\Delta t} = \frac{\Delta[COS]}{\Delta t}$$

Solve

(a) $\text{Rate} = \dfrac{\Delta[CO_2]}{\Delta t} = -\dfrac{2}{3}\dfrac{\Delta[CO]}{\Delta t}$

(b) $\text{Rate} = \dfrac{\Delta[COS]}{\Delta t} = -\dfrac{\Delta[SO_2]}{\Delta t}$

(c) $\text{Rate} = \dfrac{\Delta[CO]}{\Delta t} = 3\dfrac{\Delta[SO_2]}{\Delta t}$

Think About It

These relative rates make sense based on the stoichiometry of the reaction. For every 1 mole of SO_2 used in the reaction, 2 moles of CO_2 and 1 mole of COS are produced. So, for example, the concentration of CO_2 will increase twice as fast as the concentration of SO_2 decreases.

14.37. Collect and Organize

Using the relative rate expressions and the rate of the consumption of ClO in two reactions, we are to calculate the rate of change in the formation of the products of the two reactions.

Analyze

(a) For this reaction

$$\text{Rate} = -\frac{1}{2}\frac{\Delta[ClO]}{\Delta t} = \frac{\Delta[Cl_2]}{\Delta t} = \frac{\Delta[O_2]}{\Delta t}$$

(b) For this reaction

$$\text{Rate} = -\frac{\Delta[ClO]}{\Delta t} = -\frac{\Delta[O_3]}{\Delta t} = \frac{\Delta[O_2]}{\Delta t} = \frac{\Delta[ClO_2]}{\Delta t}$$

For each reaction we are given $\Delta[ClO]/\Delta t$, and we can use this value in the relationships above to calculate the rate of change in (a) the concentration of Cl_2 and O_2 and (b) the concentration of O_2 and ClO_2.

Solve

(a) $\dfrac{\Delta[Cl_2]}{\Delta t} = \dfrac{\Delta[O_2]}{\Delta t} = -\dfrac{1}{2}\dfrac{\Delta[ClO]}{\Delta t} = -\dfrac{1}{2}\times-2.3\times10^7 \ M/s$

$\qquad = 1.2\times10^7 \ M/s$

(b) $\dfrac{\Delta[O_2]}{\Delta t} = \dfrac{\Delta[ClO_2]}{\Delta t} = -\dfrac{\Delta[ClO]}{\Delta t} = -\left(-2.9\times10^4 \ M/s\right)$

$\qquad = 2.9\times10^4 \ M/s$

Think About It

The rates of the formation of products is positive because $[X]_f > [X]_i$ so $[X]_f - [X]_i = \Delta[X]$ is positive.

14.39. Collect and Organize

Given the $[O_3]$ over time when it reacts with NO_2^-, we are to calculate the average reaction rate for two time intervals.

Analyze

The average rate of reaction can be found according to

$$\frac{\Delta[O_3]}{\Delta t} = \frac{[O_3]_f - [O_3]_i}{t_f - t_i}$$

Solve

Between 0 and 100 μs:

$$-\frac{\Delta[O_3]}{\Delta t} = \frac{\left(9.93 \times 10^{-3} - 1.13 \times 10^{-2}\right) M}{(100 - 0)\ \mu s} = 1.4 \times 10^{-5}\ M/\mu s$$

Between 200 and 300 μs:

$$-\frac{\Delta[O_3]}{\Delta t} = \frac{\left(8.15 \times 10^{-3} - 8.70 \times 10^{-3}\right) M}{(300 - 200)\ \mu s} = 5.50 \times 10^{-6}\ M/\mu s$$

Think About It

Notice that as the reaction proceeds, the rate of consumption of ozone decreases. This is due to the decreasing reactant concentrations.

14.41. **Collect and Organize**

After we plot [ClO] versus time and [Cl$_2$O$_2$] versus time, we can determine the instantaneous rate of change at 1 s for each compound.

Analyze

The instantaneous rate is the slope of the line that is tangent to the curve at the time we are interested in. We can estimate fairly well the instantaneous rate at 1 s from the plots by choosing two points that are close to 1 s to calculate the slope. Since we are given only values for [ClO], we need to calculate [Cl$_2$O$_2$] for each time.

$$[Cl_2O_2]_t = \frac{\left([ClO]_0 - [ClO]_t\right)}{2}$$

where $[ClO]_0 = 2.60 \times 10^{11}\ M$

Time s	[ClO]$_t$ molecules/cm^3	[Cl$_2$O$_2$]$_t$ molecules/cm^3
0	2.60×10^{11}	0.00
1	1.08×10^{11}	7.60×10^{10}
2	6.83×10^{10}	9.59×10^{10}
3	4.99×10^{10}	1.05×10^{11}
4	3.93×10^{10}	1.10×10^{11}
5	3.24×10^{10}	1.14×10^{11}
6	2.76×10^{10}	1.16×10^{11}

Note that initially there is no Cl$_2$O$_2$ present, so [Cl$_2$O$_2$]$_0$ = 0 molecules/cm^3.

Solve

For the change in concentration of ClO versus time we obtain the following plot:

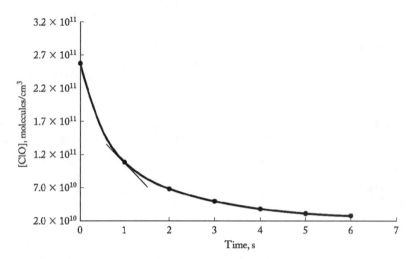

To get a fairly good estimate from the data given for the instantaneous rate, we can choose two points from the data set that surround the data point of interest and calculate $\Delta[\text{ClO}]/\Delta t$.

For $t = 1$ s, we can use the points $t = 0$ s and $t = 2$ s

$$-\frac{\Delta[\text{ClO}]}{\Delta t} = \frac{(6.83 \times 10^{10} - 2.60 \times 10^{11})\ \text{molecules/cm}^3}{(2-0)\ s} = 9.6 \times 10^{10}\ \text{molecules} \cdot \text{cm}^{-3} \cdot \text{s}^{-1}$$

Using a graphing program which calculates the slope of the tangent to the line at $t = 1$ s, we get an instantaneous rate of 8.3×10^{10} molecules $\cdot$ cm^{-3} $\cdot$ s.

For the change in concentration of Cl_2O_2 versus time we obtain the following plot:

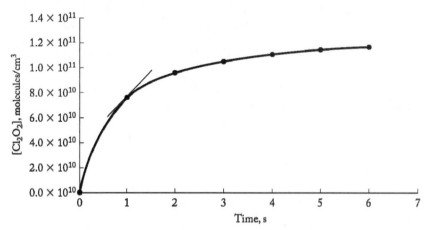

To get a fairly good estimate from the data given for the instantaneous rate, we can choose two points from the data set that surround the data point of interest and calculate $\Delta[\text{Cl}_2\text{O}_2]/\Delta t$.

For $t = 1$ s, we can use the points $t = 0$ s and $t = 2$ s

$$\frac{\Delta[\text{Cl}_2\text{O}_2]}{\Delta t} = \frac{(9.59 \times 10^{10} - 0)\ \text{molecules/cm}^3}{(2-0)\ s} = 4.8 \times 10^{10}\ \text{molecules} \cdot \text{cm}^{-3} \cdot \text{s}^{-1}$$

Using a graphing program which calculates the slope of the tangent to the line at $t = 1$ s, we get an instantaneous rate of 4.1×10^{10} molecules $\cdot$ cm^{-3} $\cdot$ s.

Think About It
Because we expect the rate of disappearance of ClO to be twice the rate of appearance of Cl_2O_2 from the balanced equation

$$2\, ClO(g) \rightarrow Cl_2O_2(g)$$

our answers make sense

$$\frac{\dfrac{\Delta[ClO]}{\Delta t}}{\dfrac{\Delta[Cl_2O_2]}{\Delta t}} = \frac{9.59 \times 10^{10}\ \text{molecules} \cdot \text{cm}^{-3} \cdot \text{s}^{-1}}{4.80 \times 10^{10}\ \text{molecules} \cdot \text{cm}^{-3} \cdot \text{s}^{-1}} = 2.0$$

14.43. Collect and Organize
We are asked to explain why the rate of a reaction generally decreases as the reaction proceeds.

Analyze
As a reaction proceeds, products are formed and the concentration of products increases, while reactants are consumed and the concentration of reactants decreases. The rate of a reaction depends on the concentration of reactants.

Solve
As the concentration of reactants decreases, so too will the number of collisions between reactants. It is thus less likely that reactants will collide with sufficient energy and the proper orientation to overcome the activation energy for the reaction, and the rate decreases.

Think About It
By the same rationale, kinetic molecular theory suggests that as the temperature increases, so does the average energy of the reactants, and so it is more likely that reactants will have sufficient energy to overcome the activation energy when colliding. For this reason, most reactions have a higher rate of reaction at higher temperatures.

14.45. Collect and Organize
We are asked if the units of the half-life for a second-order reaction are the same as those of the half-life for a first-order reaction.

Analyze
The half-life is the time it takes for the amount of reactant originally present to decrease by one-half.

Solve
Yes. Because the half-life is a time measurement, the units, no matter what the order of the reaction, are always in units of time (s, min, hr, yr, etc.).

Think About It
The half-life of a reaction depends on the value of the reaction's rate constant and, except for first-order reactions, on the initial concentration. The larger the rate constant, the faster the reaction and the shorter the half-life of the reaction.

14.47. Collect and Organize
For a second-order reaction, we are to predict the effect of doubling $[A]_0$ on the half-life.

Analyze
For a second-order reaction,

$$t_{1/2} = \frac{1}{k[A]_0}$$

Solve
From the equation for the half-life of a second-order reaction, we see that doubling $[A]_0$ halves the half-life.

Think About It
The half-life of a second-order reaction, like that of a first-order reaction, is inversely related to the rate constant.

14.49. Collect and Organize
For each rate-law expression, we are to determine the order of the reaction with respect to each reactant and the overall reaction order.

Analyze
The order of a reaction is the experimentally determined dependence of the rate of a reaction on the concentration of the reactants involved in the reaction. In the rate-law expression the order is shown as the power to which the concentration of a particular reactant is raised. The overall reaction order is the sum of the powers of the reactants in the rate-law expression.

Solve
(a) For Rate = $k[A][B]$, the reaction is first order in both A and B and second order overall.
(b) For Rate = $k[A]^2[B]$, the reaction is second order in A, first order in B, and third order overall.
(c) For Rate = $k[A][B]^3$, the reaction is first order in A, third order in B, and fourth order overall.

Think About It
The higher the order of the reaction for a particular reactant, the greater the effect of a change in concentration of that reactant on the reaction rate.

14.51. Collect and Organize
For each of the reactions described, we are to write the rate law and determine the units for k, using the units M for concentration and s for time.

Analyze
The general form of the rate law is
$$\text{Rate} = k[A]^x[B]^y$$
where k is the rate constant, A and B are the reactants, and x and y are the orders of the reaction with respect to each reactant as determined by experiment.

Solve
(a) Rate = $k[O][NO_2]$
Because rate has units of M/s and each concentration has units of M,
$$k = \frac{M/s}{M^2} = M^{-1}\,s^{-1}$$
(b) Rate = $k[NO]^2[Cl_2]$
Because rate has units of M/s and each concentration has units of M,
$$k = \frac{M/s}{M^3} = M^{-2}\,s^{-1}$$
(c) Rate = $k[CHCl_3][Cl_2]^{1/2}$
Because rate has units of M/s and each concentration has units of M,
$$k = \frac{M/s}{M^{3/2}} = M^{-\frac{1}{2}}\,s^{-1}$$
(d) Rate = $k[O_3]^2[O]^{-1}$
Because rate has units of M/s and each concentration has units of M,
$$k = \frac{M/s}{M^2/M} = s^{-1}$$

Think About It
The units of the rate constant clearly depend on the overall order of the reaction.

14.53. Collect and Organize

Given the changes in rate of the decomposition of BrO to Br_2 and O_2 when [BrO] is changed, we are to predict the rate law expression in each case.

Analyze

The general form of the rate law for the reaction is
$$\text{Rate} = k[\text{BrO}]^x$$
where x is an experimentally determined exponent.

Solve

(a) If the rate doubles when [BrO] doubles, then $x = 1$ and the rate law is
$$\text{Rate} = k[\text{BrO}]$$
(b) If the rate quadruples when [BrO] doubles, then $x = 2$ and the rate law is
$$\text{Rate} = k[\text{BrO}]^2$$
(c) If the rate is halved when [BrO] is halved, then $x = 1$ and the rate law is
$$\text{Rate} = k[\text{BrO}]$$
(c) If the rate is unchanged when [BrO] is doubled, then $x = 0$ and the rate law is
$$\text{Rate} = k[\text{BrO}]^0 = k$$

Think About It

For this reaction the relationship between the change in rate when the concentration of the reactant was changed to determine x is straightforward. To determine x for more complicated reactions, use

$$\frac{\text{rate}_2}{\text{rate}_1} = \left(\frac{[A]_2}{[A]_1}\right)^x$$

$$\ln\left(\frac{\text{rate}_2}{\text{rate}_1}\right) = x \ln\left(\frac{[A]_2}{[A]_1}\right)$$

$$x = \frac{\ln\left(\dfrac{\text{rate}_2}{\text{rate}_1}\right)}{\ln\left(\dfrac{[A]_2}{[A]_1}\right)}$$

14.55. Collect and Organize

We must predict possible reaction orders when doubling both reactant concentrations leads to a rate that is four times the initial rate.

Analyze

The reaction being considered is
$$\text{NO}(g) + O_3(g) \rightarrow NO_2(g) + O_2(g)$$
We may consider possible rate laws to account for this observation, arising from common reaction orders with respect to the reactants, NO and O_3.

Solve

At least two possible rate laws satisfy the quadrupling of rate. If the reaction is first order in each reactant, the rate is expected to quadruple.

$$\frac{\text{rate}_b}{\text{rate}_a} = \frac{[\text{NO}]_b^1 [O_3]_b^1}{[\text{NO}]_a^1 [O_3]_b^1} = \frac{(2)^1 (2)^1}{(1)^1 (1)^1} = 4$$

If the reaction is second order in one reactant and zeroth order in the other, the rate is also expected to quadruple.

$$\frac{\text{rate}_b}{\text{rate}_a} = \frac{[NO]_b^2[O_3]_b^0}{[NO]_a^2[O_3]_b^0} = \frac{(2)^2(2)^0}{(1)^2(1)^0} = 4$$

Since both situations could lead to the same observed rate behavior, it is not conclusive that the reaction is first order in both reactants.

Think About It

In order to determine the rate law for this reaction, the concentration of NO and O_3 must be manipulated independently. For example, doubling [NO] while holding [O_3] constant would have different effects on the rate:

$$\frac{\text{rate}_b}{\text{rate}_a} = \frac{[NO]_b^1[O_3]_b^1}{[NO]_a^1[O_3]_b^1} = \frac{(2)^1(1)^1}{(1)^1(1)^1} = 2$$

$$\frac{\text{rate}_b}{\text{rate}_a} = \frac{[NO]_b^2[O_3]_b^0}{[NO]_a^2[O_3]_b^0} = \frac{(2)^2(1)^0}{(1)^2(1)^0} = 4$$

14.57. **Collect and Organize**

For the reaction of NO_2 with O_3 to produce NO_3 and O_2 we are to write the rate law given that the reaction is second order overall. From the rate law and given the rate constant, we can calculate the rate of the reaction for a given [NO_2] and [O_3]. From this we can calculate the rate of appearance of NO_3 and the rate of the reaction when [O_3] is doubled.

Analyze

The general form of the rate law for this reaction is

$$\text{Rate} = k[NO_2]^x[O_3]^y$$

The rate of consumption of reactants and formation of products is

$$\text{Rate} = -\frac{\Delta[NO_2]}{\Delta t} = -\frac{\Delta[O_3]}{\Delta t} = \frac{\Delta[NO_3]}{\Delta t} = \frac{\Delta[O_2]}{\Delta t}$$

Solve

(a) Rate = $k[NO_2][O_3]$

(b) Rate = $\dfrac{1.93\times10^4}{M\cdot s}\times1.8\times10^{-8}M\times1.4\times10^{-7}M = 4.9\times10^{-11}M/s$

(c) Rate = $\dfrac{\Delta[NO_3]}{\Delta t} = 4.9\times10^{-11}M/s$

(d) When [O_3] is doubled, the rate of the reaction doubles.

Think About It

When [O_3] = 2.8×10^{-7} M (double that in part b) the rate of reaction is 9.73×10^{-11} M/s, which is twice that calculated in part (b), so our prediction in part (d) is correct.

14.59. **Collect and Organize**

By comparing the rate constants for four reactions that are all second-order, we can determine which reaction is the fastest if all the initial concentrations are the same.

Analyze

The reaction with the largest rate constant has the fastest reaction rate.

Solve

Reaction (c) has the largest value of k, so it proceeds the fastest.

Think About It
The slowest reaction is a reaction with the smallest value of k.

14.61. Collect and Organize
Given a rate constant of 3.4×10^{-5} s^{-1}, we are asked to write a rate law describing the reaction
$$2\,N_2O_5(g) \rightarrow 4\,NO_2(g) + O_2(g)$$

Analyze
The rate law for a reaction is equal to the rate constant multiplied by the concentration of each reactant raised to its order. Since the only reactant is N_2O_5, it will be the only concentration appearing in the rate law. The units of the rate constant (s^{-1}) tell us that this reaction is first order.

Solve
The rate law for this reaction is
$$\text{Rate} = 3.4 \times 10^{-5}\ s^{-1}[N_2O_5]^1$$

Think About It
The rate, expressed in terms of individual reactant and product concentrations, is
$$\text{Rate} = -\frac{1}{2}\frac{\Delta[N_2O_5]}{\Delta t} = +\frac{1}{4}\frac{\Delta[NO_2]}{\Delta t} = +\frac{\Delta[O_2]}{\Delta t}$$

14.63. Collect and Organize
In the reaction of ClO_2 with OH^-, the rate of the reaction was measured for various concentrations of both reactants. From the data, we are to determine the rate law and calculate the rate constant, k.

Analyze
To determine the dependence of the rate on a change in the concentration of a particular reactant, we can compare the reaction rates for two experiments in which the concentration of that reactant changes but the concentrations of the other reactants remain constant. Once we have the order of the reaction for each reactant, we can write the rate-law expression. To calculate the rate constant for the reaction, we can rearrange the rate law to solve for k and use the data from any of the experiments.

Solve
Using experiments 1 and 2, we find that the order of the reaction with respect to ClO_2 is 1:
$$\frac{\text{rate}_1}{\text{rate}_2} = \left(\frac{[ClO_2]_{0,1}}{[ClO_2]_{0,2}}\right)^x$$
$$\frac{0.0248\ M/s}{0.00827\ M/s} = \left(\frac{0.060\ M}{0.020\ M}\right)^x$$
$$3.00 = 3.01^x$$
$$x = 1$$
Using experiments 2 and 3 we find that the order of the reaction with respect to $[OH^-]$ is also 1:
$$\frac{\text{rate}_3}{\text{rate}_2} = \left(\frac{[OH^-]_{0,3}}{[OH^-]_{0,2}}\right)^x$$
$$\frac{0.0247\ M/s}{0.00827\ M/s} = \left(\frac{0.090\ M}{0.030\ M}\right)^x$$
$$2.99 = 3.0^x$$
$$x = 1$$
The rate law for this reaction is Rate = $k[ClO_2][OH^-]$.
Rearranging the rate-law expression to solve for k and using the data from experiment 1 gives

$$k = \frac{\text{rate}}{[\text{ClO}_2][\text{OH}^-]} = \frac{0.0248 \ M/\text{s}}{0.060 \ M \times 0.030 \ M} = 14 \ M^{-1} \ \text{s}^{-1}$$

Think About It
We may use any of the experiments in the table to calculate k. Each experiment's data give the same value of k as long as the experiments were all run at the same temperature.

14.65. Collect and Organize
In the reaction of H_2 with NO, the rate of the reaction was measured for various concentrations of both reactants. From the data we are to determine the rate law and calculate the rate constant, k.

Analyze
To determine the dependence of the rate on a change in the concentration of a particular reactant, we can compare the reaction rates for two experiments in which the concentration of that reactant changes, but the concentrations of the other reactants remain constant. Once we have the order of the reaction with respect to each reactant, we can write the rate-law expression. To calculate the rate constant for the reaction, we can rearrange the rate law to solve for k and use the data from any of the experiments.

Solve
Using experiments 1 and 2 we find that the order of the reaction with respect to NO is 2:

$$\frac{\text{rate}_2}{\text{rate}_1} = \left(\frac{[\text{NO}]_{0,2}}{[\text{NO}_2^-]_{0,1}} \right)^x$$

$$\frac{0.0991 \ M/\text{s}}{0.0248 \ M/\text{s}} = \left(\frac{0.272 \ M}{0.136 \ M} \right)^x$$

$$4.00 = 2.00^x$$

$$x = 2$$

Using experiments 3 and 4 we find that the order of the reaction with respect to H_2 is 1:

$$\frac{\text{rate}_4}{\text{rate}_3} = \left(\frac{[\text{H}_2]_{0,4}}{[\text{H}_2]_{0,3}} \right)^x$$

$$\frac{1.59 \ M/\text{s}}{0.793 \ M/\text{s}} = \left(\frac{0.848 \ M}{0.424 \ M} \right)^x$$

$$2.01 = 2.00^x$$

$$x = 1$$

The rate law for this reaction is Rate = $k[\text{NO}]^2[\text{H}_2]$.
Rearranging the rate-law expression to solve for k and using the data from experiment 1 gives

$$k = \frac{0.0248 \ M/\text{s}}{(0.136 \ M)^2 \times 0.212 \ M} = 6.32 \ M^{-2} \ \text{s}^{-1}$$

Think About It
We may use any of the experiments in the table to calculate k. Each experiment's data give the same value of k as long as the experiments were all run at the same temperature.

14.67. Collect and Organize
Using the data provided, we may calculate the average rates of consumption of HCN and acetaldehyde between 11.12 and 40.35 minutes.

Analyze
The average rate of consumption is given by the expression

$$\text{Rate} = \frac{\Delta[\text{reactant}]}{\Delta t}$$

We may calculate the individual rates by finding the difference in the concentration of each reactant at 40.35 minutes and 11.12 minutes.

Solve
(a) The rate of consumption of HCN is

$$\text{Rate} = \frac{0.0515M - 0.0619M}{40.35\,\text{min} - 11.12\,\text{min}} = -3.56 \times 10^{-4}\ M/\text{min}$$

(b) The rate of consumption of acetaldehyde is

$$\text{Rate} = \frac{0.0242M - 0.0346M}{40.35\,\text{min} - 11.12\,\text{min}} = -3.56 \times 10^{-4}\ M/\text{min}$$

Think About It
The rate of consumption of both reactants is the same, suggesting that each reactant has an equal order in the reaction.

14.69. Collect and Organize
For the first-order decomposition of H_2O_2 in the presence of Fe^{3+}, the reaction proceeds with a half-life of 17.3 minutes. Using this data, we may calculate the concentration of H_2O_2 remaining after 10.0 minutes.

Analyze
The half-life for a first-order reaction is given by the equation

$$t_{1/2} = \frac{0.693}{k}$$

Substituting in the value for the half-life will allow us to solve for the rate constant, k. Given that the initial concentration of H_2O_2 ($[H_2O_2]_0$) is 0.437 M, the rate constant may then be used to determine the concentration at 10.0 minutes ($[H_2O_2]_t$) using the integrated rate law for a first-order reaction

$$\ln\left(\frac{[H_2O_2]_t}{[H_2O_2]_0}\right) = -kt$$

Since this time is less than one half-life, it is reasonable to expect that the concentration of H_2O_2 will be slightly more than half of the original value.

Solve
The rate constant for this reaction is

$$k = \frac{0.693}{t_{1/2}} = \frac{0.693}{17.3\,\text{min}} = 0.04006\ \text{min}^{-1}$$

Solving for the concentration of H_2O_2 after 10 minutes:

$$\ln\left(\frac{[H_2O_2]_t}{0.437M}\right) = -0.04006\ \text{min}^{-1}\,(10.0\,\text{min})$$

$$\frac{[H_2O_2]_t}{0.437M} = e^{-0.4006}$$

$$[H_2O_2]_t = 0.293\ M$$

Think About It
Our initial guess seems accurate, the concentration is approximately 67% of its initial value.

14.71. Collect and Organize
For the first-order decomposition of acetoacetic acid, we must determine the rate constant, considering the half-life is 139 minutes. Given an initial concentration of acetoacetic acid, we must also calculate the concentration after 5 hours have elapsed.

Analyze

The half-life for a first-order reaction is given by the equation

$$t_{1/2} = \frac{0.693}{k}$$

Substituting in the value for the half-life will allow us to solve for the rate constant, k. Given that the initial concentration of acetoacetic acid ($[HAA]_0$) is 2.75 M, the rate constant may then be used to determine the concentration at 5 hours (18,000 s, $[HAA]_t$) using the integrated rate law for a first-order reaction

$$\ln\left(\frac{[HAA]_t}{[HAA]_0}\right) = -kt$$

Since this time is just over two half lives, it is reasonable to expect that the concentration of H_2O_2 will be slightly less than a quarter of the original value.

Solve

(a) The rate constant for this reaction is

$$k = \frac{0.693}{t_{1/2}} = \frac{0.693}{139 \text{ min}} \times \frac{1 \text{ min}}{60 \text{ s}} = 8.31 \times 10^{-5} \text{ s}^{-1}$$

(b) Solving for the concentration of acetoacetic acid after 5 hours.

$$\ln\left([X]_t\right) = -kt + \ln\left([X]_0\right)$$
$$\ln\left([X]_t\right) = -8.31 \times 10^{-5} \text{ s}^{-1} (18,000 \text{ s}) + \ln(2.75 \text{ } M)$$
$$[X]_t = e^{-0.4842}$$
$$[X]_t = 0.616 M$$

Think About It

Our initial guess was accurate: the concentration of acetoacetic acid after 5 hours will be just over 22% of the original concentration.

14.73. **Collect and Organize**

For the decomposition reaction of N_2O to N_2 and O_2, we are to write the rate law if the plot of $\ln[N_2O]$ versus time is linear and then determine the number of half-lives it would take for the concentration of N_2O to become 6.25% of its original concentration.

Analyze

The integrated rate laws for zero-, first-, and second-order reactions with their half-life equations are as follows:

$[A] = -kt + [A]_0$	zero order	$t_{1/2} = [A]_0/2k$
$\ln[A] = -kt + \ln[A]_0$	first order	$t_{1/2} = 0.693/k$
$1/[A] = kt + 1/[A]_0$	second order	$t_{1/2} = 1/k[A]_0$

Solve

(a) Given that the plot of $\ln[N_2O]$ versus time is linear, the reaction is first order with respect to N_2O and the rate law is Rate = $k[N_2O]$.

(b) The number of half-lives needed to reduce the concentration to 6.25% would be, where n = number of half-lives,

$$\frac{6.25}{100} = 0.50^n$$
$$0.0625 = 0.50^n$$
$$\ln(0.0625) = n \ln(0.50)$$
$$n = 4$$

Think About It
The number of half-lives needed to reduce the concentration of a reactant to a certain amount of product is not dependent on the order of the reaction nor on the magnitude of the rate constant.

14.75. ### Collect and Organize
From the data given for the concentration of ^{32}P over time, we are to determine the rate law, calculate the value for k, and determine the half-life for this radioactive decay.

Analyze
Radioactive decay follows first-order kinetics. This is confirmed in part (b), where we are told to determine the first-order rate constant. A plot of $\ln[^{32}P]$ versus time for this decay gives a straight line with slope $= -k$. The half-life of a first-order decay is given by

$$t_{1/2} = \frac{0.693}{k}$$

Solve
(a) First-order plot for Problem 14.75

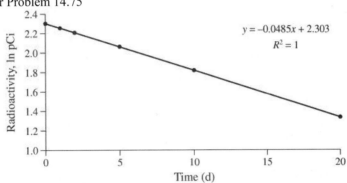

The rate law for this radioactive decay is Rate $= k[^{32}P]$.
(b) $k = -\text{slope} = 0.0485 \text{ day}^{-1}$
(c) $t_{1/2} = \dfrac{0.693}{0.0485 \text{ day}^{-1}} = 14.3 \text{ days}$

Think About It
The time needed for $[^{32}P]$ to reduce to 1.00% of its original concentration, where n = number of half-lives, is

$$\frac{1.00}{100} = 0.50^n$$

$$0.0100 = 0.50^n$$

$$\ln(0.0100) = n \ln(0.50)$$

$$n = 6.64 \text{ half-lives}$$

The number of days is $n \times t_{1/2} = 6.64 \times 14.3 \text{ days} = 95 \text{ days}$.

14.77. ### Collect and Organize
Given that the dimerization of ClO to Cl_2O_2 is second order with respect to ClO, we are to determine the value of the rate constant and calculate the half-life of this reaction.

Analyze
We are given a single data set and that the reaction is second order. For this second-order reaction, a plot of $1/[\text{ClO}]$ versus time gives a straight line with a slope equal to k, the reaction rate constant. The half-life of this second-order reaction is

$$t_{1/2} = \frac{1}{k[\text{ClO}]_0}$$

where $[\text{ClO}]_0$ is the initial concentration of ClO used in the reaction.

Solve

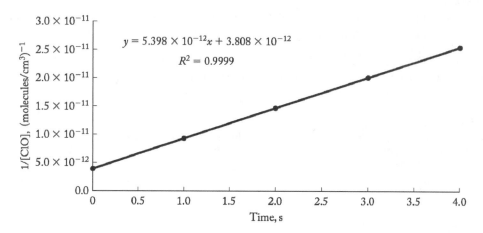

The value of k for this reaction is k = slope = 5.40×10^{-12} cm³/(molecules · s), and the half-life of the reaction is as follows:

$$t_{1/2} = \frac{1}{k[ClO]_0} = \frac{1}{\left(\dfrac{5.40 \times 10^{-12}\,\text{cm}^3}{\text{molecules} \cdot \text{s}}\right)\left(\dfrac{2.60 \times 10^{11}\,\text{molecules}}{1\,\text{cm}^3}\right)} = 0.712\ \text{s}$$

Think About It

For a second-order reaction, as we increase the initial concentration of the reactant, the half-life gets shorter.

14.79. Collect and Organize

From the data for the pseudo-first-order hydrolysis of sucrose provided, we are to write the rate law and determine the value of the pseudo-first-order rate constant, k'.

Analyze

To obtain the value of k' we plot ln[sucrose] over time. The slope of the line is equal to $-k$.

Solve

The rate law is Rate = $k[C_{12}H_{22}O_{11}][H_2O]$ = $k'[C_{12}H_{22}O_{11}]$.

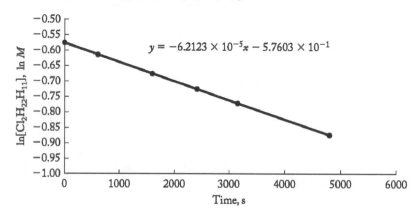

The pseudo-first-order plot gives k' = $-$slope = 6.21×10^{-5} s^{-1}.

Think About It

If we knew the concentration of water in the hydrolysis reaction we could calculate the value of k using

$$k = \frac{k'}{[H_2O]_0}$$

14.81. Collect and Organize
We are asked to identify reasons why reactions in nature have different rates of reaction.

Analyze
The rate of a reaction is dependent on a number of factors, including the temperature, concentration of reactants, order in each reactant, and the activation energy for the reaction.

Solve
The concentrations of reactions that occur in nature vary widely, as do the order with respect to each reactant and the temperature at which reactions occur. Additionally, some reactions (especially those related to biological processes) are catalyzed, and others are not. All of these factors may affect the rate of a reaction.

Think About It
Ultimately, the rate of a chemical reaction is tied to the total energy change for the process; this concept will be discussed further in Chapter 18.

14.83. Collect and Organize
We are to explain why the order of a reaction is independent of temperature, yet the value of k changes with temperature.

Analyze
We need to consider how temperature affects the motion and collision of the reactants.

Solve
An increase in temperature increases the frequency and the kinetic energy at which the reactants collide. This speeds up the reaction, changing the value of k. The activation energy of the slowest step in the reaction, however, is not affected by a change in temperature and, therefore, the order of the reaction is unaffected.

Think About It
As a general empirical observation, heating a reaction by 10°C doubles the rate of reaction.

14.85. Collect and Organize
In comparing two first-order reactions with different activation energies, we are to decide which would show a larger increase in its rate as the reaction temperature is increased.

Analyze
We can use the Arrhenius equation to mathematically determine which reaction would be most accelerated by an increase in temperature:

$$\ln k = \ln A - \frac{E_a}{RT}$$

Solve
Let's assume that $T_2 = 2T_1$. For either reaction the difference in their rate constants is as follows:

$$\ln k_{T_1} = \ln A - \frac{E_a}{RT_1} \qquad \ln k_{T_2} = \ln A - \frac{E_a}{RT_2}$$

$$\ln k_{T_2} - \ln k_{T_1} = \frac{-E_a}{RT_2} + \frac{E_a}{RT_1} = \frac{E_a}{R}\left(\frac{1}{T_1} - \frac{1}{T_2}\right)$$

But since $T_2 = 2T_1$,

$$\ln k_{T_2} - \ln k_{T_1} = \frac{E_a}{R}\left(\frac{1}{T_1} - \frac{1}{2T_1}\right) = \frac{E_a}{2RT_1}$$

For $E_a = 150$ kJ/mol,

$$\ln k_{T_2} - \ln k_{T_1} = \frac{150 \text{ kJ/mol}}{2RT_1}$$

For $E_a = 15$ kJ/mol,

$$\ln k_{T_2} - \ln k_{T_1} = \frac{15 \text{ kJ/mol}}{2RT_1}$$

Comparing these as a ratio,

$$\frac{\ln k_{T_2} - \ln k_{T_1} \text{ for } E_a = 150 \text{ kJ/mol}}{\ln k_{T_2} - \ln k_{T_1} \text{ for } E_a = 15 \text{ kJ/mol}} = \frac{\dfrac{150 \text{ kJ/mol}}{2RT_1}}{\dfrac{15 \text{ kJ/mol}}{2RT_1}} = 10$$

Therefore, the reaction with the larger activation energy (150 kJ/mol) would be accelerated more than the reaction with the lower activation energy (15 kJ/mol) when heated.

Think About It

Our derivation demonstrates that different reactions with different activation energies will not accelerate in the same way when they are heated.

14.87. ### Collect and Organize

We can use an Arrhenius plot of rate constant versus temperature for the reaction of O and O_3 to determine the activation energy (E_a) and the value of the frequency factor (A).

Analyze

The Arrhenius equation is

$$\ln k = -\frac{E_a}{R}\left(\frac{1}{T}\right) + \ln A$$

If we plot ln k (y-axis) versus $1/T$ (x-axis) we obtain a straight line with slope $m = -E_a/R$. The activation energy is therefore calculated by

$$E_a = -\text{slope} \times R$$

where $R = 8.314$ J/(mol · K). The y-intercept (b) is equal to ln A, so we can calculate the frequency factor by $b = \ln A$ or $e^b = A$.

Solve

The Arrhenius plot gives a slope of $m = -2060$ and a y-intercept of $b = 0.00160$.

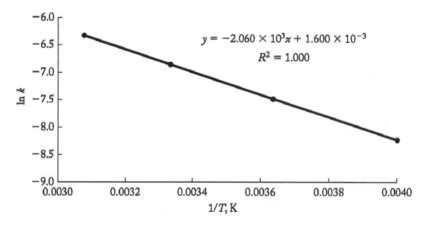

$$E_a = -[-2060 \text{ K} \times 8.314 \text{ J/(mol} \cdot \text{K)}] = 1.713 \times 10^4 \text{ J/mol or } 17.1 \text{ kJ/mol}$$
$$b = 0.00160 = \ln A$$
$$A = e^{0.00160} = 1.002$$

Think About It

Once we have the values of E_a and A from the plot, we can calculate the value of k at any temperature.

14.89. Collect and Organize

We can use an Arrhenius plot of rate constant versus temperature for the reaction of N_2 with O_2 to form NO to determine the activation energy (E_a), the frequency factor (A), and the rate constant for the reaction at 300 K.

Analyze

The Arrhenius equation is

$$\ln k = -\frac{E_a}{R}\left(\frac{1}{T}\right) + \ln A$$

If we plot ln k (y-axis) versus $1/T$ (x-axis), we obtain a straight line with slope $m = -E_a/R$. The activation energy is therefore calculated by

$$E_a = -\text{slope} \times R$$

where R = 8.314 J/(mol · K). The y-intercept (b) is equal to ln A, so we can calculate the frequency factor by $b = \ln A$ or $e^b = A$. Once E_a and A are known, we may use another form of the Arrhenius equation to calculate k at any temperature:

$$k = Ae^{\left(-\frac{E_a}{RT}\right)}$$

Solve

The Arrhenius plot gives a slope of $m = -37,758$ and a y-intercept of $b = 24.641$.

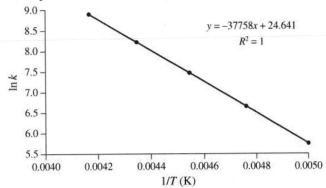

$y = -37758x + 24.641$
$R^2 = 1$

(a) $E_a = -[-37,758 \text{ K} \times 8.314 \text{ J/(mol · K)}] = 3.14 \times 10^5$ J/mol or 314 kJ/mol
(b) $A = e^{24.641} = 5.03 \times 10^{10}$
(c) $k = 5.03 \times 10^{10} e^{\left(-\frac{3.14\times10^5 \text{ J/mol}}{[8.314 \text{ J/(mol·K)}](300 \text{ K})}\right)} = 1.06 \times 10^{-44} \ M^{-1/2} \ s^{-1}$

Think About It

Alternatively, k can be calculated from the original Arrhenius equation:

$$\ln k = -\frac{E_a}{RT} + \ln A$$

$$\ln k = -\frac{3.14\times10^5 \text{ J/mol}}{[8.314 \text{ J/(mol·K)}](300 \text{ K})} + 24.641 = -101.25$$

$$k = e^{-101.25} = 1.07\times10^{-44} \ M^{-1/2} \ s^{-1}$$

14.91. Collect and Organize

We can use an Arrhenius plot of rate constant versus temperature for the reaction of ClO_2 and O_3 to determine the activation energy (E_a) and the value of the frequency factor (A).

Analyze

The Arrhenius equation is

$$\ln k = -\frac{E_a}{R}\left(\frac{1}{T}\right) + \ln A$$

If we plot ln k (y-axis) versus $1/T$ (x-axis), we obtain a straight line with slope $m = -E_a/R$. The activation energy is therefore calculated by

$$E_a = -\text{slope} \times R$$

where $R = 8.314$ J/(mol · K). The y-intercept (b) is equal to ln A, so we can calculate the frequency factor by b = ln A or $e^b = A$.

Solve

The Arrhenius plot gives a slope of $m = -4698.7$ and a y-intercept of $b = 27.872$.

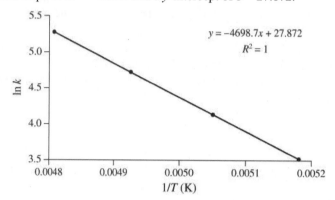

$$E_a = -[-4698.7 \text{ K} \times 8.314 \text{ J/(mol · K)}] = 3.91 \times 10^4 \text{ J/mol or } 39.1 \text{ kJ/mol}$$
$$b = 27.872 = \ln A$$
$$A = e^{27.872} = 1.27 \times 10^{12}$$

Think About It

Once we have the values of E_a and A from the plot, we can calculate the value of k at any temperature.

14.93. **Collect and Organize**

We are asked if the reaction between NO and Cl_2 could occur in one step, given that the reaction is first order in each reactant.

Analyze

If the reaction were to occur in one step, an NO molecule would have to collide with a Cl_2 molecule to form the activated complex, and onward to the products. A reaction involving multiple steps still requires that the molecularity of the rate determining (slow) step be equal to the order of the reaction.

Solve

While it is possible for the reaction to occur in one step, it is not necessarily true. It is possible for a reaction that is first order in each reactant to proceed via a multistep process in which the rate determining step is monomolecular in NO and Cl_2.

Think About It

The reaction sequence and number of steps taken to form a product are known as the reaction mechanism. Chemists will often propose a mechanism, then seek experimental confirmation of their proposal.

14.95. **Collect and Organize / Analyze**

We are to identify the conditions under which a bimolecular reaction shows behavior that is pseudo-first-order.

Solve

Pseudo-first-order kinetics occurs when one of the reactants is in a sufficiently high concentration that it does not change appreciably over the course of the reaction.

Think About It

We solved problems relating to pseudo-first-order reactions earlier in this chapter (Problems 14.79 and 14.80).

14.97. **Collect and Organize**

We are to consider a reaction in which A → B that is first order in A and first order overall. We must determine if the reaction occurs in one step, or multiple steps.

Analyze

This reaction involves a unimolecular decomposition, in which A is transformed into B. The overall rate of the reaction is dictated by the molecularity of the slowest (rate-determining) step.

Solve

These data are consistent with a reaction that occurs in one step, but they are also consistent with a reaction having a slow initial step (the decomposition of A), followed by a fast second step in which B is formed. A two-step reaction would likely involve the generation of by-products or require additional reactants. Since these are not present in the overall reaction, it is more likely that the reaction occurs in one step. It is impossible to tell with certainty from the given data which is preferred, though the direct, one-step conversion of A to B is the simplest explanation.

Think About It

It is possible for such a reaction to occur, despite having only one reactant and one product. One example is the isomerization of methylisonitrile to acetonitrile

$$CH_3N\equiv C \rightarrow CH_3C\equiv N$$

14.99. **Collect and Organize**

For each of the elementary steps given, we are to write the rate law and determine whether the step is uni-, bi- or termolecular.

Analyze

The rate law for an elementary step in a mechanism is written in the form

$$Rate = k[A]^x[B]^y[C]^z$$

where A, B, and C are the reactants involved in the elementary reaction and x, y, and z are the stoichiometric coefficients for the respective reactants in the elementary reaction.

Solve

(a) Rate = $k[SO_2Cl_2]$. Because this elementary step involves only a molecule of SO_2Cl_2, it is unimolecular.
(b) Rate = $k[NO_2][CO]$. Because this elementary step involves a molecule of NO_2 and a molecule of CO, it is bimolecular.
(c) Rate = $k[NO_2]^2$. Because this elementary step involves two molecules of NO_2, it is bimolecular.

Think About It

Termolecular elementary reactions are rare.

14.101. **Collect and Organize**

From three elementary steps that describe a reaction mechanism, we are to write the overall chemical equation.

Analyze

To write the overall chemical reaction we need to add the elementary steps, being sure to cancel the intermediates in the reaction.

Solve

$$N_2O_5(g) \rightarrow \cancel{NO_3(g)} + NO_2(g)$$
$$\cancel{NO_3(g)} \rightarrow NO_2(g) + \cancel{O(g)}$$
$$\cancel{2}\,O(g) \rightarrow O_2(g)$$
$$\overline{N_2O_5(g) + O(g) \rightarrow 2\,NO_2(g) + O_2(g)}$$

Think About It
In this reaction, NO_3 is a reaction intermediate. It is generated in the reaction but consumed in a subsequent step in the mechanism.

14.103. Collect and Organize
We are given the mechanism by which N_2 reacts with O_2 to form NO. For a given rate law of
$$Rate = k[N_2][O_2]^{1/2}$$
we are to determine which step in the mechanism is the rate-determining step.

Analyze
To determine which step in the proposed mechanism might be the slowest, we can write the rate law for the mechanism when the first, second, or third step is slow and then match the theoretical rate law to the experimental rate law.

Solve
If the first step is slow, the rate law is
$$Rate = k_1[O_2]$$
This does not match the experimental rate law, so the first step is not the slowest step in the mechanism.

If the second step is slow then the rate law is
$$Rate = k_2[O][N_2]$$
Because O is an intermediate, we use the first step to express its concentration in terms of concentrations of the reactants. For a fast step occurring before a slow step in a mechanism,
$$Rate_{forward} = Rate_{reverse}$$
$$k_1[O_2] = k_{-1}[O]^2$$
Rearranging to solve for [O],
$$[O] = \left(\frac{k_1}{k_{-1}}[O_2]\right)^{1/2}$$
Substituting this into the rate law from the second step gives
$$Rate = k_2\left(\frac{k_1}{k_{-1}}[O_2]\right)^{1/2}[N_2] = k[O_2]^{1/2}[N_2]$$
This rate law matches the experimental rate law, so the rate-determining step is the second step.
We should check to see if the mechanism of the third step is the slow step and might also give the experimental rate law. Following the logic above,
$$Rate = k_3[N][O]$$
From the second fast step in the mechanism,
$$k_2[O][N_2] = k_{-2}[NO][N]$$
Solving for [N], an intermediate, gives
$$[N] = \frac{k_2[O][N_2]}{k_{-2}[NO]}$$
From the first fast step in the mechanism,
$$k_1[O_2] = k_{-1}[O]^2$$
solving for $[O]^2$ gives
$$[O]^2 = \frac{k_1}{k_{-1}}[O_2]$$
Substituting these expressions into the rate law from the third step in the mechanism gives
$$Rate = k_3\frac{k_2}{k_{-2}}\frac{k_1}{k_{-1}}\frac{[O_2][N_2]}{[NO]}$$
$$= \frac{k[N_2][O_2]}{[NO]}$$

This rate law does not match the experimental rate law, so our earlier finding that the second step was rate-determining was correct.

Think About It
Just because the rate law for a mechanism matches the experimental rate law does not mean that mechanism is *the* mechanism. Another mechanism might also give the same experimental rate law.

14.105. **Collect and Organize**
We are given the mechanism by which NO reacts with Cl_2 to produce $NOCl_2$. For a given rate law of
$$Rate = k[NO][Cl_2]$$
we are to determine which step in the mechanism is the rate-determining step.

Analyze
To determine which step in the proposed mechanism might be the slowest, we can write the rate law for the mechanism when the first, second, or third step is slow and then match the rate law to the experimental rate law.

Solve
If the first step is slow, the rate law is
$$Rate = k_1[NO][Cl_2]$$
This matches the experimental rate law, so the first step is the rate-determining step.
We should check to see if the rate law for the mechanism with the second step as the slow step might also give the experimental rate law
$$Rate = k_2[NOCl_2][NO]$$
If the second step is slow, then
$$Rate_1 = Rate_{-1}$$
$$k_1[NO][Cl_2] = k_{-1}[NOCl_2]$$
$$[NOCl_2] = \frac{k_1}{k_{-1}}[NO][Cl_2]$$

Substituting this into the rate-law expression gives
$$Rate = k_2\left(\frac{k_1}{k_{-1}}[NO][Cl_2]\right)[NO] = k[NO]^2[Cl_2]$$

This does not match the experimental rate law, so the second step in the mechanism is not the rate-determining step.

Think About It
Just because the rate law for a mechanism matches the experimental rate law does not mean that mechanism is *the* mechanism. Another possible mechanism might also give the same experimental rate law.

14.107. **Collect and Organize**
From the mechanisms given, we are to determine which are possible for the thermal decomposition and which are possible for the photochemical decomposition of NO_2. We are given the rate laws for the thermal decomposition reaction, $Rate = k[NO_2]^2$, and for the photochemical decomposition, $Rate = k[NO_2]$.

Analyze
Using the slowest elementary step in the mechanism, we can write the rate-law expression for each of the mechanisms and then determine which is consistent with the order of the reaction given for each process.

Solve
For mechanism (a), the first step in the mechanism is slow, so the rate law is
$$Rate = k[NO_2]$$
For mechanism (b), the second step in the mechanism is slow, so the rate law is
$$Rate = k_2[N_2O_4]$$

Using the first step to express $[N_2O_4]$ in terms of the concentrations of the reactant NO_2 gives

$$k_1[NO_2]^2 = k_{-1}[N_2O_4]$$

$$[N_2O_4] = \frac{k_1}{k_{-1}}[NO_2]^2$$

Substituting into the rate expression from the second step,

$$\text{Rate} = \frac{k_2 k_1}{k_{-1}}[NO_2]^2 = k[NO_2]^2$$

For mechanism (c), the first step in the mechanism is slow so the rate law is

$$\text{Rate} = k[NO_2]^2$$

Therefore, mechanisms (b) and (c) are consistent with the thermal decomposition of NO_2, and mechanism a is consistent with the photochemical decomposition of NO_2.

Think About It

To distinguish between the two possible mechanisms for thermal decomposition, we might try to detect the different intermediates formed in each. Detection of the formation of N_2O_4 would support mechanism b over mechanism c.

14.109. Collect and Organize

We are asked if a catalyst affects both the rate and the rate constant of a reaction.

Analyze

A catalyst speeds up a reaction by providing an alternate pathway (mechanism) to the products having a lower activation energy.

Solve

Yes. Because the rate of the reaction is faster (affecting the rate) and the activation energy is lowered (affecting the value of k), a catalyst affects both the rate of the reaction and the value of the rate constant.

Think About It

A "negative" catalyst that slows down a reaction would increase E_a and decrease k for a reaction. We call these "negative catalysts" *inhibitors*.

14.111. Collect and Organize

We were asked if a substance (catalyst) that increases the rate of a reaction also increases the rate of the reverse reaction.

Analyze

A catalyst speeds up a reaction by providing an alternate pathway (mechanism) to the products having a lower activation energy.

Solve

Yes, both the reverse and forward reaction rates are increased when a catalyst is added to a reaction. The activation energies of both processes are lowered by the different pathway the catalyst provides for the reaction.

Think About It

Likewise, an inhibitor would decrease the rates of both forward and reverse reactions.

14.113. Collect and Organize

We are to explain why the concentration of a homogeneous catalyst does not appear in the rate law.

Analyze

A catalyst is used in a reaction and later regenerated.

Solve

The concentration of a homogeneous catalyst may not appear in the rate law because the catalyst itself is not involved in the rate-limiting step.

Think About It

If the catalyst is involved in the slowest step of the mechanism, however, it is involved in the rate law.

14.115. **Collect and Organize**

Given the mechanism for the reaction of NO and N_2O to form N_2 and O_2, we are to determine whether NO or N_2O is used in the reaction as the catalyst.

Analyze

A catalyst is used in a reaction and later regenerated and provides a lower energy pathway to the products by lowering the activation energy of the reaction, thereby speeding up the reaction.

Solve

We can assume that the presence of either NO or N_2O, if it is a catalyst, increases the rate of the reaction. In examining the mechanism, we see that N_2O is a reactant but does not appear as a product in either reaction. Thus, N_2O is not a catalyst. We also see that NO is present as a reactant and product, thus it is a catalyst because it is used in the reaction and then regenerated. NO is a catalyst for the decomposition of N_2O.

Think About It

If the slow step of this mechanism were the first step, the rate law would be

$$\text{Rate} = k[\text{NO}][\text{N}_2\text{O}]$$

If the second step were slow, the rate law would be

$$\text{Rate} = k \frac{[\text{NO}]^2 [\text{N}_2\text{O}]^2}{[\text{N}_2]^2}$$

14.117. **Collect and Organize**

Using the Arrhenius equation we can compute and compare the rate constants for the reaction of O_3 with O versus the reaction of O_3 with Cl.

Analyze

We are given values A and E_a for each reaction at 298 K. The Arrhenius equation is

$$\ln k = -\frac{E_a}{RT} + \ln A$$

Solve

For the reaction of O_3 with O,

$$\ln k = -\frac{17.1 \times 10^3 \text{ J/mol}}{\left[8.314 \text{ J/(mol·K)}\right](298 \text{ K})} + \ln\left(8.0 \times 10^{-12} \text{ cm}^3/\text{molecule·s}\right)$$

$$\ln k = -32.45$$

$$k = 8.05 \times 10^{-15} \text{ cm}^3/(\text{molecule·s})$$

For the reaction of O_3 with Cl,

$$\ln k = \frac{-2.16 \times 10^3 \text{ J/mol}}{8.314 \text{ J/mol·K} \times 298 \text{ K}} + \ln\left(2.9 \times 10^{-11} \text{ cm}^3/\text{molecule·s}\right)$$

$$\ln k = -25.14$$

$$k = 1.21 \times 10^{-11} \text{ cm}^3/\text{molecule·s}$$

Therefore, the reaction of O_3 with Cl has the larger rate constant.

Think About It

Our answer is consistent with a qualitative look at the activation energies and frequency factors for the two reactions. The higher activation energy and lower frequency factor for the reaction of O_3 with O give a smaller reaction rate constant.

14.119. Collect and Organize

We are to explain why a glowing wood splint burns faster in a test tube filled with O_2 than in air.

Analyze

Air is composed of about 21% O_2.

Solve

When the concentration of a reactant (O_2 for the combustion reaction) increases, the rate of reaction also increases. As we place the glowing wood in pure O_2, the rate of combustion increases.

Think About It

If the wood splint were placed in a test tube filled with argon, the combustion reaction would stop.

14.121. Collect and Organize

We are to explain why a person submerged in cold water is less likely to experience the deleterious consequences of a lack of oxygen for a given period of time than a person submerged in a warm pool.

Analyze

Chemical reactions are slower at colder temperatures than at warmer temperatures.

Solve

The bodily reactions that use O_2 are slower at colder temperatures, so the person submerged in an ice-covered lake uses less of the already dissolved oxygen in their system than the person in a warm pool.

Think About It

Rapid cooling technology is being investigated at Argonne National Laboratory for use in surgery patients and heart attack victims to reduce the damage done to cells by lack of oxygen in the blood.

14.123. Collect and Organize

In the case where rate$_{reverse}$ $\ll$ rate$_{forward}$ we are to consider whether the method to determine the rate law (initial concentrations and initial rates) would work at other times, not just at the start of the reaction. If so, we are to specify which concentrations might be used to determine the rate law.

Analyze

The method that uses initial rates and concentrations to determine the rate law is under the condition where no reverse reaction is occurring.

Solve

Yes, we could use this method at other times, not just $t = 0$, to determine the rate law if the rate of the reverse reaction is much slower than the forward reaction as long as [products] $\ll$ [reactants] at the time so that no appreciable reverse reaction is occurring.

Think About It

We will see in Chapter 15 that when the rate of the reverse reaction equals the rate of the forward reaction, the reaction is at equilibrium.

14.125. Collect and Organize

In the plot of $1/[X] - 1/[X]_0$ as a function of time, t, we are asked how the rate constant, k, can be determined.

Analyze

The plot of $1/[X] - 1/[X]_0$ as a function of time, t, is the plot for a second-order rate equation.

$$\text{Rate} = k[X]^2$$

$$\frac{1}{[X]} = kt + \frac{1}{[X]_0}$$

Solve

In this plot $1/[X] - 1/[X]_0$ divided by $t - t_0$ is the slope of the line which corresponds to k, the reaction rate constant. All we need to do to determine k from this plot is to determine the slope of the line.

Think About It

The integrated form of the rate law allows us to obtain the value of k from the concentration versus time data from a single experiment.

14.127. ### Collect and Organize

We are asked why a mechanism cannot include an elementary step with a molecularity of zero.

Analyze

The molecularity of a reaction describes the number of molecules colliding in an elementary step.

Solve

An elementary step with a molecularity of zero would have no molecules colliding, so no reaction would occur!

Think About It

A reaction may still have an order of zero with respect to a reactant if the reactant is not present in the slowest elementary step.

14.129. ### Collect and Organize

Given the balanced equation for the reaction between NO_2 and O_3 to produce N_2O_5 and O_2, we are asked to relate the rates of change in $[NO_2]$, $[O_3]$, $[N_2O_5]$, and $[O_2]$.

Analyze

From the balanced equation the rate of formation of products and consumption of reactants is

$$\text{Rate} = -\frac{1}{2}\frac{\Delta[NO_2]}{\Delta t} = -\frac{\Delta[O_3]}{\Delta t} = \frac{\Delta[N_2O_5]}{\Delta t} = \frac{\Delta[O_2]}{\Delta t}$$

Solve

The rate of consumption of O_3 is the same as the rate of formation of N_2O_5 and O_2 and one-half the rate of consumption of NO_2.

Think About It

The rate of consumption of N_2O_5 is half the rate of formation of NO_2 and twice the rate of formation of O_2.

14.131. ### Collect and Organize

We can write the rate law from the order of the decomposition reaction determined in Problem 14.130. From that we are to calculate the value of the rate constant at the experimental temperature and write the complete rate-law expression.

Analyze

From Problem 14.130 we know that the reaction is second order in $[MnO_4^-]$, first order in $[ClO_3^-]$, and half order in $[H^+]$. Once we have written a rate law for the reaction, we may use data from any of the experiments to determine the value of the rate constant.

Solve

The rate-law expression is

$$\text{Rate} = k[H^+]^{\frac{1}{2}}[ClO_3^-][MnO_4^-]^2$$

Using data from experiment 1, the rate constant is

$$5.2 \times 10^{-3} \ M/s = k(0.10 \ M)^{\frac{1}{2}}(0.10)(0.10)^2$$

$$k = 16 \ M^{-2.5} \ s^{-1}$$

The complete rate-law expression is then

$$\text{Rate} = 16 \ M^{-2.5} \ s^{-1} \ [H^+]^{\frac{1}{2}}[ClO_3^-][MnO_4^-]^2$$

Think About It

The data from experiment 2 in Problem 14.130 would give the same value of k.

14.133. Collect and Organize

For the reaction of NO with O_3 to produce NO_2 and O_2, we can use the information that the reaction is first order in both NO and O_3 along with the values of the rate constants at two different temperatures to determine whether the reaction occurs in a single or many steps and to calculate the activation energy, the rate of the reaction at different concentrations of the reactants, and the rate constants at two other temperatures.

Analyze

To answer the questions we need to use the Arrhenius equation

$$\ln k = -\frac{E_a}{RT} + \ln A$$

and the rate-law expression, which states that the reaction is first order in both NO and O_3:

$$\text{Rate} = k[NO][O_3]$$

Solve

(a) Because the rate law where the reaction is first order in both NO and O_3 is consistent with that in which the reaction would occur in a single step, this reaction might indeed occur in a single step.

(b) We can calculate the activation energy for the reaction by comparing the rate constant at the two temperatures:

$$\ln k_{298K} = -\frac{E_a}{R}\left(\frac{1}{298 \ K}\right) + \ln A$$

$$\ln k_{348K} = -\frac{E_a}{R}\left(\frac{1}{348 \ K}\right) + \ln A$$

Subtracting $\ln k$ at 25°C from $\ln k$ at 75°C gives

$$\ln k_{348K} - \ln k_{298K} = -\frac{E_a}{R}\left(\frac{1}{348 \ K} - \frac{1}{298 \ K}\right) = \ln\left(\frac{k_{348K}}{k_{298K}}\right)$$

$$\ln\left(\frac{3000 \ M^{-1} \ s^{-1}}{80 \ M^{-1} \ s^{-1}}\right) = \frac{-E_a}{8.314 \ J/(mol \cdot K)}\left(\frac{1}{348 \ K} - \frac{1}{298 \ K}\right)$$

$$E_a = 6.25 \times 10^4 \ J/mol \text{ or } 62.5 \ kJ/mol$$

(c) We can use the rate-law expression to calculate the rate of the reaction at 25°C when $[NO] = 3 \times 10^{-6} \ M$ and $[O_3] = 5 \times 10^{-9} \ M$. We are given in the statement of the problem that k at 25°C is 80 $M^{-1} \ s^{-1}$.

$$\text{Rate} = 80 \ M^{-1} \ s^{-1} \times (3 \times 10^{-6} \ M) \times (5 \times 10^{-9} \ M) = 1.2 \times 10^{-12} \ M/s$$

(d) To use the Arrhenius equation to calculate the values of k at 10°C and 35°C we have to first determine the value of the frequency factor A. To do this we can use the value for E_a and k for 25°C:

$$\ln\left(80\ M^{-1}\ s^{-1}\right) = -\frac{6.25\times10^4\ \text{J/mol}}{8.314\ \text{J/}\left(\text{mol}\cdot\text{K}\right)}\left(\frac{1}{298\ \text{K}}\right) + \ln A$$

$$\ln A = 29.608$$

$$A = 7.22\times10^{12}$$

So the value of k at 10°C (283 K) is

$$\ln k = -\frac{6.25\times10^4\ \text{J/mol}}{8.314\ \text{J/}\left(\text{mol}\cdot\text{K}\right)}\left(\frac{1}{283\ \text{K}}\right) + \ln(7.22\times10^{12})$$

$$\ln k = 3.045$$

$$k = 21\ M^{-1}\ s^{-1}$$

The value of k at 35°C (308 K) is

$$\ln k = -\frac{6.25\times10^4\ \text{J/mol}}{8.314\ \text{J/}\left(\text{mol}\cdot\text{K}\right)}\left(\frac{1}{308\ \text{K}}\right) + \ln(7.22\times10^{12})$$

$$\ln k = 5.20$$

$$k = 1.8\times10^2\ M^{-1}\ s^{-1}$$

Think About It
The values of k at 10°C and 35°C calculated above make sense because they are a little lower and a little higher, respectively, than the value of k at 25°C. Also, an Arrhenius plot using the two k values at 25°C and 75°C could be used to determine the activation energy and the frequency factor values in this problem.

14.135. Collect and Organize
We consider the mechanism for the exchange of $^{16}OH_2$ around a Na^+ cation for $^{18}OH_2$ in order to write the rate law. We also need to think about the relative energies of the reactants and products if we are to sketch the reaction-energy profile.

Analyze
We are given that the first step of the reaction is rate-determining, so this is the step from which we write the rate-law expression. In deciding which has the higher energy for the reaction profile, the reactants or products, we need to consider the relative strength of the Na^+–$^{16}OH_2$ interaction versus that of the Na^+–$^{18}OH_2$ interaction.

Solve
(a) Rate = $k[Na(H_2O)_6^+]$
(b) Neither. The ion–dipole interaction should not be significantly different for $H_2^{18}O$ versus $H_2^{16}O$, so the energy of the reactants and the products in the reaction profile will be about the same.

Think About It
In reality, there is an isotope effect in which the Na^+–$^{18}OH_2$ interaction is slightly stronger than the Na^+–$^{16}OH_2$ interaction, so the energy of the product in this reaction is slightly lower than the energy of the reactants.

14.137. Collect and Organize
For the reaction of NO with $ONOO^-$, we are to use the provided data to determine the rate law and rate constant for the reaction. We are also to draw the Lewis structure of peroxynitrite ion with resonance forms, formal charges, and an indication of which is the preferred form. We must also estimate ΔH°_{rxn} using the preferred resonance structure.

Analyze
To determine the rate law we can compare the effects of changing the concentrations of NO and $ONOO^-$ on the rate of the reaction. After drawing the resonance forms for $ONOO^-$ we can determine which is the

preferred structure by assigning formal charges to the atoms in each resonance form. The ΔH_{rxn}° may be estimated by bond energies using

$$\Delta H_{rxn}^{\circ} = \sum \Delta H_{bond\ breaking} + \sum \Delta H_{bond\ forming}$$

Solve

(a) For the order of reaction with respect to ONOO⁻ we can compare experiments 1 and 2:

$$\frac{rate_2}{rate_1} = \left(\frac{[ONOO^-]_{0,2}}{[ONOO^-]_{0,1}}\right)^x$$

$$\frac{1.02 \times 10^{-11}\ M/s}{2.03 \times 10^{-11}\ M/s} = \left(\frac{0.625 \times 10^{-4}\ M}{1.25 \times 10^{-4}\ M}\right)^x$$

$$0.502 = 0.500^x$$

$$x = 1$$

Notice that there are not two experiments in the data table for which the concentration of ONOO⁻ stays the same. For the order of reaction with respect to NO, we can compare experiments 2 and 3 as long as we take into account the knowledge that the reaction is first order in ONOO⁻. Between these two experiments we see that as the ONOO⁻ concentration is quadrupled, we would expect that the rate would be quadrupled. However, the rate of the reaction when the NO concentration is halved simultaneously with the quadrupling of the rate of the reaction, doubles:

$$\frac{rate_3}{rate_2} = \left(\frac{2.03 \times 10^{-11}\ M/s}{1.02 \times 10^{-11}\ M/s}\right) = 1.99$$

Because the $\Delta[ONOO^-] = 4$ (which would quadruple the rate of the reaction), the $\Delta[NO]$ must affect the rate by $1.99/4 = 0.498$. Therefore,

$$0.498 = \left(\frac{0.625 \times 10^{-4}\ M}{1.25 \times 10^{-4}\ M}\right)^x$$

$$0.498 = 0.500^x$$

$$x = 1$$

The rate law for this reaction is Rate = k[NO][ONOO⁻].
We can use any experiment to calculate the value of k. Using the data from experiment 1:

$$2.03 \times 10^{-11}\ M/s = k \times (1.25 \times 10^{-4}\ M) \times (1.25 \times 10^{-4}\ M)$$

$$k = 1.30 \times 10^{-3}\ M^{-1}\ s^{-1}$$

(b)

Preferred structure

(c)

Bonds broken = O—O (146 kJ/mol)
Bonds formed = N—O (201 kJ/mol)
$\Delta H_{rxn} = \{[1\ mol \times 146\ kJ/mol] + [1\ mol \times (-201\ kJ/mol)]\} = -55\ kJ$

Think About It

To solve this problem you had to draw on several concepts you have learned so far in this course. You had to review not only how to write a rate law given kinetic data, but also how to draw resonance structures and how bond energies might be used to estimate the enthalpy of a reaction.

14.139. Collect and Organize

Using data for [HNO₂] over time, we can determine the order of the isotopic exchange reaction with respect to [HNO₂]. We are also asked whether the rate of the reaction will depend on [^{18}OH₂].

Analyze

We are given a single data set. If the plot of $[HNO_2]$ versus time yields a straight line, then the reaction is zero order. If the plot of $\ln[HNO_2]$ versus time yields a straight line, then the reaction is first order. If the plot of $1/[HNO_2]$ versus time yields a straight line, then the reaction is second order.

Solve

(a)

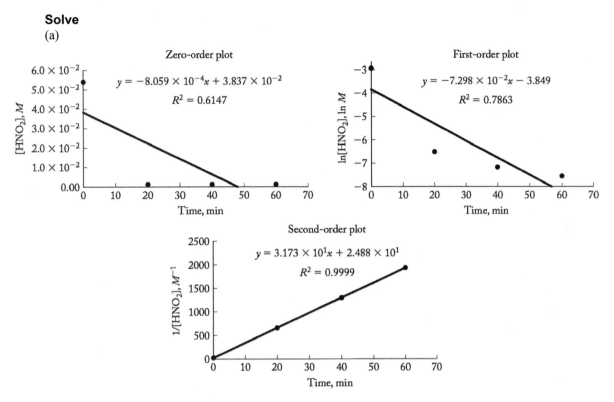

This reaction is second order in $[HNO_2]$.

(b) Because the reaction mixture has a very large $[^{18}OH_2]$, we cannot observe a change in $[^{18}OH_2]$ over time. The rate might be dependent on $[^{18}OH_2]$, but we cannot tell from the information given.

Think About It

If it were possible, we could place the reaction in a nonreactive solvent and then vary the $[^{18}OH_2]$ over time to determine the rate's dependence on its concentration.

14.141. **Collect and Organize**

Using the raw data obtained for four experiments in which the concentrations of NH_2 and NO were varied, we are to write the rate law and determine the value of k for the reaction between NH_2 and NO at 1200 K.

Analyze

To determine the rate law, we can compare the effects of changing the concentrations of NH_2 and NO on the rate of the reaction. Once we have determined the order of the reaction with respect to each reactant, we can write the rate-law expression and use any of the experiments to calculate the value of the rate constant, k.

Solve

(a) For the order of the reaction with respect to NH_2, we can use the data from experiments 1 and 2:

$$\frac{rate_2}{rate_1} = \left(\frac{[NH_2]_{0,2}}{[NH_2]_{0,1}}\right)^x$$

$$\frac{0.24 \; M/s}{0.12 \; M/s} = \left(\frac{2.00 \times 10^{-5} \; M}{1.00 \times 10^{-5} \; M}\right)^x$$

$$2.0 = 2.00^x$$

$$x = 1$$

For the order of the reaction with respect to NO, we can use the data from experiments 2 and 3:

$$\frac{rate_3}{rate_2} = \left(\frac{[NO]_{0,3}}{[NO]_{0,2}}\right)^x$$

$$\frac{0.36 \; M/s}{0.24 \; M/s} = \left(\frac{1.50 \times 10^{-5} \; M}{1.00 \times 10^{-5} \; M}\right)^x$$

$$1.5 = 1.50^x$$

$$x = 1$$

The rate-law expression is $rate = k[NH_2][NO]$.

(b) Using experiment 1 to calculate the value of k,

$$0.12 \; M/s = k \times (1.00 \times 10^{-5} \; M) \times (1.00 \times 10^{-5} \; M)$$

$$k = 1.2 \times 10^9 \; M^{-1} \; s^{-1}$$

Think About It

Any of the experiments listed in the data table would give us the same value of k in the calculation in part b.

CHAPTER 15 | Chemical Equilibrium: How Much Product Does a Reaction Really Make?

15.1. Collect and Organize

We are asked to consider the plot in Figure P15.1 and write a mass action expression, then determine the value of the equilibrium constant (K_c) for this reaction.

Analyze

The mass action expression for a reaction relates the equilibrium concentrations, raised to the power of their respective stoichiometric coefficients, to the equilibrium constant for the reaction. By substituting in the equilibrium concentrations of A_2 (0.52 M), B_2 (0.84 M), and A_3B (0.32 M) we may calculate K_c. The easiest method for determining the powers to which each concentration should be raised is to write a balanced equation from the species depicted in Figure P15.1. Here A_2 and B_2 are reactants (concentration decreases over time), and A_3B is a product (concentration increases over time). From the plot, A_2 and B_2 combine to form A_3B by the balanced equation

$$3\,A_2 + B_2 \rightleftharpoons 2\,A_3B$$

Solve

(a) Using the balanced equation above as a guide, the mass action expression is

$$K_c = \frac{[A_3B]^2}{[A_2]^3[B_2]}$$

(b) The value for K_c at equilibrium is

$$K_c = \frac{(0.32\ M)^2}{(0.52\ M)^3(0.84\ M)} = 0.87$$

Think About It

We could also use the change in concentration for each reactant to determine the relative proportions of each species in the balanced chemical equation. For example, the concentration of A_2 decreases by three times the amount that the concentration of B_2 decreases (0.48/0.16 = 3), suggesting a 3:1 ratio of A_2:B_2 in the balanced chemical equation.

15.3. Collect and Organize

From Figure P15.3, showing 26 blue spheres (product B) and 13 red spheres (reactant A), we are to write the chemical equation of the equilibrium reaction and calculate the value of K_c.

Analyze

In this reaction A is transformed into B. We represent a system at equilibrium using double-headed reaction arrows between the reactants and products. We will assume that one molecule of A produces one molecule of B in the reaction. The value of K_c is the ratio of the concentration of the products (number of B spheres) and the concentration of the reactants (number of A spheres) raised to their respective stoichiometric coefficients from the balanced chemical equation.

Solve

(a) $A \rightleftharpoons B$

(b) $K_c = \dfrac{26}{13} = 2.0$

Think About It

If the chemical equation were written as $2A \rightleftharpoons B$, then K_c would be

$$K_c = \frac{[B]}{[A]^2} = \frac{26}{(13)^2} = 0.15$$

15.5. **Collect and Organize**

By comparing the relative distributions of reactants A and B with product AB at two different temperatures as shown in Figure P15.5, we can determine whether the reaction is endothermic or exothermic.

Analyze

From the equation

$$\Delta G = -RT \ln K = \Delta H - T \Delta S$$

we see that as temperature rises for an exothermic reaction, ΔG becomes less negative and therefore K decreases. If, however, the temperature is raised on an endothermic reaction ΔG becomes more negative and K increases. Therefore, if products increase upon raising the temperature, the reaction is endothermic; if products decrease, the reaction is exothermic.

Solve

At 300 K, the equilibrium mixture is 6 A, 10 B, and 5 AB. This gives an equilibrium constant of

$$K_{300\,K} = \frac{5}{6 \times 10} = 0.083$$

At 400 K, the equilibrium mixture is 3 A, 7 B, and 8 AB. This gives an equilibrium constant of

$$K_{400\,K} = \frac{8}{3 \times 7} = 0.38$$

As temperature increases for this reaction, K increases, indicating that more products form at higher temperatures, so this reaction is endothermic.

Think About It

We assume in this problem that the difference in entropy for the two temperatures at which this reaction is run is minimal, so only ΔH contributes to the difference in ΔG at the two temperatures.

15.7. **Collect and Organize**

We are asked to define chemical equilibrium in terms of the rates of the forward and reverse reactions.

Analyze

Equilibrium is a state where the composition of the reaction is not changing over time.

Solve

The rate of the forward reaction is equal to the rate of the reverse reaction in a system at equilibrium. For a system at equilibrium, the composition of the products and reactants does not change.

Think About It

Chemical equilibrium is a dynamic process. At the molecular level the forward and reverse reactions are still occurring. Because they occur at the same rate, however, we observe no macroscopic changes in the concentrations of the reactants and products in the mixture.

15.9. **Collect and Organize**

From the graph showing the concentrations of both B and A over time (Figure P15.9), we are to determine if the reaction has reached equilibrium at 20 μs.

Analyze

At equilibrium, the rate of the reaction to produce B from A is equal to the rate of the reaction to produce A from B. The concentrations of A and B at equilibrium are constant.

Solve

No. At 20 μs the concentrations of A and B in Figure P15.9 are still changing (there is a nonzero instantaneous rate of reaction), so at this time the reaction has not yet reached equilibrium.

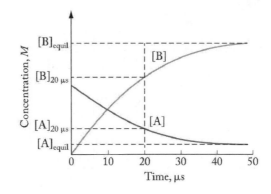

Think About It
The concentrations of both A and B level off at about 50 μs. After this time, the reaction is at equilibrium.

15.11. Collect and Organize
We are asked to describe how the rate of forward reaction (k_f) and reverse reaction (k_r) are related to the equilibrium constant, K.

Analyze
At equilibrium, the rate of forward and reverse reactions are equal. For the reaction given, the rate of forward reaction is
$$\text{Rate} = k_f[A]$$
and the rate of reverse reaction is
$$\text{Rate} = k_r[B]$$
Since the rates of forward and reverse reaction are equal at equilibrium, combining these equations gives
$$k_f[A] = k_r[B]$$
The equilibrium constant is defined as the concentration of products over the concentration of reactants raised to the power of their respective stoichiometric coefficients.

Solve
The equilibrium constant defined in terms of the rate constants for this reaction is
$$K = \frac{[A]}{[B]} = \frac{k_f}{k_r}$$

Think About It
When $K > 1$ more products are present at equilibrium than reactants.

15.13. Collect and Organize
For the decomposition of N_2O to N_2 and O_2, we are to identify the species present after one day from the given molar masses. The initial reaction mixture contained $^{15}N_2O$, N_2, and O_2.

Analyze
The only isotope in the reaction for oxygen is ^{16}O, but for nitrogen, both ^{15}N and ^{14}N are present at the beginning of the reaction. After one day, the ^{14}N will be incorporated into N_2O and ^{15}N will be incorporated into N_2.

Solve

Molar Mass	Compound	How Present
28	$^{14}N_2$	Originally present
29	$^{15}N^{14}N$	From decomposition of $^{15}N^{14}NO$
30	$^{15}N_2$	From decomposition of $^{15}N_2O$
32	O_2	Originally present
44	$^{14}N_2O$	From combination of $^{14}N_2$ and O_2
45	$^{15}N^{14}NO$	From combination of $^{15}N^{14}N$ and O_2
46	$^{15}N_2O$	Originally present

Think About It

The redistribution of ^{15}N from N_2O to N_2 and ^{14}N from N_2 to N_2O demonstrates that both forward and reverse reactions occur in a dynamic equilibrium process.

15.15. Collect and Organize

From the rate laws and rate constants for the forward and reverse reactions for
$$A \rightleftharpoons B$$
we are to calculate the value of the equilibrium constant.

Analyze

At equilibrium, the rate of the forward reaction equals the rate of the reverse reaction. The equilibrium constant is the ratio of the concentrations of the products, raised to their coefficients, to the concentrations of the reactants, raised to their respective coefficients.

Solve

For the reaction $A \rightleftharpoons B$ we are given that the forward reaction is first order in A:
$$Rate_f = k_1[A], \text{ where } k_1 = 1.50 \times 10^{-2} \text{ s}^{-1}$$

and that the reverse reaction is first order in B:
$$Rate_r = k_{-1}[B] \text{ where } k_{-1} = 4.50 \times 10^{-2} \text{ s}^{-1}$$

At equilibrium, $rate_f = rate_r$ so
$$k_1[A] = k_{-1}[B]$$

Rearranging this to give the equilibrium constant expression allows us to solve for K:
$$K_c = \frac{[B]}{[A]} = \frac{k_1}{k_{-1}} = \frac{1.50 \times 10^{-2} \text{ s}^{-1}}{4.50 \times 10^{-2} \text{ s}^{-1}} = 0.333$$

Think About It

For the reverse equilibrium
$$B \rightleftharpoons A$$
the equilibrium constant is 1/0.333 or 3.00.

15.17. Collect and Organize

From the equation relating K_c and K_p we can determine under what conditions $K_c = K_p$.

Analyze

The equation relating K_c and K_p is
$$K_p = K_c(RT)^{\Delta n}$$

Solve

K_p equals K_c when $\Delta n = 0$. This is true when the number of moles of gaseous products equals the number of moles of gaseous reactants in the balanced chemical equation.

Think About It
The value of K_p may also be less than K_c (for $\Delta n < 0$) or greater than K_c (for $\Delta n > 0$).

15.19. Collect and Organize
We are asked to write expressions for K_c and K_p for the reactions
$$C_2H_4(g) + H_2(g) \rightleftharpoons C_2H_6(g)$$
$$2\,SO_2(g) + O_2(g) \rightleftharpoons 2\,SO_3(g)$$

Analyze
K_c and K_p (law of mass action) expressions have the concentration or partial pressure of products, raised to the power of their stoichiometric coefficients, divided by the concentration or partial pressure of reactants, raised to the power of their stoichiometric coefficients.

Solve

(a) $K_c = \dfrac{[C_2H_6]}{[C_2H_4][H_2]}$ and $K_p = \dfrac{\left(P_{C_2H_6}\right)}{\left(P_{C_2H_4}\right)\left(P_{H_2}\right)}$

(b) $K_c = \dfrac{[SO_3]^2}{[SO_2]^2[O_2]}$ and $K_p = \dfrac{\left(P_{SO_3}\right)^2}{\left(P_{SO_2}\right)^2\left(P_{O_2}\right)}$

Think About It
Since all of the components in these reactions are gaseous, the K_c and K_p expressions are nearly identical.

15.21. Collect and Organize
Given a plot of the concentration versus time for the decomposition of N_2O to N_2 and O_2 (Figure P15.21), we are to estimate the value of K_c.

Analyze
The amounts of N_2O, N_2, and O_2 are given in concentration units and the form of the K_c expression is
$$K_c = \dfrac{[N_2][O_2]^{1/2}}{[N_2O]}$$

The concentrations of each species at equilibrium can be read from the graph as those concentrations that are no longer changing with time. This occurs after 5 s and gives $[N_2] = 0.00030\ M$, $[O_2] = 0.00014\ M$, and $[N_2O] = 7.10 \times 10^{-6}\ M$.

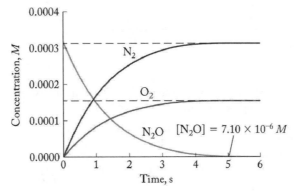

Solve

$$K_c = \frac{[N_2][O_2]^{1/2}}{[N_2O]} = \frac{(0.00030 \ M)(0.00014 \ M)^{1/2}}{(7.10 \times 10^{-6} \ M)} = 0.50$$

Think About It

Because this equilibrium constant is neither high nor low, at equilibrium the distribution of the reactants and products is expected to be roughly equal. A general rule is that if $K < 0.01$, the reactants are favored in the equilibrium and if $K > 100$ then the products are favored in the equilibrium.

15.23. **Collect and Organize**

We are given the equilibrium partial pressures for the components of the following reaction at 1200 K:

$$H_2S(g) \rightleftharpoons H_2(g) + S(g)$$

We are asked to calculate the value of K_p.

Analyze

The equilibrium constant expression for this reaction is

$$K_p = \frac{(P_S)(P_{H_2})}{(P_{H_2S})}$$

Solve

$$K_p = \frac{(0.045 \ \text{atm})(0.030 \ \text{atm})}{(0.020 \ \text{atm})} = 0.068$$

Think About It

Even at 1200 K, this reaction will favor H_2S over H_2 and S vapor.

15.25. **Collect and Organize**

Given the equilibrium molar concentrations of N_2, O_2, and NO, we are to calculate K_c for

$$N_2(g) + O_2(g) \rightleftharpoons 2 \ NO(g)$$

Analyze

The equilibrium constant expression for this reaction is

$$K_c = \frac{[NO]^2}{[N_2][O_2]}$$

Solve

$$K_c = \frac{(3.1 \times 10^{-3} \ M)^2}{(3.3 \times 10^{-3} \ M)(5.8 \times 10^{-3} \ M)} = 0.50$$

Think About It

Be careful to account for the coefficients in the mass action equation when calculating K_c. In this problem, we must be sure to square the equilibrium concentration of NO.

15.27. **Collect and Organize**

We are given the equilibrium partial pressures for the components of the following reaction at 1000 K:

$$CH_4(g) + H_2O(g) \rightleftharpoons CO(g) + 3 \ H_2(g)$$

We are asked to calculate the value of K_p.

Analyze

The equilibrium constant expression for this reaction is

$$K_p = \frac{\left(P_{CO}\right)\left(P_{H_2}\right)^3}{\left(P_{CH_4}\right)\left(P_{H_2O}\right)}$$

Solve

$$K_p = \frac{\left(1.00\ \text{atm}\right)\left(3.00\ \text{atm}\right)^3}{\left(0.71\ \text{atm}\right)\left(1.41\ \text{atm}\right)} = 27$$

Think About It

Since all species in this reaction are gases, all will be included in the mass action expression.

15.29. **Collect and Organize**

We are given $K_p = 32$ for the following reaction at 298 K:

$$A(g) + B(g) \rightleftharpoons AB(g)$$

We are to calculate the value of K_c, which we can do using the relationship

$$K_p = K_c \left(RT\right)^{\Delta n}$$

Analyze

The change in the number of moles of gas for this reaction (Δn) is

$$1\ \text{mol AB} - (1\ \text{mol A} + 1\ \text{mol B}) = -1$$

Solve

$$32 = K_c \times \left(0.08206\ \frac{\text{L} \cdot \text{atm}}{\text{mol} \cdot \text{K}} \times 298\ \text{K}\right)^{-1}$$

$$K_c = 780$$

Think About It

When Δn is positive, K_p is greater than K_c, but when Δn is negative, K_p is less than K_c. When Δn is zero, K_p equals K_c.

15.31. **Collect and Organize**

We are given $K_p = 1.45 \times 10^{-5}$ for the following reaction at 500°C:

$$N_2(g) + 3\ H_2(g) \rightleftharpoons 2\ NH_3(g)$$

We are to calculate the value of K_c.

Analyze

Rearranging the equation relating K_p and K_c to solve for K_c gives

$$K_c = \frac{K_p}{(RT)^{\Delta n}}$$

The change in the number of moles is

$$(2\ \text{mol NH}_3) - (1\ \text{mol N}_2 + 3\ \text{mol H}_2) = -2$$

The temperature in kelvins is $500 + 273 = 773$ K.

Solve

$$K_c = \frac{1.45 \times 10^{-5}}{\left(0.08206\ \frac{\text{L} \cdot \text{atm}}{\text{mol} \cdot \text{K}} \times 773\ \text{K}\right)^{-2}} = 0.0583$$

Think About It

Here, $K_p < K_c$ because Δn is negative. When $\Delta n = 0$, $K_p = K_c$.

15.33. Collect and Organize

By comparing the number of moles of gaseous reactants and products for a reaction, we may determine which reactions will have the same values for K_c and K_p.

Analyze

The relationship between K_c and K_p is

$$K_p = K_c (RT)^{\Delta n}$$

Reactions with $\Delta n = 0$ will reduce the term $(RT)^{\Delta n}$ to 1, meaning that $K_c = K_p$. We must identify which reactions will have $\Delta n = 0$, where $\Delta n = (n \text{ products}) - (n \text{ reactants})$.

Solve

(a) $\Delta n = (1 + 3) - (2 + 2) = 0$
(b) $\Delta n = (3) - (2) = 1$
(c) $\Delta n = (1) - (1) = 0$
Reactions (a) and (c) have $\Delta n = 0$, and thus $K_c = K_p$.

Think About It

Solids, liquids, and aqueous solutions are excluded from Δn calculations, as they do not have a partial pressure.

15.35. Collect and Organize

Given that $K_c = 5.0$ at 327°C (600 K) for the reaction

$$Cl_2(g) + CO(g) \rightleftharpoons COCl_2(g)$$

we are to calculate the value of K_p at 327°C. We can use the relationship

$$K_p = K_c (RT)^{\Delta n}$$

Analyze

For this reaction

$$\Delta n = (1 \text{ mol COCl}_2) - (1 \text{ mol Cl}_2 + 1 \text{ mol CO}) = -1$$

Solve

$$K_p = 5.0 \times \left(0.08206 \frac{L \cdot atm}{mol \cdot K} \times 600 \text{ K} \right)^{-1} = 0.10$$

Think About It

Because Δn is negative, the value of K_p is less than that of K_c.

15.37. Collect and Organize / Analyze

We are asked to explain why representing the same reaction with the same ratio of reactants and products but different stoichiometric coefficients results in a different value of K.

Solve

When writing a mass action expression, the concentration of each species in the reaction is raised to the power of the stoichiometric coefficient from the balanced chemical equation. By changing the stoichiometric coefficients of the equation, we are changing the exponents in the mass action expression. For the reactions provided in the problem

$$K_{p(1)} = \frac{(P_{NO_2})^2}{(P_{N_2})^2 (P_{O_2})} \quad \text{and} \quad K_{p(2)} = \frac{(P_{NO_2})}{(P_{N_2})(P_{O_2})^{\frac{1}{2}}} = (K_{p(1)})^{\frac{1}{2}}$$

Think About It
When writing a mass action expression for K_c or K_p, it is important to provide the balanced equation to avoid confusion about the stoichiometric coefficients being used.

15.39. **Collect and Organize**
Given the equilibrium constant for the reaction of 1 mole $I_2(g)$ with 1 mole $Br_2(g)$ to give 2 moles $IBr(g)$, we are to calculate the value of the equilibrium constant for the reaction of $\frac{1}{2}$ mole of I_2 and Br_2 to give 1 mole IBr.

Analyze
The K_c expressions for these reactions are

$$K_{c1} = \frac{[IBr]^2}{[I_2][Br_2]} \qquad \text{and} \qquad K_{c2} = \frac{[IBr]}{[I_2]^{1/2}[Br_2]^{1/2}}$$

K_{c2} for the second reaction is, therefore, related to that of the first by

$$K_{c2} = \left(K_{c1}\right)^{1/2}$$

Solve

$$K_{c2} = \left(K_{c1}\right)^{1/2} = (120)^{1/2} = 11.0$$

Think About It
When we multiply the coefficients of a chemical equation by a number, the new value of the equilibrium constant is the first equilibrium constant raised to that number.

15.41. **Collect and Organize**
For the reaction of NO with NO_3, we are to explain how K_c for the reverse reaction relates to K_c of the forward reaction by writing their equilibrium constant expressions.

Analyze
The form of the equilibrium constant expression for the forward reaction is

$$K_{c,forward} = \frac{[NO_2]^2}{[NO][NO_3]}$$

For the reverse reaction, the equilibrium constant expression is

$$K_{c,reverse} = \frac{[NO][NO_3]}{[NO_2]^2}$$

Solve
Examining these two expressions we see that

$$K_{c,reverse} = \frac{1}{K_{c,forward}}$$

Think About It
Another way to think about this relationship is

$$K_{c,forward} \times K_{c,reverse} = 1$$

15.43. **Collect and Organize**
Given the value of K_c for a reaction, we may calculate K_c for a reaction with stoichiometric coefficients equal to half those in the original reaction.

Analyze

When a reaction is multiplied by a factor x, the new equilibrium constant is $(K_c)^x$. For this set of reactions, x is $\frac{1}{2}$, so K_c for the second reaction is equal to $\sqrt{K_c}$ for the first reaction.

Solve

$$K_c = \sqrt{3.0 \times 10^{-4}} = 1.7 \times 10^{-2}$$

Think About It

Though the ratio of reactants and products are the same in these reactions, the value of K_c changes with the stoichiometry.

15.45. **Collect and Organize**

Given that K_c is 2.4×10^{-3} for the reaction

$$2\,SO_2(g) + O_2(g) \rightleftharpoons 2\,SO_3(g)$$

we are to calculate K_c for three other forms of this reaction at the same temperature.

Analyze

When a reaction is multiplied by a factor x, the new equilibrium constant is $(K_c)^x$. When a reaction is reversed, the new equilibrium constant is $1/K_c$.

Solve

(a) This reaction is the original equation multiplied by $\frac{1}{2}$. The new K_c is

$$\left(2.4 \times 10^{-3}\right)^{1/2} = 4.9 \times 10^{-2}$$

(b) This reaction is the reverse of the original equation. The new K_c is

$$\frac{1}{2.4 \times 10^{-3}} = 420$$

(c) This reaction is the reverse of the original equation, multiplied by $\frac{1}{2}$. The new K_c is

$$\frac{1}{\left(2.4 \times 10^{-3}\right)^{1/2}} = 20$$

Think About It

Because we can relate the equilibrium constants for different forms of a reaction, we need to only tabulate one of the equilibrium constants. All others can be calculated from that value.

15.47. **Collect and Organize**

We may calculate the value of K_p for the formation of NO_2 by summing the provided equation involving N_2, O_2, and NO_2 gas.

Analyze

We may generate the equation for the formation of NO_2 by summing the provided equations

$$N_2(g) + O_2(g) \rightleftharpoons 2\,NO(g)$$
$$\underline{2\,NO(g) + O_2(g) \rightleftharpoons 2\,NO_2(g)}$$
$$N_2(g) + 2\,O_2(g) \rightleftharpoons 2\,NO_2(g)$$

When summing equations, the value of K for the resultant equation is simply the product of the summed equations. No other manipulations to the provided equations are required, so K_p is obtained by multiplying the values of K_p provided in the problem.

Solve

$$K_p = (4.4 \times 10^{-31})(2.4 \times 10^{12}) = 1.1 \times 10^{-18}$$

Think About It

Despite the large value of K_p for the reaction of NO and O_2 to form NO_2, the overall reaction has a small K_p due to the small K_p for the reaction of N_2 and O_2 to generate NO.

15.49. **Collect and Organize / Analyze**

We are to define the term *reaction quotient*.

Solve

The reaction quotient is the ratio of the concentrations of the products, raised to their stoichiometric coefficients, to the concentrations of reactants, raised to their stoichiometric coefficients. The reaction quotient has the same form as the equilibrium constant expression, but the reaction concentrations (or partial pressures) are not necessarily at their equilibrium values.

Think About It

If the reaction quotient, Q, is greater than K, then the reaction mixture must reduce its concentration of products to attain equilibrium. When Q is less than K, the reaction mixture must increase its concentration of products to attain equilibrium.

15.51. **Collect and Organize**

We are asked what it means when $Q = K$.

Analyze

Both K and Q take the form of the ratio of the concentrations of products, raised to their stoichiometric coefficients, to the concentrations of reactants, raised to their stoichiometric coefficients.

Solve

When $Q = K$, the system is at equilibrium.

Think About It

Whenever $Q \neq K$, the reaction is not at equilibrium and it adjusts its relative concentrations of reactants and products so that $Q = K$.

15.53. **Collect and Organize**

We are asked whether the reaction

$$A(aq) \rightleftharpoons B(aq)$$

where $[A] = 0.10\ M$, $[B] = 2.0\ M$, and $K = 22$ is at equilibrium. If the reaction is not at equilibrium, we are to state in which direction the reaction will proceed to reach equilibrium.

Analyze

The reaction quotient (Q) is

$$\frac{[B]}{[A]} = Q$$ where $[B]$ and $[A]$ are the concentrations of A and B in the reaction mixture.

If $Q > K$, the reaction will proceed to the left to reach equilibrium. If $Q < K$, the reaction will proceed to the right. If $Q = K$, the reaction is at equilibrium.

Solve

$$Q = \frac{2.0\ M}{0.10\ M} = 20$$

No, the reaction is not at equilibrium. $Q < K$, so this reaction proceeds to the right to reach equilibrium.

Think About It

In this reaction, more B forms as the reaction proceeds to equilibrium.

474 | *Chapter 15*

15.55. **Collect and Organize**

Given two sets of reactant and product concentrations for the reaction of A and B to form C (where $K_p = 1.00$), we are to determine whether either reaction mixture is at equilibrium. The temperature is 300 K.

Analyze

These systems are at equilibrium when $Q = K_p = 1.00$. For the reaction where A, B, and C are expressed in terms of molarity, we must convert K_p to K_c using

$$K_p = K_c (RT)^{\Delta n}$$

where for this reaction $\Delta n = -1$.

Solve

(a) $Q_p = \dfrac{1.0 \text{ atm}}{(1.0 \text{ atm})(1.0 \text{ atm})} = 1.0$

This reaction mixture is at equilibrium.

(b) Rearranging the equation to solve for K_c gives

$$K_c = \frac{K_p}{(RT)^{\Delta n}} = \frac{1.00}{\left[\left(0.08206 \dfrac{\text{L} \cdot \text{atm}}{\text{mol} \cdot \text{K}}\right) \times 300 \text{ K}\right]^{-1}} = 24.6$$

$$Q_c = \frac{1.0 \text{ } M}{(1.0 \text{ } M)(1.0 \text{ } M)} = 1.0$$

Because $Q_c < K_c$ this reaction mixture is not at equilibrium and shifts to the right to attain equilibrium.

Think About It

Both K_p and K_c are close to 1 and so these reaction mixtures have roughly equal proportions of reactants and products when they reach equilibrium.

15.57. **Collect and Organize**

By comparing Q versus K for the reaction of N_2 with O_2 to form NO, we can determine in which direction the reaction will proceed to attain equilibrium.

Analyze

The form of the reaction quotient for this reaction is

$$Q_p = \frac{(P_{NO})^2}{(P_{N_2})(P_{O_2})}$$

Because $\Delta n = 0$ for this reaction $K_p = K_c = 1.5 \times 10^{-3}$.

Solve

$$Q_p = \frac{(1.00 \times 10^{-3})^2}{(1.00 \times 10^{-3})(1.00 \times 10^{-3})} = 1.00$$

Because $Q > K$, the system is not at equilibrium and proceeds to the left.

Think About It

Notice that for this problem $K_p = K_c$. This is not always the case, so be careful to notice which value is provided for K and in what units the amounts of the reactants and products are expressed.

15.59. **Collect and Organize**

Given the partial pressures of PCl_3, PCl_5, and Cl_2 gases, we may determine if this reaction is at equilibrium.

Analyze

We may write a mass action expression for the addition of Cl_2 and PCl_3 to form PCl_5

$$Q_p = \frac{(P_{PCl_5})}{(P_{PCl_3})(P_{Cl_2})}$$

By plugging in the provided partial pressures of PCl_3 (0.20 atm), PCl_5 (0.60 atm), and Cl_2 (0.40 atm), we obtain the reaction quotient, Q_p. Comparing Q_p and K_p (3.8) will reveal whether the reaction is at equilibrium or, alternately, to which side the reaction will shift in order to reach equilibrium. If $Q_p < K_p$, more products will be formed. If $Q_p > K_p$, more reactants will be formed, and if $Q_p = K_p$, the reaction is at equilibrium.

Solve

Solving for Q_p

$$Q_p = \frac{(0.60 \text{ atm})}{(0.20 \text{ atm})(0.40 \text{ atm})} = 7.5$$

Q_p is greater than K_p, so the reaction is not at equilibrium, and the reaction will form reactants: PCl_3 and Cl_2.

Think About It

The reaction could be brought to equilibrium by removing some of the PCl_5, thus lowering the value of Q_p.

15.61. **Collect and Organize**

We are asked to write a mass action (K_p) expression for the combustion of sulfur.

Analyze

Pure liquids and solids and compounds in aqueous solution are excluded from the mass action when expressed in terms of pressure (K_p). The partial pressure of all gaseous components are raised to the power of their stoichiometric coefficients, with products in the numerator and reactants in the denominator.

Solve

For this reaction, elemental sulfur is a solid, and so is excluded from the K_p expression

$$K_p = \frac{(P_{SO_2})}{(P_{O_2})}$$

Think About It

Elemental sulfur is a by-product of some petroleum refining processes. Modern diesel fuel, as an example, is "de-sulfured" in order to reduce SO_2 emissions when the fuel is burned in automotive engines.

15.63. **Collect and Organize**

We are asked to write a mass action expression for the reaction between iron(II) hydroxide, oxygen, and water.

Analyze

Pure liquids and solids are excluded from the mass action expression when written in terms of concentration (K_c), as their concentration does not change throughout a reaction. The concentration of gases and compounds in aqueous solution are raised to the power of their stoichiometric coefficients, with products in the numerator, and reactants in the denominator.

Solve

For this reaction

$$K_c = \frac{1}{[Fe(OH)_2]^4[O_2]}$$

Think About It

The only product of this reaction is a solid, so it is not included in the mass action expression.

15.65. **Collect and Organize**

We are asked whether the value of K increases when more reactants are added to a reaction already at equilibrium.

Analyze

The general form of the equilibrium constant expression for a reaction is

$$w\text{A} + x\text{B} \rightleftharpoons y\text{C} + z\text{D}$$

$$K_c = \frac{[\text{C}]^y [\text{D}]^z}{[\text{A}]^w [\text{B}]^x}$$

Solve

No, the equilibrium constant is not changed when the concentration of the reactants is increased. The relative concentrations of the reactants and products in that case adjust until they achieve the value of K. The value of the equilibrium constant is only affected by temperature.

Think About It

The value of the reaction quotient Q decreases below the value of K when reactants are added to a system previously at equilibrium, and the reaction shifts to the right.

15.67. **Collect and Organize**

Given that the K for the binding of CO to hemoglobin is larger than that for the binding of O_2 to hemoglobin, we are to explain how the treatment of CO poisoning by administering pure O_2 to a patient works.

Analyze

By giving a patient pure O_2 to breathe, we increase the partial pressure of O_2 to which the hemoglobin in the patient's blood is exposed. This oxygen can displace the CO bound to the hemoglobin through application of Le Châtelier's principle.

Solve

Combining the two equations, we can write the expression for the displacement of CO bound to hemoglobin by O_2:

$$\text{Hb(CO)}_4 \rightleftharpoons \text{Hb} + 4\,\text{CO}(g)$$

$$\underline{\text{Hb} + 4\,O_2(g) \rightleftharpoons \text{Hb}(O_2)_4}$$

$$\overline{\text{Hb(CO)}_4 + 4\,O_2(g) \rightleftharpoons \text{Hb}(O_2)_4 + 4\,\text{CO}(g)}$$

As the concentration of O_2 increases, the reaction shifts to the right and the CO on the hemoglobin is displaced.

Think About It

If we were given the values of the equilibrium constants for the reactions, we could calculate the new equilibrium constant for the overall reaction.

15.69. **Collect and Organize**

We are to interpret Henry's law through Le Châtelier's principle.

Analyze

The dissolution of a gas (let's use oxygen in this example) in a liquid (let's use water) can be written as a chemical equation:

$$O_2(g) \rightleftharpoons O_2(aq)$$

Solve

According to Le Châtelier's principle, an increase in the partial pressure (or concentration) of O_2 above the water shifts the equilibrium to the right so that more oxygen becomes dissolved in the water. This is consistent with Henry's law.

Think About It
The solubilities of different gases in a liquid are different from each other, but all gases are more soluble in a liquid when present at higher partial pressures.

15.71. **Collect and Organize**
Of four reactions we are to determine which will shift its equilibrium so as to form more products when the mixture is compressed to half of its original volume.

Analyze
Decreasing the volume by half doubles the partial pressures of all the gaseous reactants and products in the reactions. This would cause a shift in the position of the equilibrium toward the side of the reaction with the fewest moles of gas.

Solve
(a) This equilibrium shifts to the left, forming more reactants.
(b) This equilibrium shifts to the right, forming more products.
(c) This equilibrium shifts neither to the left nor to the right, since the number of moles of gas is the same on both sides of the equation.
(d) This equilibrium shifts to the right, forming more products.
(b) and (d) will form more products when the volume of the mixture is decreased by half.

Think About It
The opposite shifts occur in the position of the equilibrium when the volumes of the reaction mixtures are increased.

15.73. **Collect and Organize**
We are to predict the effect on the position of the equilibrium in
$$2\,O_3(g) \rightleftharpoons 3\,O_2(g)$$
with various changes in concentration and volume.

Analyze
Increasing the concentration of a reactant shifts the equilibrium to the right, whereas increasing the concentration of a product shifts the equilibrium to the left. An increase in volume decreases the pressure and shifts the equilibrium toward the side of the reaction with the greater number of moles of gas.

Solve
(a) Increasing the concentration of the reactant, O_3, shifts the equilibrium to the right, increasing the concentration of the product, O_2.
(b) Increasing the concentration of the product, O_2, shifts the equilibrium to the left, increasing the concentration of the reactant, O_3.
(c) Decreasing the volume of the reaction to 1/10 its original volume shifts the equilibrium to the left, increasing the concentration of the reactant, O_3.

Think About It
Adding O_2 and decreasing the volume create the same shift in the position of the equilibrium.

15.75. **Collect and Organize**
We are to determine how decreasing the partial pressure of O_2 affects the equilibrium of
$$2\,SO_2(g) + O_2(g) \rightleftharpoons 2\,SO_3(g)$$

Analyze
According to Le Châtelier's principle, increasing the partial pressure or concentration of a reactant shifts the equilibrium to the right. Decreasing the partial pressure of a reactant, then, shifts the equilibrium to the left.

Solve

Decreasing the partial pressure of O_2 in this reaction shifts the equilibrium to the left.

Think About It

At equilibrium then, we have less SO_3 product when we reduce the partial pressure of O_2 in the reaction mixture.

15.77. Collect and Organize

Given information about whether a reaction is endothermic or exothermic, we may determine if an increase in temperature will shift an equilibrium to the left or to the right. We are asked to identify reactions in which more product will be formed.

Analyze

An endothermic reaction has a positive value of ΔH; we may imagine heat as a reactant. An exothermic reaction has a negative value of ΔH; we may imagine heat as a product. Increasing the temperature will drive the reaction in the opposite direction, much like increasing the concentration of a reactant or product drives a reaction in the opposite direction in order to restore equilibrium. More product will be formed if heat is a reactant, and the reaction is endothermic.

Solve

(b) is the only endothermic reaction ($\Delta H > 0$); it is the only reaction that will form more product at higher temperature.

Think About It

It may be possible to form more products of reactions (a) and (c) by lowering the temperature of the reaction, but we should consider the reaction kinetics as well. Lowering the temperature at which a reaction is conducted will also lower the rate of the reaction.

15.79. Collect and Organize

We are to explain why equilibrium calculations are simpler when the value of the equilibrium constant, K, is very small.

Analyze

When K is very small, the reaction does not proceed very far to the right. The concentrations of products formed once the reaction achieves equilibrium are very small.

Solve

When K is small, the count of reactants that are transformed into products is small and so at equilibrium the concentration of the reactants is approximately equal to the initial concentrations. This means that the approximation

$$K = \frac{x}{(A-x)(B-x)} \approx \frac{x}{(A)(B)}$$

for the reaction

$$A + B \rightleftharpoons C$$

is valid. This makes our calculations easier and avoids having to use the quadratic equation.

Think About It

The assumption is considered valid if x is $< 5\%$ of the value of both A and B.

15.81. Collect and Organize

For the decomposition of PCl_5 to PCl_3 and Cl_2 ($K_p = 23.6$ at 500 K), we are to calculate the equilibrium partial pressures given the initial partial pressures of PCl_5 and PCl_3. We are also to determine how the concentration of PCl_3 and PCl_5 change when more Cl_2 is added to the system already at equilibrium.

Analyze

(a) To calculate the partial pressures of all the species present, we set up an ICE table. Because the initial partial pressure of Cl_2 is 0.0 atm, we know that the reaction proceeds to the right to attain equilibrium.

(b) To determine how the equilibrium shifts when Cl_2 is added, we apply Le Châtelier's principle.

Solve

(a)

	P_{PCl_5}	P_{PCl_3}	P_{Cl_2}
Initial	0.560 atm	0.500 atm	0.00 atm
Change	$-x$	$+x$	$+x$
Equilibrium	$0.560 - x$	$0.500 + x$	x

After placing these values into the equilibrium constant expression, we can solve for x using the quadratic formula:

$$23.6 = \frac{(0.500 + x)(x)}{(0.560 - x)}$$

$$x^2 + 24.1x - 13.216 = 0$$

$$x = 0.536 \text{ or } -24.6$$

Because $0.500 + x$ would be negative if $x = -24.6$, the actual root for this problem is $x = 0.536$. The equilibrium partial pressures of PCl_5, PCl_3, and Cl_2 are

$$P_{PCl_5} = 0.560 - 0.536 = 0.024 \text{ atm}$$

$$P_{PCl_3} = 0.500 + 0.536 = 1.036 \text{ atm}$$

$$P_{Cl_2} = 0.536 \text{ atm}$$

(b) When the partial pressure of Cl_2 is increased the partial pressure (or concentration) of PCl_3 decreases and the partial pressure of PCl_5 increases.

Think About It

Because $K > 1$, the products of this reaction are favored.

15.83. **Collect and Organize**

For the initial concentrations of H_2O and Cl_2O as 0.00432 M in the equilibrium reaction

$$H_2O(g) + Cl_2O(g) \rightleftharpoons 2\,HOCl(g)$$

where $K_c = 0.0900$ at 25°C, we are to calculate the equilibrium concentrations of H_2O, Cl_2O, and HOCl.

Analyze

We first set up an ICE table to solve this problem. We assume here that the initial concentration of HOCl = 0.00 M. The equilibrium constant expression for this reaction is

$$K_c = \frac{[HOCl]^2}{[H_2O][Cl_2O]}$$

Solve

	$[H_2O]$	$[Cl_2O]$	HOCl
Initial	0.00432 M	0.00432 M	0
Change	$-x$	$-x$	$+2x$
Equilibrium	$0.00432 - x$	$0.00432 - x$	$2x$

After placing these values into the equilibrium constant expression, we can solve for x by taking the square root of both sides:

$$0.0900 = \frac{(2x)^2}{(0.00432 - x)^2}$$

$$0.300 = \frac{2x}{0.00432 - x}$$

$$1.296 \times 10^{-3} - 0.30x = 2x$$

$$x = 5.63 \times 10^{-4}$$

The concentration of all the gases at equilibrium are

$$[H_2O] = [Cl_2O] = 0.00432 - x = 3.76 \times 10^{-3} \ M$$
$$[HOCl] = 2x = 1.13 \times 10^{-3} \ M$$

Think About It

This reaction, with its equilibrium constant being less than one, favors reactants over product at equilibrium.

15.85. Collect and Organize

Given $K_p = 1.5 \times 10^6$ at 25°C for the reaction of 1 mole of NO with $\frac{1}{2}$ mole of O_2 to give 1 mole of NO_2, we are to calculate the equilibrium ratio of the partial pressure of NO_2 to that of NO in air, where the partial pressure of oxygen is 0.21 atm.

Analyze

The K_p expression for this reaction is

$$K_p = \frac{P_{NO_2}}{P_{NO} \times \left(P_{O_2}\right)^{1/2}}$$

Solve

When $P_{O_2} = 0.21$ atm,

$$K_p = \frac{P_{NO_2}}{P_{NO} \times (0.21)^{1/2}} = 1.5 \times 10^6$$

Rearranging this equation to solve for the ratio of the partial pressure of NO_2 to the partial pressure of NO gives

$$\frac{P_{NO_2}}{P_{NO}} = \left(1.5 \times 10^6\right)(0.21)^{1/2} = 6.9 \times 10^5$$

Think About It

The high value of K indicates that the product, NO_2, is highly favored. This is consistent with the high value we calculated for the ratio of the partial pressures.

15.87. Collect and Organize

Given that $K_p = 1.5$ at 700°C, we are to calculate P_{CO_2} and P_{CO} at equilibrium for the reaction

$$CO_2(g) + C(s) \rightleftharpoons 2\ CO(g)$$

where the initial partial pressures of CO_2 and CO are 5.0 atm and 0.0 atm, respectively.

Analyze

Because carbon is a pure solid, it does not appear in the equilibrium constant expression:

$$K_p = \frac{(P_{CO})^2}{P_{CO_2}}$$

Solve

	P_{CO_2}	P_{CO}
Initial	5.0 atm	0.0 atm
Change	$-x$	$+2x$
Equilibrium	$5.0 - x$	$2x$

After placing these values into the equilibrium constant expression, we can solve for x using the quadratic formula:

$$1.5 = \frac{(2x)^2}{(5.0 - x)}$$

$$4x^2 + 1.5x - 7.5 = 0$$

$$x = 1.20 \text{ or } -1.57$$

The value of $x = -1.57$ would give a negative partial pressure for CO, so $x = 1.19$. The partial pressures of the gases at equilibrium are

$$P_{CO} = 2x = 2.4 \text{ atm}$$
$$P_{CO_2} = 5.0 - x = 3.8 \text{ atm}$$

Think About It

Checking our results should give a value close to 1.5, the equilibrium constant value:

$$\frac{(2.4)^2}{3.8} = 1.5$$

15.89. **Collect and Organize**

For the decomposition of NO_2 to NO and O_2 when $P_{O_2} = 0.136$ atm at equilibrium, we are asked to calculate the partial pressures of NO and NO_2 at equilibrium at 1000 K, where $K = 158$, and to calculate the total pressure in the flask at equilibrium.

Analyze

(a) To calculate the partial pressures of NO and NO_2, we set up an ICE table where we start with pure NO_2 (A) and end with $P_{O_2} = 0.136$ atm $= x$ at equilibrium.

(b) The total pressure is

$$P_T = P_{NO_2} + P_{NO} + P_{O_2}$$

Solve

(a)

	P_{NO_2}	P_{NO}	P_{O_2}
Initial	A atm	0.00 atm	0.00 atm
Change	$-2x$	$+2x$	$+x$
Equilibrium	$A - 2x$	$2x$	x

We know that at equilibrium $x = 0.136$ atm, so

$$K_p = 158 = \frac{(P_{NO})^2 (P_{O_2})}{(P_{NO_2})^2} = \frac{(2x)^2 (x)}{(P_{NO_2})^2} = \frac{(0.272)^2 (0.136)}{(P_{NO_2})^2}$$

$$(P_{NO_2})^2 = \frac{(0.272)^2 \times (0.136)}{158} = 6.37 \times 10^{-5}$$

$$P_{NO_2} = (6.37 \times 10^{-5})^{1/2} = 7.98 \times 10^{-3} \text{ atm}$$

At equilibrium,

$$P_{NO} = 2x = 0.272 \text{ atm}$$
$$P_{O_2} = x = 0.136 \text{ atm}$$
$$P_{NO_2} = 7.98 \times 10^{-3} \text{ atm}$$

(b) $P_T = 7.98 \times 10^{-3}$ atm + 0.272 atm + 0.136 atm = 0.416 atm

Think About It
The amount of NO_2 initially present can also be calculated:

$$A - 2x = 7.98 \times 10^{-3} \text{ atm}$$

Because $x = 0.136$ atm

$$A = 7.98 \times 10^{-3} + 2 \times 0.136 = 0.280 \text{ atm}$$

15.91. **Collect and Organize**
For the equilibrium reaction

$$N_2(g) + O_2(g) \rightleftharpoons 2\,NO(g)$$

the value of K_p is 0.050 at 2200°C. We are to calculate the partial pressures of N_2, O_2, and NO at equilibrium, given that the initial partial pressures of these gases are 0.79 atm N_2, 0.21 atm O_2, and 0.00 atm NO.

Analyze
We set up an ICE table to solve this problem. The equilibrium constant expression for this reaction is

$$K_p = \frac{\left(P_{NO}\right)^2}{\left(P_{O_2}\right)\left(P_{N_2}\right)}$$

Solve

	P_{N_2}	P_{O_2}	P_{NO}
Initial	0.79 atm	0.21 atm	0.00 atm
Change	$-x$	$-x$	$+2x$
Equilibrium	$0.79 - x$	$0.21 - x$	$2x$

After placing these values into the equilibrium constant expression, we can solve for x using the quadratic formula:

$$0.050 = \frac{\left(2x\right)^2}{(0.79 - x)(0.21 - x)}$$

$$0.050 = \frac{4x^2}{\left(0.1659 - x + x^2\right)}$$

$$3.95x^2 + 0.050x - 0.008295 = 0$$

$$x = 0.03993 \text{ or } -0.05260$$

The value of $x = -0.05260$ would give a negative partial pressure for NO, so $x = 0.03993$. The partial pressures of the gases at equilibrium are

$$P_{O_2} = 0.21 - x = 0.17 \text{ atm}$$
$$P_{N_2} = 0.79 - x = 0.75 \text{ atm}$$
$$P_{NO} = 2x = 0.080 \text{ atm}$$

Think About It
Using these equilibrium partial pressures we can check our answers, which should give $K_p = 0.050$:

$$\frac{\left(0.080\right)^2}{\left(0.75\right)\left(0.17\right)} = 0.050$$

15.93. **Collect and Organize**

For the following reaction

$$2 H_2S(g) \rightleftharpoons 2 H_2(g) + S_2(g)$$

the value of K_c is 2.2×10^{-4} at 1400 K. We are to calculate the equilibrium concentration of H_2S, given that the initial $[H_2S]$ is 6.00 M.

Analyze

We are told to assume that the initial concentrations of H_2 and S_2 are 0.00 M. Because the equilibrium constant is small we may be able to make a simplifying assumption. The equilibrium constant expression for this reaction is

$$K_c = \frac{[H_2]^2[S_2]}{[H_2S]^2}$$

Solve

	$[H_2S]$	$[H_2]$	$[S_2]$
Initial	6.00 M	0.00 M	0.00 M
Change	$-2x$	$+2x$	$+x$
Equilibrium	$6.00 - 2x$	$2x$	x

Since the equilibrium constant is small, this reaction does not proceed very far to the right. After placing these values into the equilibrium constant expression, we can solve for x:

$$2.2 \times 10^{-4} = \frac{(2x)^2(x)}{(6.00 - 2x)^2} \approx \frac{(2x)^2(x)}{(6.00)^2}$$

$$x = 0.126$$

Checking the simplifying assumption that we made shows that it is valid:

$$\frac{2(0.126)}{6.00} \times 100 = 4\%$$

The equilibrium $[H_2S]$ is $6.00 - 2x = 5.75$ M.

Think About It

Without the simplifying assumption we would have to solve a cubic equation.

15.95. **Collect and Organize**

We are asked to calculate the partial pressures of the gases at equilibrium for the following reaction, given that $K_c = 5.0$ at 600 K and the initial partial pressures of CO and Cl_2 are 0.265 atm and the initial partial pressure of $COCl_2$ is 0.000 atm:

$$CO(g) + Cl_2(g) \rightleftharpoons COCl_2(g)$$

Analyze

We must first calculate K_p from K_c where $\Delta n = -1$:

$$K_p = K_c(RT)^{\Delta n} = \frac{5.0}{[0.08206 \text{ L} \cdot \text{atm}/(\text{mol} \cdot \text{K})](600 \text{ K})} = 0.102$$

We can set up an ICE table to solve for the equilibrium partial pressures of the gases.

Solve

	P_{CO}	P_{Cl_2}	P_{COCl_2}
Initial	0.265 atm	0.265 atm	0.000 atm
Change	$-x$	$-x$	$+x$
Equilibrium	$0.265 - x$	$0.265 - x$	x

After placing these values into the equilibrium constant expression, we can solve for *x* using the quadratic formula:

$$0.102 = \frac{x}{(0.265 - x)^2}$$

$$0.102 = \frac{x}{0.0702 - 0.530x + x^2}$$

$$0.102x^2 - 1.05406x + 0.00716 = 0$$

$$x = 10.33 \text{ or } 0.00680$$

The value of *x* = 10.33 would give negative partial pressures for CO and Cl_2, so *x* = 0.00680. The partial pressures of the gases at equilibrium are

$$P_{CO} = P_{Cl_2} = 0.265 - x = 0.258 \text{ atm}$$

$$P_{COCl_2} = x = 0.00680 \text{ atm}$$

Think About It
Checking our result, we get the value of K_p:

$$\frac{0.00680}{(0.258)^2} = 0.102$$

15.97. **Collect and Organize**
We are to calculate the concentrations of all the gases at equilibrium for the reaction

$$CO(g) + H_2O(g) \rightleftharpoons CO_2(g) + H_2(g)$$

given that the initial concentrations of all the gases are 0.050 *M* and that K_c = 5.1 at 700 K.

Analyze
The equilibrium constant expression for this reaction is

$$K_c = \frac{[CO_2][H_2]}{[CO][H_2O]}$$

We have to use *Q*, the reaction quotient, to determine the direction in which the reaction goes to reach equilibrium:

$$Q = \frac{(0.050)(0.050)}{(0.050)(0.050)} = 1.0$$

Q < *K* so the reaction proceeds to the right.

Solve

	[CO]	[H₂O]	[CO₂]	[H₂]
Initial	0.050 *M*	0.050 *M*	0.050 *M*	0.050 *M*
Change	−*x*	−*x*	+*x*	+*x*
Equilibrium	0.050 − *x*	0.050 − *x*	0.050 + *x*	0.050 + *x*

$$5.1 = \frac{(0.050 + x)^2}{(0.050 - x)^2}$$

Taking the square root of both sides:

$$2.258 = \frac{(0.050 + x)}{(0.050 - x)}$$

$$0.1129 - 2.258x = 0.050 + x$$

$$0.0629 = 3.258x$$

$$0.0193 = x$$

The concentrations of the gases at equilibrium are
$$[CO] = [H_2O] = 0.050 - x = 0.031 \ M$$
$$[CO_2] = [H_2] = 0.050 + x = 0.069 \ M$$

Think About It
We did not need to use the quadratic formula here because with the concentrations of the reactants equal to each other and the concentrations of the products equal to each other we could simplify the math by taking the square root of both sides of the equation to solve for x.

15.99. **Collect and Organize**
Given the percent decomposition of CO_2 at three temperatures and that each equilibrium mixture begins with 1 atm of CO_2, we are to determine whether the reaction is endothermic and then calculate K_p at each temperature. We are also to comment on the decomposition reaction as a viable source of CO fuel and a remedy for global warming.

Analyze
The amount of CO_2 decomposed increases with increasing temperature. This means that $K_2 > K_1$ in the equation

$$\ln\left(\frac{K_2}{K_1}\right) = -\frac{\Delta H°}{R}\left(\frac{1}{T_2} - \frac{1}{T_1}\right)$$

where $T_2 > T_1$ so $(1/T_2 - 1/T_1) < 0$. To calculate K_p for each temperature, we set up an ICE table:

	P_{CO_2}	P_{CO}	P_{O_2}
Initial	1 atm	0 atm	0 atm
Change	$-2x$	$+2x$	$+x$
Equilibrium	$1 - 2x$	$2x$	x

We know that $2x$ is equal to the percent decomposition divided by 100. The form of the equilibrium expression is

$$K_p = \frac{\left(P_{CO}\right)^2 \left(P_{O_2}\right)}{\left(P_{CO_2}\right)^2}$$

Because the percent decomposition increases with temperature, the value of K_p is expected to increase as T increases.

Solve
Because the value of K increases with increasing temperature, $\ln(K_2/K_1) > 0$. With $(1/T_2 - 1/T_1) < 0$, the value of ΔH must be positive, so this reaction is endothermic.

At 1500 K, $2x = 0.00048$, $x = 0.00024$, and $1 - 2x = 0.99952$:
$$K_p = \frac{(0.00048)^2 (0.00024)}{(0.99952)^2} = 5.5 \times 10^{-11}$$

At 2500 K, $2x = 0.176$, $x = 0.088$, and $1 - 2x = 0.824$:
$$K_p = \frac{(0.176)^2 (0.088)}{(0.824)^2} = 4.0 \times 10^{-3}$$

At 3000 K, $2x = 0.548$, $x = 0.274$, and $1 - 2x = 0.452$:
$$K_p = \frac{(0.548)^2 (0.274)}{(0.452)^2} = 0.40$$

As predicted, the value of the equilibrium constant increases with increasing temperature. This reaction, however, does not favor products even at very high temperature, so this is not a viable source of CO and is not a remedy to decrease the contribution of CO_2 to global warming. Also, the process produces poisonous CO gas.

Think About It
By using values in Appendix 4 we can confirm that the reaction is endothermic.

15.101. **Collect and Organize**

By combining the two equations given, we can calculate the overall equilibrium constant, K, for the reaction and then calculate the concentration of X^{2-} in an equilibrium mixture where $[H_2X]_{eq} = 0.1$ M and $[HCl]_{eq} = [H_3O^+]_{eq} = 0.3$ M.

Analyze

When we add equations, the overall equilibrium constant is the product of the individual equilibrium constants. From the overall reaction, we can write the equilibrium constant expression and from that solve for $[X^{2-}]$ at equilibrium.

Solve

$$H_2X(aq) + H_2O(\ell) \rightleftharpoons HX^-(aq) + H_3O^+(aq) \qquad K_1 = 8.3 \times 10^{-8}$$
$$\underline{HX^-(aq) + H_2O(\ell) \rightleftharpoons X^{2-}(aq) + H_3O^+(aq) \qquad K_2 = 1 \times 10^{-14}}$$
$$H_2X(aq) + 2H_2O(\ell) \rightleftharpoons X^{2-}(aq) + 2H_3O^+(aq) \qquad K_3 = K_1 \times K_2 = 8.3 \times 10^{-22}$$

$$K_c = \frac{\left[X^{2-}\right]\left[H_3O^+\right]^2}{[H_2X]}$$

We know that $K_c = 8.3 \times 10^{-22}$, $[H_3O^+] = 0.3$ M, and $[H_2X] = 0.1$ M for the saturated solution. Rearranging the equilibrium constant expression and solving for $[X^{2-}]$ gives

$$\left[X^{2-}\right] = \frac{K_c[H_2X]}{\left[H_3O^+\right]^2} = \frac{\left(8.3 \times 10^{-22}\right)(0.1)}{(0.3)^2} = 9 \times 10^{-22}\ M$$

Think About It

Our answer makes sense because with a very small overall equilibrium constant, we expect a very small concentration of product, X^{2-}.

15.103. **Collect and Organize**

Given $K_p = 2.38 \times 10^{73}$ for the reaction in which CaO reacts with SO_2 in the presence of O_2 to form solid $CaSO_4$, we can determine P_{SO_2} in this reaction when the partial pressure of oxygen is 0.21 atm. Using this partial pressure of SO_2, we may determine how many molecules of SO_2 are in a sample containing 100.0 mol of gas.

Analyze

The equilibrium constant expression for the reaction is

$$K_p = \frac{1}{\left(P_{SO_2}\right)\left(P_{O_2}\right)^{1/2}}$$

The mole fraction of SO_2 in the sample is given by

$$\chi_{SO_2} = \frac{P_{SO_2}}{P_{Total}}$$

Multiplying the mole fraction of SO_2 by the total number of moles in the sample and Avogadro's number will yield the number of SO_2 molecules.

Solve

(a) The partial pressure of SO_2 is

$$2.38 \times 10^{73} = \frac{1}{P_{SO_2} \times (0.21)^{1/2}}$$

$$P_{SO_2} = 9.2 \times 10^{-74}\ atm$$

(b) The number of molecules of SO_2 in the sample is

$$\chi_{SO_2} = \frac{9.17 \times 10^{-74} \text{ atm}}{\left(0.21 + 9.17 \times 10^{-74} \text{ atm}\right)} = 4.37 \times 10^{-73}$$

$$n_{SO_2} = \chi_{SO_2} \cdot n_{Total} = 4.37 \times 10^{-73} \left(100.0 \text{ moles}\right) = 4.37 \times 10^{-71} \text{ mol } SO_2$$

$$4.37 \times 10^{-71} \text{ mol } SO_2 \times \frac{6.022 \times 10^{23} \text{ molecules } SO_2}{1 \text{ mol } SO_2} = 2.63 \times 10^{-47} \text{ molecules}$$

Think About It

Essentially all of the SO_2 is "scrubbed" by the CaO in this reaction, making this an efficient method to remove SO_2 from exhaust gases.

CHAPTER 16 | Acid–Base and Solubility Equilibria: Reactions in Soil and Water

16.1. Collect and Organize

In Figure P16.1 four lines are shown to describe the possible dependence of percent ionization of acetic acid on concentration. We are to choose the one that best represents the trend for this weak acid.

Analyze

The ionization of acetic acid is described by the following chemical reaction:

$$CH_3COOH(aq) \rightleftharpoons CH_3COO^-(aq) + H^+(aq)$$

The degree of ionization is the ratio of the quantity of a substance that is ionized to the concentration of the substance before ionization.

Solve

According to Figure 16.9, the change in degree of ionization of a weak acid with concentration is not linear and is best described by the red line in Figure P16.1. The degree of ionization increases with decreasing acetic acid concentration.

Think About It

The percent ionization could be calculated for each concentration if we knew the equilibrium concentration of the acetate ion in solution and the initial concentration of acetic acid dissolved.

$$\% \text{ ionization} = \frac{\left[H^+ \right]_{equilibrium}}{\left[\text{acetic acid} \right]_{initial}} \times 100$$

16.3. Collect and Organize

From Figure P16.3 we are to choose which titration curve represents a strong acid and which represents a weak acid.

Analyze

A strong acid is completely ionized in solution and has a lower initial pH than the weak acid, which is only partially ionized in solution.

Solve

The blue titration curve represents the titration of a 1 M solution of strong acid. The red titration curve represents the titration of a 1 M solution of weak acid. This is because the pH of the strong acid is expected to be much lower than that of the weak acid at the start of the titration (where no base has yet been added).

Think About It

Notice that the equivalence point of the titration of the strong acid (pH 7) does not equal that of the weak acid (pH 10).

16.5. Collect and Organize

For the red titration curve in Figure P16.3, we are to choose the indicator, according to its pK_a, that would be best for the titration.

Analyze

The best indicator is the one with a pK_a that is nearest to the end point of the titration.

Solve

The end point for the red curve is at approximately pH 10. Therefore, the best indicator is the one with a pK_a of 9.0.

Think About It
The lower pK_a indicators would show a color change before the end point of the titration was reached. Using these would underestimate the concentration of the weak acid in the original solution.

16.7. Collect and Organize
We are shown two titration curves in Figure P16.7. The blue curve has one equivalence point and the red curve has two equivalence points. We are to indicate which of these curves is Na_2CO_3 and which is $NaHCO_3$.

Analyze
Both of the bases being titrated are soluble sodium salts. The equation describing the titration of CO_3^{2-} (Na^+ is a spectator ion) shows CO_3^{2-} to be "dibasic"; it reacts in two steps to form H_2CO_3.

$$CO_3^{2-}(aq) + H^+(aq) \rightarrow HCO_3^-(aq)$$
$$HCO_3^-(aq) + H^+(aq) \rightarrow H_2CO_3(aq)$$

HCO_3^-, however, reacts with acid in one step; it is "monobasic":
$$HCO_3^-(aq) + H^+(aq) \rightarrow H_2CO_3(aq)$$

Solve
The red titration curve represents the titration of Na_2CO_3 because it shows two equivalence points. The blue titration curve represents the titration of $NaHCO_3$ because it shows one equivalence point.

Think About It
Notice that both titration curves start at high pH. This is due to the hydrolysis of CO_3^{2-} and HCO_3^- in water:
$$CO_3^{2-}(aq) + H_2O\,(\ell) \rightarrow HCO_3^-(aq) + OH^-(aq)$$
$$HCO_3^-(aq) + H_2O\,(\ell) \rightarrow H_2CO_3(aq) + OH^-(aq)$$

16.9. Collect and Organize
Figure P16.8 shows three beakers: one containing a yellow solution, one containing a very light green solution, and one containing a blue solution. Given that the bromthymol blue indicator in each beaker is yellow in acidic solutions and blue in basic solutions, we are to match solutions of the dissolved salts NH_4Cl, $NH_4C_2H_3O_2$, and $NaC_2H_3O_2$ to the correct beaker.

Analyze
We must consider the hydrolysis of the constituent cation and anion in each salt. In NH_4Cl, the NH_4^+ ion reacts with water to give an acidic solution:
$$NH_4^+(aq) + H_2O\,(\ell) \rightleftharpoons NH_3(aq) + H_3O^+(aq)$$
but Cl^- does not. In $NaC_2H_3O_2$ the Na^+ ion does not hydrolyze, but the acetate ion does to give a basic solution:
$$C_2H_3O_2^-(aq) + H_2O\,(\ell) \rightleftharpoons HC_2H_3O_2(aq) + OH^-(aq)$$
In $NH_4C_2H_3O_2$ both cation and anion hydrolyze, giving a nearly neutral solution.

Solve
The yellow solution is dissolved NH_4Cl; the blue solution is dissolved $NaC_2H_3O_2$; and the light green solution is dissolved $NH_4C_2H_3O_2$.

Think About It
The relative magnitude of the K_a of NH_4^+ (5.7×10^{-10}) compared to the K_b of $C_2H_3O_2^-$ (5.7×10^{-10}) shows that both salts hydrolyze to the same extent and that we should expect a solution of $NH_4C_2H_3O_2$ to be approximately neutral.

16.11. Collect and Organize
For $HBr(aq)$ we are to identify the Brønsted–Lowry acid and base.

Analyze
A Brønsted–Lowry acid is a proton donor. A Brønsted–Lowry base is a proton acceptor.

Solve

HBr is a strong acid in water. It acts as a Brønsted–Lowry acid, donating its proton to H_2O, the Brønsted–Lowry base:

$$HBr(aq) + H_2O\,(\ell) \rightarrow H_3O^+(aq) + Br^-(aq)$$

Think About It

Hydrobromic acid is a strong acid in water. It completely dissociates in water to H_3O^+ and Br^-.

16.13. Collect and Organize

For $NaOH(aq)$ we are to identify the Brønsted–Lowry acid and base.

Analyze

NaOH is a soluble salt that forms Na^+ and OH^- in water. Na^+ does not react with water so it is a spectator ion. We need then to consider the behavior of OH^- in water. A Brønsted–Lowry acid is a proton donor. A Brønsted–Lowry base is a proton acceptor.

Solve

OH^- is a strong base in water. It acts as a Brønsted–Lowry base, removing a proton from H_2O, the Brønsted–Lowry acid:

$$OH^-(aq) + H_2O\,(\ell) \rightleftharpoons H_2O\,(\ell) + OH^-(aq)$$

Think About It

In Problems 16.9 and 16.10, water acted as a Brønsted–Lowry base. In this problem, it acts as an acid. This dual acid–base behavior makes water *amphoteric*.

16.15. Collect and Organize

For three acid–base reactions, we are to identify which reactant is the acid and which reactant is the base.

Analyze

For all of these reactions that involve the transfer of a proton between the acid and base, we can apply the Brønsted–Lowry definitions of acid and base. A Brønsted–Lowry acid is a proton donor. A Brønsted–Lowry base is a proton acceptor.

Solve

(a) HNO_3 is the acid. It transfers H^+ to the base NaOH.
(b) HCl is the acid. It transfers H^+ to the base $CaCO_3$.
(c) HCN is the acid. It transfers H^+ to the base NH_3.

Think About It

In (a) and (b), Na^+, Ca^{2+}, and Cl^- do not get involved in the reaction. They are spectator ions. The net ionic equations are

$$HNO_3(aq) + OH^-(aq) \rightarrow NO_3^-(aq) + H_2O\,(\ell)$$
$$CO_3^{2-}(aq) + 2\,H^+(aq) \rightarrow CO_2(g) + H_2O\,(\ell)$$

16.17. Collect and Organize

For each species listed we are to write the formula for the conjugate base.

Analyze

The conjugate-base form of a species has H^+ removed from its formula.

Solve

The conjugate base of HNO_2 is NO_2^-.
The conjugate base of HClO is ClO^-.
The conjugate base of H_3PO_4 is $H_2PO_4^-$.
The conjugate base of NH_3 is NH_2^-.

Think About It
Be sure to account for the change in charge when H^+ is removed to form the conjugate base.

16.19. **Collect and Organize**
Given that the concentration of a nitric acid solution is 1.50 *M*, we are to calculate the initial concentration of H^+ ions in the solution, as well as the concentration of H^+ ions in the solution after 10 mL of 0.505 *M* NaOH is added to 100 mL of the HNO_3 solution.

Analyze
Nitric acid is a strong acid that completely dissociates in water, so the concentration of H^+ ions is stoichiometrically related to the concentration of HNO_3:
$$HNO_3(aq) \rightarrow H^+(aq) + NO_3^-(aq)$$
The reaction that occurs between HNO_3 and NaOH is
$$HNO_3(aq) + NaOH(aq) \rightarrow H_2O\,(\ell) + NaNO_3(aq)$$
The number of moles of HNO_3 in the solution is
$$100 \text{ mL HNO}_3 \times \frac{1.50 \text{ mol HNO}_3}{1000 \text{ mL HNO}_3} = 0.150 \text{ mol HNO}_3$$
The number of moles of NaOH added to the reaction is
$$10 \text{ mL NaOH} \times \frac{0.505 \text{ mol NaOH}}{1000 \text{ mL NaOH}} = 0.00505 \text{ mol NaOH}$$
We may find the number of moles of HNO_3 present after the reaction occurs using an addition table, and accounting for the final volume of solution, 100 mL + 10 mL = 110 mL.

Solve
The initial concentration of H^+ in the HNO_3 solution is 1.50 *M*. After adding NaOH to the solution, the number of moles of acid present is

	H^+	+	OH^-	$\rightarrow$	H_2O
Before	0.150 mol				
Addition			0.00505 mol		
Change	−0.00505 mol		−0.00505 mol		
Equilibrium	0.1450 mol		0 mol		

The $[H^+]$ after NaOH addition is
$$\frac{0.1450 \text{ mol H}^+}{0.110 \text{ L}} = 1.32 \text{ M H}^+$$

Think About It
The concentration of OH^- in many strong base solutions is the same as the concentration of the strong base dissolved into the solution.

16.21. **Collect and Organize**
Given that a solution is 0.0800 *M* in the strong base $Sr(OH)_2$, we are asked to calculate the concentration of OH^-, as well as the concentration of OH^- ions in the solution after 12 mL of 0.465 *M* HCl is added to 100 mL of the $Sr(OH)_2$ solution.

Analyze
$Sr(OH)_2$, being a strong base, completely dissociates according to the equation
$$Sr(OH)_2(aq) \rightarrow Sr^{2+}(aq) + 2\,OH^-(aq)$$
Therefore, the initial concentration of OH^- is equal to $2 \times [Sr(OH)_2]$.
The balanced reaction that occurs between HCl and $Sr(OH)_2$ is
$$2\,HCl(aq) + Sr(OH)_2(aq) \rightarrow 2\,H_2O\,(\ell) + SrCl_2(aq)$$

This reduces to a net ionic equation of

$$H^+(aq) + OH^-(aq) \rightarrow H_2O\,(\ell)$$

The number of moles of $Sr(OH)_2$ in the solution is

$$100 \text{ mL Sr(OH)}_2 \times \frac{0.0800 \text{ mol Sr(OH)}_2}{1000 \text{ mL Sr(OH)}_2} \times \frac{2 \text{ mol OH}^-}{1 \text{ mol Sr(OH)}_2} = 0.0160 \text{ mol OH}^-$$

The number of moles of HCl added to the reaction is

$$12 \text{ mL HCl} \times \frac{0.465 \text{ mol HCl}}{1000 \text{ mL HCl}} = 0.00558 \text{ mol NaOH}$$

We may find the number of moles of $Sr(OH)_2$ present after the reaction occurs using an addition table, and accounting for the final volume of solution, 100 mL + 12 mL = 112 mL.

Solve

The initial concentration of hydroxide ion is 2×0.0800 *M* = 0.160 *M*. After adding HCl to the solution, the number of moles of base present is

	OH⁻	+	H⁺	→	H₂O
Before	0.016 mol				
Addition			0.00558 mol		
Change	−0.00558 mol		−0.00558 mol		
Equilibrium	0.01042 mol		0 mol		

The [OH⁻] after HCl addition is

$$\frac{0.01042 \text{ mol OH}^-}{0.112 \text{ L}} = 0.093 \text{ M OH}^-$$

Think About It

Be sure to account for both OH⁻ ions in this strong base.

16.23. Collect and Organize

We are asked how to prepare 2.50 L of 0.70 *M* OH⁻ using NaOH(*s*).

Analyze

From the desired volume and concentration, we first calculate the moles of OH⁻ required for the solution. Since 1 mole of OH⁻ is produced for every 1 mole of NaOH dissolved, this is also the number of moles of NaOH required. To calculate the mass of NaOH needed, we multiply the number of moles by the molar mass of NaOH (40.00 g/mol).

Solve

Mass of NaOH needed

$$\text{Mass NaOH} = 2.50 \text{ L} \times \frac{0.70 \text{ mol OH}^-}{1 \text{ L}} \times \frac{1 \text{ mol NaOH}}{1 \text{ mol OH}^-} \times \frac{40.00 \text{ g NaOH}}{1 \text{ mol NaOH}} = 70.0 \text{ g NaOH}$$

Dissolve 70.0 g of NaOH(*s*) in water and dilute to a total volume of 2.50 L.

Think About It

Remember that volume times concentration for solutions gives us the moles of that substance in solution.

16.25. Collect and Organize

We are to explain why the pH value decreases for solutions as acidity increases.

Analyze

The pH of a solution is calculated through

$$pH = -\log[H^+]$$

Solve

Because the pH function is a –log function, as $[H^+]$ increases, the value of $-\log[H^+]$ decreases.

Think About It

The pH scale is typically 0–14 for concentrations of H^+ from 1 M to 1×10^{-14} M, but values of pH may be negative or greater than 14.

16.27. **Collect and Organize**

We are asked under what conditions the pH of a solution may be negative.

Analyze

The pH scale is often seen as 0–14. This occurs when $[H^+]$ is between 1 M and 1×10^{-14} M.

Solve

When $[H^+]$ is greater than 1 M, the pH of the solution is negative.

Think About It

For example, a 3.00 M solution of HCl has a pH of

$$pH = -\log[H^+] = -\log(3.00\ M) = -0.477$$

16.29. **Collect and Organize**

We are asked to describe how the value of pK_w changes with temperature.

Analyze

pK_w is equal to $-\log K_w$, where K_w is the autoionization constant for water. The autoionization of water is an endothermic process, in which energy must be applied to the system in order to generate H_3O^+ and OH^- ions.

Solve

As the temperature increases, so too will the value of K_w. The value of pK_w will thus decrease (become more negative) as temperature increases.

Think About It

We can imagine the autoionization of water as

$$\text{heat} + 2\ H_2O(\ell) \rightleftharpoons H_3O^+(aq) + OH^-(aq)$$

By increasing the temperature, we shift equilibrium to the right in response, thus increasing the value of K_w.

16.31. **Collect and Organize**

Given either the $[OH^-]$ or $[H^+]$ for a solution, we are asked to calculate the pH and pOH and determine whether the solution is acidic, basic, or neutral.

Analyze

To calculate the pH or the pOH from the $[H^+]$ or $[OH^-]$, respectively, we use

$$pH = -\log[H^+]$$
$$pOH = -\log[OH^-]$$

To find the pOH from the pH and vice versa, we use the relationship

$$pH + pOH = 14$$

If the pH of a solution is less than 7, the solution is acidic. If the pH is equal to 7, the solution is neutral. If the pH is greater than 7, the solution is basic.

Solve

(a) $pH = -\log(3.45 \times 10^{-8}) = 7.462$
$pOH = 14 - pH = 6.538$
This solution is basic.

(b) pH = $-\log(2.0 \times 10^{-5}) = 4.70$

pOH = $14 - pH = 9.30$

This solution is acidic.

(c) pH = $-\log(7.0 \times 10^{-8}) = 7.15$

pOH = $14 - pH = 6.85$

This solution is basic.

(d) pOH = $-\log(8.56 \times 10^{-4}) = 3.068$

pH = $14 - pOH = 10.932$

This solution is basic.

Think About It

When determining how many significant figures to include in your answers when computing the pH or pOH, remember that the first number in the pH or pOH gives the location of the decimal point. The significant digits, therefore, follow after the decimal point.

16.33. Collect and Organize

Given the concentration of a strong acid or base, we are asked to determine either the $[OH^-]$ or $[H^+]$ for a solution.

Analyze

The product of $[H^+]$ and $[OH^-]$ is equal to K_w, 1.00×10^{-14} at 25°C and 1 atm.

$$K_w = [H^+][OH^-] = 1.00 \times 10^{-14}$$

To find $[H^+]$ or $[OH^-]$ we divide K_w by the provided quantity

$$\left[H^+\right] = \frac{K_w}{\left[OH^-\right]} \quad \text{and} \quad \left[OH^-\right] = \frac{K_w}{\left[H^+\right]}$$

Solve

(a) $\left[H^+\right] = \dfrac{1.00 \times 10^{-14}}{8.42 \times 10^{-4}} = 1.19 \times 10^{-11} \ M$

(b) $[OH^-] = 2 \times [Ca(OH)_2] = 2 \times (3.97 \times 10^{-5} \ M) = 7.94 \times 10^{-5} \ M$

$\left[H^+\right] = \dfrac{1.00 \times 10^{-14}}{7.94 \times 10^{-5}} = 1.26 \times 10^{-10} \ M$

(c) $\left[OH^-\right] = \dfrac{1.00 \times 10^{-14}}{4.51 \times 10^{-3}} = 2.22 \times 10^{-12} \ M$

(d) $\left[OH^-\right] = \dfrac{1.00 \times 10^{-14}}{6.92 \times 10^{-5}} = 1.45 \times 10^{-10} \ M$

Think About It

At temperatures other than 25°C, the value of K_w changes, so the values of $[H^+]$ and $[OH^-]$ will differ from those calculated in this question.

16.35. Collect and Organize

Given the concentration of various strong acids and bases, we are asked to calculate the pH and pOH of the resulting solutions.

Analyze

To calculate the pH or the pOH from the $[H^+]$ or $[OH^-]$, respectively, we use

$$pH = -\log[H^+]$$
$$pOH = -\log[OH^-]$$

To find the pOH from the pH and vice versa, we use the relationship

$$pH + pOH = 14$$

Since the acids and bases described in the problem are all strong, we know that the acid concentration is equal to $[H^+]$ and the base concentration is equal to $[OH^-]$. In the case of the mixtures described in (c) and (d), we are to assume that acid and base will neutralize, forming a solution with a concentration equal to the difference between the molarity of the two solutions, and assuming no dilution. For (c), the reaction that occurs is

$$HCl(aq) + NaOH(aq) \rightarrow H_2O(\ell) + NaCl(aq)$$

Since the mixture is 2:1 HCl:NaOH, we can assume that $[H^+] = 2 \times (0.0125\ M) - (0.0125\ M) = 0.0125\ M$, and $[NaOH] = 0\ M$ after mixing.
For (d), the reaction that occurs is

$$H_2SO_4(aq) + 2\ KOH(aq) \rightarrow 2\ H_2O(\ell) + K_2SO_4(aq)$$

Since the mixture is 3:1 H_2SO_4:KOH, and H_2SO_4 is a diprotic acid, we can assume that $[H^+] = 6 \times (0.0125\ M) - (0.0125\ M) = 0.0625\ M$, and $[KOH] = 0\ M$ after mixing.

Solve
(a) $pH = -\log(0.155) = 0.810$
$pOH = 14 - 0.810 = 13.190$
(b) $pH = -\log(0.00500) = 2.301$
$pOH = 14 - 2.301 = 11.699$
(c) $pH = -\log(0.0125) = 1.903$
$pOH = 14 - 1.903 = 12.097$
(d) $pH = -\log(0.0625) = 1.204$
$pOH = 14 - 1.200 = 12.796$

Think About It
The acid in your stomach is fairly strong and the HCl is produced by parietal cells to break down food. Your stomach is protected in turn by epithelial cells that secrete a bicarbonate solution that neutralizes the acid and forms a coating to protect the stomach's tissues.

16.37. Collect and Organize
Given a solution that is $1.33 \times 10^{-9}\ M$ in LiOH, we are to calculate its pH.

Analyze
Since LiOH is a strong base, we know that for this solution the $[OH^-]$ from the LiOH is $1.33 \times 10^{-9}\ M$. The calculated pH from this value, however, gives a pH of less than 7, which would mean that this solution is acidic:

$$pOH = -\log(1.33 \times 10^{-9}) = 8.88$$
$$pH = 14 - 8.88 = 5.12$$

This cannot be, since we have added a base to water! This is a case where the autoionization of water (where $[H^+] = 1 \times 10^{-7}\ M$) is important. We therefore have to derive an equation to solve for the pH of this solution.

In the solution of LiOH in water, we must have an overall charge of zero:

$$[H^+] = [OH^-] - [Li^+] \quad \text{or} \quad [OH^-] = [H^+] + [Li^+]$$

This relationship puts $[OH^-]$ in terms of $[H^+]$. We also know that the concentration of Li^+ in solution is equal to the concentration of LiOH originally dissolved in the solution since LiOH is a strong base:

$$[Li^+] = [LiOH]_{initial}$$

Combining the two equations we get

$$[OH^-] = [H^+] - [LiOH]_{initial}$$

In the K_w expression we can substitute in for $[OH^-]$:

$$K_w = [H^+] \times [OH^-]$$
$$K_w = [H^+] \times ([H^+] - [LiOH]_{initial})$$

Rearranging this equation, we obtain the quadratic equation

$$[H^+]^2 - ([H^+] \times [LiOH]_{initial}) - K_w = 0$$

where for this problem $[LiOH]_{initial} = 1.33 \times 10^{-9}$ M and $K_w = 1.0 \times 10^{-14}$.

Solve

$$[H^+]^2 - ([H^+] \times 1.33 \times 10^{-9}) - 1.0 \times 10^{-14} = 0$$

Solving by the quadratic equation gives $[H^+] = 1.0067 \times 10^{-7}$ M for a pH $= -\log 1.0067 \times 10^{-7} = 6.997$.

Think About It

This answer is very close to the answer predicted using $[H^+]$ from the autoionization of water alone (pH = 7.000). This tells us that for solutions of a base approximately 100 times more dilute than the concentrations predicted for autoionization, we may approximate using K_w values alone.

16.39. Collect and Organize

For one-molar solutions of CH_3COOH, HNO_2, $HClO$, and HCl, we are to rank these in order of decreasing concentration of H^+, and increasing acid strength. For this we need the K_a values for each acid.

Analyze

The K_a values for these acids are listed in Tables 16.1 and 16.2. The greater the value of the K_a, the greater the concentration of H^+ in the solution of the acid. We note that HCl is a strong acid.

Solve

In order of decreasing $[H^+]$: $HCl > HNO_2 > CH_3COOH > HClO$.
In order of smallest K_a (weakest acid) to largest K_a (strongest acid): $HClO < CH_3COOH < HNO_2 < HCl$.

Think About It

For the weak acids in this series, there is a wide range in their acidities (about 10,000-fold comparing their K_a values).

16.41. Collect and Organize

We are to explain why the electrical conductivity of 1.0 M $NaNO_2$ is much better than that of 1.0 M HNO_2.

Analyze

Solutions with a larger concentration of dissolved ions conduct electricity better than those with lower concentrations of dissolved ions.

Solve

$NaNO_2$ is completely soluble in water, separating into Na^+ and NO_2^- ions, each in 1.0 M concentration for a total ion concentration of 2.0 M. HNO_2, however, only weakly dissociates in water:

$$HNO_2(aq) \rightleftharpoons NO_2^-(aq) + H^+(aq)$$

and so produces just slightly greater than 1.0 M ions in solution. $NaNO_2$, therefore, with more dissolved ions in solution, is a better conductor of electricity.

Think About It

NO_2^- reacts with water to a small extent:

$$NO_2^-(aq) + H_2O(\ell) \rightleftharpoons HNO_2(aq) + OH^-(aq)$$

However, this produces another ion, OH^-, and so there is no net change in the number of ions, and our analysis that a solution of 1.0 M $NaNO_2$ is 2.0 M in ions is still correct.

16.43. Collect and Organize

The formula for hydrofluoric acid is HF. From this we can write the mass action expression for this weak acid.

Analyze

The general form of the mass action expression for weak acids, based on $HA \rightleftharpoons A^- + H^+$, is

$$K_a = \frac{[H^+][A^-]}{[HA]}$$

Solve

$$HF(aq) \rightleftharpoons F^-(aq) + H^+(aq)$$

$$K_a = \frac{[H^+][F^-]}{[HF]}$$

Think About It

The hydrofluoric acid is transferring a proton to water in this equation, so an equivalent expression is

$$HF(aq) + H_2O\,(\ell) \rightleftharpoons F^-(aq) + H_3O^+(aq)$$

$$K_a = \frac{[H_3O^+][F^-]}{[HF]}$$

16.45. **Collect and Organize**

Given that the K_a of alanine is less when it is dissolved in ethanol than it is when dissolved in water, we are to determine which solvent ionizes alanine to the larger extent and which solvent is the stronger Brønsted–Lowry base.

Analyze

The larger the value of K_a, the greater the extent to which a substance has ionized. The solvent in which an acid ionizes the most must be the strongest Brønsted–Lowry base toward that acid.

Solve

(a) Because K_a for alanine is greater in water than in ethanol, alanine in water is ionized to a greater extent than in ethanol.

(b) Water is the stronger base for alanine compared to ethanol because alanine is ionized to a greater extent in water.

Think About It

This question demonstrates that acid–base strengths can depend on the basicity of the solvent.

16.47. **Collect and Organize**

Given that CH_3NH_2 is slightly basic in water we can write the equation describing its reaction with water to identify which species in the reaction is the Brønsted–Lowry acid and which is the base.

Analyze

Acting as a base, CH_3NH_2 accepts H^+ from surrounding water molecules.

Solve

The reaction describing the basicity of CH_3NH_2 is

$$CH_3NH_2(aq) + H_2O\,(\ell) \rightleftharpoons CH_3NH_3^+(aq) + OH^-(aq)$$

In this reaction, H_2O is the acid and CH_3NH_2 is the base.

Think About It

In another solvent, such as diethylamine, methylamine may act as an acid:

$$CH_3NH_2 + (CH_3CH_2)_2NH \rightleftharpoons CH_3NH^- + (CH_3CH_2)_2NH_2^+$$

This occurs because diethylamine ($K_b = 8.6 \times 10^{-4}$) is a stronger base than methylamine ($K_b = 4.4 \times 10^{-4}$).

16.49. **Collect and Organize**

Using the measured percent ionization of 1.00 M lactic acid of 2.94%, we are to calculate the K_a of this weak acid.

Analyze

The equilibrium equation from Appendix 5 that describes the ionization of lactic acid is

$$CH_3CHOHCOOH(aq) \rightleftharpoons H^+(aq) + CH_3CHOHCOO^-(aq)$$

We can set up an ICE table to solve this problem, where $x = 1.00\ M \times 0.0294 = 0.0294$. The K_a expression is

$$K_a = \frac{[H^+][CH_3CHOHCOO^-]}{[CH_3CHOHCOOH]}$$

Solve

	[CH_3CHOHCOOH]	[H^+]	[CH_3CHOHCOO^-]
Initial	1.00	0	0
Change	$-x = -0.0294$	$+x = +0.0294$	$+x = +0.0294$
Equilibrium	0.9706	0.0294	0.0294

$$K_a = \frac{(0.0294)^2}{0.9706} = 8.91 \times 10^{-4}$$

Think About It

Compare this with the K_a at 25°C value of 1.4×10^{-4} listed in Appendix 5. The difference may be attributable to a temperature difference because body temperature is 37°C.

16.51. **Collect and Organize**

We are given the $[H^+]$ for an equilibrium solution of an unknown acid with an initial concentration of 0.250 M. From this information we can calculate the degree of ionization and K_a for this weak acid.

Analyze

The equilibrium expression for the unknown acid is

$$HA(aq) \rightleftharpoons H^+(aq) + A^-(aq)$$

with a K_a expression of

$$K_a = \frac{[H^+][A^-]}{[HA]}$$

We can set up an ICE table to show how this weak acid ionizes; for this acid, $[H^+]$ at equilibrium is $4.07 \times 10^{-3}\ M$.

Solve

	[HA]	[H^+]	[A^-]
Initial	0.250	0	0
Change	$-x = -0.00407$	$+x = +0.00407$	$+x = +0.00407$
Equilibrium	0.24593	0.00407	0.00407

$$K_a = \frac{(0.00407)^2}{0.24593} = 6.74 \times 10^{-5}$$

$$\text{Degree of ionization} = \frac{[H^+]_{eq}}{[HA]_{initial}} = \frac{0.00407\ M}{0.250\ M} = 0.0163 \text{ or } 1.63\%$$

Think About It

In this problem we can use the $[H^+]$ as equivalent to x in the ICE table, enabling us to calculate both the K_a and the degree of ionization for the acid.

16.53. Collect and Organize

Given that formic acid has $K_a = 1.8 \times 10^{-4}$, we can calculate the pH of a 0.060 M aqueous solution of this weak acid.

Analyze

To solve this problem we set up an ICE table where the initial concentration of formic acid, HCOOH, is 0.060 M. We let x be the amount of formic acid that ionizes. The equilibrium and K_a expressions are

$$HCOOH(aq) \rightleftharpoons H^+(aq) + HCOO^-(aq)$$

$$K_a = \frac{[H^+][HCOO^-]}{[HCOOH]} = 1.8 \times 10^{-4}$$

The pH can be calculated through the equation $pH = -\log[H^+]$.

Solve

	[HCOOH]	[H⁺]	[HCOO⁻]
Initial	0.060	0	0
Change	$-x$	$+x$	$+x$
Equilibrium	$0.060 - x$	x	x

$$1.8 \times 10^{-4} = \frac{x^2}{0.060 - x} \approx \frac{x^2}{0.060}$$

$$x = 3.29 \times 10^{-3}$$

We should first check the simplifying assumption we made above before calculating the pH.

$$\frac{3.29 \times 10^{-3}}{0.060} \times 100 = 5.5\%$$

This is a little over 5%, so technically we should solve this by the quadratic equation (which we do below). However, if we allow this simplifying assumption to be valid, the pH of the solution would be calculated by

$$pH = -\log 3.29 \times 10^{-3} = 2.48$$

Now, solving by the quadratic equation gives

$$1.8 \times 10^{-4}(0.060 - x) = x^2$$
$$x^2 + 1.8 \times 10^{-4}x - 1.08 \times 10^{-5} = 0$$
$$x = 3.20 \times 10^{-3}$$
$$pH = -\log 3.20 \times 10^{-3} = 2.49$$

Think About It

The difference between the pH values when making the simplifying assumption and solving the equation exactly using the quadratic equation is $2.49 - 2.48 = 0.01$, which is fairly small.

16.55. Collect and Organize

By comparing the pH of rain in a weather system in the Midwest of 5.02 with the pH of the rain in that same system when it reached New England of 4.66, we can calculate how much more acidic the rain in New England was.

Analyze

We want to find the ratio

$$\frac{[H^+]_{New England}}{[H^+]_{Midwest}}$$

Because $pH = -\log[H^+]$, the $[H^+] = 1 \times 10^{-pH}$. Therefore,

$$\frac{[H^+]_{New England}}{[H^+]_{Midwest}} = \frac{1 \times 10^{-4.66}}{1 \times 10^{-5.02}}$$

Solve

$$\frac{[H^+]_{New\ England}}{[H^+]_{Midwest}} = \frac{1 \times 10^{-4.66}}{1 \times 10^{-5.02}} = \frac{2.19 \times 10^{-5}\ M}{9.55 \times 10^{-6}\ M} = 2.3$$

The rain in New England in this weather system was 2.3 times more acidic than the rain in the Midwest.

Think About It

One of the causes of the acidity of the rain in New England is the coal-burning electricity power plants in the Midwest, which expel SO_2 and SO_3 into the air, making H_2SO_3 and H_2SO_4.

16.57. Collect and Organize

We are asked to compare the basicity of aminoethanol and ethylamine, then calculate the pH and [OH⁻] for two solutions of aminoethanol.

Analyze

(a) The amine with the smaller pK_b value will be the stronger base. For aminoethanol

$$pK_b = -\log(K_b) = -\log(3.1 \times 10^{-5}) = 4.51$$

(b) To solve this problem we set up an ICE table where the initial concentration of aminoethanol is 1.67×10^{-2} *M*. Let x be the amount of aminoethanol acid that ionizes. We are asked to solve for the pH of this solution.

(c) To solve this problem we set up an ICE table where the initial concentration of aminoethanol is 4.25×10^{-4} *M*. Let x be the amount of aminoethanol acid that ionizes. We are asked to find [OH⁻] for this solution.

The equilibrium and K_b expressions for (b) and (c) are

$$H_2O\,(\ell) + HOCH_2CH_2NH_2(aq) \rightleftharpoons HOCH_2CH_2NH_3^+(aq) + OH^-(aq)$$

$$K_b = \frac{[HOCH_2CH_2NH_3^+][OH^-]}{[HOCH_2CH_2NH_2]} = 3.1 \times 10^{-5}$$

The pH can be calculated through the equations pOH = $-\log[OH^-]$ and pH = 14 – pOH. In both cases, we must check whether we can assume that x is small relative to the initial concentrations, or else we must use the quadratic equation to solve for the concentration of OH⁻ at equilibrium. The quadratic equation is

$$x = \frac{-b \pm \sqrt{b^2 - 4ac}}{2a}$$

Solve

(a) The pK_b for aminoethanol (4.51) is larger than the pK_b for ethylamine (3.36). Aminoethanol is a weaker base than ethylamine.

(b)

	$[HOCH_2CH_2NH_2]$	$[HOCH_2CH_2NH_3^+]$	$[OH^-]$
Initial	1.67×10^{-2}	0	0
Change	$-x$	$+x$	$+x$
Equilibrium	$1.67 \times 10^{-2} - x$	x	x

$$K_b = 3.1 \times 10^{-5} = \frac{x^2}{1.67 \times 10^{-2} - x} \approx \frac{x^2}{1.67 \times 10^{-2}}$$

$$x = \left[OH^-\right] = 7.195 \times 10^{-4}\ M$$

We should first check the simplifying assumption we made above before calculating the pH.

$$\frac{7.195 \times 10^{-4}}{1.67 \times 10^{-2}} \times 100\% = 4.3\%$$

The approximation is valid. The pOH and pH are

$$pOH = -\log(7.195 \times 10^{-4}\ M) = 3.143$$

$$pH = 14 - 3.143 = 10.857$$

(c)

	$[HOCH_2CH_2NH_2]$	$[HOCH_2CH_2NH_3^+]$	$[OH^-]$
Initial	4.25×10^{-4}	0	0
Change	$-x$	$+x$	$+x$
Equilibrium	$4.25 \times 10^{-4} - x$	x	x

$$K_b = 3.1 \times 10^{-5} = \frac{x^2}{4.25 \times 10^{-4} - x} \approx \frac{x^2}{4.25 \times 10^{-4}}$$

$$x = \left[OH^-\right] = 1.148 \times 10^{-4} \, M$$

We should first check the simplifying assumption we made above before calculating the pH.

$$\frac{1.148 \times 10^{-4}}{4.25 \times 10^{-4}} \times 100\% = 27\%$$

The approximation is far from valid! We will need to solve for the concentration of OH^- using the quadratic equation:

$$3.1 \times 10^{-5}(4.25 \times 10^{-4} - x) = x^2$$
$$x^2 + 3.1 \times 10^{-5}x - 1.32 \times 10^{-8} = 0$$
$$x = [OH^-] = 1.00 \times 10^{-4}$$

Think About It
If we had continued to use the approximation for (c), we would obtain a pH of 10.06, whereas when using the quadratic equation, we obtain a pH of 10.00. Though the difference in pH units is small, the distinction is important.

16.59. Collect and Organize
Given the pK_b of morphine, a weak base, we are asked to calculate the pH of a 0.115 M solution. Given the pK_a of the conjugate acid of codeine ($pK_a = 8.21$), we are to calculate the pH of a 3.42×10^{-4} M solution of codeine, a weak base.

Analyze
(a) From the pK_b we can calculate the K_b

$$K_b = 1 \times 10^{-pK_b} = 1 \times 10^{-5.79} = 1.62 \times 10^{-6}$$

The equilibrium and K_b expressions for the ionization of morphine are

$$\text{Morphine}(aq) + H_2O(\ell) \rightleftharpoons \text{morphineH}^+(aq) + OH^-(aq)$$

$$K_b = \frac{[OH^-][\text{morphineH}^+]}{[\text{morphine}]} = 1.62 \times 10^{-6}$$

(b) We first need to determine the value of the K_b from the pK_a for codeine:

$$K_a = 1 \times 10^{-pK_a} = 1 \times 10^{-8.21} = 6.17 \times 10^{-9}$$

$$K_b = \frac{K_w}{K_a} = \frac{1 \times 10^{-14}}{6.17 \times 10^{-9}} = 1.62 \times 10^{-6}$$

The equilibrium and K_b expressions for the ionization of codeine are

$$\text{Codeine}(aq) + H_2O(\ell) \rightleftharpoons \text{codeineH}^+(aq) + OH^-(aq)$$

$$K_b = \frac{[OH^-][\text{codeineH}^+]}{[\text{codeine}]} = 1.62 \times 10^{-6}$$

We can set up an ICE table to solve this problem, where x is the amount of morphine or codeine that ionizes. By solving for x we can calculate $[OH^-]$, from which we can determine the pOH and pH:

$$pOH = -\log[OH^-]$$
$$pH = 14 - pOH$$

Solve

(a)

	[Morphine]	[OH⁻]	[MorphineH⁺]
Initial	0.115	0	0
Change	−x	+x	+x
Equilibrium	0.115 − x	x	x

$$1.62 \times 10^{-6} = \frac{x^2}{0.115 - x} \approx \frac{x^2}{0.115}$$

$$x = 4.32 \times 10^{-4}$$

Checking the simplifying assumption shows that our simplifying assumption is valid:

$$\frac{4.32 \times 10^{-4}}{0.115} \times 100 = 0.38\%$$

From [OH⁻] we can calculate the pH:

$$pOH = -\log 4.32 \times 10^{-4} = 3.365$$
$$pH = 14 - 3.365 = 10.635$$

(b)

	[Codeine]	[OH⁻]	[CodeineH⁺]
Initial	3.42 × 10⁻⁴	0	0
Change	−x	+x	+x
Equilibrium	3.42 × 10⁻⁴ − x	x	x

$$1.62 \times 10^{-6} = \frac{x^2}{3.42 \times 10^{-4} - x} \approx \frac{x^2}{3.42 \times 10^{-4}}$$

$$x = 2.35 \times 10^{-5}$$

Checking the simplifying assumption shows that our simplifying assumption is not valid:

$$\frac{2.35 \times 10^{-5}}{3.42 \times 10^{-4}} \times 100 = 6.9\%$$

so we must solve this by the quadratic equation:

$$1.62 \times 10^{-6}(3.42 \times 10^{-4} - x) = x^2$$
$$x^2 + 1.62 \times 10^{-6}x - 5.54 \times 10^{-10} = 0$$
$$x = 2.27 \times 10^{-5}$$
$$pOH = -\log 2.27 \times 10^{-5} = 4.644$$
$$pH = 14 - 4.644 = 9.356$$

Think About It

Comparing the pK_b of codeine to that of morphine, we see that codeine and morphine are identical in their basicity. The difference in pH is due to the difference in the initial concentration of each base.

16.61. Collect and Organize

We are to explain why $K_{a_1} > K_{a_2} > K_{a_3}$ for H_3PO_4.

Analyze

The equations describing these acid dissociation constants are as follows:

$$H_3PO_4(aq) \rightleftharpoons H_2PO_4^-(aq) + H^+(aq) \qquad K_{a_1}$$
$$H_2PO_4^-(aq) \rightleftharpoons HPO_4^{2-}(aq) + H^+(aq) \qquad K_{a_2}$$
$$HPO_4^{2-}(aq) \rightleftharpoons PO_4^{3-}(aq) + H^+(aq) \qquad K_{a_3}$$

Solve

With each successive ionization, it becomes more difficult to remove H^+ from a species that is negatively charged. Therefore it is harder to remove H^+ from HPO_4^{2-} than from $H_2PO_4^-$, and harder to remove H^+ from $H_2PO_4^-$ than from H_3PO_4. This is reflected in decreasing K_a values as H_3PO_4 is ionized.

Think About It
From Appendix 5 we can compare the K_a values for phosphoric acid: $K_{a_1} = 7.11 \times 10^{-3}$, $K_{a_2} = 6.32 \times 10^{-8}$, $K_{a_3} = 4.5 \times 10^{-13}$. These span ten orders of magnitude.

16.63. **Collect and Organize**
We have to use the K_{a_2} of H_2SO_4 to calculate the pH of a solution of 0.300 M H_2SO_4.

Analyze
The first H^+ is completely removed from the H_2SO_4, and the initial concentrations of the species in solution before the second ionization are $[H^+] = 0.300$ M, $[H_2SO_4] = 0.0$ M, and $[HSO_4^-] = 0.300$ M. The equation describing the second ionization is

$$HSO_4^-(aq) \rightleftharpoons SO_4^{2-}(aq) + H^+(aq)$$

$$K_a = \frac{[H^+][SO_4^{2-}]}{[HSO_4^-]} = 1.2 \times 10^{-2}$$

Solve

	$[HSO_4^-]$	$[H^+]$	$[SO_4^{2-}]$
Initial	0.300	0.300	0
Change	$-x$	$+x$	$+x$
Equilibrium	$0.300 - x$	$0.300 + x$	x

Plugging equilibrium concentrations into the K_a expression gives

$$1.2 \times 10^{-2} = \frac{x(0.300 + x)}{0.300 - x}$$

Solving this by the quadratic equation gives

$$x^2 + 0.312x - 0.0036 = 0$$
$$x = 0.0111$$
$$[H^+] = 0.300 + 0.0111 = 0.311$$
$$pH = -\log 0.311 = 0.51$$

Think About It
If we did not take into consideration the second ionization of H_2SO_4 we would have underestimated the acidity of the solution by 0.016 pH units.

16.65. **Collect and Organize**
We are to calculate the pH of a 0.250 M solution of ascorbic acid. The values of K_{a_1} and K_{a_2} for this weak diprotic acid are 1.0×10^{-5} and 5×10^{-12}, respectively.

Analyze
Because the K_{a_2} is so much smaller than the K_{a_1} for ascorbic acid, we can say that the second ionization contributes little to the $[H^+]$ in the solution. Therefore, we can solve this by examining only the first ionization.

Solve

	[Ascorbic acid]	[Ascorbate$^-$]	$[H^+]$
Initial	0.250	0	0
Change	$-x$	$+x$	$+x$
Equilibrium	$0.250 - x$	x	x

$$1.0 \times 10^{-5} = \frac{x^2}{0.250 - x} \approx \frac{x^2}{0.250}$$
$$x = 1.58 \times 10^{-3}$$

Checking the simplifying assumption shows that it is valid:

$$\frac{1.58 \times 10^{-3}}{0.250} \times 100 = 0.63\%$$

The pH is found from the [H$^+$]:

$$\text{pH} = -\log 1.58 \times 10^{-3} = 2.80$$

Think About It
This solution is about as acidic as vinegar.

16.67. **Collect and Organize**
Given a 0.00100 M solution of nicotine and the K_{b_1} and K_{b_2} (from Appendix 5) for this weak dibasic compound, we are to calculate the pH.

Analyze
The K_{b_2} (1.3×10^{-11}) is so much smaller than the K_{b_1} (1.0×10^{-6}) for nicotine that we may ignore the contribution of the second ionization of nicotine to the [OH$^-$] in the solution. We can therefore solve this problem by examining only the first ionization.

Solve

	[Nicotine]	[NicotineH$^+$]	[OH$^-$]
Initial	0.00100	0	0
Change	$-x$	$+x$	$+x$
Equilibrium	$0.00100 - x$	x	x

$$1.0 \times 10^{-6} = \frac{x^2}{0.00100 - x} \approx \frac{x^2}{0.00100}$$

$$x = 3.16 \times 10^{-5}$$

Checking the simplifying assumption shows that it is valid:

$$\frac{3.16 \times 10^{-5}}{0.00100} \times 100 = 3.2\%$$

We can calculate the pOH and pH from the [OH$^-$]:

$$\text{pOH} = -\log 3.16 \times 10^{-5} = 4.500$$
$$\text{pH} = 14 - 4.500 = 9.500$$

Think About It
The relatively high pH of this dilute solution of nicotine shows that this compound is a fairly strong weak base.

16.69. **Collect and Organize**
Given a 0.01050 M solution of quinine and the K_{b_1} and K_{b_2} (from Appendix 5) for this weak dibasic compound, we are to calculate the pH.

Analyze
The K_{b_2} (1.4×10^{-9}) is so much smaller than the K_{b_1} (3.3×10^{-6}) for quinine that we may ignore the contribution of the second ionization of quinine to the [OH$^-$] in the solution. We can therefore solve this problem by examining only the first ionization.

Solve

	[Quinine]	[QuinineH$^+$]	[OH$^-$]
Initial	0.01050	0	0
Change	$-x$	$+x$	$+x$
Equilibrium	$0.01050 - x$	x	x

$$3.3 \times 10^{-6} = \frac{x^2}{0.01050 - x} \approx \frac{x^2}{0.01050}$$
$$x = 1.86 \times 10^{-4}$$

Checking the simplifying assumption shows that it is valid:

$$\frac{1.86 \times 10^{-4}}{0.01050} \times 100 = 1.8\%$$

The pOH and pH can be calculated from the $[OH^-]$:

$$pOH = -\log 1.86 \times 10^{-4} = 3.730$$
$$pH = 14 - 3.730 = 10.270$$

Think About It

Quinine has a very complicated molecular structure that includes aromatic rings, an alcohol, an amine, an alkene, and an ether as functional groups.

16.71. Collect and Organize

We are asked to explain why H_2SO_4 is a stronger acid (greater K_{a_1}) than H_2SeO_4.

Analyze

The only difference in these acids is the central atom. Sulfur and selenium belong to the same group in the periodic table. These elements differ in size and electronegativity.

Solve

Sulfur is more electronegative than selenium. This higher electronegativity on the sulfur atom stabilizes the anion HSO_4^- more than the anion $HSeO_4^-$. Therefore, H_2SO_4 is a stronger acid.

Think About It

We would expect this trend to continue, so we predict that H_2TeO_4 is a weaker acid than H_2SeO_4.

16.73. Collect and Organize

We are to predict which acid of a pair is stronger.

Analyze

The more oxygen atoms bound to the central atom and the more electronegative the central atom (X) in the acid, the more acidic is the compound.

Solve

(a) H_2SO_3 is a stronger acid than H_2SeO_3.
(b) H_2SeO_4 is a stronger acid than H_2SeO_3.

Think About It

The presence of oxygen atoms bound to the central atom in an oxyacid can have a dramatic effect on acidity.

16.75. Collect and Organize

We are asked to explain why the pK_b of ethanolamine is greater than the pK_b of ethylamine.

Analyze

A larger pK_b corresponds to a smaller K_b, meaning that ethanolamine is a weaker base than ethylamine. Drawing these two bases, we can compare structural differences that may account for the difference in basicity. The O atom of the OH group in ethanolamine is more electronegative than the corresponding H atom in ethylamine, resulting in a greater pull of electron density away from the nitrogen lone pair on ethanolamine.

Electron density pulled
away from N lone pair

Solve

Because of the decreased electron density near the nitrogen lone pair, ethanolamine is less basic than ethylamine.

Think About It

It is easy to see how this structural change could affect the basicity of the amine when considering the Lewis definition of acids and bases. A Lewis base donates an electron pair to a Lewis acid, so pulling electron density away from the nitrogen lone pair will render the amine less able to donate, and so, less basic.

16.77. **Collect and Organize**

Given the pK_a values of conjugate acids of three pyridine derivatives where methyl groups are added, we are to determine if more methyl groups increase or decrease the basicity of pyridine.

Analyze

The pK_a is a measure of the conjugate acid's acidity. From the equation

$$pK_a = -\log K_a$$

we see that the larger the pK_a, the weaker the acid. The weaker the acid, the stronger the conjugate base.

Solve

As methyl groups are added the pK_a increases, so the acidity decreases. If the acidity decreases, the basicity of the conjugate base increases. Therefore, more methyl groups on the parent pyridine increases the base strength.

Think About It

Our prediction is true. The K_b values of the pyridine bases show that adding more methyl groups leads to increased basicity.

$K_b = 1.51 \times 10^{-9}$ $K_b = 9.77 \times 10^{-8}$ $K_b = 2.69 \times 10^{-7}$

16.79. **Collect and Organize**

Of the three salts given, we are to determine which gives an acidic solution when dissolved in water.

Analyze

To give an acidic solution, the cation of the salt must donate a proton to water without the anion reacting with water, or if the anion hydrolyzes, then the pK_a of the cation must be lower than the pK_b of the anion.

Solve

Both NH_4^+ and CH_3COO^- of ammonium acetate hydrolyze:

$$NH_4^+(aq) + H_2O(\ell) \rightleftharpoons NH_3(aq) + H_3O^+(aq) \qquad pK_a = 9.25$$

$$CH_3COO^-(aq) + H_2O(\ell) \rightleftharpoons CH_3COOH(aq) + OH^-(aq) \qquad pK_b = 9.25$$

Because $pK_a = pK_b$, this salt's solution is neutral.

Only NH_4^+ of ammonium nitrate hydrolyzes. This gives an acidic solution:

$$NH_4^+(aq) + H_2O(\ell) \rightleftharpoons NH_3(aq) + H_3O^+(aq)$$

Only $HCOO^-$ of sodium formate hydrolyzes. This gives a basic solution:

$$HCOO^-(aq) + H_2O(\ell) \rightleftharpoons HCOOH(aq) + OH^-(aq)$$

Therefore, of the three salts, only ammonium nitrate dissolves to give an acidic solution.

Think About It

Remember that neither Na^+ nor NO_3^- hydrolyze because they would form either a strong base or a strong acid, which are always 100% ionized.

$$Na^+(aq) + H_2O(\ell) \not\longrightarrow NaOH(aq) + H^+(aq)$$

$$NO_3^-(aq) + H_2O(\ell) \not\longrightarrow HNO_3(aq) + OH^-(aq)$$

16.81. **Collect and Organize**

We consider why lemon juice is used to reduce the fishy odor due to the presence of $(CH_3)_3N$ in not-so-fresh seafood.

Analyze

Trimethylamine is a weak base, and lemon juice contains citric acid, which is a weak acid.

Solve

The citric acid in the lemon juice neutralizes the volatile trimethylamine to make a nonvolatile dissolved salt that neutralizes the fishy odor:

$$HOC(CH_2)_2(COOH)_3(aq) + (CH_3)_3N(aq) \rightleftharpoons HOC(CH_2)_2(COOH)_2COO^-(aq) + (CH_3)_3NH^+(aq)$$

Think About It

Because the pK_b of trimethylamine of 4.19 and the pK_a of citric acid of 3.13 are lower than the pK_a of trimethylammonium (9.81) and the pK_b of citrate (10.87), this equilibrium lies to the right, favoring the products.

16.83. **Collect and Organize**

Given $K_a = 2.1 \times 10^{-11}$ for the conjugate acid of saccharin, we are asked to calculate the value of pK_b for saccharin.

Analyze

From the K_a we can calculate the pK_a:

$$pK_a = -\log K_a$$

From the pK_a we can calculate the pK_b because $pK_w = 14.00$ at 25°C

$$pK_b = 14.00 - pK_a$$

Solve

$$pK_a = -\log(2.1 \times 10^{-11}) = 10.68$$
$$pK_b = 14.00 - 10.68 = 3.32$$

Think About It

Alternatively, we could calculate the K_b from K_a using

$$K_b = \frac{K_w}{K_a}$$

and then calculate pK_b using

$$pK_b = -\log K_b$$

16.85. Collect and Organize

From Appendix 5 we know that the K_a of HF is 6.8×10^{-4}. Using this we are to calculate the pH of a solution that is $0.00339\ M$ in NaF.

Analyze

When NaF dissolves in water the F^- ion hydrolyzes to give a basic solution:

$$F^-(aq) + H_2O\,(\ell) \rightleftharpoons HF(aq) + OH^-(aq)$$

The K_b for this reaction is

$$K_b = \frac{K_w}{K_a} = \frac{1.0 \times 10^{-14}}{6.8 \times 10^{-4}} = 1.47 \times 10^{-11}$$

We can solve for $[OH^-]$ using an ICE table and then compute the pH.

Solve

	$[F^-]$	$[HF]$	$[OH^-]$
Initial	0.00339	0	0
Change	$-x$	$+x$	$+x$
Equilibrium	$0.00339 - x$	x	x

$$1.47 \times 10^{-11} = \frac{x^2}{0.00339 - x} \approx \frac{x^2}{0.00339}$$

$$x = 2.23 \times 10^{-7}$$

Checking the simplifying assumption shows that it is valid:

$$\frac{2.23 \times 10^{-7}}{0.00339} \times 100 = 0.0066\%$$

The pH is calculated from the $[OH^-]$:

$$pOH = -\log 2.23 \times 10^{-7} = 6.65$$
$$pH = 14 - 6.65 = 7.35$$

Think About It

Because HF is a moderately strong weak acid, its conjugate base, F^-, is a fairly weak base.

16.87. Collect and Organize

We are to explain why a solution of CH_3COOH with CH_3COONa is a better pH buffer than a solution containing NaCl and HCl.

Analyze

A buffer is composed of a weak acid and its conjugate base.

Solve

Because the weak acid CH_3COOH is combined with its weak conjugate base, CH_3COO^-, this buffer can absorb added H^+ or OH^-. The other mixture, HCl with Cl^-, is a strong acid paired with its very, very weak conjugate base. This pairing cannot absorb added H^+ or OH^-.

Think About It

It is a key idea that for the acid–base pair in a buffer system both conjugates be weak.

16.89. Collect and Organize

We are asked to describe the effect of adding NaF to a buffer containing 1.0 M HF and 0.050 M F$^-$.

Analyze

The equation describing the acid ionization equilibrium for this reaction is

$$HF(aq) \rightleftharpoons F^-(aq) + H^+(aq)$$

Na$^+$ is a neutral cation that does not affect this equilibrium. The pH of this buffer may be determined using the Henderson–Hasselbalch equation

$$pH = pK_a + \log\left(\frac{[F^-]}{[HF]}\right)$$

Solve

Increasing [F$^-$] will make the logarithmic term of the Henderson–Hasselbalch equation larger and raise the pH of the buffer. The buffer capacity will also increase as F$^-$ is added to the solution until we reach [F$^-$] = 1.0 M.

Think About It

The initial pH of this buffer is 1.87, so it is not surprising that adding the conjugate base would cause the pH to rise. Without knowing the quantity of NaF added, we cannot calculate the final pH. If enough NaF were added so that [HF] = [F$^-$], the pH of the solution would be 3.17 (the pK_a for HF).

16.91. Collect and Organize

We are asked to consider two buffers prepared using different weak acids and their corresponding conjugate bases. If the solutions are all prepared using the same concentrations, we may consider the capacity of each buffer.

Analyze

The pH of a buffer is dictated by the ratio of acid and conjugate base, as well as by the pK_a of the weak acid employed. The capacity of a buffer is dictated by the concentration of acid and base components of the buffer.

Solve

Since the concentrations of acid and base components are identical between the two buffers, both should have identical capacities.

Think About It

The pH of each buffer will differ, provided the pK_a values for the two acids differ.

16.93. Collect and Organize

We are asked to choose a buffer system that would maintain a pH of 3.0, starting from equal concentrations of an acid and its conjugate base.

Analyze

A buffer will generally be effective when the desired pH lies within one unit of the pK_a of the acid employed.

Solve

Iodoacetic acid has a pK_a of 3.12. A mixture of CH$_2$ICOOH(aq) and CH$_2$ICOO$^-$(aq) will produce the desired buffer.

Think About It

The pH of this buffer will need to be adjusted by employing more acid than base. A ratio of 0.76 moles CH$_2$ICOO$^-$ to 1 mole CH$_2$ICOOH will produce a buffer with pH 3.0.

16.95. Collect and Organize

For a buffer that is 0.244 M in acetic acid and 0.122 M in sodium acetate, we can use the Henderson–Hasselbalch equation to calculate the pH of this buffer at 25°C ($K_a = 1.76 \times 10^{-5}$) and at 0°C ($K_a = 1.64 \times 10^{-5}$).

Analyze

The Henderson–Hasselbalch equation is

$$pH = pK_a + \log \frac{[base]}{[acid]}$$

For this problem, [base] = [acetate] = 0.122 *M* and [acid] = [acetic acid] = 0.244 *M*. Because the K_a values are different for the two temperatures, the pH of these solutions will differ.

Solve

At 25°C, $pH = -\log(1.76 \times 10^{-5}) + \log \dfrac{0.122}{0.244} = 4.453$

At 0°C, $pH = -\log(1.64 \times 10^{-5}) + \log \dfrac{0.122}{0.244} = 4.484$

Think About It

At the lower temperature the pH of this buffer is less acidic than at the higher temperature.

16.97. ### Collect and Organize

For a buffer that is 0.225 *M* in both HPO_4^{2-} and PO_4^{3-} (with the K_a of HPO_4^{2-} equal to 4.5×10^{-13}), we can use the Henderson–Hasselbalch equation to calculate the pH of the buffer. From the pH we can also calculate the pOH.

Analyze

The Henderson–Hasselbalch equation is

$$pH = pK_a + \log \frac{[base]}{[acid]}$$

For this problem, [base] = [acid] = 0.225 *M*.

Solve

$$pH = -\log(4.5 \times 10^{-13}) + \log \frac{0.225M}{0.225M} = 12.35$$
$$pOH = 14 - pH = 14 - 12.35 = 1.65$$

Think About It

Notice that when the concentrations of the acid and its conjugate base are equal, pH = pK_a because log 1 = 0.

16.99. ### Collect and Organize

We are asked to describe the preparation of 100 mL of a buffer at pH = 3.00 from iodoacetic acid (with a pK_a of 3.12) and sodium iodoacetate.

Analyze

Using the Henderson–Hasselbalch equation, we may determine the molar ratio of iodoacetic acid and sodium iodoacetate required to form this buffer. The Henderson–Hasselbalch equation is

$$pH = pK_a + \log \frac{[base]}{[acid]}$$

The molar masses of iodoacetic acid and sodium iodoacetate are 185.96 g/mol and 207.94 g/mol, respectively.

Solve

The ratio of base to acid is

$$3.00 = 3.12 + \log \frac{\left[CH_2ICOO^- \right]}{\left[CH_2ICOOH \right]}$$

$$\frac{\left[CH_2ICOO^- \right]}{\left[CH_2ICOOH \right]} = 0.759$$

Assuming 0.100 mol CH_2ICOOH, the quantities one could use to prepare such a buffer are

$$0.100 \text{ mol } CH_2ICOOH \times \frac{185.96 \text{ g } CH_2ICOOH}{1 \text{ mol } CH_2ICOOH} = 18.6 \text{ g } CH_2ICOOH$$

$$0.0759 \text{ mol } CH_2ICOONa \times \frac{207.94 \text{ g } CH_2ICOONa}{1 \text{ mol } CH_2ICOONa} = 15.8 \text{ g } CH_2ICOONa$$

The buffer could be prepared by mixing 18.6 g of iodoacetic acid and 15.8 g sodium iodoacetic acid, and adding water until the solution volume is 100 mL.

Think About It
Other masses of iodoacetic acid and sodium iodoacetate could be used, provided they have the same molar ratio determined above.

16.101. **Collect and Organize**
We are asked to determine which solution will have the lowest pH when buffers are prepared from formic acid, hydrofluoric acid, and hydrocyanic acid.

Analyze
We may assume that approximately equivalent concentrations of each acid and its conjugate base are used in preparing these solutions. A low pH will result when a stronger acid (more acidic compound) is used to prepare the buffer. Stronger acids have lower pK_a values. The pK_a values for formic acid, hydrofluoric acid, and hydrocyanic acid are 3.75, 3.17, and 9.21, respectively.

Solve
The weak acid with the lowest pK_a is hydrofluoric acid, so the buffer with the lowest pH is likely to be that derived from a mixture of hydrofluoric acid and the fluoride anion.

Think About It
Without knowing the exact composition of each buffer, it is impossible to tell the exact pH of each buffer solution. It is possible that a buffer derived from hydrofluoric acid could have a pH as high as 4.17 or as low as 2.17.

16.103. **Collect and Organize**
We are asked to consider the effect of diluting a buffer solution by a factor of ½ on the pH and capacity of the buffer.

Analyze
The pH of a buffer is dictated by the ratio of acid and conjugate base, as well as by the pK_a of the weak acid employed. The capacity of a buffer is dictated by the concentration of acid and base components of the buffer.

Solve
Since the ratio of acidic and basic buffer components remains the same, as does the pK_a of the acid, the pH of the buffer will remain the same. The concentrations of both the acid and base components will decrease upon dilution, reducing the buffer capacity.

Think About It
Diluting the buffer with any solution containing a spectator ion will have the same effect as diluting it with distilled water. Provided no reactions occur, only common ions (the acid or its conjugate base) will increase the buffer capacity or alter the pH of the solution.

16.105. **Collect and Organize**
We can use the Henderson–Hasselbalch equation to determine the pH of a solution prepared by mixing a volume of 0.05 M NH_3 with an equal volume of 0.025 M HCl.

Analyze

Mixing equal volumes of solutions dilutes them both. Therefore, after mixing and before reaction, $[NH_3]$ = 0.025 M and [HCl] = 0.0125 M in the combined solution. When HCl reacts with NH_3, NH_4^+ is produced according to the equation

$$NH_3(aq) + H^+(aq) \rightarrow NH_4^+(aq)$$

Stoichiometrically, this would give $[NH_3]$ = 0.0125 M and $[NH_4^+]$ = 0.0125 M after complete reaction with H^+. Because this is a solution of an acid (NH_4^+) and its conjugate base (NH_3), we can use the Henderson–Hasselbalch equation to calculate the pH of the solution. To do so we need the pK_a of NH_4^+ (9.25) from Appendix 5.

Solve

$$pH = pK_a + \log \frac{[NH_3]}{[NH_4^+]} = 9.25 + \log \frac{0.0125}{0.0125} = 9.25$$

Think About It

Because the HCl added to the NH_3 in this solution converts exactly half of the NH_3 to NH_4^+, the ratio of the concentrations equals 1 and the pH of the solution equals the pK_a of NH_4^+.

16.107. **Collect and Organize**

Given that the pH of a solution is 3.15 when 0.020 M conjugate base (A^-) and 0.080 M conjugate acid (HA) are mixed, we may determine the pK_a for the conjugate acid in question.

Analyze

Because the concentrations are within a factor of 10 of one another, we can use the Henderson–Hasselbalch equation to determine the pK_a of the weak acid in this buffer.

Solve

$$pH = pK_a + \log \frac{[A^-]}{[HA]}$$

$$pK_a = pH - \log \frac{[A^-]}{[HA]} = 3.15 - \log \frac{0.020M}{0.080M} = 3.75$$

Think About It

The pK_b of the conjugate base (A^-) of this acid is $14 - 3.75 = 10.25$.

16.109. **Collect and Organize**

We are to compare the pH of 1.00 L of a buffer that is 0.120 M in HNO_2 and 0.150 M in $NaNO_2$ before and after 1.00 mL of 12.0 M HCl is added.

Analyze

We can use the Henderson–Hasselbalch equation in both cases. After the addition of HCl, however, the amounts (calculated in moles) of HNO_2 and NO_2^- have to be adjusted before using the Henderson–Hasselbalch equation. The pK_a of HNO_2 is 3.40.

Solve

Without added HCl the pH of the buffer solution is

$$pH = 3.40 + \log \frac{0.150}{0.120} = 3.50$$

Because we have 1.00 L of the buffer solution, we originally have 0.120 mol HNO_2 and 0.150 mol NO_2^- in the solution. Adding 1.00 mL of 12.0 M HCl adds

$$1.00 \text{ mL} \times \frac{12.0 \text{ mol}}{1000 \text{ mL}} = 0.0120 \text{ mol H}^+$$

This will increase the moles of HNO_2 to 0.120 mol + 0.0120 mol = 0.132 mol and decrease the moles of NO_2^- to 0.150 mol – 0.0120 mol = 0.138 mol. Because the volume of the solution is 1.00 L, the concentrations of these species are $[HNO_2]$ = 0.132 M and $[NO_2^-]$ = 0.138 M. Using the Henderson–Hasselbalch equation to calculate the pH gives

$$pH = 3.40 + \log\frac{0.138}{0.132} = 3.42$$

Think About It
The pH of the buffer changed very little. The change in pH of 1.00 L of water after adding 0.0120 mol of H^+ would be from 7.00 to 1.92.

16.111. **Collect and Organize**
We are to describe the difference between a titration curve for a strong acid and a weak acid with a strong base.

Analyze
A strong acid is completely ionized in aqueous solution, whereas a weak acid is only partially hydrolyzed. This affects the pH of the solution at the start of the titration for equal concentrations of the acids. The equivalence, or end point, of the titration is where equal moles of OH^- have been added to the acid. The species formed at the end point for a strong acid and weak acid differ. This affects the pH of the solution at the equivalence point.

Solve
The weak acid titration curve has an initial pH that is higher (less acidic) than that of an equimolar solution of a strong acid (lower pH, more acidic). The pH at the equivalence point in the titration of a strong acid is 7.00 because the species formed in the titration are water and a nonhydrolyzing anion, such as Cl^-:
$$HCl(aq) + OH^-(aq) \rightarrow H_2O(\ell) + Cl^-(aq)$$
The pH at the equivalence point for a weak acid is basic because of the formation of a hydrolyzing anion, such as NO_2^- in the titration of HNO_2:
$$HNO_2(aq) + OH^-(aq) \rightarrow NO_2^-(aq) + H_2O(\ell)$$
$$NO_2^-(aq) + H_2O(\ell) \rightleftharpoons HNO_2(aq) + OH^-(aq)$$

Think About It
Titration of a weak acid always gives an end point of pH > 7.

16.113. **Collect and Organize**
We are asked whether the pH at the equivalence point is the same for the titration of all weak acids with strong base.

Analyze
When a weak acid is titrated, the species present in solution at the equivalence point is the conjugate base of the weak acid. This conjugate base is a weak base and will hydrolyze in water according to the equation
$$A^-(aq) + H_2O(\ell) \rightleftharpoons HA(aq) + OH^-(aq)$$

Solve
No. Because the extent to which A^- (the conjugate base) hydrolyzes depends on the base strength of A^-, the pH at the equivalence point in the titration for weak acids is not expected to be the same.

Think About It
Likewise, the pH values at the equivalence point for weak bases differ based on the strength of the conjugate acid.

16.115. **Collect and Organize**
We are to calculate the pH along various points of the titration curve when 25.0 mL of 0.100 M acetic acid ($K_a = 1.76 \times 10^{-5}$) is titrated with 0.125 M NaOH.

Analyze

For each step of the titration we have to consider the moles of NaOH added that react with the moles of CH_3COOH initially present:

$$25.0 \text{ mL} \times \frac{0.100 \text{ mol}}{1000 \text{ mL}} = 0.00250 \text{ mol } CH_3COOH$$

For each point in the titration curve, the moles of OH^- (from NaOH) added are as follows:

$$10.0 \text{ mL} \times \frac{0.125 \text{ mol}}{1000 \text{ mL}} = 0.00125 \text{ mol } OH^-$$

$$20.0 \text{ mL} \times \frac{0.125 \text{ mol}}{1000 \text{ mL}} = 0.00250 \text{ mol } OH^-$$

$$30.0 \text{ mL} \times \frac{0.125 \text{ mol}}{1000 \text{ mL}} = 0.00375 \text{ mol } OH^-$$

The OH^- reacts with the CH_3COOH in solution to give CH_3COO^-, the conjugate base of acetic acid. Our strategy for the problem is to first react as much of the added OH^- with acetic acid as possible and then use the equilibrium K_a expression to calculate the pH:

$$CH_3COOH(aq) \rightleftharpoons CH_3COO^-(aq) + H^+(aq)$$

$$K_a = \frac{[CH_3COO^-][H^+]}{[CH_3COOH]}$$

Or we can use the equivalent equilibrium K_b expression:

$$CH_3COO^-(aq) + H_2O(\ell) \rightleftharpoons CH_3COOH(aq) + OH^-(aq)$$

$$K_b = \frac{[CH_3COOH][OH^-]}{[CH_3COO^-]}$$

For that calculation we have to be careful to determine the molarity of the species in solution by remembering that the volume in a titration increases through the addition of the titrant.

Solve

When 10.0 mL of OH^- are added, the 0.00125 mol of OH^- reacts with the 0.00250 mol of CH_3COOH to produce 0.00125 mol of CH_3COO^- and leave 0.00125 mol of CH_3COOH unreacted. Since the total volume of the solution is now 25.0 + 10.0 = 35.0 mL, the molarity of these species is

$$\frac{0.00125 \text{ mol}}{0.0350 \text{ L}} = 0.0357 \text{ } M$$

Using an ICE table to calculate the pH of this solution gives the following:

	$[CH_3COOH]$	$[CH_3COO^-]$	$[H^+]$
Initial	0.0357	0.0357	0
Change	$-x$	$+x$	$+x$
Equilibrium	$0.0357 - x$	$0.0357 + x$	x

$$1.76 \times 10^{-5} = \frac{x(0.0357 + x)}{0.0357 - x} \approx \frac{x(0.0357)}{0.0357}$$

$$x = 1.76 \times 10^{-5}$$

The simplifying assumption is valid (0.05%) so

$$\text{pH} = -\log 1.76 \times 10^{-5} = 4.754$$

When 20.0 mL of OH^- are added, we have added an equal number of moles of OH^- as there are moles of CH_3COOH initially present. This reaction produces 0.00250 mol of CH_3COO^- so it makes sense here to use the K_b expression to calculate the pH of the solution. Since the total volume of the solution is now 45.0 mL, the molarity of these species is

$$\frac{0.00250 \text{ mol}}{0.0450 \text{ L}} = 0.0556 \text{ } M$$

Using an ICE table to calculate the pH of this solution gives the following:

	$[CH_3COO^-]$	$[CH_3COOH]$	$[OH^-]$
Initial	0.0556	0	0
Change	$-x$	$+x$	$+x$
Equilibrium	$0.0556 - x$	x	x

$$K_b = \frac{K_w}{K_a} = \frac{1 \times 10^{-14}}{1.76 \times 10^{-5}} = 5.68 \times 10^{-10} = \frac{x^2}{0.0556 - x} \approx \frac{x^2}{0.0556}$$

$$x = 5.62 \times 10^{-6}$$

The simplifying assumption is valid (0.01%) so

$$pOH = -\log 5.62 \times 10^{-6} = 5.250$$
$$pH = 14 - 5.250 = 8.750$$

When 30.0 mL of OH^- are added, we convert all of the CH_3COOH to 0.00250 mol CH_3COO^- and have 0.00125 mol OH^- remaining. Since the total volume of the solution is now 55.0 mL, the molarity of these species is

$$\frac{0.00250 \text{ mol}}{0.0550 \text{ L}} = 0.0455 \ M \ CH_3COO^-$$

$$\frac{0.00125 \text{ mol}}{0.0550 \text{ L}} = 0.0227 \ M \ OH^-$$

Using an ICE table to calculate the pH of this solution gives the following:

	$[CH_3COO^-]$	$[CH_3COOH]$	$[OH^-]$
Initial	0.0455	0	0.0227
Change	$-x$	$+x$	$+x$
Equilibrium	$0.0455 - x$	x	$0.0227 + x$

$$5.68 \times 10^{-10} = \frac{x(0.0227 + x)}{0.0455 - x} \approx \frac{x(0.0227)}{0.0455}$$

$$x = 1.14 \times 10^{-9}$$

The simplifying assumption is valid so

$$pOH = -\log 0.0227 = 1.644$$
$$pH = 14 - 1.644 = 12.356$$

Think About It
When exactly half the moles of strong base are added in the titration of a weak acid (as in this problem where 10 mL of the titrant were added), this point is the midpoint of the titration. Notice that at this point $pH = pK_a$ of the weak acid.

16.117. Collect and Organize
In the titration of a 100.00 mL NH_3 solution with 0.1145 M HCl, it takes 22.35 mL to reach the equivalence point. From this information we are to calculate the concentration of ammonia in the solution.

Analyze
Because at the equivalence point the moles of HCl added as a titrant equal the moles of NH_3 in the solution, we can calculate the amount of NH_3 through

$$\text{Moles } NH_3 = \text{mL HCl used as titrant} \times \text{molarity of HCl solution} \times \frac{1 \text{ mol } NH_3}{1 \text{ mol HCl}}$$

Because we know the volume of the original solution of NH_3, the molarity of the sample is

$$\frac{\text{mol } NH_3}{\text{volume of sample in L}}$$

Solve

$$\text{Moles } NH_3 = 22.35 \text{ mL HCl} \times \frac{0.1145 \text{ mol}}{1000 \text{ mL}} \times \frac{1 \text{ mol } NH_3}{1 \text{ mol HCl}} = 2.559 \times 10^{-3} \text{ mol}$$

$$\text{Molarity of NH}_3 = \frac{2.559 \times 10^{-3} \text{ mol}}{0.100 \text{ L}} = 0.02559 \ M$$

Think About It
Remember that at the equivalence point, what is equal is the moles, not the volume nor the concentration of acid and base.

16.119. **Collect and Organize**
We are to calculate the volume of 0.0100 *M* HCl required to titrate 250 mL of 0.0100 *M* Na_2CO_3 to the first equivalence point.

Analyze
Before setting out to do a lot of calculations here, let's stop and think. The concentration of the titrant (HCl) is equal to the concentration of the base we are titrating! We don't need, therefore, to do any calculations because the same volume of HCl is needed to neutralize the base.

Solve
Titration of 250 mL of 0.0100 *M* Na_2CO_3 to the first equivalence point requires 250 mL of HCl.

Think About It
To reach the second end point for the Na_2CO_3 solution, 500 mL of the HCl titrant would be required.

16.121. **Collect and Organize**
In comparing the titration of a weak acid in which the amount of NaOH titrant needed to reach the equivalence point is double that for another titration, we are asked what the pH halfway to the equivalence point is for the second titration if the pH at that point for the first titration is 4.44.

Analyze
The midpoint is where half of the weak acid has been converted into its conjugate base. It is at this point where [acid] = [conjugate base] that the pH = pK_a.

Solve
The pH at the midpoint in the second titration would be the same as it is in the first titration, 4.44.

Think About It
Certainly the volume of the added titrant at which the midpoint is reached for the second titration would be twice that as in the first titration, but the pH at those points would be the same.

16.123. **Collect and Organize**
To sketch the titration curve for the titration of 50.0 mL of a solution of the weak acid HNO_2 with 1.00 *M* NaOH, we must use the information provided to calculate the pH of the HNO_2 solution before any NaOH is added and at the equivalence point.

Analyze
The initial pH of 0.250 *M* HNO_2 is calculated from the value of $K_a = 4.0 \times 10^{-4}$ after setting up an ICE table where the initial concentrations of H^+ and NO_2^- are 0.00 *M*. To calculate the pH of the solution at the equivalence point we need to recognize that at that point all of the HNO_2 is converted to NO_2^-. We use K_b, therefore, for the ICE table calculation. We also need to take into account the added volume of the solution when NaOH titrant is added.

Solve

The initial pH of 0.250 M HNO$_2$ is found as follows:

	[HNO$_2$]	[H$^+$]	[NO$_2^-$]
Initial	0.250	0.00	0.00
Change	$-x$	$+x$	$+x$
Equilibrium	$0.250 - x$	x	x

$$K_a = 4.0 \times 10^{-4} = \frac{x^2}{0.250 - x} \approx \frac{x^2}{0.250}$$

$$x = 0.0100$$

$$pH = -\log 0.0100 = 2.00$$

At the equivalence point, moles of OH$^-$ added = moles of HNO$_2$ present in the initial solution:

$$\text{Moles HNO}_2 = 50.0 \text{ mL} \times \frac{0.250 \text{ mol}}{1000 \text{ mL}} = 0.0125 \text{ mol}$$

Because all the HNO$_2$ is converted to NO$_2^-$ at the equivalence point, the amount of NO$_2^-$ at equivalence point is initially 0.0125 mol. The NO$_2^-$ produced hydrolyzes by the equation

$$NO_2^-(aq) + H_2O(\ell) \rightleftharpoons HNO_2(aq) + OH^-(aq)$$

to give a slightly basic solution at the equivalence point.

We need to determine the volume of NaOH titrant added so that we can use the [NO$_2^-$] in an ICE table to calculate the pH at the equivalence point. The 0.0125 mol of HNO$_2$ requires 0.0125 mol of OH$^-$ to neutralize it. The volume of NaOH needed to provide this 0.0125 mol OH$^-$ is

$$0.0125 \text{ mol OH}^- \times \frac{1 \text{ mol NaOH}}{1 \text{ mol OH}^-} \times \frac{1000 \text{ mL}}{1.00 \text{ mol}} = 12.5 \text{ mL}$$

The total volume of the solution at the equivalence point is 12.5 + 50.0 mL = 62.5 mL. The molarity of NO$_2^-$ at the equivalence point is

$$\frac{0.0125 \text{ mol NO}_2^-}{0.0625 \text{ L}} = 0.200 \text{ } M$$

	[NO$_2^-$]	[HNO$_2$]	[OH$^-$]
Initial	0.200	0	0
Change	$-x$	$+x$	$+x$
Equilibrium	$0.200 - x$	x	x

This is a K_b expression so we must calculate K_b from K_a:

$$K_b = \frac{K_w}{K_a} = \frac{1 \times 10^{-14}}{4.0 \times 10^{-4}} = 2.5 \times 10^{-11}$$

Solving the equilibrium constant expression for x gives

$$K_b = 2.5 \times 10^{-11} = \frac{x^2}{0.200 - x} \approx \frac{x^2}{0.200}$$

$$x = 2.24 \times 10^{-6}$$

$$pOH = -\log 2.24 \times 10^{-6} = 5.65$$

$$pH = 14 - 5.65 = 8.35$$

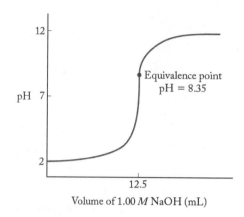

Volume of 1.00 M NaOH (mL)

Think About It

Note in the titration curve that there is a relatively flat region before the equivalence point. Here the pH does not change much despite the addition of more and more titrant. This is often called the *buffer region*.

16.125. **Collect and Organize**

For the titration of quinine, a dibasic malaria drug, we need to calculate the initial pH and the position of each of the equivalence points in order to sketch the titration curve.

Analyze

Because the second base ionization constant (K_{b_2} = 1.35 × 10⁻⁹) is much smaller than the first (K_{b_1} = 3.31 × 10⁻⁶), we can ignore it in calculating the initial pH of the solution. Because the concentration of the quinine and HCl titrant are equal (both solutions are 0.100 M), we know that the first equivalence point will be at the point where 40.0 mL of HCl have been added and the second equivalence point is at 80.0 mL of HCl added.

Solve

The initial pH is calculated as follows:

	[Quinine]	[QuinineH⁺]	[OH⁻]
Initial	0.100	0	0
Change	−x	+x	+x
Equilibrium	0.100 − x	x	x

$$K_{b_1} = 3.31 \times 10^{-6} = \frac{x^2}{0.100 - x} \approx \frac{x^2}{0.100}$$

$$x = 5.75 \times 10^{-4}$$

$$\text{pOH} = -\log 5.75 \times 10^{-4} = 3.240$$

$$\text{pH} = 14 - 3.240 = 10.760$$

The first equivalence point is where 40.0 mL of HCl have been added to give

$$40.0 \text{ mL HCl} \times \frac{0.100 \text{ mol HCl}}{1000 \text{ mL}} \times \frac{1 \text{ mol quinineH}^+}{1 \text{ mol HCl}} = 4.00 \times 10^{-3} \text{ mol quinineH}^+$$

The molarity of quinineH⁺ is

$$\frac{4.00 \times 10^{-3} \text{ mol}}{0.080 \text{ L}} = 0.0500 \ M$$

At this point quinineH⁺ can act both as an acid and as a base:

$$\text{QuinineH}^+(aq) + H_2O(\ell) \rightleftharpoons \text{quinineH}_2^{2+}(aq) + OH^-(aq) \qquad K_{b_2} = 1.35 \times 10^{-9}$$

$$\text{QuinineH}^+(aq) \rightleftharpoons \text{quinine}(aq) + H^+(aq) \qquad K_{a_2} = \frac{K_w}{K_{b_2}} = \frac{1 \times 10^{-14}}{3.31 \times 10^{-6}} = 3.02 \times 10^{-9}$$

This calculation is beyond the scope of general chemistry study, but we can see that $K_{a_2} > K_{b_1}$ and so we expect this equivalence point to be slightly acidic. In fact, it is, at approximately pH 6.8.

The second equivalence point is where 80.0 mL of HCl have been added. This solution contains 4.00×10^{-3} mol quinineH$_2{}^{2+}$ with a molarity of

$$\frac{4.00 \times 10^{-3} \text{ mol}}{0.120 \text{ L}} = 0.0333 \; M$$

because the total volume is now 40.0 mL quinine solution + 80.0 mL HCl titrant. The pH at this equivalence point is dependent on the equilibrium

$$\text{QuinineH}_2{}^{2+}(aq) \rightleftharpoons \text{quinineH}^+(aq) + \text{H}^+(aq) \quad K_{a_1} = \frac{K_w}{K_{b_2}} = \frac{1 \times 10^{-14}}{1.35 \times 10^{-9}} = 7.41 \times 10^{-6}$$

The pH at this equivalence point is as follows:

	[QuinineH$_2{}^{2+}$]	[QuinineH$^+$]	[H$^+$]
Initial	0.0333	0	0
Change	$-x$	$+x$	$+x$
Equilibrium	$0.0333 - x$	x	x

$$K_{a_1} = 7.41 \times 10^{-6} = \frac{x^2}{0.0333 - x} \approx \frac{x^2}{0.0333}$$

$$x = 4.97 \times 10^{-4}$$

$$\text{pH} = -\log 4.97 \times 10^{-4} = 3.304$$

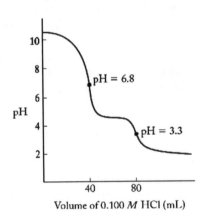

Think About It

Notice that at the two midpoints (40 mL and 80 mL HCl added) that the pH equals pK_{a_2} (8.52) and pK_{a_1} (5.13).

16.127. Collect and Organize / Analyze

We compare the terms *molar solubility* and *solubility product*.

Solve

Molar solubility is the moles of a substance that dissolves in one liter of solvent. The solubility product is the equilibrium constant for the dissolution of a substance.

Think About It

Solubility has units of grams or moles per volume of solution, but, like other equilibrium constants, the solubility product is unitless.

16.129. Collect and Organize

By comparing the K_{sp} values of $MgCO_3$, $CaCO_3$, and $SrCO_3$, we can identify which cation (Mg^{2+}, Ca^{2+}, or Sr^{2+}) precipitates first as carbonate mineral.

Analyze

From Appendix 5, the K_{sp} values are

$MgCO_3$ $K_{sp} = 6.8 \times 10^{-6}$
$CaCO_3$ $K_{sp} = 5.0 \times 10^{-9}$
$SrCO_3$ $K_{sp} = 5.6 \times 10^{-10}$

Solve

Because $SrCO_3$ has the lowest K_{sp}, the cation Sr^{2+} precipitates first as a carbonate mineral.

Think About It

The order of solubility from least to most soluble for these carbonates is $SrCO_3 < CaCO_3 < MgCO_3$.

16.131. Collect and Organize

For $SrSO_4$, whose K_{sp} increases as the temperature increases, we are to determine whether the dissolution is exothermic or endothermic.

Analyze

We can include heat as a reactant (for an endothermic reaction) or as a product (for an exothermic reaction) and apply Le Châtelier's principle:

$$SrSO_4(s) + heat \rightleftharpoons Sr^{2+}(aq) + SO_4^{2-}(aq)$$
$$SrSO_4(s) \rightleftharpoons Sr^{2+}(aq) + SO_4^{2-}(aq) + heat$$

Solve

Applying Le Châtelier's principle, we see that the reaction shifts to the right and more $SrSO_4$ dissolves as the temperature is increased. The dissolution is endothermic.

Think About It

The opposite effect of temperature occurs for an exothermic dissolution: less solid dissolves as the temperature is increased.

16.133. Collect and Organize

By writing the equation for the dissolution of hydroxyapatite, we can explain why acidic substances erode tooth enamel.

Analyze

The solubility of hydroxyapatite is described by

$$Ca_5(PO_4)_3OH(s) \rightleftharpoons OH^-(aq) + 3\ PO_4^{3-}(aq) + 5\ Ca^{2+}(aq)$$

Solve

Acidic substances react with the OH^- released upon dissolution of hydroxyapatite. The equilibrium is shifted to the right, dissolving more hydroxyapatite.

Think About It

The equilibrium would be shifted in the opposite direction (to the left) in an alkaline environment.

16.135. Collect and Organize

Given the $[Ba^{2+}]$ in a saturated solution of $BaSO_4$ ($1.04 \times 10^{-5}\ M\ Ba^{2+}$), we are to calculate the value of K_{sp} for $BaSO_4$:

$$BaSO_4(s) \rightleftharpoons Ba^{2+}(aq) + SO_4^{2-}(aq)$$

Analyze

The K_{sp} expression is

$$K_{sp} = [Ba^{2+}][SO_4^{2-}]$$

Because for every mole of $BaSO_4$ that dissolves we get 1 mole of Ba^{2+} and 1 mole of SO_4^{2-}, the molarities of Ba^{2+} and SO_4^{2-} are the same for this saturated solution of $BaSO_4$.

Solve

$$K_{sp} = (1.04 \times 10^{-5})(1.04 \times 10^{-5}) = 1.08 \times 10^{-10}$$

Think About It

Because $BaSO_4$ is quite insoluble, we can add SO_4^{2-} to a solution of dissolved Ba^{2+} to quantitatively precipitate the barium out of solution. After weighing the dried precipitate we can then calculate how much Ba^{2+} was present in the original solution.

16.137. Collect and Organize

Given that $K_{sp} = 1.02 \times 10^{-6}$, we are to calculate $[Cu^+]$ and $[Cl^-]$ for a saturated solution of CuCl.

Analyze

The solubility equation and K_{sp} expression for CuCl are

$$CuCl(s) \rightleftharpoons Cu^+(aq) + Cl^-(aq) \qquad K_{sp} = [Cu^+][Cl^-]$$

The $[Cu^+] = [Cl^-]$ in this solution because for every particle of CuCl that dissolves we get one particle of Cu^+ and one particle of Cl^-.

Solve

Let $[Cu^+] = [Cl^-] = x$. The K_{sp} expression becomes

$$K_{sp} = 1.02 \times 10^{-6} = (x)(x)$$
$$x = 1.01 \times 10^{-3}$$

Therefore, $[Cu^+] = [Cl^-] = 1.01 \times 10^{-3} \ M$.

Think About It

We do not need to know how much CuCl is originally placed into the solution because it does not appear in the K_{sp} expression, being a pure solid.

16.139. Collect and Organize

Given the K_{sp} of $CaCO_3$ (9.9×10^{-9}), we are to calculate the solubility of this substance in units of grams per milliliter.

Analyze

The solubility equation and K_{sp} expression for $CaCO_3$ are

$$CaCO_3(s) \rightleftharpoons Ca^{2+}(aq) + CO_3^{2-}(aq) \qquad K_{sp} = [Ca^{2+}][CO_3^{2-}]$$

In this solution, $[Ca^{2+}] = [CO_3^{2-}]$ because for every $CaCO_3$ that dissolves one Ca^{2+} and one CO_3^{2-} are produced. We can then say that

$$K_{sp} = x^2$$

and we can solve for x, which is the molar solubility (mol/L) of $CaCO_3$. To convert this to grams per milliliter we multiply by the molar mass of $CaCO_3$ (100.09 g/mol) and divide by 1000 mL/L.

Solve

$$K_{sp} = 9.9 \times 10^{-9} = x^2$$
$$x = 9.95 \times 10^{-5}$$

$$\frac{9.95 \times 10^{-5} \text{ mol}}{1 \text{ L}} \times \frac{100.09 \text{ g}}{1 \text{ mol}} \times \frac{1 \text{ L}}{1000 \text{ mL}} = 9.96 \times 10^{-6} \text{ g/mL}$$

Think About It
The value of x that we calculate in the K_{sp} expression is the molar solubility of the solid because it is that amount ("x") that dissolves into the solution.

16.141. **Collect and Organize**
Using the K_{sp} for the dissolution of AgOH (1.52×10^{-8}) from Appendix 5, we are to calculate the pH of a saturated solution.

Analyze
The solubility equation and K_{sp} expression for AgOH are
$$AgOH(s) \rightleftharpoons Ag^+(aq) + OH^-(aq) \qquad K_{sp} = [Ag^+][OH^-]$$
Letting $[Ag^+] = [OH^-] = x$ (because the stoichiometry is 1:1), we can solve for x using the K_{sp} expression. The pH of the solution will then be
$$pH = 14 - (-\log x)$$

Solve
$$K_{sp} = 1.52 \times 10^{-8} = x^2$$
$$x = 1.233 \times 10^{-4}$$
$$pH = 14 - \left(-\log\ 1.233 \times 10^{-4}\right) = 10.091$$

Think About It
Even though the K_{sp} of AgOH is not high, this solution is quite basic.

16.143. **Collect and Organize**
Using the common-ion effect, we can determine in which 0.1 M solution (NaCl, Na_2CO_3, NaOH, or HCl) the most $CaCO_3$ dissolves.

Analyze
Whenever a common ion is already present in the solution, the $CaCO_3$ is less soluble. Any solution, therefore, with Ca^{2+} or CO_3^{2-} would have lower solubility of $CaCO_3$ compared to water. We also should look for solutions that might react with either Ca^{2+} or CO_3^{2-} and shift the solubility equilibrium to the right.

Solve
(a) NaCl(aq) has neither an ion common to $CaCO_3$ nor ions that react with either Ca^{2+} or CO_3^{2-}. $CaCO_3$ has the same solubility in this NaCl solution as in water.
(b) The 0.1 M Na_2CO_3 solution is 0.1 M in CO_3^{2-}. This decreases the solubility of $CaCO_3$.
(c) NaOH(aq) has neither an ion common to $CaCO_3$ nor reacts with either Ca^{2+} or CO_3^{2-}. $CaCO_3$ has the same solubility in this NaOH solution as water.
(d) A solution of HCl reacts with CO_3^{2-} to form H_2CO_3 which then decomposes to H_2O and CO_2. This reaction shifts the solubility equilibrium to the right, so more $CaCO_3$ dissolves.
The solution of (d) 0.1 M HCl dissolves the most $CaCO_3$.

Think About It
A higher concentration of acid dissolves even more $CaCO_3$ as the equilibrium shifts to the right:
$$CaCO_3(s) + 2\ H^+(aq) \rightleftharpoons Ca^{2+}(aq) + H_2CO_3(aq)$$

16.145. **Collect and Organize**
Given the average concentrations of SO_4^{-2} and Sr^{2+} in seawater (0.028 M and 9×10^{-5} M, respectively) and the K_{sp} of $SrSO_4$ (3.4×10^{-7}), we are to determine if the concentration of Sr^{2+} is controlled by the relative insolubility of $SrSO_4$.

Analyze
The solubility equation and K_{sp} expression for $SrSO_4$ are
$$SrSO_4(s) \rightleftharpoons Sr^{2+}(aq) + SO_4^{2-}(aq) \qquad K_{sp} = [Sr^{2+}][SO_4^{2-}]$$

Solve

$$K_{sp} = 3.4 \times 10^{-7} = [Sr^{2+}][SO_4^{2-}]$$
$$= [Sr^{2+}](0.028 \ M)$$
$$[Sr^{2+}] = 1.21 \times 10^{-5} \ M$$

This is the expected concentration of Sr^{2+} in seawater with a known $[SO_4^{2-}]$ of 0.028 M. This $[Sr^{2+}]$ is lower than the average $[Sr^{2+}]$ of 9×10^{-5} M, so some other process must be controlling the $[Sr^{2+}]$.

Think About It
As the $[SO_4^{2-}]$ increases, the solubility of $SrSO_4$ decreases because of the common-ion effect.

16.147. **Collect and Organize**
Given 125 mL solution that is $0.375M$ in $Ca(NO_3)_2$ we are asked whether CaF_2 will precipitate when 245 mL of a 0.255 M NaF solution is added.

Analyze
The K_{sp} for CaF_2 from Appendix 5 is 3.9×10^{-11}. When the two solutions are mixed, the total volume is 370 mL and the $[Ca^{2+}]$ and $[F^-]$ in the mixed solution is

$$125 \ mL \times 0.375 M = 370 \ mL \times [Ca^{2+}]$$
$$[Ca^{2+}] = 0.127 \ M$$
$$245 \ mL \times 0.255 M = 370 \ mL \times [F^-]$$
$$[F^-] = 0.169 \ M$$

From the K_{sp} expression

$$K_{sp} = [Ca^{2+}][F^-]^2 = 3.9 \times 10^{-11}$$

If the $[Ca^{2+}]_{initial} \times [F^-]^2_{initial} > K_{sp}$, then CaF_2 will precipitate.

Solve
$[Ca^{2+}]_{initial} \times [F^-]^2_{initial} = 0.127 \times (0.169)^2 = 3.63 \times 10^{-3}$. This is greater than the value of K_{sp} for CaF_2, so it will precipitate from the mixed solution.

Think About It
Because CaF_2 has a small solubility product constant, we expect that most of the ions will precipitate as CaF_2. In this solution the Ca^{2+} ions are in excess (0.0469 mol) compared to that of F^- (0.0625 mol), so the final solution will have only a small amount of F^- in solution.

16.149. **Collect and Organize**
We can use the values of K_{sp} for $PbBr_2$ and $PbSO_4$ to determine which anion is the first to precipitate when a 0.250 M solution of $Pb^{2+}(aq)$ is added to a solution that is 0.010 M in Br^- and SO_4^{2-}. We are then asked to calculate the concentration of the anion that precipitates first at the moment that the second ion starts to precipitate. This will be when the solution is saturated in the lead salt of the first anion to precipitate.

Analyze
The K_{sp} value shows that $PbSO_4$ has a smaller solubility product constant (1.8×10^{-8}) compared to that of $PbBr_2$ (6.6×10^{-6}).

Solve
(a) $PbSO_4$, with a smaller solubility product constant, will precipitate first from the solution, so the SO_4^{2-} anion will precipitate first.
(b) When Br^- begins to precipitate, the maximum amount of Pb^{2+} that could be present that will not cause Br^- to precipitate is

$$K_{sp} = 6.6 \times 10^{-6} = [Pb^{2+}][Br^-]^2$$

$$[Pb^{2+}] = \frac{6.6 \times 10^{-6}}{(0.010)^2} = 6.6 \times 10^{-2} \ M$$

The $[SO_4^{2-}]$ in the solution when $[Pb^{2+}] = 6.6 \times 10^{-2} \ M$ is

$$K_{sp} = 1.8 \times 10^{-8} = [Pb^{2+}][SO_4^{2-}]$$

$$[Pb^{2+}] = \frac{1.8 \times 10^{-8}}{6.6 \times 10^{-2}} = 2.7 \times 10^{-7} \ M$$

Think About It
We do not need an ICE table to solve this problem.

16.151. Collect and Organize
We are asked to describe the changes in bonding and intermolecular forces when the weak base CH_3NH_2 is dissolved in water.

Analyze
Methylamine is a gas, and when dissolved in water, the individual methylamine molecules are surrounded with water. CH_3NH_2 also reacts with water according to the equation

$$CH_3NH_2(aq) + H_2O(\ell) \rightleftharpoons CH_3NH_3^+(aq) + OH^-(aq)$$

Solve
The hydrogen bonds between some of the water molecules must break and re-form around the species CH_3NH_2. Also, the amine hydrolyzes and forms $CH_3NH_3^+$ and OH^-; thus, ion–dipole forces are added when these ions are surrounded by water molecules.

Think About It
Depending on the strength of the forces formed versus those broken, this dissolution may be either exothermic or endothermic.

16.153. Collect and Organize
We are to compare the structures of H_3PO_3 and H_3PO_4 and explain why the K_{a_1} values for these acids are similar.

Analyze
After drawing the Lewis structure and identifying which H atom is ionizable on H_3PO_3, we can compare that to the structure of H_3PO_4.

Solve
(a,b) Phosphorous acid has the Lewis structure

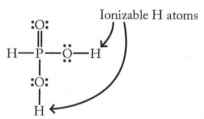

(c) Phosphoric acid has a similar structure with its ionizable H atoms also bonded to oxygen atoms, so it is not surprising that these two acids have similar values for K_{a_1}.

Think About It
Notice that the H atom bonded to P in H_3PO_3 is not ionizable. The electronegativity difference between P and H is not great enough to make this H atom acidic.

16.155. Collect and Organize

We are asked whether the pH of the solution changes when a cook adds more baking soda to water used in a recipe and to explain why or why not.

Analyze
Baking soda is a soluble sodium salt with the formula $NaHCO_3$. In solution this salt forms $Na^+(aq) + HCO_3^-$. Na^+ does not react with water but HCO_3^- does, which changes the pH of the solution.

Solve
Yes, the pH of the solution increases due to the increase in hydrolysis of HCO_3^- according to the equation

$$HCO_3^-(aq) + H_2O(\ell) \rightleftharpoons H_2CO_3(aq) + OH^-(aq)$$

Think About It
H_2CO_3 decomposes at baking temperatures to give $CO_2(g)$ and $H_2O(\ell)$.

16.157. Collect and Organize

Given the change in pH of a lake from 6.1 to 4.7 when 400 gallons of 18 M H_2SO_4 were added to the lake, we are to calculate the volume of the lake.

Analyze
To make this calculation easier, we can assume that we only ionize the first proton on H_2SO_4. First, we must calculate the moles of H_2SO_4 added to the lake and then determine the increase in the concentration of H^+ in the lake in going from pH 6.1 to pH 4.7. Knowing that we simply divide the moles of acid added by the molarity change of H^+ in the lake, we can obtain the size of the lake.

Solve
The amount of H_2SO_4 added is

$$400 \text{ gal} \times \frac{3.78 \text{ L}}{1 \text{ gal}} \times \frac{18 \text{ mol}}{L} = 27{,}216 \text{ mol}$$

The change in lake $[H^+]$ is

$$\text{Initial } [H^+] = 1 \times 10^{-6.1} = 7.94 \times 10^{-7} \, M$$
$$\text{Final } [H^+] = 1 \times 10^{-4.7} = 2.00 \times 10^{-5} \, M$$
$$\Delta[H^+] = 1.92 \times 10^{-5} \, M \text{ increase}$$

The size of the lake is

$$27{,}216 \text{ mol} \times \frac{1 \text{ L}}{1.92 \times 10^{-5} \text{ mol}} = 1.4 \times 10^9 \text{ L}$$

Think About It
This volume is equivalent to $1.4 \times 10^6 \text{ m}^3$. If the lake were 10 m deep, it would cover an area of $1.4 \times 10^5 \text{ m}^2$. If thought of as a square, that is 374 m on a side.

16.159. Collect and Organize

For the drug Zoloft we are to use Figure P16.159 to determine which form is the acid salt and whether aqueous solutions of Zoloft are acidic or basic.

Analyze
It is important to remember that this drug is sold as the HCl salt.

Solve

(a) The acid salt form has H on the amine ($R_2NH_2^+$) moiety with Cl^- as a counterion. This structure is shown on the right of Figure P16.159.

(b) Because this drug is sold as the HCl salt, solutions of this drug are acidic.

Think About It

Many drugs are sold as HCl salts to render the drugs more soluble in aqueous solution.

16.161. ## Collect and Organize

By examining the equilibrium reactions of HF in water and aqueous F^-, we are to determine the major species present at pH 7.00, calculate the equilibrium constant for the combination of the two equilibrium equations, and, finally, calculate the pH and $[HF_2^-]_{eq}$ when [HF] is 0.150 *M*.

Analyze

(a) By examining the values of the two equilibrium constants, and considering the concentrations of species in solution, we can predict which species is more likely to be present at pH 7.00, F^- or HF_2^-.

(b) The overall equation is the sum of the two equilibrium reactions, so the *K* for the combined reaction is the product of the two *K* values for the individual reactions.

(c) We can use an ICE table and the value of the overall *K* calculated in part b to determine the pH and $[HF_2^-]_{eq}$.

Solve

(a) Because the equilibrium constant of the reaction of F^- with HF is larger than the dissociation of HF, we might expect the most likely species to be HF_2^-. However, because HF is weak, the $[F^-]$ is low compared to that of water and so HF reacts with H_2O to form F^- as the major anionic species.

(b) $K_{overall} = K_a \times K = (1.1 \times 10^{-3}) \times (2.6 \times 10^{-1}) = 2.86 \times 10^{-4}$

(c) We must tackle this problem in two steps. First, we consider the hydrolysis of HF (*aq*).

	[HF]	$[H_3O^+]$	$[F^-]$
Initial	0.150	0	0
Change	$-x$	$+x$	$+x$
Equilibrium	$0.150 - x$	x	x

$$1.1 \times 10^{-3} = \frac{x^2}{(0.150 - x)}$$

$$x^2 + 1.1 \times 10^{-3} x - 1.65 \times 10^{-4} = 0$$

$$x = 0.0123$$

So $[F^-] = 0.0123$ *M* and [HF] = 0.138 *M*

Now, we consider the second equilibrium

	$[F^-]$	[HF]	$[HF_2^-]$
Initial	0.0123	0.138	0
Change	$-x$	$-x$	$+x$
Equilibrium	$0.0123 - x$	$0.138 - x$	x

$$0.26 = \frac{x}{(0.0123 - x)(0.138 - x)} \approx \frac{x}{(0.0123)(0.138)}$$

$$x = 4.4 \times 10^{-4}$$

Therefore,

$$pH = -\log(0.0123) = 1.91$$
$$[HF_2^-] = 4.4 \times 10^{-4} \ M$$

Think About It

Be careful in making simplifying assumptions. For the first equilibrium, we must solve using the quadratic equation.

16.163. Collect and Organize

Given the structure of Naproxen (Figure P16.163), we are to draw the structure of the sodium salt, explain whether a solution of the salt is acidic or basic, and explain why the salt is more soluble than Naproxen itself.

Analyze

The ionizable functional group in Naproxen is the carboxylic acid (–COOH) group.

Solve

(a)

(b) A solution of the salt of Naproxen is basic because the ionized –COO⁻ group reacts with water, giving –COOH + OH⁻:

(c) The salt is more soluble because it is charged and water molecules form stronger ion–dipole forces around the molecule compared to the dipole-induced dipole forces between the neutral molecule and water.

Think About It

Being soluble in water also helps to deliver the drug to the body.

16.165. Collect and Organize

For three reactions of nitrogen and sulfur compounds, we are to complete the equations.

Analyze

All three reactants are covalent compounds. The gases SO_3 and NO_2 are acid anhydrides, and will react with water to form mineral acids. NH_3 is oxidized in the presence of O_2 gas.

Solve

(a) $SO_3(g) + H_2O(\ell) \rightarrow H_2SO_4(\ell)$

(b) $3\,NO_2(g) + H_2O(\ell) \rightarrow 2\,HNO_3(\ell) + NO(g)$

(c) $4\,NH_3(g) + 5\,O_2(g) \rightarrow 4\,NO(g) + 6\,H_2O(g)$

Think About It

The first two reactions show how acids are produced from nonmetal oxides.

16.167. Collect and Organize

We are asked which steps of the Ostwald synthesis of nitric acid would have a higher yield at higher temperature.

Analyze

By Le Châtelier's principle the yield of a reaction increases as temperature increases for endothermic reactions.

Solve

The steps in the Ostwald process with their ΔH values calculated from the data in Appendix 4 are as follows:

$$4\,NH_3(g) + 5\,O_2(g) \rightleftharpoons 4\,NO(g) + 6\,H_2O(g)$$

$$\Delta H = [(4 \times 90.3) + (6 \times -241.8)] - [(4 \times -46.1) + (5 \times 0)]$$
$$= -905.2\ kJ$$

$$2\,NO(g) + O_2(g) \rightleftharpoons 2\,NO_2(g)$$

$$\Delta H = (2 \times 33.2) - [(2 \times 90.3) + (1 \times 0)]$$
$$= -114.2\ kJ$$

$$3\,NO_2(g) + H_2O(\ell) \rightleftharpoons 2\,HNO_3(\ell) + NO(g)$$

$$\Delta H = [(2 \times -174.1) + (1 \times 90.3)] - [(3 \times 33.2) + (1 \times -285.8)]$$
$$= -71.7\ kJ$$

All of these reactions are exothermic, so none of the steps has a higher yield at a higher temperature.

Think About It

The reactions, however, are run at 850°C, which enhances the rate of the reaction. The Ostwald process is described at the end of Chapter 16 in "The Chemistry of Two Strong Acids: Sulfuric and Nitric Acids."

16.169. **Collect and Organize**

After writing the equations corresponding to $\Delta H^{\circ}_{f,SO_2}$ and $\Delta H^{\circ}_{f,SO_3}$, we are to apply Hess's law to show how to calculate ΔH°_{rxn} for

$$2\,SO_2(g) + O_2(g) \rightarrow 2\,SO_3(g)$$

Analyze

The chemical reactions corresponding to ΔH°_f involve preparing the compound from the elements.

Solve

The balanced formation reactions for SO_2 and SO_3 are

$$\tfrac{1}{8}S_8(s) + O_2(g) \rightarrow SO_2(g) \qquad \Delta H^{\circ}_{f,SO_2}$$

$$\tfrac{1}{8}S_8(s) + \tfrac{3}{2}O_2(g) \rightarrow SO_3(g) \qquad \Delta H^{\circ}_{f,SO_3}$$

Reversing the first reaction, multiplying it by 2, and adding to the second reaction (also multiplied by 2) allow us to calculate ΔH°_{rxn}.

$$2\,SO_2(g) \rightarrow \tfrac{1}{4}S_8(s) + 2\,O_2(g) \qquad -2\,\Delta H^{\circ}_{f,SO_2}$$

$$\tfrac{1}{4}S_8(g) + 3\,O_2(g) \rightarrow 2\,SO_3(g) \qquad 2\,\Delta H^{\circ}_{f,SO_2}$$

$$\overline{2\,SO_2(g) + O_2(g) \rightarrow 2\,SO_3(g) \qquad 2\,\Delta H^{\circ}_{f,SO_3} - 2\,\Delta H^{\circ}_{f,SO_2} = \Delta H^{\circ}_{rxn}}$$

Think About It

Remember that Hess's law is applicable to other state functions, such as S and G.

16.171. **Collect and Organize**

We are asked whether we expect the Henry's law constants for SO_3 and NO_3 to be greater or less than that for CO_2 (which is 3.5×10^{-2} M/atm).

Analyze

Carbon dioxide is a linear nonpolar molecule that forms dipole-induced dipole intermolecular interactions when dissolved in water. To determine whether SO_3 and NO_2 are more or less soluble in water than CO_2 we must draw their Lewis structures and determine their molecular polarity.

Solve

SO$_3$ is trigonal planar and therefore nonpolar like CO$_2$:

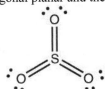

SO$_3$, like CO$_2$, is nonpolar and forms dipole-induced dipole interactions with water. These intermolecular forces, however, are stronger for SO$_3$ because SO$_3$ is a larger molecule. SO$_3$, therefore, is slightly more soluble in water than CO$_2$.

NO$_2$ is bent and therefore polar.

NO$_2$ forms dipole–dipole interactions with water. These are stronger than the water–CO$_2$ intermolecular forces, so NO$_2$ is significantly more soluble in water than CO$_2$.

Yes, the corresponding constants for SO$_3$ and NO$_2$ are expected to be greater than the Henry's law constant for CO$_2$.

Think About It

Of the three substances, NO$_2$ is the most soluble in water.

16.173. **Collect and Organize**

After drawing the Lewis structure for H$_2$S$_2$O$_3$, we are to consider the acid properties of thiosulfuric acid compared to H$_2$SO$_4$.

Analyze

We are given that H$_2$S$_2$O$_3$ is isostructural with H$_2$SO$_4$. This means that their atoms are arranged in the same way, with an S atom in H$_2$S$_2$O$_3$ taking the place of one of the O atoms in H$_2$SO$_4$.

Solve

(a)

$$\text{H}-\overset{..}{\underset{..}{\text{O}}}-\overset{\overset{\overset{..}{\text{O}}}{\|}}{\underset{\underset{|}{\underset{\text{H}}{\overset{..}{\text{O}}}}}{\text{S}}}=\overset{.}{\underset{..}{\text{S}}}:$$

(b) When a less electronegative sulfur atom replaces an oxygen atom in the acid, the acidity decreases. Therefore, H$_2$S$_2$O$_3$ is less acidic than H$_2$SO$_4$.

Think About It

Thiosulfuric acid is indeed less acidic than sulfuric acid. H$_2$S$_2$O$_3$ has a pK_{a_1} = 0.6 and pK_{a_2} = 1.6 whereas H$_2$SO$_4$ has a pK_{a_1} < 0 and a pK_{a_2} = 1.92. Although the pK_{a_2} for H$_2$S$_2$O$_3$ is higher than that of H$_2$SO$_4$, remember that the first ionization constant in diprotic acids dominates, so H$_2$SO$_4$ is a stronger acid than H$_2$S$_2$O$_3$.

CHAPTER 17 | Metal Ions: Colorful and Essential

17.1. Collect and Organize
From the highlighted elements in Figure P17.1, we are to choose those whose chlorides are colored.

Analyze
Chloride compounds are colored for the transition elements that have incomplete d shells. The chlorides of the highlighted elements and the electron configurations of the transition metal ions are

$$
\begin{array}{lll}
CaCl_2 & Ca^{2+} & [Ar] \\
CrCl_2 & Cr^{2+} & [Ar]3d^4 \\
CrCl_3 & Cr^{3+} & [Ar]3d^3 \\
CoCl_2 & Co^{2+} & [Ar]3d^7 \\
CoCl_3 & Co^{3+} & [Ar]3d^6 \\
ZnCl_2 & Zn^{2+} & [Ar]3d^{10}
\end{array}
$$

Solve
Chromium (green) and cobalt (yellow) have colored chloride salts.

Think About It
Remember that we remove the s electrons first in forming transition metal cations.

17.3. Collect and Organize
Of the elements highlighted in Figure P17.3, we are to identify which have M^{2+} ions that form colorless tetrahedral complex ions.

Analyze
Transition metal ions that are colorless have either filled or empty d orbitals. The electron configurations for the M^{2+} cations are

$$
\begin{array}{lll}
Red & V^{2+} & [Ar]d^3 \\
Purple & Mn^{2+} & [Ar]d^5 \\
Yellow & Co^{2+} & [Ar]d^7 \\
Blue & Zn^{2+} & [Ar]d^{10}
\end{array}
$$

Solve
Zinc (blue) forms colorless tetrahedral complex ions.

Think About It
Because nearly all tetrahedral complex ions are high-spin with the d orbital splitting diagram shown below,

the numbers of unpaired electrons for the other ions in this problem are as follows: V^{2+}, three unpaired e^-; Mn^{2+}, five unpaired e^-; and Co^{2+}, three unpaired e^-.

17.5. Collect and Organize
From Figure P17.5 showing different possible d-orbital splitting diagrams, we are to choose the one that best represents the Mn^{2+} ions in a tetrahedral crystal field in amethyst.

Analyze
Mn^{2+} has five electrons in its d orbitals so the correct diagram will have five electrons in it. A tetrahedral crystal field splits the energy levels so that there are two low-lying d orbitals and three high-lying d orbitals.

Solve

Only (b) and (d) show five electrons in the *d*-orbital splitting diagrams, and only (d) is for a tetrahedral crystal field (diagram (b) shows an octahedral field), so Mn^{2+} is represented by diagram (d).

Think About It

The magnitude of the tetrahedral crystal field splitting is about half that of the octahedral field splitting. As a consequence, transition metals in a tetrahedral field are nearly always high-spin.

17.7. **Collect and Organize**

The spectrum of $Ti(H_2O)_6^{3+}$ (Figure P17.7) shows an absorption maximum of 512 nm. We are asked to predict the color of this solution based on the absorption spectrum.

Analyze

The observed color of a solution is the complementary color of the absorbed wavelength.

Solve

The absorption maximum in the spectrum is around 512 nm. This corresponds to green in the visible spectrum. The complementary color is red, so this solution appears red.

Think About It

Because the color that this complex absorbs is of a short wavelength, we can say that H_2O on Ti^{3+} is a fairly strong-field ligand.

17.9. **Collect and Organize**

By analyzing the absorbance spectrum for a gemstone, we may determine the color of the sample.

Analyze

The color we perceive for an object is the color of light the object transmits. Unlike a transmittance spectrum, absorption spectra illustrate the wavelengths of light absorbed by the sample. The wavelength of light with the highest absorption will correspond to the color of light complementary to the color of the sample.

Solve

The sample has a large peak around 550 nm, continuing to increase as the wavelength of light increases. This corresponds to absorption in the green to red range, meaning that purple/blue light will be transmitted. The gemstone will have a blue to purple appearance.

Think About It

By consulting the wavelengths of *lowest* absorption, we would also arrive at the conclusion that blue and purple light are transmitted rather than absorbed.

17.11. **Collect and Organize**

We are asked if all Lewis bases are also Brønsted–Lowry bases.

Analyze

In the Lewis definition of acids and bases, a base is a species that donates an electron pair to a Lewis acid. In the Brønsted–Lowry definition of acids and bases, a base is a species that accepts a proton (H^+ ion) from a Brønsted–Lowry acid. If we can imagine a scenario in which a species is donating an electron pair, but not accepting a proton, the statement that all Lewis bases must also be Brønsted–Lowry bases is false.

Solve

The statement is false, since it is possible for a base to donate electrons without accepting a proton. One example is the acid–base adduct formed between trimethylamine $(CH_3)_3N$ and borane, BH_3. In this case, trimethylamine is acting as a Lewis base (donating an electron pair to borane), but is not accepting a proton (i.e., forming an N–H bond).

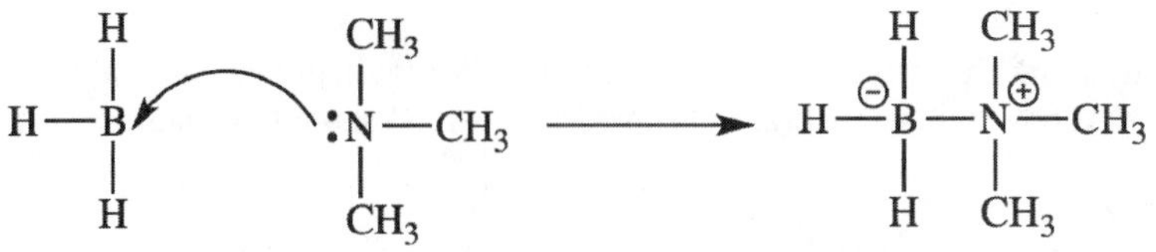

Think About It

Although trimethylamine is not serving as a Brønsted–Lowry base in this case, it is capable of doing so. Consider the reaction of trimethylamine with HCl, in which trimethylamine serves as both a Lewis base and a Brønsted–Lowry base, forming trimethylammonium chloride.

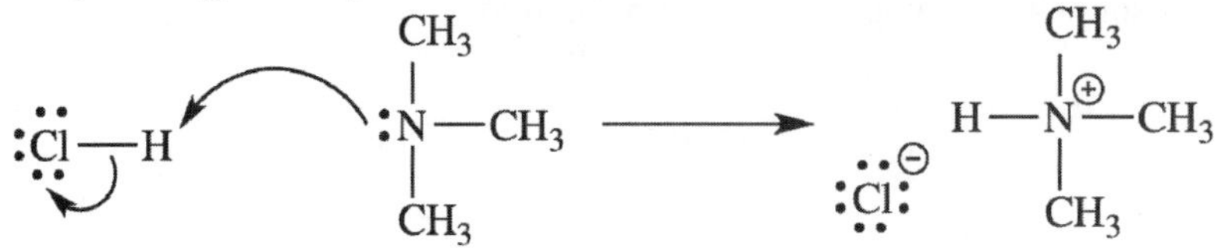

17.13. Collect and Organize

We are given that BF_3 is a Lewis acid. We are to explain why it is not a Brønsted–Lowry acid.

Analyze

A Brønsted–Lowry acid is a proton donor. A Lewis acid is an electron-pair acceptor.

Solve

BF_3 is a Lewis acid because it accepts electron pairs from Lewis bases to complete the octet around boron. It is not a Brønsted–Lowry acid, however, because is does not have any H atoms to donate as H^+.

Think About It

The reaction of NH_3 with BF_3 is typical of the Lewis acid–base reactions of BF_3:

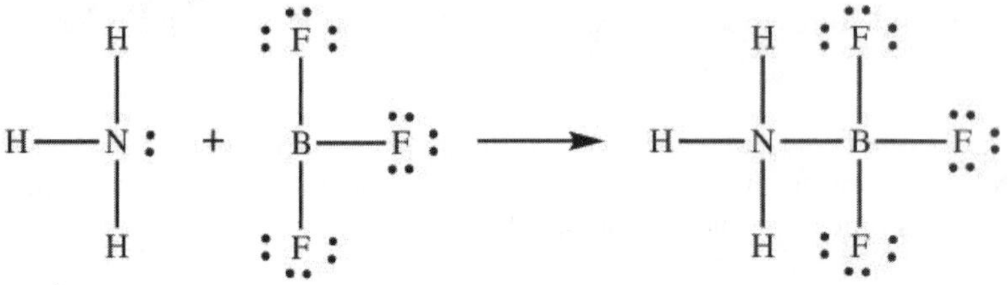

17.15. Collect and Organize

After drawing the Lewis structures for NH_3 and $B(CH_3)_3$, we can show how the electron pairs move and bonds form in the formation of $NH_3 \cdot B(CH_3)_3$, and identify the Lewis acid and Lewis base in the reaction.

Analyze

NH_3 has a full octet in its Lewis structure. $B(CH_3)_3$ has only six electrons around B in its Lewis structure.

Solve

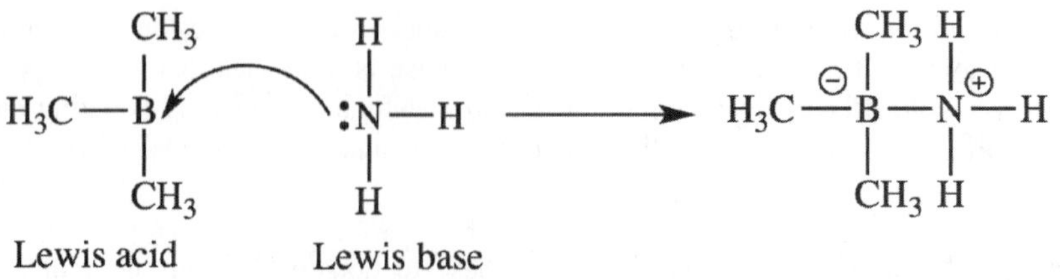

The Lewis base is NH_3 and the Lewis acid is $B(CH_3)_3$.

Think About It

$B(CH_3)_3$ is often called an *electron-deficient* compound and is a strong Lewis acid.

17.17. Collect and Organize

After drawing the Lewis structures for SO_2, H_2O, and H_2SO_3, we can show how the electron pairs move and bonds form and break in the formation of H_2SO_3 and identify the Lewis acid and Lewis base in the reaction.

Analyze

All the compounds have covalent bonding. A Lewis acid accepts an electron pair in an acid–base reaction, whereas a Lewis base donates an electron pair.

Solve

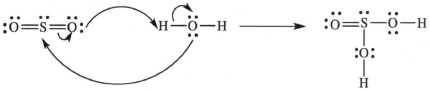

The sulfur atom in SO_2 and the hydrogen atom in H_2O are acting as Lewis acids, while the oxygen atoms in SO_2 and H_2O are acting as Lewis bases.

Think About It

Notice that one of the O—H bonds in water breaks in this reaction so that two O—H bonds are present in the product, H_2SO_3.

17.19. Collect and Organize

After drawing the Lewis structures for $B(OH)_3$, H_2O, $B(OH)_4^-$, and H^+, we can show how the electron pairs move and bonds form and break in the formation of $B(OH)_4^-$ and identify the Lewis acid and Lewis base in the reaction.

Analyze

All the compounds have covalent bonding. A Lewis acid accepts an electron pair in an acid–base reaction, whereas a Lewis base donates an electron pair.

Solve

$B(OH)_3$ is the Lewis acid, and H_2O is the Lewis base.

Think About It

Notice that one of the O—H bonds in water breaks in this reaction so that an additional –OH group on B is formed.

17.21. Collect and Organize

We are asked which molecules or ions (H_2O or Cl^-) surround Ca^{2+} when $CaCl_2$ is dissolved in water.

Analyze

$CaCl_2$ is completely soluble in water; the Ca^{2+} and Cl^- ions are 100% dissociated.

Solve

Water molecules occupy the inner coordination sphere of Ca^{2+} ions.

Think About It

The oxygen atoms, which carry partial negative charge, are pointed toward the Ca^{2+} ion in the coordination sphere.

17.23. Collect and Organize

We are to consider a solution formed when $AgNO_3$ is dissolved in aqueous solution, in the presence of ammonia, NH_3. We are asked to identify the molecules or ions found in the inner coordination sphere of Ag^+.

Analyze

Silver will form a complex with ammonia according to the equation

$$Ag^+(aq) + 2\,NH_3(aq) \rightleftharpoons Ag(NH_3)_2^+(aq)$$

Solve

Ag^+ ions in solution are bonded to the ammonia molecules, forming the adduct $Ag(NH_3)_2^+$. NH_3 occupies the inner coordination sphere.

Think About It

Though H_2O molecules will surround ions in solution, they will be found in the secondary coordination sphere, due to the presence of the ammonia ligand.

17.25. Collect and Organize

We are to identify the counter ion present in $Na_3[Fe(CN)_6]$.

Analyze

A counter ion is the ion of opposite charge to the complex ion and is not directly bonded to the metal cation.

Solve

Na^+, sodium ion

Think About It

The counter ion balances the charge to give a neutral salt. In this case we know that the charge on the complex ion is 3– because there are three Na^+ counter ions.

17.27. Collect and Organize

We are asked to describe the rationale behind chelation therapy for the treatment of lead poisoning.

Analyze

If we assume the lead is present as Pb^{2+}, the formation of a complex with a chelating ligand such as EDTA is described by the following equation

$$Pb^{2+}(aq) + EDTA^{4-}(aq) \rightleftharpoons [Pb(EDTA)]^{2-}(aq)$$

Solve

Complexes of lead are less toxic than Pb^{2+}. Adding a chelating ligand shifts the equilibrium in favor of the complex ion, rather than free Pb^{2+}.

Think About It

Aqueous solutions of metal cations such as $[Pb(EDTA)]^{2-}$ are more readily excreted from the body than are the metal cations themselves.

17.29. Collect and Organize

We are to explain why AgCl dissolves easily in a solution of NH_3 but not in water.

Analyze

The K_{sp} reaction for AgCl is

$$AgCl(s) \rightleftharpoons Ag^+(aq) + Cl^-(aq) \qquad K_{sp} = 1.8 \times 10^{-10}$$

The K_f reaction for $Ag(NH_3)_2^+$ is

$$Ag^+(aq) + 2\ NH_3(aq) \rightleftharpoons Ag(NH_3)_2^+(aq) \qquad K_f = 1.7 \times 10^7$$

Solve

Ag^+ forms a soluble complex with NH_3, removing Ag^+ from solution and shifting the equilibrium for the dissolution of AgCl to the right.

Think About It

The overall reaction for the dissolution of AgCl in aqueous ammonia is

$$AgCl(s) + 2\ NH_3(aq) \rightleftharpoons Ag(NH_3)_2^+(aq) + Cl^-(aq) \qquad K_{overall} = K_{sp} \times K_f = 3.1 \times 10^{-3}$$

17.31. **Collect and Organize**

Given the amount of $Ni(NO_3)_2$ (0.00100 mol) that is dissolved in 1.00 L of a 0.500 M aqueous solution of NH_3 we can calculate the concentration of $Ni(NO_3)_2$ dissolved. Using the formation constant of $Ni(NH_3)_6^{2+}$ ($K_f = 5.5 \times 10^8$) we can then calculate the concentration of uncomplexed $Ni^{2+}(aq)$ in the solution at equilibrium. The relevant equation is

$$Ni^{2+}(aq) + 6\ NH_3(aq) \rightleftharpoons Ni(NH_3)_6^{2+}(aq)$$

Analyze

Because the K_f for this complex is very large we can expect that all of the Ni^{2+} dissolved [from 1 mmol of $Ni(NO_3)_2$] is converted into $Ni(NH_3)_6^{2+}$. We can use the K_f to calculate the amount of Ni^{2+} that remains in solution at equilibrium. The amount of $Ni(NO_3)_2$ and Ni^{2+} initially in the solution is

$$\frac{1.0 \times 10^{-3}\ \text{mol}}{1.00\ \text{L}} = 1.0 \times 10^{-3}\ M$$

At equilibrium then, if $[Ni^{2+}] = x$, the $[Ni(NH_3)_6^{2+}] = 1.0 \times 10^{-3} - x$. The $[NH_3]$ at equilibrium will be $[0.500 - (6 \times 1.0 \times 10^{-3})] + 6x$.

Solve

(a) The amount of $Ni(NO_3)_2$ initially dissolved in the solution is

$$\frac{1.0 \times 10^{-3}\ \text{mol}}{1.00\ \text{L}} = 1.0 \times 10^{-3}\ M$$

(b) To determine the concentration of free Ni^{2+} in the solution, we set up an ICE table and use the K_f expression:

	$[Ni^{2+}]$	$[NH_3]$	$[Ni(NH_3)_6^{2+}]$
Initial	1.0×10^{-3}	0.500	0
Change	$-(1.0 \times 10^{-3} - x)$	$-6(1.0 \times 10^{-3} - x)$	$1.0 \times 10^{-3} - x$
Equilibrium	x	$0.494 + 6x$	$1.0 \times 10^{-3} - x$

$$5.5 \times 10^8 = \frac{1.0 \times 10^{-3} - x}{x(0.494 + 6x)^6} \approx \frac{1.0 \times 10^{-3}}{x(0.494)^6}$$

$$x = 1.25 \times 10^{-10}\ M$$

Think About It

Indeed, only a very small amount of nickel is present as uncomplexed Ni^{2+} in this solution.

17.33. Collect and Organize

Given the initial molar amounts of $Co(NO_3)_2$, NH_3, and ethylenediamine (en) in a 250 mL solution (1.00 mmol, 100 mmol, and 100 mmol, respectively), we can calculate the concentration of $Co^{2+}(aq)$ in the solution at equilibrium.

Analyze

The complex ion formation equations are

$$Co^{2+}(aq) + 6\ NH_3(aq) \rightleftharpoons Co(NH_3)_6^{2+} \qquad K_f = 7.7 \times 10^4$$

$$Co^{2+}(aq) + 3\ en(aq) \rightleftharpoons Co(en)_3^{2+} \qquad K_f = 8.7 \times 10^{13}$$

We can reverse the first equation and add it to the second equation to obtain an equilibrium equation for the conversion of $Co(NH_3)_6^{2+}$ into $Co(en)_3^{2+}$:

$$Co(NH_3)_6^{2+}(aq) + 3\ en(aq) \rightleftharpoons Co(en)_3^{2+}(aq) + 6\ NH_3(aq)$$

$$K = \frac{8.7 \times 10^{13}}{7.7 \times 10^4} = 1.13 \times 10^9$$

Because the K_f for $Co(en)_3^{2+}$ is so large we can assume that initially all Co^{2+} is converted into $Co(en)_3^{2+}$. Therefore $[Co(en)_3^{2+}]_{initial} = 4.00 \times 10^{-3}\ M$, $[en]_{initial} = (0.400\ M - 3 \times 4.00 \times 10^{-3})$, and $[Co(NH_3)_6^{2+}]_{initial} = 0$. Once this equilibrium is established we can calculate $[Co^{2+}]$, knowing $[Co(NH_3)_6^{2+}]$ and using the first equilibrium equation above.

Solve

	$[Co(NH_3)_6^{2+}]$	$[en]$	$[Co(en)_3]^{2+}$	$[NH_3]$
Initial	0	0.388	4.00×10^{-3}	0.400
Change	$+x$	$+3x$	$-x$	$-6x$
Equilibrium	x	$0.388 + 3x$	$4.00 \times 10^{-3} - x$	$0.400 - 6x$

$$1.13 \times 10^9 = \frac{\left(4.00 \times 10^{-3} - x\right)\left(0.400 - 6x\right)^6}{x\left(0.388 + 3x\right)^3} \approx \frac{\left(4.00 \times 10^{-3}\right)\left(0.400\right)^6}{x\left(0.388\right)^3}$$

$$x = 2.48 \times 10^{-13}$$

This result can now be used in the first equilibrium equation where $[Co(NH_3)_6^{2+}] = 2.48 \times 10^{-13}\ M$, $[NH_3] \approx 0.400\ M$, and $[Co^{2+}] = 0$ initially:

	$[Co^{2+}]$	$[NH_3]$	$[Co(NH_3)_6^{2+}]$
Initial	0	0.400	2.48×10^{-13}
Change	x	$+6x$	$-x$
Equilibrium	x	$0.400 + 6x$	$2.48 \times 10^{-13} - x$

$$7.7 \times 10^4 = \frac{\left(2.48 \times 10^{-13} - x\right)}{x\left(0.400 + 6x\right)^6} \approx \frac{\left(2.48 \times 10^{-13} - x\right)}{x\left(0.400\right)^6}$$

$$315.4x = \left(2.48 \times 10^{-13} - x\right)$$

$$316.4x = 2.48 \times 10^{-13}$$

$$x = 7.8 \times 10^{-16}\ M = [Co^{2+}]$$

Think About It

In this 250 mL solution there are only

$$0.250\ L \times \frac{7.8 \times 10^{-16}\ mol}{1\ L} \times \frac{6.022 \times 10^{23}\ Co^{2+}}{1\ mol} = 1.2 \times 10^8\ Co^{2+}\ ions$$

compared to the initial Co^{2+} dissolved in solution

$$1.00 \times 10^{-3} \text{ mol} \times \frac{6.022 \times 10^{23} \text{ Co}^{2+}}{1 \text{ mol}} = 6.022 \times 10^{20} \text{ Co}^{2+} \text{ ions}$$

for a % of Co^{2+} at equilibrium of

$$\frac{1.2 \times 10^8}{6.022 \times 10^{20}} \times 100 = 2.0 \times 10^{-11}\%$$

17.35. Collect and Organize

We can use the conventions for naming given in Section 17.4 to name three transition metal complex ions: $Ag(NH_3)_2^+$, $Co(H_2O)_6^{3+}$, and $[Fe(NH_3)_5Br]^{2+}$.

Analyze

For each cation, we first name the ligands in alphabetical order, indicating with a prefix how many of each ligand are bonded to the metal ion. Then, we add the name of the metal, indicating the charge on the metal ion with a Roman numeral.

Solve

(a) Diamminesilver(I)
(b) Hexaaquacobalt(III)
(c) Pentaamminebromoiron(III)

Think About It

Be sure to correctly account for the charge on the metal ion by considering the charge on the ligands and the charge on the overall complex. In part (c) the bromide ligand has a 1– charge. With an overall charge on the complex of 2+, the iron ion must have a 3+ charge.

17.37. Collect and Organize

We can use the conventions for naming given in Section 17.4 to name and indicate the coordination numbers (CN) of three transition metal complex ions: $CoBr_4^{2-}$, $Zn(H_2O)(OH)_3^-$, and $Ni(CN)_5^{3-}$.

Analyze

For each anion, we first name the ligands in alphabetical order, indicating with a prefix how many of each ligand are bonded to the metal ion. Then we add the name of the metal, using *-ate* as the ending and indicate the charge on the metal ion with a Roman numeral.

Solve

(a) Tetrabromocolbaltate(II), CN = 4.
(b) Aquatrihydroxozincate(II), CN = 4.
(c) Pentacyanonickelate(II), CN = 5.

Think About It

Be sure to correctly account for the charge on the metal ion by considering the charge on the ligands and the charge on the overall complex. In part (c), the cyanide ligands have a 1– charge each. With an overall charge on the complex of 3–, the Ni metal ion must have a 2+ charge.

17.39. Collect and Organize

We can use the conventions for naming given in Section 17.4 to name three transition metal coordination compounds: $[Zn(en)_2]SO_4$, $[Ni(NH_3)_5(H_2O)]Cl_2$, and $K_4Fe(CN)_6$. We are also asked to identify the charge on the complex ion and the oxidation state and coordination number (CN) of the metal in the complex ion.

Analyze

To name these coordination compounds, we separately name the cation and the anion, with the name of the cation being written first. The charge on the complex ion may be determined by considering the charge and quantity of counterions present. The oxidation state of the metal may be determined by subtracting the number of anionic ligands from the charge on the complex ion.

Solve

(a) i. Bis(ethylenediamine)zinc(II) sulfate
ii. Pentaammineaquanickel(II) chloride
iii. Potassium hexacyanoferrate(II)
(b) i. Charge on complex ion = +2; oxidation state of Zn = 2+; CN = 4.
ii. Charge on complex ion = +2; oxidation state of Ni = 2+; CN = 6.
iii. Charge on complex ion = −4; oxidation state of Fe = 2+; CN = 6.

Think About It

We need not indicate the number of sulfate, chloride, or potassium ions in the name. They are understood to be counter ions. When writing the formulas we would indicate how many are needed to balance the charge to make a neutral compound.

17.41. Collect and Organize

We are to identify which chloride salts would produce an acidic solution.

Analyze

All the salts are soluble. The Cl^- does not hydrolyze but the cations might.

Solve

(a) $Ca^{2+}(aq)$ does not hydrolyze because it would form the strong base $Ca(OH)_2$, which is 100% ionized in solution.
(b) $Cr^{3+}(aq)$ does hydrolyze to form an acidic solution according to the equation

$$Cr^{3+}(aq) + 3\ H_2O\ (\ell) \rightleftharpoons Cr(OH)_3(s) + 3\ H^+(aq)$$

(c) $Na^+(aq)$ does not hydrolyze because it would form the strong base NaOH, which is 100% ionized in solution.
(d) $Fe^{2+}(aq)$ does hydrolyze to form an acidic solution according to the equation

$$Fe^{2+}(aq) + 2\ H_2O\ (\ell) \rightleftharpoons Fe(OH)_2(s) + 2\ H^+(aq)$$

Both (b) $CrCl_3$ and (d) $FeCl_2$ produce an acidic solution.

Think About It

According to Table 17.3, Cr^{3+} is among the most acidic of the hydrated metal ions.

17.43. Collect and Organize

We are asked how the oxidation of Fe^{2+} to Fe^{3+} affects the acidity of the solution. To answer this we are to compare the acidity of $Fe^{2+}(aq)$ versus $Fe^{3+}(aq)$.

Analyze

Both $Fe^{2+}(aq)$ and $Fe^{3+}(aq)$ hydrolyze according to the equations

$$Fe^{2+}(aq) + 2\ H_2O(\ell) \rightleftharpoons Fe(OH)_2(s) + 2\ H^+(aq)$$
$$Fe^{3+}(aq) + 3\ H_2O(\ell) \rightleftharpoons Fe(OH)_3(s) + 3\ H^+(aq)$$

Solve

An aqueous solution of Fe^{3+} is more acidic and has a lower pH because Fe^{3+} hydrolyzes to a greater extent than Fe^{2+}. The higher ionic charge of Fe^{3+} polarizes the O—H bonds in water bound to Fe^{3+} to a greater extent than Fe^{2+} does.

Think About It

Notice too that Fe^{3+} could hydrolyze to produce three moles of H^+ rather than two moles of H^+ from Fe^{2+} hydrolysis.

17.45. Collect and Organize

For the amphiprotic $Cr(OH)_3$ we are to write chemical equations to describe this property.

Analyze

Amphiprotic compounds react with both acids and bases.

Solve

In basic solution $Cr(OH)_3$ adds OH^- to form the soluble $Cr(OH)_4^-$ ion:

$$Cr(OH)_3(s) + OH^-(aq) \rightleftharpoons Cr(OH)_4^-(aq)$$

In acidic solution $Cr(OH)_3$ reacts with H^+ to form Cr^{3+} and water:

$$Cr(OH)_3(s) + 3\,H^+(aq) \rightleftharpoons Cr^{3+}(aq) + 3\,H_2O(\ell)$$

Think About It

Other amphiprotic hydroxide compounds that behave similarly to $Cr(OH)_3$ are $Al(OH)_3$ and $Zn(OH)_2$.

17.47. Collect and Organize

We are to explain why Ca^{2+}, Mg^{2+}, and Fe^{3+} combined with strong base (OH^-) are insoluble, but Al^{3+} is soluble.

Analyze

All of these metal cations form insoluble hydroxide compounds (see Appendix 5 for K_{sp} values, which range from 1.9×10^{-33} for Al^{3+} to 4.7×10^{-6} for Ca^{2+}). $Al(OH)_3$, however, is amphoteric and reacts with OH^- to form a complex ion.

Solve

Although $Al(OH)_3$ is insoluble, it reacts with more OH^- in solution to form the soluble $Al(OH)_4^-$. The other ions do not form this type of soluble complex ion and therefore remain insoluble as $Mg(OH)_2$, $Ca(OH)_2$, and $Fe(OH)_3$ in strongly basic solution.

Think About It

If either Cr^{3+} or Zn^{2+} contaminated the ore, they too would be soluble as the complex ions $Cr(OH)_4^-$ and $Zn(OH)_4^{2-}$.

17.49. Collect and Organize

From the K_a of $Al^{3+}(aq)$ we can calculate the pH of a solution that is 0.25 M in $Al(NO_3)_3$.

Analyze

The nitrate ions of $Al(NO_3)_3$ do not react with water. The reaction of Al^{3+} with water, however, gives an acidic solution with $K_a = 1 \times 10^{-5}$:

$$Al^{3+}(aq) + 2\,H_2O(\ell) \rightleftharpoons Al(OH)^{2+}(aq) + H_3O^+(aq)$$

Solve

	$[Al^{3+}]$	$[Al(OH)^{2+}]$	$[H_3O^+]$
Initial	0.25	0	0
Change	$-x$	$+x$	$+x$
Equilibrium	$0.25 - x$	x	x

$$1 \times 10^{-5} = \frac{(x)(x)}{(0.25 - x)} \approx \frac{x^2}{0.25}$$

$$x = 1.58 \times 10^{-3}$$

The assumption that $x \ll 0.500$ is valid:

$$\frac{1.58 \times 10^{-3}}{0.25} \times 100 = 0.63\%$$

The pH of the solution is, therefore

$$pH = -\log(1.58 \times 10^{-3}) = 2.80$$

Think About It
Aluminum, with its small size and high positive charge as Al^{3+}, gives fairly acidic aqueous solutions.

17.51. Collect and Organize
From the K_a of $Fe^{3+}(aq)$ we can calculate the pH of a solution that is 0.100 M in $Fe(NO_3)_3$.

Analyze
The nitrate ions of $Fe(NO_3)_3$ do not react with water. The reaction of Fe^{3+} with water, however, gives an acidic solution with $K_a = 3 \times 10^{-3}$:

$$Fe^{3+}(aq) + 2\ H_2O(\ell) \rightleftharpoons Fe(OH)^{2+}(aq) + H_3O^+(aq)$$

Solve

	$[Fe^{3+}]$	$[Fe(OH)^{2+}]$	$[H_3O^+]$
Initial	0.100	0	0
Change	$-x$	$+x$	$+x$
Equilibrium	$0.100 - x$	x	x

$$3\times10^{-3} = \frac{(x)(x)}{(0.100-x)} \approx \frac{x^2}{0.100}$$

$$x = 1.73\times10^{-2}$$

The assumption that x $\ll$ 0.100 is not valid:

$$\frac{1.73\times10^{-2}}{0.100}\times100 = 17\%$$

so we must solve by the quadratic equation:

$$3\times10^{-3}(0.100-x) = x^2$$
$$3\times10^{-4} - 3\times10^{-3}\ x = x^2$$
$$x^2 + 3\times10^{-3}x - 3\times10^{-4} = 0$$
$$x = 0.0159 \text{ or } -0.0189$$

The concentration of H_3O^+ and $Fe(OH)^{2+}$ must be positive, so $x = 0.0159$ and the pH of the solution is

$$pH = -\log(0.0159) = 1.80$$

Think About It
Fe^{3+} is the most acidic hydrated metal ion listed in Table 17.3, so we expect this solution to be quite acidic.

17.53. Collect and Organize
We are to sketch the titration curve for the titration of 25 mL of 0.25 M $FeCl_3$ with 0.50 M NaOH.

Analyze
The Fe^{3+} ions react with OH^- to form, in steps, $Fe(OH)_3$:

$$Fe^{3+}(aq) + OH^-(aq) \rightleftharpoons Fe(OH)^{2+}(aq)$$
$$Fe(OH)^{2+}(aq) + OH^-(aq) \rightleftharpoons Fe(OH)_2^+(aq)$$
$$Fe(OH)_2^+(aq) + OH^-(aq) \rightleftharpoons Fe(OH)_3(s)$$

The titration curve shows three equivalence points. Notice that the concentration of OH^- in the titrant is twice the concentration of Fe^{3+} in the solution. The equivalence points will, therefore, occur at 12.5 mL, 25.0 mL, and 37.5 mL.

Solve

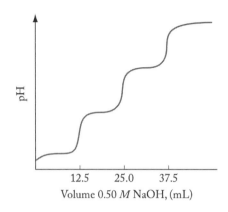

Think About It

Remember that $Fe(OH)_3$ does not further react with OH^-.

17.55. Collect and Organize / Analyze

After first defining *chelating agent*, we are to describe the properties that make a compound an effective chelating agent.

Solve

A chelating agent is a multidentate ligand that separates metal ions from other substances so that they can no longer react. Properties that make a chelating agent effective include strong bonds formed between the metal and the ligand and large formation constants.

Think About It

EDTA is an example of a good chelating agent.

17.57. Collect and Organize

By examining the structure of an aminocarboxylate ligand, we can predict how its chelating ability changes as pH changes.

Analyze

The general structure of an aminocarboxylate is

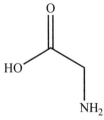

Solve

At low pH the aminocarboxylate ligand contains $-COOH$ and $-NH_3^+$ groups. As the pH increases, the chelating ability increases because OH^- removes the H on the amine and carboxylic acid groups, providing additional sites for binding to the metal cation.

Think About It

EDTA is an example of an aminocarboxylate ligand.

17.59. Collect and Organize

For a square planar crystal field geometry, we are to explain why d_{xy} is higher in energy than either the d_{xz} or d_{yz} orbitals.

Analyze

In a crystal field, d orbitals that are pointed directly at the ligands are raised in energy and those that are not pointed directly at ligands are lowered in energy.

Solve

The repulsions due to the ligands in a square planar crystal field are highest for the d_{xy} orbital and so it is raised in energy because this orbital lies in the plane of the ligands. The d_{xz} and d_{yz} orbitals, however, are perpendicular to the plane of the four ligands, and therefore these orbitals are lower in energy compared to the energy of the d_{xy} orbital.

Think About It

In a square planar geometry the $d_{x^2-y^2}$ orbital points directly at the four ligands. It has the most repulsions and therefore is the highest orbital in energy.

17.61. **Collect and Organize**

We can explain why compounds of Ti^{4+} are colorless, whereas those of Ti^{3+} are violet, by examining the d-electron configuration of the ion.

Analyze

When the transition metals bond to ligands, the d orbitals split in energy. If there is a d to d transition possible for the ion, the compound is likely to be colored.

Solve

Ti^{4+} has an electron configuration of $[Ar]4s^03d^0$. Because Ti^{4+} has no electrons in its d orbitals, no d–d transitions can occur, and the compounds of Ti^{4+} are colorless. Ti^{3+} has an electron configuration of $[Ar]4s^03d^1$. The violet color of Ti^{3+} tells us the magnitude of Δ_o for this Ti^{3+} ion.

Think About It

For the same reason, Sc^{3+} compounds are colorless.

17.63. **Collect and Organize**

Considering solutions of $Cr(H_2O)_6^{3+}$ and $Cr(NH_3)_6^{3+}$, we are to determine which solution is violet and which is yellow.

Analyze

From the spectrochemical series we know that NH_3 splits the d orbitals to a greater extent than H_2O (Table 17.5). $Cr(NH_3)_6^{3+}$ therefore absorbs a higher energy (shorter wavelength) of light than $Cr(H_2O)_6^{3+}$.

Solve

The wavelength (color) absorbed is complementary to the wavelength (color) observed. The yellow aqueous solution absorbs violet light and the violet aqueous solution absorbs yellow light. Because violet light has a higher energy and shorter wavelength than yellow light, the yellow solution contains (b) $Cr(NH_3)_6^{3+}$. The violet solution contains (a) $Cr(H_2O)_6^{3+}$.

Think About It

A Cr^{3+} compound giving a red solution would have a d-orbital splitting energy between that of $Cr(NH_3)_6^{3+}$ and $Cr(H_2O)_6^{3+}$.

17.65. **Collect and Organize**

Given the octahedral crystal field splitting for $Co(phen)_3^{3+}$ (5.21×10^{-19} J/ion), we can determine the color of a solution of $Co(phen)_3^{3+}$ by calculating the wavelength of light absorbed and correlating that to the color of light reflected or transmitted.

Analyze

To calculate the wavelength of light absorbed we rearrange $E = hc/\lambda$ to solve for the wavelength of light:

$$\lambda = \frac{hc}{E}$$

We can use the visible spectrum (400–700 nm) to determine the color of the absorbed wavelength and then the color wheel (Figure 17.16) to choose the complementary color (the color we observe).

Solve

$$\lambda = \frac{6.626 \times 10^{-34}\ \text{J} \cdot \text{s} \times 3.00 \times 10^{8}\ \text{m/s}}{5.21 \times 10^{-19}\ \text{J/ion}} = 3.82 \times 10^{-7}\ \text{m or 382 nm}$$

This wavelength is in the UV region, so the solution is colorless.

Think About It

If a complex absorbs in the UV, sometimes it appears slightly yellow because its absorption "tails" into the violet region of the visible spectrum.

17.67. **Collect and Organize**

For the complexes $NiCl_4^{2-}$ and $NiBr_4^{2-}$, we are asked which complex absorbs longer wavelengths of visible light.

Analyze

$NiCl_4^{2-}$ and $NiBr_4^{2-}$ solutions absorb light at 702 and 756 nm, respectively. The larger the split of the d-orbital energies, Δ_o, the shorter the wavelength absorbed.

Solve

Because $NiCl_4^{2-}$ has a greater split of the d-orbital energies, $NiCl_4^{2-}$ absorbs at a shorter wavelength than $NiBr_4^{2-}$.

Think About It

A solution of $NiCl_4^{2-}$ appears blue-green, and a solution of $NiBr_4^{2-}$ appears yellow-green.

17.69. **Collect and Organize**

From the list of metal ions provided, we are asked to determine how many d-electrons each metal has in its valence shell, and which ions have the possibility of high-spin and low-spin configurations.

Analyze

The octahedral field diagram is

and may hold 10 e^-. The possibility for high-spin or low-spin configurations occurs when, depending on the magnitude of Δ_o, electrons may be placed in either the low-lying orbital set or the high-lying orbital set after the d^3 configuration. These situations may occur for metal ions that have 4, 5, 6, or 7 d electrons.

Solve

(a) Co^{2+} has the electron configuration $[Ar]4s^0 3d^7$, so it has 7 electrons in its $3d$ orbitals.
Cr^{2+} has the electron configuration $[Ar]4s^0 3d^4$, so it has 4 electrons in its $3d$ orbitals.
Ni^{2+} has the electron configuration $[Ar]4s^0 3d^8$, so it has 8 electrons in its $3d$ orbitals.
Cu^{2+} has the electron configuration $[Ar]4s^0 3d^9$, so it has 9 electrons in its $3d$ orbitals.
(b) Both Co^{2+} and Cr^{2+} may have high-spin and low-spin states.

Think About It

The high-spin and low-spin configurations for Co^{2+} (a d^7 ion) are

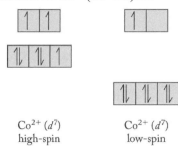

Co^{2+} (d^7)
high-spin

Co^{2+} (d^7)
low-spin

17.71. Collect and Organize

From the list of metal ions provided, we are asked to determine how many unpaired electrons each metal has in its valence shell in the hexachloro complex ion.

Analyze

Chloride is a weak field ligand, meaning that the hexachloro complex ions will all be high-spin, with an octahedral geometry. The octahedral field diagram is

Solve

Fe^{2+} has the electron configuration $[Ar]4s^0 3d^6$, so it has 6 electrons in its $3d$ orbitals. In the high-spin case, this corresponds to 4 unpaired electrons.

Cu^{2+} has the electron configuration $[Ar]4s^0 3d^9$, so it has 9 electrons in its $3d$ orbitals. There is only one arrangement of nine electrons in an octahedral field, corresponding to 1 unpaired electron.

Co^{2+} has the electron configuration $[Ar]4s^0 3d^7$, so it has 7 electrons in its $3d$ orbitals. In the high-spin case, this corresponds to 3 unpaired electrons.

Mn^{2+} has the electron configuration $[Ar]4s^0 3d^5$, so it has 5 electrons in its $3d$ orbitals. In the high-spin case, this corresponds to 5 unpaired electrons.

The high-spin electron configurations for these ions are

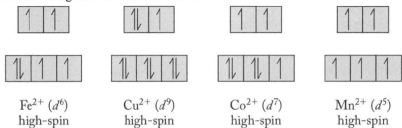

Fe^{2+} (d^6)
high-spin

Cu^{2+} (d^9)
high-spin

Co^{2+} (d^7)
high-spin

Mn^{2+} (d^5)
high-spin

Think About It

Despite having the greatest number of d-electrons (9), Cu^{2+} has the fewest number of unpaired electrons (1). It is important to consider the arrangement of electrons in orbitals, instead of just the number of electrons.

17.73. Collect and Organize

For the minerals MnO_2 and Mn_3O_4, where the Mn ions are surrounded by six O^{2-} ions (and are therefore in an octahedral crystal field), we are to determine the charges of the Mn ions in each mineral and which mineral might have a possibility of high-spin and low-spin Mn ions.

Analyze
High-spin and low-spin complexes are possible in an octahedral field when there are 4, 5, 6, or 7 electrons occupying the d orbitals.

Solve
(a) Mn^{4+} in MnO_2.
Two Mn^{3+} and one Mn^{2+} in Mn_3O_4.
(b) Both low-spin and high-spin configurations are possible in Mn_3O_4 (d^4 and d^5) but not in MnO_2 (d^3).

Think About It
It is not unusual for minerals to have a metal ion present in two different oxidation states, as is the case here for Mn_3O_4.

17.75. Collect and Organize
We are asked whether tetrahedral $CoCl_4^{2-}$ is paramagnetic or diamagnetic.

Analyze
The electron configuration of Co^{2+} in $CoCl_4^{2-}$ is $[Ar]4s^03d^7$.

Solve
In a tetrahedral field, the Co^{2+} ions in $CoCl_4^{2-}$ have a d^7 configuration and have three unpaired electrons in the d_{xy}, d_{xz}, and d_{yz} orbitals. This complex is paramagnetic.

Think About It
Because of its partially filled d orbitals, we also expect this compound to be colored.

17.77. Collect and Organize / Analyze
We are asked to differentiate between *cis*- and *trans*- for an octahedral complex ion.

Solve
For an octahedral geometry, *cis*- means that two ligands are side by side and have a 90° bond angle between them. Ligands that are *trans*- to each other have a 180° bond angle between them.

Cis   Trans

Think About It
Complexes that have *cis*- and *trans*-placed ligands are isomers of each other.

17.79. Collect and Organize
We are to determine how many different ligands are necessary to give stereoisomers of a square planar complex.

Analyze
There are four coordination sites on a square planar complex.

Solve
To have stereoisomers we must have at least two ligands that are different from the others.

Think About It
Some stereoisomer possibilities for square planar complexes are as follows:

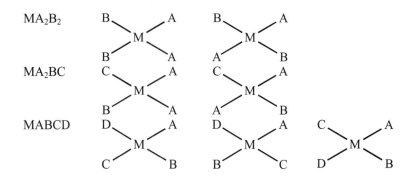

MA_2B_2

MA_2BC

$MABCD$

17.81. Collect and Organize

By looking at the structure of $Co(en)(H_2O)_2Cl_2$, we can determine if it can have stereoisomers.

Analyze

Ethylenediamine (en) must bind in a cis fashion to the Co^{2+} metal ion.

Solve

Yes, $Co(en)(H_2O)_2Cl_2$ has geometric isomers because the H_2O (or Cl) ligands may be either cis or trans to each other and relative to the ethylenediamine ligand:

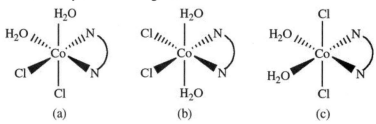

Think About It

Isomer a also has an optical isomer (nonsuperimposable mirror image) as a possibility.

17.83. Collect and Organize

For the square planar complex ion $CuCl_2Br_2^{2-}$, we are asked to sketch and name the possible stereoisomers and to indicate if any of the isomers are chiral.

Analyze

The possible arrangements of the two different ligands on a square planar metal cation are cis and trans.

Solve

$$\begin{bmatrix} Br & Cl \\ & Cu & \\ Br & Cl \end{bmatrix}^{2-} \quad \begin{bmatrix} Cl & Br \\ & Cu & \\ Br & Cl \end{bmatrix}^{2-}$$

        Cis   Trans

No, neither isomer is chiral because both are superimposable on their mirror images.

Think About It

If this complex were tetrahedral, it also would not have any stereoisomers and would not be chiral.

17.85. Collect and Organize

Given that the K_f for the following equation is 5×10^{13}:

$$Ag^+(aq) + 2\ S_2O_3^{2-}(aq) \rightleftharpoons Ag(S_2O_3)_2^{3-}(aq)$$

we are to calculate the ratio of $[Ag^+]$ to $[Ag(S_2O_3)_2^{3-}]$, where $[S_2O_3^{2-}] = 0.250\ M$.

Analyze

The equilibrium constant expression for the reaction is

$$K_f = \frac{\left[Ag(S_2O_3)_2^{3-}\right]}{\left[Ag^+\right]\left[S_2O_3^{2-}\right]^2}$$

Rearranging this to solve for the desired concentration ratio gives

$$\frac{\left[Ag^+\right]}{\left[Ag(S_2O_3)_2^{3-}\right]} = \frac{1}{K_f \times \left[S_2O_3^{2-}\right]^2}$$

Solve

$$\frac{\left[Ag^+\right]}{\left[Ag(S_2O_3)_2^{3-}\right]} = \frac{1}{\left(5 \times 10^{13}\right) \times \left(0.250\right)^2} = 3.2 \times 10^{-13}$$

Think About It

Our answer, which indicates that there is very little Ag^+ compared to $Ag(S_2O_3)_2^{3-}$ in solution, is consistent with the large K_f value for this reaction.

17.87. Collect and Organize

Given the observed colors and magnetic properties of two cobalt complexes, we are asked which has the largest Δ_o.

Analyze

We can presume that both complexes have octahedral geometry. Oxidation converts Co^{2+} to Co^{3+}. Co^{2+} has a d^7 configuration. Because this complex is observed to be purple, it is absorbing relatively low-energy yellow light. For this complex to have three unpaired e^- (high-spin), Δ_o must be small. Co^{3+} has a d^6 configuration. Because this complex is observed to be yellow, it is absorbing relatively high-energy purple light. For this complex to have no unpaired e^- (low-spin), Δ_o must be large.

Solve

The yellow complex containing Co^{3+} in aqueous ammonia has the larger Δ_o.

Think About It

Remember that the spin of a complex is a result of the magnitude of Δ_o.

17.89. Collect and Organize

We are to explain why square planar Ag^{2+} is paramagnetic but square planar AgO, Ag^+, and Ag^{3+} are diamagnetic.

Analyze

The square planar crystal field diagram is

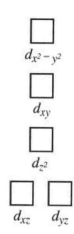

$d_{x^2-y^2}$

d_{xy}

d_{z^2}

d_{xz} d_{yz}

Solve

Ag^{2+} has 9 d electrons, leaving an unpaired electron in the $d_{x^2-y^2}$ orbital to make it paramagnetic. Ag^{3+} has 8 d electrons and Ag^+ has 10 d electrons. Both have all electrons paired, so these silver ions are diamagnetic.

Think About It

Because AgO is diamagnetic we know that Ag^{2+} is not in the compound.

17.91. **Collect and Organize**

As we replace H_2O ligands on $Cu(H_2O)_6{}^{2+}$ with NH_3 ligands, we can use the ligand's relative placement in the spectrochemical series (Table 17.5) to predict whether the color of the series of complexes will shift to longer or shorter wavelengths.

Analyze

In the spectrochemical series, NH_3 is a stronger-field ligand than H_2O. We must keep in mind that the color we see is complementary to the color absorbed.

Solve

As we replace weak-field H_2O ligands with stronger-field NH_3 ligands, the Δ_o for the complex increases. This means that the complex absorbs at shorter wavelengths. The color we see, therefore, shifts to longer wavelengths.

Think About It

Water and ammonia are not far apart in the spectrochemical series, however, so we may not expect the shift in wavelength to be significant.

CHAPTER 18 | Thermodynamics: Spontaneous and Nonspontaneous Reactions and Processes

18.1. Collect and Organize

We are asked to compare the internal pressure and entropy for two balloons of equal volume at the same temperature.

Analyze

Since the volume and temperature are the same, the only thing different between the balloons is the number of particles depicted. We can assume that the pressures will be different based on the difference in the number of particles. Pressure and the number of moles are directly related; as the number of moles increases, so too does the pressure. The entropy of a system (S) is defined using Equation 18.21, and tells us that as the number of microstates (W) increases, so too does the entropy.

Solve

The balloon on the right has a greater number of gas particles, and so it must have the greater pressure. Similarly, the presence of a greater number of gas particles in the balloon at right implies that there are a greater number of microstates (arrangements of those particles), so it also has a higher entropy.

Think About It

We can guess that the balloon on the left requires fewer gas particles to reach the same volume because the external pressure is lower than that for the balloon on the right.

18.3. Collect and Organize

For the system in Figure P18.3 showing a random distribution of two gases, we are to assess the probability that one bulb will collect gas A and the other gas B. We are also asked to determine if the system is isolated.

Analyze

The system is already at high entropy because both gases can randomly move throughout the entire volume.

Solve

The probability that gases A and B will separate in the apparatus so as to occupy separate bulbs is very low. Each gas would then be confined to a smaller volume.

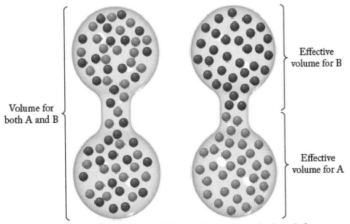

This change would involve a decrease in entropy. The bulbs are not isolated from one another, but they are isolated from the universe outside the vessel.

Think About It

Systems tend to move spontaneously to maximum randomness. In that way we can think of spontaneous changes as tending toward increasing entropy.

18.5. **Collect and Organize**

We are asked to comment on the spontaneity of the process depicted in Figure P18.4 at different temperatures.

Analyze

B_2 (blue) changes very little, whereas A_2 (red) condenses in Figure P18.4. Condensation is an exothermic process, in which molecules move from the gas phase to the liquid phase, releasing heat in the process.

$$A_2(g) \rightarrow A_2(\ell) + \text{heat}$$

Solve

Lowering the temperature will drive this process in the forward direction by Le Châtelier's principle. Condensation will become more spontaneous at lower temperatures.

Think About It

Our solution makes sense; ice cubes are formed spontaneously in your freezer, but not at the warmer temperatures in the rest of your kitchen. Both condensation (gas to liquid) and freezing (liquid to solid) are exothermic phase transitions.

18.7. **Collect and Organize**

We are asked to identify phase changes from those listed that fit the thermodynamic pattern depicted in Figure P18.6.

Analyze

Because $T\Delta S$ decreases with increasing temperature on the graph, ΔS must be negative in Figure P18.6. ΔH has a consistent value over the temperature range depicted, though it is not clear if this value is positive or negative. When the sign of ΔS is negative the process is not favored by entropy, and entropy is decreasing. When the sign of ΔS is positive, the process is favored by entropy, and entropy is increasing. We may categorize the phase changes by their increase or decrease in entropy; those with a negative entropy may correspond to the plot in Figure P18.6.

Solve

The sign of the change in entropy for each phase change is

Melting	Vaporization	Condensation	Freezing	Sublimation	Deposition
S increases (+)	S increases (+)	S decreases (−)	S decreases (−)	S increases (+)	S decreases (−)

Condensation, freezing, and deposition could fit the profile.

Think About It

Freezing could correspond to the plot in Figure P18.6. The point at which the ΔH and $T\Delta S$ lines cross would correspond to the freezing point of this compound.

18.9. **Collect and Organize**

We consider what happens to the sign of ΔS when we reverse a process.

Analyze

When the sign of ΔS is negative, the process is not favored by entropy, and entropy is decreasing. When the sign of ΔS is positive, the process is favored by entropy, and entropy is increasing.

Solve

If a process that is favored by entropy (+ΔS) is reversed, the reverse process will not be favored (−ΔS). Therefore, the ΔS when a process is reversed has its sign reversed.

Think About It

In terms of order and disorder in a chemical or physical process, a process that has increasing disorder (ΔS positive) has increasing order (ΔS negative) when the process is reversed.

18.11. **Collect and Organize**

Given ice cubes (the system) in a glass of lemonade (the surroundings), we are to determine the signs of ΔS for the system and for the surroundings as the ice cools the lemonade from 10.0°C to 0.0°C.

Analyze

Cooling is the result of decreased molecular motion, which is correlated to a decrease in entropy.

Solve

The sign of ΔS_{surr} is negative because the lemonade is cooling, which decreases the motion of the lemonade molecules. The sign of ΔS_{sys} is positive because the ice is melting. When a phase change from solid to liquid occurs in the lemonade, disorder, and therefore entropy, of the ice (the system) increases.

Think About It

In this case, as long as $\Delta S_{sys} > -\Delta S_{surr}$, the process is spontaneous.

18.13. **Collect and Organize**

Given three combinations of signs for changes in entropy for a system, the surroundings, and the universe, we are to determine which combinations are possible.

Analyze

In order for a process to occur, the second law of thermodynamics must be obeyed. This law states that the combination of the change in entropy for a system and the change in entropy for the surroundings must be greater than zero. Therefore, ΔS_{univ}, which is equal to $\Delta S_{sys} + \Delta S_{surr}$, must be greater than zero for a process to occur and

$$\Delta S_{sys} + \Delta S_{surr} > 0$$

Solve

(a) If $\Delta S_{sys} > 0$ and $\Delta S_{surr} > 0$, then $\Delta S_{univ} > 0$ under all conditions. This combination of entropy changes is possible.
(b) If $\Delta S_{sys} > 0$ and $\Delta S_{surr} < 0$, then $\Delta S_{univ} > 0$ when $\Delta S_{sys} > -\Delta S_{surr}$. This combination of entropy changes is possible.
(c) If $\Delta S_{sys} > 0$ and $\Delta S_{surr} > 0$, then ΔS_{univ} must be > 0. ΔS_{univ} cannot be < 0 for these changes in entropy for the system and the surroundings, so this combination of entropy changes is not possible.

Think About It

In part (c), ΔS_{univ} would be calculated as < 0 if ΔS_{sys} and ΔS_{surr} were both < 0 or if $\Delta S_{sys} + \Delta S_{surr} < 0$. This process, however, would not be spontaneous.

18.15. **Collect and Organize**

We are asked to determine the signs of ΔS for the system and for the universe as glucose is formed from carbon dioxide and water by photosynthesis.

Analyze

For a spontaneous reaction, ΔS_{univ} will be positive; a nonspontaneous process will have a negative value of ΔS_{univ}. The sign of ΔS_{sys} may be determined by considering the quantity of reactants and products in the reaction. For a reaction in which many molecules are combined to form a smaller number of more complicated species, the sign of ΔS_{sys} will be negative. The equation for photosynthesis described in Chapter 5 is

$$\text{Sunlight} + 6\ CO_2(g) + 6\ H_2O(\ell) \rightarrow C_6H_{12}O_6(aq) + 6\ O_2(g)$$

Solve

From the equation, small molecules such as CO_2 and H_2O are being combined to form a more ordered species, glucose. The sign of ΔS_{sys} for this reaction will be negative. This reaction is nonspontaneous, as it requires an input of sunlight to drive the reaction to completion. The sign of ΔS_{univ} will be negative.

Think About It

The metabolism of glucose to produce energy (with the by-products CO_2 and H_2O) is a spontaneous reaction for which ΔS_{univ} will be positive.

18.17. Collect and Organize

Given that a chemical reaction has $\Delta S_{sys} = -66.0$ J/K, we are to determine the maximum value of ΔS_{surr} needed in order for the reaction to be nonspontaneous.

Analyze

For a spontaneous process,

$$\Delta S_{univ} = \Delta S_{sys} + \Delta S_{surr} > 0$$

Solve

In order for the reaction to be nonspontaneous, ΔS_{surr} must be less (more negative) than +66.0 J/K.

Think About It

If ΔS_{surr} is more positive than 66.0 J/K, then $\Delta S_{univ} > 0$ and the reaction would be spontaneous.

18.19. Collect and Organize

From the list of equimolar ionic compounds, we are asked to determine which will experience the greatest increase in entropy upon dissolution in water.

Analyze

The entropy of a system will increase when the number of particles increases. For ionic compounds, more moles of ions are liberated than moles of the compound dissolved. The greater the number of ions released when the ionic compound dissolves in solution, the greater the predicted increase in entropy for the system. The balanced equations for the dissolution of these salts are

$$CaCl_2(s) + H_2O(\ell) \rightarrow Ca^{2+}(aq) + 2\ Cl^-(aq)$$
$$NaBr(s) + H_2O(\ell) \rightarrow Na^+(aq) + Br^-(aq)$$
$$KCl(s) + H_2O(\ell) \rightarrow K^+(aq) + Cl^-(aq)$$
$$Cr(NO_3)_3(s) + H_2O(\ell) \rightarrow Cr^{3+}(aq) + 3\ NO_3^-(aq)$$
$$LiOH(s) + H_2O(\ell) \rightarrow Li^+(aq) + OH^-(aq)$$

Solve

The greatest increase in entropy will be observed when (d), $Cr(NO_3)_3$, is dissolved.

Think About It

An even greater increase in entropy would be expected if a species such as aluminum sulfate, $Al_2(SO_4)_3$, were dissolved in solution. Each mole of $Al_2(SO_4)_3$ would be expected to release five moles of ions into solution.

18.21. Collect and Organize

For each of the pairs of species provided, we are asked to identify the molecule expected to have the greater entropy.

Analyze

Entropy increases with the number of atoms, as well as when moving from the solid or liquid phase to the gas phase. It is important to consider the number of atoms as well as the arrangement of atoms in a molecule.

Solve

(a) 1 mole of $S_8(g)$ will have greater entropy than 1 mole of $S_2(g)$. The greater number of S atoms in S_8 will lead to a greater entropy for the system.
(b) 1 mole of $S_2(g)$ will have greater entropy than 1 mole of $S_8(s)$. The more ordered arrangement of S_8 molecules in the solid state will have a greater effect on the entropy than the increase in the number of atoms.

(c) 1 mole of $O_3(g)$ will have greater entropy than 1 mole of $O_2(g)$. The greater number of O atoms in O_3 will lead to a greater entropy for the system.
(d) 1 gram of $O_2(g)$ will have greater entropy than 1 gram of $O_3(g)$. Though both have the same number of O atoms, O_2 is less ordered than O_3, so it has a greater number of microstates available and a higher entropy.

Think About It
We can confirm some of our choices by consulting Appendix 4. As an example, O_2 has $S° = 205.0$ J/K, while O_3 has $S° = 238.8$ J/K.

18.23. **Collect and Organize**
We are to predict which has the higher standard molar entropy, diamond or the fullerenes.

Analyze
Diamond is a three-dimensional, highly ordered network of covalently bonded carbon atoms (see Chapter 5, Carbon: Diamonds, Graphite, and the Molecules of Life). Fullerenes are also made up of covalently bonded carbon atoms, but they form discrete structures instead of an extended network.

Solve
Fullerenes, with less extensive bonding, have a higher standard molar entropy than diamond.

Think About It
We would expect the standard molar entropy of graphite to be in between that of diamond and the fullerenes because of its intermediate structure.

18.25. **Collect and Organize**
By considering the phase and size of a compound, we can rank the compounds in each series in order of increasing standard molar entropy.

Analyze
Generally speaking, compounds that are larger in size have greater $S°$.

Solve
(a) $CH_4(g) < CF_4(g) < CCl_4(g)$
(b) $CH_2O(g) < CH_3CHO(g) < CH_3CH_2CHO(g)$
(c) $HF(g) < H_2O(g) < NH_3(g)$

Think About It
Larger molecules have greater $S°$ because they have more opportunities for internal motion.

18.27. **Collect and Organize**
We are to determine the sign of the entropy change for the overall process when the products of a process have a greater entropy than the reactants.

Analyze
The entropy change for a process is calculated by
$$\Delta S_{rxn} = \sum nS_{products} - \sum mS_{reactants}$$

Solve
If $nS_{products} > mS_{reactants}$, then ΔS_{rxn} for the process is positive.

Think About It
A positive ΔS_{rxn} means that the process resulted in less order in going to products.

18.29. **Collect and Organize**
We are asked to predict whether precipitation reactions tend to have positive or negative ΔS_{rxn} values.

Analyze

A precipitation reaction is one in which a solid is formed from soluble aqueous ions, such as in the generic net ionic equation

$$m\,A^{n+}(aq) + n\,B^{m-}(aq) \rightarrow A_mB_n(s)$$

In the absence of any information about the exact stoichiometry of the reaction, we may generalize based on the knowledge that species in solution combine in some proportion to form a solid.

Solve

Two factors suggest that ΔS_{rxn} will be negative for a precipitation reaction. First, solids have lower entropy than soluble ions due to their decreased vibrational energy. Secondly, two or more ions are forming one solid, so a decrease in entropy is expected.

Think About It

In part, the value of ΔS_{rxn} for a precipitation reaction will depend on the number of ions assembling to form the solid; the greater the number of ions combined, the greater (more negative) the change in entropy.

18.31. **Collect and Organize**

Using values of $S°$ for the reactants and products for four atmospheric reactions, we are to calculate $\Delta S°_{rxn}$. Standard molar entropies are in Appendix 4.

Analyze

A change in entropy for a reaction is

$$\Delta S°_{rxn} = \sum nS°_{products} - \sum mS°_{reactants}$$

Solve

(a) $\Delta S°_{rxn} = \left(2\text{ mol NO} \times \dfrac{210.7\text{ J}}{\text{mol}\cdot\text{K}}\right)$

$\qquad\qquad -\left[\left(1\text{ mol N}_2 \times \dfrac{191.5\text{ J}}{\text{mol}\cdot\text{K}}\right) + \left(1\text{ mol O}_2 \times \dfrac{205.0\text{ J}}{\text{mol}\cdot\text{K}}\right)\right]$

$\qquad = 24.9\text{ J/K}$

(b) $\Delta S°_{rxn} = \left(2\text{ mol NO}_2 \times \dfrac{240.0\text{ J}}{\text{mol}\cdot\text{K}}\right)$

$\qquad\qquad -\left[\left(2\text{ mol NO} \times \dfrac{210.7\text{ J}}{\text{mol}\cdot\text{K}}\right) + \left(1\text{ mol O}_2 \times \dfrac{205.0\text{ J}}{\text{mol}\cdot\text{K}}\right)\right]$

$\qquad = -146.4\text{ J/K}$

(c) $\Delta S°_{rxn} = \left(1\text{ mol NO}_2 \times \dfrac{240.0\text{ J}}{\text{mol}\cdot\text{K}}\right)$

$\qquad\qquad -\left[\left(1\text{ mol NO} \times \dfrac{210.7\text{ J}}{\text{mol}\cdot\text{K}}\right) + \left(\tfrac{1}{2}\text{ mol O}_2 \times \dfrac{205.0\text{ J}}{\text{mol}\cdot\text{K}}\right)\right]$

$\qquad = -73.2\text{ J/K}$

(d) $\Delta S°_{rxn} = \left(1\text{ mol N}_2O_4 \times \dfrac{304.2\text{ J}}{\text{mol}\cdot\text{K}}\right) - \left(2\text{ mol NO}_2 \times \dfrac{240.0\text{ J}}{\text{mol}\cdot\text{K}}\right)$

$\qquad = -175.8\text{ J/K}$

Think About It

Notice that because the balanced equation in part (b) is twice that in part (c) the value of the entropy change for the reaction is doubled.

18.33. Collect and Organize

We are asked to calculate ΔS_{surr} for the melting of ice at 0°C on a granite surface at 12°C.

Analyze

To calculate ΔS_{surr} under reversible conditions, we must consider the heat flow between the system and the surroundings, and the temperature of the system or surroundings. Under reversible conditions, the relationship between ΔS_{surr} and heat is

$$\Delta S_{surr} = \frac{q_{surr}}{T}$$

We are given the enthalpy of fusion of ice (6.01 kJ/mol) and the initial temperature of the granite block (12°C, 285 K). The number of moles of ice present are

$$n_{H_2O} = 456 \text{ g } H_2O \times \frac{1 \text{ mol } H_2O}{18.02 \text{ g } H_2O} = 25.31 \text{ mol } H_2O$$

Since heat is flowing from the granite block into the ice, we should expect that the sign of ΔS_{surr} will be negative.

Solve

$$\Delta S_{surr} = \frac{25.31 \text{ mol } H_2O \left(-6.01 \times 10^3 \text{ J/mol}\right)}{285 \text{ K}} = -534 \text{ J/K}$$

Think About It

Our prediction was correct; the sign of ΔS_{surr} is negative. The sign of ΔS_{sys} is thus positive, and since the process is spontaneous, the sign of ΔS_{univ} is also positive.

18.35 Collect and Organize

We are asked to determine the relationship between the sign of ΔG and spontaneity.

Analyze

Spontaneous reactions have positive values of ΔS_{univ}. By definition
$$\Delta G_{sys} = -T \Delta S_{univ}$$

Solve

For a spontaneous reaction, ΔS_{univ} will be positive, leading to a negative value of ΔG_{sys}.

Think About It

A spontaneous reaction may be either exothermic or endothermic according to the equation
$$\Delta G = \Delta H - T \Delta S$$

18.37. Collect and Organize

We are to discuss a possible explanation for the historical belief that all exothermic reactions are spontaneous.

Analyze

A reaction is spontaneous when its free-energy change is negative according to the equation
$$\Delta G = \Delta H - T \Delta S$$
When H is negative, the reaction is exothermic.

Solve

Though the assumption that all exothermic reactions are spontaneous is incorrect, most of the exothermic reactions observed by 19th-century scientists were spontaneous at the temperatures they were being observed at. As an example, most hydrocarbon combustion reactions are exothermic, and result in an increase in the entropy of the system. In this case, ΔS_{sys} is positive, and ΔS_{surr} is positive, so ΔS_{univ} will always be positive, leading to a spontaneous reaction.

Think About It

A reaction that is exothermic is nonspontaneous when its entropy change is negative and the reaction is run at "high temperature." If the value of $T\Delta S$ is more negative than the value of ΔH, then ΔG would be positive and the reaction would be nonspontaneous.

18.39. **Collect and Organize**

For the sublimation of dry ice at room temperature, we are to determine the signs of ΔS, ΔH, and ΔG.

Analyze

The reaction describing the sublimation is

$$CO_2(s) \rightarrow CO_2(g)$$

Solve

ΔS is positive because the solid is subliming to a gas, which has much greater entropy.
ΔH is positive because heat is required to effect the phase change.
ΔG is negative because at room temperature (25°C) this process spontaneously occurs.

Think About It

The spontaneity of this process is temperature dependent. At lower temperatures the value of $T\Delta S$ in the equation

$$\Delta G = \Delta H - T\Delta S$$

would not be great enough to give a negative value for ΔG.

18.41. **Collect and Organize**

For the processes described, we are to determine whether they are spontaneous.

Analyze

Spontaneous processes occur without outside intervention once they are started.

Solve

(a) A tornado does not form spontaneously, it requires certain weather conditions.
(b) Unfortunately, a broken cell phone does not spontaneously mend itself.
(c) You will have to put effort in to get an A in this course—this is a nonspontaneous process!
(d) Provided the room is cooler than the soup, it will cool spontaneously as heat escapes the bowl.

Think About It

Spontaneous processes are favored either by enthalpy, entropy, or both according to the equation
$$\Delta G = \Delta H - T\Delta S$$

18.43. **Collect and Organize**

For the dissolution of NaBr and NaI in water, we are to calculate the value of $\Delta G°$ knowing the $\Delta H°_{sol}$ and $\Delta S°_{sol}$ for each of these soluble salts.

Analyze

We can calculate the value of $\Delta G°$ from the standard state enthalpy and entropy of the reaction using
$$\Delta G°_{rxn} = \Delta H°_{rxn} - T\Delta S°_{rxn}$$
where $T = 298$ K.

Solve

For NaBr

$$\Delta G^{\circ}_{rxn} = -1 \text{ kJ/mol} - \left(298 \text{ K} \times \frac{0.057 \text{ kJ}}{\text{mol} \cdot \text{K}}\right) = -18 \text{ kJ/mol}$$

For NaI

$$\Delta G^{\circ}_{rxn} = -7 \text{ kJ/mol} - \left(298 \text{ K} \times \frac{0.074 \text{ kJ}}{\text{mol} \cdot \text{K}}\right) = -29 \text{ kJ/mol}$$

Think About It

Be sure to use consistent units for ΔH° and ΔS° in calculations. In this problem, we have to change the ΔS° values given in J/(mol $\cdot$ K) to kJ/(mol $\cdot$ K).

18.45. **Collect and Organize**

For the reaction of C(*s*) with H_2O(*g*) to produce H_2(*g*) and CO(*g*), we are to calculate the value of ΔG°_{rxn} from the data listed in Appendix 4:

$$\Delta G^{\circ}_{rxn} = \sum n\Delta G^{\circ}_{f,products} - \sum m\Delta G^{\circ}_{f,reactants}$$

Analyze

To calculate the value of ΔG°_{rxn}, we may use tabulated values for ΔG°_f found in Appendix 4. For a pure species, $\Delta G^{\circ}_f = 0$ and for a spontaneous reaction ΔG is negative.

Solve

$$\Delta G^{\circ}_{rxn} = \left[\left(1 \text{ mol } H_2 \times 0.0 \text{ kJ/mol}\right) + \left(1 \text{ mol CO} \times -137.2 \text{ kJ/mol}\right)\right]$$
$$- \left[\left(1 \text{ mol } H_2O \times -228.6 \text{ kJ/mol}\right) + \left(1 \text{ mol C} \times 0.0 \text{ kJ/mol}\right)\right]$$
$$= 91.4 \text{ kJ}$$

Think About It

For this calculation we used carbon in the form of graphite. We would not want to use diamond as a reactant!

18.47. **Collect and Organize**

From tabulated values of ΔH° and ΔS° for the listed species, we are asked to calculate ΔG° for the reduction of iron oxide at 25°C.

Analyze

We can calculate ΔG° from the standard state enthalpy and entropy of the reaction using

$$\Delta G^{\circ}_{rxn} = \Delta H^{\circ}_{rxn} - T\Delta S^{\circ}_{rxn}$$

where ΔH°_{rxn} and ΔS°_{rxn} may be determined by

$$\Delta H^{\circ}_{rxn} = \sum n\Delta H^{\circ}_{f,products} - \sum n\Delta H^{\circ}_{f,reactants}$$
$$\Delta S^{\circ}_{rxn} = \sum n\Delta S^{\circ}_{f,products} - \sum n\Delta S^{\circ}_{f,reactants}$$

and the temperature is 298 K.

Solve

$$\Delta H^{\circ}_{rxn} = \left[\left(2 \text{ mol Fe} \times 0 \text{ kJ/mol}\right) + \left(3 \text{ mol H}_2\text{O} \times \frac{-241.8 \text{ kJ}}{\text{mol}}\right) \right]$$

$$- \left[\left(1 \text{ mol Fe}_2\text{O}_3 \times \frac{-824.2 \text{ kJ}}{\text{mol}}\right) + \left(3 \text{ mol H}_2 \times 0 \text{ kJ/mol}\right) \right]$$

$$= +98.8 \text{ kJ}$$

$$\Delta S^{\circ}_{rxn} = \left[\left(2 \text{ mol Fe} \times \frac{27.3 \text{ J}}{\text{K} \cdot \text{mol}}\right) + \left(3 \text{ mol H}_2\text{O} \times \frac{188.8 \text{ J}}{\text{K} \cdot \text{mol}}\right) \right]$$

$$- \left[\left(1 \text{ mol Fe}_2\text{O}_3 \times \frac{87.4 \text{ J}}{\text{K} \cdot \text{mol}}\right) + \left(3 \text{ mol H}_2 \times \frac{130.6 \text{ J}}{\text{K} \cdot \text{mol}}\right) \right]$$

$$= +141.8 \text{ J/K}$$

Using these values for ΔH° and ΔS°, we may determine ΔG°

$$\Delta G^{\circ}_{rxn} = 98.8 \text{ kJ} - 298 \text{ K} \left(0.1418 \text{ kJ/K}\right) = +56.5 \text{ kJ}$$

Think About It

At 298 K, this reaction is nonspontaneous. As the temperature is increased, $T\Delta S$ will be larger than ΔH, leading to a spontaneous reaction.

18.49. **Collect and Organize**

We are asked to calculate the value of ΔG°_{rxn} for the formation of sulfuric acid in rain droplets using standard free energies of formation.

Analyze

From the values of ΔG°_f in Appendix 4, we can calculate ΔG°_{rxn} by

$$\Delta G^{\circ}_{rxn} = \sum n \, \Delta G^{\circ}_{f,\text{products}} - \sum m \Delta G^{\circ}_{f,\text{reactants}}$$

Solve

$$\Delta G^{\circ}_{rxn} = \left(1 \text{ mol H}_2\text{SO}_4 \times \frac{-690.0 \text{ kJ}}{\text{mol}}\right)$$

$$- \left[\left(1 \text{ mol SO}_3 \times \frac{-371.1 \text{ kJ}}{\text{mol}}\right) + \left(1 \text{ mol H}_2\text{O} \times \frac{-237.2 \text{ kJ}}{\text{mol}}\right) \right]$$

$$= -81.7 \text{ kJ}$$

Think About It

This reaction is exothermic, with decreasing entropy. Therefore, as the temperature decreases from 298 K, the spontaneity of the reaction will increase (more negative value of ΔG).

18.51. **Collect and Organize**

We consider whether exothermic reactions are spontaneous only at low temperature.

Analyze

Spontaneity is shown by a negative free-energy value (ΔG) in the equation

$$\Delta G = \Delta H - T\Delta S$$

For exothermic reactions, we also know that ΔH is negative.

Solve

No. As we can see from the equation, if ΔS is positive then the value of $T\Delta S$ would be positive. In that case ΔG would always be negative, regardless of the temperature, and the exothermic reaction would be spontaneous at all temperatures.

Think About It
If ΔS is negative for an exothermic reaction, then it is true that the reaction would be spontaneous only at low temperatures.

18.53. **Collect and Organize**
For the reaction of C(s) with $H_2O(g)$ to produce $H_2(g)$ and CO(g), we are to calculate the lowest temperature at which the reaction is spontaneous. Using

$$\Delta G^\circ_{rxn} = \Delta H^\circ_{rxn} - T\Delta S^\circ_{rxn}$$

we can predict the lowest temperature at which the reaction is spontaneous.

Analyze
To calculate the lowest temperature at which CO and H_2 form from C and steam, we must first calculate ΔH°_{rxn} and ΔS°_{rxn} using Appendix 4. For a spontaneous reaction, ΔG is negative. Therefore, if we set $\Delta G = 0$ and solve for T, that would give the temperature at which the reaction changes spontaneity.

$$\Delta G = 0 = \Delta H - T\Delta S$$

$$T = \frac{\Delta H}{\Delta S}$$

Be sure to have consistent units for ΔH and ΔS for this calculation.

Solve
To determine the lowest temperature at which this reaction is spontaneous, we must first calculate ΔH° and ΔS°:

$$\Delta H^\circ_{rxn} = \left[\left(1 \text{ mol } H_2 \times 0.0 \text{ kJ/mol}\right) + \left(1 \text{ mol CO} \times \frac{-110.5 \text{ kJ}}{\text{mol}}\right) \right]$$
$$- \left[\left(1 \text{ mol } H_2O \times \frac{-241.8 \text{ kJ}}{\text{mol}}\right) + \left(1 \text{ mol C} \times 0.0 \text{ kJ/mol}\right) \right]$$
$$= 131.3 \text{ kJ}$$

$$\Delta S^\circ_{rxn} = \left[\left(1 \text{ mol } H_2 \times \frac{130.6 \text{ J}}{\text{mol} \cdot K}\right) + \left(1 \text{ mol CO} \times \frac{197.7 \text{ J}}{\text{mol} \cdot K}\right) \right]$$
$$- \left[\left(1 \text{ mol } H_2O \times \frac{188.8 \text{ J}}{\text{mol} \cdot K}\right) + \left(1 \text{ mol C} \times \frac{5.7 \text{ J}}{\text{mol} \cdot K}\right) \right]$$
$$= 133.8 \text{ J/K}$$

Solving for T when $\Delta G^\circ_{rxn} = 0$ gives

$$T = \frac{131.3 \text{ kJ}}{0.1338 \text{ kJ/K}} = 981.3 \text{ K or } 708.2°C$$

The lowest temperature at which this reaction is spontaneous is just above 981.3 K or 708.2°C.

Think About It
The combustion of graphite is not spontaneous under standard conditions, making it a reasonable choice for industrial applications where the temperature will not exceed 708.2°C.

18.55. **Collect and Organize**
For the vaporization of hydrogen peroxide at 1.00 atm we are to calculate ΔH° and ΔS°. Assuming that calculated quantities (using data listed in Appendix 4 at 298 K) are not temperature dependent, we are to calculate the boiling point of hydrogen peroxide at 1.00 atm.

Analyze
To calculate both ΔH° and ΔS° for the phase change

$$H_2O_2(\ell) \rightarrow H_2O_2(g)$$

we need the following (from Appendix 4): ΔH_f° of $H_2O_2(\ell)$ = −187.8 kJ/mol, ΔH_f° of $H_2O_2(g)$ = −136.3 kJ/mol, S° of $H_2O_2(\ell)$ = 109.6 J/(mol·K), and S° of $H_2O_2(g)$ = 232.7 J/(mol·K).

To calculate the boiling point, we can rearrange the Gibbs free-energy equation to solve for the temperature where $\Delta G = 0$ because at the boiling point $H_2O_2(\ell)$ and $H_2O_2(g)$ are in equilibrium

$$\Delta G_{rxn} = 0 = \Delta H_{rxn}^\circ - T\Delta S_{rxn}^\circ$$

$$T_b = \frac{\Delta H_{rxn}^\circ}{\Delta S_{rxn}^\circ}$$

Solve

$$\Delta H_{rxn}^\circ = \left(1 \text{ mol } H_2O_2(g) \times \frac{-136.3 \text{ kJ}}{\text{mol}}\right) - \left[1 \text{ mol } H_2O_2(\ell) \times \frac{-187.8 \text{ kJ}}{\text{mol}}\right]$$

$$= 51.5 \text{ kJ}$$

$$\Delta S_{rxn}^\circ = \left[1 \text{ mol } H_2O_2(g) \times 232.7 \text{ J/(mol·K)}\right] - \left[1 \text{ mol } H_2O_2(\ell) \times 109.6 \text{ J/(mol·K)}\right]$$

$$= 123.1 \text{ J/K or } 0.1231 \text{ kJ/K}$$

$$T_b = \frac{51.5 \text{ kJ}}{0.1231 \text{ kJ/K}} = 418.4 \text{ K or } 145.2°C$$

Think About It

There is some temperature dependence of ΔH and ΔS, as the calculated value for the boiling point is lower than the actual value of 150°C.

18.57. **Collect and Organize**

For the three listed reactions, we are asked to determine which are spontaneous at low temperatures, which are spontaneous at high temperatures, and which are spontaneous at all temperatures.

Analyze

To calculate the spontaneity of the reaction at a temperature other than 298 K, we consider the equation

$$\Delta G^\circ = \Delta H^\circ - T\Delta S^\circ$$

We use data in Appendix 4 to also calculate ΔH_{rxn}° (and ΔS_{rxn}° if necessary).

The spontaneity of a reaction as a function of ΔH_{rxn}° and ΔS_{rxn}° may be summarized as follows:

	ΔH_{rxn}°	ΔS_{rxn}°
Spontaneous at all T	−	+
Spontaneous only at low T	−	−
Spontaneous only at high T	+	+

Solve

(a) ΔS will be negative for this process, as three moles of gas are converted to two moles of gas. The increase in molecular complexity renders ΔS negative. ΔH° for this reaction is

$$\Delta H_{rxn}^\circ = \left[\left(2 \text{ mol } NO_2 \times \frac{+33.2 \text{ kJ}}{\text{mol}}\right)\right]$$

$$- \left[\left(1 \text{ mol } O_2 \times 0 \text{ kJ/mol}\right) + \left(2 \text{ mol } NO \times \frac{+90.3 \text{ kJ}}{\text{mol}}\right)\right]$$

$$= -114.2 \text{ kJ}$$

Both ΔH and ΔS for this reaction are negative, so the reaction will be spontaneous only at low temperatures.

(b) The value of ΔS for this reaction will be close to zero because neither the number of moles nor the phase changes; we will have to calculate both ΔH and ΔS.

$$\Delta H^{\circ}_{rxn} = \left[\left(1 \text{ mol N}_2\text{O} \times \frac{82.1 \text{ kJ}}{\text{mol}}\right) + \left(3 \text{ mol H}_2\text{O} \times \frac{-241.8 \text{ kJ}}{\text{mol}}\right)\right]$$
$$- \left[\left(2 \text{ mol NH}_3 \times \frac{-46.1 \text{ kJ}}{\text{mol}}\right) + \left(2 \text{ mol O}_2 \times 0 \text{ kJ/mol}\right)\right]$$
$$= -551.1 \text{ kJ}$$

$$\Delta S^{\circ}_{rxn} = \left[\left(1 \text{ mol N}_2\text{O} \times \frac{219.9 \text{ J}}{\text{mol} \cdot \text{K}}\right) + \left(3 \text{ mol H}_2\text{O} \times \frac{188.8 \text{ J}}{\text{mol} \cdot \text{K}}\right)\right]$$
$$- \left[\left(2 \text{ mol NH}_3 \times \frac{192.5 \text{ J}}{\text{mol} \cdot \text{K}}\right) + \left(2 \text{ mol O}_2 \times \frac{205.0 \text{ J}}{\text{mol} \cdot \text{K}}\right)\right]$$
$$= -8.7 \text{ J/K}$$

Both ΔH and ΔS for this reaction are negative, so the reaction will be spontaneous only at low temperatures.
(c) ΔS will be positive for this process, as three moles of gas are generated for each mole of solid that decomposes. The increase in the number of molecules, as well as the change from the solid phase to the gas phase, lead to a positive value of ΔS. ΔH° for this reaction is

$$\Delta H^{\circ}_{rxn} = \left[\left(2 \text{ mol H}_2\text{O} \times \frac{-241.8 \text{ kJ}}{\text{mol}}\right) + \left(1 \text{ mol N}_2\text{O} \times \frac{+82.1 \text{ kJ}}{\text{mol}}\right)\right]$$
$$- \left[\left(1 \text{ mol NH}_4\text{NO}_3 \times \frac{-365.6 \text{ kJ}}{\text{mol}}\right)\right]$$
$$= -35.9 \text{ kJ}$$

For this reaction ΔH is negative and ΔS is positive, so the reaction will be spontaneous at all temperatures.

Think About It
Though ΔS for (b) is negative, the value is close to zero. This tells us that the reaction is nearly spontaneous at all temperatures. By substituting these values into the equation below and setting ΔG equal to zero, we may determine the temperature below which the reaction is spontaneous

$$\Delta G_{rxn} = \Delta H^{\circ}_{rxn} - T\Delta S^{\circ}_{rxn} = 0$$

$$T = \frac{\Delta H^{\circ}_{rxn}}{\Delta S^{\circ}_{rxn}} = \frac{-551.1 \text{ kJ}}{-0.0087 \text{ kJ/K}} = 63,345 \text{ K}$$

The reaction will be spontaneous at temperatures below 63,345 K; this is essentially always spontaneous!

18.59. Collect and Organize
We are to determine the sign of ΔG° when $K < 1$.

Analyze
The equation relating ΔG° to K is

$$\Delta G^{\circ} = -RT \ln K$$

Solve
The value of ΔG° is positive when $K < 1$. Because $\ln K$ is negative for $K < 1$ and because $\Delta G^{\circ} = -RT \ln K$, the value of ΔG° is positive.

Think About It
This means that reactions with $K < 1$ are not spontaneous.

18.61. **Collect and Organize**

If a reaction starts with 100% reactants, we are to determine if the reaction will proceed even if $\Delta G° > 0$.

Analyze

The equation relating $\Delta G°$ to K is

$$\Delta G° = -RT \ln K$$

Solve

When $\Delta G° > 0$, then $\ln K$ of the reaction is negative, giving $K < 1$. The reaction favors the formation of reactants, so the reaction will not shift to the right.

Think About It

The reaction described in this problem is not likely to ever form products under these conditions! A reaction consisting of pure reactants with $\Delta G° > 0$ will remain stable with respect to this reaction indefinitely.

18.63. **Collect and Organize**

Given three reactions with their associated $\Delta G°$ values, we are asked to determine which reaction has the largest value of K_p at 25°C.

Analyze

The equilibrium constant of a reaction is related to the Gibbs free energy through the equation
$$\Delta G° = -RT \ln K$$
Rearranging this equation gives

$$\ln K = \frac{-\Delta G°}{RT}$$

From this we see that the more negative the value of ΔG, the larger K will be.

Solve

Two of the reactions have negative values of $\Delta G°$ and are spontaneous (b and c); they will have $K > 1$. The reaction with the more negative free energy (c) has the largest equilibrium constant.

Think About It

The value of K for reaction c is

$$\ln K = \frac{-\Delta G°}{RT} = \frac{27.9 \text{ kJ}}{\left[8.314 \times 10^{-3} \text{ kJ/(mol} \cdot \text{K)}\right](298 \text{ K})} = 11.26$$

$$K = e^{11.26} = 7.8 \times 10^4$$

18.65. **Collect and Organize**

By consulting Appendix 5 for tabulated values of K, we are to determine the value of $\Delta G°_{rxn}$ at 298 K for

$$NH_3(g) + H_2O(\ell) \rightleftharpoons NH_4^+(aq) + OH^-(aq)$$

Analyze

By definition, the reaction depicted in this problem is the base hydrolysis reaction of ammonia, with an associated K_b value of 1.76×10^{-5}. The equilibrium constant of a reaction is related to the Gibbs free energy through the equation

$$\Delta G° = -RT \ln K$$

where the temperature is 298 K, and R is 8.314 J/(mol $\cdot$ K).

Solve

$$\Delta G^{\circ}_{rxn} = -[8.314 \text{ J/(K} \cdot \text{mol)}](298 \text{ K}) \ln(1.76 \times 10^{-5})$$

$$= 27{,}124 \text{ J/mol} \times \frac{1 \text{ kJ}}{1000 \text{ J}} = +27.1 \text{ kJ/mol}$$

Think About It

The value of K for this reaction is less than one, in keeping with the positive value of ΔG°. Both of these values correspond to a reaction that favors reactants over products under the listed conditions.

18.67. **Collect and Organize**

Using the relationship

$$\ln\left(\frac{K_2}{K_1}\right) = -\frac{\Delta H^{\circ}}{R}\left(\frac{1}{T_2} - \frac{1}{T_1}\right)$$

we can determine whether a reaction is endothermic or exothermic, considering the value of K decreases with increasing T.

Analyze

In the equation, $K_2 < K_1$ so $\ln(K_2/K_1) < 0$ and $T_2 > T_1$ so $(1/T_2 - 1/T_1) < 0$.

Solve

In order for $\ln(K_2/K_1)$ to be negative when $(1/T_2 - 1/T_1)$ is also negative, the value of ΔH must be negative and the reaction must be exothermic.

Think About It

An endothermic reaction, on the other hand, shows an increased K with increasing T because $\Delta H > 0$ and thus $\ln(K_2/K_1) > 0$, so $K_2/K_1 > 0$.

18.69. **Collect and Organize**

Given that K_p for the reaction of CO with H_2O increases as the temperature decreases, we can use the relationship

$$\ln\left(\frac{K_2}{K_1}\right) = -\frac{\Delta H^{\circ}}{R}\left(\frac{1}{T_2} - \frac{1}{T_1}\right)$$

to determine whether the reaction is endothermic or exothermic.

Analyze

In the equation $K_2 > K_1$ so $\ln(K_2/K_1) > 0$. Since $T_2 < T_1$, $(1/T_2 - 1/T_1) > 0$.

Solve

In order for $\ln(K_2/K_1)$ to be positive when $(1/T_2 - 1/T_1)$ is also positive, ΔH must be negative and the reaction must be exothermic.

Think About It

An endothermic reaction, on the other hand, shows a decreased K with a decrease in temperature.

18.71. **Collect and Organize**

Given $\Delta H^{\circ} = 180.6$ kJ and $K_c = 4.10 \times 10^{-4}$ for the conversion of N_2 and O_2 to NO at 2000°C, we are to calculate the value of K_c at 25°C.

Analyze

The relationship between equilibrium constants at two different temperatures is given by

$$\ln\left(\frac{K_2}{K_1}\right) = -\frac{\Delta H^\circ}{R}\left(\frac{1}{T_2} - \frac{1}{T_1}\right)$$

where ΔH° is the enthalpy of the reaction (in J/mol), R is the gas constant [in J/(mol·K)], and T_1 and T_2 are the temperatures (in Kelvin).

Solve

$$\ln\left(\frac{4.1\times10^{-4}}{K_2}\right) = -\frac{1.806\times10^5 \text{ J/mol}}{8.314 \text{ J/(mol·K)}}\left(\frac{1}{2273 \text{ K}} - \frac{1}{298 \text{ K}}\right)$$

$$\ln\left(\frac{4.1\times10^{-4}}{K_2}\right) = 63.337$$

$$\frac{4.1\times10^{-4}}{K_2} = e^{63.337} = 3.214\times10^{27}$$

$$K_2 = 1.3\times10^{-31}$$

Think About It

As we would expect, the equilibrium constant for this endothermic reaction decreases as the reaction is cooled.

18.73. Collect and Organize

Given that $K_1 = 1.5 \times 10^5$ at 430°C and $K_2 = 23$ at 1000°C, we are to calculate the standard enthalpy of the reaction between NO and O_2 to produce NO_2.

Analyze

We can use the following equation to solve this problem:

$$\ln\left(\frac{K_2}{K_1}\right) = -\frac{\Delta H^\circ}{R}\left(\frac{1}{T_2} - \frac{1}{T_1}\right)$$

Solve

$$\ln\left(\frac{23}{1.5\times10^5}\right) = -\frac{\Delta H^\circ}{8.314 \text{ J/(mol·K)}}\left(\frac{1}{1273 \text{ K}} - \frac{1}{703 \text{ K}}\right)$$

$$\Delta H^\circ = -1.15\times10^5 \text{ J/mol or } -115 \text{ kJ/mol}$$

Think About It

The result that this reaction, where K decreases with increasing T, is exothermic is consistent with our answer to Problem 18.72.

18.75. Collect and Organize

We are asked to describe the conditions required for a spontaneous reaction to couple with a nonspontaneous reaction so that the overall process is spontaneous.

Analyze

For a spontaneous reaction, ΔG is less than zero, whereas for a nonspontaneous reaction, ΔG is greater than zero.

Solve

For the overall reaction to be spontaneous, the sum of ΔG for each component reaction must also be negative. To truly consider the reactions as a pair, some of the products of one reaction must be reactants for the second reaction.

Think About It

Spontaneous reactions also have $K > 1$, whereas nonspontaneous reactions have $K < 1$. For the sum of the two reactions to be spontaneous, $K_{\text{spontaneous}} > \dfrac{1}{K_{\text{nonspontaneous}}}$

18.77. Collect and Organize

By examining the structures of glucose 6-phosphate and fructose 6-phosphate in Figure P18.77 we can explain why the conversion between these two sugars has a $\Delta G°$ close to zero.

Analyze

Changes in free energy result from changes in enthalpy (related to the number and types of chemical bonds) and changes in entropy (disorder in the system).

Solve

In the conversion of glucose 6-phosphate to fructose 6-phosphate the six-membered ring becomes a five-membered ring. The bond arrangements are only slightly different between the two structures and so the enthalpies and entropies of the product and reactant are very close in value.

Think About It

Glucose 6-phosphate and fructose 6-phosphate are structural isomers of each other.

18.79. Collect and Organize

By consulting Appendix 4 for tabulated values of $\Delta G_f°$, we are to determine the value of $\Delta G_{\text{rxn}}°$ for the direct conversion of methane into hydrogen gas and carbon, as well as for the oxidation of carbon to form carbon dioxide at 298 K. We are then asked to combine these two equations, calculate $\Delta G_{\text{rxn}}°$ for the resultant equation, and determine if the reaction is spontaneous under standard conditions.

Analyze

From the values of $\Delta G_f°$ in Appendix 4, we can calculate $\Delta G_{\text{rxn}}°$ by

$$\Delta G_{\text{rxn}}° = \sum n\,\Delta G_{f,\text{products}}° - \sum m\Delta G_{f,\text{reactants}}°$$

where the temperature is 298 K, and R is 8.314 J/(mol·K). For both reactions, we may assume that elemental carbon is present as graphite. When reactions are summed, the values for $\Delta G_{\text{rxn}}°$ for each reaction are also summed.

Solve

(a)

$$\Delta G_{\text{rxn}}° = \left[\left(1 \text{ mol C} \times 0 \text{ kJ/mol}\right) + \left(2 \text{ mol H}_2 \times 0 \text{ kJ/mol}\right) \right]$$
$$- \left[\left(1 \text{ mol CH}_4 \times \frac{-50.8 \text{ kJ}}{\text{mol}}\right) \right]$$
$$= +50.8 \text{ kJ}$$

$$\Delta G_{\text{rxn}}° = \left[\left(1 \text{ mol CO}_2 \times \frac{-394.4 \text{ kJ}}{\text{mol}}\right) \right]$$
$$- \left[\left(1 \text{ mol C} \times 0 \text{ kJ/mol}\right) + \left(1 \text{ mol O}_2 \times 0 \text{ kJ/mol}\right) \right]$$
$$= -394.4 \text{ kJ}$$

(b)

$$CH_4(g) \rightarrow C(s,\text{graphite}) + 2\,H_2(g)$$

$$\underline{C(s,\text{graphite}) + O_2(g) \rightarrow CO_2(g)}$$

$$CH_4(g) + O_2(g) \rightarrow CO_2(g) + 2\,H_2(g)$$

For the overall reaction

$$\Delta G^\circ_{\text{rxn}} = +50.8 \text{ kJ} + (-394.4 \text{ kJ}) = -343.6 \text{ kJ}$$

The reaction is spontaneous under standard conditions.

Think About It

If the experimental details were worked out for this reaction in such a way that it could be run on an industrial scale, it would be energetically favorable to produce hydrogen by this process rather than the water–gas shift reaction and steam reformation described in Problem 18.78.

18.81. **Collect and Organize**

We can use the equation

$$\Delta G^\circ = -RT \ln K$$

to calculate the value of K at 298 K for the phosphorylation of glucose, which has $\Delta G^\circ = 13.8$ kJ/mol.

Analyze

Rearranging the equation to solve for $\ln K$ gives

$$\ln K = \frac{-\Delta G^\circ}{RT}$$

where $R = 8.314 \times 10^{-3}$ kJ/(mol·K).

Solve

$$\ln K = \frac{-13.8 \text{ kJ/mol}}{\left[8.314 \times 10^{-3} \text{ kJ/(mol·K)}\right] \times 298 \text{ K}} = -5.57$$

$$K = e^{-5.57} = 3.81 \times 10^{-3}$$

Think About It

When ΔG is positive the value of K is less than 1 and the reactants are favored at equilibrium over the products.

18.83. **Collect and Organize**

From the value of K_c for the hydrolysis of sucrose ($K_c = 5.3 \times 10^{12}$ at 298 K), we are to calculate the value of ΔG°.

Analyze

The value of ΔG° is calculated from K using the equation

$$\Delta G^\circ = -RT \ln K$$

where $R = 8.314$ J/(mol·K) and T is the temperature in Kelvin.

Solve

$$\Delta G^\circ = -8.314 \frac{\text{J}}{\text{mol·K}} \times 298 \text{ K} \times \ln(5.3 \times 10^{12}) = -7.3 \times 10^4 \text{ J/mol or } -73 \text{ kJ/mol}$$

Think About It

When the value of K is larger than 1, the Gibbs free energy of the reaction is negative.

18.85. **Collect and Organize**

We are asked to determine the number of microstates when arranging four chairs in four possible positions.

Analyze

Each microstate represents a distinct arrangement of the identical chairs in separate positions. One such arrangement of chairs places the first chair on the first level, the second chair on the second level, etc.

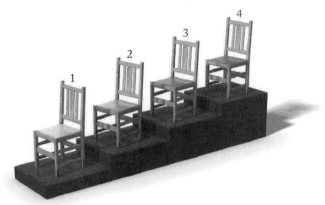

We are looking at a permutation of four chairs over four spaces, symbolized $_4P_4$.

Solve

The permutation $_4P_4$ is equal to

$$\frac{4!}{0!} = 4 \times 3 \times 2 \times 1 = 24 \text{ permutations of four chairs on four levels}$$

Think About It

If we instead had five different chairs to distribute over four levels (with one chair remaining unused), the number of microstates would be

$$\frac{5!}{1!} = 120 \text{ permutations}$$

The number of microstates increases quickly with the number of items being considered. It is unreasonable to simply count all of the possible arrangements by hand.

18.87. **Collect and Organize**

We are asked to calculate the number of microstates available to one H_2O molecule at 298 K.

Analyze

Entropy and the number of microstates are related by the Boltzmann equation

$$S = k_B \left(\ln W \right)$$

where the Boltzmann constant is 1.38×10^{-23} J/K. We may solve for the entropy of a single molecule by dividing the molar entropy of H_2O (69.9 J/K·mol) by Avogadro's number, 6.022×10^{23} molecules/mol.

Solve

$$S = \frac{69.9 \text{ J}}{K \cdot mol} \times \frac{1 \text{ mol}}{6.022 \times 10^{23} \text{ H}_2\text{O molecules}} = 1.161 \times 10^{-22} \text{ J/K}$$

$$W = e^{\frac{S}{k_B}} = e^{\left(\frac{1.161 \times 10^{-22} \text{ J/K}}{1.38 \times 10^{-23} \text{ J/K}} \right)} = 4.47 \times 10^3 \text{ microstates}$$

Think About It

A single molecule of H_2O at 298 K has over four thousand microstates!

18.89. Collect and Organize

Given the boiling point and the molar enthalpy of vaporization $\left(\Delta H^{\circ}_{vap}\right)$ for trichlorofluorobenzene, we are asked to calculate the molar entropy of vaporization $\left(\Delta S^{\circ}_{vap}\right)$.

Analyze

At the boiling point (23.8°C = 297.0 K), $\Delta G^{\circ}_{vap} = 0$. By substituting the provided values into the equation below and setting ΔG equal to zero, we may determine ΔS°_{vap}.

$$\Delta G_{vap} = \Delta H^{\circ}_{vap} - T\Delta S^{\circ}_{vap} = 0$$

Rearranging gives

$$\Delta S^{\circ}_{vap} = \frac{\Delta H^{\circ}_{vap}}{T}$$

Solve

$$\Delta S^{\circ}_{vap} = \frac{24.8 \text{ kJ/mol}}{297.0 \text{ K}} \times \frac{1000 \text{ J}}{1 \text{ kJ}} = 83.5 \text{ J/(K} \cdot \text{mol)}$$

Think About It

Our value for the molar entropy of vaporization is within the expected range for other similar compounds. The molar entropies of formation for most species are between a few tens to a few hundred J/K·mol.

18.91. Collect and Organize

For the decomposition of solid NH_4Cl into gaseous NH_3 and HCl, we are to calculate the temperature at which $\Delta G^{\circ}_{rxn} = 0$.

Analyze

For this we need to first calculate ΔH°_{rxn} and ΔS°_{rxn} using values in Appendix 4. Then we can calculate T by rearranging the free-energy equation:

$$\Delta G^{\circ}_{rxn} = \Delta H^{\circ}_{rxn} - T\Delta S^{\circ}_{rxn} = 0$$

$$T = \frac{\Delta H^{\circ}_{rxn}}{\Delta S^{\circ}_{rxn}}$$

Solve

$$\Delta H^{\circ}_{rxn} = \left[\left(1 \text{ mol NH}_3 \times \frac{-46.1 \text{ kJ}}{\text{mol}}\right) + \left(1 \text{ mol HCl} \times \frac{-92.3 \text{ kJ}}{\text{mol}}\right)\right]$$
$$- \left(1 \text{ mol NH}_4Cl \times \frac{-314.4 \text{ kJ}}{\text{mol}}\right)$$
$$= 176.0 \text{ kJ}$$

$$\Delta S^{\circ}_{rxn} = \left[\left(1 \text{ mol NH}_3 \times \frac{192.5 \text{ J}}{\text{mol} \cdot \text{K}}\right) + \left(1 \text{ mol HCl} \times \frac{186.9 \text{ J}}{\text{mol} \cdot \text{K}}\right)\right]$$
$$- \left(1 \text{ mol NH}_4Cl \times \frac{94.6 \text{ J}}{\text{mol} \cdot \text{K}}\right)$$
$$= 284.8 \text{ J/K}$$

$$T = \frac{176.0 \text{ kJ}}{0.2848 \text{ kJ/K}} = 618.0 \text{ K or } 344.8°C$$

Think About It

This reaction is favored by entropy but not by enthalpy. It is spontaneous at high temperature (above 618 K).

18.93. Collect and Organize

For the gas-phase reaction of NO with H_2 to form N_2 and H_2O we can use the values of ΔG_f° in Appendix 4 to calculate ΔG_{rxn}° and then determine if the reaction is spontaneous.

Analyze

We can use the values for ΔG_f° of the products and reactants to calculate ΔG_{rxn}°:

$$\Delta G_{rxn}^\circ = \sum n\Delta G_{f,products}^\circ - \sum m\Delta G_{f,reactants}^\circ$$

If ΔG_{rxn}° is negative then the reaction is spontaneous.

Solve

$$\Delta G_{rxn}^\circ = \left[\left(1\ \text{mol}\ N_2 \times 0.0\ \text{kJ/mol}\right) + \left(2\ \text{mol}\ H_2O \times \frac{-228.6\ \text{kJ}}{\text{mol}}\right)\right]$$
$$- \left[\left(2\ \text{mol}\ NO \times \frac{86.6\ \text{kJ}}{\text{mol}}\right) + \left(2\ \text{mol}\ H_2 \times 0.0\ \text{kJ/mol}\right)\right]$$
$$= -630.4\ \text{kJ}$$

Yes, the reaction is spontaneous at standard temperature and pressure.

Think About It

Because 4 moles of gas combine as reactants and form 3 moles of gas as products, we predict that ΔS_{rxn}° is negative. From the data in Appendix 4, we see that this is indeed the case:

$$\Delta S_{rxn}^\circ = \left[\left(1\ \text{mol}\ N_2 \times \frac{191.5\ \text{J}}{\text{mol}\cdot K}\right) + \left(2\ \text{mol}\ H_2O \times \frac{188.8\ \text{J}}{\text{mol}\cdot K}\right)\right]$$
$$- \left[\left(2\ \text{mol}\ NO \times \frac{210.7\ \text{J}}{\text{mol}\cdot K}\right) + \left(2\ \text{mol}\ H_2 \times \frac{130.6\ \text{J}}{\text{mol}\cdot K}\right)\right]$$
$$= -113.5\ \text{J/K}$$

18.95. Collect and Organize

For HCN we are to estimate the normal boiling point (T_b) given ΔH_f° and S_f° values for HCN(ℓ) and HCN(g).

Analyze

At the boiling point $\Delta G = 0$ because the system is at equilibrium. Therefore,
$$\Delta G = 0 = \Delta H_{vap} - T_b \Delta S_{vap}$$
$$T = \frac{\Delta H_{vap}}{\Delta S_{vap}}$$

Solve

For the vaporization process
$$HCN(\ell) \rightarrow HCN(g)$$
$$\Delta H_{vap} = \left(1\ \text{mol}\ HCN(g) \times 135.1\ \text{kJ/mol}\right) - \left(1\ \text{mol}\ HCN(\ell) \times 108.9\ \text{kJ/mol}\right)$$
$$= 26.2\ \text{kJ}$$
$$\Delta S_{vap} = \left[1\ \text{mol}\ HCN(g) \times 202\ \text{J/(mol}\cdot K)\right] - \left[1\ \text{mol}\ HCN(\ell) \times 113\ \text{J/(mol}\cdot K)\right]$$
$$= 89\ \text{J/K}$$
$$T_b = \frac{26.2\ \text{kJ}}{0.089\ \text{kJ/K}} = 294\ K$$

Think About It
The actual boiling point of $HCN(\ell)$ is 299 K, just about room temperature, so our calculation is approximately correct.

18.97. **Collect and Organize**
Given the melting point ($T_m = 3422°C$) and ΔH_{fus} (35.4 kJ/mol) of tungsten, we are to calculate ΔS_{fus}.

Analyze
The reaction for this process is

$$W(s) \rightarrow W(\ell)$$

At the melting point $\Delta G° = 0$ because the system is at equilibrium. Therefore,

$$\Delta G = 0 = \Delta H_{fus} - T_m \, \Delta S_{fus}$$

$$\Delta S_{fus} = \frac{\Delta H_{fus}}{T_m}$$

Solve

$$\Delta S_{fus} = \frac{35.4 \text{ kJ/mol}}{3695 \text{ K}} = 0.00958 \text{ kJ/(mol} \cdot \text{K) or } 9.58 \text{ J/(mol} \cdot \text{K)}$$

Think About It
Tungsten has the highest melting point of all the metals, making it useful as a filament in incandescent light bulbs.

18.99. **Collect and Organize**
Given the transition temperature (369 K) and the enthalpy change for the interconversion of two allotropes of S_8 (297 J/mol), we are to calculate the entropy change for the transition.

Analyze
At the transition temperature, $\Delta G = 0$ because the system is at equilibrium. Therefore,

$$\Delta G = 0 = \Delta H_{trans} - T\Delta S_{trans}$$

$$\Delta S_{trans} = \frac{\Delta H_{trans}}{T}$$

Solve

$$\Delta S_{trans} = \frac{297 \text{ J/mol}}{369 \text{ K}} = 0.805 \text{ J/(mol} \cdot \text{K)}$$

Think About It
This transition, not favored by enthalpy (ΔH positive), is favored by entropy (ΔS positive).

18.101. **Collect and Organize**
Using the values given for $\Delta H_f°$ and $S°$ for $CaCO_3$, CaO, and CO_2, we are to explain why $S°$ of $CaCO_3$ is higher than that of CaO and calculate the temperature at which the pressure of CO_2 over $CaCO_3$ is 1.0 atm.

Analyze
By considering the phase and size of each compound, we can rank the compounds in order of increasing standard molar entropy. The reaction involved is

$$CaCO_3(s) \rightarrow CaO(s) + CO_2(g)$$

To calculate the temperature at which the partial pressure of CO_2 is 1.0 atm, we must recognize that at that temperature, the reaction will be at equilibrium, $\Delta G = 0$.

Solve

$S°$ for $CaCO_3$ is greater than $S°$ for CaO because there are more atoms in $CaCO_3$.

To calculate the temperature at which the pressure of CO_2 is 1 atm, we must first calculate $\Delta H°_{rxn}$ and $\Delta S°_{rxn}$.

$$\Delta H°_{rxn} = \left[\left(1 \text{ mol CaO} \times \frac{-636 \text{ kJ}}{\text{mol}}\right) + \left(1 \text{ mol CO}_2 \times \frac{-394 \text{ kJ}}{\text{mol}}\right)\right]$$
$$- \left(1 \text{ mol CaCO}_3 \times \frac{-1207 \text{ kJ}}{\text{mol}}\right)$$
$$= 177 \text{ kJ}$$

$$\Delta S°_{rxn} = \left[\left(1 \text{ mol CaO} \times \frac{40 \text{ J}}{\text{mol} \cdot \text{K}}\right) + \left(1 \text{ mol CO}_2 \times \frac{214 \text{ J}}{\text{mol} \cdot \text{K}}\right)\right]$$
$$- \left(1 \text{ mol CaCO}_3 \times \frac{93 \text{ J}}{\text{mol} \cdot \text{K}}\right)$$
$$= 161 \text{ J/K}$$
$$\Delta G = 0 = 177 \text{ kJ} - T \times 0.161 \text{ kJ/K}$$
$$T = 1099 \text{ K or } 826°C$$

Think About It

Although this reaction is endothermic, it is favored by entropy and so is spontaneous at high temperature (above 1099 K).

18.103. Collect and Organize

We consider what changes occur in ΔS for the heating and cooling of DNA, and then we are to write an equation that relates the melting temperature of DNA to ΔH and ΔS.

Analyze

(a and b) Entropy is a measure of disorder in the system. As DNA unwinds into its two single strands, more disorder is present in the system.

(c) At the melting point, ΔG is equal to 0. We can rearrange the free-energy equation to solve for T.
$$\Delta G = \Delta H - T\Delta S = 0$$

Solve

(a) The sign of ΔS for the process of DNA separating into two strands through heating is positive.

(b) ΔS for the re-formation of the double helix of DNA is negative.

(c) $\Delta H - T\Delta S = 0$

$$T = \frac{\Delta H}{\Delta S}$$

Think About It

When $\Delta G = 0$, the system is at equilibrium, as we see in the study of chemical equilibria in Chapter 16.

CHAPTER 19 | Electrochemistry: The Quest for Clean Energy

19.1. Collect and Organize
For the voltaic cell shown in Figure P19.1, we are to explain why a porous separator is not required.

Analyze
The porous separator serves to keep the reduction and oxidation half-reactions separate so that electrons are passed through the external circuit.

Solve
Because of the careful layering, each half-cell has its metal in contact with its cation solution. The solutions are not mixing, but nevertheless the layers allow the ions needed to balance the charge in each half-cell to pass.

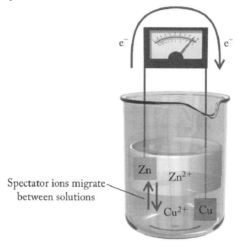

Think About It
The half-reactions and overall reaction for this voltaic cell are

$$Zn(s) \rightarrow Zn^{2+}(aq) + 2\ e^- \qquad E°_{anode} = -0.7618\ V$$
$$Cu^{2+}(aq) + 2\ e^- \rightarrow Cu(s) \qquad E°_{cathode} = 0.3419\ V$$
$$\overline{Zn(s) + Cu^{2+}(aq) \rightarrow Zn^{2+}(aq) + Cu(s) \qquad \begin{array}{l} E°_{cell} = E°_{cathode} - E°_{anode} \\ = 0.3419\ V - (-0.7618\ V) = 1.1037\ V \end{array}}$$

19.3. Collect and Organize
For the voltaic cell shown in Figure P19.3 in which an Ag^+/Ag cell is connected to a standard hydrogen electrode (SHE), we are to determine which electrode is the anode and which is the cathode and indicate in which direction the electrons flow in the outside circuit.

Analyze
A voltaic cell runs spontaneously when E_{cell} is positive. By comparing the reduction potentials of each half-cell, we can write the reaction that is spontaneous for the cell. The half-cell where reduction (gain of electrons) occurs contains the cathode and the half-cell where oxidation occurs contains the anode. Electrons flow from the anode, where they are produced by oxidation, toward the cathode, where they are required for reduction.

Solve

The spontaneous reaction for this cell is

$$2 \times (Ag^+ + e^- \rightarrow Ag) \qquad\qquad E^\circ_{cathode} = 0.7996 \text{ V}$$

$$\underline{H_2 \rightarrow 2\,H^+ + 2\,e^- \qquad\qquad E^\circ_{anode} = 0.000 \text{ V}}$$

$$2\,Ag^+(aq) + H_2(g) \rightarrow 2\,H^+(aq) + 2\,Ag(s) \qquad E^\circ_{cell} = E^\circ_{cathode} - E^\circ_{anode}$$

$$= 0.7996 \text{ V} - 0.000 \text{ V} = 0.7996 \text{ V}$$

Thus, Ag is the cathode, Pt in the SHE is the anode, and electrons flow from the SHE to Ag (to the left in the circuit shown in Figure P19.3).

Think About It

The shorthand notation for this cell would be

$$Pt(s) \mid H_2(g) \mid H^+(aq) \parallel Ag^+(aq) \mid Ag(s)$$

where Pt is an inert electrode used in the SHE.

19.5. **Collect and Organize**

The graph of cell potential versus $[H_2SO_4]$ (Figure P19.5) shows four lines, some curved, some linear, some increasing, and some decreasing, as the concentration of H_2SO_4 decreases. From the shape of the curves and their trends, we are to choose the line that best represents the trend in potential versus $[H_2SO_4]$ in a lead–acid battery.

Analyze

The scale for $[H_2SO_4]$ is logarithmic and voltage in the battery varies with the $[H_2SO_4]$ according to the Nernst equation:

$$E_{cell} = 2.04 \text{ V} - \frac{0.0592}{2} \log \frac{1}{[H_2SO_4]^2}$$

Solve

From the Nernst equation we see that the cell potential drops as the $\log 1/[H_2SO_4]^2$ decreases. So as $[H_2SO_4]$ decreases, the cell potential also decreases. The red line on the graph shows the opposite trend: The voltage increases as $[H_2SO_4]$ decreases. In considering which of the remaining lines might describe the lead–acid battery, we must consider that because the cell voltage drops as $\log 1/[H_2SO_4]^2$, we expect that the decrease in potential is linear. Therefore, the blue line best describes the potential as a function of $[H_2SO_4]$ concentration.

Think About It

Another characteristic of lead–acid batteries is that their cell voltage does not drop substantially until over 90% of the battery has been discharged (see Figure 19.13).

19.7. **Collect and Organize**

For the electrolysis of water in which the two product gases, H_2 and O_2, are collected in burettes (Figure P19.7), we are to write the half-reactions and standard potentials occurring at each electrode and discuss why a small amount of acid was added to the water to speed up the reaction.

Analyze

In the electrolysis of water, electricity is supplied to make the nonspontaneous oxidation and reduction reactions occur. From Figure P19.7 we notice that the left burette has collected twice the volume of gas as the right burette. From the overall balanced equation

$$2\,H_2O(\ell) \rightarrow 2\,H_2(g) + O_2(g)$$

we can identify the left burette as containing H_2 and the right burette as containing O_2. E°_{red} values are given in Appendix 6.

Solve

(a) The left electrode is the cathode where reduction is occurring:

$$2 H_2O(\ell) + 2 e^- \rightarrow H_2(g) + 2 OH^-(aq) \qquad E^\circ_{cathode} = -0.8277 \text{ V}$$

The right electrode is the anode where oxidation is occurring:

$$2 H_2O(\ell) \rightarrow O_2(g) + 4 H^+(aq) + 4 e^- \qquad E^\circ_{anode} = 1.229 \text{ V}$$

(b) A small amount of H_2SO_4 is added to the water to increase the conductivity of the solution.

Think About It

The electrochemical potential for the overall process is

$$E^\circ_{cell} = E^\circ_{cathode} - E^\circ_{anode} = -0.8277 \text{ V} - 1.229 \text{ V} = -2.057 \text{ V}$$

19.9. Collect and Organize

We are asked to account for the observation that metal posts corrode at the waterline rather than above or below the waterline.

Analyze

Corrosion is a redox reaction involving the metal of the post. In this redox reaction, O_2 serves as the oxidizing agent, and water is the solvent for the reaction.

Solve

Water on the post provides a medium for the reaction to occur. At the waterline, both O_2 and water are present, so oxidation may occur. In the air, O_2 is present, but water is not. Below the waterline, water is present, but the concentration of dissolved oxygen is low enough that the oxidation of metal is much lower than at the waterline.

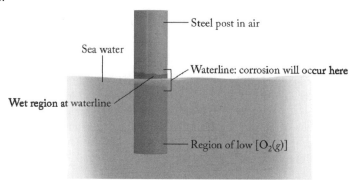

Think About It

The dissolved ions in saltwater accelerate the redox reaction by facilitating the transfer of electrons.

19.11. Collect and Organize

We are to explain the function of the porous separator in an electrochemical cell and why a wire would not function the same.

Analyze

The porous separator is situated between the half-reactions in the electrochemical cell.

Solve

(a) The porous separator allows nonreactive ions to pass through the separator so that electrical neutrality is maintained in each of the half-cells.

(b) A wire would allow electrons to flow between halves of the cell, but not ions. Ions must migrate between halves of the cell to maintain electrical neutrality as electrons flow through the circuit.

Think About It

Without the flow of ions through the separator, the flow of electrons stops.

19.13. Collect and Organize

Given the descriptions of the redox reactions provided, we are to choose appropriate half-reactions from Appendix 6, and write a net ionic equation for the reaction that occurs.

Analyze

(a) In this reaction, aluminum is oxidized to Al^{3+}, and Fe^{3+} is reduced to Fe^{2+}.

(b) In this reaction, I_2 is reduced to I^-, and NO_2^- is oxidized to NO_3^-.

(c) In this reaction, MnO_4^- is reduced to Mn^{2+}, and Cr^{3+} is oxidized to $Cr_2O_7^{2-}$.

Solve

(a)

$$Fe^{3+}(aq) + e^- \rightarrow Fe^{2+}(aq)$$
$$\underline{Al(s) \rightarrow Al^{3+}(aq) + 3\ e^-}$$
$$Al(s) + 3\ Fe^{3+}(aq) \rightarrow Al^{3+}(aq) + 3\ Fe^{2+}(aq)$$

(b)

$$I_2(s) + 2\ e^- \rightarrow 2\ I^-(aq)$$
$$\underline{NO_2^-(aq) + 2\ OH^-(aq) \rightarrow NO_3^-(aq) + H_2O(\ell) + 2\ e^-}$$
$$NO_2^-(aq) + 2\ OH^-(aq) + I_2(s) \rightarrow NO_3^-(aq) + H_2O(\ell) + 2\ I^-(aq)$$

(c)

$$MnO_4^-(aq) + 8\ H^+(aq) + 5\ e^- \rightarrow Mn^{2+}(aq) + 4\ H_2O(\ell)$$
$$\underline{2\ Cr^{3+}(aq) + 7\ H_2O(\ell) \rightarrow Cr_2O_7^{2-}(aq) + 14\ H^+(aq) + 6\ e^-}$$
$$6\ MnO_4^-(aq) + 10\ Cr^{3+}(aq) + 11\ H_2O(\ell) \rightarrow 6\ Mn^{2+}(aq) + 5\ Cr_2O_7^{2-}(aq) + 22\ H^+(aq)$$

Think About It

When balancing redox reactions in aqueous solution, it is important to note if the solution is acid or basic.

19.15. Collect and Organize

For the Cd/Cd^{2+} and Ni^{2+}/Ni voltaic cell in which we are told that the products are $Cd^{2+}(aq)$ and $Ni(s)$, we are to write the appropriate half-reactions, write the balanced overall cell reaction, and diagram the cell in aqueous solution.

Analyze

Because we know that $Cd^{2+}(aq)$ and $Ni(s)$ are produced, Ni^{2+} is reduced and Cd is oxidized in this process. In balancing the overall reaction, we must be sure to cancel all the electrons produced by oxidation with those used in reduction. To do so, we might have to multiply either half-reaction, or both, by some factor. Finally, to diagram the cell we use the following convention:

Anode | oxidation half-reaction species || reduction half-reaction species | cathode

making sure to indicate the phases of the species involved and to use vertical lines to separate phases.

Solve

(a) $Ni^{2+}(aq) + 2\ e^- \rightarrow Ni(s)$ cathode, reduction

     $Cd(s) \rightarrow Cd^{2+}(aq) + 2\ e^-$ anode, oxidation

(b) To cancel the electrons, neither reaction needs to be multiplied by any factor:

$$Ni^{2+}(aq) + 2\ e^- \rightarrow Ni(s)$$
$$\underline{Cd(s) \rightarrow Cd^{2+}(aq) + 2\ e^-}$$
$$Cd(s) + Ni^{2+}(aq) \rightarrow Cd^{2+}(aq) + Ni(s)$$

(c) $Cd(s) \mid Cd^{2+}(aq) \parallel Ni^{2+}(aq) \mid Ni(s)$

Think About It
Depicting this electrochemical cell

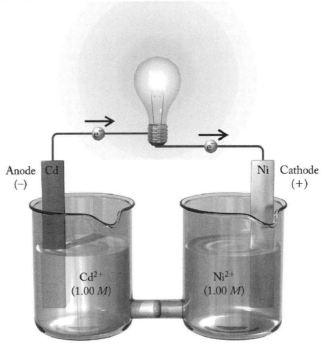

19.17. Collect and Organize
For the $Cd/Cd(OH)_2$ and MnO_4^-/MnO_2 voltaic cell in which we are told that the products are $Cd(OH)_2(s)$ and $MnO_2(s)$, we are to write the appropriate half-reactions, write the balanced overall cell reaction, and diagram the cell. The electrolyte is an alkaline aqueous solution.

Analyze
Because we know that $Cd(OH)_2$ and MnO_2 are produced, MnO_4^- is reduced and Cd is oxidized in this process. In balancing the overall reaction, we must be sure to cancel all the electrons produced by oxidation with those used in reduction. To do so, we might have to multiply either half-reaction, or both, by some factor. Finally, to diagram the cell we use the following convention:

Anode | oxidation half-reaction species || reduction half-reaction species | cathode

making sure to indicate the phases of the species involved and to use vertical lines to separate phases.

Solve
(a) $MnO_4^-(aq) + 2\,H_2O(\ell) + 3\,e^- \rightarrow MnO_2(s) + 4\,OH^-(aq)$ cathode, reduction

   $Cd(s) + 2\,OH^-(aq) \rightarrow Cd(OH)_2(s) + 2\,e^-$ anode, oxidation

(b) To cancel the electrons, we multiply the cathode reaction by 2 and the anode reaction by 3:

$$2\,MnO_4^-(aq) + 4\,H_2O(\ell) + 6\,e^- \rightarrow 2\,MnO_2(s) + 8\,OH^-(aq)$$

$$3\,Cd(s) + 6\,OH^-(aq) \rightarrow 3\,Cd(OH)_2(s) + 6\,e^-$$

$$\overline{2\,MnO_4^-(aq) + 4\,H_2O(\ell) + 3\,Cd(s) \rightarrow 2\,MnO_2(s) + 3\,Cd(OH)_2(s) + 2\,OH^-(aq)}$$

(c) $Cd(s) \mid Cd(OH)_2(s) \parallel MnO_4^-(aq) \mid MnO_2(s) \mid Pt(s)$

Think About It
Because we would find it difficult to make an electrode out of $MnO_2(s)$, this cell requires the use of an inert electrode such as Pt.

19.19. Collect and Organize

For the "super iron" battery, we are to determine the number of electrons transferred in the spontaneous reaction and the oxidation states for Fe and Zn in the reactants and products. Finally, we are to diagram the cell.

Analyze

(a) By writing the half-reactions involved in the redox reaction and balancing them, we can determine the number of electrons transferred.

(b) Knowing the typical oxidation states of oxygen (–2) and potassium (+1) helps us determine the oxidation states of Fe in K_2FeO_4 and Zn in ZnO and K_2ZnO_2.

(c) To diagram the cell we use the following cell notation:

Anode | oxidation half-reaction species || reduction half-reaction species | cathode

making sure to indicate the phases of the species involved and to use vertical lines to separate phases.

Solve

(a) The half-reactions are as follows (K^+ is a spectator ion):
$$5\ H_2O(\ell) + 3\ Zn(s) \rightarrow ZnO(s) + 2\ ZnO_2^{2-}(aq) + 10\ OH^+(aq) + 6\ e^-$$
$$6\ e^- + 10\ H^+(aq) + 2\ FeO_4^{2-}(aq) \rightarrow Fe_2O_3(s) + 5\ H_2O\ (\ell)$$

Six electrons are transferred in this reaction.

(b) FeO_4^{2-} has Fe^{6+}

Fe_2O_3 has Fe^{3+}

Zn has Zn^0

ZnO and ZnO_2^{2-} have Zn^{2+}

(c) $Zn(s) | ZnO(s) | ZnO_2^{2-}(aq) \| FeO_4^{2-}(aq) | Fe_2O_3(s) | Pt(s)$

Think About It

Because we would find it difficult to make an electrode out of $Fe_2O_3(s)$, this cell requires the use of an inert electrode such as Pt.

19.21. Collect and Organize / Analyze

We are to describe the function of platinum in the standard hydrogen electrode.

Solve

The platinum electrode transfers electrons to the half-cell; because it is inert, it is not involved in the reaction.

Think About It

The inert electrode simply serves as a place for the oxidation or reduction reaction to occur and as a conduit for electrons for the reaction.

19.23. Collect and Organize

We are asked to equate two different ways to calculate E°_{cell}.

Analyze

Equation 19.2 is
$$E^\circ_{cell} = E^\circ_{cathode} - E^\circ_{anode}$$

where both potentials are understood to be the reduction at the cathode and the anode. Other textbooks express E°_{cell} in terms of both E°_{red} and E°_{ox}:
$$E^\circ_{cell} = E^\circ_{red}(\text{cathode}) + E^\circ_{ox}(\text{anode})$$

Solve

Because $E^\circ_{red}(\text{anode}) = -E^\circ_{ox}(\text{anode})$ we can substitute $-E^\circ_{ox}(\text{anode})$ for $E^\circ_{red}(\text{anode})$ in the expression

$$E^\circ_{cell} = E^\circ_{red}(\text{cathode}) + E^\circ_{ox}(\text{anode})$$

to obtain

$$E^\circ_{cell} = E^\circ_{red}(\text{cathode}) + \left(-E^\circ_{red}(\text{anode})\right)$$

$$= E^\circ_{red}(\text{cathode}) - E^\circ_{red}(\text{anode})$$

This is equivalent to the expression for E°_{cell} in Equation 19.2.

Think About It

Either expression to calculate E°_{cell} is valid.

19.25. Collect and Organize

By consulting the list of standard reduction potentials in Appendix 6, we are asked to determine which of Cl_2 or O_2 is a stronger oxidizing agent.

Analyze

An oxidizing agent is itself reduced in the process. The larger the value of E°_{red}, the more readily a species is reduced, and so the stronger it will be as an oxidizer. The half-reactions for O_2 and Cl_2 are

$$O_2(g) + 4\ H^+(aq) + 4\ e^- \rightarrow 2\ H_2O(\ell) \qquad E^\circ_{red} = 1.229\ \text{V}$$

$$Cl_2(g) + 2\ e^- \rightarrow 2\ Cl^-(aq) \qquad E^\circ_{red} = 1.3583\ \text{V}$$

Solve

The reduction potential for Cl_2 is larger, so Cl_2 is expected to be a stronger oxidizer than O_2.

Think About It

Sorting redox half-reactions in order of their reduction potential allows us to quickly assess the oxidizing and reducing power of listed species.

19.27. Collect and Organize

For each redox reaction given, we are to calculate the cell potential after calculating the value of free energy.

Analyze

We calculate ΔG° for each reaction using G°_f values for the products and reactants from Appendix 4. The cell potential can then be calculated using

$$\Delta G^\circ = -nFE^\circ_{cell}$$

$$E^\circ_{cell} = \frac{-\Delta G^\circ}{nF}$$

Solve

(a) $\Delta G^\circ = \left[(1\ \text{mol}\ Cu^{2+} \times 65.5\ \text{kJ/mol}) + (1\ \text{mol}\ Cu \times 0.0\ \text{kJ/mol})\right] - (2\ \text{mol}\ Cu^+ \times 50.0\ \text{kJ/mol}) = -34.5\ \text{kJ}$

$$E^\circ_{cell} = \frac{+34,500\ \text{J}}{1\ \text{mol} \times 9.65 \times 10^4\ \text{C/mol}} = +0.358\ \text{V}$$

(b)

$$\Delta G^\circ = \left[(1\ \text{mol}\ Cu^{2+} \times 65.5\ \text{kJ/mol}) + (2\ \text{mol}\ Fe^{2+} \times -78.9\ \text{kJ/mol})\right] -$$

$$\left[(1\ \text{mol}\ Cu \times 0\ \text{kJ/mol}) + (2\ \text{mol}\ Fe^{3+} \times -4.7\ \text{kJ/mol})\right] = -82.9\ \text{kJ}$$

$$E^\circ_{cell} = \frac{-(-82,900\ \text{J})}{2\ \text{mol} \times (9.65 \times 10^4\ \text{C/mol})} = +0.430\ \text{V}$$

Think About It
Reactions (a) and (b) are both spontaneous, as we can see from the sign of $\Delta G°$ (<0) and $E°_{cell}$ (>0).

19.29. Collect and Organize
By calculating E_{cell} for the possible redox reaction between Ag and Cu^{2+}, we can determine if the reaction is spontaneous.

Analyze
We are told that $[Ag^+] = [Cu^{2+}] = 1.00$ *M*, and we can assume this reaction occurs under standard conditions, so we can calculate $E°_{cell}$ from values of $E°_{red}$ in Appendix 6. If $E°_{cell}$ is calculated as positive, the reaction is spontaneous.

Solve
The half-reactions, $E°_{cathode}$ and $E°_{anode}$, and the overall reaction and cell potential are

$$2\ Ag(s) \rightarrow 2\ Ag^+(aq) + 2\ e^- \qquad E°_{anode} = 0.7996\ V$$
$$\underline{Cu^{2+}(aq) + 2\ e^- \rightarrow Cu(s) \qquad E°_{cathode} = 0.3419\ V}$$
$$2\ Ag(s) + Cu^{2+}(aq) \rightarrow 2\ Ag^+(aq) + Cu(s) \qquad E°_{cell} = E°_{cathode} - E°_{anode} = -0.4577\ V$$

No, the reaction is not spontaneous.

Think About It
The reverse reaction, Cu placed into Ag^+ to dissolve copper and deposit silver, is spontaneous.

19.31. Collect and Organize
We are asked to write the overall cell reaction for the reaction of Zn with O_2 in the zinc–air battery to form $Zn(OH)_4^{2-}$.

Analyze
For each half-reaction, we need to have OH^- or H_2O as a reactant. We balance each for both atoms and charge before adding the two half-reactions together.

Solve

Cathode	$O_2(g) + 2\ H_2O(\ell) + 4\ e^- \rightarrow 4\ OH^-(aq)$
Anode	$[Zn(s) + 4\ OH^-(aq) \rightarrow Zn(OH)_4^{2-}(aq) + 2\ e^-] \times 2$

$$O_2(g) + 2\ H_2O(\ell) + 2\ Zn(s) + 4\ OH^-(aq) \rightarrow 2\ Zn(OH)_4^{2-}(aq)$$

Think About It
This differs from the other overall reaction for the zinc–air battery shown in Figure 19.7, in that water and OH^- are reactants to give soluble $Zn(OH)_4^{2-}$ instead of solid ZnO.

19.33. Collect and Organize
We are asked to consider the $E°_{cell}$ of a Ni–Zn cell, similar to the Cu–Zn cell in Figure 19.2, and indicate if it will have an $E°_{cell}$ greater than, less than, or equal to 1.10 V.

Analyze
Using $E°_{red}$ values in Appendix 6, we can identify the spontaneous reaction (having positive $E°_{cell}$) and calculate $E°_{cell}$.

Solve

The spontaneous reaction in the Ni–Zn cell is

$$Ni^{2+}(aq) + 2\,e^- \rightarrow Ni(s) \qquad E^\circ_{cathode} = -0.257\ V$$

$$Zn(s) \rightarrow Zn^{2+}(aq) + 2\,e^- \qquad E^\circ_{anode} = -0.7618\ V$$

$$E^\circ_{cell} = E^\circ_{cathode} - E^\circ_{anode} = 0.505\ V$$

The E°_{cell} for a Ni–Zn cell is less than 1.10 V.

Think About It

Because E°_{cell} for the Ni–Zn cell is less positive, we also know that the cell reaction has a less negative ΔG.

19.35. Collect and Organize

We can break up each reaction into its appropriate half-reactions and, using E°_{red} values from Appendix 6, calculate E°_{cell}. From this cell potential we can then calculate the free energy for each reaction.

Analyze

To calculate the cell potential we use

$$E^\circ_{cell} = E^\circ_{cathode} - E^\circ_{anode}$$

Once we have calculated E°_{cell} we use

$$\Delta G^\circ = -nFE^\circ_{cell}$$

to calculate the free energy of the reaction. Here, we have to remember that n is the number of moles of electrons transferred in the overall balanced equation, and that 1 V = 1 J/C.

Solve

(a)

$$2\,Br^-(aq) \rightarrow Br_2(\ell) + 2\,e^- \qquad E^\circ_{anode} = 1.066\ V$$

$$Cl_2(g) + 2\,e^- \rightarrow 2\,Cl^-(aq) \qquad E^\circ_{cathode} = 1.3583\ V$$

$$Cl_2(g) + 2\,Br^-(aq) \rightarrow 2\,Cl^-(aq) + Br_2(\ell) \quad E^\circ_{cell} = E^\circ_{cathode} - E^\circ_{anode} = 0.292\ V$$

$$\Delta G^\circ = -2\ mol\ e^- \times \frac{9.65\times10^4\ C}{1\ mol\ e^-} \times \frac{0.292\ J}{1\ C} \times \frac{1\ kJ}{1000\ J} = -56.4\ kJ$$

(b)

$$Zn(s) \rightarrow Zn^{2+}(aq) + 2\,e^- \qquad E^\circ_{anode} = -0.7618\ V$$

$$Ni^{2+}(aq) + 2\,e^- \rightarrow Ni(s) \qquad E^\circ_{cathode} = -0.257\ V$$

$$Zn(s) + Ni^{2+}(aq) \rightarrow Zn^{2+}(aq) + Ni(s) \qquad E^\circ_{cell} = E^\circ_{cathode} - E^\circ_{anode} = 0.5048\ V$$

$$\Delta G^\circ = -2\ mol\ e^- \times \frac{9.65\times10^4\ C}{1\ mol\ e^-} \times \frac{0.5048\ J}{1\ C} \times \frac{1\ kJ}{1000\ J} = -97.4\ kJ$$

Think About It

Reaction (b) is spontaneous, so we may conclude that zinc metal in contact with Ni^{2+} solution would oxidize.

19.37. Collect and Organize

We are asked if H_2 gas can reduce the listed ions (Ag^+, Mg^{2+}, Cu^{2+}, and Cd^{2+}) to their corresponding metals.

Analyze

H_2 gas is capable of reducing anything more easily reduced than the standard hydrogen electrode (at $E^\circ_{red} = 0.00\ V$). Any reduction half-reactions with $E^\circ_{red} > 0.00\ V$ will be reduced by H_2.

Solve

(a) Ag^+ and (c) Cu^{2+} have reduction potentials greater than 0.00 V, so they will be reduced by $H_2(g)$.

Think About It
Other metal cations such as Hg^{2+} and Au^{3+} could also be reduced by $H_2(g)$.

19.39. Collect and Organize
We can write the overall cell reaction for the battery by adding the equations together. The cell potential is $E°_{cell} = E°_{cathode} - E°_{anode}$.

Analyze
Because the reduction reaction and the oxidation reaction each involve the exchange of one electron, we do not need to multiply either reaction in order to write the overall reaction.

Solve
(a) $NiO(OH)(s) + TiZr_2H(s) \rightarrow TiZr_2(s) + Ni(OH)_2(s)$
(b) $E°_{cell} = 1.32\ V - 0.00\ V = 1.32\ V$

Think About It
This reaction is spontaneous with a ΔG value of
$$\Delta G° = -1\ mol\ e^- \times \frac{9.65 \times 10^4\ C}{1\ mol\ e^-} \times \frac{1.32\ J}{1\ C} \times \frac{1\ kJ}{1000\ J} = -127\ kJ$$

19.41. Collect and Organize
Using the relationship between free energy and cell potential
$$\Delta G_{cell} = -nFE_{cell}$$
we can calculate ΔG_{cell} for a Zn–MnO_2 battery generating 1.50 V.

Analyze
To use the equation, we need the value of n, the number of electrons transferred in the overall balanced equation. From the cell reaction provided, we see that Zn is oxidized to Zn^{2+}, and two moles of Mn^{4+} (in MnO_2) are reduced Mn^{3+} (in Mn_2O_3), so two moles of electrons are transferred in the reaction. We should also be aware of the units for the quantities in this equation. The value of F is 9.65×10^4 C/mol and E is in volts, which, being equivalent to joules per coulomb (J/C), means that ΔG is calculated in joules, which we can convert to kilojoules, the usual units for free energy.

Solve
$$\Delta G_{cell} = -2\ mol\ e^- \times \frac{9.65 \times 10^4\ C}{1\ mol\ e^-} \times \frac{1.50\ J}{1\ C} \times \frac{1\ kJ}{1000\ J} = -290\ kJ$$

Think About It
This reaction is spontaneous because the cell potential is positive, making ΔG_{cell} negative.

19.43. Collect and Organize
Using the relationship between free energy and cell potential
$$\Delta G_{cell} = -nFE_{cell}$$
we can calculate ΔG_{cell} for a nickel–metal hydride battery cell generating 1.20 V.

Analyze
To use the equation, we need the value of n, the number of electrons transferred in the overall balanced equation. From the cell reaction provided, we see that Ni^{3+} [in NiO(OH)] is reduced to Ni^{2+} [in $Ni(OH)_2$] and M^- (in MH) is oxidized to M, so one mole of electrons are transferred in the reaction. We should also be aware of the units for the quantities in this equation. The value of F is 9.65×10^4 C/mol and E is in volts, which, being equivalent to joules per coulomb (J/C), means that ΔG is calculated in joules, which we can convert to kilojoules, the usual units for free energy.

Solve

$$\Delta G_{cell} = -1 \text{ mol } e^- \times \frac{9.65 \times 10^4 \text{ C}}{1 \text{ mol } e^-} \times \frac{1.20 \text{ J}}{1 \text{ C}} \times \frac{1 \text{ kJ}}{1000 \text{ J}} = -116 \text{ kJ}$$

Think About It

This reaction is spontaneous because the cell potential is positive, making ΔG_{cell} negative.

19.45. Collect and Organize

We are to explain why the voltage of most batteries changes little until the battery is almost discharged, at which time the voltage drops significantly.

Analyze

Voltage of a battery (a voltaic cell) is governed by the Nernst equation:

$$E_{cell} = E_{cell}^\circ - \frac{RT}{nF} \ln Q$$

As a battery discharges, the value of Q, the reaction quotient, changes:

$$Q = \frac{[\text{products}]^x}{[\text{reactants}]^y}$$

Solve

At the start of the reaction, Q is very small because [reactants] >> [products]. As the reaction proceeds [products] grows and Q increases but does not increase significantly until significant amounts of products form, that is, when the battery is nearly discharged.

Think About It

As an example, Figure 19.13 shows the cell potential of a lead–acid battery as a function of discharge. Notice that the voltage is relatively constant until the battery is approximately 90% discharged.

19.47. Collect and Organize

We can use the Nernst equation to calculate the cell potential when Fe^{3+} is combined with Cu^{2+} at nonstandard conditions.

Analyze

Because $T = 298$ K, we can use the following form of the Nernst equation

$$E_{cell} = E_{cell}^\circ - \frac{0.0592 \text{ V}}{n} \log Q$$

where E_{cell}° is the potential of the cell under standard conditions (calculated from tabulated E_{red}° values), n is the moles of electrons transferred in the overall balanced redox equation, and Q is the reaction quotient.

Solve

First, we need to calculate E_{cell}° and determine the value of n. The half-reactions and overall cell reaction are

$$Fe^{3+}(aq) + e^- \rightarrow Fe^{2+}(aq) \qquad E_{cathode}^\circ = 0.770 \text{ V}$$
$$Cu^+(aq) \rightarrow Cu^{2+}(aq) + e^- \qquad E_{anode}^\circ = 0.153 \text{ V}$$
$$\overline{Fe^{3+}(aq) + Cu^+(aq) \rightarrow Fe^{2+}(aq) + Cu^{2+}(aq) \qquad E_{cell}^\circ = E_{cathode}^\circ - E_{anode}^\circ = 0.617 \text{ V}}$$

We see that $n = 1$. Now we can use the Nernst equation to calculate E_{cell} when $[Fe^{3+}] = [Cu^+] = 1.50 \times 10^{-3}$ M and $[Fe^{2+}] = [Cu^{2+}]$ 2.5×10^{-4} M.

$$E_{cell} = 0.617 \text{ V} - \frac{0.0592}{1} \log \frac{\left(2.5 \times 10^{-4}\right)^2}{\left(1.50 \times 10^{-3}\right)^2} = 0.709 \text{ V}$$

Think About It

This reaction became more spontaneous (higher cell potential) under these conditions.

19.49. Collect and Organize

We can use the following equation to calculate the equilibrium constant for the given redox reaction:

$$\log K = \frac{nE^{\circ}_{cell}}{0.0592 \text{ V}}$$

Analyze

First, we have to determine the standard cell potential, E°_{cell}, using the E°_{red} values in Appendix 6 and also determine the value of n, the moles of electrons transferred in the overall balanced equation.

Solve

$$Fe^{3+}(aq) + e^- \rightarrow Fe^{2+}(aq) \qquad E^{\circ}_{cathode} = 0.770 \text{ V}$$
$$\underline{Cr^{2+}(aq) \rightarrow Cr^{3+}(aq) + e^- \qquad E^{\circ}_{anode} = -0.41 \text{ V}}$$
$$Fe^{3+}(aq) + Cr^{2+}(aq) \rightarrow Fe^{2+}(aq) + Cr^{3+}(aq) \quad E^{\circ}_{cell} = E^{\circ}_{cathode} - E^{\circ}_{anode} = 1.18 \text{ V}$$
$$n = 1$$

$$\log K = \frac{1 \times 1.18 \text{ V}}{0.0592} = 19.9324$$
$$K = 1 \times 10^{19.9324} = 8.56 \times 10^{19}$$

Think About It

Because K for this reaction is very large, the reaction goes very far to the right.

19.51. Collect and Organize

We can use the Nernst equation to calculate the potential of the hydrogen electrode at pH = 7.00.

Analyze

We are reminded that under standard conditions [1 atm $H_2(g)$ and 1.00 M H^+ = pH = 0.00] the voltage of the hydrogen cell is zero. This is E°_{cell}. The overall reaction for the cell (against SHE) is

$$2 H^+(aq) + 2 e^- \rightarrow H_2(g) \qquad \text{(where } [H^+] = 1 \times 10^{-7} M)$$
$$\underline{H_2(g) \rightarrow 2 H^+(aq) + 2 e^- \qquad \text{(SHE)}}$$
$$2 H^+(aq) (1 \times 10^{-7} M) + H_2(g) (1 \text{ atm}) \rightarrow 2 H^+(aq) (1.00 M) + H_2(g) (1 \text{ atm})$$

The form of Q for the Nernst equation is

$$Q = \frac{(1.00 \ M)^2 \times (1 \text{ atm})}{(1 \times 10^{-7} \ M)^2 \times (1 \text{ atm})} = \frac{1}{(1 \times 10^{-7})^2}$$

Solve

$$E_{cell} = 0.000 \text{ V} - \frac{0.0592 \text{ V}}{2} \log \frac{1}{(1 \times 10^{-7})^2} = -0.414 \text{ V}$$

Think About It

The spontaneous reaction actually is the reverse reaction:

$$2 H^+(aq) (1.00 M) + H_2(g) (1 \text{ atm}) \rightarrow 2 H^+(aq) (1 \times 10^{-7} M) + H_2(g) (1 \text{ atm}) \qquad E^{\circ}_{cell} = 0.414 \text{ V}$$

In this redox cell, acid in the SHE will be reduced and H_2, in the cell where pH = 7.00, will be oxidized.

584 | *Chapter 19*

19.53. Collect and Organize

We can use the Nernst equation to calculate the potential for the reduction of MnO_4^- to MnO_2 in the presence of SO_3^{2-} when $[MnO_4^-] = 0.250\ M$, $[SO_3^{2-}] = 0.425\ M$, $[SO_4^{2-}] = 0.075\ M$, and $[OH^-] = 0.0200\ M$. We are also to assess whether the potential increases or decreases as reactants become products.

Analyze

To use the Nernst equation we need to know the value of E°_{cell} and n. We can determine both of these by writing out the half-reactions and balancing the redox reaction.

Solve

$$2\ MnO_4^-(aq) + 4\ H_2O(\ell) + 6\ e^- \rightarrow 2\ MnO_2(s) + 8\ OH^-(aq) \qquad E^\circ_{cathode} = 0.59\ V$$

$$3\ SO_3^{2-}(aq) + 6\ OH^-(aq) \rightarrow 3\ SO_4^{2-}(aq) + 3\ H_2O(\ell) + 6\ e^- \qquad E^\circ_{anode} = -0.92\ V$$

$$2\ MnO_4^-(aq) + 3\ SO_3^{2-}(aq) + H_2O(\ell) \rightarrow 2\ MnO_2(s) + 3\ SO_4^{2-}(aq)\ 2\ OH^-(aq) \quad E^\circ_{cell} = E^\circ_{cathode} - E^\circ_{anode} = 1.51\ V$$

$$n = 6$$

$$E_{cell} = 1.51\ V - \frac{0.0592}{6} \log \frac{(0.075\ M)^3(0.0200\ M)^2}{(0.250\ M)^2(0.425\ M)^3}$$

$$E_{cell} = 1.55\ V$$

As the reaction proceeds, the concentrations of the reactants decrease and the concentrations of the products increase, so Q increases and $\log Q$ becomes more positive. When a more positive $\log Q$ is multiplied by $0.0592/6$ and then subtracted from the E°_{cell}, E_{cell} decreases.

Think About It

Be sure to use half-reactions to determine the correct value of n.

19.55. Collect and Organize

For the reaction of copper pennies with nitric acid, we are to calculate E°_{cell} and then E_{cell} when $[H^+] = 0.500\ M$, $[NO_3^-] = 0.0550\ M$, $[Cu^{2+}] = 0.0500\ M$, and $P_{NO} = 0.00250$ atm.

Analyze

E°_{cell} is calculated by adding E°_{anode} and $E^\circ_{cathode}$. The reaction is spontaneous, so our calculated E°_{cell} must be positive. In balancing the reaction, we can also determine the value of n for the Nernst equation.

Solve
(a)

$$3\ Cu(s) \rightarrow 3\ Cu^{2+}(aq) + 6\ e^- \qquad E^\circ_{anode} = 0.3419\ V$$

$$2\ NO_3^-(aq) + 8\ H^+(aq) + 6\ e^- \rightarrow 2\ NO(g) + 4\ H_2O(\ell) \qquad E^\circ_{cathode} = 0.96\ V$$

$$3\ Cu(s) + 2\ NO_3^-(aq) + 8\ H^+(aq) \rightarrow 3\ Cu^{2+}(aq) + 2\ NO(g) + 4\ H_2O(\ell) \quad E^\circ_{cell} = E^\circ_{cathode} - E^\circ_{anode} = 0.62\ V$$

(b)
$$E_{cell} = 0.6181\ V - \frac{0.0592\ V}{6} \log \frac{(0.0500\ M)^3(0.00250\ \text{atm})^2}{(0.0550\ M)^2(0.500\ M)^8} = 0.66\ V$$

Think About It

We may mix concentration units of atmospheres and molarity as we do in this calculation of Q.

19.57. Collect and Organize

For the reaction of NH_4^+ with O_2 in water, we are to calculate E°_{cell}. We can then use the Nernst equation to determine $[NO_3^-]/[NH_4^+]$ for $P_{O_2} = 0.21$ atm and pH = 7.00 at 298 K.

Analyze

E°_{cell} is calculated by adding E°_{anode} and $E^{\circ}_{cathode}$. The reaction is spontaneous, so our calculated E°_{cell} must be positive. In balancing the reaction, we can determine the value of n. Because the system is at equilibrium in part (b), we know that $E_{cell} = 0$. Therefore, $[NO_3^-]/[NH_4^+]$ can be determined through the equation

$$0 = E^{\circ}_{cell} - \frac{0.0592 \text{ V}}{n} \log \frac{[NO_3^-][H^+]^2}{[NH_4^+]\left(P_{O_2}\right)^2}$$

Solve

(a)

$$NH_4^+(aq) + 3\ H_2O(\ell) \rightarrow NO_3^-(aq) + 10\ H^+ + 8\ e^- \qquad\qquad E^{\circ}_{anode} = 0.88 \text{ V}$$

$$2\ O_2(g) + 8\ H^+(aq) + 8\ e^- \rightarrow 4\ H_2O(\ell) \qquad\qquad E^{\circ}_{cathode} = 1.229 \text{ V}$$

$$\overline{NH_4^+(aq) + 2\ O_2(g) \rightarrow NO_3^-(aq) + H_2O(\ell) + 2\ H^+(aq) \quad E^{\circ}_{cell} = E^{\circ}_{cathode} - E^{\circ}_{anode} = 0.349 \text{ V}}$$

(b)

$$0 = 0.349 \text{ V} - \frac{0.0592}{8} \log \frac{[NO_3^-]\left(1.00 \times 10^{-7} \text{ M}\right)^2}{[NH_4^+]\left(0.21 \text{ atm}\right)^2}$$

$$-0.349 \text{ V} = -\frac{0.0592}{8} \log \frac{[NO_3^-]\left(1.00 \times 10^{-7} \text{ M}\right)^2}{[NH_4^+]\left(0.21 \text{ atm}\right)^2}$$

$$47.16 = \log\left(\left(2.268 \times 10^{-13}\right) \times \frac{[NO_3^-]}{[NH_4^+]}\right)$$

$$1 \times 10^{47.16} = 1.445 \times 10^{47} = 2.268 \times 10^{-13} \times \frac{[NO_3^-]}{[NH_4^+]}$$

$$\frac{[NO_3^-]}{[NH_4^+]} = 6.37 \times 10^{59}$$

Think About It

This ratio is consistent with a spontaneous reaction, as indicated by the positive E°_{cell}.

19.59. Collect and Organize

We are to compare two 12-volt lead–acid batteries, one of which has a lower ampere-hour rating.

Analyze

An ampere-hour is a unit of electrical charge and is defined as the electric charge transferred by 1 A of current for 1 hr. It is used to describe the life of a battery.

Solve

The total masses of the electrode materials (c) and the combined surface areas of the electrodes (f) are likely to be different.

Think About It

Both batteries use the same components (b and e) and have the same voltage (a and d).

19.61. Collect and Organize

We are to compare two voltaic cells to determine which produces more charge per gram of anode material.

Analyze

For each cell we must first identify which species is the anode and the number of electrons transferred when one mole of anode is consumed in the reaction. The charge generated by the reaction is

$$C = nF$$

where C = charge in coulombs, n = moles of electrons transferred in balanced equation, and $F = 9.65 \times 10^4$ C/mol e⁻. This gives the charge per mol of anode. To convert this into charge per gram we use the molar mass:

$$\frac{\text{charge}}{\text{gram}} = \frac{\text{coulombs/mol}}{\text{molar mass of anode material}}$$

Solve
For the Ni–Cd voltaic cell, Cd is the anode material:

$$\frac{C}{g} = \frac{2 \text{ mol e}^- \times 9.65 \times 10^4 \text{C / mol e}^-}{112.41 \text{ g/mol e}^-} = 1.72 \times 10^3 \text{ C/g Cd}$$

For the Al–O_2 voltaic cell, Al is the anode material:

$$\frac{C}{g} = \frac{12 \text{ mol e}^- \times 9.65 \times 10^4 \text{C / mol e}^-}{26.98 \text{ g/mol e}^-} = 4.29 \times 10^4 \text{ C/g Al}$$

Therefore, the Al–O_2 cell produces a greater charge per gram.

Think About It
Notice that the number of electrons transferred in the oxidation of Al to Al^{3+} is 4 mol × 3 e⁻/mol = 12 e⁻.

19.63. Collect and Organize
We are to compare two voltaic cells to determine which produces more energy per gram of anode material.

Analyze
The energy of a voltaic cell is the force to move electrons from the anode to the cathode. The unit of volts is energy per unit charge, so

$$\text{Energy} = \text{volts} \times \text{charge (in units of V·C)}$$

where 1 V = 1 J/C. The charge in the cell generated by 1 g of anode material is

$$\text{Charge} = 1 \text{ g} \times \text{molar mass of anode material} \times \frac{\text{mol e}^-}{\text{mol anode}} \times \frac{9.65 \times 10^4 \text{C}}{1 \text{ mol e}^-}$$

Solve
For the Zn–$Ni(OH)_2$ cell, Zn is the anode material:

$$\text{Charge} = 1 \text{ g} \times \frac{1 \text{ mol}}{65.38 \text{ g}} \times \frac{2 \text{ mol e}^-}{1 \text{ mol Zn}} \times \frac{9.65 \times 10^4 \text{ C}}{1 \text{ mol e}^-} = 2.952 \times 10^3 \text{ C}$$

$$\text{Energy} = 1.20 \frac{\text{J}}{\text{C}} \times 2.952 \times 10^3 \text{ C} = 3.54 \times 10^3 \text{ J}$$

For the Li–MnO_2 cell, Li is the anode material:

$$\text{Charge} = 1 \text{ g} \times \frac{1 \text{ mol}}{6.941 \text{ g}} \times \frac{1 \text{ mol e}^-}{1 \text{ mol Li}} \times \frac{9.65 \times 10^4 \text{ C}}{1 \text{ mol e}^-} = 1.390 \times 10^4 \text{ C}$$

$$\text{Energy} = 3.15 \frac{\text{J}}{\text{C}} \times 1.390 \times 10^4 \text{C} = 4.38 \times 10^4 \text{ J}$$

Therefore, the Li–MnO_2 cell generates more energy per gram of anode.

Think About It
Notice that although the charge generated per mole of Li versus that of a mole of Zn in these cells is lower, the high voltage of the Li–MnO_2 cell means that this cell generates more energy.

19.65. Collect and Organize
We are asked to comment on why Teflon was used to replace the asbestos mats separating the skeleton and outer surface of the Statue of Liberty.

Analyze

Corrosion of the Statue of Liberty occurred due to the electrochemical reaction that occurred between the steel frame and the copper surface of the statue. Ideally, the mat separating these two materials should prevent such an electrochemical reaction from occurring. Teflon is a polymeric material composed of poly(tetrafluoroethylene). Teflon is water-repellant, and does not ionize due to the extremely stable C–F bonds present in the polymer chain.

poly(tetrafluoroethylene)

Solve

Since no metals or ionizable bonds are present in Teflon, it is not expected to react with either steel or the copper surface of the Statue of Liberty. The water-repellency of the material also prevents water from coming into contact with both surfaces and setting up an electrochemical cell, using the mats as a substrate.

Think About It

Electrochemical reactions between dissimilar materials could lead to serious problems with the structural integrity of buildings or monuments. When possible, it is best to avoid direct contact between dissimilar metals.

19.67. **Collect and Organize**

We are asked to choose a material that would make a good sacrificial anode for the aluminum frames of a window in contact with salt water in an aquarium.

Analyze

Aluminum is itself relatively resistant to redox activity once it forms a protective oxide layer of $Al_2O_3(s)$. A sacrificial anode material must be more readily oxidized than aluminum metal. When consulting Appendix 6, any material with $E^°_{red}$ more negative than -1.662 V would serve.

Solve

Magnesium metal is the most obvious choice. It forms a protective oxide, MgO, and has $E^°_{red} = -2.37$ V, meaning that it would be oxidized rather than the aluminum frame.

Think About It

Though several alkali metals such as Na and K are nearby in Appendix 6, they would react with the water in the aquarium to generate NaOH(*aq*) and hydrogen gas!

19.69. **Collect and Organize**

We are to explain the differences in the signs of the cathode in a voltaic versus an electrolytic cell.

Analyze

The signs of the electrodes in a cell indicate the direction of electron flow.

Solve

In a voltaic cell, the electrons are produced at the anode so a positive (+) charge builds up there; in an electrolytic cell, electrons are being forced onto the cathode so that it builds up negative (–) charge. The flow of electrons in the outside circuit is reversed in an electrolytic cell compared to the flow in a voltaic cell.

Think About It

An electrolytic cell uses an outside source of electrical energy to cause a nonspontaneous reaction to occur.

19.71. Collect and Organize

In a mixture of molten salts containing Br^- and Cl^- ions, we are to predict which product, Br_2 or Cl_2, forms first in an electrolytic cell as the voltage is increased.

Analyze

The oxidations of Br^- and Cl^- are expressed as

$$2\,Br^-(\ell) \rightarrow Br_2(\ell) + 2\,e^-$$
$$2\,Cl^-(\ell) \rightarrow Cl_2(g) + 2\,e^-$$

Solve

The halide that is first to be oxidized is the one with the lowest ionization energy. Br^-, being larger and less electronegative than Cl^-, loses its electron more readily and therefore Br_2 forms first in the cell as the voltage is increased.

Think About It

If the molten salt also contains F^-, F_2 would form after Br_2 and Cl_2.

19.73. Collect and Organize

For the electrolysis of a 1.0 M Cu^{2+} solution, we are to determine whether the potential at the cathode where the reduction of Cu^{2+} occurs needs to be more negative or less negative than 0.34 V in order to quantitatively reduce the Cu^{2+} in solution to Cu.

Analyze

We are given that $E°_{red}$ for Cu^{2+} is 0.34 V. This is under standard conditions when $[Cu^{2+}]$ = 1.0 M, the concentration of the solution at the start of the electrolysis. As $[Cu^{2+}]$ decreases as Cu is deposited on the electrode, E_{cell} can be calculated using the Nernst equation:

$$E_{cell} = 0.34\ V - \frac{0.0592\ V}{2} \log \frac{1}{[Cu^{2+}]}$$

Solve

As the reaction proceeds and $[Cu^{2+}]$ decreases, the value of log $(1/[Cu^{2+}])$ becomes more positive. As a result E_{cell} decreases, so the cathode potential must be more negative than 0.34 V to complete the reduction.

Think About It

There might be a slight overpotential required to accomplish the electrolysis, however, because of a kinetic barrier to the reduction reaction.

19.75. Collect and Organize

In an electroplating process, we are to calculate the mass of silver deposited on an object when 1.3 A · hr of charge is delivered from a battery. To determine this we need to relate the ampere-hours to the total number of electrons generated. Then we can relate the number of electrons to the mass of Ag deposited from a solution of Ag^+.

Analyze

We can convert ampere-hours to coulombs:

$$A \cdot hr \times \frac{1\ C}{A \cdot s} \times \frac{3600\ s}{1\ hr}$$

The moles of electrons used in the process can be calculated from this result knowing that 1 mol e^- = 9.65 × 10^4 C. To calculate the mass of Ag deposited, we also need to know that the reduction of Ag^+ to Ag is a one-electron process.

Solve

$$1.3 \text{ A} \cdot \text{hr} \times \frac{1 \text{ C}}{\text{A} \cdot \text{s}} \times \frac{3600 \text{ s}}{\text{hr}} = 4680 \text{ C}$$

$$4680 \text{ C} \times \frac{1 \text{ mol e}^-}{9.65 \times 10^4 \text{ C}} = 4.850 \times 10^{-2} \text{ mol e}^-$$

$$4.850 \times 10^{-2} \text{ mol e}^- \times \frac{1 \text{ mol Ag}}{1 \text{ mol e}^-} \times \frac{107.87 \text{ g Ag}}{1 \text{ mol}} = 5.2 \text{ g Ag}$$

Think About It

The higher the amps for the battery, the faster an object can be electroplated.

19.77. Collect and Organize

We are to calculate how long it will take to recharge a battery that contains 4.10 g of NiO(OH) and is 75% discharged. This means that 3.075 g of NiO(OH) has been depleted from the battery. The charger for the battery operates at 2.00 A and 1.3 V.

Analyze

We first need to calculate the moles of electrons needed to recover 3.075 g of NiO(OH), which in the NiMH battery forms $Ni(OH)_2$ in a 1-electron process. Next, we will convert the moles of electrons to coulombs. Because 1 C = 1 A·s and we know the amperes at which the charger operates, we can then calculate the time it takes the charger to deliver the electrons to recharge the battery.

Solve

$$3.075 \text{ g NiO(OH)} \times \frac{1 \text{ mol NiO(OH)}}{91.70 \text{ g NiO(OH)}} \times \frac{1 \text{ mol e}^-}{1 \text{ mol NiO(OH)}} = 0.0335 \text{ mol e}^-$$

$$0.0335 \text{ mol e}^- \times \frac{9.65 \times 10^4 \text{C}}{1 \text{ mol e}^-} \times \frac{1 \text{ A} \cdot \text{s}}{1 \text{ C}} \times \frac{1}{2.00 \text{ A}} = 1618 \text{ s}$$

In minutes this is 1618 s × 1 min/60 s = 27.0 min.

Think About It

As the size of the battery increases, so does the amount of reactant that needs to be regenerated, and thus the time it takes to recharge it.

19.79. Collect and Organize

We are to calculate the amount of O_2 that could be generated in one hour on a submarine using electrolysis, and then consider the practicality of using seawater as the source of oxygen for the submarine.

Analyze

We can calculate the moles of O_2 produced by the electrolytic cell through

$$\text{Moles O}_2 = \text{time in seconds} \times \text{amperes} \times \frac{1 \text{ C}}{\text{A} \cdot \text{s}} \times \frac{1 \text{ mol e}^-}{9.65 \times 10^4 \text{C}} \times \frac{1 \text{ mol O}_2}{4 \text{ mol e}^-}$$

Notice that the oxidation of water to O_2 is a 4 e$^-$ process. We can then use the ideal gas law to calculate the volume of O_2 produced.

Solve

(a)

$$\text{Moles O}_2 = 1 \text{ hr} \times \frac{3600 \text{ s}}{1 \text{ hr}} \times 0.025 \text{ A} \times \frac{1 \text{ C}}{\text{A} \cdot \text{s}} \times \frac{1 \text{ mol e}^-}{9.65 \times 10^4 \text{C}} \times \frac{1 \text{ mol O}_2}{4 \text{ mol e}^-} = 2.332 \times 10^{-4} \text{ mol O}_2$$

$$V = \frac{2.332 \times 10^{-4} \text{ mol} \times 0.08206 \text{ L} \cdot \text{atm} / \text{mol} \cdot \text{K} \times 298 \text{ K}}{0.98692 \text{ atm}/1 \text{ bar}} = 5.78 \times 10^{-3} \text{ L or 5.8 mL}$$

(b) Seawater contains a fairly high concentration of Cl$^-$ and Br$^-$ that can be oxidized, so the direct electrolysis of seawater would not be useful as an oxygen source.

Think About It
Submarines probably purify their water, perhaps through a reverse osmosis process, to remove the chloride and bromide and other ions before the electrolysis process.

19.81. **Collect and Organize**
For the process that electroplates nickel, we are to calculate the lowest potential required to deposit Ni onto a piece of iron using a 0.35 M Ni^{2+} solution.

Analyze
To solve this problem we need to use the Nernst equation:

$$E_{cell} = E^{\circ}_{cell} - \frac{0.0592 \text{ V}}{n} \log Q$$

where E°_{cell} is the reduction potential of Ni^{2+} versus the SHE (−0.257 V), n is the number of electrons needed to reduce Ni^{2+} to Ni (2), and Q is $1/[Ni^{2+}]$ based on the reduction half reaction
$$Ni^{2+}(aq) + 2 \text{ e}^- \rightarrow Ni(s)$$

Solve

$$E_{cell} = -0.257 \text{ V} - \frac{0.0592 \text{ V}}{2} \log \frac{1}{0.35 \text{ } M} = -0.270 \text{ V}$$

Think About It
For this electrolysis reaction, using a more dilute solution of the metal cation necessitates an increase in the potential needed to cause Ni^{2+} to deposit on the iron.

19.83. **Collect and Organize**
We are to consider the advantages and disadvantages of hybrid power systems versus all-electric fuel-cell systems.

Analyze
A parallel hybrid power system uses traditional petroleum-based fuel for high power demands and an electric motor for lower power demands. An all-electric fuel cell system uses only electrochemical power based on combustion half-reactions to supply power.

Solve
A hybrid vehicle uses a relatively inexpensive fuel (gasoline) in the internal combustion engine and has good fuel economy but still gives off emissions. A fuel-cell vehicle does not give off emissions (the reaction produces H_2O) but requires a more expensive and explosive fuel (hydrogen); moreover, current battery technologies in all-electric and hybrid vehicles incorporate materials that are still very expensive and bulky.

Think About It
The determining factor in whether alternate fuels and power systems get used will be the cost of petroleum used to produce gasoline, which traditionally has been much less expensive than alternative energy sources.

19.85. **Collect and Organize**
We are to consider why methane fuel cells are likely to produce less CO_2 emissions per mile than an internal combustion engine fueled by methane.

Analyze
Both the methane fuel cell and the combustion of methane in a combustion reaction have the balanced equation
$$CH_4(g) + 2 \text{ } O_2(g) \rightarrow 2 \text{ } H_2O(g) + CO_2(g)$$

Solve
Fuel cells burn methane fuel more efficiently. Electric engines are more efficient because they convert more of the energy into motion instead of losing it as heat. Therefore, less CO_2 is produced per mile with fuel cells.

Think About It
The bulkiness and short range of fuel cells currently limit the use of fuel cells in transportation.

19.87. **Collect and Organize**
For the reactions of CH_4 and CO with water, we are to assign oxidation numbers to the C and H atoms in all the species in the reactions and calculate ΔG°_{rxn} for each and for the overall reaction

$$CH_4(g) + 2\ H_2O(g) \rightarrow 4\ H_2(g) + CO_2(g)$$

Analyze
We can use the usual rules of assigning oxidation states from Chapter 4. To calculate ΔG°_{rxn} we use ΔG°_f values from Appendix 4 in the following equation:

$$\Delta G^\circ_{rxn} = \sum n\Delta G^\circ_{f,products} - \sum m\Delta G^\circ_{f,reactants}$$

Solve
(a) $\overset{-4}{C}\overset{+1}{H_4}(g) + \overset{+1}{H_2}O(g) \rightarrow \overset{+2}{C}O(g) + 3\overset{0}{H_2}(g)$

$\overset{+2}{C}O(g) + \overset{+1}{H_2}O(g) \rightarrow \overset{0}{H_2}(g) + \overset{+4}{C}O_2(g)$

(b) For the reaction of CH_4 with H_2O
$$\Delta G^\circ_{rxn} = \left[(1\ mol\ CO)(-137.2\ kJ/mol) + (3\ mol\ H_2 \times 0.0\ kJ/mol)\right] -$$
$$\left[(1\ mol\ CH_4)(-50.8\ kJ/mol) + (1\ mol\ H_2O)(-228.6\ kJ/mol)\right] = 142.2\ kJ$$

For the reaction of CO with H_2O
$$\Delta G^\circ_{rxn} = \left[(1\ mol\ CO_2)(-394.4\ kJ/mol) + (1\ mol\ H_2 \times 0.0\ kJ/mol)\right] -$$
$$\left[(1\ mol\ CO)(-137.2\ kJ/mol) + (1\ mol\ H_2O)(-228.6\ kJ/mol)\right] = -28.6\ kJ$$

For the overall reaction
$$\Delta G^\circ_{overall} = \Delta G^\circ_{rxn_1} + \Delta G^\circ_{rxn_2} = 113.6\ kJ$$

Think About It
The spontaneity of the second reaction is not enough to overcome the positive free energy of the first reaction, so the overall reaction is nonspontaneous.

19.89. **Collect and Organize**
We consider electrolysis of a molten salt containing the Mg^{2+} ion from evaporated seawater (so it may also contain NaCl).

Analyze
The possible reactions are (with E° values when listed in Appendix 6)

$$Mg^{2+}(\ell) + 2\ e^- \rightarrow Mg(s)$$
$$Na^+(\ell) + e^- \rightarrow Na(s)$$
$$2\ H_2O(\ell) + 2\ e^- \rightarrow H_2(g) + 2\ OH^-(aq) \qquad E^\circ_{red} = -0.8277\ V$$
$$2\ H_2O(\ell) \rightarrow O_2(g) + 4\ H^+(aq) + 4\ e^- \qquad E^\circ_{ox} = 1.229\ V$$
$$Mg^{2+}(aq) + 2\ e^- \rightarrow Mg(s) \qquad E^\circ_{red} = -2.37\ V$$
$$Na^+(aq) + e^- \rightarrow Na(s) \qquad E^\circ_{red} = -2.71\ V$$
$$2\ Cl^-(aq) + 2\ e^- \rightarrow Cl_2(g) \qquad E^\circ_{ox} = 1.3583\ V$$

Solve
(a) Mg^{2+} undergoes a reduction reaction that occurs at the cathode. Mg forms at the cathode.
(b) No. Mg^{2+}, with a higher positive charge, has a lower (less negative) reduction potential than Na^+, so the Mg^{2+} would not need to be separated from the NaCl in seawater first.

(c) No. The electrolysis of $MgCl_2(aq)$ would not produce $Mg(s)$ because water, with a less negative reduction potential, would be electrolyzed.
(d) H_2 and O_2 gases would be produced.

Think About It

Because different components are reduced at different potentials, electrolysis of molten salts is one way to separate components (e.g., metals) from each other.

19.91. **Collect and Organize**

For a Mg–Mo_3S_4 battery for which we are given the half-reaction potential of the anode reaction (–2.37 V) and the overall cell potential (1.50 V), we are to calculate E°_{red} of Mo_3S_4. We are asked to give the apparent oxidation states of Mo in Mo_3S_4 and $MgMo_3S_4$. We are also to consider why Mg^{2+} is added to the battery's electrolyte and determine the oxidation states and electron configurations of Mo in Mo_3S_4 and $MgMo_3S_4$.

Analyze

Because $E^{\circ}_{cell} = E^{\circ}_{cathode} - E^{\circ}_{anode}$, the reduction potential for Mo_3S_4 will be $E^{\circ}_{cell} + E^{\circ}_{anode}$.

Solve

(a) $E^{\circ}_{cell} = 1.50 \text{ V} + (-2.37 \text{ V}) = -0.87 \text{ V}$

(b) Sulfur usually carries a 2– charge so each Mo atom in Mo_3S_4 has a calculated charge of 2.67+. This, therefore, is a mixed oxidation state compound where it is likely that two of the Mo atoms have a 3+ charge and one Mo atom has a 2+ charge. In $MgMo_3S_4$, the Mg atom has a 2+ charge, so the Mo atoms have a 2+ charge. The electron configurations for the two oxidation states of Mo are

Mo in +2 oxidation state $[Kr]4d^4$
Mo in +3 oxidation state $[Kr]4d^3$

(c) Mg^{2+} is added to the electrolyte to better carry the charge in the cell. This cation is produced at the anode and consumed at the cathode.

Think About It

This battery resembles the lithium–ion battery in that the migration of a cation in the cell generates the electrical current.

CHAPTER 20 | Biochemistry: The Compounds of Life

20.1. Collect and Organize
We are asked to draw the Lewis structure of the hydrogen phosphite ion, a reactant with formaldehyde to produce p-(hydroxymethyl)-phosphonic acid.

Analyze
We learned how to draw Lewis structures in Chapter 8. We must remember here that phosphorus may expand its octet so that formal charges on all of the atoms are minimized.

Solve

Think About It
Indeed, in this compound phosphorus has expanded its octet so that the formal charge on phosphorus is zero. The oxygen atoms that have three lone pairs in this structure each carry a formal charge of -1.

20.3. Collect and Organize
For the triglycerides shown in Figure P20.3, we are to identify which of the fatty acids that make up the triglycerides are saturated.

Analyze
A saturated fatty acid has no $C=C$ double bonds in its structure. Common fatty acids are shown in Table 20.3.

Solve
(a) This triglyceride contains palmitic acid (saturated), oleic acid (unsaturated), and linoleic acid (unsaturated).
(b) This triglyceride contains oleic acid (unsaturated) and stearic acid (saturated).
The saturated fatty acids in these triglycerides are palmitic acid and stearic acid.

Think About It
The triglycerides are formed through a condensation reaction between glycerol and three fatty acids.

20.5. Collect and Organize
For the structure of enkephalin shown in Figure P20.5, we are to name the five amino acids that make up the polypetide.

Analyze
The 20 common amino acids are shown in Table 20.1. A peptide bond is formed between two amino acids in an acid–base reaction:

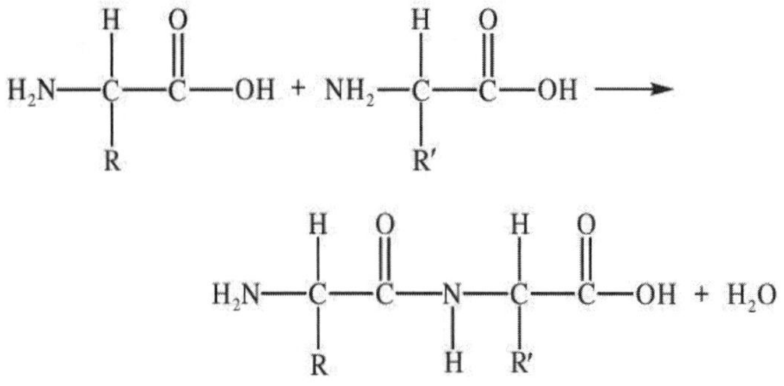

Therefore, the $\overset{O}{\overset{\|}{C}}$—NH linkage is where two amino acids are joined.

Solve

The amino acids that make up the structure are (from left to right in the molecule): tyrosine, glycine, glycine, phenylalanine, and methionine.

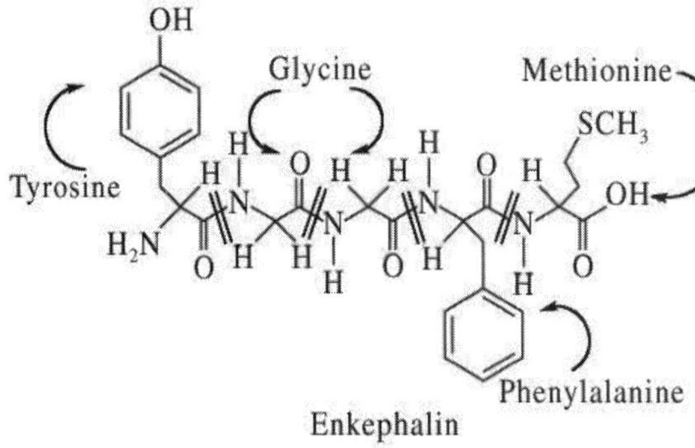

Think About It

Although there are only 20 common amino acids, they can join in seemingly endless combinations to make a huge variety of peptides.

20.7. **Collect and Organize**

We are to describe the type of isomerism that is implied by the term *trans fats* and identify which molecules shown in Figure P20.7 are trans fats.

Analyze

Trans isomerism occurs around a C=C double bond.

Solve

Trans fats exhibit geometric isomerism around the C=C bond where similar groups (such as the H atoms) on the two carbon atoms are situated on opposite sides of the double bond. Both (a) and (c) are trans fats.

Think About It

Most trans fats are made through partially hydrogenating the double bonds in plant oils to give them higher melting points and a longer shelf life.

20.9. **Collect and Organize**

From the structure given for sucralose in Figure P20.9, we are to identify the sugar from which this sweetener is prepared and then comment on the implications that this sweetener is "natural."

Analyze

Some of the common natural sugars (glucose, fructose, and sucrose) are shown in Figures 20.20, 20.22, and 20.23.

Solve

The structure of sucralose appears to be derived from sucrose. The difference in the structures is that in sucralose, three OH groups on sucrose have been replaced by Cl atoms.

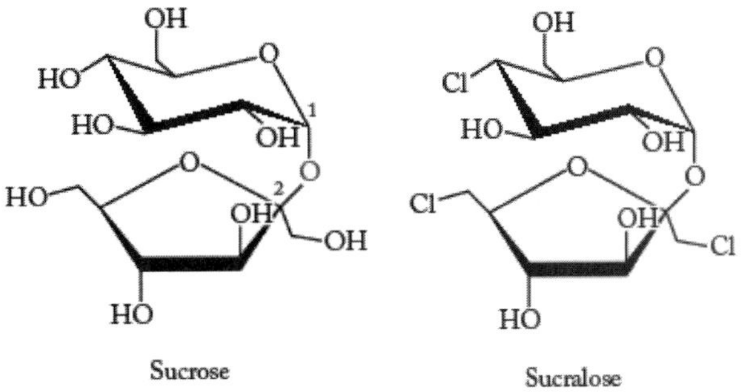

Sucrose   Sucralose

Being derived from sucrose implies that the sugar is natural, but the presence of Cl atoms on sugars is not natural.

Think About It

Sucralose is different than other artificial sweeteners like aspartame or saccharin in that it is heat stable and therefore can be used in baking.

20.11. **Collect and Organize**

We are asked whether the assembly of small molecules into larger molecules in a cell is accompanied by an increase or a decrease in entropy.

Analyze

An increase in entropy is an increase in the number of particles in a process, whereas a decrease in entropy is a decrease in the number of particles in a process.

Solve

Because the number of particles (molecules) is decreasing as small molecules are bonded together to form a larger molecule, entropy decreases.

Think About It

Remember that a decrease in entropy is not favorable for a process. For the process with negative ΔS to be spontaneous, the enthalpy of the reaction must be negative so that the free energy is negative for the process.

20.13. **Collect and Organize**

We are to explain the meaning of alpha in "α-amino acid."

Analyze

The α refers to the placement of the –NH$_2$ and –COOH groups on the structure of the amino acid.

Solve

The α refers to the single carbon atom in amino acids to which both –NH$_2$ and –COOH groups are bonded.

Think About It

A β-amino acid has its amino group bonded to the β-carbon instead of the α-carbon. An example is β-alanine.

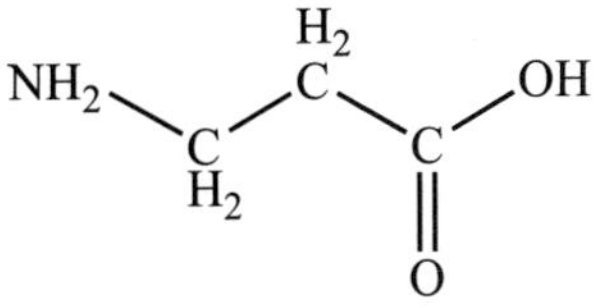

20.15. **Collect and Organize / Analyze**

We are asked what the prefixes D- and L- indicate.

Solve

D- and L- refer to how the four groups on a chiral carbon atom are oriented.

Think About It

Very few biological systems on Earth contain D-amino acids; they have been found in bacterial cell walls, in fungal toxins, and in some exotic sea animals.

20.17. **Collect and Organize**

Of the structures shown in Figure P20.17, we are asked which is not an α-amino acid.

Analyze

An α-amino acid has both the amine and carboxylic acid groups bonded to the same carbon.

Solve

Compounds (a) and (c) have their amine and carboxylic acid groups bonded to different carbons and so they are not α-amino acids. Only (b) has the amine group bonded to the same carbon as the carboxylic acid is bonded, and so it is the only α-amino acid in Figure P20.17.

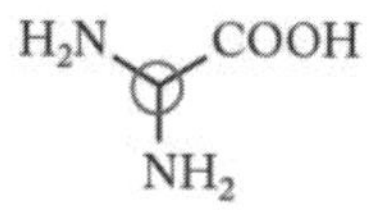

$$H_2N \quad COOH$$
$$NH_2$$

Think About It

Compound (b) actually has two amine groups in its structure.

20.19. **Collect and Organize / Analyze**

We are asked to explain why most amino acids exist in their zwitterionic form (contain both a positive and a negative group on the same molecule) at a slightly basic pH.

Solve

On an amino acid, the amine group is basic (becomes protonated) and the carboxylic acid group is acidic (deprotonates) to form the zwitterion at pH 7.4. This near-neutral pH is basic enough to deprotonate the carboxylic acid and yet acidic enough to protonate the amine group.

Think About It

For example, the zwitterion of alanine is

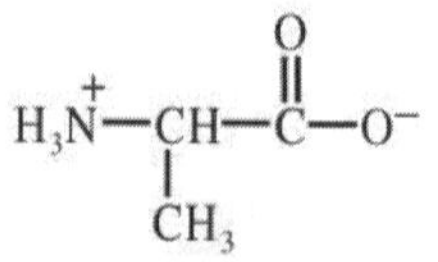

$$H_3\overset{+}{N}-CH-\overset{\overset{\displaystyle O}{\|}}{C}-O^-$$
$$CH_3$$

20.21. **Collect and Organize**

In the folding of proteins, we are asked why lysine often pairs with glutamic acid.

Analyze

The structures of the two amino acids appear in Table 20.1.

Solve

Lysine contains two amino groups, one of which is on a long carbon tail. This can react with the carboxylic acid on the carbon tail of glutamic acid to form a salt bridge.

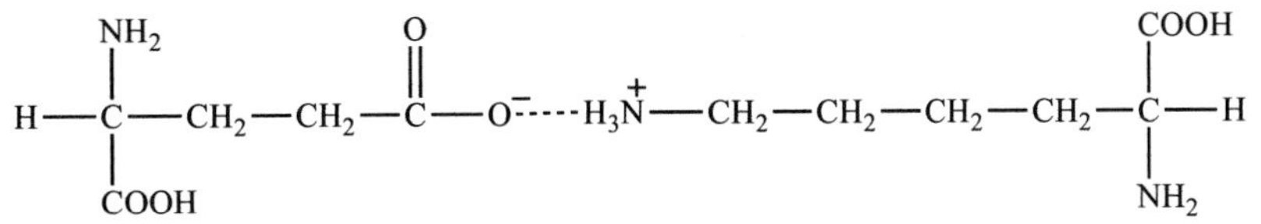

Think About It
This interaction is formed without much crowding of the amino acids and thus is facile.

20.23. **Collect and Organize**
We are asked to draw and name the dipeptides that form when two amino acids undergo condensation.

Analyze
The condensation reaction between amino acids is the result of the carboxylic acid of one amino acid reacting with the amine of another to give the peptide bond (Figure 20.7) and release water.

Solve
(a) Alanine + serine = Alaser (or AS)

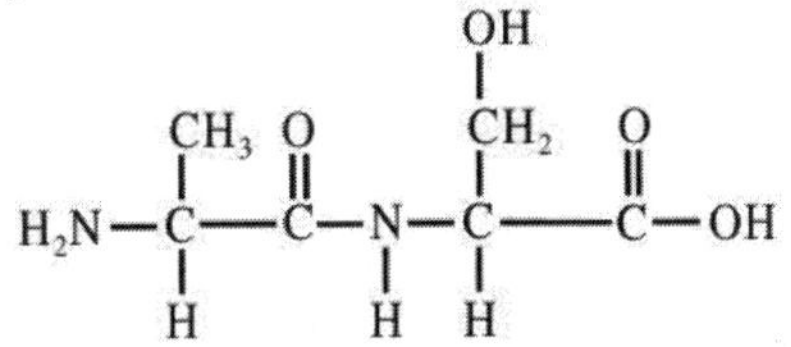

(b) Alanine + phenylalanine = Alaphe (or AF)

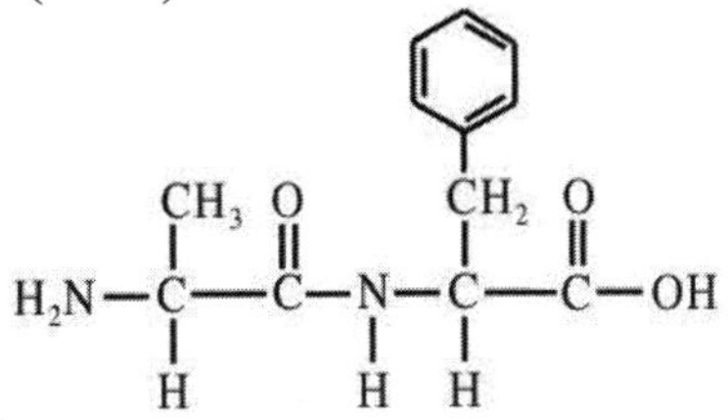

(c) Alanine + valine = Alaval (or AV)

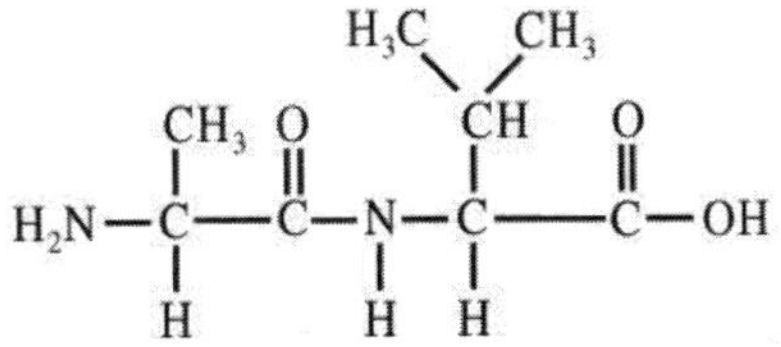

Think About It
Using the naming rules described in the textbook, these dipeptides may also be named (a) alanylserine, (b) alanylphenylalanine, and (c) alanylvaline.

20.25. **Collect and Organize**
For each dipeptide structure shown in Figure P20.25, we are to identify the parent amino acids.

Analyze
The peptide bond formed between two amino acids is

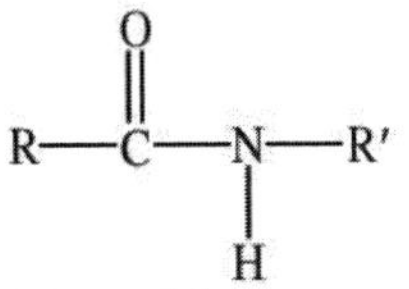

If we imagine the C—N bond rehydrated, we would regenerate the carboxylic acid containing R and the amine containing R'. From the structures of R and R', we can identify the amino acids.

Solve

(a) Alanine + glycine

(b) Leucine + leucine

(c) Tyrosine + phenylalanine

Think About It

For longer chain peptides that make up proteins, the order in which the amino acids are linked is very important to their structure, which is key to their function in the cell.

20.27. **Collect and Organize**

We are asked to identify the second product in the reaction shown in Figure P20.27.

Analyze

The only difference in the structures of the products and the reactants is the replacement of the $-NH_2$ group in the reactant with $-OH$ in the product.

Solve

Upon reaction with water the amide group forms a carboxylic acid group and ammonia. The second product in this reaction is simply NH_3.

Think About It

If we think of the reverse reaction as the condensation of NH_3 with the carboxylic acid, we see that we have the same condensation reaction that produces peptides.

20.29. **Collect and Organize / Analyze**

By looking at the structures in Figure 20.24, we can describe the structural differences between starch and cellulose.

Solve

Starch has α-glycosidic bonds and cellulose has β-glycosidic bonds. Starch coils into granules, and cellulose forms linear molecules.

Think About It

The subtle difference in how the glucose chain is oriented gives starch and cellulose very different properties. We can easily digest starch, but we cannot digest cellulose.

20.31. **Collect and Organize**

We are asked if the fuel value of glucose in its cyclic form is the same as that in its linear form.

Analyze

In the cyclic form of glucose we have an additional C—O bond that is not present in the linear form. We also have the formation of a C—O single bond from a C=O.

Solve

No. Because the bonding is different in the cyclic form of glucose, we do not expect the fuel values of the cyclic and linear forms to be exactly the same.

Think About It

We might expect, though, that the fuel values are very close to each other since the structures differ only by a few bonds.

20.33. **Collect and Organize**

We are to explain why the free-energy change for the conversion of glucose 6-phosphate into fructose 6-phosphate is close to zero.

Analyze
Glucose and fructose are structural isomers of each other and differ only in the location of the carbonyl group.

Solve
The bonding in fructose and glucose is nearly the same, so we expect that the free energy change from one to the other would be close to zero.

Think About It
Because the bonding is nearly identical, the enthalpy change is close to zero. Likewise, the structures of these two sugars are very similar, and so the entropy change is also close to zero.

20.35. **Collect and Organize**
We can recall from Chapter 18 how to calculate ΔG for an overall process that is made up of two steps.

Analyze
ΔG is a state function, and therefore we can apply Hess's law to calculate the free-energy change for a multistep process.

Solve
To calculate the free-energy change for a two-step process, we need only to sum the individual ΔG values for each reaction.

Think About It
Other state functions include enthalpy and entropy.

20.37. **Collect and Organize**
For the sugar galactose (Figure P20.37), we are to use the flat Lewis structure to draw the isomers that form when this linear structure forms a ring, describing the similarities and differences between the α or β structures formed in each case.

Analyze
Galactose is a 6-carbon sugar, and the OH group on the C-5 carbon reacts with the aldehyde group at C-1 to form a six-membered ring. Depending on the orientation of the OH group attached relative to the aldehyde, we obtain either the α or β isomer.

Solve
Using Figure 20.22 as a reference, the α and β forms of galactose are

β-Galactose   α-Galactose

The position of the hydroxy group on carbon 1 differs between the α and β forms of galactose. The relative positions of the hydroxy groups on carbons 2, 3, and 4 are the same on both isomers.

Think About It
Galactose is not as water soluble as glucose and is less sweet.

20.39. Collect and Organize
Given the cyclic structures of sugars shown in Figure P20.39, we are to identify those that are β isomers.

Analyze
Structures that are β forms have the OH group at C-1 pointing up.

Solve
(c) shows a β isomer.

Think About It
Because the orientation of the OH group on the carbon-3 atom in structure (b) is different from the others, we can say that (b) is a different sugar.

20.41. Collect and Organize
Given the cyclic structures shown in Figure P20.41, we are to determine which are α isomers.

Analyze
Structures that are α forms have the OH group at C-1 pointing down.

Solve
(a) shows an α isomer.

Think About It
Both structures (b) and (c) are β isomers.

20.43. Collect and Organize
Given the structures shown in Figure P20.43, we are to choose the saccharides that are digestible by humans.

Analyze
Humans can digest α-glycosidic linkages between sugars, but not β linkages. Structures that are α forms have the OH group at C-1 pointing down.

Solve
Only structure (b) is fully digestible by humans.

Think About It
Structure (c) has both α and β linkages, so it is only partially digestible.

20.45. Collect and Organize
Given values for the free energy of formation of maltose, glucose, and water, we are to calculate the free-energy change for the reaction of maltose with water to produce glucose.

Analyze
To calculate the free-energy change we use the equation
$$\Delta G^{\circ}_{rxn} = \sum n \Delta G^{\circ}_{f,products} - \sum m \Delta G^{\circ}_{f,reactants}$$

Solve
$$\Delta G^{\circ}_{rxn} = \left[(2 \text{ mol})(-1274.4 \text{ kJ/mol})\right] - \left[(1 \text{ mol})(-2246.6 \text{ kJ/mol}) + (1 \text{ mol})(-285.8 \text{ kJ/mol})\right] = -16.4 \text{ kJ}$$

Think About It
This reaction is spontaneous under standard conditions, as indicated by the negative free-energy change for the reaction.

20.47. Collect and Organize / Analyze

We can use definitions to discriminate between a saturated and unsaturated fatty acid.

Solve

Saturated fatty acids have all C—C single bonds in their structure; unsaturated fatty acids have C═C double bonds in their structure.

Think About It

The common fatty acids are shown in Table 20.3. Of those listed, oleic, linoleic, and linolenic acids have C═C double bonds in their structures.

20.49. Collect and Organize

We consider why Arctic explorers would eat sticks of butter.

Analyze

Butter is composed mostly of fatty acids.

Solve

Fatty acids have a high fuel value (see Problem 20.48) and eating sticks of butter affords Arctic explorers with more energy per gram of food compared to carbohydrates or proteins.

Think About It

Arctic explorers need many more calories, not only to move around the environment, but also to keep warm.

20.51. Collect and Organize

By examining the structure of a triglyceride, we can determine whether these molecules have a chiral center.

Analyze

A triglyceride is formed when glycerol reacts with long-chain fatty acids to form esters.

Solve

If the two fatty acids linked to the glycerol at C-1 and C-3 are different then, yes, the triglyceride has a chiral center. This will give a structure in which the central carbon on the triglyceride has four different groups attached and the fatty acid is therefore chiral.

Think About It

If all three fatty acid chains are different, the triglyceride must be chiral.

20.53. Collect and Organize

We are asked to compare oleic acid, $CH_3(CH_2)_7CH=CH(CH_2)_7COOH$, and α-linoleic acid, $CH_3CH_2CH=CHCH_2CH=CHCH_2CH=CH(CH_2)_7COOH$, and determine which would consume a greater quantity of hydrogen when hydrogenated. We are also asked to determine if we could distinguish between the two fatty acids from the hydrogenation products of each fatty acid.

Analyze

Each mole of oleic acid ($\mathcal{M} = 282.52$ g/mol) contains one mole of "unsaturation" (i.e., one C=C double bond), whereas each mole of α-linoleic acid ($\mathcal{M} = 278.48$ g/mol) contains three moles of unsaturated species. Upon hydrogenation, the molar ratio of $H_2(g)$ required to fully hydrogenate each species is 1:3. Upon hydrogenation, both species will be fully saturated, forming stearic acid, $CH_3(CH_2)_{16}COOH$.

Solve

(a) For each of the listed masses of fatty acid, the quantity of $H_2(g)$ required is

$$1.0 \text{ kg oleic acid} \times \frac{1000 \text{ g}}{1 \text{ kg}} \times \frac{1 \text{ mol oleic acid}}{282.52 \text{ g oleic acid}} \times \frac{1 \text{ mol C=C}}{1 \text{ mol oleic acid}} \times \frac{1 \text{ mol H}_2}{1 \text{ mol C=C}} = 3.54 \text{ mol H}_2(g)$$

$$0.5 \text{ kg } \alpha\text{-linoleic acid} \times \frac{1000 \text{ g}}{1 \text{ kg}} \times \frac{1 \text{ mol } \alpha\text{-linoleic acid}}{278.48 \text{ g } \alpha\text{-linoleic acid}} \times \frac{3 \text{ mol C=C}}{1 \text{ mol } \alpha\text{-linoleic acid}} \times \frac{1 \text{ mol H}_2}{1 \text{ mol C=C}} = 5.39 \text{ mol H}_2(g)$$

α-linoleic acid will consume more $H_2(g)$ than oleic acid.

(b) No, we would not be able to distinguish the hydrogenation products; upon hydrogenation, both species will be fully saturated, forming stearic acid, $CH_3(CH_2)_{16}COOH$.

Think About It

Because of the high degree of unsaturation, α-linoleic acid is a liquid under standard conditions.

20.55. ### Collect and Organize

For the reaction of glycerol with octanoic acid, decanoic acid, and dodecanoic acid, we are to draw the structures of the resulting triglycerides.

Analyze

The reaction of glycerol with the carboxylic acid functional group of the fatty acid makes an ester bond and releases water as a by-product.

Solve

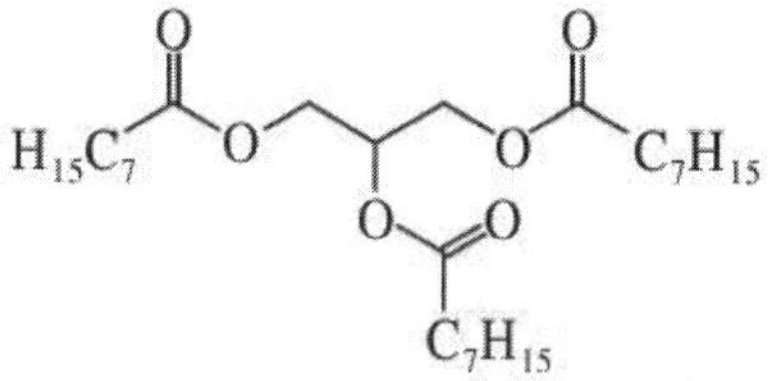

Glycerol with octanoic acid

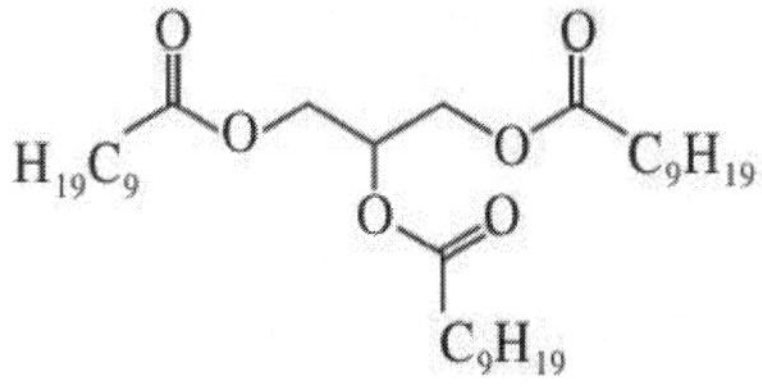

Glycerol with decanoic acid

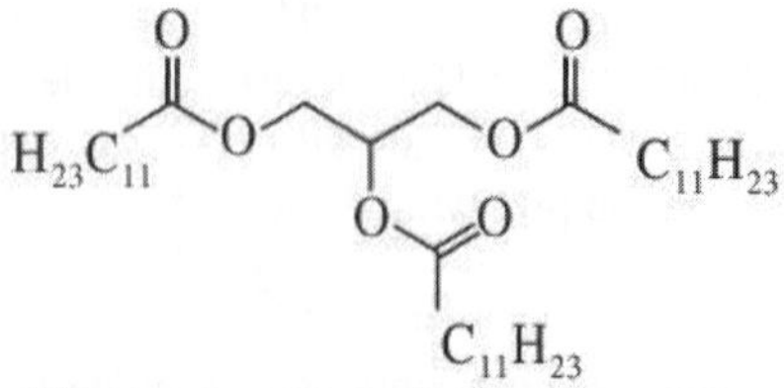

Glycerol with dodecanoic acid

Think About It

You may have read this problem as asking for the reaction of glycerol with one equivalent each of the fatty acids. In that case you would have the following triglycerides:

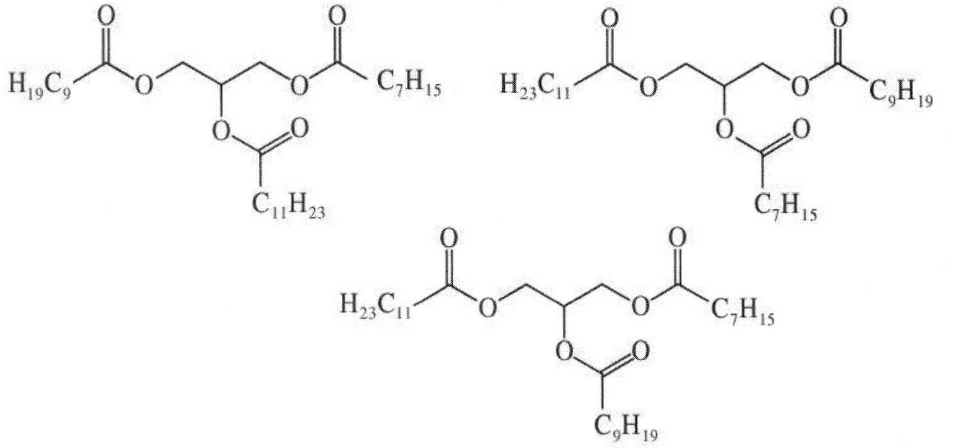

20.57. Collect and Organize / Analyze

We are to name the three molecular subunits present in DNA and identify the two subunits that form the backbone of DNA strands.

Solve

The three subunits of the DNA structure are a phosphate group, a five-carbon sugar, and a nitrogen base. As shown in Figure 20.35, the backbone of DNA is composed of alternating sugar residues and phosphate groups.

Think About It

The entire unit of the three subunits is called a *nucleotide*.

20.59. Collect and Organize

We are to identify the kind of intermolecular force that holds together the strands of DNA and forms the double-helix configuration.

Analyze

The intermolecular force between the base pairs is illustrated in Figure 20.36.

Solve

Hydrogen bonds hold the base pairs together to form the double helix of the DNA molecule.

Think About It

Recall that hydrogen bonds are quite strong, and so the integrity of the DNA molecule is fairly secure.

20.61. Collect and Organize

We are to draw the structure of a ribonucleotide of RNA, namely, adenosine 5′-monophosphate.

Analyze

A ribonucleotide of RNA consists of the five-carbon sugar ribose, a phosphate group, and a nitrogen base with the general form

The structure of adenine is

Solve

Think About It
This nucleoside plays an important role in energy transfer as ATP and ADP.

20.63. **Collect and Organize**
In the replication of DNA, we are to determine the sequence of the strand that is formed when the original strand has the sequence T-C-G and draw the section of the double helix corresponding to this sequence.

Analyze
In replication, DNA copies information from a DNA strand by making a sequence of bases complementary to that on the DNA segment. For DNA, thymine (T) complements adenine (A), and guanine (G) complements cytosine (C).

Solve
(a) The sequence of the double-stranded helix containing the complementary base sequence of T-C-G is
A-G-C

(b)

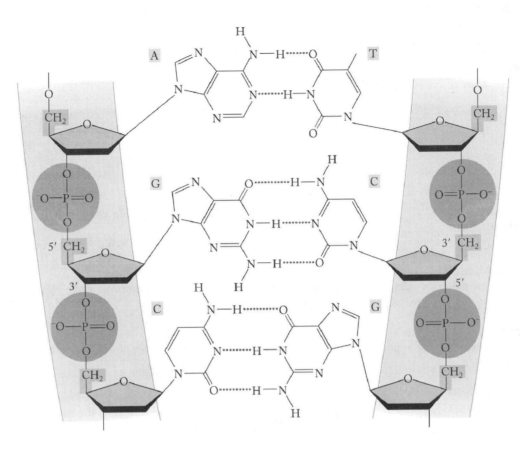

Think About It

DNA is a very important molecule for life on our planet. It contains the instructions used in the development and functioning of all known living organisms and even of some viruses.

20.65. **Collect and Organize**

We consider the structure of olestra shown in Figure P20.65.

Analyze

Olestra is a disaccharide in which the –OH groups on the sugars have been converted to esters through reaction with long-chain carboxylic acids.

Solve

(a) The disaccharide core of olestra is composed of sucrose.
(b) Esters have replaced the hydroxyl groups on the disaccharide as functional groups.
(c) The carboxylic acid used to make olestra is $C_{15}H_{31}COOH$.

Think About It

The fatty acid that reacted with the disaccharide in this problem is palmitic acid.

20.67. **Collect and Organize**

We are to compare the structure of cysteine and with that of homocysteine shown in Figure P20.67. We are also asked if homocysteine is chiral.

Analyze

The structure of cysteine from Table 20.1 is

$$
\begin{array}{c}
SH \\
| \\
CH_2 \\
| \\
H_2N - C - COOH \\
| \\
H
\end{array}
$$

Solve

(a) There is an extra $-CH_2-$ group in homocysteine's sulfur-containing side chain that is not present in cysteine.
(b) Yes, homocysteine is chiral because it contains a carbon that is bonded to four different groups.

Think About It

Because of the extra $-CH_2-$ group in homocysteine, it can form a five-membered ring. This does not allow this amino acid to form stable peptide bonds.

20.69. **Collect and Organize**

Given the structure of hypoglycin shown in Figure P20.69, we are to determine whether this compound is an α-amino acid.

Analyze

An α-amino acid has both the carboxylic acid group and the amine group bonded to the same carbon atom.

Solve

Yes, hypoglycin is an α-amino acid. The carbon to which the carboxylic acid and amine group are bonded is indicated below.

Think About It

Hypoglycin is chemically related to lysine, and its toxic effects may include coma and death.

20.71. **Collect and Organize**

We are asked to identify and draw the amino acids listed in Table 20.1 that include a branched carbon chain in the R group.

Analyze

Branched chains are those R groups that feature an aliphatic carbon atom bound to two other carbon atoms. Aromatics groups and carbon atoms bound to non-carbon atoms are not counted as "branched."

Solve

The amino acids featuring a branched aliphatic side chain are valine (Val), leucine (Leu), and isoleucine (Ile).

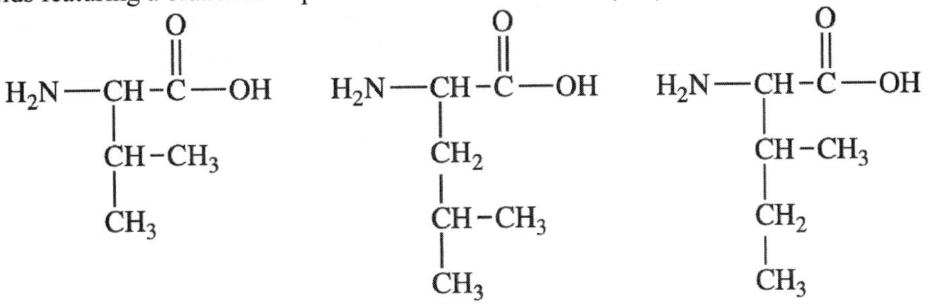

$$
\begin{array}{ccc}
\text{H}_2\text{N}-\underset{\underset{\displaystyle\text{CH}_3}{|}}{\underset{\underset{}{|}}{\underset{\text{CH}-\text{CH}_3}{}}}\overset{\overset{\displaystyle\text{O}}{\|}}{\text{CH}-\text{C}}-\text{OH} &
\text{H}_2\text{N}-\text{CH}-\text{C}-\text{OH} &
\text{H}_2\text{N}-\text{CH}-\text{C}-\text{OH}
\end{array}
$$

Valine   Leucine   Isoleucine

Think About It

Phenylalanine, tryptophan, and tyrosine are not branched chain amino acids because the substituents are aromatic rather than aliphatic.

20.73. Collect and Organize

From the structure of glutathione (Figure P20.73), we are to identify the three amino acids that make up this essential molecule.

Analyze

The 20 common naturally occurring amino acids are shown in Table 20.1. Glutathione is prepared from the condensation of three amino acids to make the two peptide bonds that are in the structure.

Solve

Glutamic acid, cysteine, and glycine are the amino acids that make up glutathione.

Think About It

Glutathione is an antioxidant and, as such, protects cells from damage due to free radicals.

20.75. Collect and Organize

By considering only the average bond energies, we are asked whether the relative fuel values of leucine and isoleucine should be the same. We are also to explain why calorimetric measurements show isoleucine has a lower fuel value than leucine.

Analyze

The structures of leucine and isoleucine show that they are isomers with very similar structures and bonding between the atoms. They differ only in location of the methyl ($-\text{CH}_3$) group on the C_3 side chain.

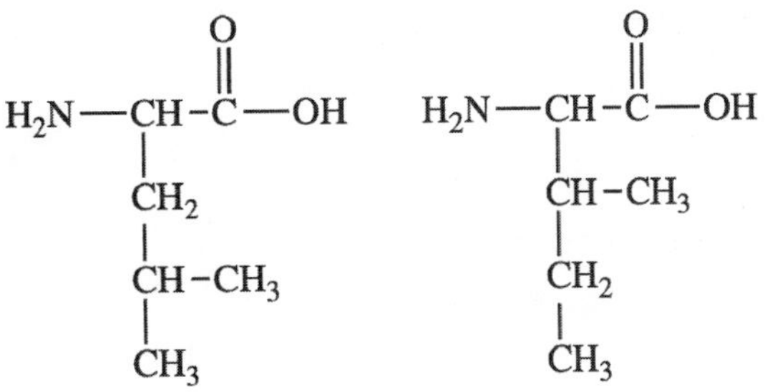

Leucine   Isoleucine
(Leu)   (Ile)

Solve

Yes, we would expect the fuel values to be the same. Because there is no difference in the number of C—C, C—H, C=O, C—O, or N—H bonds between the two compounds, we expect on the basis of average bond energies that the fuel values of leucine and isoleucine should be identical. Experimentally, isoleucine might have a lower fuel value because the CH_3 group is closer to the COOH and NH_2 groups; this difference in shape must contribute to the slightly different fuel values.

Think About It

Isoleucine and leucine are not enantiomers of each other but are structural isomers.

CHAPTER 21 | Nuclear Chemistry: Applications to Energy and Medicine

21.1. Collect and Organize

From the elements depicted in Figure P21.1, we are asked to identify which plays a key role in the controlled fusion of hydrogen.

Analyze

The elements depicted in Figure P21.1 are:

Purple – Li; orange – Mg; blue – Sc; green – Zr; yellow – La; red – Ra

Solve

Purple (Li). Lithium atoms in the walls of a tokamak reactor are bombarded with ^{4}He nuclei during the controlled fusion of hydrogen.

Think About It

In a nuclear fusion reaction, small elements are built up into larger elements, so heavy radioactive nuclides are not required as they are for nuclear fission reactions.

21.3. Collect and Organize

From the elements depicted in Figure P21.1, we are asked to identify which is formed when ^{238}U decays.

Analyze

The elements depicted in Figure P21.1 are:

Purple – Li; orange – Mg; blue – Sc; green – Zr; yellow – La; red – Ra

Solve

Red (Ra).

Think About It

Radium is formed when ^{238}U decays through a series of α and β emissions, ultimately producing ^{206}Pb.

21.5. Collect and Organize

From the two graphs shown in Figure P21.5, we are to determine which one describes β decay.

Analyze

In β decay a neutron is changed into a proton and an electron is produced. In β decay, then, the number of neutrons decreases and the number of protons increases.

Solve

Because β decay results in the emission of an electron from the nucleus, which increases the atomic number of the nucleus, graph a illustrates β decay.

Think About It

Figure P21.5(b) represents the process of proton emission from the nucleus, where the number of protons is reduced by one but the number of neutrons remains the same.

21.7. Collect and Organize

Given the five curves shown in Figure P21.7, we are to choose the one that represents a nuclear process with $t_{1/2} = 2.0$ days.

Analyze

The half-life of a nuclear decay process is the time it takes for half of the original concentration to decay. Starting from the original isotope quantity of 100%, it is the time it takes for that concentration to decrease to 50%.

Solve

The blue line (b) represents a decay process with $t_{1/2}$ = 2.0 days. The quantity of the isotope is 50% after 2 days after starting at 100%.

Think About It

Notice that the value of $t_{1/2}$ does not change. On line (b) the time it takes for the quantity of the isotope to drop from 50% to 25% is also 2.0 days.

21.9. Collect and Organize

From the processes depicted in Figure P21.9, we are to assign each as either fission or fusion.

Analyze

In a fission process, the nucleus splits into smaller nuclei. In a fusion process, smaller nuclei combine to form a heavier nucleus.

Solve

Process 1 represents fission, in which smaller nuclei are generated (along with some nuclear particles shown as gray spheres). Process 2 represents fusion, in which lighter nuclei are fused into a heavier nucleus.

Think About It

Note that in fusion, some nuclear particles (neutrons, α particles, or β particles) may be released to stabilize the larger nucleus formed.

21.11. Collect and Organize / Analyze

We can use the definitions of *mass defect* and *binding energy* to discriminate between these two terms.

Solve

The *mass defect* is the difference between the mass of the nucleus of an isotope and the sum of the masses of the individual nuclear particles that make up that isotope. The *binding energy* is the energy released when individual nucleons combine to form the nucleus of an isotope.

Think About It

The mass defect and binding energy are related to each other through Einstein's equation:

$$\Delta E = \Delta mc^2$$

21.13. Collect and Organize

We are asked to calculate the binding energy of ^{2}H.

Analyze

The binding energy is calculated using

$$\Delta E = \Delta mc^2$$

where c is the speed of light and Δm is the mass defect. Because the nucleus of ^{2}H is made up of one proton (1.67262×10^{-27} kg each) and one neutron (1.67493×10^{-27} kg each), the mass defect is the sum of the masses of those particles subtracted from the actual mass of ^{2}H (given as 3.34370×10^{-27} kg).

Solve

$$\Delta m = 3.34370 \times 10^{-27} \text{ kg} - \left[\left(1 \times 1.67262 \times 10^{-27} \text{ kg}\right) + \left(1 \times 1.67493 \times 10^{-27} \text{ kg}\right)\right] = -3.85 \times 10^{-30} \text{ kg}$$

$$\Delta E = 3.85 \times 10^{-30} \text{ kg} \times \left(3.00 \times 10^8 \text{ m/s}\right)^2 = 3.46 \times 10^{-13} \text{ J}$$

Think About It

Though less stable than ^{1}H, deuterium (^{2}H) is found in reasonable quantities in seawater and other hydrogen-containing molecules.

21.15. **Collect and Organize**

We are asked to compute the binding energy (BE) per nucleon in a nucleus of ^{4}He.

Analyze

We can calculate the binding energy using Einstein's equation:

$$BE = \Delta E = \Delta mc^2$$

where Δm is the difference in the mass between the ^{4}He nucleus and the sum of the masses of the two protons and two neutrons that compose the nucleus. Because there are four total nucleons, we divide the calculated value of ΔE by four.

Solve

$$\Delta m = 4.00260 \text{ amu} - \left[(2 \times 1.00728 \text{ amu}) + (2 \times 1.00867 \text{ amu}) \right] = -0.0293 \text{ amu}$$

$$BE \text{ per nucleon} = \frac{\left(0.0293 \text{ amu} \times \dfrac{1.6605402 \times 10^{-27} \text{ kg}}{1 \text{ amu}} \right) \times (3.00 \times 10^8 \text{ m/s})^2}{4 \text{ nucleons}} = 1.09 \times 10^{-12} \text{ J/nucleon}$$

Think About It

As shown in Figure 21.1, ^{56}Fe has the highest binding energy per nucleon and therefore has the highest nuclear stability.

21.17. **Collect and Organize**

We are to describe how the belt of stability (shown in Figure 21.2) can be used to predict the possible decay modes of radioactive nuclides.

Analyze

The belt of stability in Figure 21.2 plots the number of neutrons versus the number of protons.

Solve

If the nuclide lies in the belt of stability (green dots on the plot in Figure 21.2), it is not radioactive and is stable. If it lies above the belt of stability, then it is neutron rich and tends to undergo β decay to increase the number of protons and reduce the number of neutrons in its nucleus. If it lies below the belt of stability, it is neutron poor and tends to undergo positron emission or electron capture to increase the number of neutrons and reduce the number of protons in its nucleus.

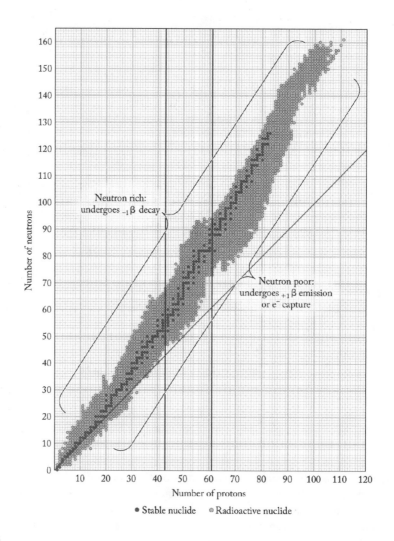

● Stable nuclide　　○ Radioactive nuclide

Think About It
From Figure 21.2 we can see that there are often several known isotopes for an element that are radioactive (orange dots).

21.19. Collect and Organize
Using the decay series for ^{238}U shown in Figure 21.5, we can explain why several α decays are often followed by β decay.

Analyze
We know from the statement of the problem that the neutron-to-proton ratio of decay products must decrease to form stable fission products.

Solve
Alpha decay increases the neutron-to-proton ratio to produce less stable isotopes, which can then be made more stable through β emission to decrease the neutron-to-proton ratio.

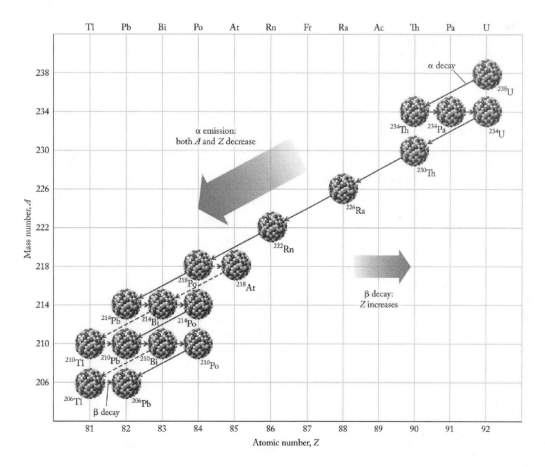

Think About It
Ultimately, ^{238}U decays to form stable ^{206}Pb.

21.21. Collect and Organize
For the nuclear decay reactions described, we are to identify the modes of decay.

Analyze
By writing balanced equations for each decay, we can identify the mode of decay.

Solve

$$^{137}_{53}I \rightarrow {}^{137}_{54}Xe + {}^{0}_{-1}\beta$$
$$^{137}_{54}Xe \rightarrow {}^{137}_{55}Cs + {}^{0}_{-1}\beta$$

Both of these processes are β decays.

Think About It
The effect of the β emission process is that the atomic number increases by one, leaving the mass number unchanged.

21.23. Collect and Organize
We consider the neutron-to-proton ratio for an isotope with a mass number, A, that is more than two times the atomic number, Z.

Analyze

When the neutron-to-proton ratio is 1.00, there are equal numbers of protons and neutrons in the nucleus. When the ratio is greater than 1, there are more neutrons than protons. When the ratio is less than 1, there are more protons than neutrons.

Solve

When $A > 2Z$, the neutron-to-proton ratio is greater than 1.

Think About It

For light elements the neutron-to-proton ratio is about 1.0, but for heavier stable nuclei the ratio is greater than 1.0 and up to 1.5 for the heaviest elements.

21.25. Collect and Organize

For the decay of ^{26}Al to ^{26}Mg, we are to write a balanced nuclear equation.

Analyze

Since the mass number in this nuclear process stays the same but the atomic number decreases, ^{26}Al undergoes positron decay.

Solve

$$^{26}_{13}\text{Al} \rightarrow \, ^{0}_{1}\beta + \, ^{26}_{12}\text{Mg}$$

Think About It

We might predict that this process would occur, since the neutron-to-proton ratio for ^{26}Al is 1.0, but should be slightly higher according to the belt of stability shown in Figure 21.2.

21.27. Collect and Organize

For the isotopes listed, we can use the belt of stability shown in Figure 21.2 to predict the modes of decay.

Analyze

If the nuclide lies in the belt of stability (green dot on the plot), it is not radioactive and is stable. If it lies above the belt of stability, then it is neutron rich and tends to undergo β decay to increase the number of protons and reduce the number of neutrons in its nucleus. If it lies below the belt of stability, it is neutron poor and tends to undergo positron emission or electron capture to increase the number of neutrons and reduce the number of protons in its nucleus.

Solve

(a) ^{10}C has 6 protons and 4 neutrons and is neutron poor; it may undergo electron capture or positron emission.
(b) ^{19}Ne has 10 protons and 9 neutrons and is neutron poor; it may undergo electron capture or positron emission.
(c) ^{50}Ti has 22 protons and 28 neutrons and is stable.

Think About It

Notice that all of these isotopes are known, as they appear as orange or green dots in Figure 21.2.

21.29. Collect and Organize

For ^{56}Co and ^{44}Ti, we are to predict the mode of decay.

Analyze

If the nuclide lies in the belt of stability (green dot in the plot in Figure 21.2), it is not radioactive and is stable. If it lies above the belt of stability, then it is neutron rich and tends to undergo β decay to increase the number of protons and reduce the number of neutrons in its nucleus. If it lies below the belt of stability, it is neutron poor and tends to undergo positron emission or electron capture to increase the number of neutrons and reduce the number of protons in its nucleus.

Solve

^{56}Co has 27 protons and 29 neutrons and is neutron poor; it may undergo electron capture or positron emission.

^{44}Ti has 22 protons and 22 neutrons and is neutron poor; it may undergo electron capture or positron emission.

Think About It

Both of these radioisotopes, then, are expected to be positron emitters.

21.31. **Collect and Organize**

Based on the number of protons and neutrons present in the isotopes of chlorine, we may determine which will decay by beta emission, positron emission, or both.

Analyze

Neutron rich nuclides will decay by emitting a β particle, whereas neutron poor nuclides will decay by positron emission. The isotopes of chlorine are

Neutron poor ← ^{32}Cl ^{33}Cl ^{34}Cl [^{35}Cl] ^{36}Cl [^{37}Cl] ^{38}Cl ^{39}Cl → Neutron rich

Neither

^{35}Cl Stable ^{37}Cl Stable

Solve

(a) ^{32}Cl, ^{33}Cl, and ^{34}Cl will undergo positron emission.

(b) ^{38}Cl and ^{39}Cl will emit a β particle.

(c) ^{36}Cl may emit either a β particle or a positron.

Think About It

^{36}Cl will form either ^{36}S or ^{36}Ar depending on whether it emits a β particle or a positron:

$$^{36}_{17}Cl \rightarrow ^{36}_{18}Ar + ^{0}_{-1}\beta$$

$$^{36}_{17}Cl \rightarrow ^{36}_{16}S + ^{0}_{+1}\beta$$

21.33. **Collect and Organize**

We are to determine what percentage of radioactivity of a sample remains after two half-lives. A half-life is defined as the time at which half of the radioactivity remains.

Analyze

After each half-life, there is a 50% decrease in radioactivity. We can start with 100% and decrease by 50% for each half-life.

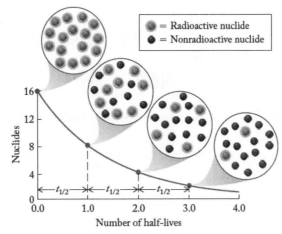

Solve

$$100\% \xrightarrow[\text{half-life}]{\text{first}} 50\% \xrightarrow[\text{half-life}]{\text{second}} 25\%$$

Think About It

Alternatively, we can use Equation 21.2

$$\frac{N_t}{N_0} = 0.5^n$$

where n is the number of half-lives:

$$\frac{N_t}{N_0} = 0.5^2 = 0.25 \text{ or } 25\%$$

21.35. **Collect and Organize**

We may determine the half-life of the radionuclide using the percentage decayed in the specified time period.

Analyze

The amount remaining after 6.6 days is 100% − 87.5% = 12.5%. The equation describing the amount of radionuclide, the time elapsed, and the half-life is

$$\ln\left(\frac{N_t}{N_0}\right) = -0.693\frac{t}{t_{\frac{1}{2}}}$$

Since more than 50% of the sample has decayed, the half-life should be less than 6.6 days.

Solve

$$\ln\left(\frac{12.5}{100}\right) = -0.693\frac{(6.6 \text{ days})}{t_{\frac{1}{2}}}$$

$$t_{\frac{1}{2}} = 2.2 \text{ days}$$

Think About It

Our prediction was correct, the half-life was less than 6.6 days. We could also have solved this problem by recognizing that 12.5% is equal to three half-lives

$$100\% \xrightarrow[\text{half-life}]{\text{first}} 50\% \xrightarrow[\text{half-life}]{\text{second}} 25\% \xrightarrow[\text{half-life}]{\text{third}} 12.5\%$$

21.37. **Collect and Organize**

We are asked to determine how long it will take for ^{137}Cs to decay to 5% of its initial concentration.

Analyze

The half-life of ^{137}Cs is 30.2 years. The equation describing the amount of radionuclide, the time elapsed, and the half-life is

$$\ln\left(\frac{N_t}{N_0}\right) = -0.693\frac{t}{t_{\frac{1}{2}}}$$

The concentration will be less than that expected after four half-lives, so the time elapsed will be at least four half-lives

$$100\% \xrightarrow[\text{half-life}]{\text{first}} 50\% \xrightarrow[\text{half-life}]{\text{second}} 25\% \xrightarrow[\text{half-life}]{\text{third}} 12.5\% \xrightarrow[\text{half-life}]{\text{fourth}} 6.25\%$$

Solve

$$\ln\left(\frac{5}{100}\right) = -0.693\frac{t}{30.2 \text{ years}}$$

$$t = 131 \text{ years}$$

It will take 131 years for the radiation level to decrease to 5% of the value in 2011.

Think About It
Our prediction was correct. Four half-lives would correspond to 121 years, and our calculated value is in fact slightly larger, as expected.

21.39. **Collect and Organize**
We are to explain why radiocarbon dating is reliable only for artifacts younger than 50,000 years.

Analyze
The half life of ^{14}C is 5730 years. The number of half-lives that 50,000 years represents is 50,000/5730 = 8.726 half-lives.

Solve
After 8.726 half-lives, the ratio of ^{14}C present to that originally in an artifact is

$$\frac{N_t}{N_0} = 0.50^{8.726} = 0.00236 \text{ or } 0.236\%$$

This is too little to detect.

Think About It
Longer-lived isotopes would be better suited for dating old artifacts. For example, ^{40}K (Problem 21.41) is useful for objects older than 300,000 years!

21.41. **Collect and Organize**
For this question we consider why ^{40}K is a useful isotope only for dating objects older than 300,000 years.

Analyze
The half-life of ^{40}K is 1.28×10^9 yr. The number of half-lives that 300,000 years represents is $300,000/1.28 \times 10^9 = 0.00023$ half-lives.

Solve
After 0.00023 half-lives the ratio of ^{40}K present to that originally in a sample is

$$\frac{N_t}{N_0} = 0.50^{0.00023} = 0.9998 \text{ or } 99.98\%$$

This level is just when we can detect the difference in amounts of ^{40}K.

Think About It
For objects younger than 300,000 years, an isotope with a shorter half-life must be used.

21.43. **Collect and Organize**
For a piece of charcoal that is 8700 years old, we are to calculate the fraction of ^{14}C remaining.

Analyze
We can use Equation 21.4 to solve for N_t/N_0:

$$t = -\frac{t_{1/2}}{0.693} \ln \frac{N_t}{N_0}$$

Solve

$$8700 \text{ yr} = -\frac{5730 \text{ yr}}{0.693} \ln \frac{N_t}{N_0}$$

$$\frac{N_t}{N_0} = 0.35 \text{ or } 35\%$$

Think About It

Alternatively, this problem can be solved by first calculating n, the number of half-lives:

$$n = \frac{8700}{5730} = 1.518 \text{ half-lives}$$

$$\frac{N_t}{N_0} = 0.5^{1.518} = 0.35 \text{ or } 35\%$$

21.45. Collect and Organize

For a wood sample from a giant sequoia tree that in 1891 was 1342 years old, we are to compare the fraction of ^{14}C in the innermost ring with that in the outermost ring.

Analyze

We can use Equation 21.4 to solve for N_t/N_0:

$$t = -\frac{t_{1/2}}{0.693} \ln \frac{N_t}{N_0}$$

Solve

$$1342 \text{ yr} = -\frac{5730 \text{ yr}}{0.693} \ln \frac{N_t}{N_0}$$

$$\frac{N_t}{N_0} = 0.850 \text{ or } 85.0\%$$

Think About It

Alternatively, this problem can be solved by first calculating n, the number of half-lives:

$$n = \frac{1342}{5730} = 0.2342 \text{ half-lives}$$

$$\frac{N_t}{N_0} = 0.5^{0.2342} = 0.850 \text{ or } 85.0\%$$

21.47. Collect and Organize

Given that the ^{14}C to ^{12}C ratio is only 1.19% for the ancient mammoth tusk compared to the ratio in elephants today, we are to calculate the age of the mammoth tusk.

Analyze

We can use Equation 21.4 to solve for t:

$$t = -\frac{t_{1/2}}{0.693} \ln \frac{N_t}{N_0}$$

Solve

$$t = -\frac{5730 \text{ yr}}{0.693} \ln \frac{1.19}{100} = 36,640 \text{ yr}$$

Think About It

The number of half-lives this represents for ^{14}C is

$$n = \frac{36,640}{5730} = 6.39 \text{ half-lives}$$

21.49. Collect and Organize / Analyze

We can use the definitions of the terms to describe the difference between the *level* of radioactivity and the *dose* of radioactivity.

Solve

The level of radioactivity is the amount of radioactive particles present in a given instant of time. The dose is the accumulation of exposure over a length of time.

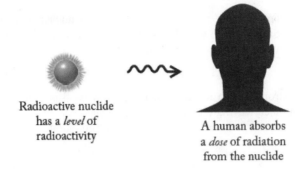

Radioactive nuclide
has a *level* of
radioactivity

A human absorbs
a *dose* of radiation
from the nuclide

Think About It

A person can get a high dose of radioactivity either from a brief exposure to a highly intense radioactive source or through prolonged exposure to low levels of radiation.

21.51. **Collect and Organize**

We are to describe the dangers of ^{222}Rn.

Analyze

Radon is a colorless, odorless gas that results from the natural decay of uranium in the earth. It is an α emitter that decays to ^{218}Po, also an α emitter.

Solve

When radon-222 decays to polonium-218 while in the lungs, the ^{218}Po, a reactive solid that is chemically similar to oxygen, lodges in the lung tissue where it continues to emit α radiation. Alpha radiation is one of the most damaging kinds of radiation when in contact with biological tissues. The result of exposure to high levels of radon is an increased risk for lung cancer.

Think About It

Most hardware stores sell radon detection kits that help homeowners to determine whether the level of radon-222 in their homes is unusually high.

21.53. **Collect and Organize**

We are asked to calculate the grays that are equal to 5 μSv and how much energy 5 μSv corresponds to for a 50 kg person.

Analyze

From Table 21.4, we see that

$$1 \text{ Sv} = 1 \text{ Gy} \times \text{RBE}$$

The relative biological effectiveness (RBE) of X-rays is given in the problem as 1, so for dental X-rays

$$1 \text{ Sv} = 1 \text{ Gy} \times 1 \text{ or } 1 \text{ μSv} = 1 \text{ μGy} \times 1$$

Also from Table 21.4, we see that

$$1 \text{ Gy} = 1 \text{ J/kg of tissue mass or } 1 \text{ μGy} = 1 \text{ μJ/kg}$$

Solve

$$5 \text{ μSv} = 5 \text{ μGy}$$

$$5 \text{ μGy} \times \frac{1 \text{ μJ/kg}}{1 \text{ μGy}} \times 50 \text{ kg} = 250 \text{ μJ}$$

Think About It
This dosage of 5 μSv is well below the dose that may cause toxic effects (Table 21.5).

21.55. **Collect and Organize**

For the radioactive isotope ^{90}Sr, we are to write a balanced nuclear equation corresponding to its decay, calculate the atoms of ^{90}Sr in 200 mL of milk that has 1.25 Bq/L of ^{90}Sr radioactivity, and give a reason why ^{90}Sr would be more concentrated in milk than other foods.

Analyze

(a) Strontium-90 is neutron rich, so it is expected to decay by β particle emission. The process of β decay increases the atomic number, leaving the mass number unchanged.

(b) To calculate the atoms of ^{90}Sr in the milk we can use the equation

$$\text{mL of milk} \times \frac{1.25 \text{ Bq}}{1000 \text{ mL}} \times \frac{1 \text{ disintegration/s}}{1 \text{ Bq}} = \text{disintegrations/s}$$

This is the rate of the first-order decay of ^{90}Sr that follows the rate law

$$\text{Rate} = k[^{90}\text{Sr}]$$

where $k = 0.693/t_{1/2}$. Given that the $t_{1/2}$ for ^{90}Sr is 28.8 yr, we need to convert from years to seconds.

Solve

(a) $^{90}_{38}\text{Sr} \rightarrow {}^{0}_{-1}\beta + {}^{90}_{39}\text{Y}$

(b) The number of disintegrations in 200 mL of milk is

$$200 \text{ mL} \times \frac{1.25 \text{ Bq}}{1000 \text{ mL}} \times \frac{1 \text{ disintegration/s}}{1 \text{ Bq}} = 0.250 \text{ disintegration/s}$$

The rate constant k in reciprocal seconds is

$$k = \frac{0.693}{28.8 \text{ yr}} \times \frac{1 \text{ yr}}{365 \text{ d}} \times \frac{1 \text{ d}}{24 \text{ hr}} \times \frac{1 \text{ hr}}{60 \text{ min}} \times \frac{1 \text{ min}}{60 \text{ s}} = 7.63 \times 10^{-10} \text{ s}^{-1}$$

The concentration of ^{90}Sr in the milk from the first-order rate law is

$$\left[^{90}\text{Sr}\right] = \frac{0.250 \text{ disintegrations/s}}{7.63 \times 10^{-10} \text{ s}^{-1}} = 3.28 \times 10^{8} \text{ }^{90}\text{Sr atoms}$$

(c) Strontium-90 is found in milk and not other foods because it is chemically similar to calcium, and milk is rich in calcium.

Think About It
In more familiar chemical concentration terms, the concentration of ^{90}Sr in these samples is

$$3.28 \times 10^{8} \text{ }^{90}\text{Sr atoms} \times \frac{1 \text{ mol}}{6.022 \times 10^{23} \text{ atoms}} \times \frac{1}{0.200 \text{ L}} = 2.72 \times 10^{-15} \text{ } M$$

21.57. **Collect and Organize**

For drinking water, we are to calculate the number of decay events per second in 1.0 mL with a radon level of 4.0 pCi/mL. We are then to calculate the number of Rn atoms in the 1.0 mL, given that $t_{1/2}$ of ^{222}Rn = 3.8 d.

Analyze

(a) To calculate the number of decay events per second from picocuries, we need the conversions

$$1 \text{ pCi} = 1 \times 10^{-12} \text{ Ci}$$
$$1 \text{ Ci} = 3.70 \times 10^{10} \text{ Bq}$$
$$1 \text{ Bq} = 1 \text{ disintegration/s}$$

(b) The decay of ^{222}Rn follows the first-order rate law

$$\text{Rate} = k[^{222}\text{Rn}]$$

where the rate is the number of decay events per second and k, the rate constant, is

$$k = \frac{0.693}{t_{1/2}}$$

Here $t_{1/2} = 3.8$ d, must be converted to seconds.

Solve

(a) $4.0 \text{ pCi} \times \dfrac{1 \times 10^{-12} \text{ Ci}}{1 \text{ pCi}} \times \dfrac{3.70 \times 10^{10} \text{ Bq}}{1 \text{ Ci}} \times \dfrac{1 \text{ decay/s}}{1 \text{ Bq}} = 0.15 \dfrac{\text{decays}}{\text{s}}$

(b) $[^{222}\text{Rn}] = \dfrac{0.148 \text{ decays/s}}{2.11 \times 10^{-6} \text{ s}^{-1}} = 7.0 \times 10^4 \text{ atoms}$

Think About It

Even though this seems like a large number of ^{222}Rn atoms, the percentage of ^{222}Rn atoms in 1.0 mL of water is very low:

$$1.0 \text{ mL} \times \frac{1 \text{ g}}{\text{mL}} \times \frac{1 \text{ mol}}{18 \text{ g}} \times \frac{6.022 \times 10^{23} \text{ H}_2\text{O molecules}}{\text{mole}} = 3.35 \times 10^{22} \text{ molecules of H}_2\text{O}$$

$$\% \ ^{222}\text{Rn atoms} = \frac{7.0 \times 10^4 \text{ atoms of } ^{222}\text{Rn}}{3.35 \times 10^{22} \text{ molecules of H}_2\text{O}} \times 100$$

$$= 2.1 \times 10^{-16} \ \%$$

21.59. Collect and Organize / Analyze

We are asked to consider how a radioactive isotope for radiotherapy is selected, considering its half-life, decay mode, and properties of its products.

Solve

(a) The half-life should be long enough to effect treatment of the cancerous cells but not so long as to cause damage to healthy tissues.

(b) Because α radiation does not penetrate far beyond a tumor, the α decay mode is best.

(c) Products should be nonradioactive, if possible, or have short half-lives and be able to be flushed from the body by normal cellular and biological processes.

Think About It

The investigation and development of new radioisotopes are very active areas of research.

21.61. Collect and Organize

For each of the isotopes given, we can use the belt of stability (Figure 21.2) to predict the mode of decay.

Analyze

If the nuclide has a neutron-to-proton ratio that places it below the belt of stability, either electron capture or positron emission is likely. If a nuclide lies above the belt of stability, β emission is likely.

Solve

(a) $^{197}_{80}$Hg has 80 protons and 117 neutrons. This nuclide lies below the belt of stability and so is likely to decay by positron emission or electron capture.

(b) $^{75}_{34}$Se has 34 protons and 41 neutrons. This nuclide lies below the belt of stability and so is likely to decay by positron emission or electron capture.

(c) $^{18}_{9}$F has 9 protons and 9 neutrons. This nuclide lies below the belt of stability and so is likely to decay by positron emission or electron capture.

Think About It

These imaging agents have relatively short half-lives: ^{197}Hg, 64 hr; ^{75}Se, 120 days; ^{18}F, 110 min.

21.63. **Collect and Organize**

For a 1.00 mg sample of ^{192}Ir, 0.756 mg remains after 30 days. From this information we are to calculate the half-life of ^{192}Ir.

Analyze

We need to rearrange Equation 21.4 to solve this problem:

$$t = -\frac{t_{1/2}}{0.693} \ln \frac{N_t}{N_0}$$

$$t_{1/2} = -\frac{0.693t}{\ln(N_t/N_0)}$$

Solve

$$t_{1/2} = -\frac{0.693 \times 30 \text{ d}}{\ln(0.756 \text{ mg}/1.00 \text{ mg})} = 74.3 \text{ d}$$

Think About It

After 30 days, then, this isotope has decayed through not even one half-life (Equation 21.3):

$$\frac{0.756}{1.00} = 0.5^n$$

$$0.756 = 0.5^n$$

$$n = 0.404 \text{ half-lives}$$

21.65. **Collect and Organize**

Using the information given about the half-life of ^{131}I ($t_{1/2} = 8.1$ d) and the initial and residual activity of the isotope after 30 days (108 and 4.1 counts/min, respectively), we are to determine whether the brain cells took up any of the ^{131}I.

Analyze

We need to calculate the expected activity of the ^{131}I if it were not taken up by the cells. This is found through the use of Equation 21.4 to solve for N_t.

$$t = -\frac{t_{1/2}}{0.693} \ln \frac{N_t}{N_0}$$

If the activity of the sample after 30 days is less than the calculated activity, then the cell did take up ^{131}I.

Solve

$$30 \text{ days} = -\frac{8.1 \text{ days}}{0.693} \ln \frac{N_t}{108}$$

$$-2.567 = \ln \frac{N_t}{108}$$

$$0.07679 = \frac{N_t}{108}$$

$$N_t = 8.29 \text{ counts/min}$$

Because this is more than the counts/min in the sample, yes, the brain cells must have incorporated some of the ^{131}I.

Think About It

The sample was taken after the ^{131}I had decayed through 3.7 half-lives:

$$\frac{8.29}{108} = 0.5^n$$

$$n = 3.7 \text{ half-lives}$$

21.67. Collect and Organize

Given that the half-life of ^{11}C is 20.4 min, we are to calculate the time it takes for 99% of the injected ^{11}C to decay.

Analyze

We can use Equation 21.4 to solve this problem:

$$t = -\frac{t_{1/2}}{0.693} \ln \frac{N_t}{N_0}$$

where $N_t = 1$ and $N_0 = 100$ for the 99% decay of the isotope.

Solve

$$t = -\frac{20.4 \text{ min}}{0.693} \ln \frac{1}{100} = 136 \text{ min}$$

Think About It

This time is the equivalent of over 6 half lives:

$$\frac{1}{100} = 0.0100 = 0.5^n$$

$$n = 6.64 \text{ half-lives}$$

21.69. Collect and Organize

For the decay of ^{11}B, formed during boron neutron-capture therapy (BNCT) for cancers, which decays to ^7Li, we are to write balanced nuclear equations for the neutron absorption and α decay; then calculate the energy released in the process, using Einstein's equation; and finally consider why this process, in which an α emitter is produced by neutron capture, is an effective cancer treatment.

Analyze

(a) In writing the balanced equation, we need to balance the sum of the mass numbers of the products with that of the reactants. Likewise, the sum of the atomic numbers of the products must equal that of the reactants.
(b) In the Einstein equation

$$\Delta E = \Delta mc^2$$

c is the speed of light and Δm is the mass defect. The mass defect is the mass lost in the reaction expressed in kilograms. Since the masses in this problem are expressed in amu we need to use the conversion factor
$$1 \text{ amu} = 1.6605402 \times 10^{-27} \text{ kg}$$

Solve

(a) $^{10}_{5}B + ^{1}_{0}n \rightarrow ^{11}_{5}B$

$^{11}_{5}B \rightarrow ^{7}_{3}Li + ^{4}_{2}\alpha$

Overall process: $^{10}_{5}B + ^{1}_{0}n \rightarrow ^{7}_{3}Li + ^{4}_{2}\alpha$

(b) $\Delta m = (7.01600 \text{ amu} + 4.00260 \text{ amu}) - (10.0129 \text{ amu} + 1.008665 \text{ amu}) = -0.002965 \text{ amu}$

$$0.002965 \text{ amu} \times \frac{1.6605402 \times 10^{-27} \text{ kg}}{1 \text{ amu}} = 4.9235 \times 10^{-30} \text{ kg}$$

$$\Delta E = 4.9235 \times 10^{-30} \text{ kg} \times (3.00 \times 10^8 \text{ m/s})^2 = 4.43 \times 10^{-13} \text{ J}$$

(c) Alpha particles have a high RBE, and they do not penetrate into healthy tissue if the radionuclide is placed inside a tumor.

Think About It

BNCT is especially useful in the treatment of brain tumors. However, the only source of neutrons is from nuclear reactors and there are only a few sites worldwide that have this treatment available.

21.71. Collect and Organize
We are asked to describe how the rate of energy release is controlled in nuclear reactors.

Analyze
Nuclear reactions that power the reactors release neutrons, which promote more nuclear fission processes as shown in Figure 21.21. The reaction can be controlled by absorbing some of the neutrons.

Solve
Control rods made of boron or cadmium are used to absorb the excess neutrons to control the rate of energy release in a nuclear reactor.

Think About It
When the control rods are removed, the reactor core may go critical. This is what occurred in the reactor at Chernobyl in the Soviet Union on April 26, 1986.

21.73. Collect and Organize
Figure 21.2 shows the belt of stability of the radionuclides. Using this we are to explain why neutrons are by-products in fission reactions.

Analyze
In a fission reaction, a heavier, unstable nucleus splits into two lighter nuclei.

Solve
The neutron-to-proton ratio for heavy nuclei is high and when the nuclide undergoes fission to form smaller nuclides, it must emit neutrons because the fission products require a lower neutron-to-proton ratio for stability.

Think About It
In fusing nuclei, more neutrons are needed for the heavier nuclei, and so we might expect that β decay often accompanies those processes.

21.75. Collect and Organize
For the incomplete fission reactions given, we are to determine the missing nuclides.

Analyze
To balance these reactions, the sum of the mass numbers of the reactants must equal that of the products. Likewise, the sum of the atomic numbers of the reactants must equal that of the products.

Solve
(a) $^{235}_{92}\text{U} + ^{1}_{0}\text{n} \rightarrow ^{96}_{40}\text{Zr} + ^{138}_{52}\text{Te} + 2\ ^{1}_{0}\text{n}$

(b) $^{235}_{92}\text{U} + ^{1}_{0}\text{n} \rightarrow ^{99}_{41}\text{Nb} + ^{133}_{51}\text{Sb} + 4\ ^{1}_{0}\text{n}$

(c) $^{235}_{92}\text{U} + ^{1}_{0}\text{n} \rightarrow ^{90}_{37}\text{Rb} + ^{143}_{55}\text{Cs} + 3\ ^{1}_{0}\text{n}$

Think About It
Notice that all of these fission products (^{96}Zr, ^{138}Te, ^{99}Nb, ^{133}Sb, ^{90}Rb, and ^{143}Cs) come from the fission of ^{235}U.

21.77. Collect and Organize
For the incomplete fission reactions given, we are to determine the missing nuclides.

Analyze
To balance these reactions, the sum of the mass numbers of the reactants must equal that of the products. Likewise, the sum of the atomic numbers of the reactants must equal that of the products.

Solve

(a) $^{235}_{92}U + ^1_0n \rightarrow ^{131}_{53}I + ^{103}_{39}\underline{Y} + 2\,^1_0n$

(b) $^{235}_{92}U + ^1_0n \rightarrow ^{103}_{44}Cs + ^{130}_{48}Cd + 3\,^1_0n$

(c) $^{235}_{92}U + ^1_0n \rightarrow ^{95}_{40}Ce + ^{138}_{52}\underline{Se} + 3\,^1_0n$

Think About It

Notice that all of these fission products (^{131}I, ^{103}Y, ^{103}Cs, ^{130}Cd, ^{95}Ce, and ^{138}Se) come from the fission of ^{235}U.

21.79. Collect and Organize

We are to comment on the differences between the reactions that form alpha particles in the primordial Sun and those that fuel today's Sun.

Analyze

The primordial Sun contained a far greater concentration of free neutrons that the Sun does today.

Solve

In the primordial Sun, direct fusion of a proton and a neutron formed deuterons, which combined to generate alpha particles

$$2\,^1_1p + 2\,^1_0n \rightarrow 2\,^2_1H \rightarrow ^4_2He$$

In the contemporary Sun, protons fuse to form a positron and a deuteron, which may then react with another proton to generate helium-3. Two helium-3 nuclides may fuse to generate an alpha particle and free protons.

$$2\,^1_1p \rightarrow ^2_1H + ^0_{+1}\beta$$

$$2\,^2_1H + 2\,^1_1p \rightarrow 2\,^3_2He \rightarrow ^4_2He + 2\,^1_1p$$

Think About It

As the Sun ages, its supply of free neutrons decreases, and it begins to emit positrons.

21.81. Collect and Organize

For the four fusion reactions given, we can use Einstein's equation to calculate the energy released in each.

Analyze

In the equation

$$\Delta E = \Delta mc^2$$

c is the speed of light and Δm is the mass defect. The mass defect is the mass lost in the reaction and must be expressed in kilograms. Since the masses in this problem are expressed in atomic mass units (amu) we need to use the conversion factor

$$1\text{ amu} = 1.6605402 \times 10^{-27}\text{ kg}$$

Solve

(a) For the production of ^{28}Si from $^{14}N + ^{14}N$

$$\Delta m = 27.97693\text{ amu} - (2 \times 14.00307\text{ amu}) = -0.02921\text{ amu}$$

$$0.02921\text{ amu} \times \frac{1.6605402 \times 10^{-27}\text{ kg}}{1\text{ amu}} = 4.8504379 \times 10^{-29}\text{ kg}$$

$$\Delta E = 4.8504379 \times 10^{-29}\text{ kg} \times (3.00 \times 10^8\text{ m/s})^2 = 4.37 \times 10^{-12}\text{ J}$$

(b) For the production of ^{28}Si from $^{10}B + ^{16}O + ^2H$

$$\Delta m = 27.97693\text{ amu} - (10.0129\text{ amu} + 15.99491\text{ amu} + 2.0146\text{ amu}) = -0.04548\text{ amu}$$

$$0.04548\text{ amu} \times \frac{1.6605402 \times 10^{-27}\text{ kg}}{1\text{ amu}} = 7.552 \times 10^{-29}\text{ kg}$$

$$\Delta E = 7.552 \times 10^{-29} \text{ kg} \times \left(3.00 \times 10^8 \text{ m/s}\right)^2 = 6.80 \times 10^{-12} \text{ J}$$

(c) For the production of ^{28}Si from ^{16}O + ^{12}C

$$\Delta m = 27.97693 \text{ amu} - \left(15.994915 \text{ amu} + 12.000 \text{ amu}\right) = -0.01798 \text{ amu}$$

$$0.01798 \text{ amu} \times \frac{1.6605402 \times 10^{-27} \text{ kg}}{1 \text{ amu}} = 2.98565 \times 10^{-29} \text{ kg}$$

$$\Delta E = 2.98565 \times 10^{-29} \text{ kg} \times \left(3.00 \times 10^8 \text{ m/s}\right)^2 = 2.69 \times 10^{-12} \text{ J}$$

(d) For the production of ^{28}Si from ^{24}Mg + ^{4}He

$$\Delta m = 27.97693 \text{ amu} - \left(23.98504 \text{ amu} + 4.00260 \text{ amu}\right) = -0.01071 \text{ amu}$$

$$0.01071 \text{ amu} \times \frac{1.6605402 \times 10^{-27} \text{ kg}}{1 \text{ amu}} = 1.77844 \times 10^{-29} \text{ kg}$$

$$\Delta E = 1.77843 \times 10^{-29} \text{ kg} \times \left(3.00 \times 10^8 \text{ m/s}\right)^2 = 1.60 \times 10^{-12} \text{ J}$$

Think About It

These reactions all release energy (there is a negative mass defect in proceeding from reactants to products). Notice that in solving for the *energy released* we have used the positive value of the mass defect.

21.83. Collect and Organize

For the two fusion reactions given, we can use Einstein's equation to calculate the energy released in each.

Analyze

In the equation

$$\Delta E = \Delta mc^2$$

c is the speed of light and Δm is the mass defect. The mass defect is the mass lost in the reaction and must be expressed in kilograms. Since the masses in this problem are expressed in atomic mass units (amu), we need to use the conversion factor

$$1 \text{ amu} = 1.6605389 \times 10^{-27} \text{ kg}$$

Solve

(a) For the reaction ^{6}Li

$$\Delta m = \left(4.002603 \text{ amu} + 3.01605 \text{ amu}\right) - \left(1.00867 \text{ amu} + 6.015121 \text{ amu}\right) = -0.005138 \text{ amu}$$

$$0.005138 \text{ amu} \times \frac{1.6605389 \times 10^{-27} \text{ kg}}{1 \text{ amu}} = 8.532 \times 10^{-30} \text{ kg}$$

$$\Delta E = 8.532 \times 10^{-30} \text{ kg} \times \left(3.00 \times 10^8 \text{ m/s}\right)^2 = 7.67 \times 10^{-13} \text{ J}$$

(b) For the reaction ^{7}Li

$$\Delta m = \left(4.002603 \text{ amu} + 3.01605 \text{ amu} + 1.00867 \text{ amu}\right) - \left(1.00867 \text{ amu} + 7.016003 \text{ amu}\right) = +0.00265 \text{ amu}$$

$$-0.00265 \text{ amu} \times \frac{1.6605389 \times 10^{-27} \text{ kg}}{1 \text{ amu}} = -4.400 \times 10^{-30} \text{ kg}$$

$$\Delta E = -4.400 \times 10^{-30} \text{ kg} \times \left(3.00 \times 10^8 \text{ m/s}\right)^2 = -3.96 \times 10^{-13} \text{ J}$$

Think About It

Energy is released in (a), whereas energy must be consumed in (b). Notice that in solving for the *energy released* in part (a) we have used the positive value of the mass defect.

21.85. Collect and Organize

We are asked to explain why antihydrogen would have been a suitable fuel for the starship *Enterprise*, and to comment on any challenges of using this fuel.

Analyze

A large amount of energy (1.813×10^{14} J) is released when one mole of hydrogen collides with one mole of antihydrogen.

Solve

(a) Besides releasing a large amount of energy to power the starship *Enterprise*, hydrogen is an abundant fuel in the universe and therefore could easily react with any antihydrogen produced.

(b) If any of the antihydrogen came in contact with conventional matter such as the reactor wall, the ship, or the crew members, the two would annihilate each other, resulting in a large explosion.

Think About It

Antihelium, although much more difficult to produce, might also have been a good choice.

21.87. Collect and Organize

We need to use Einstein's equation to calculate the energy released in the fusion of four hydrogen atoms compared that to the energy released by the fission of ^{235}U. Because this fusion reaction produces positrons, which are immediately annihilated, we include their energy in the calculation.

Analyze

To determine the energy released in the fusion reaction, we need to first calculate the mass defect for the reactions and then solve Einstein's equation.

Solve

The mass defect for the fusion reaction is

$$\Delta m = \left[(2 \times 5.485799 \times 10^{-4}\,\text{amu}) + 4.00260\,\text{amu} \right] - \left[(4 \times 1.00728\,\text{amu}) + (2 \times 5.485799 \times 10^{-4}\,\text{amu}) \right]$$

$$= -0.02652\,\text{amu}$$

$$0.02652\,\text{amu} \times \frac{1.6605402 \times 10^{-27}\,\text{kg}}{1\,\text{amu}} = 4.404 \times 10^{-29}\,\text{kg}$$

The energy released in the fusion reaction is

$$\Delta E = 4.404 \times 10^{-29}\,\text{kg} \times (3.00 \times 10^{8}\,\text{m/s})^2 = 3.9633 \times 10^{-12}\,\text{J/atom}\,{}^{4}\text{He}$$

$$\Delta E \text{ per nucleon} = \frac{3.9633 \times 10^{-12}\,\text{J}}{4\,\text{nucleons}} = 9.91 \times 10^{-13}\,\text{J/nucleon}$$

The energy released in the fission reaction is

$$\Delta E = 3.2 \times 10^{-11}\,\text{J/atom} \times 1\,\text{atom}/235\,\text{nucleons} = 1.4 \times 10^{-13}\,\text{J/nucleon}$$

On a per nucleon basis, the fusion reaction generates more energy.

Think About It

On a per gram basis, the fusion reaction also provides more energy.

21.89. Collect and Organize

We consider the ^{241}Am isotope in smoke detectors, which decays by α emission.

Analyze

(a) A Geiger counter functions by conducting electricity when there are ions in the gas due to the presence of ionizing radiation. A scintillation counter detects radiation through the detection of light caused by the fluorescence of the material when ionizing radiation strikes it.

(b) To calculate the time it takes for the activity of ^{241}Am to decay to 1% of its original activity, we can use Equation 21.4:

$$t = -\frac{t_{1/2}}{0.693} \ln \frac{N_t}{N_0}$$

Solve

(a) Because the ^{241}Am ionizes the air and the smoke detector registers a change in current, a smoke detector resembles a Geiger counter in its operation.

(b)

$$t = -\frac{433 \text{ yr}}{0.693} \ln \frac{1}{100} = 2877 \text{ yr}$$

(c) Smoke detectors are safe to handle because the ^{241}Am is an α emitter and α particles do not travel more than a few inches in air and cannot penetrate the first layer of skin.

Think About It

Even though the half-life of ^{241}Am is very long, the useful life of a smoke detector is 10 years.

21.91. **Collect and Organize**

For the synthesis of ^{294}Uuo, we are to describe its decay reactions with balanced equations and, by looking at its position in the periodic table, select another element that has similar properties to Uuo.

Analyze

To balance these reactions, the sum of the mass numbers of the reactants must equal that of the products. Likewise, the sum of the atomic numbers of the reactants must equal that of the products. By balancing the equations, we can identify the nuclides in parts b, c, and d of this problem.

Solve

(a) $^{249}_{98}\text{Cf} + ^{48}_{20}\text{Ca} \rightarrow ^{294}_{118}\text{Uuo} + 3\,^{1}_{0}\text{n}$

(b) $^{294}_{118}\text{Uuo} \rightarrow ^{4}_{2}\alpha + ^{290}_{116}\text{Uuh}$

(c) $^{290}_{116}\text{Uuh} \rightarrow ^{4}_{2}\alpha + ^{286}_{114}\text{Uuq}$

(d) $^{286}_{114}\text{Uuq} \rightarrow ^{4}_{2}\alpha + ^{282}_{112}\text{Cn}$

(e) Because ^{294}Uuo is a member of the noble gas family, it has chemical and physical properties similar to naturally occurring radon.

Think About It

Even though these superheavy elements are short-lived, their half-lives are sufficiently long (milliseconds) to allow for some of their chemical and physical properties to be experimentally determined.

21.93. **Collect and Organize**

By examining the half-lives for nuclides in a decay series for ^{214}Bi, we can identify which element will be present in the highest amount in a sample after one year.

Analyze

The decay process can only go as fast as the slowest step. This means that there will be a buildup of the isotope with the longest half-life.

Solve

Because ^{210}Pb has the longest half-life in the decay series, it will be the most abundant of the isotopes after one year.

Think About It

In the language of kinetics, we can think of the decay of ^{210}Pb as the rate-limiting step.

21.95. **Collect and Organize**

For the synthesis of Pt by two fusion reactions, we can determine which isotopes of Pt are formed by writing the balanced nuclear reactions for their formation.

Analyze
To balance these reactions, the sum of the mass numbers of the reactants must equal that of the products. Likewise, the sum of the atomic numbers of the reactants must equal that of the products.

Solve
(a) $^{64}_{28}\text{Ni} + ^{124}_{50}\text{Sn} \rightarrow ^{188}_{78}\text{Pt}$

(b) $^{64}_{28}\text{Ni} + ^{132}_{50}\text{Sn} \rightarrow ^{196}_{78}\text{Pt}$

Think About It
Both of these isotopes of Pt lie well above the belt of stability (Figure 21.2), so we would expect that they might decay by β or α emission.

21.97. **Collect and Organize**
We are to determine the ratio of ^{14}C present in a bone sample that is 15,000 years old to another sample that is 25,000 years old.

Analyze
We can determine the ratio using

$$\frac{N_t/N_0 (15,000 \text{ years old})}{N_t/N_0 (25,000 \text{ years old})} = \frac{0.5^n}{0.5^n}$$

where n for 15,000 years ago is 15,000 yr/5730 yr and n for 25,000 years ago is 25,000 yr/5730 yr.

Solve

$$\frac{N_t/N_0 (15,000 \text{ years ago})}{N_t/N_0 (25,000 \text{ years ago})} = \frac{0.5^{15,000/5730}}{0.5^{25,000/5730}} = 3.35$$

Think About It
The number of half-lives of ^{14}C that have passed for the 25,000-year-old sample is 25,000 yr/5730 yr = 4.4 half-lives.

21.99. **Collect and Organize**
For the dating of ancient human skulls in Ethiopia based on the amount of ^{40}Ar, we are to propose a decay mechanism for this nuclide to form from ^{40}K and explain why researchers used ^{40}Ar to date the skulls and not ^{14}C.

Analyze
(a) The mass number in the decay of ^{40}K to ^{40}Ar does not change, which indicates either a β decay or positron emission.
(b) To explain why researchers used ^{40}Ar instead of ^{14}C dating, we must compare the half-lives of the parent isotopes: ^{14}C, 5730 yr; ^{40}K, 1.28×10^9 yr.

Solve
(a) $^{40}_{19}\text{K} \rightarrow ^{40}_{18}\text{Ar} + ^{0}_{1}\beta$
(b) Because the half-life of ^{40}K is so much longer than that of ^{14}C, ^{40}Ar can be used to date much, much older objects.

Think About It
The amount of ^{14}C remaining after 154,000 years would be too small to measure

$$\frac{N_t}{N_0} = 0.5^{154,000/5730} = 8.1 \times 10^{-9}$$

CHAPTER 22 | Life and the Periodic Table

22.1. Collect and Organize

Given the figures showing periodic trends, we are to determine which describes the periodic trend in monatomic cation radii.

Analyze

Moving across a period, cation size decreases because the number of protons in the cations increases, but the added electrons are located in the same valence shell. The electrons, therefore, feel greater effective nuclear charge and the cations decrease in size. Moving down a group, the cation size increases because electrons are added to larger valence shells.

Solve

(d) best describes the increasing trend in size in monatomic cations as we descend a group in the periodic table and the decreasing size as we traverse a period.

Think About It

Figure (b) would best describe the periodic trends in effective nuclear charge of the elements.

22.3. Collect and Organize

In Figure P22.3 group 1 (green) and group 16 (pink) are highlighted. Given the two groups, we are to determine which forms ions of larger radii than the neutral atoms.

Analyze

When an atom forms a cation, the cation is smaller than the neutral atom. When an atom forms an anion, the anion is larger than the neutral atom.

Solve

Because group 1 elements are metallic and typically form cations whereas group 16 elements are nonmetallic and typically form anions, group 16 (lavender) forms ions (anions) that have radii larger than their neutral atoms.

Think About It

An anion that is isoelectronic with a cation (for example, Br^- and Rb^+, each with 36 electrons) is also larger than the cation.

22.5. Collect and Organize

For a system in which two solutions (one at 150 mM Na^+ and the other at 10 mM Na^+) are separated by a semipermeable membrane, we can use the Nernst equation to calculate the cell potential due to the concentration difference and then calculate ΔG from E_{cell}.

Analyze

The Nernst equation is

$$E_{cell} = E_{cell}^{\circ} - \frac{0.0592 \text{ V}}{n} \log Q$$

where $E_{cell}^{\circ} = 0$ because both cells are composed of Na^+, $Q = (10 \text{ m}M/150 \text{ m}M)$ for a spontaneous reaction, and $n = 1$. Once we have calculated E_{cell}, ΔG may be found through

$$\Delta G = -nFE$$

Solve

$$E_{cell} = 0 \text{ V} - \frac{0.0592 \text{ V}}{1} \log\left(\frac{10 \text{ m}M}{150 \text{ m}M}\right) = 0.0696 \text{ V}$$

$$\Delta G = -(1)(9.65 \times 10^4 \text{ C})(0.0696 \text{ V}) = -6.72 \times 10^3 \text{ J or } -6.72 \text{ kJ}$$

Think About It

This reaction would be nonspontaneous in the reverse direction—transport of Na^+ from the lower concentration side to the higher concentration side of the semipermeable membrane.

22.7. **Collect and Organize**

For each germanium (Ge) atom in the structure shown in Figure P22.7, we are to identify the molecular geometry.

Analyze

The geometry around an atom is based on the steric number (or number of lone pairs and bonding pairs). If the steric number is 2, the geometry is linear. If the steric number is 3, the geometry is trigonal planar. If the steric number is 4, the geometry is tetrahedral.

Solve

Both Ge atoms in the structure have a steric number equal to 3 (three bonding pairs including one double bond that is counted as one steric bonding pair), so the geometry around the Ge atoms is trigonal planar.

Think About It

From our knowledge of orbital hybridization, we could also assign the Ge atoms as sp^2 hybridized.

22.9. **Collect and Organize**

Given the three unit cells shown in Figure P22.9, we are to choose the correct structure for NiTi.

Analyze

The stoichiometry of the unit cell must match that of the empirical formula of the shape memory alloy and must also reflect its stated similarity to that of the CsCl crystal, which has a body-centered cubic structure. The stoichiometry and descriptions of the unit cells shown in Figure P22.9 are

(a) Stoichiometry = (4 Ni at corners $\times \frac{1}{8}$) + (4 Ti at corners $\times \frac{1}{8}$) = $\frac{1}{2} : \frac{1}{2}$, Ni:Ti = 1:1

This structure is based on a simple cube with Ti and Ni atoms at the corners of the cube.

(b) Stoichiometry = (8 Ni at corners $\times \frac{1}{8}$) + (6 Ti on faces $\times \frac{1}{2}$) + (1 Ti in center $\times$ 1) = 1:4, Ni:Ti

This structure is based on a face-centered cube with Ni atoms at the corners of the cube and Ti atoms in the center of the unit cell and located on each face.

(c) Stoichiometry = (8 Ni at corners $\times \frac{1}{8}$) + (1 Ti in center $\times$ 1) = 1:1, Ni:Ti

This structure is based on a simple cube with Ni atoms at the corners of the cube and a Ti atom in the center.

Solve

Both structures (a) and (c) have the correct stoichiometry of 1:1, Ni:Ti, but only structure (c) is like the body-centered cubic structure of CsCl.

Think About It

Using the radius ratio rule for NiTi, we see that the Ti does indeed fit into the cubic hole of Ni atoms just as Cs^+ fits into a cubic lattice of Cl^-.

$$\frac{r_{Ni}}{r_{Ti}} = \frac{124 \text{ pm}}{147 \text{ pm}} = 0.844$$

$$\frac{r_{Cs^+}}{r_{Cl^-}} = \frac{170 \text{ pm}}{181 \text{ pm}} = 0.939$$

22.11. **Collect and Organize / Analyze**

We can use their functions and properties to differentiate between an essential and a nonessential element.

Solve

Without an essential element, biological processes that rely on that element would shut down or deteriorate. If a nonessential element is missing, there would not be any severe deleterious effects.

Think About It

Nonessential elements, however, may aid in processes such as growth.

22.13. **Collect and Organize**

We can distinguish between major, trace, and ultratrace elements by defining each of these categories.

Analyze

The difference between these element categories is the amount in which they are present in the body.

Solve

Major elements are present in milligram per gram quantities in the body, whereas trace elements are present in microgram to milligram per gram quantities, and ultratrace elements are present in nanogram to microgram per gram quantities.

Think About It

All of these elements are still *essential* and therefore are important in the body.

22.15. **Collect and Organize**

In this question we are to explain why the methylmercury cation is more toxic than mercury metal.

Analyze

Mercury metal is uncharged and has no bonds to a carbon group, whereas the methylmercury cation is charged and is bonded to a methyl group.

Solve

The methyl group on CH_3Hg^+ is relatively nonpolar and this helps the cation be soluble in nonpolar environments. Its charge, on the other hand, allows it to be soluble in polar environments. Mercury metal, however, is neutral and has no nonpolar substituents and so is less soluble in biological systems (both polar, like water, and nonpolar, like fats).

Think About It

The methyl mercury cation is dangerously toxic because it can pass through the blood (water)–brain (lipid) barrier.

22.17. **Collect and Organize**

We are to explain why Be^{2+} is more likely than Ca^{2+} to replace Mg^{2+} in biomolecules.

Analyze

All of these ions belong to group 2. Their sizes increase in order $Be^{2+} < Mg^{2+} < Ca^{2+}$.

Solve

The difference in behavior must be due to size. Ca^{2+} must simply be too large to fit into the biomolecules where Mg^{2+} is important.

Think About It

Ca^{2+}, however, is easily replaced in bone by Sr^{2+}. This cation is being considered (as its nonradioactive isotope) as a treatment and preventative for osteoporosis.

22.19. Collect and Organize

By using periodic trends in sizes of cations, we can determine which is larger, the potassium ion channel or the sodium ion channel.

Analyze

Moving down a group, the cation size increases because electrons are added to larger valence shells.

Solve

K^+ is the larger cation, so the potassium ion channel is larger than the sodium ion channel.

Think About It

The alkali metal cations increase in size in the order $Li^+ < Na^+ < K^+ < Rb^+ < Cs^+$.

22.21. Collect and Organize

For the amounts of elements present as essential elements given in grams or milligrams per kilogram, we are to express them in parts per million (ppm).

Analyze

First, we have to express both masses in each concentration in a common unit. Grams is convenient here. The concentrations in parts per million then are

$$\frac{\text{grams of element}}{\text{total grams of matter}} \times 10^6 = \text{ppm}$$

Solve

(a) $\dfrac{0.110 \text{ g}}{70 \times 10^3 \text{ g}} \times 10^6 = 1.6 \text{ ppm Cu}$

(b) $\dfrac{0.033 \text{ g}}{1000 \text{ g}} \times 10^6 = 33 \text{ ppm Zn}$

(c) $\dfrac{0.043 \text{ g}}{100 \times 10^3 \text{ g}} \times 10^6 = 0.43 \text{ ppm I}$

Think About It

We could also have chosen kilograms as the common units for mass. The result would be the same, as it is the ratio of masses that is important in the ppm unit.

22.23. Collect and Organize

Given three pairs of elements, we are to determine which one of the pair is more abundant in the human body.

Analyze

The abundance of the elements in our bodies is given in Table 22.1. Listed in the table are the compositions in terms of the numbers of atoms of the elements in the human body. Elements without a percent entry or those not listed in the table are presumed to be less than 0.01% abundant.

Solve

(a) Oxygen is more abundant than silicon.
(b) Oxygen is more abundant than iron.
(c) Carbon is more abundant than aluminum.

Think About It

From Figure 22.6 we see that both iron and silicon are essential trace elements, whereas aluminum is a nonessential element.

22.25. Collect and Organize

We are to list three ways in which essential element ions such as Na^+ and K^+ can enter and exit cells.

Analyze

The discussion about ion transport across a cell membrane is explained in the text in Section 22.2.

Solve

The three mechanisms for ion transport across a cell membrane are osmosis, transport through ion channels, and ion pumps.

Think About It

It is indeed difficult to get charged particles into and out of the cell because the cell membrane is composed of nonpolar molecules.

22.27. Collect and Organize

We are asked what makes it difficult for (charged) ions to simply diffuse across a cell membrane.

Analyze

Figure 22.3 shows the structure of the cell membrane. It is composed of a phospholipid bilayer with the polar head groups facing the interior and exterior of the cell.

Solve

The hydrophobic interior of the cell membrane (the long-chain nonpolar groups that make up the phospholipids) make it difficult to transport charged ions through the cell membrane.

Think About It

The polar head groups of the cell membrane, however, are attractive to ions such as Na^+ and K^+.

22.29. Collect and Organize

We can use similarities in size of the alkali metal cations to predict which ion Rb^+ is most likely to substitute for.

Analyze

The sizes of the alkali metal ions are shown in Figure 10.2.

Solve

Because Rb^+ (radius 149 pm) is closest in size to K^+ (138 pm), it is most likely to substitute for K^+ in the body.

Think About It

Rubidium in low concentrations is not particularly toxic and is excreted from the body readily in sweat and urine. The other ion we should consider is Na^+; with a radius of only 102 pm, it is unlikely than the larger Rb^+ ion would substitute for Na^+.

22.31. Collect and Organize

We need to consider the relative solubilities of $CaSO_4$ and $CaCO_3$ in water. The least soluble salt would be more readily used by nature as the structural material of shells.

Analyze

From Appendix 5 we see that the K_{sp} for $CaSO_4$ at 25°C is 7.1×10^{-5}, which is higher than the K_{sp} for $CaCO_3$ (5.0×10^{-9}).

Solve

The greater insolubility of $CaCO_3$ compared to $CaSO_4$ makes calcium carbonate a better structural material. Also, the partial pressure of CO_2 in the atmosphere is higher than SO_3, so the carbonate solubility equilibrium is shifted more to the left by Le Châtelier's principle than the sulfate equilibrium.

Think About It
Calcium phosphate, with an even lower solubility, would make an even better structural material. Indeed, calcium and phosphate ions make up hydroxyapatite, tooth enamel.

22.33. **Collect and Organize**
Given that the concentration of NaCl in a cell at 37°C is 11 mM, we are to calculate the osmotic pressure exerted across the cell membrane using Equation 11.14:

$$\Pi = iMRT$$

Analyze
In the equation, Π is the osmotic pressure, i is the van't Hoff factor (2 for NaCl), M is the molarity of the solution (11×10^{-3} M), R is the gas constant [0.0821 L · atm/(mol · K)], and T is the temperature (310 K).

Solve

$$\Pi = 2 \times (11 \times 10^{-3}\,M) \times \left(0.0821 \frac{L \cdot atm}{mol \cdot K}\right) \times 310\ K = 0.56\ atm$$

Think About It
As either the concentration of NaCl or the temperature increases, the osmotic pressure increases.

22.35. **Collect and Organize**
From the difference in concentration of NaCl between red blood cells and the plasma, we can use the Nernst equation to calculate the electrochemical potential created by this concentration difference.

Analyze
The Nernst equation is

$$E = E^{\circ}_{cell} - \frac{0.0592\ V}{n} \log Q$$

where $E^{\circ}_{cell} = 0.00$ V, $n = 1$, and $Q = [Na^+]_{cell}/[Na^+]_{plasma}$

Solve

$$E = 0.00\ V - \frac{0.0592\ V}{1} \log\left(\frac{11\ mM}{160\ mM}\right) = 0.0688\ V$$

Think About It
As the $[Na^+]$ in the plasma decreases, the electrochemical potential decreases because log Q becomes less negative.

22.37. **Collect and Organize**
Given that $\Delta G^{\circ} = -34.5$ kJ for the reaction of ATP^{4-} with water, we are to calculate the moles of ATP^{4-} that must be hydrolyzed to provide the 5 kJ/mol to transport K^+ across a cell membrane.

Analyze
To solve this problem we need only to divide the energy needed by the energy released by the hydrolysis of ATP^{4-}.

Solve

$$\frac{5\ kJ}{mol\ K^+} \times \frac{1\ mol\ K^+}{1\ mol\ ATP^{4-}} \times \frac{1\ mol\ ATP^{4-}}{34.5\ kJ} = 0.15\ mol\ ATP^{4-}$$

Think About It
We used +34.5 kJ/mol ATP^{4-} to solve this problem because that amount of energy would be needed for the endergonic transport of K^+ across the cell membrane.

22.39. Collect and Organize

Given the solubility product constant for strontium sulfate ($K_{sp} = 3.4 \times 10^{-7}$), we are asked to calculate the solubility (in mol/L) of $SrSO_4$.

Analyze

The K_{sp} expression for strontium sulfate is

$$K_{sp} = 3.4 \times 10^{-7} = [Sr^{2+}][SO_4^{2-}] = s^2$$

Solve

$$K_{sp} = 3.4 \times 10^{-7} = [Sr^{2+}][SO_4^{2-}] = s^2$$
$$s = 5.8 \times 10^{-4} \text{ mol/L}$$

Think About It

In terms of grams per liter the solubility of $SrSO_4$ at this temperature is

$$\frac{5.83 \times 10^{-4} \text{ mol}}{L} \times \frac{183.68 \text{ g}}{mol} = 0.107 \text{ g/L}$$

22.41. Collect and Organize

We are to explain the dangers of ^{137}Cs.

Analyze

Cesium is a member of the alkali metal family and thus is present in the form of Cs^+. Its radius as a cation (170 pm) is slightly larger than that of K^+ (138 pm). From Appendix 3 we find that ^{137}Cs is a beta emitter.

Solve

Because of their similar sizes $^{137}Cs^+$ may substitute for K^+ in cells, and as a β emitter with a relatively long half-life it may cause cancer.

Think About It

^{137}Cs is produced in nuclear bomb explosions and was released into the atmosphere during the Chernobyl accident.

22.43. Collect and Organize

We are asked to determine the signs of ΔS and ΔG for the decay of tooth enamel.

Analyze

Tooth enamel is comprised of $Ca_3(PO_4)_2$ and $Ca_8(HPO_4)_2(PO_4)_4 \cdot 6 H_2O$. Decay occurs in the presence of acid to form Ca^{2+}, HPO_4^{2-}, and $H_2PO_4^-$ ions.

Solve

ΔS will be positive, as matter is transferred from the solid phase to the aqueous phase. ΔG will be negative, as the reaction occurs spontaneously in acidic solution.

Think About It

Although this reaction is spontaneous, it occurs at a slow rate.

22.45. Collect and Organize

We are to describe the function of enzymes, and determine whether all proteins are enzymes.

Analyze

Enzymes are proteins that serve a specific role in biochemical reactions.

Solve

(a) Enzymes function as catalysts for biochemical reactions involving small molecules. Enzymes serve as a template for reactions, by creating cavities of the appropriate size and shape. Small molecules are brought in close proximity or with the appropriate orientation to react in the desired manner.

(b) No, not all proteins are enzymes. Other examples of proteins that are not enzymes include muscle tissue and ion channels.

Think About It

It may be helpful to think of enzymes as biochemical catalysts. Much like a metal catalyst, they promote a reaction occurring under mild conditions at a faster rate than the uncatalyzed process.

22.47. **Collect and Organize / Analyze**

We can use the definition of an enzyme to describe the effect on the activation energy of a biochemical process that uses an enzyme as a catalyst.

Solve

Because an enzyme is a catalyst, it lowers the activation energy of the biochemical process by providing an alternate, lower energy pathway for the reactants to form products.

Think About It

The activity of enzymes is affected by reaction conditions such as temperature, pH, and the presence of chemical inhibitors.

22.49. **Collect and Organize**

We are to predict whether the reaction between a transition metal cation like Cu^{2+} and a protein to produce a metalloenzyme has an equilibrium constant much greater than or much less than 1.

Analyze

A protein has many sites to which the metal ion may bind, so a protein is a multidentate ligand. The formation of metal complexes with multidentate ligands is highly favored.

Solve

The K of the reaction of a transition metal ion with a protein to form a metalloenzyme has a value much greater than 1.

Think About It

The large value of K means that there is very little free transition metal ion present in a solution of protein.

22.51. **Collect and Organize**

We are asked to predict the sign of ΔS for the reaction catalyzed by carboxypeptidase, described in Table 22.6.

Analyze

The reaction catalyzed by carboxypeptidase in Table 22.6 is

In this reaction, an amide bond is broken to form the corresponding carboxylic acid and amine.

Solve

One molecule is split into two molecules, so the entropy of this reaction increases, and ΔS is positive.

Think About It

This reaction is the reverse of a peptide bond formation.

22.53. Collect and Organize

We are to write a balanced equation for the nuclear decay of radioactive ^{137}Cs.

Analyze

We learned in the text that ^{137}Cs is a β emitter. We expect a product with a greater number of protons in its nucleus as a neutron transforms into a proton with release of $^{0}_{-1}$e.

Solve

$$^{137}_{55}Cs \rightarrow \, ^{0}_{-1}\beta + \, ^{137}_{56}Ba$$

Think About It

The ^{137}Ba isotope is stable and is the final product of ^{137}Cs decay.

22.55. Collect and Organize

Given that $pK_{a_1} = 2.21$ and $pK_{a_2} = 5.43$ for selenocysteine ($C_3H_7NO_2Se$), we are to calculate the pH of a 1.00×10^{-3} M solution.

Analyze

Because the second ionization constant ($pK_{a_2} = 3.72 \times 10^{-6}$) is so much less than the first ionization constant ($pK_{a_1} = 6.17 \times 10^{-3}$), we can calculate the pH of the solution using just the first ionization constant.

Solve

	$[C_3H_7NO_2Se]$	$[H^+]$	$[C_3H_6NO_2Se^-]$
Initial	1.00×10^{-3}	0	0
Change	$-x$	$+x$	$+x$
Equilibrium	$1.00 \times 10^{-3} - x$	x	x

Substituting values into the mass action expression gives

$$6.17 \times 10^{-3} = \frac{x^2}{1.00 \times 10^{-3} - x}$$

Expanding and solving by the quadratic equation gives

$$x^2 + 6.17 \times 10^{-3} x - 6.17 \times 10^{-6} = 0$$
$$x = 8.76 \times 10^{-4}$$

The pH of the solution is

$$pH = -\log(8.76 \times 10^{-4}) = 3.06$$

Think About It

The K_{a_1} of cysteine is 2×10^{-2}. This means that an equimolar solution of cysteine is more acidic than selenocysteine.

22.57. Collect and Organize

For the reaction describing the replacement of OH$^-$ in hydroxyapatite with F$^-$, we are to write the equilibrium constant expression and, using the value of the equilibrium constant given ($K = 8.48$), determine whether the reaction favors products (lies to the right) or reactants (lies to the left).

Analyze

The equilibrium constant expression takes the form

$$wA + xB \rightleftharpoons yC + zD$$

$$K_c = \frac{[C]^y[D]^z}{[A]^w[B]^x}$$

where [C], [D], [A], and [B] are the concentrations of the soluble species in moles per liter. Remember that pure solids and liquids do not appear in the equilibrium constant expression. To determine whether products or reactants are favored in this reaction, we remember from Chapter 15 that when $K > 1$ products are favored in the reaction and when $K < 1$ reactants are favored.

Solve
The only soluble species in the equilibrium reaction are F^- and OH^-. The equilibrium constant expression, therefore, is

$$K = \frac{[OH^-]}{[F^-]} = 8.48$$

Because $K = 8.48$, which is greater than 1, the equilibrium favors the products and lies to the right.

Think About It
This equilibrium does not lie overwhelmingly to the right, however. For products to be greatly favored in a reaction, K values should be above 1000.

22.59. ## Collect and Organize
By comparing the values of the equilibrium constants for the dissolution of hydroxyapatite and the calcium phosphate that together form tooth enamel, we can determine whether the calcium phosphate is more or less soluble in water than hydroxyapatite. Using the pH of the solution, we may use $[OH^-]$ to calculate the molar solubility of hydroxyapatite.

Analyze
(a) The larger the K_{sp} of a solid, the more soluble it is, but that would only be strictly true for molecules of the same formula type. To answer this problem, we should solve for s in the solubility product expression.
(b) The chemical equation and K_{sp} expression for the dissolution of hydroxyapatite is

$$Ca_5(PO_4)_3OH(s) \rightleftharpoons 5\ Ca^{2+}(aq) + 3\ PO_4^{3-}(aq) + OH^-(aq)$$
$$K_{sp} = [Ca^{2+}]^5[PO_4^{3-}]^3[OH^-] = 2.3 \times 10^{-59}$$

When pH = 7.0, pOH = 7.0 and $[OH^-] = 1 \times 10^{-7}$ M. For every mole of $Ca_5(PO_4)_3OH$ or $Ca_5(PO_4)_3F$ that dissolves, 5 moles of Ca^{2+} and 3 moles of PO_4^{3-} are released into the solution along with 1 mole of OH^- or F^-. The molar solubility of hydroxyapatite at pH = 7.0 is thus given by the equation

$$K_{sp} = (5s)^5(3s)^3(1 \times 10^{-7})$$

(c) When pH = 5.0, pOH = 9.0 and $[OH^-] = 1 \times 10^{-9}$ M. For every mole of $Ca_5(PO_4)OH$ or $Ca_5(PO_4)F$ that dissolves, 5 moles of Ca^{2+} and 3 moles of PO_4^{3-} are released into the solution along with 1 mole of OH^- or F^-. The molar solubility of hydroxyapatite at pH = 5.0 is thus given by the equation
$$K_{sp} = (5s)^5(3s)^3(1 \times 10^{-9})$$

Solve
(a) For the solubility of $Ca_8(HPO_4)_2(PO_4)_4 \cdot 6\ H_2O$ we have the following equation and K_{sp} expression

$$Ca_8(HPO_4)_2(PO_4)_4 \cdot 6\ H_2O(s) \rightleftharpoons 6\ H_2O(\ell) + 8\ Ca^{2+}(aq) + 4\ PO_4^{3-}(aq) + 2\ HPO_4^{2-}(aq)$$
$$K_{sp} = [Ca^{2+}]^8[PO_4^{3-}]^4[HPO_4^{2-}]^2$$
$$1.1 \times 10^{-47} = (8s)^8(4s)^4(2s)^2 = 1.718 \times 10^{10} s^{14}$$
$$s^{14} = 6.403 \times 10^{-58}$$
$$s = 8.2 \times 10^{-5}\ M$$

For the solubility of $Ca_5(PO_4)_3(OH)$ we have the following equation and K_{sp} expression

$$Ca_5(PO_4)_3(OH)(s) \rightleftharpoons 5\ Ca^{2+}(aq) + 3\ PO_4^{3-}(aq) + OH^-(aq)$$
$$K_{sp} = [Ca^{2+}]^5[PO_4^{3-}]^3[OH^-]$$
$$2.3 \times 10^{-59} = (5s)^5(3s)^3 s = 8.4375 \times 10^4 s^9$$
$$s^9 = 2.726 \times 10^{-64}$$
$$s = 8.7 \times 10^{-8}\ M$$

Therefore, the calcium mineral $Ca_8(HPO_4)_2(PO_4)_4 \cdot 6\ H_2O$ is more soluble than hydroxyapatite.

(b) For $Ca_5(PO_4)_3OH$ at pH = 7.0, the solubility (s) is
$$K_{sp} = 2.3 \times 10^{-59} = (5s)^5(3s)^3(1 \times 10^{-7}) = 3125s^5 \times 27s^3 \times (1 \times 10^{-7})$$
$$s = 8.5 \times 10^{-8} \ M$$
(c) For $Ca_5(PO_4)_3OH$ at pH = 5.0, the solubility (s) is
$$K_{sp} = 2.3 \times 10^{-59} = (5s)^5(3s)^3(1 \times 10^{-9}) = 3125s^5 \times 27s^3 \times (1 \times 10^{-9})$$
$$s = 1.5 \times 10^{-7} \ M$$

Think About It
Notice that both substances are still relatively insoluble compounds, with very small K_{sp} values. In grams per liter the solubility of hydroxyapatite at this temperature and pH = 5.0 is
$$\frac{1.512 \times 10^{-7} \ \text{mol}}{1 \ \text{L}} \times \frac{502.31 \ \text{g}}{1 \ \text{mol}} = 7.59 \times 10^{-5} \ \text{g/L}$$

22.61. Collect and Organize
We relate the K_{sp} of $Ca_5(PO_4)_3OH$ to that of $Ca_{10}(PO_4)_6(OH)_2$.

Analyze
The difference in the two formulas is that in $Ca_{10}(PO_4)_6(OH)_2$ the atoms have all been doubled compared to the formula $Ca_5(PO_4)_3OH$. Therefore, the K_{sp} of $Ca_{10}(PO_4)_6(OH)_2$ is the K_{sp} of $Ca_5(PO_4)_3OH$ squared.

Solve
$$K_{sp,Ca_{10}(PO_4)_6(OH)_2} = \left(K_{sp,Ca_5(PO_4)_3OH}\right)^2 = \left(2.3 \times 10^{-59}\right)^2 = 5.3 \times 10^{-118}$$

Think About It
For $Ca_5(PO_4)_3OH$ the K_{sp} expression is $K_{sp} = [Ca^{2+}]^5[PO_4^{3-}]^3[OH^-]$. For $Ca_{10}(PO_4)_6(OH)_2$ the K_{sp} expression is $K_{sp} = [Ca^{2+}]^{10}[PO_4^{3-}]^6[OH^-]^2$.

22.63. Collect and Organize
Using the equation
$$RT \ln\left(\frac{k_1}{k_2}\right) = E_{a_2} - E_{a_1}$$
we can calculate the effect a catalyst has on the decomposition of H_2O_2, given the values of the activation energies for the catalyzed and the uncatalyzed reactions (29.3 kJ/mol and 75.3 kJ/mol, respectively).

Analyze
We can solve for the value of k_1/k_2 from the given form of the Arrhenius equation through
$$\frac{k_1}{k_2} = e^{\left(\frac{E_{a_2} - E_{a_1}}{RT}\right)}$$

Solve
$$\frac{k_1}{k_2} = e^{\left(\frac{75.3 \ \text{kJ/mol} - 29.3 \ \text{kJ/mol}}{8.314 \times 10^{-3} \ \text{kJ/(mol·K)} \times 293 \ \text{K}}\right)} = 1.59 \times 10^8$$
The catalyzed reaction will be 1.59×10^8 times faster than the uncatalyzed reaction.

Think About It
Remember from chemical kinetics that a catalyst provides a lower energy pathway to products in a reaction but does not change the amount of products formed once the reaction achieves equilibrium.

22.65. Collect and Organize
We are asked to describe the apparent reaction order at the far right side of the plot in Figure P22.65.

Analyze

We may estimate the order of a reaction at a given time by drawing a tangent to the instantaneous rate. The slope of this tangent should tell us about the instantaneous order of the reaction at that time. At the far right side of the plot, there is essentially no change in the rate of reaction as a function of concentration.

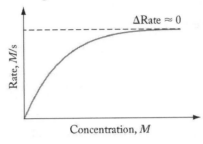

Solve

At the far right of this plot, the rate of reaction is independent of the substrate concentration, as would be expected for a zero order reaction.

Think About It

The reaction rate appears to be closer to a first or second order dependence at the outset of the reaction, slowing as the reaction proceeds.

22.67. **Collect and Organize**

We are asked to comment on why both decay mode and half-life are important considerations in choosing a medical isotope for imaging the human body.

Analyze

The main decay modes for radionuclides are α emission, β emission, and γ emission. α particles have a large mass and are unable to penetrate deeply into materials and tissue. However, due to their large mass, α particles may cause significant damage to organs and tissues if directly adjacent. γ rays, while more ionizing, can easily penetrate the body, including the skin.

Solve

α emitters are much more damaging as radioisotopes than γ emitters. While γ rays have a greater ionizing power, they can also leave the body by passing through organs and tissues. α particles cannot escape the body, so they stay resident, damaging organs and tissues in the process.

Think About It

Radionuclides are often chosen so that their half-life is long enough for diagnostic imaging, but short enough that excessive levels of radiation in the patient are avoided.

22.69. **Collect and Organize**

We consider the characteristics of α emitters that make them good choices for chemotherapy.

Analyze

An α particle is the same as a nucleus of helium-4 and consists of two protons and two neutrons with a charge of 2+.

Solve

Because α particles are both charged and heavy compared to other particles emitted from the nucleus (β particles, neutrons, positrons, and γ rays), they do not penetrate deeply into tissues. Also, of all the particles, α particles have the highest RBE (relative biological effectiveness).

Think About It

An α emitter can be implanted into a tumor to destroy the tumor and since the particles do not penetrate far, healthy tissue can be left unaffected.

22.71. **Collect and Organize**

We consider the radiation released by the decay of ^{153}Gd.

Analyze

We are given that ^{153}Gd decays by electron capture. In this process a proton is changed into a neutron and there is an accompanying loss of energy as γ radiation. X-rays are also emitted.

Solve

The radiation produced by the decay of ^{153}Gd is γ radiation and X-rays.

Think About It

The nuclide that is formed in the electron capture decay of ^{153}Gd is ^{153}Eu:

$$^{153}_{64}\text{Gd} \xrightarrow{\text{electron capture}} {}^{153}_{63}\text{Eu}$$

22.73. **Collect and Organize**

We are to describe how platinum- and ruthenium-containing drugs fight cancer.

Analyze

The biological activity of these compounds is described in the text in Section 22.5.

Solve

These drugs fight cancer by binding to the nitrogen atoms in DNA to stop the division of cells so that the tumors cannot grow.

Think About It

Platinum drugs are used widely to treat a variety of cancers including lymphomas, lung cancer, and colorectal cancer.

22.75. **Collect and Organize**

From our knowledge of the magnetic properties of transition metal complexes (Chapter 17), we can determine whether the chromium(III) ion in the glucose tolerance factor is paramagnetic or diamagnetic.

Analyze

The geometry of the chromium(III) ion in the glucose tolerance factor is octahedral, as shown in Figure 22.20. When surrounded by an octahedral field of ligands, the *d* orbitals split into two levels as shown below. A paramagnetic substance has unpaired electrons. A diamagnetic substance has no unpaired electrons.

Solve

The Cr(III) ion has three *d* electrons that will fill the octahedral field diagram as follows:

The electrons are unpaired, so this complex is paramagnetic.

Think About It
The number of unpaired electrons in this compound is the same (3) whether the ligands are strong-field or weak-field ligands.

22.77. **Collect and Organize**
We describe why mercury amalgams used in dentistry are not toxic.

Analyze
Dental alloys are described in Section 22.5.

Solve
Making an amalgam composed of silver, tin, and mercury renders the mercury chemically unreactive and no longer toxic.

Think About It
Mercury amalgams have the advantages of low cost, strength, and durability over other dental materials.

22.79. **Collect and Organize**
We are asked to determine the electron configuration and number of unpaired electrons for a Gd^{3+} ion.

Analyze
A gadolinium atom has 10 valence electrons, and the electron configuration $[Xe]\ 6s^2\ 4f^8$. We must remove three electrons to form the 3+ cation. Recall that electrons are removed from the s orbital prior to the f or d orbitals.

Solve
The electron configuration for Gd^{3+} is $[Xe]\ 6s^0\ 4f^7$. By preparing a condensed orbital box diagram, we can see that all seven electrons in the $4f$ orbital are unpaired

Think About It
The $6s$ electrons will be removed first because they are shielded by the $4f$ electrons and thus experience a lower Z_{eff} than the $4f$ electrons.

22.81. **Collect and Organize**
We are asked to determine the electron configuration for the iron cation in sodium nitroprusside, $Na_2[Fe(NO)(CN)_5]$. We may also predict whether this complex will be high spin or low spin.

Analyze
The two sodium cations imply that the complex nitroprusside anion has a charge of 2–. Nitrosyl (NO) bears a formal charge of +1, and each cyanide (CN^-) bears a formal charge of –1, so the oxidation state of iron (x) is
$$(x) + (+1) + (5 \times -1) = -2$$
$$x = +2$$
The oxidation state of iron is +2. We must remove two electrons from the electron configuration of a neutral iron atom to generate the 2+ cation. The oxidation state of an iron atom is $[Ar]\ 4s^2\ 3d^8$. Complexes featuring a strong-field ligand will adopt a low-spin (more electrons paired) configuration, and complexes featuring a weak-field ligand will adopt a high-spin (fewer electrons paired) configuration.

Solve
The electron configuration for Fe^{2+} is $[Ar]\ 4s^0\ 3d^6$. Cyanide is a strong-field ligand, so this complex is likely to be low spin.

Think About It
The crystal field splitting diagram for the Fe^{2+} ion in a low-spin configuration is

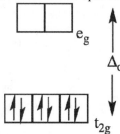

This complex should be diamagnetic.

22.83. **Collect and Organize**
Using the equation

$$\frac{N_t}{N_0} = 0.5^n \qquad \text{where } n = \frac{t}{t_{1/2}}$$

we can calculate the time (t) at which the activity of ^{68}Ga ($t_{1/2} = 9.4$ hr) drops to 5% of its initial value.

Analyze
Combining the equations above we get

$$\frac{N_t}{N_0} = 0.5^{t/t_{1/2}}$$

Rearranging to solve for t we obtain the equation

$$t = \frac{\ln(N_t/N_0) \times t_{1/2}}{\ln 0.5}$$

Solve

$$t = \frac{\ln(0.05) \times 9.4 \text{ hr}}{\ln 0.5} = 41 \text{ hr}$$

Think About It
This loss of activity over a short period of time is characteristic of many radioactive imaging agents.

22.85. **Collect and Organize**
We are asked to draw the Lewis structure of BiO^+.

Analyze
Bismuth has 5 valence electrons and oxygen has 6 valence electrons for a total of 11 valence electrons. We have to subtract one electron for the Lewis structure because the overall charge on the ion is 1+. This Lewis structure, then, has 10 valence electrons.

Solve

$$\left[:Bi \equiv O: \right]^+$$

Think About It
Notice that in order for both atoms to have a complete octet, a triple bond must be drawn between Bi and O.

22.87. **Collect and Organize**
Given the log of the formation constants for $Hg(\text{methionine})^{2+}$ and $Hg(\text{penicillamine})^{2+}$, we are asked to calculate the K for the reaction of $Hg(\text{methionine})^{2+}$ with penicillamine.

Analyze

The value of the K_f for each of the reactions is

$$K_f \text{ of Hg(methionine)}^{2+} = 1 \times 10^{14.2} = 1.58 \times 10^{14}$$
$$K_f \text{ of Hg(penicillamine)}^{2+} = 1 \times 10^{16.3} = 2.00 \times 10^{16}$$

The two complex formation reactions can be added together to obtain the desired overall equation:

$$\text{Hg(methionine)}^{2+} \rightleftharpoons \text{Hg}^{2+} + \text{methonine}$$
$$\text{Hg}^{2+} + \text{penicillamine} \rightleftharpoons \text{Hg(penicillamine)}^{2+}$$

Notice that the first reaction is reversed and the overall value of K for the reaction is

$$K = \frac{K_{f,\text{Hg(penicillamine)}^{2+}}}{K_{f,\text{Hg(methionine)}^{2+}}}$$

Solve

$$K = \frac{2.00 \times 10^{16}}{1.58 \times 10^{14}} = 127$$

Think About It

The replacement of methionine by penicillamine complexed to Hg^{2+} does proceed to the right, and the reaction is spontaneous since ΔG is negative when $K > 1$.

22.89. Collect and Organize

Given that $K = 0.633$ for the reaction in which cysteine replaces penicillamine on the methylmercury cation (CH_3Hg^+), we are to calculate the concentration of cysteine and penicillamine at equilibrium when the initial concentrations are 1.00 M cysteine and 1.00 mM methylmercury–penicillamine complex.

Analyze

To solve this problem we set up an ICE table.

Solve

	$[\text{CH}_3\text{Hg(penicillamine)}^+]$	[Cysteine]	$[\text{CH}_3\text{Hg(cysteine)}^+]$	[Penicillamine]
Initial	1.00×10^{-3}	1.00	0	0
Change	$-x$	$-x$	$+x$	$+x$
Equilibrium	$1.00 \times 10^{-3} - x$	$1.00 - x$	x	x

Substituting into the equilibrium constant expression gives

$$0.633 = \frac{x^2}{\left(1.00 \times 10^{-3} - x\right)\left(1.00 - x\right)} \approx \frac{x^2}{\left(1.00 \times 10^{-3} - x\right)\left(1.00\right)}$$

Expanding and solving by the quadratic equation gives

$$0.367\,x^2 + 0.6336x - 6.33 \times 10^{-4} = 0$$
$$x = 1.0 \times 10^{-4}\,M$$

Therefore, $[\text{penicillamine}]_{eq} = 1.0 \times 10^{-4}\,M$ and $[\text{cysteine}]_{eq} = 1.00 \times 10^{-3} - 1.0 \times 10^{-4}\,M = 9.0 \times 10^{-4}\,M$.

Think About It

With the equilibrium constant close to 1, we can say that both penicillamine and cysteine bind to the methylmercury cation about equally.

22.91. Collect and Organize

We are to draw the Lewis structure of the citrate anion.

Analyze

We are given the skeletal structure of citrate in Figure P22.91. Each carbon atom in the structure brings four valence electrons, each hydrogen atom brings one valence electron, and each oxygen atom brings six valence electrons. The total number of electrons in bonding pairs and lone pairs in the structure, including counting of the three additional electrons to give the anion its 3– charge, is

$$(6\ C \times 4\ e^-) + (5\ H \times 1\ e^-) + (7\ O \times 6\ e^-) + 3\ e^- = 74\ e^-$$

Solve

Think About It

Citrate is an intermediate in the Krebs cycle for cellular respiration.

22.93. Collect and Organize

We are asked to write a balanced net ionic equation describing the reaction between $Al(OH)_3$ and HCl.

Analyze

In this reaction solid aluminum hydroxide acts as a base and HCl acts as an acid to produce Al^{3+} and water. HCl is a strong acid and thus is completely dissociated in aqueous solution. Cl^- is a spectator ion in this reaction.

Solve

$$Al(OH)_3(s) + 3\ H^+(aq) \rightarrow 3\ H_2O(\ell) + Al^{3+}(aq)$$

Think About It

$Al(OH)_3$ also reacts with OH^- to give the soluble species $Al(OH)_4^-$. Because it reacts with both acids and bases, $Al(OH)_3$ is *amphoteric*.

22.95. Collect and Organize

Given the reduction potentials of ZnO and Ag_2O, we are to determine the overall reaction and $E°_{cell}$ for a silver/zinc battery.

Analyze

A battery must have a spontaneous electrochemical reaction and therefore $E°_{cell}$ must be positive. To determine the overall reaction, therefore, we reverse the equation for the reaction that has the most negative reduction potential.

Solve

$$
\begin{aligned}
Zn(s) + 2\ OH^-(aq) &\rightarrow ZnO(s) + H_2O(\ell) + 2\ e^- & E°_{anode} &= -1.258\ V\\
Ag_2O(s) + H_2O(\ell) + 2\ e^- &\rightarrow 2\ Ag(s) + 2\ OH^-(aq) & E°_{cathode} &= 0.342\ V\\
\hline
Zn(s) + Ag_2O(s) &\rightarrow ZnO(s) + 2\ Ag(s) & E°_{cell} &= 1.600\ V
\end{aligned}
$$

Think About It

Notice that in this battery, the hydroxide produced by the cathode is used by the anode.

22.97. Collect and Organize

Given the density of a sample of ^{192}Ir and knowing that it crystallizes in an fcc unit cell, we may calculate the radius of an iridium-192 atom.

Analyze

We must first calculate the mass of each iridium-192 atom, accounting for the different mass of this isotope compared to the average mass found in the periodic table. ^{192}Ir contains 115 neutrons (1.67493×10^{-27} kg), 77 protons (1.67263×10^{-27} kg), and 77 electrons (9.10939×10^{-31} kg). The mass of ^{192}Ir is

$$115(1.67493 \times 10^{-27} \text{ kg}) + 77(1.67263 \times 10^{-27} \text{ kg}) + 77(9.10939 \times 10^{-31} \text{ kg}) = 3.214796 \times 10^{-25} \text{ kg}$$

The fcc unit cell contains four atoms. The volume of the unit cell (V) is related to the edge length of the unit cell (ℓ) and the radius of the ^{192}Ir atom (r) by

$$\ell = \sqrt[3]{V}$$
$$r = 0.3536\ell$$

Solve

The volume of the unit cell is

$$V = 1 \text{ unit cell} \times \frac{4 \ ^{192}Ir \text{ atoms}}{1 \text{ unit cell}} \times \frac{3.214796 \times 10^{-25} \text{ kg}}{1 \ ^{192}Ir \text{ atom}} \times \frac{1000 \text{ g}}{1 \text{ kg}} \times \frac{\text{cm}^3}{22.42 \text{ g}} = 5.736 \times 10^{-23} \text{ cm}^3$$

The radius of an iridium-192 atom is

$$\ell = \sqrt[3]{5.736 \times 10^{-23} \text{ cm}^3} = 3.857 \times 10^{-8} \text{ cm}$$
$$r = 0.3536 \left(3.857 \times 10^{-8} \text{ cm} \right)$$

$$= 1.364 \times 10^{-8} \text{ cm} \times \frac{1 \text{ m}}{100 \text{ cm}} \times \frac{10^{12} \text{ pm}}{1 \text{ m}} = 136.4 \text{ pm}$$

Think About It

We must account for the mass of the ^{192}Ir isotope, rather than using the average atomic mass 192.22 g/mol.

22.99. **Collect and Organize**

We may calculate the molar composition of an alloy that is 6% Al, 4% V, and 90% Ti by weight. We are asked to compare this composition with that of a bcc arrangement of Ti atoms, with two corner atoms substituted by Al and V, shown here

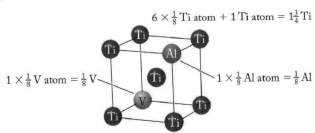

$$6 \times \tfrac{1}{8} \text{ Ti atom} + 1 \text{ Ti atom} = 1\tfrac{3}{4} \text{ Ti}$$

$$1 \times \tfrac{1}{8} \text{ V atom} = \tfrac{1}{8} \text{ V} \qquad 1 \times \tfrac{1}{8} \text{ Al atom} = \tfrac{1}{8} \text{ Al}$$

Analyze

The molar masses of the alloy components are Al (26.98 g/mol), V (50.94 g/mol), and Ti (47.87 g/mol). Assuming 100 g of the alloy, we may calculate the molar quantity of each element and divide by the smallest value to arrive at a molar ratio.

Solve

(a) Assuming 100 g of the alloy

$$6 \text{ g Al} \times \frac{1 \text{ mol Al}}{26.98 \text{ g Al}} = 0.2224 \text{ mol Al}/0.07852 = 2.83$$

$$4 \text{ g V} \times \frac{1 \text{ mol V}}{50.94 \text{ g V}} = 0.07852 \text{ mol V}/0.07852 = 1$$

$$90 \text{ g Ti} \times \frac{1 \text{ mol Ti}}{47.87 \text{ g Ti}} = 1.8801 \text{ mol Ti}/0.07852 = 23.94$$

This reduces to approximately 3:1:24, or Al_3VTi_{24}.

(b) From the bcc unit cell described, the Al:V:Ti ratio would be 0.125:0.125:1.75, which reduces to AlVTi$_{14}$. The unit cell described does not match the experimental composition of Al$_3$VTi$_{24}$ determined in (a).

Think About It

2.83 is on the low side for rounding up to 3. We could instead multiply each coefficient by a factor of 18, to arrive at a ratio of rougly Al$_{51}$V$_{18}$Ti$_{431}$. Because these numbers become unwieldy, it is common to report decimals in the subscripts for substitutional alloys.